Table of Atomic Masses

NAME	SYMBOL	ATOMIC NUMBER	ATOMIC MASS*
Actinium	Ac	89	[227.0278]
Aluminum	Al	13	26.981538
Americium	Am	95	[243.0614]
Antimony	Sb	51	121.757
Argon	Ar	18	39.948
Arsenic	As	33	74.92160
Astatine	At	85	[209.9871]
Barium	Ba	56	137.327
Berkelium	Bk	97	[247.0703]
Beryllium	Be	4	9.012182
Bismuth	Bi	83	208.98038
Bohrium	Bh	107	[262.1231]
Boron	B	5	10.811
Bromine	Br	35	79.904
Cadmium	Cd	48	112.411
Calcium	Ca	20	40.078
Californium	Cf	98	[251.0796]
Carbon	C	6	12.0107
Cerium	Ce	58	140.116
Cesium	Cs	55	132.90545
Chlorine	Cl	17	35.4527
Chromium	Cr	24	51.9961
Cobalt	Co	27	58.933200
Copper	Cu	29	63.546
Curium	Cm	96	[247.0703]
Dubnium	Db	105	[262.1144]
Dysprosium	Dy	66	162.50
Einsteinium	Es	99	[252.0830]
Erbium	Er	68	167.26
Europium	Eu	63	151.964
Fermium	Fm	100	[257.0951]
Fluorine	F	9	18.9984032
Francium	Fr	87	[223.0197]
Gadolinium	Gd	64	157.25
Gallium	Ga	31	69.723
Germanium	Ge	32	72.61
Gold	Au	79	196.96655
Hafnium	Hf	72	178.49
Hassium	Hs	108	[265.1306]
Helium	He	2	4.002602
Holmium	Ho	67	164.93032
Hydrogen	H	1	1.00794
Indium	In	49	114.818
Iodine	I	53	126.90447
Iridium	Ir	77	192.22
Iron	Fe	26	55.847
Krypton	Kr	36	83.80
Lanthanum	La	57	138.9055
Lawrencium	Lr	103	[262.1098]
Lead	Pb	82	207.2
Lithium	Li	3	6.941
Lutetium	Lu	71	174.967
Magnesium	Mg	12	24.3050
Manganese	Mn	25	54.938049
Meitnerium	Mt	109	[266.1378]
Mendelevium	Md	101	[258.0984]
Mercury	Hg	80	200.59
Molybdenum	Mo	42	95.94
Neodymium	Nd	60	144.24
Neon	Ne	10	20.1797
Neptunium	Np	93	[237.0482]
Nickel	Ni	28	58.6934
Niobium	Nb	41	92.90638
Nitrogen	N	7	14.00674
Nobelium	No	102	[259.1011]
Osmium	Os	76	190.23
Oxygen	O	8	15.9994
Palladium	Pd	46	106.42
Phosphorus	P	15	30.973761
Platinum	Pt	78	195.078
Plutonium	Pu	94	[244.0642]
Polonium	Po	84	[208.9824]
Potassium	K	19	39.0983
Praseodymium	Pr	59	140.90765
Promethium	Pm	61	[144.9127]
Protactinium	Pa	91	231.03588
Radium	Ra	88	[226.0254]
Radon	Rn	86	[222.0176]
Rhenium	Re	75	186.207
Rhodium	Rh	45	102.90550
Rubidium	Rb	37	85.4678
Ruthenium	Ru	44	101.07
Rutherfordium	Rf	104	[261.1089]
Samarium	Sm	62	150.36
Scandium	Sc	21	44.955910
Seaborgium	Sg	106	[263.1186]
Selenium	Se	34	78.96
Silicon	Si	14	28.0855
Silver	Ag	47	107.8682
Sodium	Na	11	22.989770
Strontium	Sr	38	87.62
Sulfur	S	16	32.066
Tantalum	Ta	73	180.9479
Technetium	Tc	43	[97.9072]
Tellurium	Te	52	127.60
Terbium	Tb	65	158.92534
Thallium	Tl	81	204.3833
Thorium	Th	90	232.0381
Thulium	Tm	69	168.93421
Tin	Sn	50	118.710
Titanium	Ti	22	47.88
Tungsten	W	74	183.84
Uranium	U	92	238.0289
Vanadium	V	23	50.9415
Xenon	Xe	54	131.29
Ytterbium	Yb	70	173.04
Yttrium	Y	39	88.90585
Zinc	Zn	30	65.39
Zirconium	Zr	40	91.224

*Atomic masses, to greatest known significant figures, are taken from *Pure and Appl. Chem. 68* 2340 (1996). A bracketed value denotes the mass of the longest-lived isotope.

Essentials of

GENERAL, ORGANIC, & BIOLOGICAL CHEMISTRY

MELVIN T. ARMOLD
Adams State College

HARCOURT COLLEGE PUBLISHERS

Fort Worth Philadelphia San Diego New York Orlando Austin San Antonio
Toronto Montreal London Sydney Tokyo

Publisher: Emily Barrosse
Publisher/Acquisitions Editor: John Vondeling
Marketing Strategist: Pauline Mula
Developmental Editor: Sandra Kiselica
Project Editor: Theodore Lewis
Production Manager: Charlene Catlett Squibb
Art Director: Paul Fry

Cover legend: Digitalis, an extract of the leaves of the foxglove plant, has long been used in medicine to treat congestive heart failure. The structure shown is that of digitalin, one of several active ingredients in digitalis. *(Larry Lefever and Jane Grushow/Grant Heilman Photography, Inc.)*

Essentials of General, Organic, & Biological Chemistry
ISBN: 0-03-005648-9
Library of Congress Card Number: 00-106151

Address for domestic orders:
Saunders College Publishing, 6277 Sea Harbor Drive, Orlando, FL 32887-6777
1-800-782-4479
e-mail collegesales@harcourt.com

Address for international orders:
International Customer Service, Harcourt, Inc.
6277 Sea Harbor Drive, Orlando, FL 32887-6777
Phone (407) 345-3800   Fax (407) 345-4060
e-mail hbintl@harcourt.com

Address for editorial correspondence:
Saunders College Publishing,
Public Ledger Building, Suite 1250, 150 S. Independence Mall West,
Philadelphia, PA 19106-3412

Web Site address: http://www.harcourtcollege.com

Printed in the United States of America

9012345678   048   10 987654321

What Your Colleagues Are Saying About

ARMOLD'S

Essentials of

GENERAL, ORGANIC, & BIOLOGICAL CHEMISTRY

DONALD WILLIAMS, Hope College

"The coverage is great. The order and the coverage are what I want in my one-semester course."

"The end-of-chapter questions are one of the strengths. There are enough and some of them are great."

"I would advertise that the book is direct and to the point and that it assumes little of the students but moves to a good discussion of metabolism."

SHARON KAPICA, County College of Morris

"The coverage seems well planned. It covers the topics students need as background in allied health fields, without covering every topic a general chem or organic chemistry student needs. The length is appropriate."

"The writing style is very simple—very clear throughout."

"Overall the material is up to date and accurate. The chapter on metabolism was wonderful for me to read—cogent, clear, nothing extraneous."

"I very much like the large number of problems, their variety and types. This is a real strength of the text."

KEVIN GRATTON, Johnson County Community College

"This book will compete with Timberlake, which is the text we now use. It is appropriate for our course. I would consider using this book."

OWEN McDOUGAL, Southern Oregon University

"I think the author does a nice job of relating the content of the chapters to topics that students can relate to in their everyday lives."

"The end-of-chapter summaries do a nice job of incorporating many of the key terms into a clear and understandable overview. The summaries of reactions are kept brief and to the point. I think this is an attribute."

"Students of variable scientific background will not only be able to read this book, they will be able to relate to it."

"There are sufficient problems covering each section of the text so that even the best students will have no problem keeping busy."

"There are connections to topics that are current and understandable."

LINDA WILSON, Middle Tennessee State University

"The level is appropriate for my introductory chemistry student. The material is written in an understandable fashion and good examples are given."

"The writing style is very readable, almost conversational. The students should have no trouble understanding it."

This textbook is dedicated to the memory of

Professor Veryl Keen—

colleague, educator, gentleman.

About the Author

Melvin T. Armold is a professor at Adams State College in Colorado, where he is well known for his high expectations in introductory and general chemistry courses. Dr. Armold received his PhD in biochemistry from Purdue University in 1974. He then taught both biology and chemistry and conducted biochemical research at several colleges and universities before settling at Adams State College in 1985. At ASC, he also has served in several administrative roles, including interim dean; however, his primary professional interest is chemical education. He, his wife Anita (a high school chemistry teacher), and their sons Adrian and Jeremiah enjoy many outdoor activities. Mel and Anita take an annual trip to northern Canada to experience wilderness fishing and canoeing.

Mel Armold at his favorite sport catches a grayling.

Preface

Audience

Essentials was written for students taking a broad introductory course in general, organic, and biological chemistry in a limited time. These students are often preparing for careers in allied health, such as nursing and various fields of therapy, but students in many other fields of study are often enrolled in this introductory chemistry course as well. This book assumes no background in chemistry, and students taking this course need only basic math skills.

(Charles D. Winters)

Objectives

This textbook has three primary objectives: (1) to present only that chemistry that is *essential* to the students' needs. The choice of material and depth of coverage was greatly influenced by the comments and suggestions of reviewers who have considerable experience in teaching general, organic, and biological chemistry; (2) to provide students with many examples of chemistry in health care and other applications to real life; and (3) to contribute to the broad educational experience of students by fostering analytical thinking and problem-solving skills.

Content

The book includes 17 chapters of general, organic, and biological chemistry. Chapters 1 through 7 provide an introduction to general principles of chemistry, including measurements, atomic structure, chemical bonding, naming compounds, states of matter, solutions, and chemical reactions. These chapters include applications in health care and build a strong foundation for organic and biological chemistry. Chapters 8 through 11 cover those aspects of organic chemistry that both directly affect health care and develop a solid base for understanding biochemistry. These chapters include a study of saturated and unsaturated hydrocarbons, as well as ten other common organic functional groups. Chapters 12 through 17 cover the basics of biochemistry that are so important in health care and that provide an understanding of how our bodies work. In these chapters biomolecules, such as carbohydrates, lipids, proteins, enzymes, and nucleic acids are covered. Students are also introduced to the metabolic pathways that provide the energy and molecules needed for life.

Features

Because no chemistry background is assumed in *Essentials*, the textbook includes many features designed to facilitate student learning.

- **Chapter Outline and Objectives** initially focuses the students' attention on the organization and content of the chapters. The outline gives an overview of what is to come and the objectives are keyed to each section of a chapter.
- **Examples** and **Self-Tests,** which facilitate learning and understanding, give students the opportunity to practice solving problems and check their mastery of the material. The worked examples have complete solutions to the problems, whereas the self-tests have answers only.
- **Connections** are essays on special topics, which extend the student's understanding of chemistry from the classroom and science lab to health care and everyday life.
- **Marginal Notes** highlight topics covered in the textbook or provide interesting asides.
- **Key Terms**—boldfaced in the book and listed at the end of the chapter—emphasize the words and expressions that are an essential part of chemistry.
- **Chapter Summaries** review, in order of presentation, the main concepts and many of the key terms of the chapter.
- **Summary of Organic Reactions** provides a quick review of the organic reactions at the ends of Chapters 8 to 11.
- **Exercises** at the end of each chapter provide many additional opportunities to complete and master understanding. Answers to the odd-numbered problems can be found at the end of the book.
- **CD-ROM icon** alerts students to use the enclosed **Harcourt Interactive General, Organic, & Biological Chemistry** CD-ROM to enhance and extend their knowledge of key topics.
- **Appendix on Scientific Notation** is provided for students needing a refresher on expressing numbers in this format.
- **Glossary and Index** are included for the student's convenience and are located at the back of the book.

(Charles D. Winters)

Although many features have been included in the text, the size of the text has been deliberately kept small because of the limited time the student and instructor has for this introductory course.

Ancillaries

A complete integrated learning package is provided with this book:

- ***Student Study Guide and Solutions Manual,*** by Owen McDougal (Southern Oregon University), contains complete solutions to one half of the end-of-chapter problems, as well as review of all key concepts in the chapter and a list of all definitions of key terms. Most important, this guide contains a set of new problems with answers for each chapter so that students can more actively gauge their mastery of the material.
- ***Laboratory Experiments for General, Organic, and Biochemistry,*** 4th edition, by Frederick Bettelheim and Joseph Landesberg (Adelphi University), contains 52 experiments to illustrate the important concepts and principles taught in this course. Many experiments in this manual have been revised to miniscale use of chemicals for environmental concerns and economic reasons. The large number of experiments allows sufficient flexibility for the instructor.

- **Harcourt Interactive General, Organic, and Biological Chemistry CD-ROM** is a high-quality, interactive, dual-platform CD-ROM created by William Vining (University of Massachusetts). It features tutorials in general chemistry, practice in visualizing organic molecules, and biochemical animations. This CD comes free of charge with each textbook.
- **Harcourt GOB Web Site** is an interactive site of extra practice problems and teaching and learning exercises.
- ***Instructor's Manual and Test Bank,*** by Melvin T. Armold, contains answers to the even-numbered end-of-chapter exercises, chapter highlights, suggested course outlines, and numerous test questions to use as quizzes, homework assignments, or exams.
- **ExaMaster™ Computerized Test Bank** is the software version of the printed test bank. Instructors can generate many versions of tests and quizzes in multiple-choice format. A command reformats the multiple-choice question into a short-answer question. New problems can be added, existing problems modified, and graphics can be incorporated. ExaMaster™ has grade-book capabilities for recording and graphing students' grades.
- **Instructor's Resource CD-ROM** provides imagery from the text. Available as a presentation tool, this CD-ROM can be used in conjunction with commercial presentation packages, such as PowerPoint™, Persuasion™, and Podium™, as well as with Saunders LectureActive™ presentation software. Available in both Macintosh and Windows platforms.
- **Overhead Transparency Acetates:** A set of 125 full-color transparency acetates of key figures and tables from this text.

(Greg Vaughn/Tony Stone Images)

Saunders College Publishing, a division of Harcourt College Publishers, may provide complimentary instructional aids and supplements or supplement packages to those adopters qualified under our adoption policy. Please contact your sales representative for more information. If as an adopter or potential user you receive supplements you do not need, please return them to your sales representative or send them to

Attn: Return Department
Troy Warehouse
465 South Lincoln Drive
Troy, MO 63379

Acknowledgments

I want to thank all of the reviewers who helped with the development of *Essentials*. Their broad experience in teaching introductory chemistry and their commitment to teaching excellence honed the comments, suggestions, and constructive criticism they provided me. A special thanks goes to Professors Deborah and Thomas Nycz for reading all galleys and page proofs with an eye for accuracy and to Professor Brian Lipscomb for checking the accuracy of the artwork.

Tim Burch, *Milwaukee Area Technical College*
Ana Ciereszko, *Miami Dade Community College, Kendall Campus*
Csilla Duneczky, *Johnson County Community College*
Kevin Fujita, *Santa Rosa Junior College*
Kevin Gratton, *Johnson County Community College*
Martin Jones, *Adams State College*
Sharon Kapica, *County College of Morris*
Joanne Lin, *Houston Community College, Central College*
N. T. Lipscomb, *University of Louisville*
William Masterton, *University of Connecticut*
Owen McDougal, *Southern Oregon University*
Deborah Nycz, *Broward Community College, Central Campus*
Thomas Nycz, *Broward Community College, North Campus*
Robert Smith, *Skyline College*
Donald Williams, *Hope College*
Linda Wilson, *Middle Tennessee State University*
Pamela Zelmer, *Miami Dade Community College, North Campus*

(Charles D. Winters)

The staff at Saunders College Publishing and our consultants have been most helpful and supportive during the development of this textbook. Special thanks to vice president and publisher John Vondeling for his acceptance and support of this project; senior developmental editors Sandi Kiselica and Beth Rosato for their commitment and professional expertise; consulting editor Irene Nunes for her fine editing on an early version of the manuscript; Professor Peter Krieger for his help on the Connections and Exercises; project editors Ted Lewis and Bonnie Boehme for their professional and cheerful management of the production phase of the project; art and layout staff Paul Fry and Karen Gloyd; photoresearcher Dena Betz; and photographer Charlie Winters for his many fine photographs.

Special thanks to Professors John Amend and Brad Mundy for getting me started in all this. Thanks to the Adams State College chemistry faculty and students for their help and patience. And finally, thanks to the family—Anita, Adrian, and Jeremiah—for their patience and understanding. As I look back on all this, I realize that Jeremiah probably does not remember a time when his dad wasn't working on a book.

Melvin T. Armold
September, 2000

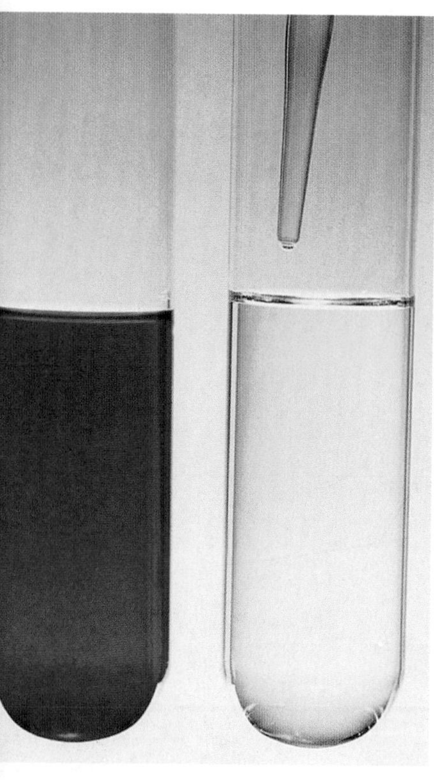

(Charles D. Winters)

Contents Overview

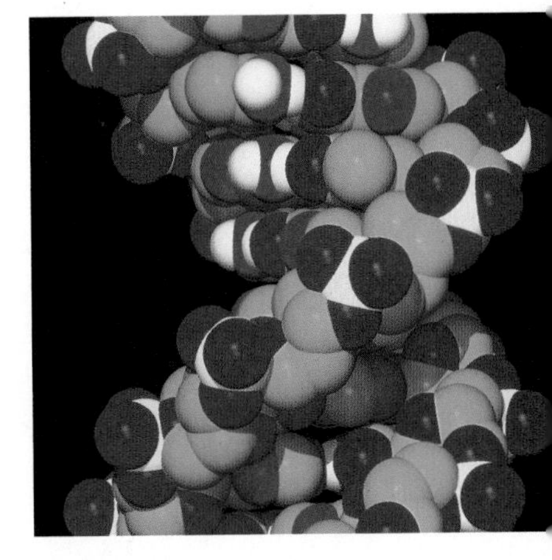

(Professor K. Seddon and Dr. T. Evans, Science Photo Library/Photo Researchers, Inc.)

Contents

(Dr. Jeremy Burgess/Science Photo Library/Photo Researchers, Inc.)

(Paul Boisvert, FPG International)

Essentials of

GENERAL, ORGANIC, & BIOLOGICAL CHEMISTRY

1 The Principles and Tools of Science

SCIENTIFIC STUDIES PROVIDE US WITH AN UNDERSTANDING OF THE WORLD AND UNIVERSE.

THE UNIVERSE consists of matter and energy. Earth is a tiny part of the universe, and except for energy from the sun, all of the matter and energy that is available to us exists on Earth. Science provides the basic knowledge of the world and universe, and technology uses this knowledge to provide useful tools, materials, and methods. Health care relies heavily on science for knowledge about human health and on technology for the modern medical devices and procedures that are now so common in health care. This chapter begins with an introduction to some methods and measurements that are used in science and technology and then explores some basic properties of matter and energy.

Outline

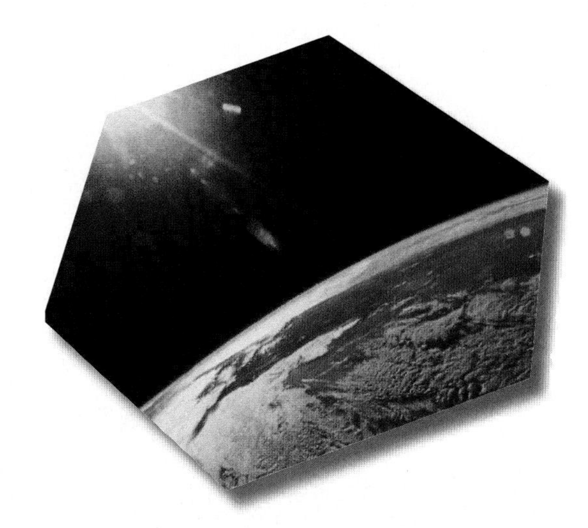

Except for the energy that arrives on Earth as sunlight, all of the matter and energy available to us is already on Earth. It is important that we use our limited resources wisely. *(NASA/Science Source/Photo Researchers)*

1.1 The Scientific Method

There are many ways to view or study the world around us. One very useful way is the **scientific method,** which uses known information, statements of explanation, and experiments to gain understanding. First, the scientific method *begins* with a thorough (ideally, complete) search for all known information about the topic of interest. Specific factual information is called **data,** and data gathering is often very time-consuming. The data are then studied to find patterns and gain understanding.

Second, a scientist attempts to provide an *explanation* for the topic of interest that is valid for the gathered data. A statement that explains a known set of data is called a **hypothesis.** In addition to explaining known facts, a useful hypothesis also makes predictions that can be tested by experimentation.

Third, a scientist uses *experiments* to test the hypothesis. If the results of these experiments are consistent with the predictions of the hypothesis, then the hypothesis gains support. If the experimental data are not consistent with the hypothesis, then the hypothesis must be either rejected or modified. If modified, additional experiments are carried out to test the modified hypothesis. With time, either a hypothesis becomes an accepted explanation of a phenomenon or it is rejected. A hypothesis can never be proved correct; experimentation can only show it to be acceptable or invalid. Consider this example: A physician and microbiologist hypothesize that a certain set of symptoms in a patient are *caused* by a specific bacterium. If each additional experiment shows that the bacterium is present whenever

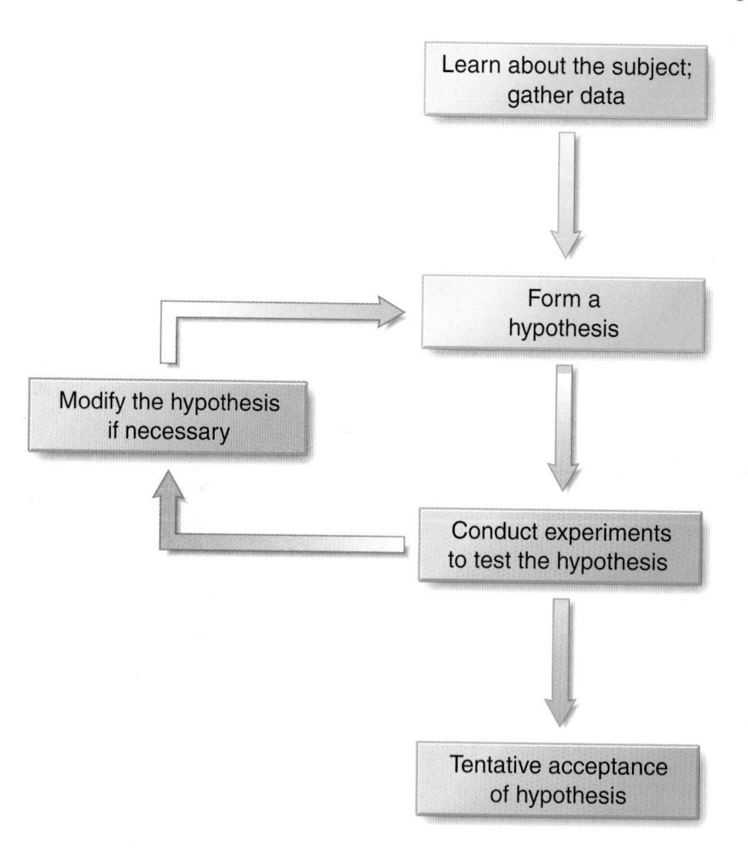

Figure 1.1 The steps of the scientific method. Although this diagram suggests the steps are independent of each other, there is often overlap.

a patient has this set of symptoms, then the hypothesis gains support. But if a patient shows these symptoms and the bacterium is absent, then the hypothesis is invalid. Because a hypothesis can never be proved correct, a scientist never *believes* a hypothesis; instead a hypothesis is treated as an accepted explanation until a better one (perhaps) replaces it. Figure 1.1 summarizes the steps of the scientific method.

The scientific method is not limited to scientists in research laboratories; we use it in everyday life as well. A weather forecaster uses existing data and trends to predict tomorrow's weather. (Tomorrow, observations of the weather will provide a test of the forecaster's hypothesis.) An experienced baseball pitcher decides which pitch or sequence of pitches to use (forms a hypothesis) based on a hitter's past performance and tendencies. The hitter's performance on that pitch either reinforces the pitcher's hypothesis or causes a change in strategy. The scientific method is also an integral part of health care, as is shown by the similarities between a hypothesis and a diagnosis (see Connections below).

There are some related scientific terms that you may read or hear. A **theory**, like a hypothesis, explains a set of data. However, theories have more established credibility and are more broadly accepted than hypotheses. A theory generally emerges from one or more accepted hypotheses. The theory of evolution, for example, has broad (though no universal) acceptance, and it explains many aspects of biology, geology, and biochemistry. A **law** is a generalization or summary that is based on consistent experimental results or experiences. A law summarizes a set of data, but it does not explain why or how. Consider gravity. When an object is dropped, it always falls to the ground. Three centuries ago, Sir Isaac Newton developed several laws that summarize gravity and the attraction of objects. These laws accurately summarize the gravitational attraction between two objects, but they do not explain how or why gravity works.

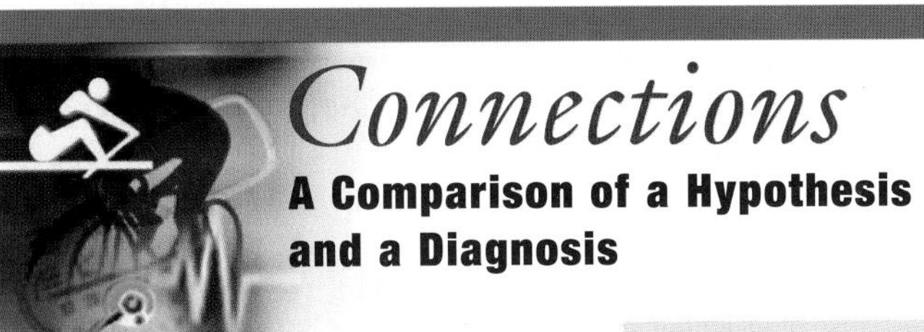

Connections

A Comparison of a Hypothesis and a Diagnosis

The scientific method is used in health care as well as in science. A physician begins with a review of the patient's history, an examination, and perhaps some laboratory tests. All of this information (data) allows the physician to evaluate the patient's past and present medical condition. Based on these data, the physician forms an explanation for the condition. A scientist would call this a hypothesis, but in medicine it is called a **diagnosis.** Just as a hypothesis predicts the results of experiments, a diagnosis provides the basis for (predicts) the best course of treatment.

Once the diagnosis is made and the treatment administered, the patient's medical progress is noted. If the patient gets better, then the diagnosis was probably correct. If the patient does not respond to the treatment, then the diagnosis was probably incorrect. If the diagnosis was incorrect, the results of the treatment are included with the previous data about the patient, new laboratory tests are performed if necessary, and the diagnosis is modified or abandoned for a new diagnosis. A new treatment is administered, and the results are again evaluated.

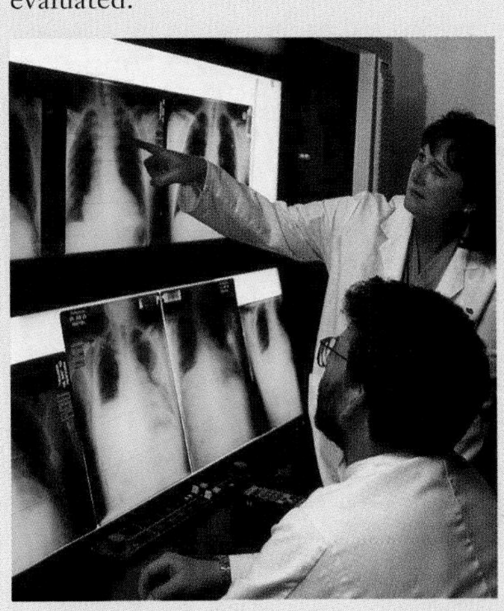

Ben Edwards/Tony Stone Images

1.2 Measurements and Units

Measurements are observations that provide numerical information about a subject or topic. When you measure something, you gather quantitative data, which generally provide greater understanding than qualitative data. A patient may look overweight (a qualitative assessment), but you will know this for certain only after you measure the patient's weight, height, and percentage of body fat (quantitative assessments). Scientists and health care professionals rely heavily on measurements (Figure 1.2).

Figure 1.2 A nurse measures out the proper dose of a drug. Note that the flow of intravenous solution in the background is also being delivered to the patient in measured quantities. *(Larry Mulvehill/Science Source/Photo Researchers).*

Measurements describe a variety of physical properties, including length, volume, mass, and time. A measurement contains a *number* and a *unit*. A **unit** specifies the physical property and size of the measurement, and the number indicates how many units are present (Figure 1.3). For example, quarts and gallons are units for volume, but they specify different sizes. When you read "three gallons", you get a very different mental picture of volume than when you read "three quarts".

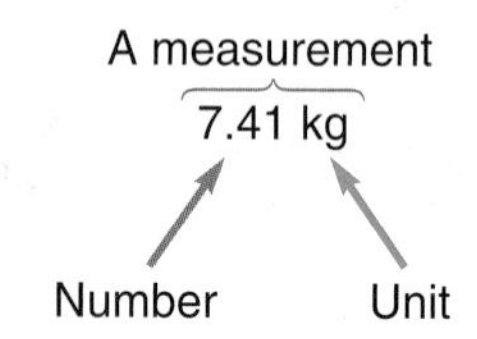

Figure 1.3 A measurement contains a number and a unit.

In the past, many systems of units have been used in the world. At the present time, three systems are in common use. The United States primarily uses the English (British) system; the English units of pounds, feet, and quarts are familiar to you. Most of the world uses the metric system; the gram, meter, and liter are common metric units. For the most part, the scientific community uses the **Système International (SI),** which is a modification of the metric system. Table 1.1 summarizes the standard units of SI and the metric system and compares them with units of the English system.

The term International System is also used.

Units for Mass

Mass is a measure of how much material (matter) is present in an object. The standard SI unit for mass is the **kilogram (kg),** which is equivalent to 2.205 lb (Figure 1.4). The metric system uses the **gram (g)** as its standard unit for mass. The gram is 1/1000 of a kilogram, and there are 28.4 g to an ounce. Both the gram and the kilogram are commonly used in science and health care.

"Mass" is often confused with "weight". Mass is a measure of the amount of material present in an object, which does not change with location. Thus an object's mass is the same everywhere. Balances are devices that are used to measure mass. A balance compares the unknown mass of an object to known masses, as

Table 1.1 Some Common Units of Measurement

MEASUREMENT	SYSTÈME INTERNATIONAL (SI)	METRIC SYSTEM	APPROXIMATE EQUIVALENTS IN THE ENGLISH SYSTEM
Mass	kilogram (kg)	gram (g)	1 kg = 2.205 lb; 28.4 g = 1 oz
Length	meter (m)	meter (m)	1 m = 1.094 yd
Volume	cubic meter (m^3)	liter (L)	1 m^3 = 264 gal; 1 L = 1.057 qt
Time	second (s)	second (s)	—
Temperature	kelvin (K)	Celsius (°C)	Fahrenheit (°F)

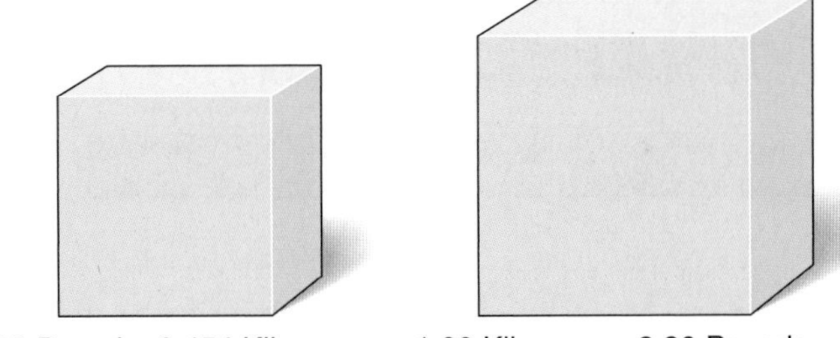

Figure 1.4 Two volumes of water are used to compare the size of one pound and one kilogram.

shown in Figure 1.5 (a, b). **Weight** is a measure of the gravitational pull on an object. A man weighing 168 lb on Earth would weigh 28 lb on the moon because gravity is about one sixth as large on the moon as on Earth. Scientists use mass rather than weight because mass does not change with location.

Weight is often measured with a spring scale, such as those used in some grocery stores [Figure 1.5(c)]. The gravitational force of attraction between the object and Earth pulls the spring down. Although they are different, the terms "mass" and "weight" are often used interchangeably. This interchanging works because mass and weight are directly proportional to each other if gravity does not change. (By the way, the verb "weighing" is commonly used to describe either the act of measuring the mass or the weight of an object.)

Units for Length and Volume

The **meter (m)** is both the SI and metric standard unit for length. A meter is approximately 1.1 yd (Figure 1.6). The English system uses several standard units; inches, feet, yards, and miles are common ones.

A commonly used unit for length is the centimeter, which is 1/100 of a meter; 2.54 cm = 1 in.

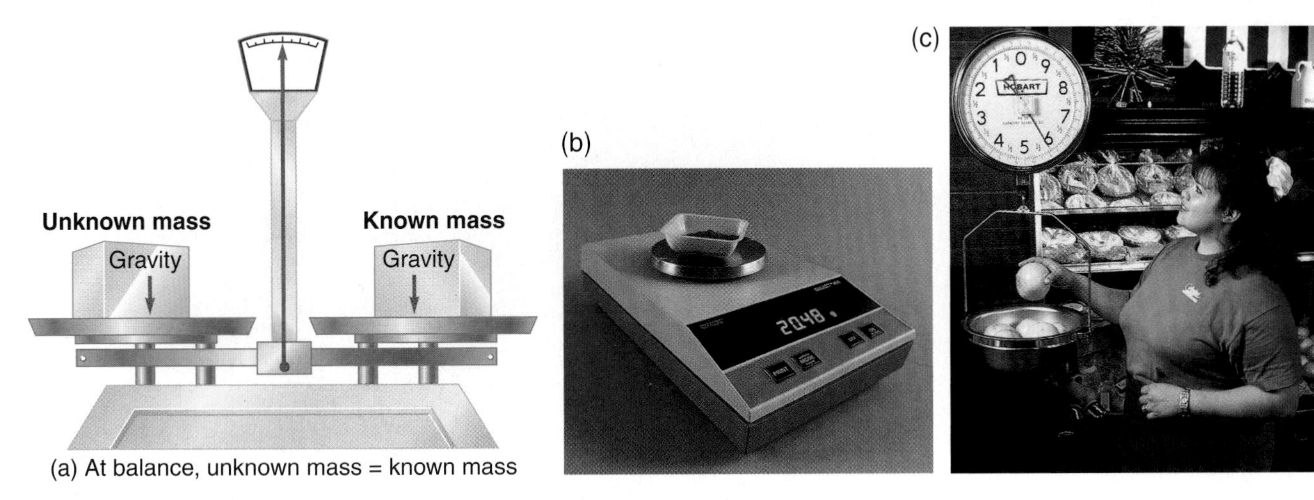

Figure 1.5 (a) Mass is determined by comparing the unknown mass of an object to a known mass. (b) Modern balances are fast, accurate, and easy to use. (c) Spring scales can only measure weight. [(b) © *1986 by SIU/Peter Arnold, Inc.*; (c) *Charles D. Winters*].

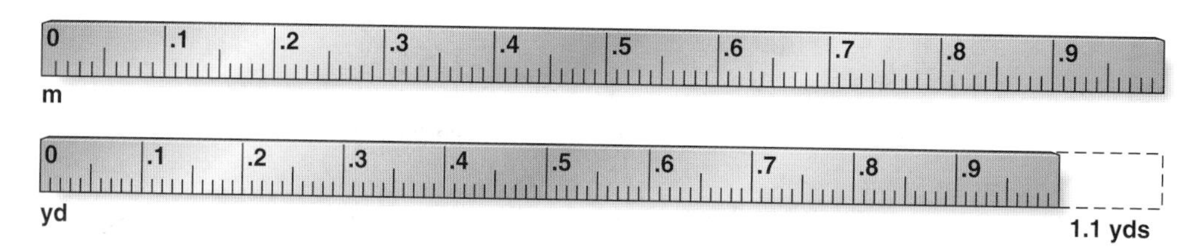

Figure 1.6 One meter is equal to 1.09 yards.

These devices are commonly used to measure volumes and properties of liquids. *(Charles D. Winters)*

The units for volume are derived from the length units in the metric system and SI. Volume is the cube of length, for example: 1 m × 1 m × 1 m = 1 m^3 (1 cubic meter). In SI, the cubic meter is the standard unit for volume. This is a very large volume, roughly 264 gal. Because the cubic meter is too large for many scientific uses, the standard metric unit for volume, the **liter (L),** is commonly used. One liter is 1/1000 of a m^3 and a bit larger than a quart (Figure 1.7). Volume units in the English system are arbitrary units; thus there is no relationship between fluid ounces, quarts, and gallons and the English length units. To refresh your memory, there are 32 fl oz per quart; 4 qt per gal.

Units for Time and Temperature

The **second** is the standard unit of time in all three systems. Of course, for longer periods of time, other units, such as minutes, hours, days, and years, are used.

The units for temperature are different in all three systems. The Celsius degree is the metric unit, the kelvin is the SI unit, and the Fahrenheit degree is the English unit (Table 1.1). You will learn more about temperature scales in Section 1.7.

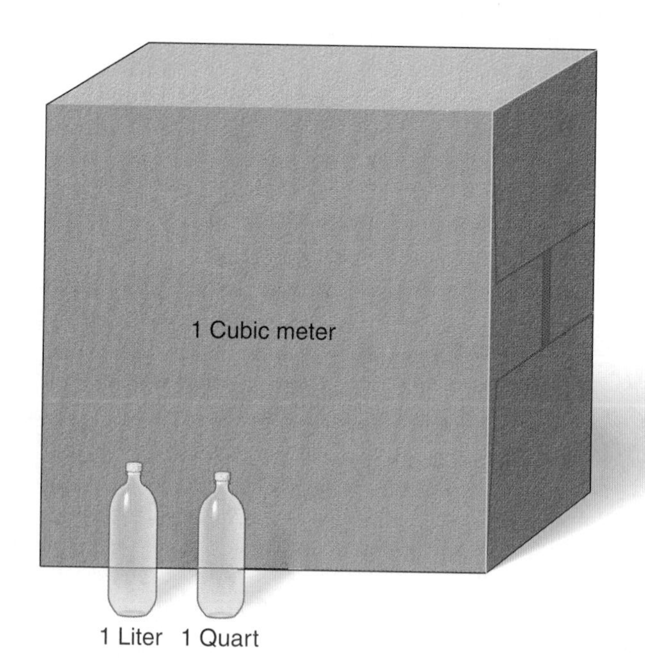

Figure 1.7 One liter equals 1.06 quarts. The cubic meter is a much larger unit of volume (1000 L or 264 gal).

Example 1.1 Common Units in Science and Health Care

1. What is the standard unit for volume in SI? In the metric system?
2. The liter is most like what unit in the English system?
3. A baby has a mass of 3.0 kg. This mass corresponds to a weight of how many pounds?

SOLUTIONS

1. The standard units for volume are the cubic meter in SI and the liter in the metric system.
2. The liter is just a bit bigger than a quart (1 L = 1.06 qt).
3. Because a kilogram is about 2.2 times greater than a pound, the baby would weigh $2.2 \times 3.0 = 6.6$ lb.

   Note: A more general method for solving problems that have measurements will be explained in Section 1.5.

Self-Test

1. What is the standard unit for length in SI and in the metric system?
2. What is the temperature unit in SI?

ANSWERS

1. The meter is used in both SI and the metric system. 2. Kelvin

Combining Prefixes with Units

Measurements can vary greatly in size. An astronomer measures distances between planets or between galaxies, microbiologists deal with the sizes of viruses and bacteria, and a nurse may measure the length of a newborn baby. If a single unit were used for each of these measurements, some would require a very large number, and others would be very small (Figure 1.8). When measurements are very large or very small, the standard units become inconvenient. We could use many different units for a measurement (fluid ounce, cup, quart, gallon, and barrel are examples from the English system), but a simpler method is used with metric and SI units. The value of a standard unit is changed with *prefixes*.

You were exposed to the use of prefixes when you learned that the kilogram is the standard SI unit of mass. The word "kilogram" consists of a prefix, *kilo-*, and a unit, *-gram*. A **prefix** changes a unit by a fixed value. "Kilo-" means "1000". When "kilo-" and "-gram" are combined, they form "kilogram", which means "1000 grams". Another example is the word "kilometer", which is 1000 meters. *Milli-* is a prefix that means 1/1000 (0.001). Thus a "milligram" (mg) is 0.001 gram. *Milli-* plus *liter* yields **milliliter (mL),** which is 0.001 liter. By the way, health care professionals often use the unit **cubic centimeter (cc** or **cm**^{3}) for volume, and 1 cc equals 1 mL. Table 1.2 includes several prefixes that you will see in common use.

There are two ways to use the information contained in prefixes. Consider the term "centimeter". Because *centi-* means 1/100, 1 cm equals 0.01 m. This relationship can also be viewed in another way. A centimeter is 1/100 of a meter; therefore, there must be 100 cm in a meter: 1 m = 100 cm. Both statements are equivalent, and when you do calculations that involve prefixes, you can use either one. Table 1.2 contains examples that express prefixes in both ways.

(a)

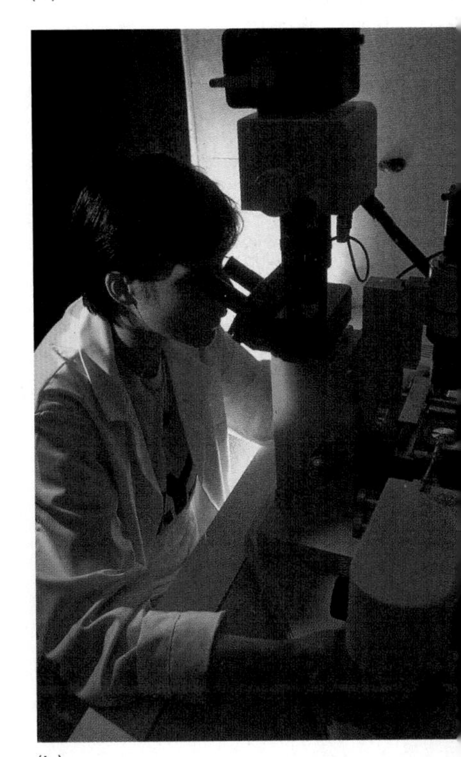

(b)

Figure 1.8 Large differences in size require standard units that are modified with prefixes. (a) A galaxy; (b) Scientist implanting DNA into a cell. [(a) *NASA/ Science Source, Photo Researchers;* (b) *Richard Nowitz/Phototake]*

Table 1.2 Some Common Prefixes for Measurements

PREFIX	ABBREVIATION	VALUE OF THE PREFIX[a]	EXAMPLES
mega (million)	M	1,000,000 (1×10^6)	1 Mg = 1×10^6 g or 1 g = 1×10^{-6} Mg
kilo (thousand)	k	1000 (1×10^3)	1 km = 1000 m or 1 m = 0.001 km
deci (tenth)	d	0.1 (1×10^{-1})	1 dL = 0.1 L or 1 L = 10 dL
centi (hundredth)	c	0.01 (1×10^{-2})	1 cm = 0.01 m or 1 m = 100 cm
milli (thousandth)	m	0.001 (1×10^{-3})	1 mg = 0.001 g or 1 g = 1000 mg
micro (millionth)	μ	0.000001 (1×10^{-6})	1 μm = 0.000001 m or 1 m = 1,000,000 μm
nano (billionth)	n	0.000000001 (1×10^{-9})	1 ns = 1×10^{-9} s or 1 s = 1×10^9 ns

[a]Scientific notation, which is another way to express large and small numbers, will be introduced in Section 1.4.

Example 1.2 Using Prefixes with Units

1. How many grams equal 1 kilogram?
2. One meter will contain how many centimeters?
3. Express one second in milliseconds.

SOLUTIONS

1. Because *kilo-* means "1000", a kilogram contains 1000 grams.
2. *Centi-* means "1/100", thus there are 100 centimeters in one meter.
3. *Milli-* means "1/1000", so one second could be written as 1000 milliseconds.

Self-Test

1. How many liters are in a deciliter?
2. How many micrograms are in a gram?
3. Compare a nanosecond with a second.

ANSWERS

1. 0.1 L 2. 1,000,000 μg 3. 1 ns = 1/1,000,000,000 s

1.3 Making Measurements

Errors and Variations

All measurements are subject to variation and error. One source of error lies in instruments and their use. If an instrument is not adjusted correctly or if it is used improperly, then the measurement it provides will be wrong. This type of error, which is called either a *systematic error* or a *determinate error*, is avoidable and should not happen. Trained professionals know how to use and adjust their instruments.

A second kind of error, an *indeterminate error*, is a result of random variation. Consider this example: A nurse may measure a patient's blood pressure as 120/80, but when the same nurse repeats the measurement a few moments later, the blood pressure is 118/82. The patient's blood pressure could have changed slightly, or the nurse may simply have observed (measured) it as a different value. Variations of this type are always present and cannot be eliminated. However, well-trained personnel make careful measurements that minimize these variations.

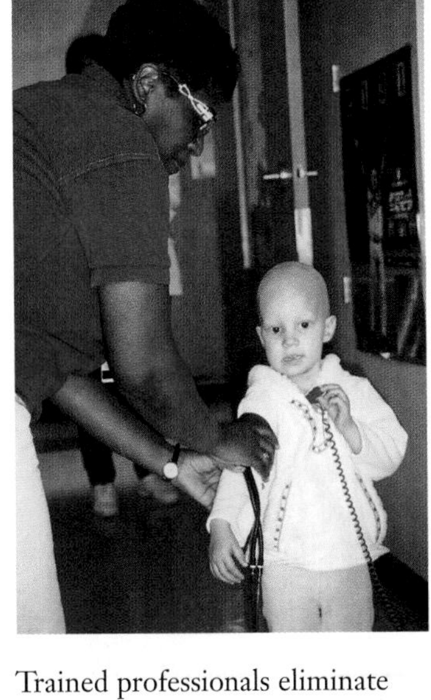

Trained professionals eliminate systematic errors and minimize random errors. Here a nurse takes the blood pressure of a young cancer patient. *(Brian DeWitt)*

Measuring Samples and Objects

Measurements are made with measuring devices, which, no matter how well-made, have limitations. Consider the graduated cylinders shown in Figure 1.9. The smaller (10-mL) one has marks for each 0.1 mL, but the larger (100-mL) one is marked only to the nearest 1 mL. Whenever scientists or technicians make a measurement, they estimate the measurement to one place beyond the marks on the

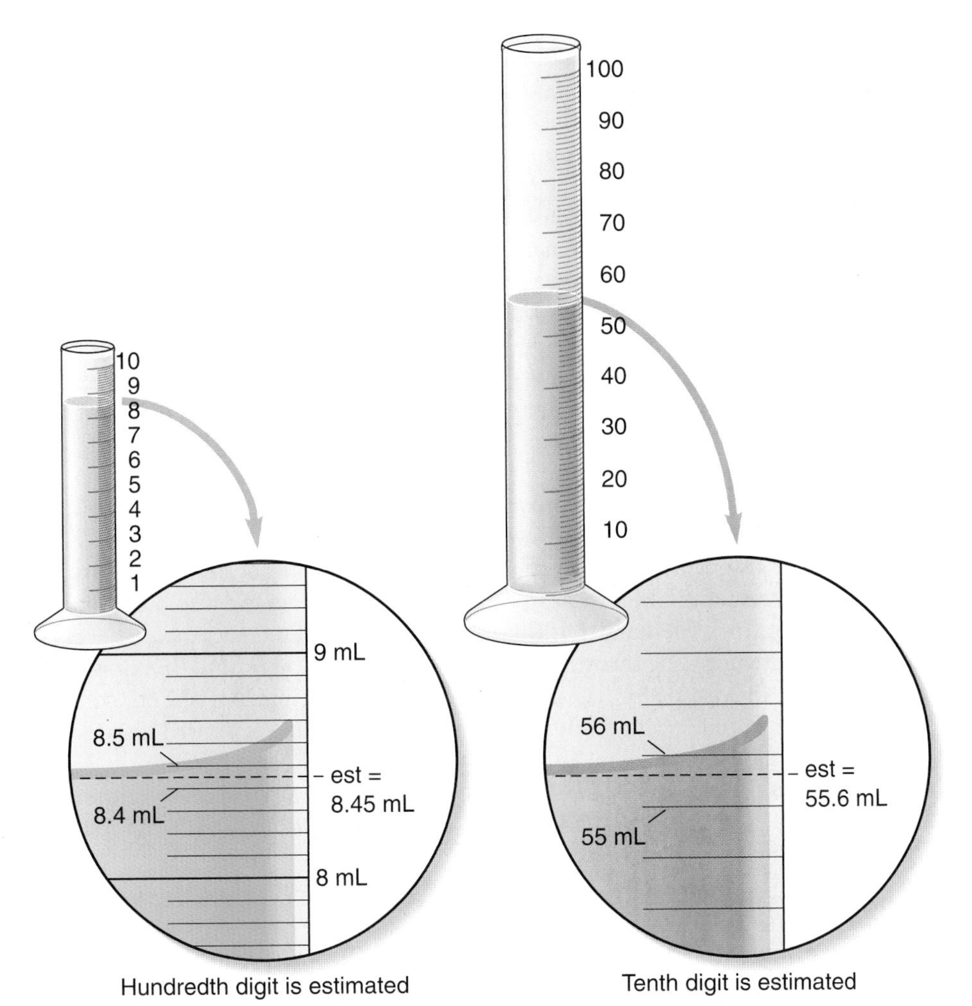

Figure 1.9 Graduated cylinders are used to measure volume. The volume in the smaller cylinder can be estimated to the hundredths place and the volume in the larger cylinder can be estimated to the tenths place. The measurement from the smaller (10-mL) cylinder is more precise.

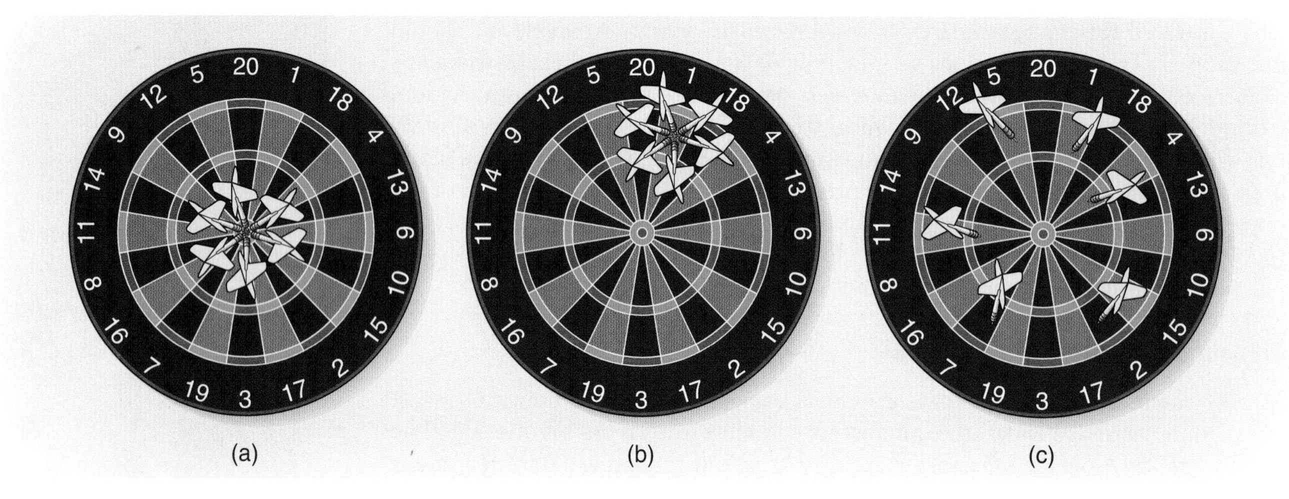

Figure 1.10 (a) The dart thrower was both accurate (darts are close to the center—that is, the "true value") and precise (darts are close together). (b) The dart thrower was precise but not accurate. (c) The thrower was neither precise nor accurate.

device. Thus the 10-mL graduated cylinder would be read (estimated) to the hundredths place (10.00), and the 100-mL cylinder would be read (estimated) to the tenths place (100.0). We do this because in most measurements the material being measured does not fall exactly on a mark. By estimating the last place, the measurement more closely reflects the true value.

Because the last place of a measurement is estimated, and because variation occurs during measuring, measurements are always uncertain. This uncertainty can be expressed in different ways. **Accuracy** is the term used to describe how close a measurement is to the true value of the property being measured. Consider a person aiming darts at the bull's-eye of a dartboard (Figure 1.10). The darts are all in the bull's-eye in Figure 1.10(a); thus for this set of darts, the thrower was accurate. If no systematic errors are made, and if random errors are kept to a minimum, a measurement should be an accurate reflection of the true value. Unfortunately, in real measurements the true value is usually not known (if it were known, why bother to measure it again?). As a consequence, we can usually say little about the accuracy of a measurement.

Precision also deals with uncertainty. **Precision** is an indicator of similarity in a set of measurements—that is, how closely the measurements agree. Look at Figure 1.10 again. In 1.10(a) and 1.10(b), the darts are clustered together, whereas in 1.10(c), they are not. The thrower showed good precision in (a) and (b) but was imprecise in (c). Precision also has another meaning in measurements. Let's consider the graduated cylinders again (Figure 1.9). If a small volume is poured into the 10-mL cylinder, the volume is measured to the hundredths place because the tenths place is marked on the cylinder. Three measurements might yield these values: 8.03 mL, 8.01 mL, and 8.00 mL. If the same volume is measured with the 100-mL cylinder, the measurements would be to the tenth place. Three trials might yield 8.0 mL, 8.1 mL, and 8.2 mL. The measurements from the 10-mL cylinder are said to be *more precise* than those from the 100-mL cylinder because the measurements from the 10-mL cylinder do not vary until the hundredths place, whereas those from the 100-mL cylinder vary in the tenths place.

Example 1.3 Making Measurements

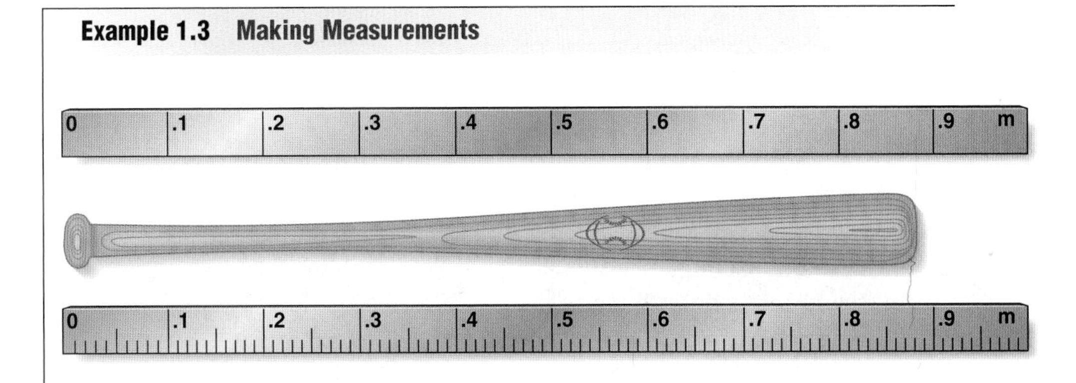

Determine the length of this bat using each ruler, starting with the top ruler.

SOLUTION

The top ruler is marked in 0.1-m intervals. The bat is longer than 0.8 m but shorter than 0.9 m. When making measurements, estimate one more place than the place marked on the device. Therefore, when the top ruler is used, the bat is about 0.88 m in length. (You may have estimated 0.87 or 0.89.) The lower ruler is calibrated to 0.01 m (1 cm), and the bat extends just beyond the 0.88 mark. Estimate one more place than shown on the device. Thus the bat is about 0.883 m (88.3 cm). (Again, your estimate may vary slightly from this value.)

Self-Test

1. Use each ruler in this figure to measure the length from the handle end of the bat to the top of the trademark.
2. Use a metric ruler to measure the length and width of your textbook cover. Express the answer in centimeters.

ANSWERS

1. Top ruler: 0.60 m; bottom ruler: 0.598 m   2. Approximately 21.6 cm × 25.4 cm

Note that the number of figures you include in your measurement will depend on the markings on your ruler. If the ruler is calibrated in millimeters, then your answer should be expressed to the nearest 0.1 mm.

Significant Figures in Measurements

When a measurement is made, all of the digits shown by the device plus the estimated digit are included in the measurement. All of these numbers are said to be **significant figures.** In Example 1.3, the top ruler gave measurements that contain two significant figures (0.88 m), and the bottom ruler yielded measurements that contain three significant figures (0.883 m). Note that the *number of significant figures* indicates the *precision of the measurement.* A measurement should not be reported with more or fewer numbers than were measured. If the mass of a newborn baby was measured as 3.42 kg, that mass should be reported in the baby's records to the nearest hundredth of a kilogram. It should not be reported as 3.4 kg or 3.420 kg.

Table 1.3 Rules for Determining Significant Figures in a Measurement

RULE 1 All nonzero digits are significant.	Example: 6.439 cm has four significant figures.
RULE 2 All zeros between two nonzero digits are significant.	Example: 104 s has three significant figures.
RULE 3 All zeros that follow a digit located to the right of the decimal point are significant. Zeros to the right of a decimal point that precede the first digit are not significant.	Example: 2.600 mL has four significant figures, but 0.00560 g has only three.
RULE 4 Zeros such as those in 10 or 200 may or may not be significant. (In this book, always assume the smallest number of significant figures in such cases; thus 400 has one significant figure.)	Example: 400 g may have one, two or three significant figures.

Unfortunately, because of our numbering system the number of digits shown in a measurement may not be the same as the number of significant figures in the measurement. This is because zeros in measurements may or may not be significant. A zero (a) may be a part of a measurement or (b) may merely be used as a placeholder in the number to give the correct size. Consider the measurement 100 cm. Were the zeros measured, or are they only indicating size and place? Put another way, does 100 mean that the measurement was made to the nearest centimeter (telling us that the length is closer to 100 cm than to 99 or 101), or to the nearest 10 cm (closer to 100 cm than 90 or 110), or the nearest 100 cm (closer to 100 cm than 0 or 200 cm)? Because significance in measurements is very important, rules for determining the number of significant figures have been established (Table 1.3).

Example 1.4 Determining the Number of Significant Figures

Use Table 1.3 to determine the number of significant figures in these measurements.

1. 487 cm 2. 3.20 kg 3. 1100 ms 4. 11.056 g 5. 0.00045 L

SOLUTIONS

1. Because all of the digits are nonzero, they are all significant; three.
2. The zero follows a number and the decimal point, therefore it is significant; three.
3. The zeros in this large number may or may not be significant. We assume the minimum; two.
4. Because the zero is between nonzero digits, it is significant; five.
5. Because the zeros precede the two digits, the zeros are not significant; two.

Self-Test

Determine the number of significant figures in these measurements:

1. 0.0406 L
2. 340,000 mg
3. 0.1450 kg

ANSWERS

1. Three
2. Two
3. Four

1.4 Calculations Involving Measurements

Significant Figures in Calculations

Measurements are used in calculations, thus significant figures must be considered in calculations as well as measurements. The rules for determining significant figures in calculations depend on the type of calculation. *For multiplication and division, the measurement containing the fewest significant figures determines the number of significant figures in the answer.* Consider this multiplication:

$$1.4\ \text{cm} \times 1.31\ \text{cm} \times 0.570\ \text{cm} = 1.04538\ \text{cm}^3 = 1.0\ \text{cm}^3$$

The answer before rounding contains six digits, but the measurement 1.4 cm has only two significant figures (the others each have three significant figures). Therefore, the answer should have only two significant figures, 1.0 cm^3.

Now consider this division:

$$\frac{245\ \text{m}}{26.03\ \text{s}} = 9.412216673\ \text{m/s} = 9.41\ \text{m/s}$$

The measurement 245 m has three significant figures, and 26.03 s has four. The answer must therefore be expressed only to three significant figures: 9.41 m/s.

For addition and subtraction, the calculated answer should end at the same place as the measurement that has the least precision. Look at this example for addition:

$$\begin{array}{r} 2.92\ \text{g} \\ +\ 21.5\ \text{g} \\ \hline 24.42\ \text{g} \end{array} = 24.4\ \text{g}$$

The number 2.92 is expressed to the hundredths place (precise to the hundredths place), but 21.5 is expressed (precise) only to the tenths place. Therefore, the answer can be expressed only to the tenths place, 24.4 g.

Subtractions are treated the same way:

$$\begin{array}{r} 1.908\ \text{cm} \\ -\ 0.55\ \text{cm} \\ \hline 1.358\ \text{cm} \end{array} = 1.36\ \text{cm}$$

The number 1.908 is expressed to the thousandths place, and the number 0.55 is expressed to the hundredths place. The answer must therefore be expressed to the hundredths place: 1.36 cm.

Table 1.4 Guide for Rounding Numbers

RULE 1 If the first digit beyond the last significant digit is a 4 or less, just drop it and all digits that follow it.	**EXAMPLE** If 23.7839 mL should be expressed to four significant figures, drop the 3 and 9 to get 23.78 mL.
RULE 2 If the first number beyond the last significant figure is a 5 or greater, increase the last significant digit by 1 and drop the rest.	**EXAMPLE** To three significant figures, the mass 12.1866 g would be expressed as 12.2 g.

When you do calculations, you often must round your answer to get the correct number of significant figures. (Note that the four calculations shown previously all involve rounding.) When rounding numbers use Table 1.4 as a guide; however, other rounding rules exist and your instructor may suggest a different set of rules.

Example 1.5 Significant Figures and Rounding

Perform the following calculations and determine the number of significant figures in each answer. Then round each answer to the proper number of significant figures.

1. 354.62 g ÷ 27.02 mL =
2. $\begin{array}{r} 11.88 \text{ mL} \\ +\ 21.7\ \ \text{mL} \\ \hline \end{array}$

SOLUTIONS

1. The division yields **13.12**435233 g/mL. Because the volume 27.02 mL has the fewer number of significant figures (four), the answer must also have four (shaded part of answer). The first digit beyond the fourth place is a 4. Therefore the 4 and all digits beyond it are dropped. The correct answer is 13.12 g/mL.
2. The volume 21.7 mL is measured only to the tenths place; thus the answer must be to the tenths place. This gives three significant figures: **33.5**8 mL. The digit beyond the third significant figure is an 8; thus the last significant digit must be rounded up and any remaining numbers dropped. The answer is 33.6 mL.

Self-Test

1. 44.56 cm × 0.0507 cm =
2. $\begin{array}{r} 5.277\ \text{g} \\ -\ 2.4938\ \text{g} \\ \hline \end{array}$

ANSWERS

1. 2.26 cm^2 2. 2.783 g

We have said that the numbers in measurements have some degree of uncertainty in them. This is not true for all numbers. Some numbers, called **exact numbers,** have no uncertainty. They are assumed to have an unlimited number of significant figures. A number that results from a count is exact. These are examples of counts that are exact numbers: there are 31 students in lab section 1; the woman in delivery just had triplets; there are $11 in his wallet. By the way, not all *counts* are really counts. The established population number of any city is always an estimate rather than an actual count.

Definitions within a system of units are another example of exact numbers. In the expressions "1 foot = 12 inches" and "1 m = 100 cm", the 1s, the 12, and the 100 are exact. Exact numbers are also used in many equations and expressions. When 100 is used to determine percentage, it is an exact number. *When an exact number is used in a calculation, it is assumed to have an unlimited number of significant figures.* It is our measurements, not exact numbers, that determine the number of significant figures in the answer to a calculation.

Scientific Notation

Often in science and technology a person must work with very large or very small numbers. Prefixes are one way to make these numbers more convenient to use (85,000 m = 85 km), but there is another way as well. **Scientific (exponential) notation** expresses numbers as a small number multiplied by a power of ten. One advantage of scientific notation is that large and small numbers are easily expressed and read (with a little experience, it is easier to read exponents than count zeros). The number 125 can be expressed as 1.25×10^2 (1.25×100 or $1.25 \times 10 \times 10$). Because 125 is quite easily read, scientific notation may or may not be a useful way to express it. On the other hand, consider Avogadro's number (more on this later), which is 602,000,000,000,000,000,000,000. This number is hard to understand written in the normal way, and it takes up lots of space. In scientific notation, it is written 6.02×10^{23}.

A second advantage of scientific notation relates to significant figures. The first part of any number written in scientific notation (the part to the left of the multiplication sign) is always written to the correct number of significant figures. The power-of-ten part of the notation indicates size but not significance.

This shows the number of significant figures.

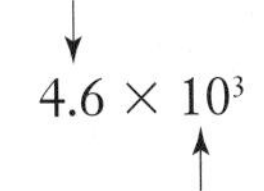

This shows size.

Consider the number 2470. Written in this way, we do not know whether the zero is or is not significant. But when the number is written in scientific notation, the number of significant figures must be shown in the first part of the number. The number 2.470×10^3 has four significant figures. With scientific notation, there is no need to guess how many digits are significant. Because the zero is shown, it must be significant. Appendix A contains a review of writing and reading scientific notation. If you are not familiar with this notation, please review it before continuing.

Example 1.6 Numbers and Scientific Notation

Review Appendix A if you are not familiar with scientific notation.

1. Convert to scientific notation: (a) 4500 (b) 0.00006703
2. Convert to regular numerical format: (a) 3.40×10^{-3} (b) 2.921×10^{4}

SOLUTIONS

1. (a) Using the assumption we make in this book, 4500 has two significant figures and is the same as 4.5×1000. Therefore $4500 = 4.5 \times 10^{3}$

   (b) 0.00006703 has four significant figures and is equal to 6.703×0.00001. Therefore, $0.00006703 = 6.703 \times 10^{-5}$
2. (a) 3.40×10^{-3} has three significant figures and is the same as 3.40×0.001, therefore it equals 0.00340 (b) 2.921×10^{4} has four significant figures and is the same as $2.921 \times 10{,}000$. It can be written as 29,210.

Self-Test

1. Write in scientific notation: 0.0042; 671,000
2. Change to regular format: 3.4×10^{5}; 2.779×10^{-4}

ANSWERS

1. 4.2×10^{-3}; 6.71×10^{5} 2. 340,000; 0.0002779

1.5 Dimensional Analysis

As we learned earlier, measurements consist of a number and a unit. There are occasions when the units of a measurement must be converted to a different unit, such as the conversion of milliliters to liters or the conversion of liters to quarts. There are other times when measurements are used in a calculation and the answer has different units than the original measurement. **Dimensional analysis** is a systematic approach for solving both conversions and calculations. In dimensional analysis, the number and unit of a measurement are multiplied by one or more conversion factors, yielding an answer that is correct for both the number and unit. The **conversion factors** used in dimensional analysis are ratios or fractions derived from definitions or equalities. The definition 1.000 kg = 2.205 lb can be written as two different conversion factors:

$$\frac{1.000 \text{ kg}}{2.205 \text{ lb}} \quad \text{and} \quad \frac{2.205 \text{ lb}}{1.000 \text{ kg}}$$

The first conversion factor says that there is 1.000 kg per 2.205 lb, and the second conversion factor says there are 2.205 lb per 1.000 kg. Although these two conversion factors contain the same information, they are not equivalent and they are not used interchangeably in dimensional analysis.

Let's illustrate the use of conversion factors in dimensional analysis by converting 7.50 lb (the weight of a really good trout or a typical baby) to kilograms.

Begin the conversion by writing down the given information and the units needed in the answer:

$$\text{Step 1} \quad 7.50 \text{ lb} \longrightarrow \text{kg}$$

Next write down the original measurement that we want to convert:

$$\text{Step 2} \quad \frac{7.50 \text{ lb}}{1}$$

To emphasize that "7.50 lb" is in the numerator of our original measurement, the "7.50 lb" is divided by 1 (any number divided by 1 has the same numerical value).

Next select the appropriate definition and write it as a conversion factor. For this conversion we need the definition 1.000 kg = 2.205 lb, which can be written as these two conversion factors:

$$\text{Step 3} \quad \frac{2.205 \text{ lb}}{1.000 \text{ kg}} \quad \text{or} \quad \frac{1.000 \text{ kg}}{2.205 \text{ lb}}$$

In the last step, multiply the original measurement by the conversion factor that has the units of lb in the denominator of the fraction:

$$\text{Step 4} \quad \frac{7.50 \text{ lb}}{1} \times \frac{1.000 \text{ kg}}{2.205 \text{ lb}} = 3.40 \text{ kg}$$

The lb in the numerator of the measurement cancels with the lb in the denominator of the conversion factor, leaving kg as the unit. When the numbers are multiplied and divided, the answer is 3.40 to the correct number of significant figures. Our conversion is complete: 7.50 lb converts to 3.40 kg.

Now consider what would happen if the incorrect conversion factor were used in the calculations:

$$\frac{7.50 \text{ lb}}{1} \times \frac{2.205 \text{ lb}}{1.000 \text{ kg}} = 16.5 \text{ lb}^2/\text{kg}$$

The desired unit is kg, but the units in the answer are lb^2/kg. This combination of units makes no sense, and our original units did not cancel out. Clearly the conversion was set up incorrectly, and the answer, both number and units, is wrong.

Dimensional analysis is excellent for conversions, but it also works well for calculations. Let's use dimensional analysis to solve this problem: A patient requires 1.5 L of an intravenous fluid over the course of a day. How many milliliters should be delivered each hour? Using Step 1 we write down the information we know and the units needed in the answer:

$$\text{Step 1} \quad 1.5 \text{ L per day} \longrightarrow \text{mL per hr}$$

Next (Step 2) we write down what we want to convert:

$$\text{Step 2} \quad \frac{1.5 \text{ L}}{1 \text{ day}}$$

In this calculation the numerator changes from L to mL, and the denominator changes from days to hours. Thus we must convert both liters to milliliters and days to hours. In Step 3 we need two conversion factors, one to convert L to mL and one to convert days to hours:

$$\text{Step 3} \quad \frac{1000 \text{ mL}}{1 \text{ L}} \quad \frac{1 \text{ day}}{24 \text{ hr}}$$

Only the correct conversion factors for this problem are shown in Step 3. In the final step we multiply our original term by the conversion factors to get the correct answer:

$$\text{Step 4} \quad \frac{1.5 \not{\text{L}}}{1 \not{\text{day}}} \times \frac{1000 \text{ mL}}{1 \not{\text{L}}} \times \frac{1 \not{\text{day}}}{24 \text{ hr}} = \frac{63 \text{ mL}}{1 \text{ hr}}$$

Because the units we wanted (mL/hr) are present in the answer, the solution has been set up correctly. The IV should deliver 63 mL of the medication per hour.

When you do your calculations by dimensional analysis, just remember to include both the number and units for all terms and to perform the appropriate calculations on both the numbers and the units. If you use the correct conversion factors, your answer will not only be numerically correct (unless you make a calculational error), it will also be expressed in the correct units.

Example 1.7 Dimensional Analysis

1. A patient has received 6.00 pt of blood. This is how many quarts and how many liters of blood?
2. Use dimensional analysis to calculate the volume of this cube in cm^3: 1.4 cm by 1.7 cm by 2.2 cm.

SOLUTIONS

1. To convert pints to quarts, first write down what is known and the desired units of the answer:

$$\text{Step 1} \quad 6.00 \text{ pt} \rightarrow \text{qt}$$

Now write the original volume over the number 1 to emphasize the fact that 6.00 pints is in the numerator:

$$\text{Step 2} \quad \frac{6.00 \text{ pt}}{1}$$

Next select the appropriate definition—in this case from experience and memory, 2 pt = 1 qt. (Remember, these are exact numbers, thus they have an infinite number of significant figures in a calculation.) Rearrange the definition into a conversion factor that cancels pints and leaves quarts:

$$\text{Step 3} \quad \frac{1 \text{ qt}}{2 \text{ pt}}$$

Now combine the original measurement and conversion factor and carry out the calculation:

$$\text{Step 4} \quad \frac{6.00 \not{\text{pt}}}{1} \times \frac{1 \text{ qt}}{2 \not{\text{pt}}} = 3.00 \text{ qt}$$

A volume of 6.00 pt equals 3.00 qt. How many liters is this?

$$\text{Step 1} \quad 3.00 \text{ qt} \rightarrow \text{L}$$

$$\text{Step 2} \quad \frac{3.00 \text{ qt}}{1}$$

From Table 1.1, 1.000 L = 1.057 qt. Rearrange to get the appropriate conversion factor:

$$\text{Step 3} \quad \frac{1.000 \text{ L}}{1.057 \text{ qt}}$$

Now combine the original volume and the conversion factor and do the calculation:

$$\text{Step 4} \quad \frac{3.00 \cancel{\text{qt}}}{1} \times \frac{1.000 \text{ L}}{1.057 \cancel{\text{qt}}} = 2.84 \text{ L}$$

A volume of 6.00 pt equals 3.00 qt, which equals 2.84 L.

The two conversion steps used to convert pints to liters could have been combined into a single solution: pints → quarts → liters:

$$\frac{6.00 \cancel{\text{pt}}}{1} \times \frac{1 \cancel{\text{qt}}}{2 \cancel{\text{pt}}} \times \frac{1.000 \text{ L}}{1.057 \cancel{\text{qt}}} = 2.84 \text{ L}$$

Note that pints and quarts cancel, leaving only liters as the units.

2. Because volume is length times width times height, multiply these measurements times each other (be sure to include the units):

$$1.4 \text{ cm} \times 1.7 \text{ cm} \times 2.2 \text{ cm} = 5.2 \text{ cm}^3$$

Multiplication of the numbers yields 5.2 to the correct number of significant figures, and multiplication of the units (cm × cm × cm) yields cm^3. (Note that we began with units for length—cm—and ended with units for volume—cm^3).

Self-Test

Convert these measurements to the desired units: 1. 4.73 m to cm
2. 4683 ms to s 3. 17.4 in to m

ANSWERS

1. 473 cm 2. 4.683 s 3. 0.442 m (Hint: use the definition 2.54 cm = 1.00 in to convert inches to centimeters, then convert centimeters to meters.)

1.6 Matter

As mentioned at the beginning of this chapter, the universe consists of matter and energy. **Matter** is anything that has mass and occupies space. This book, you, your chair, air, water, and all the objects around you are examples of matter. Although matter can be quite varied in form, most of us have no trouble visualizing the concept of matter. The absence of matter is a *vacuum*.

Matter on Earth exists in three states or phases, called solids, liquids, and gases (Figure 1.11). **Gases** have no definite volume or shape; they expand to fill any space

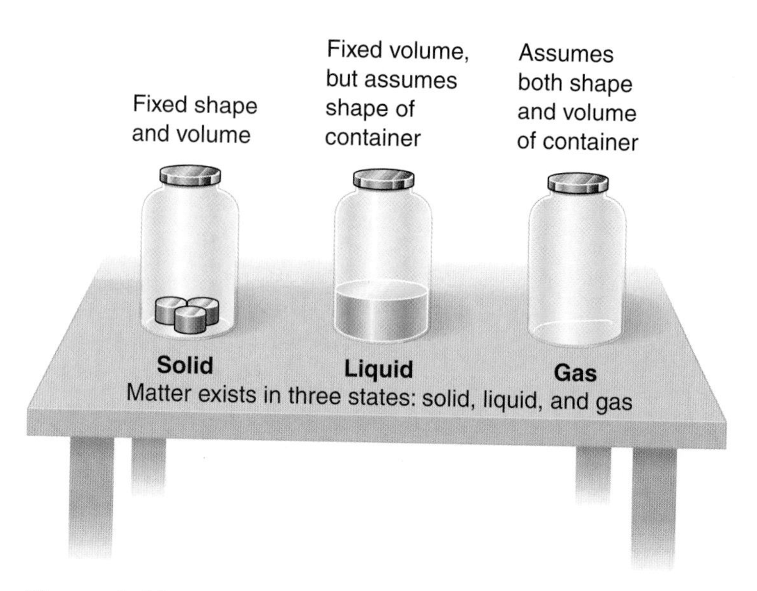

Figure 1.11 Gases, liquids, and solids differ in their ability to assume shapes and volumes.

On Earth, everything around us consists of matter.
(John Kieffer/Peter Arnold, Inc.)

they occupy. When we cook, for instance, aromas and smells (which are gases) expand from the kitchen and soon fill the other rooms of the house. Air, oxygen, and natural gas (methane) are common examples of gases. **Liquids** have a definite volume but no definite shape. When a liquid is poured into a container, it assumes the shape of the container, but it retains its original volume. Unlike a gas, a liquid does not expand to fill its container. Water, gasoline, and intravenous fluids are examples of liquids. **Solids** have both definite shape and definite volume. A solid retains its original volume and shape regardless of the size or shape of its container. This book, a coffee cup, a table, and salt are examples of solids.

Composition is a term describing the contents of a sample of matter. For example, water is composed of hydrogen and oxygen.

Matter can undergo two kinds of changes. A *physical change* alters the state or form of matter but not its content or composition. When ice (solid water) melts or liquid water boils, the water is undergoing a physical change. Whether water is a solid such as ice or snow, a liquid, or a gas such as steam or vapor, the composition of water is unchanged. On the other hand, when wood burns, it is no longer wood. Ashes, smoke, and gases are all that remain. This is a *chemical change*, which is a change in the composition of the matter. Chemical changes are also called **chemical reactions.** Changes in matter nearly always involve changes in energy, and chemical changes usually have larger energy changes than physical changes. Burning a kilogram of wood (chemical change) releases much more energy than the freezing of a kilogram of water (physical change).

Most samples of matter that we encounter are mixtures. Wood, coffee, salt water, and air are examples. A **mixture** has variable composition and consists of two or more materials that can be separated from each other by physical changes (Figure 1.12). Milk is a mixture, and butterfat, one of milk's components, is easily separated from the rest of the milk because it is less dense. The butterfat can be skimmed from the top or separated by centrifugation. Note that skimming and centrifugation are both physical processes because neither the butterfat nor the rest of the milk is changed in composition. Blood is also a mixture, and proteins can be filtered from the smaller components in blood serum. This is a physical change

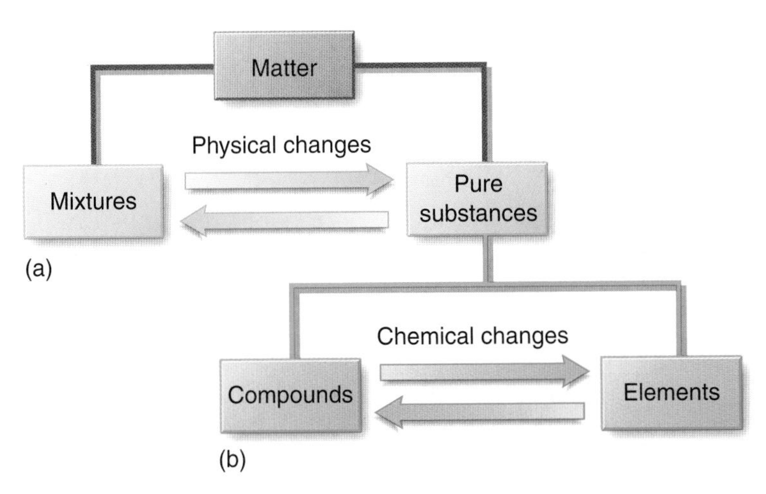

Figure 1.12 (a) Mixtures can be separated by physical methods into substances. (b) Compounds and elements are substances that can be changed into each other by chemical methods.

because the composition of the proteins and the other components is not altered.

Unlike mixtures, **substances,** sometimes called *pure substances,* are samples of matter that always have constant composition. Substances cannot be separated into components by any physical changes. Table salt and gold are substances. Pure substances are not common in nature but are found as components in mixtures.

You will learn more about elements and their symbols in Chapter 2.

Based on their composition, substances can be classified into two groups: elements and compounds. An **element** is a substance that cannot be broken down into any simpler substance. Gold is 100% gold (but your gold jewelry probably isn't), and oxygen is 100% oxygen. Both are elements. There are roughly 90 elements that occur naturally, and about 20 have been made in the laboratory. For the remainder of this course, you will deal with elements regularly. The inside front cover of this book has a table that contains information about them. Each element is represented by an abbreviation called its **chemical symbol.** Examples of chemical symbols include H for hydrogen, O for oxygen, and Na for sodium. Neither physical nor chemical changes can convert elements into simpler substances. However, through chemical change, two or more elements can combine to form compounds (see Figure 1.12).

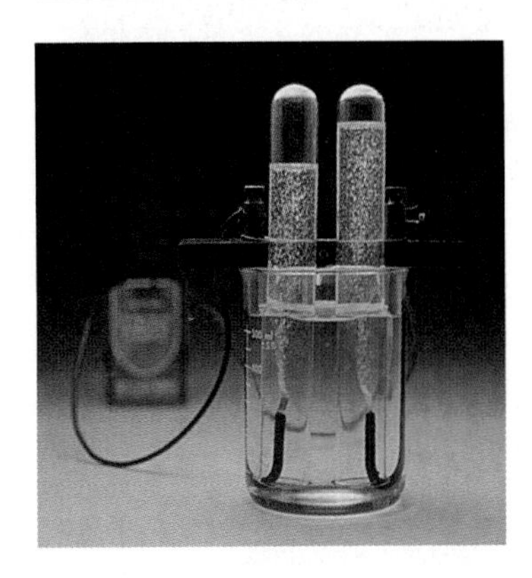

Figure 1.13 Water can be broken down into the elements oxygen and hydrogen. Electricity provides the energy for this chemical change. *(Charles D. Winters)*

A **compound** is a pure substance that contains two or more elements in fixed proportions. Compounds can be broken down into simpler substances (either simpler compounds or elements) by chemical changes (Figure 1.13). Water (H_2O) is made up of hydrogen and oxygen. When water is broken down to its elements, the mass of the hydrogen always makes up 11.19% of the original mass of the water, and oxygen makes up 88.81%. Table salt (NaCl) is a compound that always contains 39.3% of the element sodium and 60.7% of the element chlorine.

The symbols H_2O after water and NaCl after salt are chemical formulas, which are formally defined in Chapter 4.

Density and Specific Gravity

All matter has mass and occupies space. Any sample of matter has a particular volume and mass. **Density** is a measure of the mass present in a unit volume of

Figure 1.14 Objects float in air or water because they are less dense than air or water. Hot air balloons soar near Snowmass, Colorado. *(Doug Lee/Peter Arnold, Inc.)*

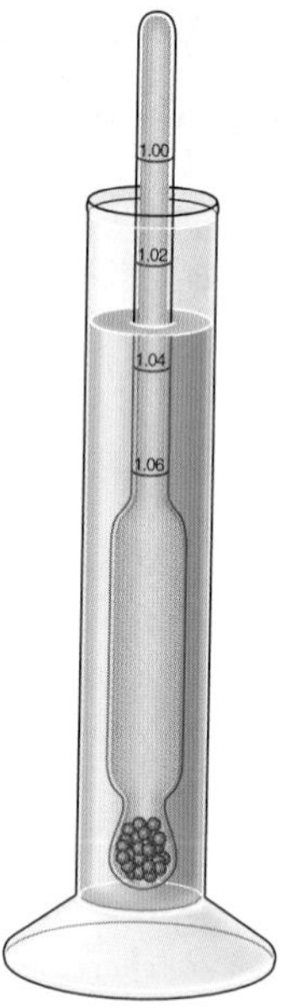

Figure 1.15 One way in which the density or specific gravity of a liquid can be measured directly is with a hydrometer.

any substance and is often expressed as grams per milliliter. The density of an object can be determined with this formula:

$$\text{Density} = \frac{\text{mass of object}}{\text{volume of object}} = \frac{\text{grams (g)}}{\text{milliliters (mL)}}$$

For example, if a 8.3-mL sample of a liquid had a mass of 11.43 g, its density would be

$$\text{Density} = \frac{11.43 \text{ g}}{8.3 \text{ mL}} = 1.4 \frac{\text{g}}{\text{mL}}$$

Densities vary widely—for example, the water in a glass of ice water has a density of 1.00 g/mL, the density of gold is 19.3 g/mL, and air has a density of 0.00129 g/mL. The densities of several substances and materials are shown in Table 1.5. You have considerable experience with densities of matter because differences in densities explain why objects sink or float. Wood floats because it is less dense than water; glass sinks because it is more dense than water. Why does a helium or hot air balloon rise in air? Because the gas in the balloon is less dense than air (Figure 1.14).

Specific gravity is the density of a sample divided by the density of water. If, for instance, a urine sample has a density of 1.017 g/mL, its specific gravity is

$$\text{Specific gravity of urine} = \frac{\text{density of urine}}{\text{density of water}} = \frac{1.017 \cancel{\frac{\text{g}}{\text{mL}}}}{1.000 \cancel{\frac{\text{g}}{\text{mL}}}} = 1.017$$

Note that although density always has units, specific gravity, because it is a ratio, has no units. Specific gravity shows how dense a substance is compared to water. Traditionally, health professionals use specific gravity for body fluids, because unlike density, specific gravity varies little with temperature for these fluids. Density and specific gravity can be measured directly with a device called a hydrometer (Figure 1.15).

Table 1.5 Densities of Some Common Materials at 25°C[a]

SOLIDS		LIQUIDS		GASES	
Substance	**Density (g/mL)**	**Substance**	**Density (g/mL)**	**Substance**	**Density (g/mL)**
Balsa wood[b]	0.12–0.20	Gasoline[b]	0.66–0.69	Hydrogen	0.000090
Aluminum	2.70	Ethyl alcohol	0.789	Helium	0.000179
Iron	7.8	Water	1.00	Nitrogen	0.00125
Lead	11.3	Ethylene glycol (antifreeze)	1.12	Oxygen	0.00143
Uranium	18.7	Mercury	13.6	Xenon	0.00586

[a]Density varies with temperature.

[b]The density varies because this is a mixture.

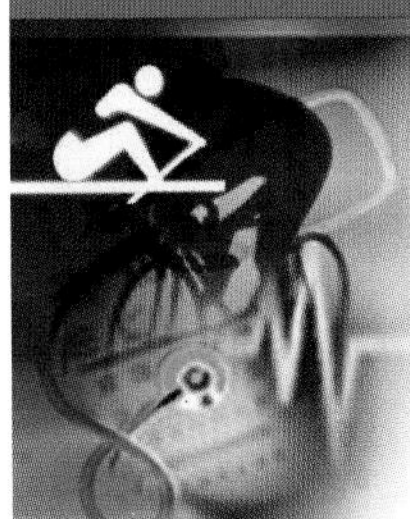

Connections

Using Density to Determine Percentage of Body Fat

Our body fat serves as an energy reserve, as insulation against heat loss, and as a protective cushion for internal organs. Body fat is often expressed as a percentage of body weight. A very slender person, such as a distance runner, has only a few percent body fat, whereas an obese person has more than 30% body fat. The normal range of body fat for males is around 12 to 18%, and the range for females is 18 to 24%. Because obesity is linked to several health risks, including cardiovascular disease, the percentage of body fat of some patients is measured.

There are several methods for determining body fat in general use, and each has its advantages and limitations. The simplest test is the *pinch test,* which uses a set of calipers at specific points on the body to measure the thickness of skin folds. This measures the amount of fat beneath the skin, and these measurements are then compared to standard tables to obtain an estimate of the percentage of body fat. Although this test is less accurate than some others, it is used because the equipment required is both inexpensive and portable.

A more accurate measure of percentage of body fat involves *hydrostatic immersion.* The patient is first weighed normally on a balance or scale. Then the patient is weighed again while submerged in a tank of water. The patient's weight in water is much less than on land because the displaced water buoys up the body. The difference between the two weighings, after applying several correction factors, is then used to determine the volume of the body. Once the patient's mass and volume have been determined, the patient's density can be calculated using the expression for density. Consider this example for a patient weighing 57.2 kg with a body volume of 53.8 L:

$$\text{Density} = \frac{\text{body mass}}{\text{body volume}} = \frac{57.2\text{ kg}}{53.8\text{ L}} = \frac{57{,}200\text{ g}}{53{,}800\text{ mL}} = 1.06\text{ g/mL}$$

Fat (adipose tissue) is less dense than other parts of the body such as muscle and bone. Fatty tissue has a density of about 0.92 g/mL, and lean tissue has a density of about 1.10 g/mL. The patient's density depends on both the proportion of lean tissue and body fat. The patient's body density is then compared to standard tables to get an estimate of percentage of body fat. For example, the patient with a body density of 1.06 g/mL has 21% body fat according to such a table.

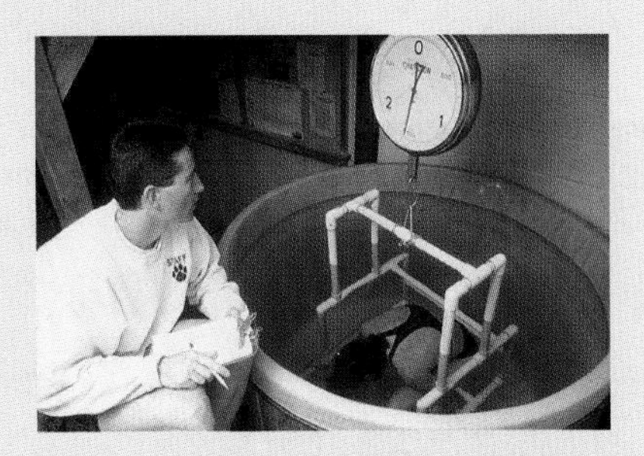

Determining percentage of body fat by hydrostatic immersion. *(Tom Pantages)*

Example 1.8 Density and Specific Gravity

A 7.41-mL sample of urine has a mass of 7.561 g. Determine the density and specific gravity of this sample.

SOLUTION

Density is calculated with this formula:

$$\text{Density} = \frac{\text{mass}}{\text{volume}} = \frac{7.561\text{ g}}{7.41\text{ mL}} = 1.02\ \frac{\text{g}}{\text{mL}}$$

The specific gravity of a sample is the density of the sample divided by the density of water:

$$\text{Specific gravity} = \frac{\text{sample density}}{\text{density of water}} = \frac{1.02\ \cancel{\text{g/mL}}}{1.00\ \cancel{\text{g/mL}}} = 1.02$$

By the way, the specific gravity of this urine sample falls within the normal range of 1.003 to 1.03.

Self-Test

A certain liquid sample has a mass of 4.921 g and a volume of 5.77 mL. Determine the density and specific gravity of this sample.

ANSWERS

Density is 0.853 g/mL; specific gravity is 0.853.

1.7 Energy

We have defined matter as anything that has mass and occupies space. That, at first glance, is nearly everything familiar to us, but the universe also contains energy. **Energy** is the capacity to cause a change, or, stated another way, the ability to do work. The wind blows dust in the alley behind the house and blows a sailboat through the water. The positions of the dust and sailboat have changed; thus we conclude that moving air (wind) possesses energy (Figure 1.16).

Figure 1.16 Moving air has kinetic energy. Wind provides the energy to propel this sailboat near Victoria, Australia. *(Bill Bachman/Photo Researchers)*

There are several forms of energy. The energy associated with motion is called **kinetic energy.** The energy of wind, flowing water, and a moving car are examples of kinetic energy. **Heat** (thermal energy) is a form of kinetic energy that involves the motion or vibration of particles in matter. The particles making up a hot object are moving or vibrating rapidly. The particles in a cold object are moving or vibrating more slowly.

Objects can also possess energy called **potential energy,** which is stored energy. Water stored behind a dam has potential energy that can be used to generate electricity, and food has potential energy that is released by chemical reactions in the body.

The amount of energy in any object can be measured, and a number of different units are commonly used. The standard SI unit for energy is the **joule (J),** and kilojoules (kJ) are commonly used for larger quantities of energy. Other common units are the **calorie (cal), kilocalorie (kcal),** and **Calorie (Cal).** You are probably most familiar with the Calorie (capital C), which is the same size as a kilocalorie (1 Cal = 1 kcal). This is the unit used to measure energy in foods. Table 1.6 summarizes the common units of energy.

Table 1.6 Some Common Units for Energy[a]

UNIT	SYSTEM	COMPARISON
joule	SI	1 J = 0.2390 cal
calorie	Metric	1 cal = 4.184 J
Calorie = 1000 cal = 1 kcal	Metric	1 Cal = 4,184 J

[a]The unit for energy in the English System is the British Thermal Unit (BTU). This unit is rarely used in science and health care, but your air conditioner or furnace may be rated in BTUs.

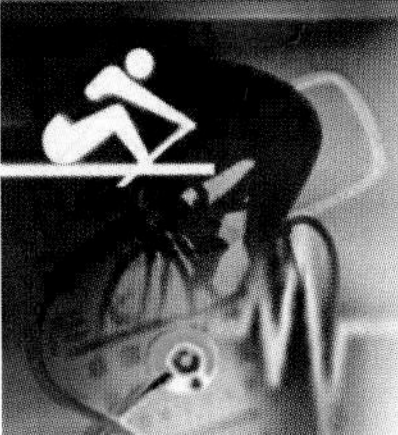

Connections
Energy in Foods

Foods provide the energy needed by the body for growth, motion, and internal (metabolic) activities. The energy content of foods varies greatly. Fruits and vegetables contain relatively few Calories, whereas meats, seeds, and foods made from grains are generally higher in Calories. Butter and margarine are high in Calories. The number of Calories in a particular food depends on the amounts of fats, carbohydrates, and proteins present. Butter and margarine are high in Calories because they contain mostly fats and oils that provide 9 Cal (kcal) per gram. Lean meats and many grains are lower in Calories because they contain primarily proteins and digestible carbohydrates that provide about 4 Cal per gram. Many fruits and vegetables contain mostly water and fiber (indigestible carbohydrate); thus they contain few Calories.

Although daily caloric needs vary with age, gender, and activity level, the United States Department of Agriculture bases its daily guidelines for nutrients on 2000 Calories per day for a typical adult. Growing teenagers and very active people such as athletes require more energy, and infirm persons or couch potatoes need less. Maintaining a balance between energy needs and energy intake contributes to good health.

Approximate Energy Content of Some Common Foods and Beverages

FOOD OR BEVERAGE	CALORIES	KILOJOULES
Fruits		
Apple, medium	80	335
Banana, 1 medium	105	440
Grapefruit, 1/2	40	165
Vegetables		
Carrot, 1 cup	45	190
Celery, raw, 1 stalk	5	20
Potato, 1 medium	105	440
Grain Products		
Bagel, plain	200	835
Oil, corn, 1 Tbl	125	525
Popcorn, airpopped, 1 cup	30	125
Meats		
Beef, lean, 3 oz	130	545
Chicken, skinless, 3 oz	110	460
Salmon, 3 oz	130	545
Dairy Products		
Butter, 1 Tbl	100	440
Cheese, cheddar, 1 oz	115	480
Egg, 1 large	80	335
Milk, 2%, 8 oz	120	500

Food and drink provide the energy needed by the body. *(Charles D. Winters)*

Temperature Scales

We use thermometers regularly, and we know from this experience that they are used to measure how hot or cold something is. Heat causes the liquid in the thermometer to expand and cold causes it to contract. **Temperature** is a measure of the kinetic energy in matter—that is, it is related to the amount of motion or vibration in a sample of matter. As temperature increases, the motion of the particles increases (and that is what causes the liquid in the thermometer to expand).

Temperature can be used to indicate direction of heat flow; heat always moves from a hotter object to a colder one. Temperature is not, however, a measure of the total amount of heat energy in a sample of matter. Let's compare temperature and total energy with this example. Consider a spoon sitting in a pot of boiling water. Both the spoon and the water are at the same temperature (they have the same amount of motion and vibration in their particles), but the water possesses much more heat energy than the spoon does. Which would cause the greater burn: the spoon pressed against your skin or the boiling water poured over your hand? The water would damage far more tissue because it possesses far more energy. Temperature is a measure of kinetic energy (motion) in matter, not total energy content.

Several scales are used in the measurement of temperature (Figure 1.17). In the United States, the common scale is the **Fahrenheit scale.** In this scale, water freezes at 32°F and boils at 212°F, and normal human body temperature is 98.6°F. In the metric system's **Celsius scale,** water freezes at 0°C and boils at 100°C, and normal body temperature is 37°C. The SI temperature scale is called the **Kelvin scale;** the units on the Kelvin scale are called kelvins (K) instead of degrees. In the Kelvin scale, water freezes at 273 K and boils at 373 K, and normal human body temperature is 310 K.

Fahrenheit degrees are smaller than Celsius degrees. For water, the range from the freezing point to the boiling point on the Fahrenheit scale is 180°; the same range in Celsius is 100° (Figure 1.17). A kelvin is the same size as the Celsius degree (there are 100°C and 100 kelvins between the freezing points and boiling

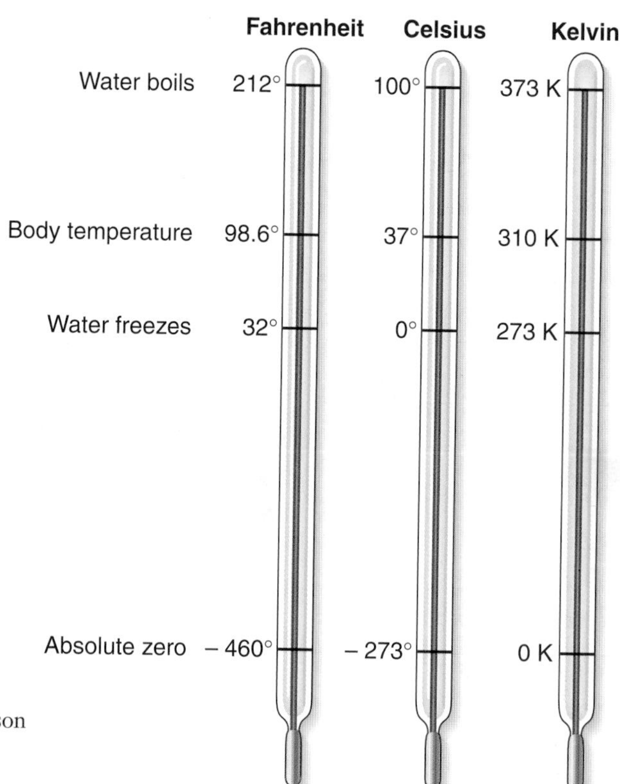

Figure 1.17 A comparison of the three common temperature scales.

points of water), but the kelvin temperature is 273 units larger than the temperature in Celsius. Water freezes at 0°C and 273 K and water boils at 100°C and 373 K (Figure 1.17).

The zero temperature on the Kelvin scale is called **absolute zero** because it is not possible to have a lower temperature. At absolute zero, the particles in matter are motionless.

It is sometimes necessary to change from one temperature scale to another. These formulas are useful for these conversions:

$$T_F = 1.8(T_C) + 32$$

$$T_C = \frac{(T_F - 32)}{1.8}$$

$$T_K = T_C + 273$$

In these equations, T_F, T_C, and T_K are Fahrenheit, Celsius, and Kelvin temperatures, respectively.

Let's do a temperature conversion to illustrate the use of these equations. If the temperature is 71°F, what is the temperature in degrees Celsius? We want °C, so we choose the middle equation and substitute the Fahrenheit temperature into it:

$$T_C = \frac{(T_F - 32)}{1.8}$$

$$T_C = \frac{(71 - 32)}{1.8} = 22°C$$

A temperature of 71°F corresponds to 22°C.

Example 1.9 Converting Temperatures Between Scales

A nurse from Japan is visiting your clinic and says that one of her patients had a temperature of 40°C. What was the patient's temperature in °F?

SOLUTION

To convert from one system to another, use the appropriate equation. We were given a temperature in degrees Celsius and want a temperature in degrees Fahrenheit. Using the appropriate expression and appropriate substitution yields

$$T_F = 1.8(T_C) + 32 = 1.8(40) + 32 = 72 + 32 = 104°F$$

The patient's temperature was well above normal and thus the patient had a fever.

Self-Test

1. Convert −35°F (a cold winter morning in Alamosa, Colorado) to degrees Celsius.
2. Convert 155°C to degrees Fahrenheit.
3. Convert 84°C to kelvins.

ANSWERS

1. −37°C 2. 311°F 3. 357 K

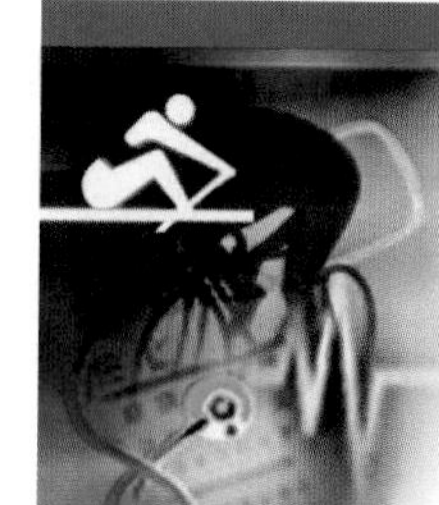

Connections

Fahrenheit and Normal Human Body Temperature

Temperature scales are set using specific reference points. For example, the Celsius temperature scale uses the freezing point of water (0°C) and the boiling point of water (100°C) as its reference points. The temperature range between these points was divided into 100 equal parts (degrees), and this 100-degree range is why the Celsius scale was formerly called the centigrade scale.

Although most of the world uses the Celsius scale, we use the Fahrenheit scale in the United States. What did Gabriel Daniel Fahrenheit use as his reference points in 1714 when he devised the Fahrenheit temperature scale? He used an equal mixture of ice and salt as the lower reference point and human body temperature as the higher one. He divided the temperature range into 96 equal parts. He then adjusted the scale so the freezing point and boiling point of water came out as whole numbers (32°F and 212°F, respectively). On the Fahrenheit scale, human body temperature is often said to be 98.6°F. But is it?

Actually it is very difficult to assign a "normal" body temperature, because body temperature varies with activity, time of day, and various physiological states such as sleep and the menstrual cycle. It is easier and more accurate to talk about a normal body temperature range rather than a single temperature. For many healthy people, the typical temperature range is from 98.2 to 98.8°F (36.8 to 37.1°C).

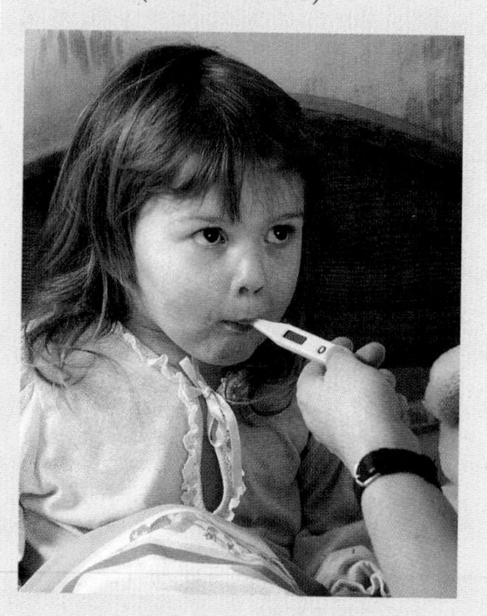

(Bob Thomas/Tony Stone Images)

CHAPTER SUMMARY

The **scientific method** uses **data,** hypotheses, and experiments to test and explain the world around us. **Hypotheses** may evolve into well-accepted **theories,** and scientific **laws** are summarizing statements rather than explanations. A **diagnosis** in medicine is similar to a hypothesis. **Measurements** are observations that provide numerical information and consist of a number and a unit. Standard units are defined in the metric system and the **Système International (SI)** for several properties, including mass, length, time, and temperature. These standard **units** are often combined with **prefixes** to provide units that vary greatly in size. Measurements are always subject to error. When measurements are close to the true value of what is being measured, the measurements are **accurate.** When measurements are quite similar to each other, they are **precise.** Measurements are always estimated to one place beyond the markings on a measuring device. All of the numbers in a measurement are significant. In contrast, some numbers, such as definitions and counts, are **exact numbers** and have an infinite number of **significant figures. Scientific notation** is a way of writing small and large numbers in a more compact form. **Dimensional analysis** is a problem-solving method that uses both the numbers and units of measurements to provide the correct numerical answer and correct units. **Matter** has **mass** and occupies space; thus it has a **density.** Matter can exist as a **gas, liquid,** or **solid.** Matter can undergo chemical and physical changes and can exist as an **element, compound,** or **mixture. Energy** is the ability to do work. Energy involving motion is **kinetic energy,** and **heat** is one form of kinetic energy. Stored energy is **potential energy. Temperature** is a measure of kinetic energy (motion) in matter, and in science and health care it is usually measured in degrees **Celsius** or **kelvins.**

KEY TERMS

Section 1.1
data
diagnosis
hypothesis
law
scientific method
theory

Section 1.2
cubic centimeter
gram
kilogram
liter
mass
measurements
meter

milliliter
prefix
second
Système International (SI)
unit
weight

Section 1.3
accuracy
precision
significant figures

Section 1.4
exact numbers
scientific (exponential) notation

Section 1.5
conversion factors
dimensional analysis

Section 1.6
chemical reactions
chemical symbol
compound
density
element
gases
liquids
matter
mixture
solids
specific gravity
substances

Section 1.7
absolute zero
calorie
Calorie
Celsius scale
energy
Fahrenheit scale
heat
joule
Kelvin scale
kilocalorie
kinetic energy
potential energy
temperature

EXERCISES

The Scientific Method (*Section 1.1*)

1. **(a)** What are the three steps of the scientific method? **(b)** Explain how a scientist would use these steps.
2. What is a hypothesis? Compare a hypothesis and a theory.
3. **(a)** What is a law? **(b)** What is the difference between a law and a hypothesis or theory?
4. A classmate says a recent archeological find in China proves the theory of evolution. Given the nature of the scientific method, how would you respond to this statement?
5. How many times must experimental data support a hypothesis to prove a hypothesis correct?

Measurements and Units (*Section 1.2*)

6. We often make measurements at work or at home. Just what is a measurement?
7. **(a)** What are the two components of a measurement? **(b)** Why are both components necessary to provide useful information?
8. **(a)** What is the system of units used by scientists? **(b)** Is the same system used by a physician?
9. What is the standard SI unit for
   (a) Length **(b)** Mass **(c)** Volume **(d)** Time
10. **(a)** What are the metric units for mass and volume? **(b)** Why are they commonly used in science and health care instead of the standard SI units?
11. Provide the full unit name for these abbreviations and indicate what is being measured (length, volume, time, etc.).
    (a) 14 g **(b)** 16.3 s **(c)** 47.3 L **(d)** 385 m
12. The units in these measurements use standard abbreviations, just as in Exercise 11. Provide the full unit name and indicate what is being measured (volume, time, mass, etc.).
    (a) 61 kg **(b)** 1.3 m **(c)** 13 m^3 **(d)** 257 g
13. Although we often consider mass and weight to be the same thing, they are not. What is the difference between them?
14. To help you become more familiar with units used in science, make these comparisons between SI, metric, and English units.
    (a) 1.0 kilogram is equal to how many pounds?
    (b) 1.0 liter is equal to how many quarts?
    (c) 1.0 meter is equal to how many feet?
    (d) 1.0 cubic meter is equal to how many gallons?
15. Express the meaning of each prefix as a decimal number.
    (a) *centi-* **(b)** *milli-* **(c)** *kilo-*
16. Express the meaning of each prefix as a decimal number.
    (a) *nano-* **(b)** *micro-* **(c)** *deci-*
17. What is the abbreviation for each prefix?
    (a) *centi-* **(b)** *milli-* **(c)** *kilo-*
18. What is the abbreviation for each prefix?
    (a) *nano-* **(b)** *micro-* **(c)** *deci-*
19. You have learned about the abbreviations of both units and prefixes. Provide the full unit name and indicate what is being measured (length, volume, etc.) for these units containing a prefix.
    (a) 64.1 mL **(b)** 4.3 ns **(c)** 23 Mg **(d)** 34.32 cm

20. Provide the full unit name and indicate what is being measured (length, volume, etc.) for these units containing a prefix.
(a) 324 km **(b)** 2.5 cm^3 **(c)** 10.0 dL
(d) 2.0 μL

21. Syringes are often marked in cc's, which is the abbreviation for cubic centimeter.
(a) A cc is equal to how many milliliters?
(b) How many cc's are in one liter?

22. Provide the answer to these statements.
(a) 100 cm = _____ m **(b)** 1 L = _____ mL
(c) 1000 ms = _____ s

23. Complete these statements.
(a) 1000 m = _____ km **(b)** 1 s = _____ μs
(c) 1,000,000 g = _____ Mg

Making Measurements (Section 1.3)

24. (a) What is the difference between a determinate error and an indeterminate error? **(b)** Give an example of each.

25. Barometers are instruments used to measure atmospheric pressure. Your laboratory has a barometer marked to the nearest millimeter of mercury. When you read this barometer, your measurement should be expressed to which decimal place?

These barometers measure atmospheric pressure. *(C. Steele)*

26. Define accuracy and give an example.

27. (a) What is the meaning of precision? **(b)** How would you compare the precision of a sundial and a fine watch?

28. You are deciding between two bathroom scales, one of which weighs to the nearest pound, the other to the nearest tenth of a pound. Which is more precise?

29. Significant figures are important in measurements and calculations involving measurements. What is a significant figure?

30. How many significant figures are in each of these measurements? (Review Table 1.3 if necessary.)
(a) 3.1 kg **(b)** 123.3 cm **(c)** 417 mL

31. How many significant figures are in each of these measurements?
(a) 17 km **(b)** 12.417 g **(c)** 68.4 s

32. How many significant figures are in each of these measurements?
(a) 1.2 m **(b)** 0.005023 s **(c)** 640 m^3
(d) 0.040 lb **(e)** 11.05 g

33. How many significant figures are in each of these measurements?
(a) 83,000 mi **(b)** 138.006 μm
(c) 0.000005 L **(d)** 340.2100 cm
(e) 1,000,000 km

Calculations Involving Measurements (Section 1.4)

34. (a) The numbers in a definition linking two units have how many significant figures? **(b)** In the English system, one furlong equals 220 yd. When converting 555 yd to furlongs, the answer in furlongs will have how many significant figures?

35. Compare how you determine the number of significant figures in the answer for a multiplication or division problem to how you determine the number of significant figures in an addition or subtraction problem.

36. Work these multiplication and division problems and express each answer to the correct number of significant figures. Remember to round the answers correctly.
(a) $345 \times 11 =$ **(b)** $468 \div 53.22 =$
(c) $0.02331 \div 4 =$ **(d)** $123.668 \times 3.004 =$

37. Work these multiplication and division problems and express each answer to the correct number of significant figures. Remember to round the answers correctly.
(a) $93{,}800 \times 0.0022800 =$ **(b)** $11{,}428 \times 6 =$
(c) $11 \div 0.040063 =$ **(d)** $5280 \div 11.09 =$

38. Do these addition and subtraction problems and express each answer to the correct number of significant figures. Don't forget to round properly.
(a) $428 + 5212.22 =$ **(b)** $11.411 + 0.66445 =$
(c) $345.02 - 125 =$ **(d)** $1.0099 - 0.44 =$

39. Do these addition and subtraction problems and express each answer to the correct number of significant figures. Don't forget to round properly.
(a) $43008 + 0.22 =$ **(b)** $11.400 - 9.55523 =$
(c) $23.40 - 27.699 =$ **(d)** $0.002 + 11.34411 =$

40. Express these numbers in scientific notation:
(a) 321.01 **(b)** 458,000
(c) 0.0000033 **(d)** 0.0004

41. Express these numbers in scientific notation:
(a) 32,110 **(b)** 0.0000000005598
(c) 1008 **(d)** 0.000200

42. Change these numbers from scientific notation to regular format:
(a) 4.662×10^5 **(b)** 2.01×10^{-4}
(c) 1.3×10^3 **(d)** 1.930×10^{-2}

43. Change these numbers from scientific notation to regular format:
(a) 1.77×10^{11} **(b)** 2.92×10^4
(c) 6.58×10^{-3} **(d)** 9.477×10^{-7}

Dimensional Analysis *(Section 1.5)*

44. Conversion factors are used in dimensional analysis to convert from one unit to another. What are the two conversion factors that can be derived from each of these equalities?
(a) 5280 ft = 1 mi **(b)** 1000 mL = 1 L
(c) 100 cm = 1 m

45. Use these equalities to write the two different conversion factors that are derived from each of them.
(a) 1×10^9 ns = 1 s **(b)** 1000 g = 1 kg
(c) 1 L = 1.057 qt

46. What are the equalities that were used to write these conversion factors?
(a) 1 gal/4 qt **(b)** 10 dL/1 L
(c) 1 mi/5280 ft

47. What equalities were used to derive these conversion factors?
(a) 1 in/2.54 cm **(b)** 2.205 lb/1 kg
(c) 1000 m/1 km

48. You want to buy a carpet runner in a shop in Tijuana. You need an 11.25-ft length of runner, and you know that 3.3 ft = 1.0 m. When you convert feet to meters, which of the two conversion factors from this equality would give you the correct answer? What would be the units in your answer if you used the incorrect conversion factor?

49. Use dimensional analysis to convert these English system measurements to the desired units:
(a) 33.5 in to ft (1 ft = 12 in)
(b) 7.3 pt to qt (1 qt = 2 pt)
(c) 3.770 lb to oz (1 lb = 16 oz)

50. Use dimensional analysis to convert these English system measurements:
(a) 750 yd to mi (1 mi = 1760 yd)
(b) 3.5 cups to pt (1 pt = 2 cups)
(c) 1150 lb to tn (1 tn = 2000 lb)

51. Use dimensional analysis to convert each measurement to the desired metric or SI units:
(a) 807 cm to m **(b)** 43 g to kg
(c) 2.31 L to mL

52. Use dimensional analysis to convert each measurement to the desired metric or SI units:
(a) 4.81 kg to g **(b)** 150 μL to L
(c) 2850 m to km

53. Use dimensional analysis to make these conversions between English and metric units:
(a) 2.7 kg to lb (1 kg = 2.2 lb)
(b) 100 yd to m (1 m = 1.094 yd)
(c) 37.4 L to qt (1 L = 1.057 qt)

54. Use dimensional analysis to make these conversions between English and metric units:
(a) 880 yd to m (1 m = 1.094 yd)
(b) 55 gal to m^3 (1 m^3 = 264 gal)
(c) 10.04 g to oz (1 oz = 28.4 g)

55. This set of conversions may require more than a single conversion factor. Show all steps that you use in your conversion.
(a) 126 μL to mL
(b) 3.5 cups to qt (1 qt = 2 pt, 1 pt = 2 cups)
(c) 163 μm to in (1 in = 2.54 cm)

56. Make the indicated conversions showing all the steps that are involved.
(a) 37.5 L to gal (1 gal = 4 qt, 1 L = 1.057 qt)
(b) 241 km to mi (1 m = 1.094 yd, 1 mi = 1760 yd)
(c) 2.7 oz to mg (1 oz = 28.4 g)

57. A premature baby had a mass of 2.25 kg at birth. Use dimensional analysis to determine the baby's weight in pounds.

58. The common barrel used in the United States for storing liquids such as gasoline, diesel, and motor oil contains 55 gal. What is the barrel's volume in cubic meters?

59. A small bottle contains 3.0 oz of medicine by weight. Use dimensional analysis to determine the medicine's mass in grams.

60. Suppose you were to buy an over-the-counter medication for 47.30 pesos in Mexico City. When you made the purchase the exchange rate was 9.55 pesos per U.S. dollar. Use dimensional analysis to determine the cost of the medicine in U.S. dollars.

61. A patient must take 1000 mg of a prescription medication three times a day. **(a)** How many 500 mg tablets are required to fill a one-month (30-day) prescription? **(b)** If the cost of a tablet is 17 cents, how much will the patient have to pay for the prescription?

Matter *(Section 1.6)*

62. **(a)** What is the formal definition of matter? **(b)** Write a definition for matter using your own words.

63. **(a)** What is the definition of a vacuum? **(b)** Where would a perfect vacuum be found in nature on Earth? **(c)** Is outer space a perfect vacuum? Justify your answer.

64. Name the three states of matter, beginning with the state that would exist at a low temperature and proceeding to the states that exist as the temperature progressively increases.

65. A substance has a definite volume but indefinite shape. This substance exists in the ____________ state.

66. A substance has both indefinite volume and shape. This substance exists in the ________________ state.

67. How do you describe a solid in terms of shape and volume?

68. **(a)** What is the definition of a mixture? **(b)** Make a list of three or four mixtures commonly found in the kitchen or medicine cabinet.

69. **(a)** A mixture can be separated into its components by what type of process? **(b)** While camping, you make coffee "cowboy style" by boiling water and coffee grounds together. When the coffee is ready, how can you separate this mixture of grounds and coffee? **(c)** Is the process involved in the separation you chose in part (b) consistent with your answer in part (a)?

70. **(a)** What is a chemical change (reaction)? **(b)** How would you distinguish between a chemical change and a physical change?

71. **(a)** What is a substance? **(b)** Give three examples of a substance.

72. **(a)** Define an element in both formal terms and in your own words. **(b)** Give three examples of an element.

73. **(a)** What is a compound? **(b)** Give three examples of common compounds (you will probably find several compounds in the kitchen or medicine cabinet).

74. A cup of saltwater is heated to drive off the water, leaving only a solid residue. **(a)** Did the heating result in a physical or chemical change? **(b)** Is saltwater a mixture, compound, or element?

75. What is the chemical symbol of each of these elements? First try to answer this question from your experiences and memory, then if necessary refer to the inside front cover of this book for the information. **(a)** Oxygen **(b)** Gold **(c)** Sulfur **(d)** Uranium **(e)** Sodium

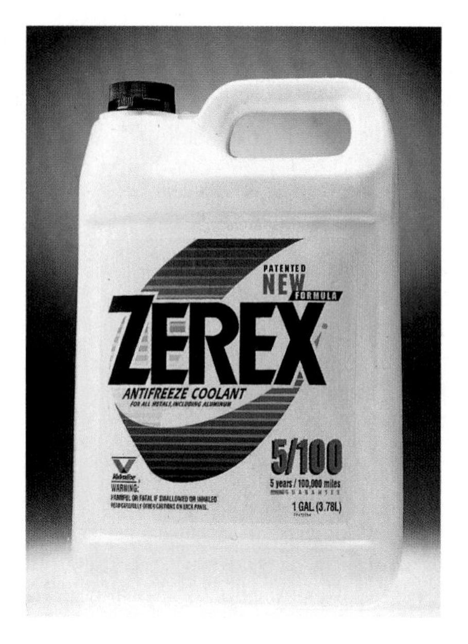

Commercial antifreeze contains ethylene glycol. *(Charles D. Winters)*

76. What is the chemical symbol of each of these elements? **(a)** Magnesium **(b)** Bromine **(c)** Phosphorus **(d)** Boron **(e)** Tungsten

77. What is the name of each of these elements? (Try answering first from memory, then use the inside front cover of this book.) **(a)** K **(b)** Cu **(c)** Ca **(d)** N **(e)** Cl

78. What is the name of each of these elements? **(a)** Fe **(b)** Li **(c)** I **(d)** Si **(e)** Ag

79. Define density and describe how it differs from specific gravity.

80. A sample of a liquid has a mass of 17.329 g and a volume of 19.41 mL. **(a)** What is the density of this liquid? **(b)** Does density depend on the state of the matter? Would a solid of the same volume and mass as the liquid have the same density?

81. A small statue has a mass of 723.4 g, and displaces 68.6 mL of water when added to a container of water. **(a)** What is the density of the solid used to make the statue? **(b)** You need to determine the density of a solid cube. Although you could determine the cube's density the same way the density of the statue was determined, describe an alternative method you could use.

82. Pure copper has a density of 8.9 g/cm^3. Calculate the volume of a sample of copper that has a mass of 255 g.

83. Ethylene glycol is used in automotive antifreeze. A 150.0-mL sample of ethylene glycol (density = 1.12 g/mL) would have what mass?

84. A liquid has a density of 0.97 g/mL. **(a)** What is the specific gravity of this liquid? **(b)** If this liquid does not mix with water, will the liquid float or sink in water?

Energy (*Section 1.7*)

85. What is the definition of energy?

86. **(a)** Compare kinetic energy and potential energy. **(b)** Use parts of your car to give examples of kinetic and potential energy.

87. **(a)** What is heat? **(b)** What effect does heat have on the particles (atoms, molecules) of matter?

88. **(a)** What is the SI unit for energy? **(b)** What units of energy are commonly used in the metric system?

89. **(a)** Express 1.00×10^2 J as kilojoules, calories, kilocalories, and Calories. **(b)** Which of these units is used in the U.S. to express energy in foods?

90. Convert 75 Calories to calories and to joules.

91. **(a)** What is the unit of temperature for each of the three common systems of units? **(b)** Of the three units, which is the smallest? (Hint: How many units of temperature exist between freezing and boiling water in the three scales?) **(c)** Are any of these units the same size?

92. Convert these Fahrenheit temperatures to Celsius.
(a) 74°F **(b)** 247°F **(c)** −11.2°F

93. Express these temperatures in degrees Celsius.
(a) 1132°F **(b)** 103°F **(c)** −317°F

94. Convert these Celsius temperatures to Fahrenheit.
(a) 13°C **(b)** 142.8°C **(c)** −40°C

95. Express these temperatures in degrees Fahrenheit.
(a) 106°C **(b)** 4.3°C **(c)** −273.15°C

96. Convert these Celsius temperatures to the Kelvin scale.
(a) 47.3°C **(b)** 845°C **(c)** −115°C

97. Convert these Celsius temperatures to the Kelvin scale.
(a) 56°C **(b)** 263°C **(c)** −83°C

98. Provide the Celsius equivalent of these temperatures.
(a) 296 K **(b)** 481 K **(c)** 27 K

99. Convert these Kelvin temperatures to Celsius.
(a) 373 K **(b)** 829 K **(c)** 116 K

Challenge Exercises

100. An athlete has a vertical jump of 21 in. **(a)** How high can this athlete jump in meters? **(b)** How high could she jump in inches on the moon? **(c)** Express the height of this jump on the moon in meters.

101. Most balances that measure to the tenth of a gram (0.1 g) cost a few hundred dollars. Balances that measure to a ten-thousandth of a gram (0.0001 g) cost a few thousand dollars. **(a)** Why is there such a large difference in price? **(b)** You are asked to purchase a new balance for the laboratory of the small clinic where you work. What would you need to know about the measurements taken on the current lab balances to determine whether you should buy the more or less expensive balance?

102. A report on a running back who plays football in Canada lists his speed as 4.9 s for 40 m. How fast can he run 40 yd?

103. A pharmacist is preparing a prescription for a patient who must take 1 teaspoon (tsp), as needed up to a maximum of four times a day. The prescription is written for a three-week supply. How many mL must the pharmacist dispense to ensure that the patient will have enough doses? If you solve this by dimensional analysis, the units in your answer will be mL per prescription. [3 tsp = 1 tablespoon (tbl); 0.5 fluid ounces (fl oz) = 1 tbl; 29.6 mL = 1 fl oz.]

104. OK, let's go for the whole nine yards . . . literally. In the good old days a typical dump truck carried nine cubic yards of gravel (9.00 yd^3). **(a)** What was the capacity of the truck in m^3? **(b)** If the delivered cost of the gravel was $8.00 U.S. per cubic yard, what would be the cost in Canadian dollars for a m^3? (Assume $1.00 U.S. = $1.45 Canadian.)

The Atom

2

ATOMIC STRUCTURE PROVIDES CLUES TO UNDERSTANDING THE PROPERTIES OF MATTER AND RADIATION.

IN CHAPTER 1 you learned about the scientific method, measurements, calculations, matter, and energy. These basic tools and concepts are needed in any branch of science. In this chapter you will learn about matter, which is the principal subject of chemistry. **Chemistry** is the study of matter and of the changes that occur in matter. This chapter begins with a discussion of the particulate nature of matter. The structure of these particles, which are called atoms, is then covered. The relationship between the structure of atoms and the properties of matter follows. The chapter concludes with an introduction to radiation that originates in atoms. Although this chapter covers some very basic ideas, you will also see topics that have direct applications to the health sciences.

Outline

2.1 The Atomic Nature of Matter
Recognize and explain the atomic nature of matter.

2.2 The Structure of Atoms
Differentiate among the subatomic particles and explain the composition of the nucleus and the atom.

2.3 Electron Shells and Valence Electrons
Recognize the arrangement of electrons in shells, and distinguish between core shells and the valence shell.

2.4 The Periodic Table
Explain the basis for the arrangement of the elements in the periodic table.

2.5 Electron Subshells and Orbitals
Describe and fill electron subshells and orbitals (optional material for some courses).

2.6 Nuclear Radiation
Describe nuclear radiations and distinguish among alpha, beta, and gamma decays.

2.7 Detecting and Measuring Radiation
Distinguish among the common detectors and units for radiation, and describe the relative penetration of nuclear radiations.

2.8 Induced Nuclear Reactions
Explain fission, fusion, and nuclear transformation.

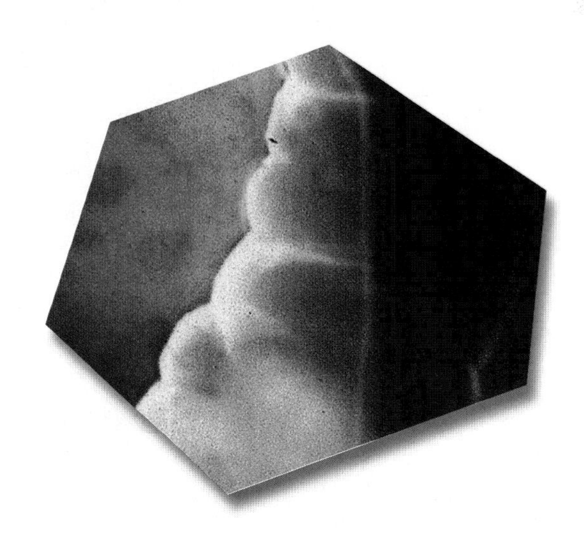

False-color x-ray image of a human abdomen. The patient drank a preparation of barium sulfate. Because the barium absorbs x-rays better than soft body tissues do, the intestines appear in sharp contrast to the rest of the body. ***(CNRI/SPL/Photo Researchers)***

2.1 The Atomic Nature of Matter

In Chapter 1 you learned that an element cannot be broken down into a simpler substance. Let's do a mental experiment. Take a piece of some element. Let's use gold (if we're going to experiment, let's do it in style). Mentally cut the sample in half. We now have two smaller pieces of gold. Push one piece aside, and cut the other in half. Again, we get two smaller pieces of gold. Repeat the cut again and again (Figure 2.1); because this is a mental exercise, we will assume we can continue to cut the gold pieces even after they are too small to see. Can we keep cutting the remaining piece of gold forever? Would each cut simply yield a smaller piece of gold? Or eventually will a cut yield a piece of matter that is something other than gold? Eventually we get an extremely small piece of gold that, *when cut*, is no longer gold. This smallest piece of gold is the fundamental particle of gold. It is a gold atom. An **atom** is the smallest particle of an element that possesses the properties of that element (Figure 2.2). We can also turn this definition around to define an element in another way: A sample of matter is an *element* if all of its atoms have the same properties.

Although our mental exercise is similar to the thinking of some ancient Greek philosophers, the idea of atoms was not accepted until the nineteenth century. In the early 1800s, John Dalton proposed that matter consists of atoms, and he hypothesized how atoms are involved in compounds and chemical changes. For the most part, our modern atomic theory (Table 2.1) is the same as that proposed by Dalton.

Self-Test

Matter Consists of Atoms

1. What is an atom?
2. Compare the definition of an element given in Section 1.6 with that given in Section 2.1. Are the two definitions consistent with each other?

ANSWERS

1. An atom is the smallest particle of an element that possesses the properties of the element. 2. In Section 1.6 an element is defined as a substance that cannot be broken down into simpler substances. In Section 2.1 an element is said to possess only one kind of atom. These definitions are consistent. Because atoms cannot be created or destroyed under normal conditions, the atoms of an element cannot be changed into other kinds of atoms to form a new substance.

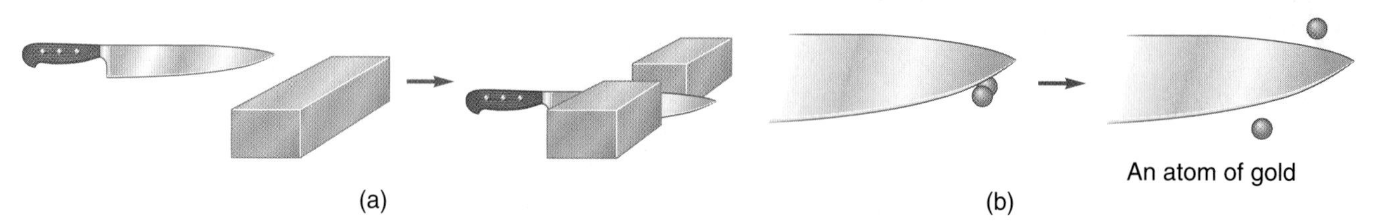

Figure 2.1 (a) A piece of gold can be cut into smaller pieces. (b) Eventually, a cut yields a piece of gold that cannot be cut and still be gold. This piece is an atom.

Table 2.1 Modern Atomic Theory

1. Elements consist of tiny particles called atoms.
2. *With the exception of mass*,[a] all of the atoms of an element are the same. The atoms of an element differ from the atoms of all other elements.
3. Compounds form when atoms of different elements combine. In any compound, the ratio of the different atoms present is always the same (e.g., water always has two hydrogen atoms per one oxygen atom).
4. In chemical changes (reactions), atoms combine in various ways to yield different substances. During chemical reactions, atoms are never created or destroyed.

[a]In Dalton's original hypothesis, the atoms of an element were the same in all properties, including mass.

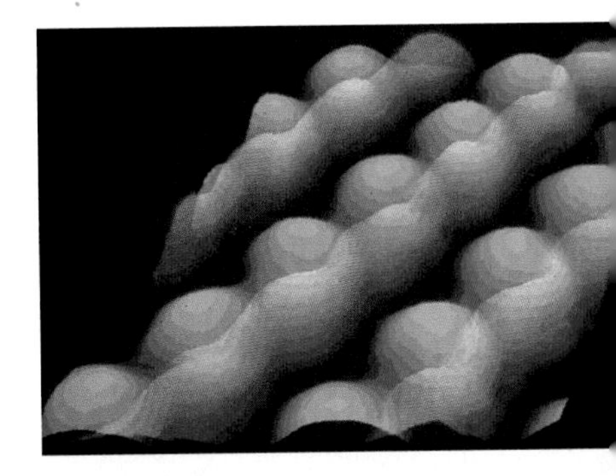

Figure 2.2 An image of individual atoms in a sample of gallium arsenide. Although there were no techniques that could produce images of individual atoms until recently, the existence of atoms has been supported by experimental evidence for nearly two centuries. This photograph was produced by a scanning-tunneling electron microscope. *(IBM Research)*

Symbols for Elements and Atoms

Because we will be studying elements and atoms throughout this course, it is worthwhile to learn some conventions and symbols that relate to them. In Section 1.6 you learned that each element is assigned an abbreviation that is the *symbol* for that element. The first letter of the element symbol is always capitalized, but the rest of the symbol (when there is more than one letter) is never capitalized. These examples illustrate this point: N is the symbol for the element nitrogen and Cu is the symbol for the element copper. Table 2.2 contains some common elements and symbols that you will encounter routinely in your chemistry studies and in health care.

The element symbol and additional information about elements can be found in the *periodic table of the elements*, a copy of which is on the inside cover of your textbook. In this table, each element is represented by a box, which always includes the element symbol. (For your convenience, the element name is also included in this textbook's periodic table.)

In chemistry, element symbols are used not only to represent an element but also an atom or group of atoms of that element. You will be introduced to this convention in Chapter 3, and will see it repeatedly throughout your chemistry course.

For some elements the symbol is not an obvious abbreviation of the English name. This is because some symbols were derived from the element name in other languages such as Latin and Greek. The symbol for sodium is derived from *natrium*

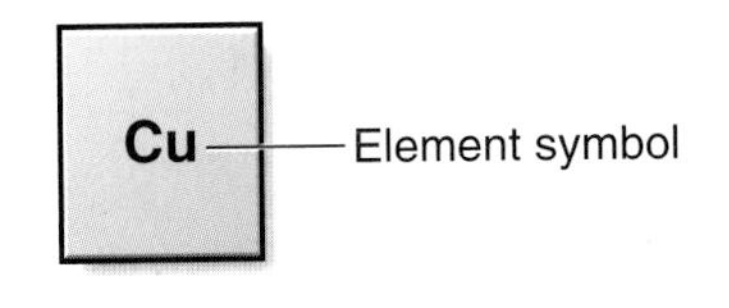

Copper

Table 2.2 Chemical Symbols of Some Common Elements[a]

ELEMENT	SYMBOL	ELEMENT	SYMBOL
Aluminum	Al	Iron	Fe
Bromine	Br	Magnesium	Mg
Calcium	Ca	Nitrogen	N
Carbon	C	Oxygen	O
Chlorine	Cl	Phosphorus	P
Copper	Cu	Potassium	K
Hydrogen	H	Sodium	Na
Iodine	I	Sulfur	S

[a]Because you will see these symbols often, it is worth your time to learn them.

(Na), and potassium's symbol comes from *kalium* (K). Some medical terms are also derived from these words. *Hyponatremia*, for example, means decreased concentrations of sodium in the blood.

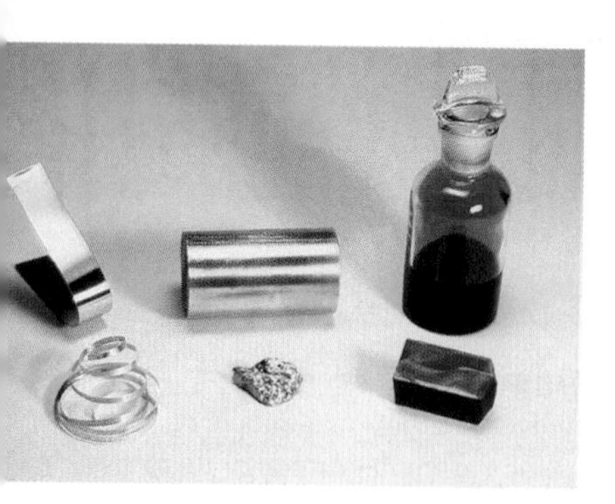

Elements vary greatly in their appearance. In this picture are shown the metals copper (wide strip), magnesium (narrow strips), and aluminum (cylinder); the nonmetals bromine (liquid) and carbon (gray-black block); and the metalloid silicon (shiny chunk). *(Charles D. Winters)*

Self-Test

Elements and Their Symbols

Table 2.2 and the periodic table of the elements found on the inside cover of this textbook provide the information you need to answer these questions.

1. What are the symbols for (a) calcium, (b) phosphorus, and (c) lead?
2. Cl, K, and Zn are the symbols for which elements?

ANSWERS

1. Ca, P, Pb 2. Chlorine, potassium, zinc

All matter is made up of atoms. In a few cases the atoms are alone and independent, but much more often atoms are combined with other atoms in larger particles called molecules. Chemists study atoms and molecules to develop a basic understanding of matter and to provide ideas that can be used in a practical way. For now, let's put molecules aside and concentrate on the structure of atoms, because their structure reveals why matter has its characteristic properties.

2.2 The Structure of Atoms

Subatomic Particles

Atoms consist of three smaller particles, called *subatomic particles*. An **electron** is a lightweight particle that has a mass of 9.109×10^{-28} g and has an electric charge of -1. A **proton** is a particle that has a mass of 1.673×10^{-24} g and an electric charge of $+1$. A **neutron** is a particle that has about the same mass as a proton (1.675×10^{-24} g), but a neutron has no electric charge; it is electrically *neutral.*

The mass of a subatomic particle expressed in grams is very small. Because numbers this small are inconvenient, another mass unit is commonly used for atoms and subatomic particles: **atomic mass unit (amu),** where 1 amu is now defined as 1/12 of the mass of a carbon-12 atom (the meaning of "carbon-12" is discussed shortly). A proton and neutron each have a mass of about 1 amu. The much smaller electron has a mass of 1/1837 amu. This is so much smaller than protons and neutrons that the mass of an electron in an atom is often considered to be zero (Table 2.3).

Arrangement of Subatomic Particles in Atoms

The protons and neutrons in an atom are found in a tiny core called the **nucleus** (plural: *nuclei*). Because protons and neutrons are found in the nucleus, they are collectively called *nucleons.* The nucleus of an atom has very little volume, but it contains most of the mass of an atom. The nucleus has a positive electric charge that is equal to the number of protons that it contains. Outside the nucleus and

Table 2.3 Subatomic Particles

NAME	CHARGE	MASS
Electron	−1	9.109×10^{-28} g; 0.0005486 amu ≈ 0 amu
Proton	+1	1.673×10^{-24} g; 1.007277 amu ≈ 1 amu
Neutron	0	1.675×10^{-24} g; 1.008665 amu ≈ 1 amu

moving around it are the electrons (Figure 2.3). In an atom, the number of protons equals the number of electrons, which means an atom is electrically neutral. If an atom gains or loses an electron it is no longer electrically neutral (the number of protons no longer equals the number of electrons), and it is now called an *ion*. For example, if a sodium atom loses an electron it becomes a sodium ion, and if a sulfur atom gains two electrons it becomes a sulfide ion. The arrangement of electrons in atoms varies with the number of electrons that are present. Ultimately, as we shall see, the arrangement of electrons in atoms accounts for many of the properties of the elements.

Atoms are very small, but just how small are they? The atoms of different elements have different sizes, but a reasonable average atomic diameter is 2 to 3×10^{-10} m (0.2 to 0.3 nm). Atoms are indeed very small, but they are much larger than the nucleus that is in them. An average nucleus is about 1×10^{-14} m. If the nucleus of an atom were the size of a basketball, the atom would have a four-mile diameter. Most of the space surrounding the nucleus is empty except for the tiny electrons that pass through this space.

Self-Test

Subatomic Particles

1. Name the three particles found in an atom.
2. Which of the particles has the smallest mass?
3. Which of the particles has a positive charge?
4. Which of the particles are found in the nucleus?

ANSWERS

1. Electron, neutron, proton 2. Electrons 3. Protons 4. Neutrons and protons

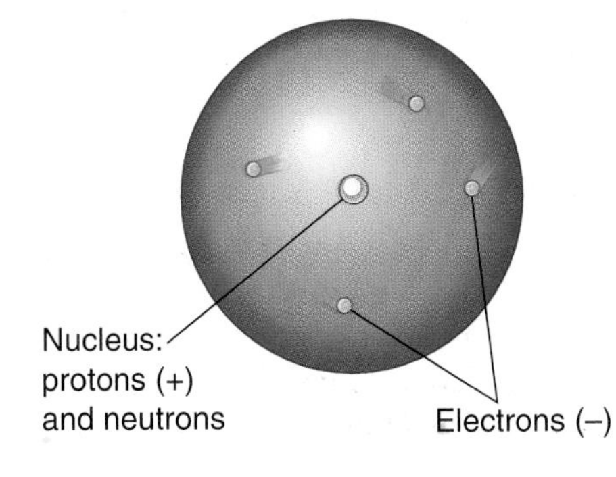

Figure 2.3 A simplified view of an atom. Most of the mass of the atom is in the nucleus. Electrons move around in the space surrounding the nucleus.

Atomic Number

Atoms differ in the number of protons, electrons, and neutrons they contain. Let's consider protons first. All of the atoms of an *element* have the same number of protons. Consider these examples: All hydrogen atoms have a single proton, all carbon atoms have 6 protons, and all uranium atoms have 92 protons. This concept can be viewed the other way around: Any atom with 1 proton must be a hydrogen atom; any atom with 6 protons must be a carbon atom.

The **atomic number** of an element is the number of protons found in any atom of that element:

29 — Atomic number
Cu

Copper atoms have 29 protons in their nucleus.

$$\text{Atomic number} = \text{number of protons in an atom}$$

Because atoms contain the same number of protons and electrons, the atomic number is also equal to the number of electrons in the atom. The atomic number of an element can always be found in a periodic table of the elements. It is the integer number in the element box and is usually shown above the element symbol.

Self-Test

Protons and Electrons in Elements

Use the periodic table to determine the number of protons and electrons in an atom of (a) magnesium, (b) N, (c) tin, and (d) Au.

ANSWERS

Because atoms are electrically neutral, each atom contains the same number of protons and electrons: (a) 12, (b) 7, (c) 50, (d) 79

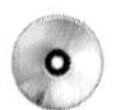

Isotopes and Mass Number

Although all atoms of an element have the same number of protons and the same number of electrons, the number of neutrons varies. Let's use carbon as an example. Carbon atoms contain six protons and six electrons, but they do not all contain the same number of neutrons. Most carbon atoms have six neutrons, fewer have seven neutrons, and still fewer have eight. Atoms of an element that have different numbers of neutrons are **isotopes** of each other. There are three naturally occurring isotopes of carbon. The mass of an atom is almost entirely in the nucleus and is due to the number of protons and neutrons *(nucleons)* in the nucleus. The **mass number** of an atom is its number of nucleons, which is the sum of the neutrons and protons.

$$\text{Mass number} = \text{total number of protons and neutrons in an atom}$$

Thus the mass numbers of the three isotopes of carbon are 12, 13, and 14.

The isotopes of an element are sometimes represented by the element name followed by the mass number. Carbon-12, carbon-13, and carbon-14, represent the isotopes of carbon. The isotopes can also be represented by combining the element symbol, the mass number of the isotope as a superscript, and the atomic number as a subscript. Carbon-12 is used as an example:

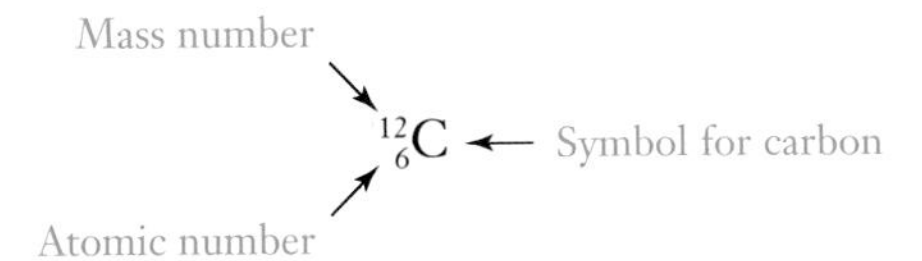

The three isotopes of carbon would be represented this way: $^{12}_{6}C$, $^{13}_{6}C$, $^{14}_{6}C$.

In a neutral atom both the number of protons and the number of electrons is equal to the atomic number of the element. The number of neutrons is determined by subtracting the atomic number from the mass number of an isotope. The number of neutrons in an atom of carbon-14 would be

$$\text{Mass number} - \text{atomic number} = 14 - 6 = 8 \text{ neutrons}$$

Example 2.1 Atomic Numbers and Mass Numbers

Use the atomic number and mass number to determine the number of neutrons in these isotopes:

1. $^{23}_{11}Na$ 2. chlorine-37

SOLUTION

The number of neutrons is determined by subtracting the atomic number (number of protons) from the mass number (total number of neutrons and protons).

1. The isotope representation provides both numbers, thus mass number minus atomic number equals number of neutrons: 23 − 11 = 12 neutrons.
2. This isotope representation provides the mass number (37), but the atomic number (in this case 17) is found in the element box of chlorine in the periodic table. The number of neutrons is 37 − 17 = 20 neutrons.

Self-Test

1. How many protons and neutrons are present in these isotopes? (a) iron-56, (b) bromine-79, (c) $^{19}_{9}F$
2. How many electrons would be present in the isotopes in question 1?

ANSWERS

1. (a) 26 protons and 30 neutrons, (b) 35 protons and 44 neutrons, (c) 9 protons and 10 neutrons   2. (a) 26, (b) 35, (c) 9

Atomic Mass

The decimal number in an element box in the periodic table is the **atomic mass** of the element (sometimes called atomic weight). This is the average mass of the naturally occurring isotopes of an element expressed in atomic mass units. For example, the atomic mass of copper is 63.55 amu. This means that the average mass of the copper atoms in a sample of copper is 63.55 amu, even though no individual atom of copper has this mass. There are two common isotopes of copper, copper-63 with a mass of about 63 amu, and copper-65, with a mass of about 65 amu. In a sample of copper, about 69% of the atoms are copper-63, and 31% are copper-65.

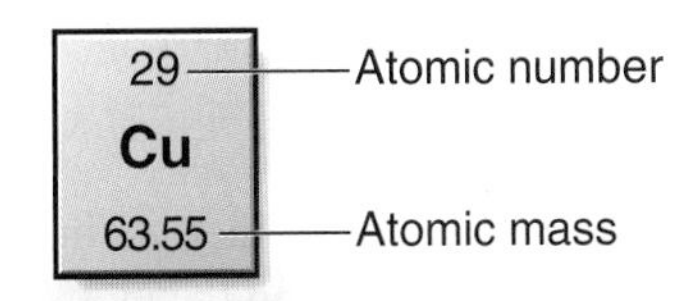

The average mass of a copper atom is 63.55 amu.

The atomic mass of copper, 63.55 amu, is closer to that of copper-63 than to copper-65 because copper-63 is more abundant than copper-65.

Atoms are very small, thus any natural sample of an element will always contain many atoms. (A paper clip with a mass of 0.44 g has 4.7×10^{21} atoms of iron.) Because of this, the atomic mass (the average mass of the atoms of the element) is far more important in normal use than the masses of the individual isotopes. Whenever the mass of an element is needed in clinical and chemistry laboratories, it is the atomic mass of an element that is used, rather than individual isotope masses.

Self-Test

Atomic Mass

Use the periodic table to determine the atomic mass of these elements: (a) silver, (b) Cl, and (c) nitrogen.

ANSWERS

(a) 107.87 amu (b) 35.45 amu (c) 14.01 amu

2.3 Electron Shells and Valence Electrons

So far we have described the atom as a nucleus of protons and neutrons surrounded by electrons. The electrons are not randomly arranged, however, but instead are arranged in layers around the nucleus. These layers of electrons are called either **shells** or *energy levels* (Figure 2.4). The shell closest to the nucleus has electrons with the lowest energy, and electron energy increases the further a shell is from the nucleus. The innermost shell, the *first shell*, can hold a maximum of two electrons. The single electron of a hydrogen (H) atom and the two electrons of a helium (He)

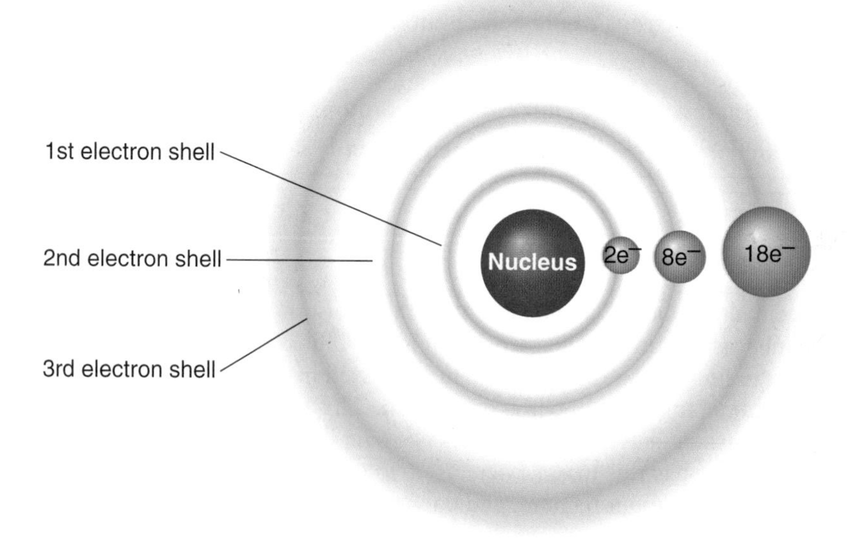

Figure 2.4 The electrons of an atom are layered around the nucleus in shells. The first three shells are labeled with the maximum number of electrons they can contain. Electrons have the least energy when they are in the innermost shell and the most energy when they are in the outermost shell.

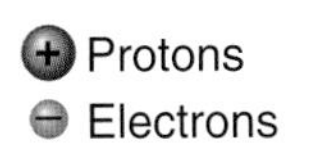

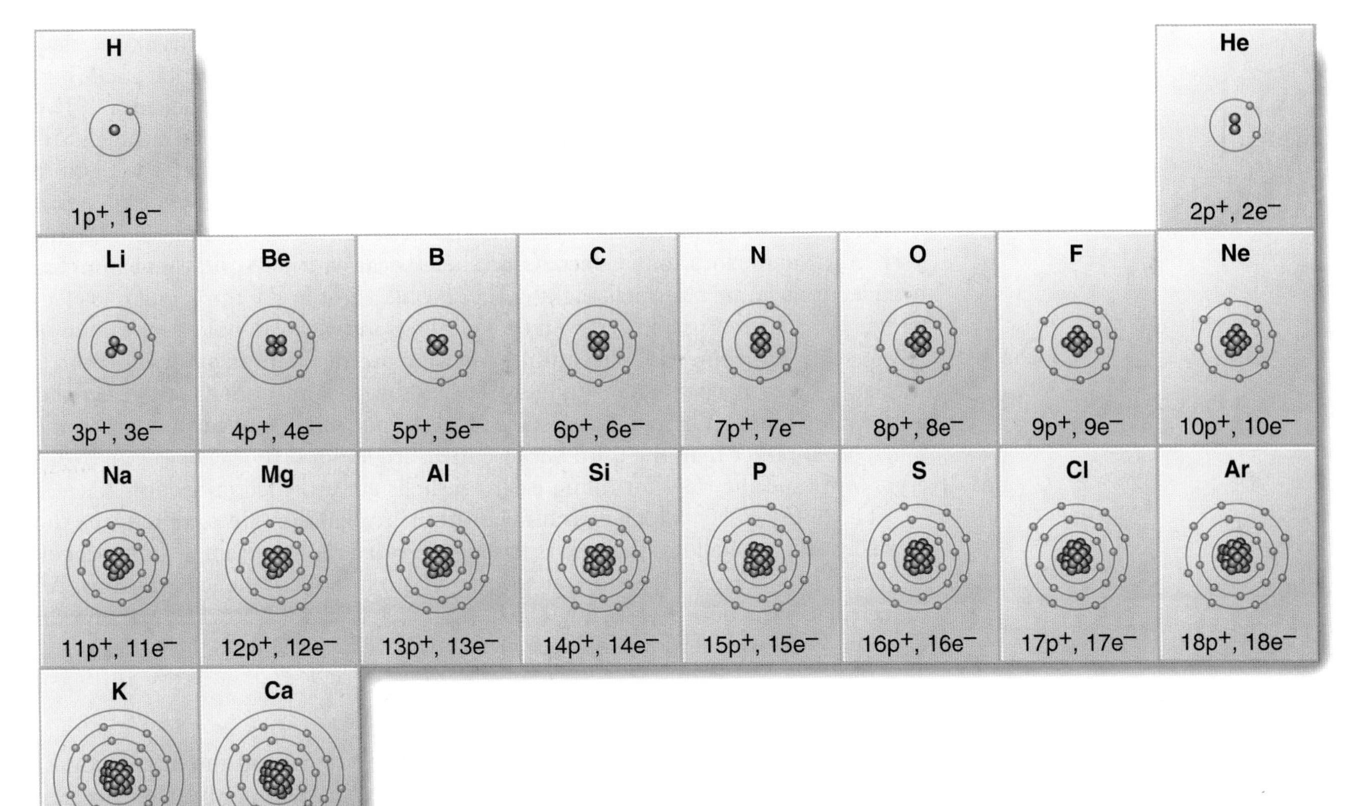

Figure 2.5 The arrangement of electrons in shells for the first 20 elements. The protons are shown in the nucleus to emphasize the electrical neutrality of an atom.

atom are in the first shell (Figure 2.5). If an atom contains more than two electrons, these additional electrons will be in shells further from the nucleus. The *second shell* can hold up to eight electrons, thus an oxygen atom that contains a total of eight electrons will have two of them in the first shell and the remaining six in the second shell (Figure 2.5). The *third* shell can hold up to 18 electrons, the *fourth* up to 32, and so on (Table 2.4). The largest atoms, like those of uranium (U), have seven shells of electrons.

The filling of shells with electrons is straightforward for the elements whose atoms contain 1 to 18 electrons. When a shell becomes filled, the next electron

Table 2.4 Shells of Electrons in Atoms

SHELL	MAXIMUM NUMBER OF ELECTRONS IN A SHELL[a]
1	2
2	8
3	18
4	32

[a]The maximum number of electrons in a shell can be calculated with the formula $2n^2$, where n is the number of the shell. Example: The maximum number of electrons in the third shell is $2 \times 3^2 = 18$.

goes into the next higher energy shell. For example, a lithium atom has three electrons, with two of them in the first shell and the third one in the second shell (Figure 2.5). In what shells would we find the seven electrons in a nitrogen atom? Two would be in the first shell, and the remaining five would be in the second shell. Argon, the element with 18 electrons, has two electrons in the first shell and eight in each of the second and third shells.

But beginning with potassium atoms, which have 19 electrons, this simple rule for predicting the location of electrons in shells breaks down. A potassium atom has two electrons in the first shell, eight in the second, eight in the third, and one electron in the fourth (Figure 2.5), even though the third shell can hold a maximum of 18 electrons. Calcium has 20 electrons, and their arrangement in a calcium atom is similar to those in a potassium atom except that the calcium atom has two electrons in the fourth shell. These outermost electrons in potassium and calcium are in the fourth shell rather than the third because the third and fourth shells overlap in energy. More specific rules for filling electron shells are provided in Section 2.5.

The outermost shell of electrons in an atom is called the **valence shell,** and the electrons in this shell are the atom's **valence electrons.** For example, a hydrogen atom has one valence electron and an oxygen atom has six valence electrons. All other electrons, those in the inner shells, are referred to as **core electrons.** In Chapter 3 you will learn that valence electrons have a special role in chemistry.

Light and Matter

We have just learned that electrons are in shells in atoms, and the interaction of light and matter provides evidence for this. Before we examine this evidence, we need to learn a little bit about light. Visible light is a part of the electromagnetic spectrum which also includes x-rays, ultraviolet light, infrared light, radio waves, and other radiations (Figure 2.6). Light, and all other *electromagnetic radiation,* is

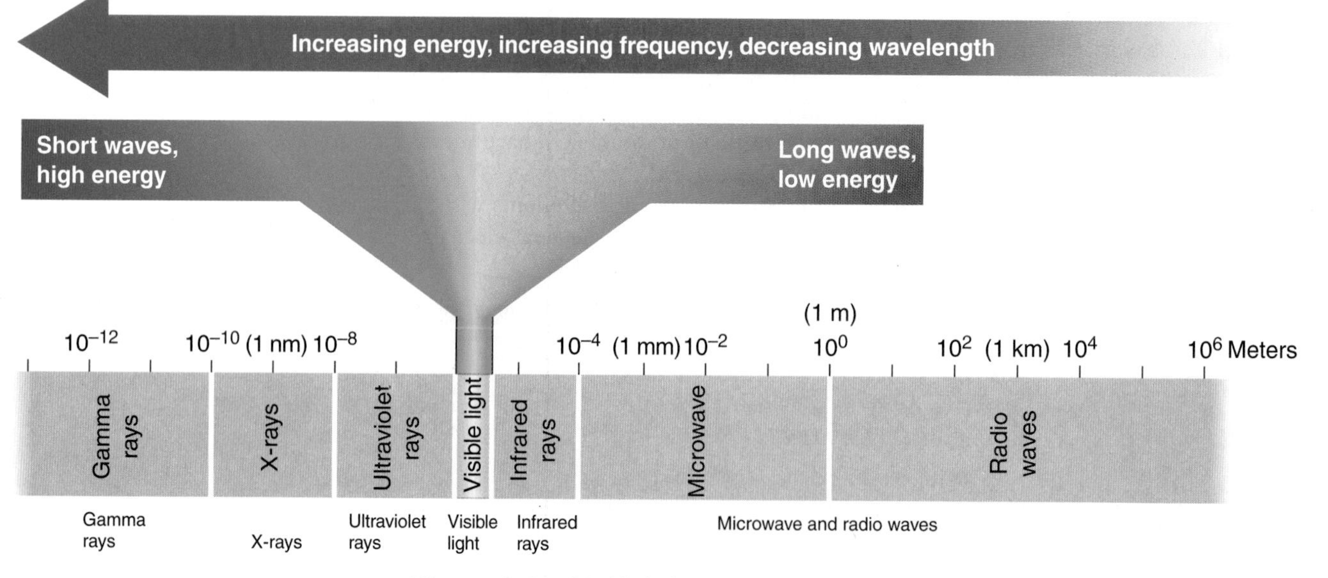

Figure 2.6 Visible light is a part of the electromagnetic spectrum. The amount of energy possessed by electromagnetic radiation depends on the wavelength of the radiation.

energy that is emitted and propagated in the form of rays or waves. Thus light behaves like a wave, and the energy of light varies with the *wavelength*—the length of the wave—and *frequency* of the light—number of waves that pass a point in a second (Figure 2.7). The shorter the wavelength of light (the higher the frequency), the higher the energy. Gamma rays and x-rays are high-energy radiation, radio waves are low-energy radiation, and visible light is intermediate in energy. Matter can either absorb or emit light, but when it does this, the matter gains or loses the exact amount of energy that corresponds to the energy of that light. This emission of light by matter is the basis for several common light sources, including neon lights and sodium-vapor lights.

Atoms normally exist in a lowest energy, most-stable state. If an atom gains energy (absorption of light is one way to do this), it becomes less stable and emits light to return to the more stable state. The atoms of each element emit very specific light. In fact, the emission of light by the elements is so distinctive that light is often used to determine which elements are present in a sample. The distinctive light emitted by hydrogen atoms is shown in Figure 2.8 as an example.

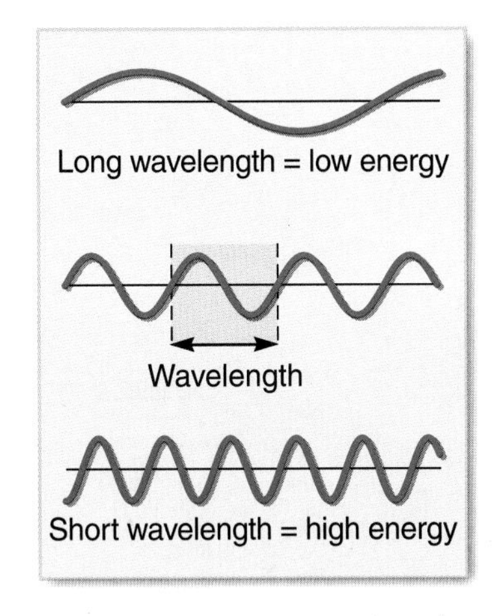

Figure 2.7 One wavelength is the length of the wave. The energy possessed by radiation depends on the wavelength of the radiation.

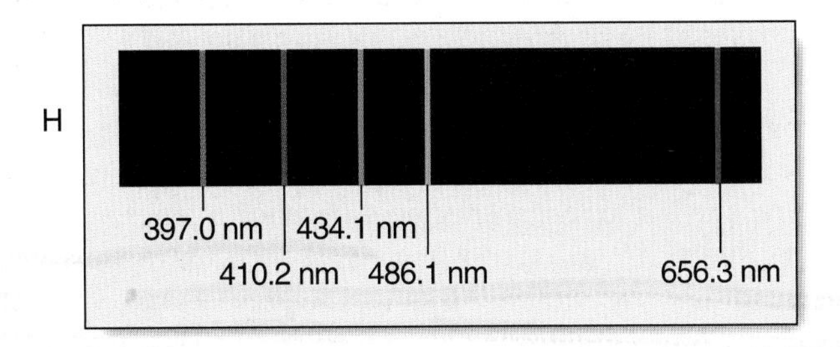

Figure 2.8 Hydrogen, like all other elements when they are gases, emits lines of visible light of very specific wavelengths.

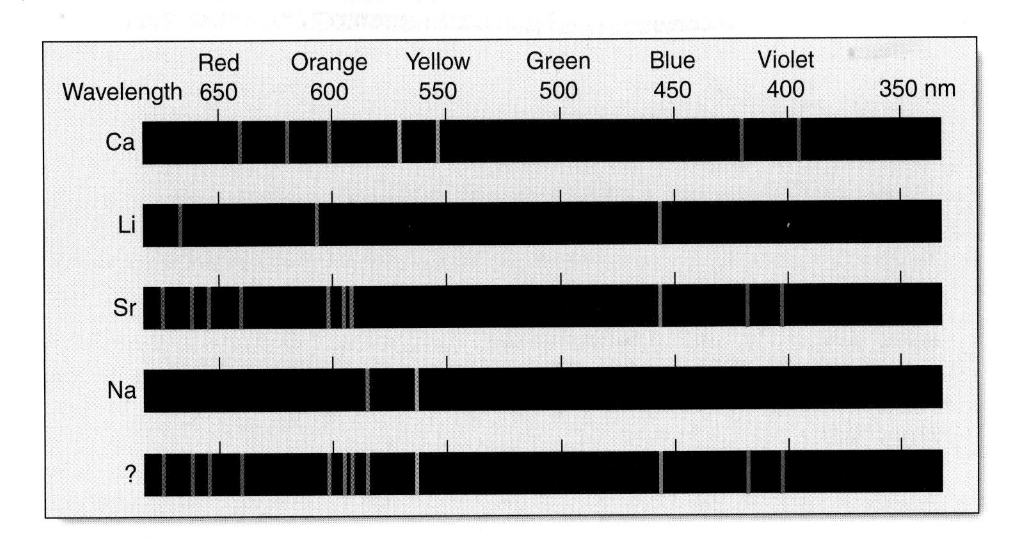

These four elements have the spectra shown here. What element or combination of elements produced the spectrum at the bottom? (Strontium, Sr, and Sodium, Na)

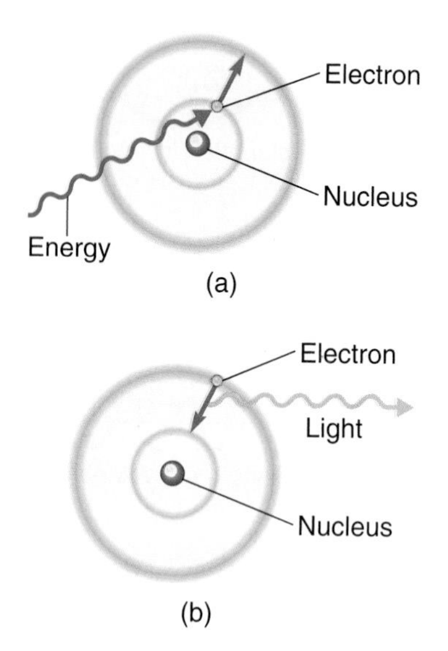

Figure 2.9 (a) When an atom gains a specific amount of energy, an electron moves from an inner (lower energy) shell to an outer (higher energy) one. (b) When the electron falls back into an inner shell, light of a specific energy and wavelength is emitted.

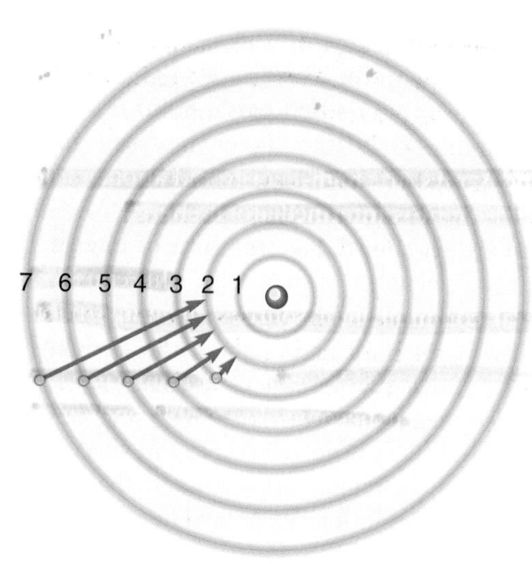

Figure 2.10 Electron transitions between shells produce the visible-light spectrum of hydrogen. The five transitions shown correspond to the five lines in the visible spectrum of hydrogen shown in Figure 2.8. Can you guess which orbital change corresponds to which line? (Hint: The farther the shell is from the nucleus, the higher the energy and the shorter the wavelength of the light.)

Let's consider the emission of light by hydrogen in a bit more detail. As first hypothesized by Niels Bohr in 1913, the electron in a hydrogen atom is normally located in the first shell. There is a specific difference in energy between any two shells in an atom [Figure 2.9 (a)], thus when a hydrogen atom gains energy, an electron moves from a lower energy shell to one of higher energy. The energy gained by the atom is exactly the amount of energy needed to move the electron from the lower energy shell to the higher energy one. With its electron in a higher energy shell, the atom is now in an unstable, higher energy, excited state. Remember, though, the tendency is for electrons to occupy the lowest energy shell available. So shortly after gaining the energy, the electron returns to a lower energy shell [Figure 2.9(b)]. Energy must be given up when this occurs, and the energy is lost as light. Each line in the emitted light of hydrogen represents the movement of an electron from a specific outer (higher energy) shell to a specific inner (lower energy) one (Figure 2.10).

2.4 The Periodic Table

We have just learned that the electrons of atoms are arranged into shells. The periodic table can be used to determine the arrangement of electrons in shells in an atom, but interestingly, the periodic table was not originally made to explain the location of electrons. It was set up to organize the elements according to their properties.

The properties of the elements have been studied for many years. From these studies, it was clear that each element has its own set of properties, but in addition, some elements have properties *similar* to those of some other elements. More than a century ago, a Russian chemist named *Dmitri Mendeleev* arranged the elements by atomic mass into a table, with elements having similar properties placed in the same column. Mendeleev's hypothesis organized the known elements well, but there were holes in his table where no known element fit. Mendeleev predicted that these holes represented elements that were not yet found, and he predicted the properties for these elements. Over the years these elements were found, and their properties were very similar to those predicted by Mendeleev. The accurate predictions made by Mendeleev demonstrated that the particular arrangement he had made was a useful one. (This is also a good example of the utility of the scientific method.)

The modern version of Mendeleev's table is called the **periodic table of the elements,** and scientists use it repeatedly to get information about the elements (Figure 2.11). Although this table is useful, for many years no one could provide any explanation—any physical basis—for why the elements should be arranged this

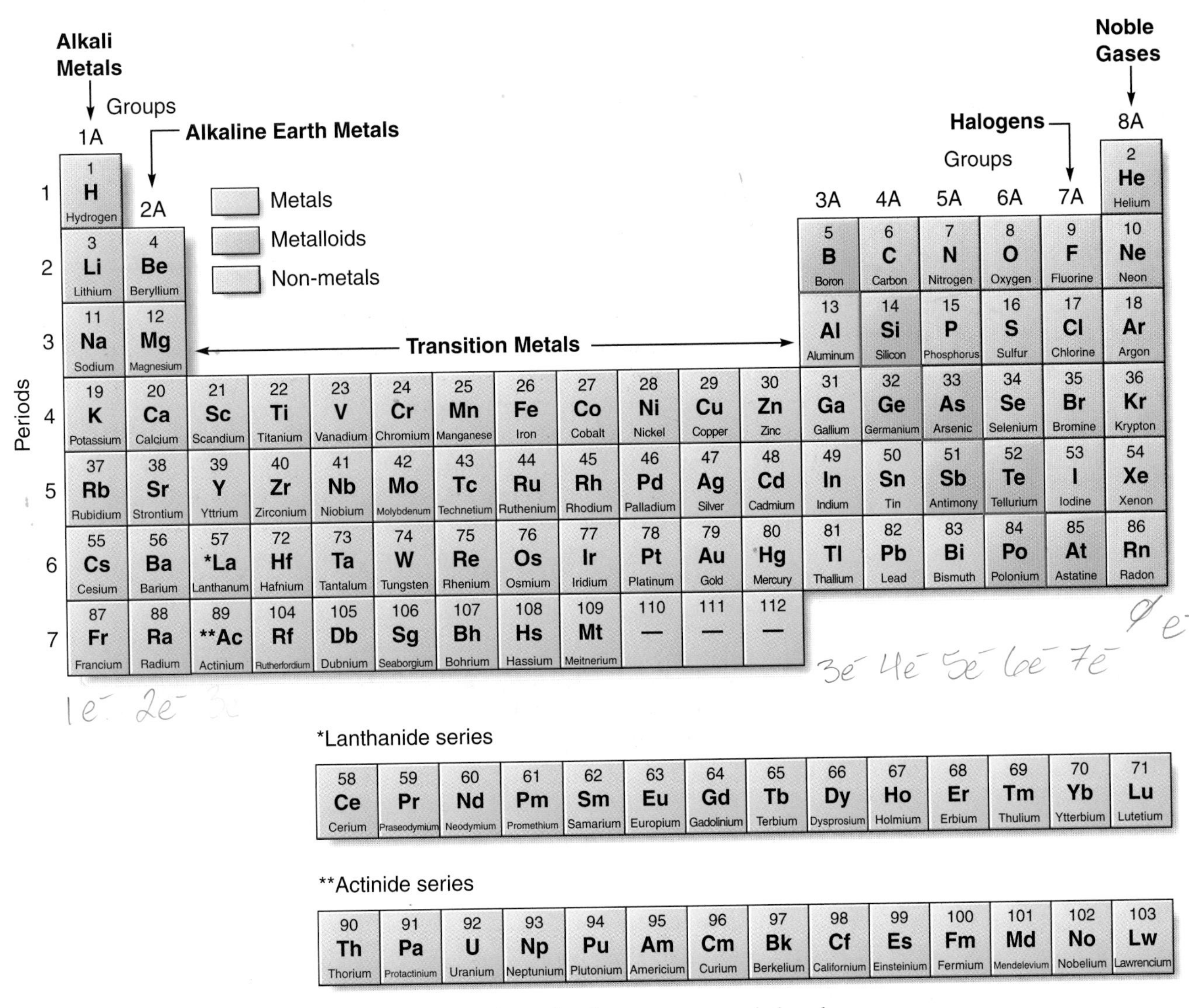

Figure 2.11 This version of the periodic table shows the element name, symbol, and atomic number.

These metals are examples of transition metals found in the fourth period. From left to right: Ti, V, Cr, Mn, Fe, Co, Ni, Cu *(Charles D. Winters)*

Some fourth period transition metals: *left to right*, Ti, V, Cr, Mn, Fe, Co, Ni, Cu.

21	22	23	24	25	26	27	28	29	30
Sc	**Ti**	**V**	**Cr**	**Mn**	**Fe**	**Co**	**Ni**	**Cu**	**Zn**
Scandium	Titanium	Vanadium	Chromium	Manganese	Iron	Cobalt	Nickel	Copper	Zinc

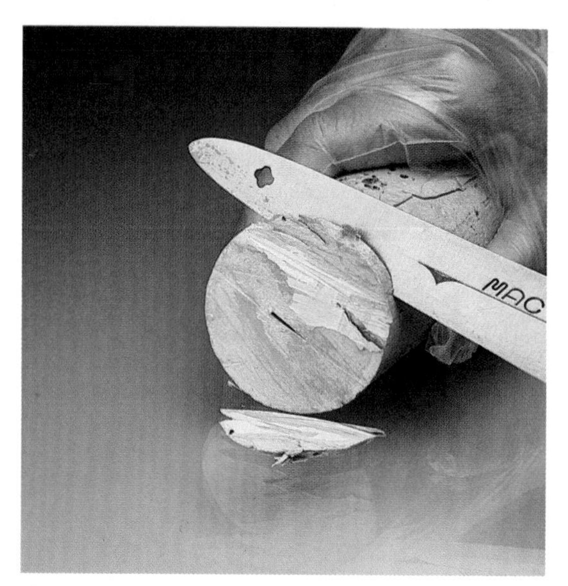

Group 1A: Sodium (Na).

1A		
1	**H**	Hydrogen
3	**Li**	Lithium
11	**Na**	Sodium
19	**K**	Potassium
37	**Rb**	Rubidium
55	**Cs**	Cesium
87	**Fr**	Francium

Sodium is an example of an alkali metal (Group 1A). *(Charles D. Winters)*

way. However, when the arrangement of electrons in atoms was discovered in the early twentieth century, the basis for the location of elements in the periodic table became known: (1) the elements were ordered in the table by their atomic number (number of protons and electrons) and (2) they were arranged into columns containing elements that had the same number of electrons in the valence shell. This regular and periodic arrangement of the elements is summarized by the **periodic law:** The properties of the elements vary in a systematic way according to outer electronic arrangement.

Columns in the periodic table are called either **groups** or **families.** Groups are often identified by a number and letter at the top of the column (Figure 2.11), although in this textbook some groups are simply lumped together and called the transition metals. An "A" at the end of a group number means the elements in the group are *representative* elements. For us, this means that the number of valence electrons are readily predicted because the group number is the same as the number of valence (outer-shell) electrons.

Look at the Group 1A elements lithium (Li), sodium (Na), and potassium (K) shown in Figure 2.11. These elements all contain a single electron in their outer shell—that is, they contain a single valence electron. The Group 2A elements, such as beryllium and magnesium, have two valence electrons. For columns 1A through 8A the outer electronic arrangement is the same for each element in that column. Elements in the same column have the same general properties because they have the same number of valence electrons.

In Chapter 3, you will learn that helium (He) differs from the outer electronic arrangement of the other 8A elements.

Group 2A: *left*, Magnesium (Mg); *right*, Calcium (Ca).

2A
4 **Be** Beryllium
12 **Mg** Magnesium
20 **Ca** Calcium
38 **Sr** Strontium
56 **Ba** Barium
88 **Ra** Radium

Magnesium and calcium are the most common alkaline earth metals (Group 2A). *(Charles D. Winters)*

7A
9 F Fluorine
17 Cl Chlorine
35 Br Bromine
53 I Iodine
85 At Astatine

Group 7A: *left*, Bromine (Br_2); *right*, Iodine (I_2).

Bromine (liquid) and iodine (solid) are halogens (Group 7A). *(Charles D. Winters)*

Some of the representative groups have common names. The Group 1A *metals* are called the *alkali metals* because they react with water to make it basic (alkaline). Note that hydrogen, which is not a metal, is also in Group 1A. Group 2A elements are called the *alkaline earth metals* because they are common in the rocks and minerals of the Earth and some of their compounds also make water basic. The *halogens* (salt formers) are found in Group 7A. Group 8A elements are called the *noble gases* because they are much less reactive than other elements.

Self-Test

The Periodic Table

1. Which elements are in Group 6A?
2. The elements lead (Pb), tin (Sn), and carbon (C) are in which group?
3. How many valence electrons are present in an atom of a Group 3A element?
4. A calcium atom has how many valence electrons?

ANSWERS

1. Oxygen (O), sulfur (S), selenium (Se), tellurium (Te), and polonium (Po)
2. 4A 3. Three 4. Two

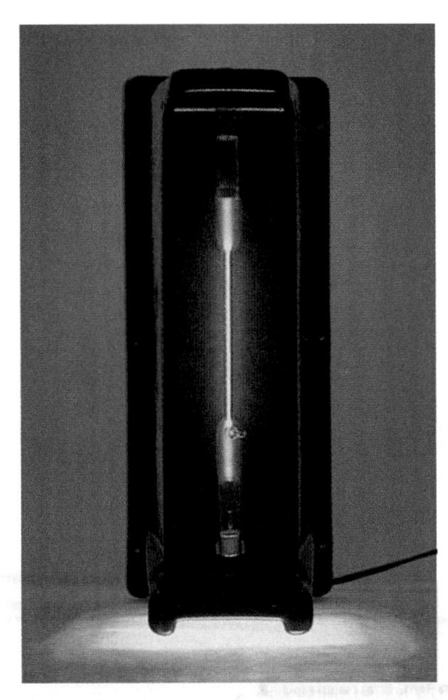
Group 8A: Neon (Ne).

8A
2 **He** Helium
10 **Ne** Neon
18 **Ar** Argon
36 **Kr** Krypton
54 **Xe** Xenon
86 **Rn** Radon

This neon light contains the noble gas neon (Group 8A). *(Charles D. Winters)*

The elements are often classified as metals or nonmetals. **Metals** are elements that are good conductors of heat and electricity. Metals also tend to be shiny, easily hammered into thin sheets (malleable) or pulled into wires (ductile). Metals tend to lose electrons in chemical reactions. Aluminum, copper, gold, and iron are metals. **Nonmetals** have opposite properties. They tend to be electrical insulators, they are not as shiny when solid, and they are brittle. In chemical reactions, nonmetal atoms tend to either gain or share electrons. Sulfur, iodine, oxygen, carbon, and phosphorus are examples of nonmetals.

Metals are located together in the periodic table because they have similar properties and related electronic arrangements. They are found at the left and bottom of the table (shown in blue in Figure 2.11). The nonmetals are mostly on the right and toward the top of the table (tan in Figure 2.11). The elements located at the boundary between the metals and nonmetals possess properties intermediate between those of metals and nonmetals (green in Figure 2.11). These elements are called either **metalloids** or *semimetals*. Metalloids are semiconductors (they conduct electricity somewhat), and the metalloid silicon is the basis for much of the semiconductor industry that has given us computers and other electronic devices.

Self-Test

Metals and Nonmetals

1. Which of these elements is an electrical insulator: Y, Hg, S, Al?
2. Which of these elements gives up electrons in a chemical reaction: Br, Mg, C?
3. Which element might be a semiconductor like silicon?

ANSWERS

1. Because S (sulfur) is a nonmetal, it should be an insulator.
2. Mg (magnesium) is a metal and therefore should give up electrons during a reaction.
3. Any of the metalloids should have intermediate properties and could be a semiconductor. [Germanium (Ge), like silicon, has applications in the semiconductor industry.]

Elements are also arranged into rows in the periodic table. Each row is called a **period**. All the representative elements in a period have their valence electrons in the same shell. For example, hydrogen and helium have their valence electrons in the first shell, and the second-period elements (the eight elements lithium through neon) have their valence electrons in the second shell (Figure 2.11). To understand the arrangement of electrons in shells for Period 4 and beyond you should read Section 2.5.

Example 2.2 Groups, Periods, and Valence Shell Electrons

1. The valence electrons of aluminum (Al) and iodine (I) are found in what shell?
2. How many valence electrons are present in aluminum and iodine?

SOLUTIONS

1. The valence shell (i.e., the outer shell) has the same number as the period number for the element. Because aluminum is in Period 3, the third shell is the outer shell. Iodine is in the fifth period, thus the fifth shell is the outer shell.
2. Use the group number to predict the number of electrons. Because aluminum is in Group 3A, there are three electrons in the outer shell. Iodine is in Group 7A, thus the number of valence electrons is seven.

Self-Test

1. The valence electrons are found in what shell in atoms of C, Ca, sulfur, and Na?
2. Determine the number of valence electrons for an atom of the elements listed in Question 1.

ANSWERS

1. 2, 4, 3, 3 2. Four, two, six, one

The universe consists of mostly hydrogen and helium atoms, with smaller amounts of the other naturally occurring elements. In contrast, the Earth's crust has a larger proportion of heavier elements such as oxygen, silicon, aluminum, and iron. The chemical composition of your body is still different because living organisms take up and retain elements selectively from their environment. The elements taken in and required by the body are called *minerals* by nutritionists.

Recommended Dietary Allowance (RDA) for Some Minerals

ELEMENT	AMOUNT PER DAY	
	Young Female Adults	Young Male Adults
Calcium	1.2 g	1.2 g
Copper	1.5–3.0 mg	1.5–3.0 mg
Iodine	150 μg	150 μg
Iron	15 mg	10 mg
Magnesium	280 mg	350 mg
Phosphorus	1.2 g	1.2 g
Zinc	12 mg	15 mg

Foods are our sources of minerals, but some people's food selection results in inadequate amounts of some minerals in the diet. To help make people aware of our mineral needs, the Food and Drug Administration (FDA) provides recommendations for daily intake of minerals (see table). In addition, the food industry enriches some foods by adding one or more minerals and vitamins to the food. For example, iron is added to flour, which is then used in many baked products such as bread. Dietary supplements can also be used to provide minerals. Be careful when using dietary supplements, however. If you don't need the minerals, the supplements are an unnecessary expense. Furthermore, approximately one person in 200 has the genetic disease hemochromatosis and accumulates iron in the body, sometimes to potentially harmful or fatal levels. If these people take dietary supplements containing iron, they increase the risk of accumulated iron.

2.5 Electron Subshells and Orbitals (Optional in some courses)

Although an understanding of electron shells and valence electrons provides a solid basis for understanding much of chemistry, sometimes additional details about electronic structure are needed. Electron shells can be further divided into **subshells** that are labeled *s*, *p*, *d*, and *f* (Table 2.5). The number of subshells in a shell is equal to the shell number; thus the first shell has one subshell and the second shell has two subshells. The first shell contains only an *s* subshell, the second shell contains an *s* and a *p* subshell, the third shell an *s*, *p*, and *d* subshell, and the fourth shell an *s*, *p*, *d*, and *f* subshell. Subshells are distinguished from each other by a number indicating the shell and a letter indicating the type of subshell. For instance, the *s* subshell in the first shell is called the 1*s* subshell, and the two subshells in the second shell are designated 2*s* and 2*p*.

Subshells differ in the number of electrons they can hold. An *s* subshell holds a maximum of two electrons—that is, it may contain zero, one, or two electrons. A *p* subshell holds a maximum of six electrons, a *d* subshell a maximum of ten electrons, and an *f* subshell a maximum of 14 electrons.

Table 2.5 Subshells of Electrons in the First Three Shells of Atoms

SHELL	SUBSHELL	MAXIMUM NUMBER OF ELECTRONS IN THE SUBSHELL	MAXIMUM NUMBER OF ELECTRONS IN THE SHELL
1	1*s*	2	2
2	2*s*	2	
	2*p*	6	8
3	3*s*	2	
	3*p*	6	
	3*d*	10	18

Example 2.3 Shells and Subshells of Electrons

1. How many subshells are found in the fourth shell?
2. What subshells are found in the third shell?
3. How many electrons can a *d* subshell hold?

SOLUTIONS

1. The number of subshells in a shell equals the shell number. Therefore, the fourth shell has four subshells.
2. The third shell has three subshells: 3*s*, 3*p*, and 3*d*.
3. Any *d* subshell can hold up to ten electrons.

Self-Test

1. How many subshells are in the second shell?
2. The *p* subshells hold a maximum of how many electrons?
3. Could a *p* subshell contain four electrons?
4. Could a *d* subshell hold 11 electrons?

ANSWERS

1. Two 2. Six 3. Yes 4. No

Filling Subshells with Electrons

There is an organized way to predict the shells and subshells that contain electrons in an atom. The subshells—like the shells—differ in energy, and electrons fill the lowest energy subshells available to them. Figure 2.12 shows this order of filling. Consider a hydrogen atom. Because the atomic number of hydrogen is one, there is one electron in a hydrogen atom. The lowest energy subshell is the 1*s* subshell, and therefore the single electron of hydrogen must be in the 1*s* subshell. The arrangement of electrons in an atom is called electronic configuration, and the electronic configuration of hydrogen is

$$\text{H} \quad 1s^1$$

which means one electron (the superscript 1) is found in the 1*s* subshell of a hydrogen (H) atom.

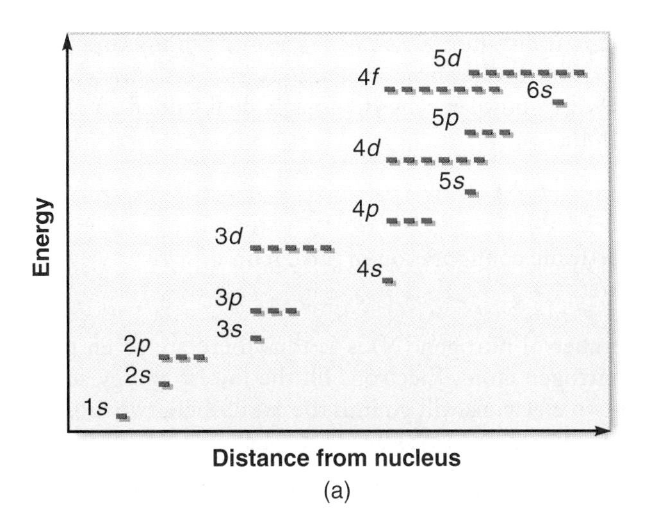

(a)

Subshell	Maximum number of electrons
5s	2
4p	6
3d	10
4s	2
3p	6
3s	2
2p	6
2s	2
1s	2

(b)

Figure 2.12 (a) The subshells are arranged in the order of their energies. (b) When you need to fill subshells with electrons, begin by filling the bottom subshell first and then fill them in the order shown here. Remember, each orbital—shown as a short red line—can hold zero, one, or two electrons.

The atoms of helium (He) are the next smallest atoms (atomic number = 2), and they contain two electrons. Where are these two electrons located? Remember, subshells can hold more than one electron, and electrons go into the lowest energy subshell that is available. For this reason, both the electrons of helium are in the 1*s* subshell:

$$\text{He} \quad 1s^2$$

Lithium (Li) has three electrons. A quick glance at the electronic configuration of helium suggests that two of these electrons must be in the first shell (in the 1*s* subshell), but the third electron cannot fit into the 1*s* subshell because that subshell already has its maximum of two electrons. The third electron of lithium must go into the next-lowest energy subshell, which is the 2*s* (Figure 2.12). The electronic configuration of lithium is Li $1s^2 2s^1$. Beryllium (Be) has four electrons, and its electronic configuration is Be $1s^2 2s^2$. (Do you understand why the fourth electron was put into the 2*s* subshell?) Boron (B) has an atomic number of 5, and thus has five electrons per atom. A boron atom has the same electronic configuration as beryllium plus an electron in the 2*p* subshell: B $1s^2 2s^2 2p^1$. The electronic configuration of all the remaining elements can be determined in the same manner, and Table 2.6 shows the electronic configuration of several elements. When determining

Table 2.6 Electronic Configuration of Some Elements

ELEMENT	ATOMIC NUMBER	ELECTRONIC CONFIGURATION
Hydrogen	1	H $1s^1$
Helium	2	He $1s^2$
Lithium	3	Li $1s^2\ 2s^1$
Carbon	6	C $1s^2\ 2s^2\ 2p^2$
Oxygen	8	O $1s^2\ 2s^2\ 2p^4$
Sodium	11	Na $1s^2\ 2s^2\ 2p^6\ 3s^1$
Phosphorus	15	P $1s^2\ 2s^2\ 2p^6\ 3s^2\ 3p^3$
Calcium	20	Ca $1s^2\ 2s^2\ 2p^6\ 3s^2\ 3p^6\ 4s^2$

electronic configurations, just remember to use the filling order shown in Figure 2.12. Consider again the location of electrons in the shells of a potassium atom (Section 2.3). Do you now see why the fourth shell gained an electron before the third shell was full?

Example 2.4 Electronic Configuration of Atoms

What is the electronic configuration of nitrogen?

SOLUTION

The atomic number of nitrogen (N) is 7; thus there are seven protons and seven electrons in a nitrogen atom. Electrons fill the lowest energy subshells first (Figure 2.12); thus two electrons will go into the 1*s* subshell, two more into the 2*s* subshell, and the remaining three into the 2*p* subshell. The electronic configuration is N $1s^22s^22p^3$.

Self-Test

What is the electron configuration for (a) fluorine, (b) sulfur, and (c) potassium?

ANSWERS

(a) F $1s^22s^22p^5$ (b) S $1s^22s^22p^63s^23p^4$ (c) K $1s^22s^22p^63s^23p^64s^1$

Orbitals

Electrons are found in shells and subshells of atoms, but there is still one more level of organization for electrons. Within any subshell are specific regions in space where the electrons can be found. These regions in space are called **orbitals.** Each orbital can hold a maximum of two electrons. A single, spherical orbital called the *1s orbital* is found in the first shell (Figure 2.13). Note that the name of an orbital indicates the subshell where the orbital is located: a 2*s* orbital is in the *s* subshell of the second shell. Each *s* subshell consists of a single *s* orbital. Figure 2.13 shows the shapes and relative sizes of the 1*s*, 2*s*, and 3*s* orbitals.

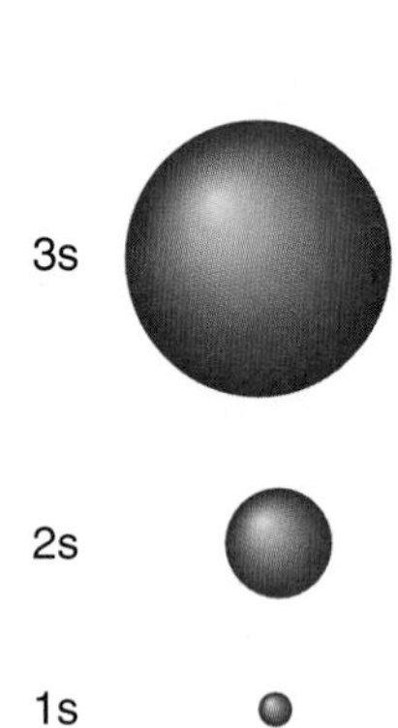

Figure 2.13 Each shell of electrons contains an *s* orbital that is spherical. The *s* subshell, which is an *s* orbital, increases in size as the shell number increases.

A second type of orbital is the *p* orbital. This orbital is dumbbell, or propeller, shaped (Figure 2.14). There are no *p* orbitals in the first shell, but each higher numbered shell contains three *p* orbitals located in the *p* subshell. Thus the 2*p* subshell contains three *p* orbitals that are designated the *2p orbitals.* The *p* orbitals in a given subshell point in different directions. Subscripts are sometimes added to the orbital symbol to indicate these directional differences (Figure 2.14).

There are two more types of orbitals found in atoms: *d* orbitals and *f* orbitals. These orbitals are more varied in shape than *s* and *p* orbitals. The third shell and all higher numbered shells have five *d* orbitals. The fourth shell and all higher numbered shells have seven *f* orbitals.

An atom consists of a nucleus surrounded by electrons located in orbitals. If we merge the orbitals of the first two shells of an atom (the 1*s*, 2*s*, and 2*p* orbitals), we get the image shown in Figure 2.15. Note that two or more orbitals often share the same space.

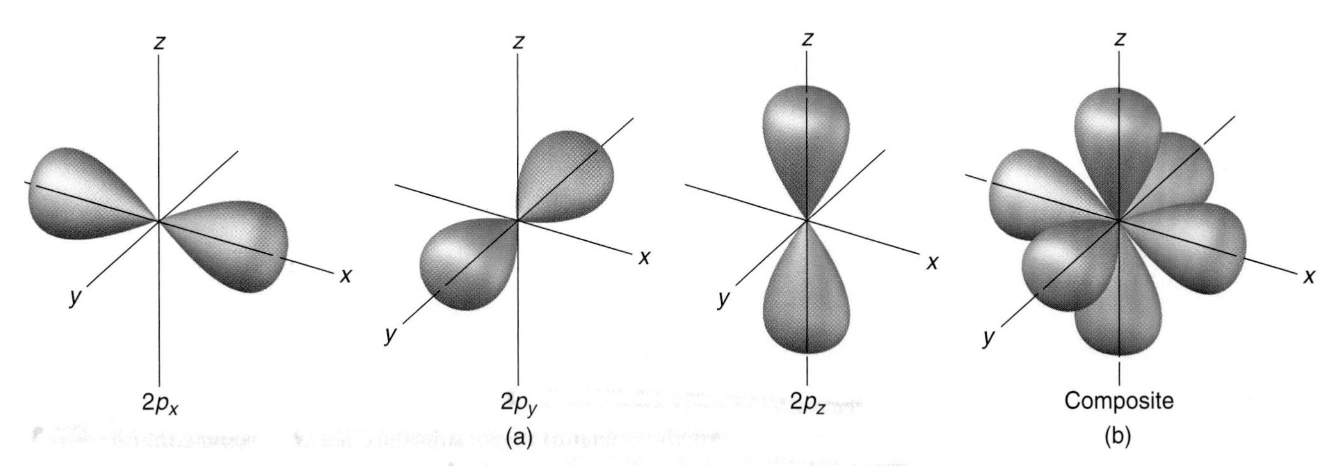

Figure 2.14 (a) Except for the first shell, each shell contains three *p* orbitals. The three *p* orbitals are identical in shape, but they point in different directions. The subscripts *x*, *y*, and *z* are sometimes used to show that the orbitals point along the *x*, *y*, and *z* axes of a three-dimensional graph. (b) This is the arrangement of the three 2*p* orbitals in the second shell.

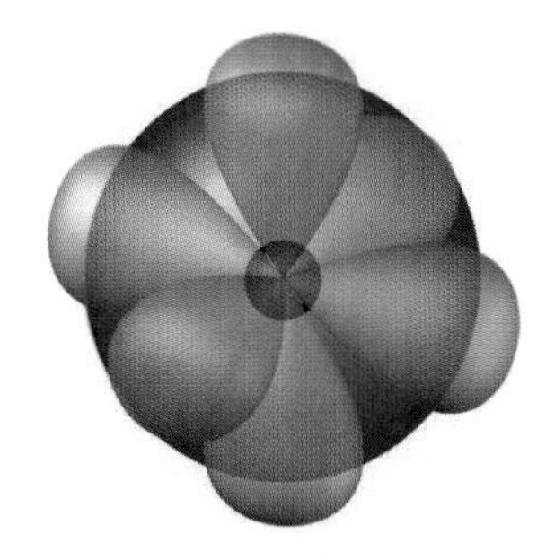

Figure 2.15 The orbitals of an atom overlap in space. This diagram shows the four orbitals found in the second shell (one *s* orbital and three *p* orbitals), and the 1*s* orbital of the first shell.

Self-Test

Atomic Orbitals

1. What does the term *3d orbital* mean?
2. How many orbitals are found in the second shell? What are these orbitals called?
3. How many electrons are found in a full 3*d* orbital?

ANSWERS

1. This means the orbital is located in the d subshell of the third shell.
2. Four; the 2s orbital and the three 2*p* orbitals.
3. Two electrons.

2.6 Nuclear Radiation

Radiation is a topic that is often in the news. You are probably aware that nuclear weapons produce radiation and that radiation is used in health care in a number of ways. As we learned earlier, light is a part of the electromagnetic spectrum and thus is one kind of electromagnetic radiation. Radiation also can involve high-speed particles. In this section we concentrate on **nuclear radiation,** which is radiation that originates in the nuclei of atoms.

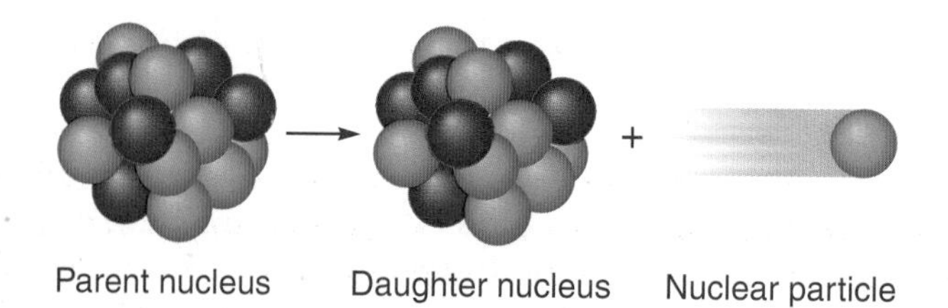

Figure 2.16 An unstable nucleus (parent) disintegrates to give a daughter nucleus and a nuclear particle.

The nuclei of most atoms are *stable*, which means they remain in their original form indefinitely, but the nuclei of some atoms are *unstable*. Atoms that have unstable nuclei are called **radioactive nuclides** (sometimes simply called *nuclides*). A radioactive nuclide (the parent nuclide) changes with time into an atom containing a different nucleus (the daughter nuclide), and this process is called **radioactive decay** (Figure 2.16). Unstable nuclei decay by emitting a particle that changes either the mass number or atomic number of the nucleus. Through one or more decays, the nucleus changes to a more stable one.

Half-Life

The concept of half-life is related to many other important topics; for example, in your studies you will probably hear about the half-life of a drug in a patient.

The instability of a radioactive nuclide is expressed as a **half-life,** which is the time required for one half of a sample of the nuclide to decay. In Figure 2.17, 200 atoms of a radioisotope exist initially, but only 100 exist after three years—one half-life—and 50 atoms exist after six years—two half-lives. The half-lives of radioactive nuclides vary greatly: carbon-14 = 5730 years; uranium-238 = 4.5 billion years; polonium-214 = 1 millisecond. Some nuclides are used in health care, and a short half-life is a desirable factor when choosing these nuclides. For example, nitrogen-13, which is used to image organs in the body, has a half-life of 10 min.

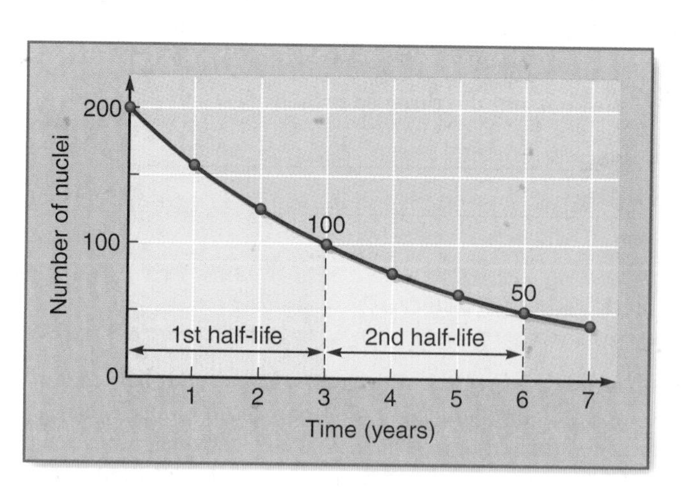

Figure 2.17 One half-life is the amount of time needed for one-half of a radioactive sample to decay. In this example, half of the sample has decayed every three years. The half-life is therefore three years.

Although the number of atoms of a radioactive nuclide that remain after any period of time can be calculated, we will restrict our calculations to times involving whole numbers of half-lives. Let's use a sample of phosphorus-32 as an example. Phosphorus-32 has a half-life of 14 days, and is used both in biochemical research and to treat bone pain in some breast cancer or prostate cancer patients. If we begin with a 64-μg sample of this isotope, half of it will decay in one half-life—in 14 days—leaving 32 μg of the sample. In another 14 days, another half of the sample will decay leaving 16 μg. This decay can be represented in this way:

$$64\ \mu\text{g phosphorus-32} \xrightarrow[\text{14 days}]{\text{1 half-life}} 32\ \mu\text{g phosphorus-32} \xrightarrow[\text{14 days}]{\text{1 half-life}} 16\ \mu\text{g phosphorus-32}$$

The labeled arrow indicates that one half-life—14 days—has passed. How much of the phosphorus-32 will remain after a third half-life? One-half of the 16 μg will remain, which is 8 μg.

Determining the amount of a radioactive sample that remains after the passage of one or more half-lives can be done with multiplication. Take the original amount of the sample and multiply it by 0.5 (½) for each half-life that has passed. To determine how much of our original phosphorus-32 sample would remain after four half-lives, we would multiply the original mass by 0.5 four times:

$$64\ \mu\text{g phosphorus-32} \times 0.5 \times 0.5 \times 0.5 \times 0.5 = 4\ \mu\text{g phosphorus-32}$$

Another way to solve this problem would be to begin with the original mass and divide the mass by 2 for each half-life:

$$\overset{\text{1st half-life}}{\frac{64\ \mu\text{g}}{2}} \rightarrow \overset{\text{2nd half-life}}{\frac{32\ \mu\text{g}}{2}} \rightarrow \overset{\text{3rd half-life}}{\frac{16\ \mu\text{g}}{2}} \rightarrow \overset{\text{4th half-life}}{\frac{8\ \mu\text{g}}{2}} \rightarrow 4\ \mu\text{g phosphorus-32}$$

Example 2.5 Half-Life

The radioactive nuclide iodine-131 is used in the diagnosis and treatment of thyroid conditions. The patient drinks a liquid containing sodium iodide that has a small amount of iodine-131. The iodide, and thus the iodine-131, is taken up by the thyroid gland and the amount in the gland is measured. Use iodine-131's half-life of 8.0 days to answer the following questions:

(a) How much of a 96-μg sample of iodine-131 will remain after one half-life?
(b) How much of the original sample would remain after 16 days?
(c) How much of the original sample would remain after five half-lives?

SOLUTIONS

(a) By definition, one-half of a sample will decay in one half-life, and so we can calculate the amount remaining by multiplying the original mass by 0.5 (½): 96 μg $\times$ 0.5 = 48 μg of iodine-131.

(b) For this question we must first convert time into half-lives, then use the number of half-lives to determine the mass of the remaining iodine-131. To determine the number of half-lives in 16 days use the original time and a conversion factor derived from the definition of half-life for iodine-131:

$$16 \text{ days} \times \frac{1 \text{ half-life}}{8.0 \text{ days}} = 2 \text{ half-lives}$$

Now multiply the original mass by 0.5 twice (once for each half-life) to determine the amount of the sample that remains:

$$96\ \mu\text{g iodine-131} \times 0.5 \times 0.5 = 24\ \mu\text{g iodine-131}$$

(c) Multiply the original mass of the sample by 0.5 five times:

$$96\ \mu\text{g iodine-131} \times 0.5 \times 0.5 \times 0.5 \times 0.5 \times 0.5 = 3.0\ \mu\text{g iodine-131}$$

Remember, you could also have divided the masses by 2 five times:

$$\frac{96\ \mu\text{g}}{2} \rightarrow \frac{48\ \mu\text{g}}{2} \rightarrow \frac{24\ \mu\text{g}}{2} \rightarrow \frac{12\ \mu\text{g}}{2} \rightarrow \frac{6\ \mu\text{g}}{2} \rightarrow 3.0\ \mu\text{g iodine-131}$$

Self-Test

1. The radioactive nuclide cobalt-60 is used to destroy or reduce tumors. How much of a 2.4-μg sample of cobalt-60 will remain after two half-lives?
2. The half-life of strontium-90 is 28.8 years. How much of a 0.36-μg sample of strontium-90 will remain after 86.4 years?

ANSWERS

1. 0.60 μg of cobalt-60 2. 0.045 μg of strontium-90

Figure 2.18 Legend says that the Shroud of Turin is the burial shroud of Christ. Recent dating using carbon-14 indicates that the shroud is less than 1000 years old; however, some of the procedures used in this study have been questioned. Thus the age of the shroud remains in dispute. The origin of the image on the shroud is not yet understood. *(Shroud of Turin Project)*

Nuclear Dating

Carbon-14, which has a half-life of 5730 years, can be used to estimate the age of objects containing material from plants or animals. This is because some carbon-14 containing carbon dioxide becomes incorporated into plants. In living plants there is enough carbon-14 to provide about 14 disintegrations per minute (dpm) for each gram of carbon in the sample. When a tree or other plant dies, it no longer takes in any carbon dioxide, and so the amount of carbon-14 in the sample decreases with time. If the radioactivity of fresh plant material and that of an old sample are compared, the age of the sample can be estimated (Figure 2.18).

Example 2.6 Carbon-14 Dating

A piece of wood from a carved bowl has been found buried in a cave. The sample has a measured activity of 7.0 dpm from carbon-14, per gram of carbon in the sample. What is the approximate age of the wood?

SOLUTION

A piece of living wood has an activity of 14 dpm. The wood sample has half the activity (7 dpm/14 dpm) that is in fresh wood. This means that one half-life has passed. Because the half-life for carbon-14 is 5730 years, the sample is about 5730 years old.

Self-Test

What if the wood sample from the cave in Example 2.6 had an activity of 3.5 dpm per gram. What would its age be?

ANSWER

Two half-lives have passed (14 dpm × ½ × ½ = 3.5 dpm); the sample is around 11,460 years old (5730 yr/half life × 2 half-lives).

Types of Radiation

Radioactive nuclides become more stable by emitting energy and mass. These emissions are called *nuclear radiation*, and in this section we consider the three most common: alpha, beta, and gamma radiation. **Alpha (α)** radiation consists of *alpha particles* that have two protons and two neutrons. Compared with other radiations, alpha particles are very heavy. They have the same composition as the nucleus of a helium-4 atom, and alpha particles are represented with the symbol for the helium-4 nucleus: ${}^{4}_{2}He$. **Beta (β)** radiation consists of *beta particles*, which are electrons. Compared to alpha particles, beta particles have an insignificant mass and are represented with the symbol ${}^{0}_{-1}e$, where the "e" is a symbol for an electron. **Gamma (γ)** radiation is similar to x-rays. These high-energy radiations have no measurable mass. A symbol for gamma rays is ${}^{0}_{0}\gamma$.

NUCLEAR RADIATION	SYMBOL	RELATIVE MASS
Alpha (α)	${}^{4}_{2}He$	Heavy
Beta (β)	${}^{0}_{-1}e$	Light
Gamma (γ)	${}^{0}_{0}\gamma$	None

Alpha Decay

Radioactive decay, the spontaneous emission of radiation, can be represented by *nuclear equations*, which resemble mathematical equations.

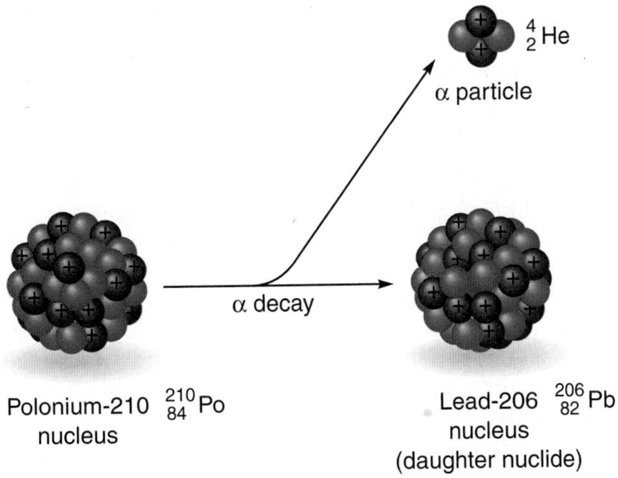

$${}^{210}_{84}Po \xrightarrow{\alpha} {}^{206}_{82}Pb + {}^{4}_{2}He$$

polonium-210 → lead-206 + alpha particle

Radioactively unstable nuclide → Daughter nuclide + Alpha radiation

The equation begins with the unstable nuclide, which is followed by an arrow. The arrow and α signify that the unstable nuclide undergoes alpha decay and changes to a new nuclide, commonly referred to as the daughter nuclide. The daughter nuclide is shown to the right of the arrow, as is the radiation produced by the decay. This nuclear equation is read this way: the polonium-210 nuclide undergoes alpha decay to yield a *daughter nuclide* of lead-206 and an alpha particle. These equations, like the chemical equations you will learn about later, must be balanced so that the mass is the same on each side of the equation. A nuclear equation is balanced when both sides have the same number of protons and the same number of nucleons (protons + neutrons).

Because a nuclear equation is balanced, the daughter nuclide can be determined by examining the atomic numbers and mass numbers on both sides of the equation. An alpha particle (helium-4 nucleus) has an atomic number of 2 and a mass number of 4. If the atomic number of the alpha particle is subtracted from the atomic number of the parent nuclide, the atomic number of the daughter nuclide can be determined: $84 - 2 = 82$, which means the daughter nuclide contains 82 protons. A look at the periodic table shows lead is the element that has 82 protons in its nucleus, thus the daughter nuclide must be lead. The mass number of the daughter nuclide is determined by subtracting 4 (the mass number of an alpha particle) from the mass number of the parent nuclide: $210 - 4 = 206$. The daughter nuclide is lead-206.

Is the nuclear equation balanced? The atomic number of polonium is 84, and the sum of the atomic numbers of the alpha particle and lead is also 84 (2 + 82). There are 210 nucleons in the polonium nucleus, and the sum of the mass numbers of the alpha particle and lead is also 210 (4 + 206). The equation is balanced.

Example 2.7 Predicting the Products of Alpha Decay

Determine the products of this alpha decay of thorium:

$$^{232}_{90}Th \xrightarrow{\alpha} ?$$

SOLUTION

Because this is alpha decay, one product is an alpha particle (a helium-4 nucleus). The daughter nuclide will have two fewer protons and four fewer nucleons that the parent nuclide. Radium (Ra) is the element that has 88 protons. The nuclear equation is therefore

$$\underset{\text{Unstable nuclide}}{^{232}_{90}Th} \xrightarrow{\alpha} \underset{\text{Alpha particle}}{^{4}_{2}He} + \underset{\text{Daughter nuclide}}{^{228}_{88}Ra}$$

Is the equation balanced? The mass number is 232 on the left side of the arrow, and 232 (4 + 228) on the right side of the arrow. The atomic number is 90 on the left, and 90 (2 + 88) on the right. The equation is balanced.

Self-Test

1. $^{222}_{86}Rn \xrightarrow{\alpha} ?$
2. $^{238}_{92}U \xrightarrow{\alpha} ?$

ANSWERS

1. $^{222}_{86}\text{Rn} \xrightarrow{\alpha} {}^{4}_{2}\text{He} + {}^{218}_{84}\text{Po}$
2. $^{238}_{92}\text{U} \xrightarrow{\alpha} {}^{4}_{2}\text{He} + {}^{234}_{90}\text{Th}$

Beta Decay

Carbon-14 undergoes beta decay. This nuclide emits a beta particle—that is, an electron. This seems odd because you learned that the nucleus contains only protons and neutrons. A neutron can be viewed, perhaps simplistically, as a particle made up of a proton and an electron. The charges of the proton and electron cancel, leaving no net charge on the neutron. Because the mass of an electron is so small, the mass of a neutron is about the same as that of a proton. If a carbon-14 nucleus emits a beta particle, then a neutron in the nucleus must have been converted to a proton. Beta decays are shown by equations like this one:

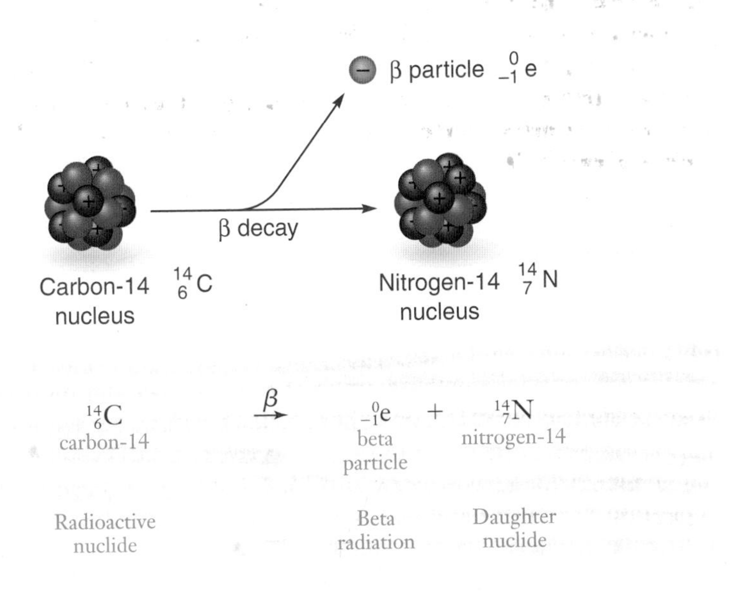

For beta decay, the atomic number of the daughter nuclide is determined by subtracting −1 from the atomic number of the parent nuclide. Remember, when you subtract a negative number, it is the same as adding. The atomic number for the daughter nuclide that is produced by beta decay of carbon-14 will be: $6 - (-1) = 7$. The atomic number of nitrogen is 7, thus the daughter nuclide is a nitrogen atom. For beta decay, the number of nucleons in the daughter nuclide is always equal to the number of nucleons in the parent nuclide. Because carbon-14 has 14 nucleons, the daughter nuclide is nitrogen-14.

Example 2.8 Balancing Equations for Beta Decay

Determine the daughter nuclide for the beta decay of phosphorus-32, which is used in treating malignant tumors, and write the complete balanced equation:

$$^{32}_{15}\text{P} \xrightarrow{\beta} ?$$

SOLUTION

In beta decay, an electron, ${}^{0}_{-1}e$, is emitted. The atomic number (number of protons) in the daughter nuclide is determined by subtracting the −1 of the beta particle from the atomic number of the parent nuclide: 15 − (−1) = 16. Because there are 16 protons in the daughter nuclide, it is sulfur. For all beta decays, the number of nucleons in the product is the same as the number of nucleons in the parent—in this case, 32. The daughter nuclide is sulfur-32, and the complete balanced equation is

$${}^{32}_{15}P \xrightarrow{\beta} {}^{0}_{-1}e + {}^{32}_{16}S$$

Self-Test

Determine the daughter nuclide and write the complete balanced equations for these beta decays, both of which are used in treating cancer:

1. ${}^{131}_{53}I \xrightarrow{\beta} ?$
2. ${}^{60}_{27}Co \xrightarrow{\beta} ?$

ANSWERS

1. ${}^{131}_{53}I \xrightarrow{\beta} {}^{0}_{-1}e + {}^{131}_{54}Xe$
2. ${}^{60}_{27}Co \xrightarrow{\beta} {}^{0}_{-1}e + {}^{60}_{28}Ni$

Gamma Decay

Alpha decays and many beta decays simultaneously yield gamma radiation. There are, however, a few examples of radioactive decay that emit only gamma radiation. Because gamma radiation has no charge and no mass, the daughter nuclide is the same as the parent, but it has lost energy, and so is more stable. One of the most commonly used gamma emitters in nuclear medicine is technetium (Tc). The unstable form is said to be metastable, thus it is symbolized as technetium-99m or ${}^{99m}_{43}Tc$. This is an example of a gamma decay:

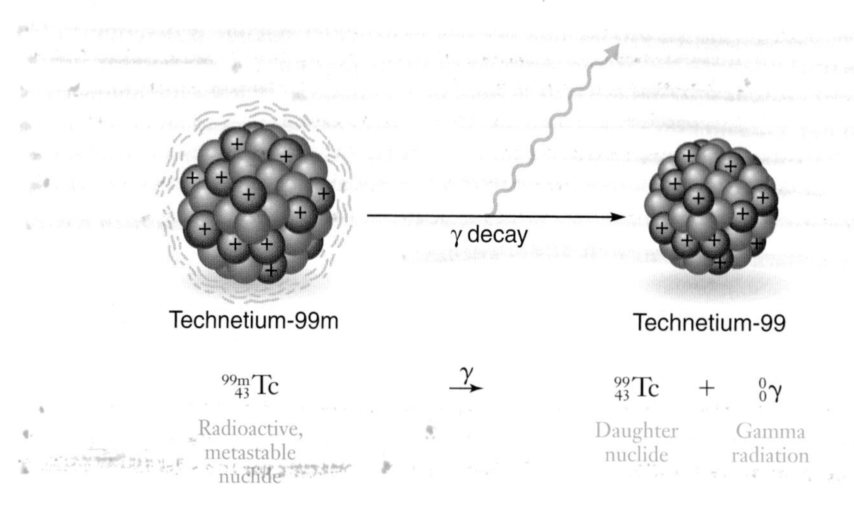

$${}^{99m}_{43}Tc \xrightarrow{\gamma} {}^{99}_{43}Tc + {}^{0}_{0}\gamma$$

Radioactive, metastable nuclide — Daughter nuclide — Gamma radiation

Connections

Positron Emission Tomography (PET)

Several imaging techniques allow a physician to examine an organ's function without the risks of exploratory surgery. One of these imaging techniques, positron emission tomography (PET), uses nuclear decay involving a positron. A positron is an electron-sized particle that has a positive charge, and positron decay is illustrated with this equation for the emission of a positron by fluorine-18:

$$^{18}_{9}F \rightarrow \underset{\text{Positron}}{^{0}_{+1}e} + ^{18}_{8}O$$

A positron is the antimatter equivalent of matter's electron, and any positron produced by positron decay reacts very quickly with an electron in the surrounding tissue. The products of the reaction of a positron and electron are two gamma rays that travel in opposite directions:

$$\underset{\text{Positron}}{^{0}_{+1}e} + \underset{\text{Electron}}{^{0}_{-1}e} \rightarrow \underset{\text{Gamma rays}}{2^{0}_{0}\gamma}$$

Before a patient is given a PET scan, the patient is given a positron-emitting substance that is used (metabolized) by the organ to be studied. After the nuclide enters the organ, the patient undergoes the PET scan. When the nuclide decays and the positron reacts with an electron, the resulting gamma rays pass through the patient and strike a surrounding detector. This generates an electrical signal that is then stored in a computer. After the scan is complete, the computer uses the collected data to generate a three-dimensional image of the organ in the patient. More gamma radiation will be emitted from the most active parts of the organ. Analysis of the scan can find either overactive regions (perhaps a tumor) or underactive regions (degeneration) in the organ.

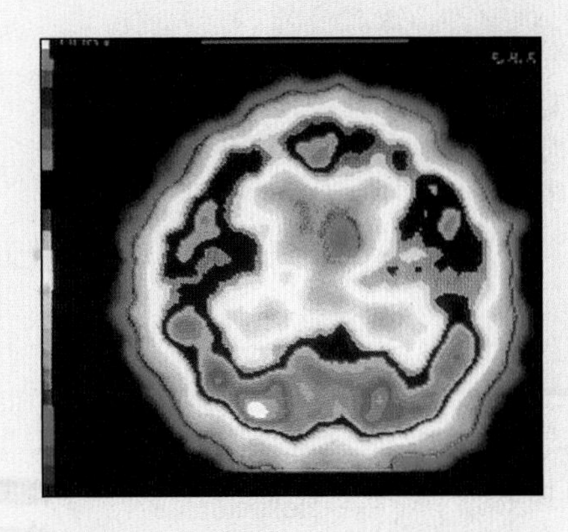

Positron emission tomography (PET) is a powerful diagnostic tool for health care. This PET scan indicates the brain of this patient is normal. *(CEA-ORSAY/CNRI/Science Photo Library/Photo Researchers)*

2.7 Detecting and Measuring Radiation

There are many forms of radiation all around us. Some radiation is helpful, some is harmful. Understanding radiation and knowing how to detect it are important and useful information. Detecting radiation involves the interaction of the radiation with matter. When radiation strikes an atom, that atom absorbs energy. If the radiation has enough energy, an electron can be knocked from the atom (Figure 2.19). The atom is no longer neutral, and it is now an ion (you will learn more about ions in Chapter 3). These radiations are sometimes called *ionizing radiations*. Photographic film is perhaps the simplest detector for ionizing radiation. The radiation exposes the film because the electrons formed during ionization combine with silver in the film, causing the silver to appear dark when the film is developed. The amount of radiation is proportional to the darkness of the exposed film. X-ray technologists and dentists wear a badge that contains film to monitor their exposure to ionizing radiation.

Another common detector of ionizing radiation is the *Geiger-Müller tube*, which is a key component of a Geiger counter (Figure 2.20). This detector has a gas-filled tube whose walls are electrically charged. When ionizing radiation passes

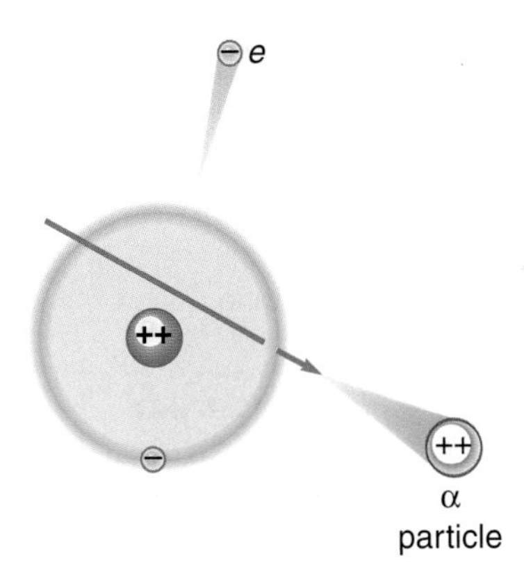

Figure 2.19 As an alpha particle passes through an atom, it may knock an electron from the atom forming an ion. It is the ions formed by ionizing radiation that cause tissue damage.

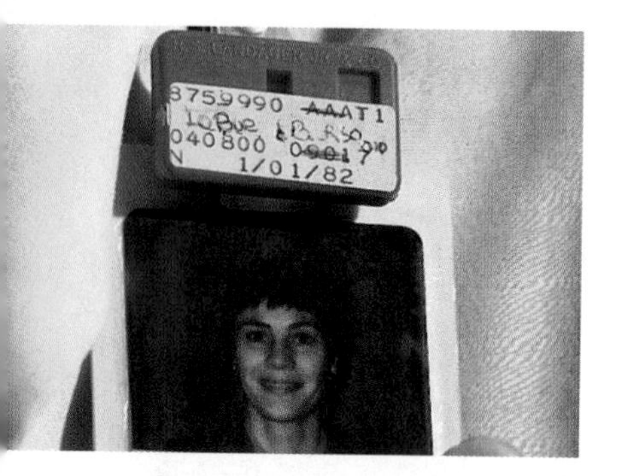
This worker wears a badge to detect exposure to radiation. *(Yoav Levy, Phototake)*

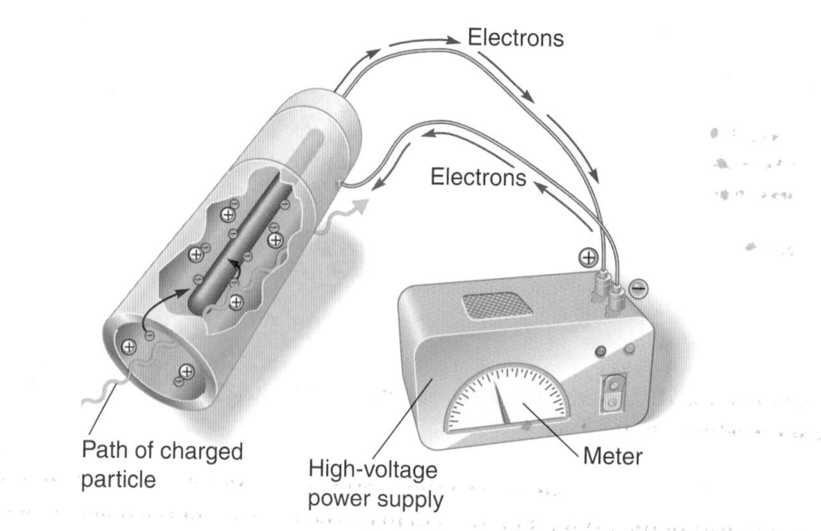

Figure 2.20 A Geiger-Müller tube is the radiation detector of a Geiger counter.

through the tube, it strikes atoms of the gas and knocks electrons from them. The electrons and ions that are produced are attracted to the walls of the tube, and when they strike the walls an electric signal is generated. The strength of the electric signal increases with the amount of ionizing radiation.

Another type of radiation detector involves the production of light by the interaction of radiation and matter. A *scintillation detector* contains a substance that releases light when it is struck by radiation (Figure 2.21). The detector measures the amount of light produced, which is proportional to the amount of radiation.

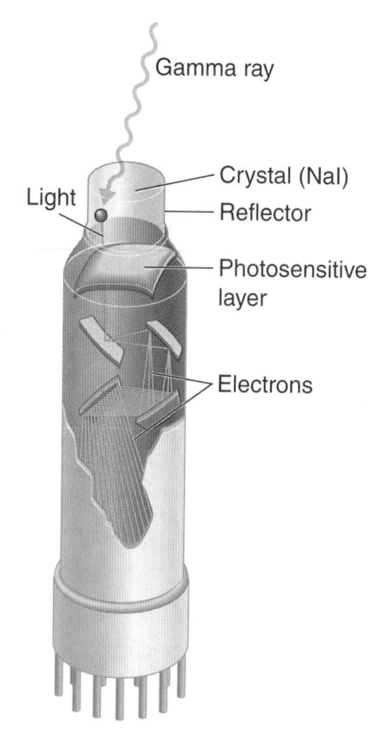

Figure 2.21 In a scintillation detector, radiation strikes a substance that emits light when struck. The light then generates an electric current that indicates the presence and amount of radiation.

Units of Radiation

Once radiation is detected and measured, the amount is expressed in some particular unit. The **curie (Ci)** (named for Marie and Pierre Curie, who discovered the radioactive elements radium and polonium) expresses the number of disintegrations per second (dps), where a *disintegration* is one nuclear decay. One curie equals 3.7×10^{10} dps.

The curie is fine for measuring counting rates, but the number of counts per unit time does not adequately express the danger the radiation poses to a living organism. The greater the amount of energy in a radiation, the greater the risk to health. This is because energy from radiation either destroys or alters proteins and genetic material in cells. A second unit, the **radiation absorbed dose (rad),** expresses the amount of energy associated with the radiation. Tissues exposed to 1 rad absorb 1×10^{-2} J of energy per kg of tissue.

There is still one more factor that must be considered regarding radiation and health. Not all radiations are equally penetrating nor equal in their ability to cause ionization. Thus they vary in their ability to damage tissues. Because of this, still another unit of radiation, the **radiation equivalent for man (rem),** is used to compensate for these differences. One rem is one rad multiplied by a factor called *relative biological effectiveness*, RBE:

$$1 \text{ rem} = 1 \text{ rad} \times \text{RBE}$$

Of course, "man" should now be human or person; should we call it a rep?

Table 2.7 Biological Effects of Exposure to Radiation

DOSE IN rem	EFFECT
0–25	No observable effect; genetic damage possible
25–100	Temporary decrease in the number of white blood cells
100–200	Vomiting, diarrhea (mild radiation sickness); strong decrease in number of white blood cells
500	Death in half the population (this is the LD_{50} value for humans)

The value for RBE is 1 for beta and gamma radiations and about 20 for alpha particles produced in the body.

How much exposure to radiation is safe? Table 2.7 summaries the effects of radiation exposure. For a single exposure, 500 rem is the dose of radiation that is lethal to half of the exposed humans (lethal dose for 50% of the population is symbolized LD_{50}). Most normal exposures to radiation are much smaller. The typical annual exposure to diagnostic x-rays, for instance, corresponds to a radiation dose of about 0.05 rem (50 mrem). In radiation treatment of a tumor, intense radiation is focused on the tumor, exposing it to lethal doses of radiation while surrounding tissues receive nonlethal doses.

High-energy radiation is dangerous. As the rays pass through tissue, they ionize molecules in the body by knocking electrons from the molecules. The molecules may be permanently damaged, or they may react with other molecules and destroy them. The result is damage, temporary or permanent, to important molecules such as proteins and genetic material. If the dose is very large, so much damage will occur that death or serious injury will result. If the dose is small, no immediate observable effects will be seen, but damage to genetic material could lead to birth defects in offspring or cancer in the recipient.

The best protection against ionizing radiation is either to avoid it or to provide shielding materials, such as lead, between the radiation source and the workers. Ionizing radiations vary in their ability to penetrate matter (Figure 2.22). Workers require only modest protection from alpha particles because they penetrate skin poorly (ingested or inhaled sources of alpha particles are a much more serious

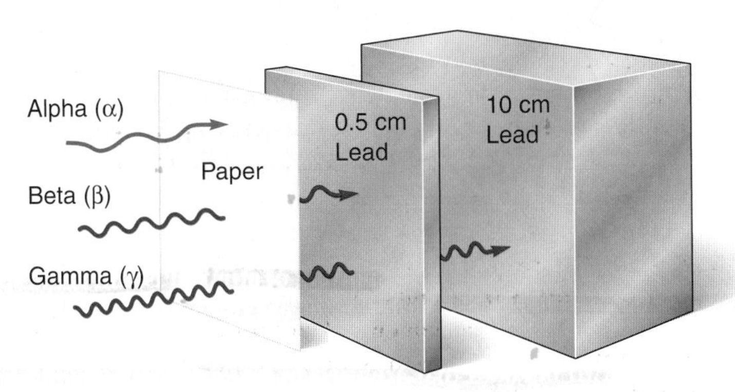

Figure 2.22 Radiations vary in their ability to penetrate matter. The penetration of x-rays is similar to that of gamma rays.

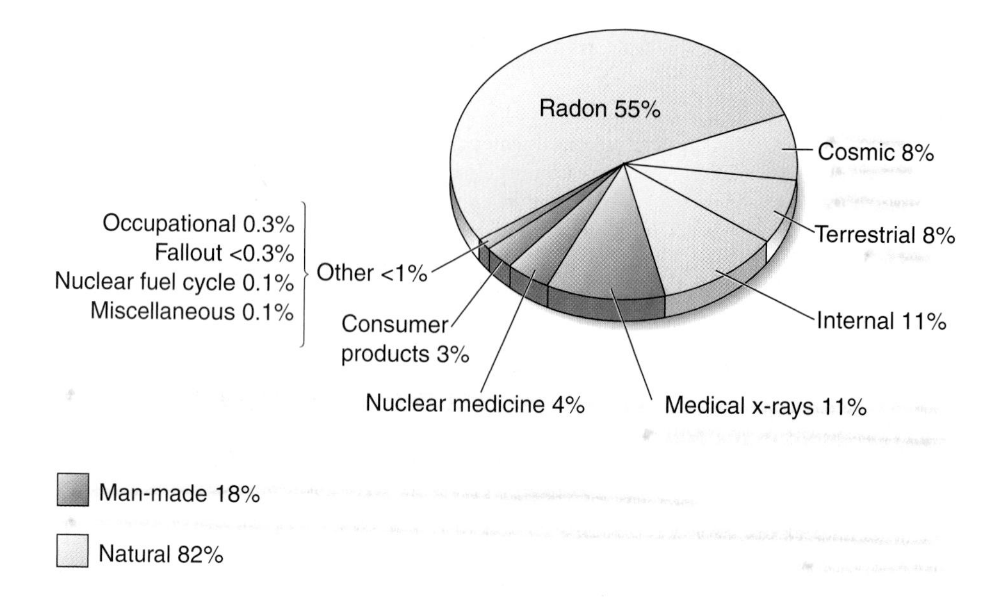

Average exposure to radiation of a U.S. resident.

matter). Beta particles are intermediate in their ability to penetrate, and gamma radiation is the most penetrating. Health professionals and others who work with radiation should always minimize their exposure by following safety procedures that establish proper shielding and distance between the radiation and worker and adequate time between exposures. When radioactive materials are used properly, there is little increased risk.

It is not possible for anyone to avoid radiation entirely. This is because *background radiation* from natural sources such as soil and cosmic rays is always present in our environment. In addition, the mineral potassium contains small amounts of the radioactive nuclide potassium-40, and because potassium is present in food and is concentrated in cells, the body is constantly exposed to the radiation from potassium-40 decay.

2.8 Induced Nuclear Reactions

Many nuclear reactions occur spontaneously, but there are other nuclear reactions that occur only under certain conditions. These latter nuclear reactions are used for a variety of purposes. Some serve a positive role, such as the production of isotopes used in health care and the production of energy in nuclear power plants. Others have a potentially more ominous role in nuclear weapons.

Fission

One kind of nuclear reaction involves the splitting of atoms. **Fission** is the splitting of a large nucleus into smaller nuclei with the release of enormous amounts of energy. The energy is formed as tiny amounts of the mass of the atoms are changed to energy. There are two nuclides that are used in fission reactions: uranium-235 and plutonium-239. When hit by a neutron, a uranium-235 nucleus absorbs the neutron to become uranium-236. This isotope then splits into two daughter nuclei and several neutrons, with the release of large amounts of energy. There are many possible daughter nuclei, and the number of neutrons produced varies as well. One particular fission of uranium-235 is:

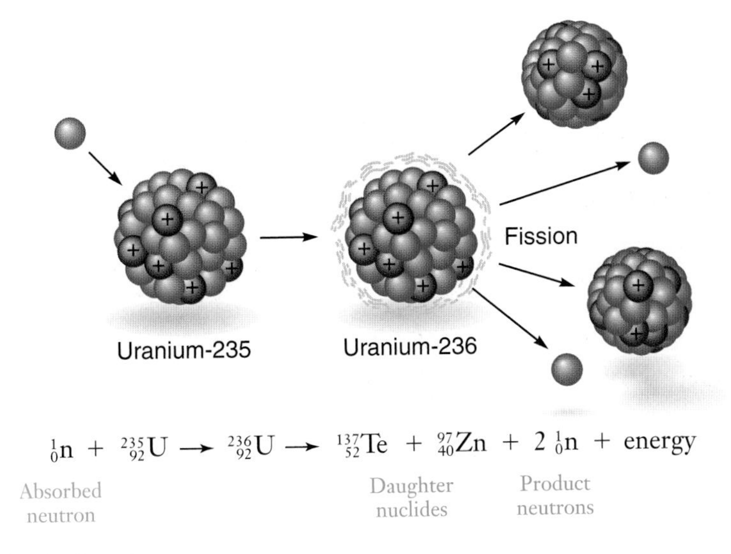

Fission Chain Reactions

Fission reactions produce neutrons as well as energy and smaller daughter nuclei. As these product neutrons fly through a uranium sample, they collide with other uranium-235 nuclei, producing additional fission events, which in turn produce more neutrons. If, on average, fewer than one of the product neutrons from a fission event causes an additional fission event, then the process dies out. The process is said to be *subcritical.* But if, on average, one neutron from each fission event produces another fission event, then a controlled *chain reaction* occurs, and fission is self-sustaining. The process is *critical.* This is what occurs in nuclear power plants. Finally, if, on average, more than one neutron from each fission event causes another fission event, then the number of fission events increases very rapidly, and an uncontrolled chain reaction occurs. This process is *supercritical*, and this is what occurs in the explosion of an atomic bomb (Figure 2.23).

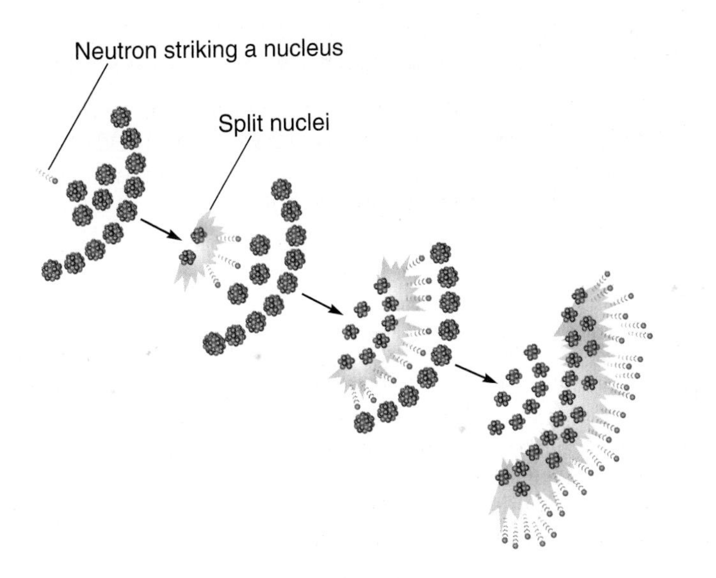

Figure 2.23 A nuclear chain reaction occurs when the neutrons produced in a fission event strike other nuclei, causing them to split. In this example, the number of nuclei that are split is constantly increasing. If this continues, an uncontrolled explosion (an atomic bomb) is the result.

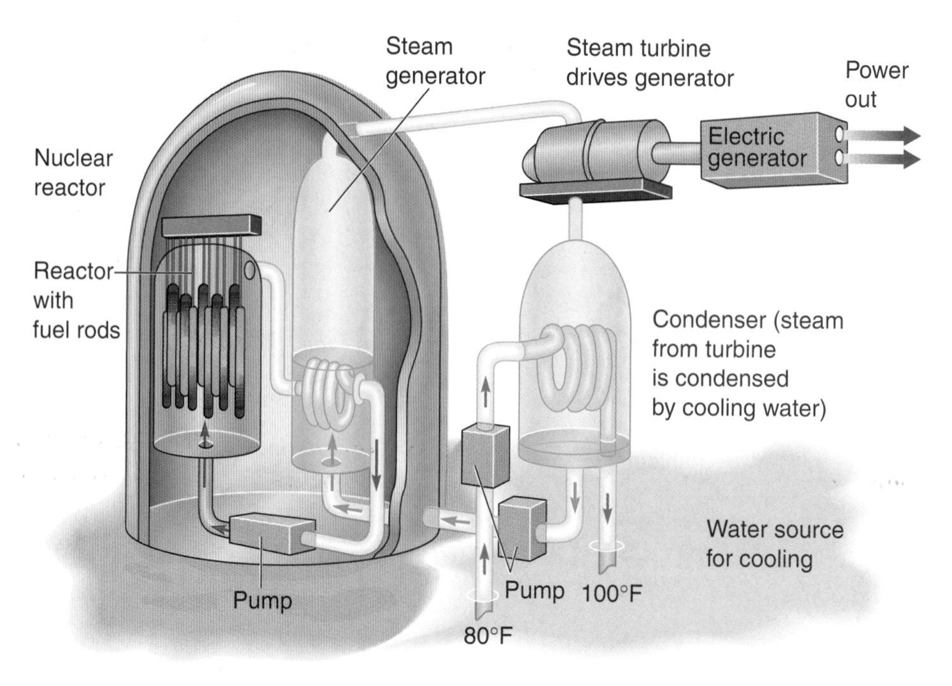

Figure 2.24 This diagram shows the operation of a nuclear reactor and its generator.

Connections

Radiation in the Food Web

We are constantly exposed to very small quantities of radiation from natural sources such as potassium-40 in our diets or carbon-14 in carbon dioxide in the air. Unfortunately, our exposure to radiation has increased because of radiation produced by nuclear reactions resulting from human activity. Consider the radiation produced by atmospheric testing of atomic bombs in the 1940s and 1950s. When the uranium-235 or plutonium-239 in these bombs split, smaller atoms were formed and scattered throughout the atmosphere by the explosion and winds. Many of these daughter nuclides were radioactive, and as they fell to Earth, living organisms became exposed to radiation from them. Some of these radioactive nuclides were potentially more dangerous than others, however, because they were taken up by living organisms and became concentrated in the food web (food chain).

One of the radioactive isotopes formed in atmospheric testing of nuclear weapons that became concentrated in the food web was strontium-90. As grazing animals fed on plants, they ingested some of the fallen strontium-90. Because strontium and calcium are both in Group 2A, they have similar properties, and strontium-90 became incorporated into calcium-rich materials in these animals. The strontium-90 that became concentrated in the milk of cattle, goats, reindeer, and other domestic milk-producing animals became concentrated in the bones of the people who drank the milk.

More recently, in 1986 a reactor meltdown occurred in a nuclear-powered electricity-generating plant at Chernobyl in Ukraine. This accident released huge quantities of radioactive materials that spread for many kilometers around the plant. Among the released materials was radioactive cesium, which like potassium is in Group 1A. Cesium thus resembles potassium, and some of the cesium was absorbed by living organisms. The radioactive cesium became incorporated into the organisms living in this contaminated area and became more concentrated in animals higher in the food web.

Nuclear power plants generate significant amounts of electrical energy using fission reactors (Figure 2.24). Nuclear power provides more than 7% of the total energy needs in the United States. This may not sound like much until you think about the huge amounts of energy used each year. Some nations rely much more on nuclear power than the United States. In France around 70% of the electricity comes from nuclear power, and in Japan, it is nearly 30%. At the present time, there are more than 350 nuclear power plants in the world.

Concerns about safe operation of nuclear plants and about the disposal of nuclear wastes have limited development of the nuclear industry in the United States. The United States has around 100 nuclear plants, but no new ones are being built.

Fusion

Fission *splits* large nuclei into smaller ones. **Fusion** *combines (fuses)* smaller nuclei into larger ones, and in this way also enormous amounts of energy are released. The energy from the sun is generated by fusion reactions in the center of the sun. A common fusion reaction in the sun involves the joining of two hydrogen atoms (hydrogen-2, which is called *deuterium*) to form a helium nucleus:

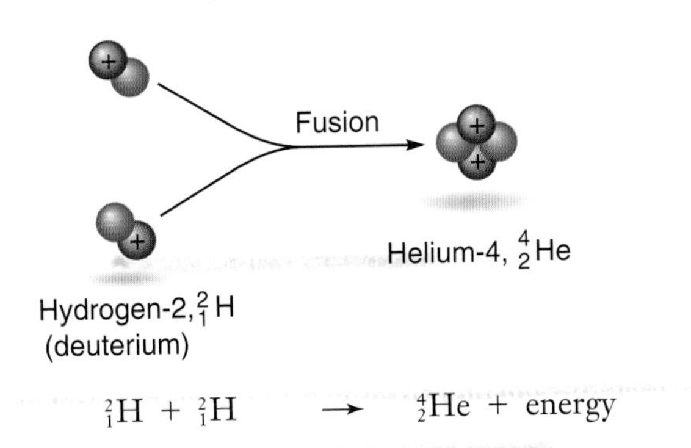

$$^{2}_{1}H + ^{2}_{1}H \longrightarrow ^{4}_{2}He + \text{energy}$$

Fusion occurs only at very high temperatures like those in the sun. A hydrogen bomb is made by surrounding a fission (atomic) bomb with small nuclei that can undergo fusion. The atomic bomb serves as a trigger. When it goes off, huge amounts of energy are released. This creates the extremely high temperatures that are needed to fuse the light nuclei. Because such high temperatures are required, fusion research has progressed very slowly. If controlled fusion can be achieved, then an almost endless supply of energy will be available from the small atoms such as hydrogen found on Earth.

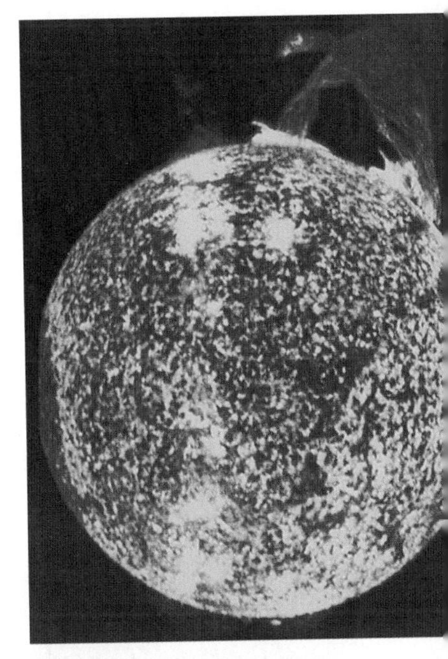

The energy from the sun is produced by nuclear fusion. *(NASA)*

Nuclear Transformation

Nuclei repel particles such as neutrons and protons unless these particles collide with the nucleus with very high energy. Fusion of small nuclei occurs because the extremely high temperatures needed for fusion provide this energy. **Nuclear transformation** involves the conversion of one nucleus into another by high-energy bombardment with a particle. When the nucleus and particle collide, they fuse into a new nucleus. Many radioactive nuclides are formed in this way, including all those with an atomic number greater than 92 (the *transuranium elements*). Uranium is the largest naturally occurring element. Since 1940, around 20 transuranium elements have been made by transformation. The first to be made was neptunium-239:

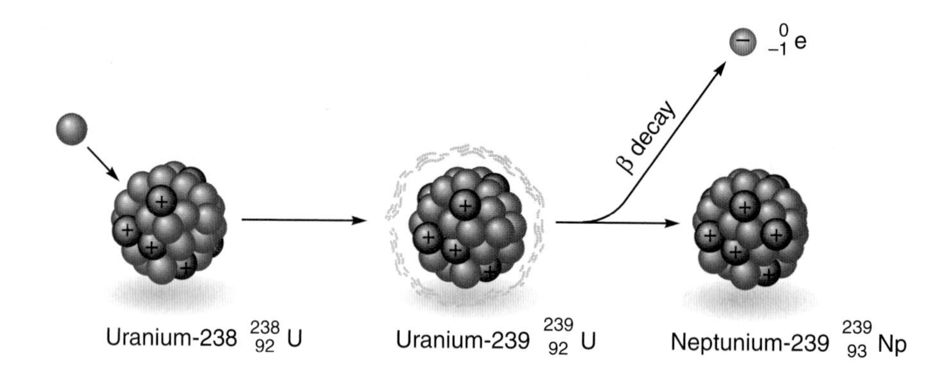

$$^{1}_{0}n + ^{238}_{92}U \xrightarrow{\text{fusion}} ^{239}_{92}U \xrightarrow{\beta\text{-decay}} ^{239}_{93}Np + ^{0}_{-1}e$$

In this nuclear transformation, a uranium-238 nucleus fuses with a neutron to form uranium-239. Uranium-239 does not fission the way uranium-235 does; instead it undergoes beta decay to form neptunium-239. Neptunium-239 in turn undergoes beta decay to form plutonium-239:

$$^{239}_{93}Np \xrightarrow{\beta\text{-decay}} ^{239}_{94}Pu + ^{0}_{-1}e$$

Nuclear transformations are also used to make radioisotopes smaller than the transuranium elements, and some of these nuclides are used in health care. Cobalt-60, which is used for radiation treatment of cancer, is formed in this way:

$$^{59}_{27}Co + ^{1}_{0}n \rightarrow ^{60}_{27}Co + ^{0}_{0}\gamma$$

Uses of Radioactive Nuclides

There are several significant applications for radioactive nuclides, including food processing and health care (you saw the use of uranium in energy production earlier).

Food Processing

A developing use of radiation is the irradiation of foods to improve shelf life (Figure 2.25). Much of the spoilage of foods is a result of bacteria and other microorganisms rather than internal breakdown of materials in the foods. Irradiation kills many of the microorganisms that promote spoilage, which greatly increases the shelf life of the food. Irradiation can also be used to reduce insect damage to food and retard sprouting in root crops.

Public acceptance of food irradiation has not been rapid nor is it universal. The gamma radiation used in food processing passes through the food and *does not* make foods radioactive. Some people, however, remain concerned that irradiation may decrease the nutritional value of food or may leave dangerous or damaging materials in the food.

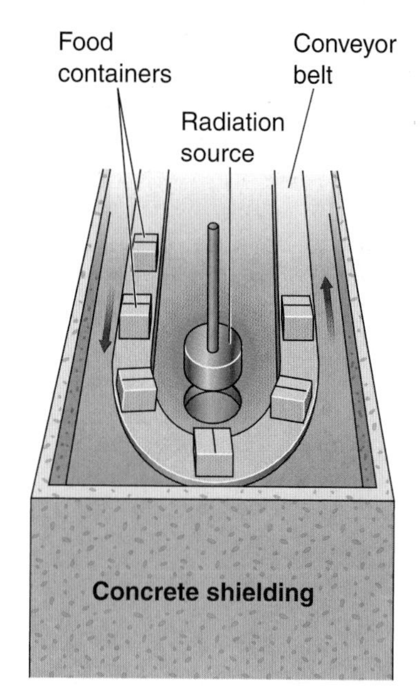

Figure 2.25 Packages of food pass by a source of radiation and become irradiated. The use of radiation to improve food storage is increasing.

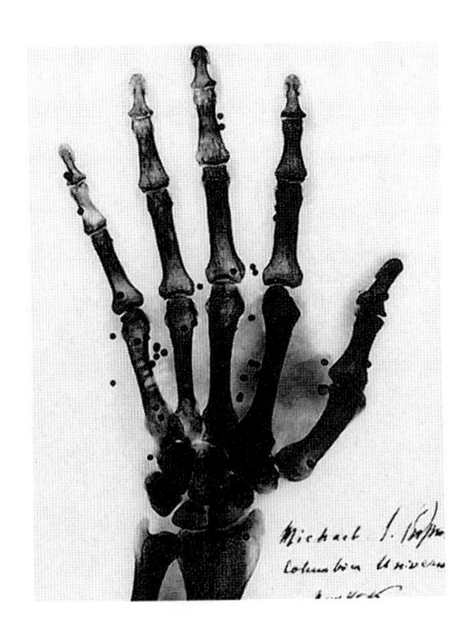

Figure 2.26 Only months after their discovery, x-rays were used in health care. This x-ray of a patient's hand, taken in 1896, shows the lead shot from a shooting accident. *(Columbia University)*

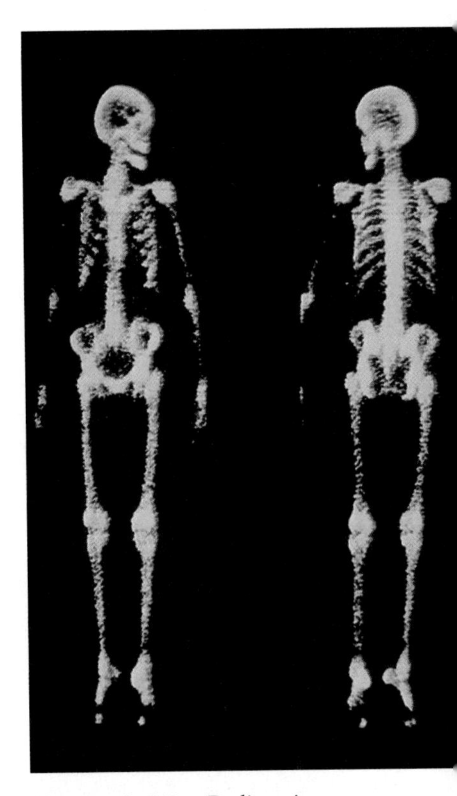

Figure 2.27 Radioactive nuclides, such as technetium-99, can be used to obtain a whole-body scan of patients. Cancerous growths absorb the isotope more readily, with the result that a tumor appears as a bright spot on the scan. This individual is healthy, with no tumors present. *(CNRI/SPL/Photo Researchers)*

Nuclear Medicine

Nuclear medicine is another major application of radioactive materials and radiation. Radiation is used for both diagnosis and treatment. X-rays, which are non-nuclear in origin but similar to gamma rays, have been used for diagnosis since their discovery in the late 1800s (Figure 2.26). Several radioactive nuclides are used as *tracers*. If a substance is taken up by a tissue or organ preferentially, then a radioactive nuclide of that substance will accumulate there. For example, when technetium-99 is complexed with a specific organic compound called methylene diphosphonate, it is absorbed by bones. After a patient has been given this complexed isotope, a whole-body scan can be obtained (Figure 2.27). The amount of technetium-99 taken up by bone associated with tumor cells is generally greater than the amount taken up by normal bone. Thus bone associated with tumors contrasts with normal bone in these scans. Table 2.8 lists several radioactive nuclides that are commonly used for diagnosis.

Table 2.8 Some Common Radioactive Nuclides Used in Diagnosis[a]

ISOTOPE	HALF-LIFE	USE
Iodine-123	13.1 hr	Diagnosis of thyroid function
Sodium-24	15 hr	Diagnosis of blood flow and volume
Technetium-99m	6 hr	Diagnosis of bone, bone marrow, and many other tissues and organs. This is the most commonly used radiopharmaceutical.
Fluorine-18	110 min	Used in positron emission tomography (PET) to provide detailed images of organs

[a]These substances are also known as *radiopharmaceuticals*.

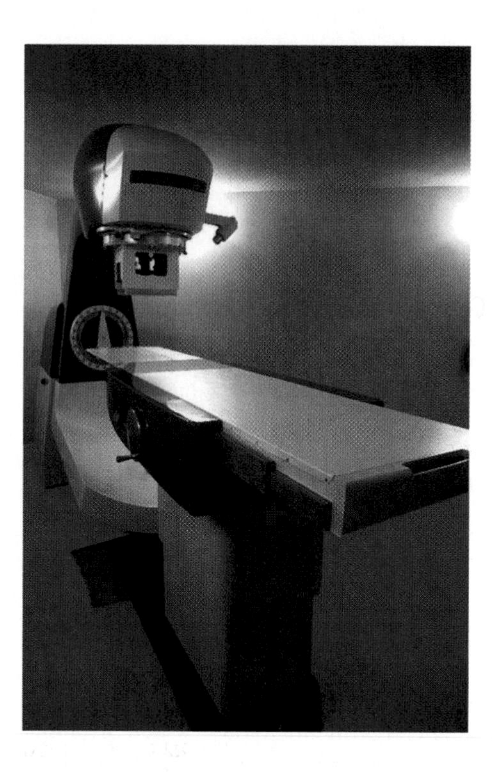

Figure 2.28 This device is used to treat tumors. A beam of penetrating radiation from cobalt-60 is focused on the tumor to destroy it. *(Earl Roberge/Photo Researchers)*

Radioactive nuclides are also used in treatment of disease. Cobalt-60 releases gamma radiation. An intense beam of gamma rays from cobalt-60 is focused on tumors to destroy them (Figure 2.28). Iodine-131 is used to treat hyperthyroid conditions (*hyper* means above normal). The iodine-131 accumulates in the thyroid, and as it decays, some thyroid cells are destroyed. The goal is to lower thyroid activity to normal or near-normal levels.

CHAPTER SUMMARY

Chemistry is the study of matter and changes in matter. Matter is made up of atoms, which are the smallest particles that possess the properties of an element. **Atoms** contain subatomic particles called **electrons, protons,** and **neutrons.** Protons and neutrons are found in the **nucleus** of an atom, and the number of protons in an atom is the **atomic number** of that element. Atoms of an element that differ in mass are called **isotopes,** and the **mass number** of the element is the total number of protons and neutrons in an atom. The weighted average mass of the naturally occurring isotopes of an element is the element's **atomic mass.** Around the atom's nucleus are layers of electrons called **shells,** which differ in energy. Electrons in the outer shell are called **valence electrons,** and the electrons in the other shells are called **core electrons.** Studies of the interaction of light and matter have provided data that has led to our modern view of the atom. The **periodic table** arranges the elements by increasing atomic number. In this table, elements with similar properties are in columns labeled **groups** or families, and elements with valence electrons in the same shell are in the same rows, called **periods.** Elements in the lower left of the periodic table are **metals** that readily lose electrons in reactions; those in the upper right are **nonmetals** that gain or share electrons; and those between them are **metalloids** whose properties are intermediate between metals and nonmetals. Electron shells can be divided into **subshells,** and these subshells contain **orbitals** that differ in shape, orientation in space, and energy. When **radioactive nuclides** undergo **radioactive decay** they emit either **alpha, beta,** or **gamma nuclear radiation.** Ionizing radiations generate ions as they pass through matter. These radiations can be detected by several devices, and exposure to radiation can be expressed in **curies, rads,** or **rems. Fission** is the splitting of a large nucleus into small nuclei, and **fusion** is the joining of smaller nuclei to form larger ones. Some nuclei can be transformed into other nuclei by bombardment with high-energy particles. Nuclear reactions produce a significant amount of our electrical energy and have several other applications, including use in health care.

KEY TERMS

Introduction
chemistry

Section 2.1
atom

Section 2.2
atomic mass
atomic mass unit (amu)
atomic number
electron
isotopes
mass number
neutron
nucleus
proton

Section 2.3
core electrons
shells
valence electrons
valence shell

Section 2.4
families
groups
metalloids
metals
nonmetals
periods
periodic law
periodic table of the elements

Section 2.5
subshells
orbitals

Section 2.6
alpha (α) radiation
beta (β) radiation
gamma (γ) radiation
half-life
nuclear radiation
radioactive decay
radioactive nuclides

Section 2.7
curie (Ci)
radiation absorbed dose (rad)
radiation equivalent for man (rem)

Section 2.8
fission
fusion
nuclear transformation

EXERCISES

The Atomic Nature of Matter (*Section 2.1*)

1. A sample of an element can be separated into smaller pieces until what particle remains?

2. Now that you know that matter is made up of atoms, use this idea to define an element.

3. Modern atomic theory differs from Dalton's original hypothesis in what single feature?

4. A sample of pure chalk is analyzed and found to contain calcium, carbon, and oxygen. Is chalk an element? Why or why not?

5. What is the element symbol for these elements?
(a) Boron **(b)** Sodium
(c) Aluminum **(d)** Platinum

6. What is the element symbol for these elements?
(a) Iron **(b)** Iodine
(c) Uranium **(d)** Strontium

7. These symbols represent what elements?
(a) S **(b)** Sn **(c)** Br **(d)** Cr

8. These symbols represent what elements?
(a) H **(b)** Mg **(c)** Co **(d)** Ba

The Structure of Atoms (*Section 2.2*)

9. What three subatomic particles are found in an atom?

10. What subatomic particles are found in the nucleus?

11. Identify the subatomic particle that has these features.
(a) Located in nucleus; has a positive charge
(b) Has the smallest mass of the subatomic particles
(c) Has no electric charge

12. Identify the subatomic particle that has these features.
(a) Has a mass similar to a proton
(b) Is located around the nucleus; has a negative charge
(c) Is in the nucleus; has a charge

13. What is the term we use to identify all of the subatomic particles found in the nucleus of an atom?

14. What electric charge is associated with each of the subatomic particles? Why is an atom electrically neutral?
15. What is an atomic mass unit (amu)? We use atomic mass units to describe atomic-sized particles because these particles are so small. Express 1 amu as a mass in grams.
16. Compare the mass of the subatomic particles in grams and amu.
17. Each element has a unique atomic number. What does the atomic number tell us about an element?
18. From information in the periodic table, how can you determine the number of protons and electrons in an atom of an element?
19. Provide the atomic number of these elements.
    (a) Li (b) P (c) Br (d) N
    (e) Magnesium (f) Sulfur
20. What is the atomic number of these elements?
    (a) C (b) Ca (c) I (d) Fe
    (e) Neon (f) Lead
21. What are the names of the elements with these atomic numbers?
    (a) 1 (b) 8 (c) 47 (d) 80
22. What are the names of the elements with these atomic numbers?
    (a) 11 (b) 30 (c) 50 (d) 28
23. Determine the number of protons and electrons in an atom of these elements:
    (a) Potassium (b) Au (c) Boron
24. Determine the number of protons and electrons in an atom of these elements:
    (a) Silicon (b) Bromine (c) Ba
25. A Zn atom has how many electrons?
26. What is an isotope? What are the isotopes of carbon?
27. To what does the mass number of an atom correspond?
28. What is the mass number of a carbon-14 atom?
29. Consider this isotope: $^{23}_{11}Na$. What is the atomic number of this isotope? What is the mass number of this isotope?
30. How many neutrons and protons are in an atom of these isotopes?
    (a) $^{11}_{5}B$ (b) $^{60}_{27}Co$
    (c) Aluminum-27 (d) Uranium-235
31. How many neutrons and protons are in an atom of these isotopes?
    (a) $^{81}_{35}Br$ (b) $^{40}_{19}K$
    (c) Oxygen-18 (d) Lithium-7
32. Complete this table for a neutral atom of each element:

ELEMENT NAME	ELEMENT SYMBOL	ATOMIC NUMBER	MASS NUMBER	NUMBER OF PROTONS AND ELECTRONS	NUMBER OF NEUTRONS
Oxygen			16		
	Sr				52
		6			8
		30	64		

33. Complete this table for a neutral atom of each element:

ELEMENT NAME	ELEMENT SYMBOL	ATOMIC NUMBER	MASS NUMBER	NUMBER OF PROTONS AND ELECTRONS	NUMBER OF NEUTRONS
	S		32		
Uranium			238		
Chlorine					20
		19			22

34. Explain the meaning of the atomic mass of an element.
35. The element bromine has two isotopes of approximately equal abundance: bromine-79 and bromine-81, yet its atomic mass is 79.9. Explain.
36. What is the atomic mass of these elements?
    (a) F (b) Cobalt (c) Th
    (d) Carbon (e) Potassium
37. What is the atomic mass of these elements?
    (a) P (b) Na (c) Beryllium
    (d) Iodine (e) Argon
38. Which element has this atomic mass?
    (a) 114.82 (b) 195.08 (c) 14.01
39. Which element has this atomic mass?
    (a) 183.85 (b) 10.81 (c) 58.69

Electron Shells and Valence Electrons (Section 2.3)

40. Electron shells, or just shells, are found in atoms. What is an electron shell?
41. Which shell in an atom has the lowest energy?

42. Each electron shell can contain from zero to a maximum number of electrons. How many electrons are in each of these shells when they are completely filled?
(a) First (b) Third (c) Sixth
43. How many electrons are found in each of these shells when they are completely filled?
(a) Second (b) Fourth (c) Fifth
44. Indicate how many electrons are in each shell of an atom of these elements:
(a) H (b) S (c) Carbon
45. Indicate how many electrons are in each shell of an atom of these elements:
(a) Al (b) Nitrogen (c) Calcium
46. Identify the elements having these arrangements of electrons in an atom.
(a) Two electrons in the first shell, six electrons in the second shell
(b) Two electrons in the first shell
(c) Two electrons in the first shell, eight electrons in the second shell, five electrons in the third shell
47. Which elements have these arrangements of electrons in an atom?
(a) One electron in the first shell
(b) Two electrons in the first shell, eight electrons in the second shell, two electrons in the third shell
(c) Two electrons in the first shell, three electrons in the second shell
48. Which shell is the valence shell of an atom?
49. Where are valence electrons located?
50. What are core electrons?
51. How many valence and core electrons are in an atom of
(a) Li (b) Nitrogen (c) Ca
52. How many valence and core electrons are in an atom of
(a) Na (b) He (c) Silicon
53. If an atom gains specific amounts of energy, what change takes place in the atom?
54. After an atom gains energy, what change will take place?
55. Consider the spectrum of light emitted by hydrogen. What does each of these lines represent?

The Periodic Table *(Section 2.4)*

56. In a periodic table of the elements, the rows are called what?
57. In a periodic table of the elements, the columns are called what?
58. What do all of the elements in a group have in common?
59. What do all of the elements in a period have in common?
60. These elements are found in what group?
(a) Al (b) Oxygen (c) Sodium
61. These elements are found in what period?
(a) Calcium (b) P (c) Carbon
62. Identify the elements in these groups or periods.
(a) Group 6A (b) Group 3A
(c) Second period (d) First period
63. Identify the elements in these groups or periods.
(a) Third period (b) Group 1A
(c) Group 4A (d) Period four and Group 8A
64. How many valence electrons are found in an atom of these elements?
(a) Na (b) Calcium (c) O
65. How many valence electrons are found in an atom of these elements?
(a) Carbon (b) Al (c) Br
66. Name the alkali metals, and explain why they were given this name.
67. What are the halogens?
68. List the properties associated with metals.
69. Describe the location of metals in the periodic table.
70. List the properties associated with nonmetals.
71. Describe the location of the nonmetals in the periodic table.
72. What is a metalloid or semimetal?
73. Describe the location of the metalloids in the periodic table.

Electron Subshells and Orbitals *(Section 2.5)*

74. Subshells of electrons are designated by symbols consisting of a number and a letter. What letters are used to designate subshells?
75. What is the maximum number of electrons that each of the subshells can hold?
76. How many subshells are in these shells?
(a) Second (b) Fourth
(c) First (d) Third
77. Give the electron configurations of these elements.
(a) He (b) Boron (c) Ar
78. Give the electron configurations of these elements.
(a) Li (b) F (c) Magnesium
79. What is an orbital?

80. How many electrons can be found in an orbital?

81. How many electrons could be found in each of these subshells?

(a) $2p$ (b) $3s$ (c) $4f$ (d) $3d$

Nuclear Radiation (Section 2.6)

82. Provide a definition of radiation.

83. What is nuclear radiation?

84. What occurs when an atom undergoes nuclear decay?

85. What name is given to an atom that has an unstable nucleus?

86. Define the term half-life and give an example.

87. How many half-lives must pass for 12 mg of the radioactive nuclide nickel-66 to decay to 3.0 mg?

88. How many half-lives must pass for 36 ng of the radioactive nuclide technetium-99m to decay to 4.5 ng?

89. Nickel-66 has a half-life of 56 hr. How much time must pass for a 6.0-μg sample of nickel-66 to decay to 1.5 μg?

90. Technetium-99m has a half-life of 5.9 hr. How much time must pass for a 128-μg sample to decay to 4.0 μg?

91. Hydrogen-3 is called tritium and has a half-life of 12.5 years. How much of a 2.50 g sample of tritium would remain after 50 years?

92. Fluorine-17 has a half-life of 66 s. How much of a 36-μg sample would remain after 3.30 min?

93. A sample of cotton cloth is found during an archeological excavation and is thought to be around 5000 years old. Carbon-14 radiodating is used to estimate the age of some archeological specimens. How is a carbon-14 study performed? Would a carbon-14 study of this cotton cloth provide a very accurate estimate of its age? Why or why not?

94. What is alpha radiation? What effect does the emission of an alpha particle have on the parent nuclide?

95. What is beta radiation? What effect does beta emission have on the parent nuclide?

96. What is gamma radiation? What effect does gamma emission have on the parent nuclide?

97. Predict the products of these nuclear decays:

(a) ${}^{222}_{86}\text{Rn} \xrightarrow{\alpha} ?$

(b) ${}^{210}_{82}\text{Pb} \xrightarrow{\beta} ?$

(c) ${}^{45}_{19}\text{K} \xrightarrow{\beta} ?$

(d) ${}^{238}_{92}\text{U} \xrightarrow{\alpha} ?$

(e) ${}^{3}_{1}\text{H} \xrightarrow{\beta} ?$

98. Predict the products of these nuclear decays:

(a) ${}^{234}_{91}\text{Pa} \xrightarrow{\beta} ?$

(b) ${}^{214}_{83}\text{Bi} \xrightarrow{\alpha} ?$

(c) ${}^{226}_{88}\text{Ra} \xrightarrow{\alpha} ?$

(d) ${}^{17}_{7}\text{N} \xrightarrow{\beta} ?$

(e) ${}^{24}_{11}\text{Na} \xrightarrow{\beta} ?$

Detecting and Measuring Radiation (Section 2.7)

99. The detection of radiation is an important factor in protecting people who may be exposed to radiation. Name three common methods of detecting radiation.

100. What is a curie?

101. What is a radiation absorbed dose (rad)?

102. How does a radiation equivalent of man (rem) differ from a rad?

103. Explain the meaning of LD_{50}.

104. Very intense radiation is used to kill tumor cells. Explain why the patient is not killed by this radiation dose.

Induced Nuclear Reactions (Section 2.8)

105. Briefly describe what happens during nuclear fission and name an isotope that undergoes fission.

106. Fission and fusion are nuclear processes that both release large amounts of energy. **(a)** Define the process of fusion. **(b)** Identify an element that undergoes fusion.

107. What is needed besides light nuclei for fusion to occur?

108. For a fission reaction, what is meant by the term critical mass?

109. Explain what will happen if a mass of uranium-235 is supercritical.

110. What is a nuclear transformation?

111. List some common uses for nuclear reactions and nuclear energy.

112. Several radioisotopes are now commonly used in health care. Identify three of these radioactive nuclides and describe their uses.

Challenge Exercises

113. The element chlorine consists of two naturally occurring isotopes. Chlorine-35 has a mass of 34.97867 amu and a natural abundance of 75.4%. Chlorine-37 has a mass of 36.97750 amu and an abundance of 24.6%. Determine the atomic mass of chlorine from these data. Is your calculation consistent with the value shown in the periodic table?

114. The atomic mass of an iron atom is 55.85 amu, and iron's density is 7.86 g/cm^3. Use this information and the conversion factor for atomic mass units to grams to determine the number of iron atoms in 1.00 cubic centimeter of iron.

115. A representative atom has a diameter of approximately 1×10^{-8} cm, and its nucleus is about 1×10^{-13} cm in diameter. Determine the percentage of an atom's volume that is occupied by its nucleus. Assume the atom and nucleus are spherical. The volume of a sphere is $4/3\pi r^3$.

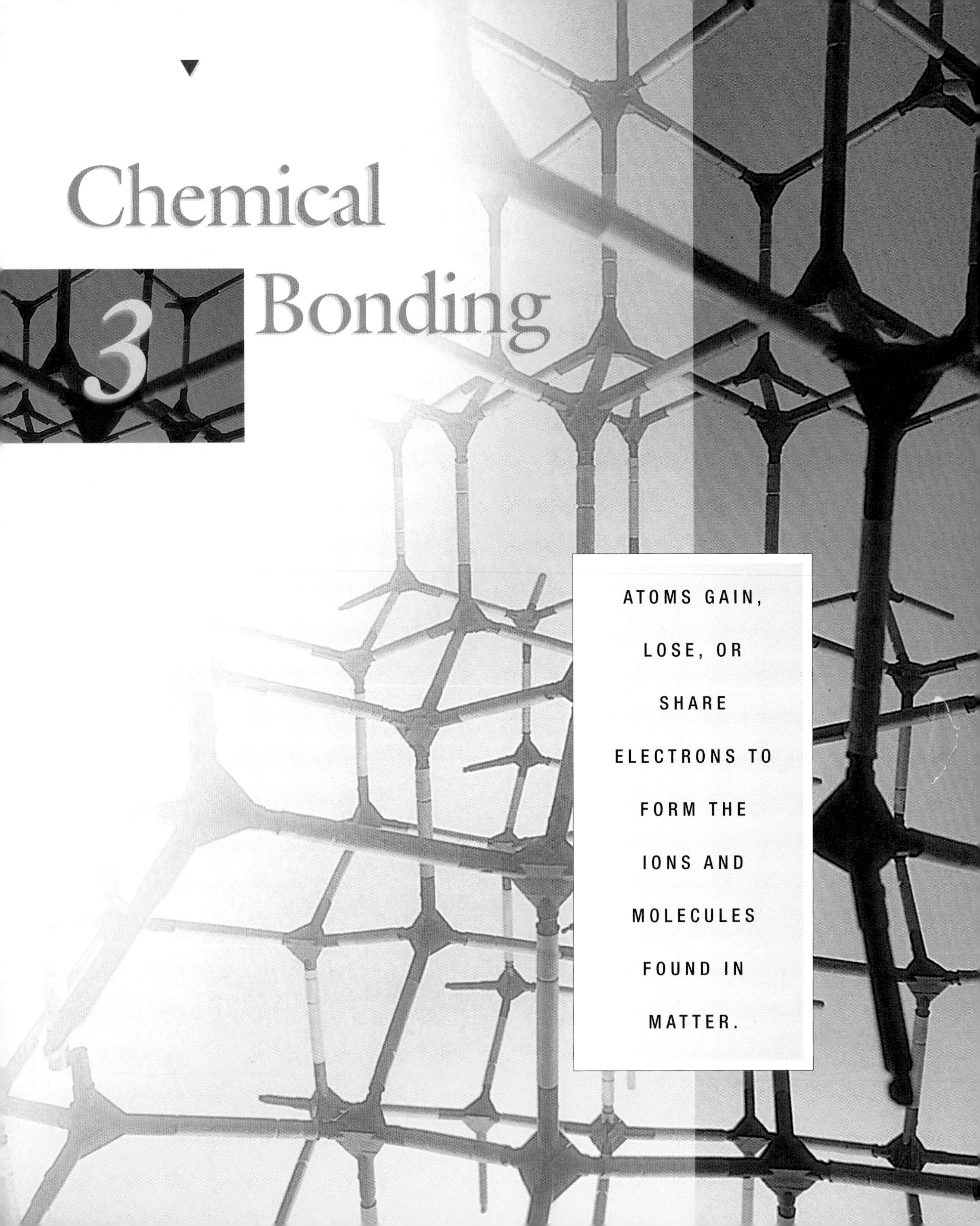

Chemical Bonding

3

ATOMS GAIN, LOSE, OR SHARE ELECTRONS TO FORM THE IONS AND MOLECULES FOUND IN MATTER.

On Earth, only the atoms of the Group 8A elements, the noble gases such as helium and neon, exist as free uncombined atoms. The atoms of all other elements are combined with other atoms. In a piece of iron or gold, the iron or gold atoms are bound to each other. In water (H_2O), two hydrogen atoms are bound to an oxygen atom. Even oxygen (O_2) in the air consists of two atoms of oxygen bound to each other. Why are there so few free atoms in nature? How do atoms bind to each other? Is there anything about the structure of atoms that provides clues about why and how they bind to each other? Is there anything special about the structure of noble gas atoms that might give us a hint about their unique independent existence? This chapter explores the answers to these questions and then discusses the properties of the ions and molecules that form when atoms bond.

Outline

3.1 Octet Rule and Lewis Structures

Explain how atoms gain, lose, or share electrons to satisfy the octet rule, and use Lewis structures to represent atoms, molecules, and monatomic ions.

3.2 Ions

Predict the charge on monatomic ions formed from atoms of the representative elements.

3.3 Ionic Bonds and Ionic Compounds

Describe ionic bonds and ionic compounds.

3.4 Covalent Bonding: Atoms Share Electrons in Molecules

Define a molecule and describe how covalent bonds link atoms in a molecule.

3.5 Predicting Bond Type

Predict the type of bond that will form between two atoms.

3.6 Shapes of Molecules

Recognize the shapes of simple molecules.

3.7 Polarity of Molecules

Use bond polarity and molecular shape to predict molecular polarity.

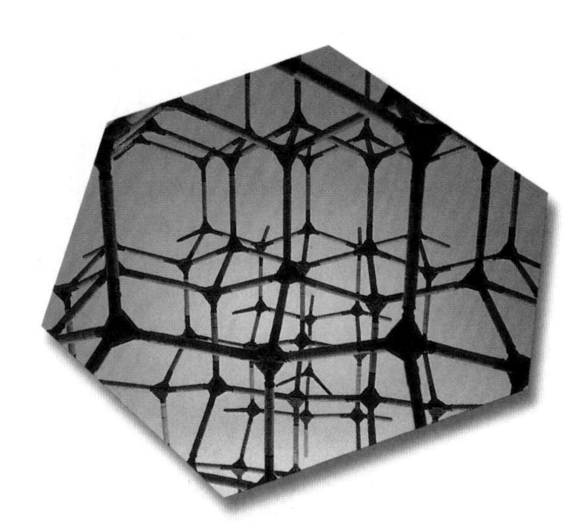

Most atoms on earth are bonded to other atoms. Diamond is one form of pure carbon. In this model of its structure, we see that each carbon is surrounded by and bonded to four other carbon atoms at the corner of a tetrahedron. ***(Charles D. Winters)***

3.1 Octet Rule and Lewis Structures

These eight electrons fill the *s* and *p* subshells of the valence shell (see Section 2.5).

The atoms of the noble gases exist as free atoms because they have little tendency to combine with other atoms (Figure 3.1). A chemist would say these elements are *unreactive*, whereas the atoms of all other elements are much more *reactive*. What do the noble gases have in common that makes them unreactive? Their atoms (except helium) all have eight electrons in their valence shell (Table 3.1), whereas the atoms of no other elements have this structural feature. It is these eight valence electrons that make the noble gas atoms stable and unreactive. All other atoms are more reactive because they have a different number of electrons in their valence shell. Atoms combine with other atoms to get eight electrons in their outer shell and thus gain stability.

There are two ways an atom can get eight valence electrons when it combines with another atom. One, through electron transfer, an atom can either gain electrons from another atom or lose electrons to another atom. Two, the atom can share electrons with another atom. The tendency for an atom to seek eight electrons for their valence shell is summarized in the **octet rule:** Atoms will gain, lose, or share electrons to obtain eight electrons in their valence shell. Note that the octet rule does not apply to the smallest elements such as hydrogen and helium because their outer shell—the first shell—contains only two electrons. After atoms combine with other atoms they have different, and usually reduced, reactivities.

Because the valence shell of an atom is so important in chemistry, chemists use a model called a **Lewis structure (electron dot structure)** to help visualize it. In

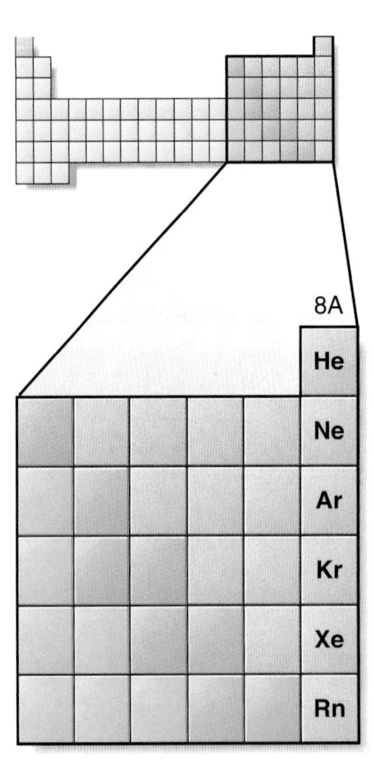

Figure 3.1 The noble gases (Group 8A) are unreactive and exist as free atoms.

a Lewis structure the atom's nucleus and core electrons are represented by the element's symbol, and the valence electrons are represented by dots around the element's symbol. Table 3.2 shows the Lewis structures for the elements in Period 2 of the periodic table. For comparison, the Lewis structures of sodium and fluorine atoms and their electronic shells are shown in Figure 3.2.

To draw Lewis structures you need to know the number of valence electrons in an atom, which for the representative elements—Group 1A through 8A—is the group number (Section 2.4).

Example 3.1 Valence Shell Electrons and Lewis Structures

1. How many electrons are in the valence shell of aluminum?
2. What is the Lewis structure of aluminum?

SOLUTIONS

1. Because aluminum is in Group 3A there are three electrons in its valence shell.
2. Each of the three valence electrons of the atom is represented by a dot around the symbol for aluminum. This is an electron dot structure for aluminum:

·Al·

Self-Test

Determine the number of valence electrons and draw the Lewis structure for

1. Sodium (Na) 2. Iodine (I) 3. Arsenic (As)

ANSWERS

1. One; Na· 2. Seven; :I:
3. Five; ·As:

Table 3.1 Number of Electrons in the Valence Shell of the Noble Gases (Group 8A)

ELEMENT	NUMBER OF ELECTRONS IN SHELLS SHELL					NUMBER OF ELECTRONS IN THE VALENCE SHELL
	1st	2nd	3rd	4th	5th	
He	2					2
Ne	2	8				8
Ar	2	8	8			8
Kr	2	8	18	8		8
Xe	2	8	18	18	8	8

Table 3.2 Lewis (Electron Dot) Structures of the Period 2 Elements

ELEMENT	GROUP NUMBER	NUMBER OF VALENCE ELECTRONS	LEWIS STRUCTURE
Lithium	1A	1	Li·
Beryllium	2A	2	·Be·
Boron	3A	3	·B·
Carbon	4A	4	·C·
Nitrogen	5A	5	:N·
Oxygen	6A	6	:O:
Fluorine	7A	7	:F:
Neon	8A	8	:Ne:

3.2 Ions

Atoms are electrically neutral because they contain an equal number of positively charged protons and negatively charged electrons. Some atoms gain or lose valence electrons in chemical reactions, and when they do this the atom is no longer electrically neutral. It is now an **ion,** which is an atom or group of atoms that carries an electrical charge. If an atom *loses* one or more electrons, it becomes a positively

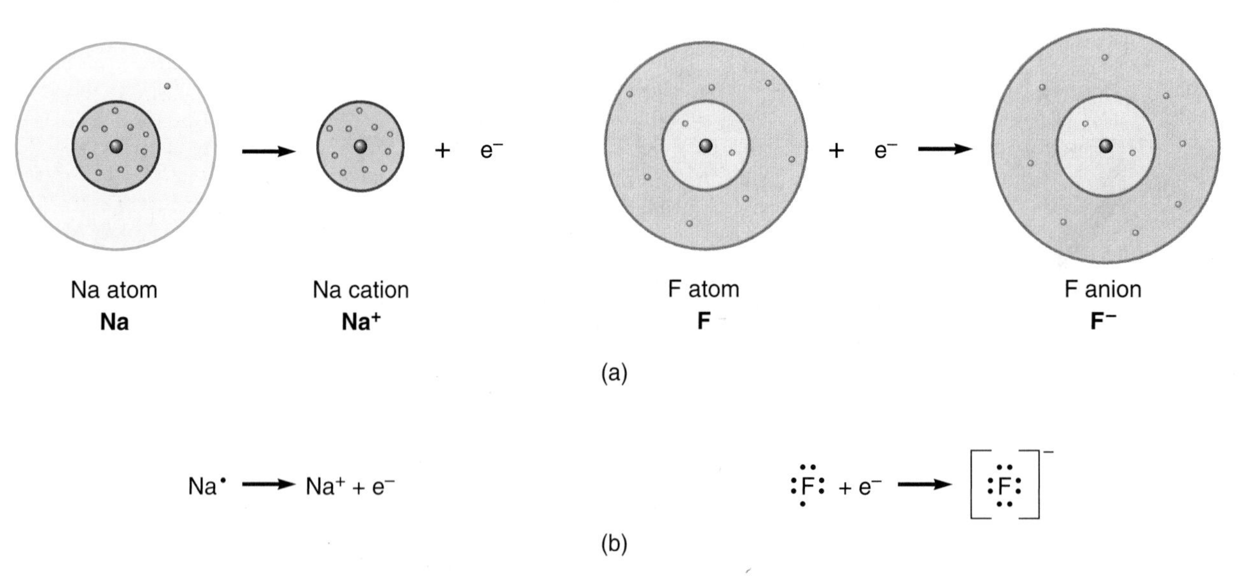

Figure 3.2 (a) These electron shell representations of a sodium atom and ion (blue) and a fluorine atom and ion (tan) show all the electrons of the atoms. (b) Lewis structures of the same atoms and ions emphasize the valence shell electrons; there is one in the sodium atom and seven in the fluorine atom and zero in the sodium ion and eight in the fluoride ion.

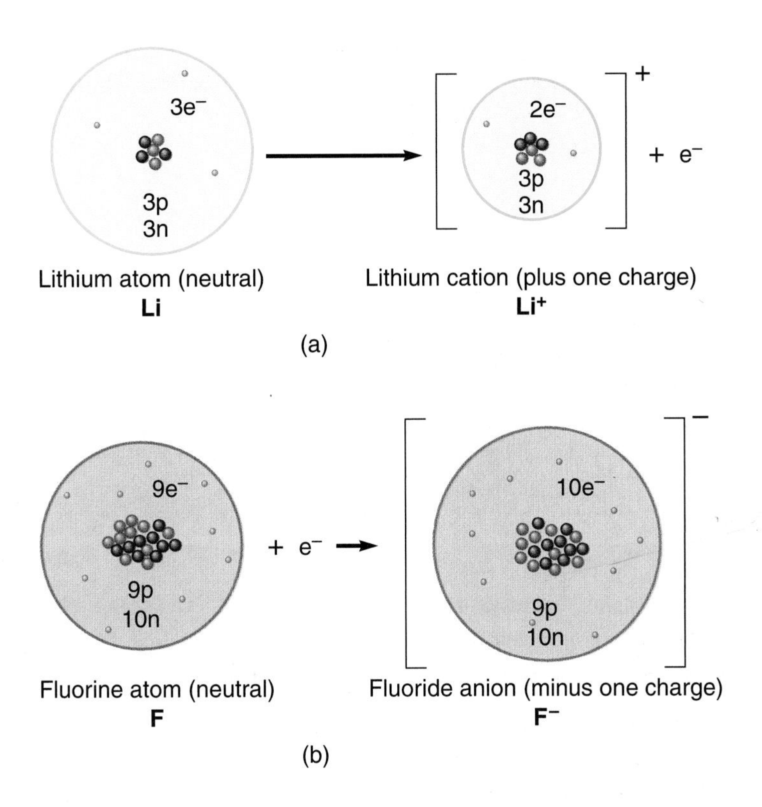

Figure 3.3 (a) Metal atoms lose electrons to become positively charged ions called cations. The lithium ion contains one fewer electron than the lithium atom, but they contain the same number of protons. (b) Nonmetal atoms can gain electrons to become negatively charged ions called anions. The fluoride ion contains one more electron than the fluorine atom, but they contain the same number of protons.

charged ion because it now has fewer electrons than protons. A positively charged ion is called a **cation.** If an atom gains one or more electrons, it becomes an **anion,** a negatively charged ion (Figure 3.3).

A simple and convenient way to represent ions uses the element symbol plus a superscript to indicate the charge on an ion. A charge of plus one or minus one is shown by a plus sign or a minus sign. For example, the sodium ion is a cation that has a charge of plus one, and is written Na^+, and the chloride ion, an anion with a minus one charge, is shown as Cl^-. All charges other than plus one or minus one are shown as a number followed by a plus or minus sign. The magnesium ion has a two plus charge and is represented as Mg^{2+}. The oxide ion has a charge of two minus and so is written O^{2-}.

Several chemical names, which may be unfamiliar to you, are used here; naming compounds will be covered in more detail in Chapter 4.

The ions Na^+, Cl^-, Mg^{2+}, and O^{2-} are **monatomic ions** because they contain only one atom. Some ions have more than one atom, and these ions are called **polyatomic ions.** The charge on a polyatomic ion is indicated in the same way as that on a monatomic ion. The ammonium ion is polyatomic (it is made up of one nitrogen atom and four hydrogen atoms) and has a charge of plus one. Ammonium ion is symbolized this way: NH_4^+.

There is an easy way to predict the charge on many monatomic ions that form by electron transfer between a metal and nonmetal. The periodic table provides the information. The atoms of Group 1A elements, such as sodium (Na), lose one electron when they react and become cations with a plus one charge. This makes sense; it is easier for these atoms to give up one electron to get eight valence electrons than it is to gain seven electrons. The atoms of the Group 2A elements lose two electrons. Except for boron (B), the elements in 3A are metals. The atoms of Group 3A metals usually lose three electrons. The atoms of the transition metals

Table 3.3 Rules for Predicting Charge on Monatomic Ions

1. For the representative metals in Groups 1A, 2A, and 3A, the charge on the ion is simply the group number.
   (a) For Group 1A metals the charge will be plus one. Example: The potassium ion is K^+.
   (b) For Group 2A metals the charge will be plus two. Example: The calcium ion is Ca^{2+}.
   (c) For Group 3A metals the charge will be plus three. Example: The aluminum ion is Al^{3+}.
2. For the representative nonmetals in Groups 5A, 6A, and 7A, the charge is the group number minus 8.
   (a) For Group 5 nonmetals the charge will be minus three ($5 - 8 = -3$). Example: The nitride ion is N^{3-}.
   (b) For Group 6 nonmetals the charge will be minus two ($6 - 8 = -2$). Example: The sulfide ion is S^{2-}.
   (c) For Group 7 nonmetals the charge will be minus one ($7 - 8 = -1$). Example: The bromide ion is Br^-.

also tend to lose electrons, but the charge on the resulting cation is not easily predicted by the periodic table. Furthermore, some transition metals can form more than one cation. For the transition metals, it is easier to just memorize the common charge or charges on the ions.

Atoms of Group 7A elements gain one electron. Again this makes sense, these atoms can gain one electron more easily than they can lose seven. Atoms of Group 6A gain two electrons. Atoms of Group 5A elements, such as Nitrogen (N), can gain three electrons.

Table 3.3 contains these simple rules to predict the charge on monatomic ions.

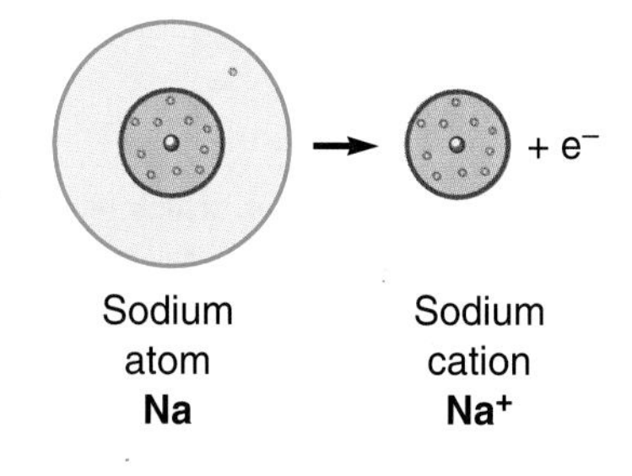

Brackets are placed around a charged Lewis structure to show that the charge is associated with the entire Lewis structure.

Atoms become ions when they gain or lose electrons. Let's use Lewis structures to look at two examples. Consider first a sodium atom and a fluorine atom. A sodium (Na) atom has one valence electron and can satisfy the octet rule most easily by losing that electron:

$$Na\cdot \longrightarrow Na^+ + e^-$$

This is consistent with what we know about metals: They lose electrons to satisfy the octet rule.

A fluorine (F) atom has seven electrons in its outer shell. It can satisfy the octet rule by gaining an electron:

$$:\ddot{\underset{\cdot}{F}}: + e^- \longrightarrow \left[:\ddot{\underset{\cdot\cdot}{F}}:\right]^-$$

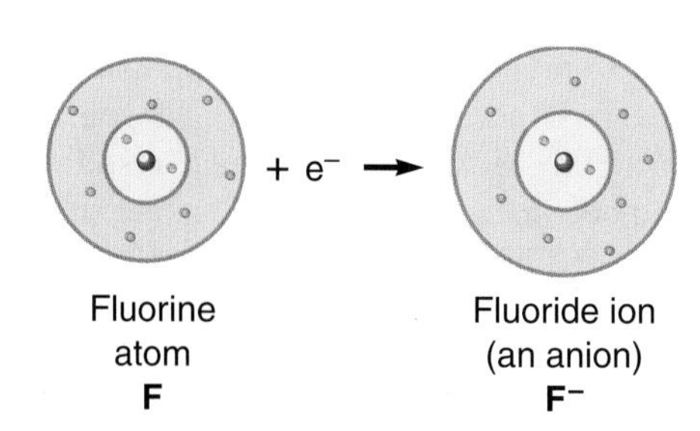

This is consistent with what we know about nonmetals: They may gain electrons to get eight valence electrons, thus satisfying the octet rule.

A sodium atom loses an electron to become a sodium ion, and the fluorine atom gains an electron to become a fluoride ion. Because the sodium ion lost an electron, its outer shell is now the second shell, which contains eight electrons. The fluoride ion has gained an electron, which puts eight electrons in its second shell. F^- and Na^+ each have filled valence shells, and each contains eight valence electrons, thus the octet rule is satisfied for both of them.

Gain of an electron by a fluorine atom or loss of one by a sodium atom normally occurs when contact is made between them:

$$\underset{\text{Sodium atom}}{Na\cdot} + \underset{\text{Fluorine atom}}{:\ddot{F}:} \longrightarrow \underset{\text{Cation}}{Na^+} + \underset{\text{Anion}}{[:\ddot{F}:]^-}$$

Sodium atom Na + Fluorine atom F → Sodium cation Na^+ + Fluoride anion F^-

The sodium and fluorine atoms have chemically reacted with each other to form ions. When atoms react with each other to form ions, the number of electrons gained and the number of electrons lost must be the same. Thus the net charge before and after the transfer of electrons must be the same. The sodium and fluorine atoms have a net charge of zero ($0 + 0 = 0$), as do the sodium and fluoride ions ($+1 + -1 = 0$).

Now consider a different electron-transfer reaction, one that occurs between a calcium atom and two chlorine atoms, that yields a calcium ion and two chloride ions:

$$\cdot Ca \cdot + 2:\ddot{Cl}: \longrightarrow Ca^{2+} + 2[:\ddot{Cl}:]^-$$

As in the reaction involving sodium and fluorine, a neutral metal atom (the calcium atom) gave up electrons and became a cation, Ca^{2+}. *When metals and nonmetals react, the metal atoms lose electrons.* The neutral nonmetal atoms, the chlorine atoms, accepted electrons and became anions, Cl^-. *When metals and nonmetals react, the nonmetal atoms gain electrons.* Note that the net charge is the same before ($0 + 0 + 0 = 0$) and after ($+2 + -1 + -1 = 0$) the reaction.

Example 3.2 Predicting Charges and Numbers of Ions When Electron Transfer Occurs

1. (a) When sodium reacts with oxygen, what are the charges on the ions that form?
   (b) How many ions of each element are formed?
   (c) Use electron dot structures to show the transfer of electrons.

SOLUTIONS

1. (a) Sodium is in Group 1A; thus it forms a +1 ion (Na^+). Oxygen is in Group 6A; thus the charge on the oxide ion will be $6 - 8 = -2$ (O^{2-}).

   (b) The number of electrons gained and lost must be the same. From the charges on the ions, it can be seen that the oxygen atom gains two electrons and a sodium atom loses only one. There must therefore be two sodium ions formed for each oxide ion.

   (c) $2\text{Na}\cdot + :\dot{\underset{\cdot}{\text{O}}}: \longrightarrow 2\text{Na}^+ + \left[:\ddot{\underset{\cdot\cdot}{\text{O}}}:\right]^{2-}$

   Note that the net charge is the same (zero) for the reacting atoms and for the resulting ions.

Self-Test

The elements shown here react together to form ions. Determine the charge on each ion and the number of ions that will form from each element. Show the reactions with electron dot structures.

1. Magnesium (Mg) and sulfur (S)
2. Aluminum (Al) and oxygen (O)

ANSWERS

1. Magnesium forms Mg^{2+} ions and sulfur forms S^{2-} ions. There will be one of each ion formed.

$$\cdot\text{Mg}\cdot + :\dot{\underset{\cdot}{\text{S}}}: \longrightarrow \text{Mg}^{2+} + \left[:\ddot{\underset{\cdot\cdot}{\text{S}}}:\right]^{2-}$$

2. Aluminum forms Al^{3+} ions and oxygen forms O^{2-} ions. Two Al^{3+} ions will form and three O^{2-} ions will form.

$$2\cdot\underset{\cdot}{\text{Al}}\cdot + 3:\dot{\underset{\cdot}{\text{O}}}: \longrightarrow 2\text{Al}^{3+} + 3\left[:\ddot{\underset{\cdot\cdot}{\text{O}}}:\right]^{2-}$$

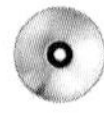

3.3 Ionic Bonds and Ionic Compounds

When fluorine atoms react with sodium atoms, anions and cations form. These cations and anions are of opposite charge, and oppositely charged particles attract each other. Because of this attraction, these ions form solids that are usually crystalline (crystal-like). The force of attraction between oppositely charged ions is an **ionic bond.** The strength of an ionic bond depends partly on the distance between the charges and partly on the size of the charges on the ions. The closer the charges and the larger the charges on the ions, the stronger the attraction between them.

When sodium and fluorine react, they form a new substance, a compound. Any compound that consists of ions held together by ionic bonds is called an **ionic compound** (Figure 3.4). To determine if a compound is ionic, first dissolve the

Connections

Heavy Metal Toxicity

Although we often refer to heavy metal toxicity, in reality it is usually not the metal atoms but the ions of heavy metals that are toxic. Let's take lead and mercury as examples. When ions of these metals are ingested or breathed in, they are taken up by cells. In the cells these ions bind to sulfur-containing groups in proteins (see Chapter 14) and alter or eliminate the protein's normal function. This alters the cell's function or kills the cell, and if the cell is part of a vital organ like the brain, the normal function of the organ is impaired.

Most commonly, lead poisoning in North America happens when children ingest lead-containing paint flakes or dust. Lead-containing compounds were used in paints as coloring agents (pigments), and the risk to children is highest in older buildings and homes where paint peels from the walls. Lead poisoning in adults more commonly comes from occupational exposure. Lead poisoning is more easily prevented than cured, but substances that bind (chelate) lead ions can be given to patients, and the bound lead is then excreted from the body.

Mercury poisoning most commonly arises from environmental sources. Mercury ions are toxic, but the most toxic mercury compounds are organic compounds such as methyl mercury (organic compounds are discussed later in the book). Although very small quantities of mercury are found throughout the environment from both natural sources and human activity, some localities are contaminated by significant amounts of mercury from mine or factory pollution. One example from the 1950s is Minamata Bay in Japan. A company discharged untreated, mercury-containing wastes into the bay and the mercury became concentrated in the food web. Fish and other seafood became heavily contaminated and were then harvested and eaten. Many people in the area developed mercury poisoning from eating the contaminated seafood.

The yellow solid formed in this reaction is lead chromate, which was the pigment in some older yellow paints. Because lead salts have a variety of colors, lead-based paints were available in several colors. *(Charles D. Winters)*

compound in water (if it's water soluble) or melt it, then determine whether the mixture or melt will conduct electricity. Electric current is flowing charge, and ions or free electrons must be present for a current to exist. Ionic compounds conduct electricity when dissolved in water. Compounds that conduct electricity are also known as **electrolytes.** Ionic compounds are not the only compounds that conduct electricity when they are dissolved in water. Acids and bases do also (Chapter 7).

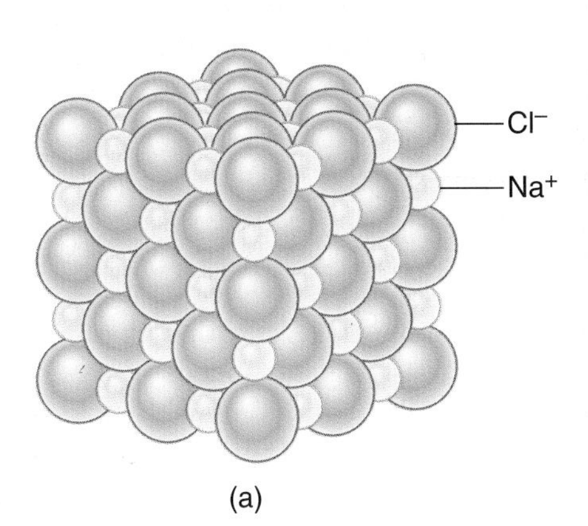

Figure 3.4 The ions of an ionic compound are held together by ionic bonds. (a) The sodium ions and chloride ions of sodium chloride are arranged systematically in a crystal to maximize bonding between ions of opposite charge. (b) Many compounds form beautiful or unusual crystals (sodium chloride is shown here). *(b, Charles D. Winters)*

Connections

Electrolytes in Body Fluids

Electrolytes are ions found in blood, interstitial fluids (fluid around cells), and intracellular fluids (fluid in cells). Although the concentrations of electrolytes vary from one body fluid to another, the amounts of electrolytes in any given fluid are normally maintained within narrow ranges. The amounts of these electrolytes are often expressed in milliequivalents per liter (meq/L), which are discussed in Chapter 7. Because solutions are electrically neutral, the total milliequivalents of the cations and anions are equal.

Electrolytes serve several roles in the body. Sodium, potassium, and chloride ions are involved in the electrical potential of cell membranes. Calcium ions are involved in muscle contraction, and sodium ions play a role in transmission of nerve impulses. Several conditions can result in loss of electrolytes from the body, including severe diarrhea, vomiting, or kidney dysfunction. To replenish lost electrolytes, patients are given electrolyte-containing intravenous fluids.

Some athletes use sports drinks to restore electrolytes lost in perspiration. However, unless the exercise session is long, replacing electrolytes with sports drinks during exercise is not generally needed. The relatively small amounts of electrolytes lost through exercise are easily replaced by the foods in a normal diet.

Electrolytes Found in Blood and Cells

ELECTROLYTE	BLOOD PLASMA (mEQ/L)	INTRACELLULAR FLUID (mEQ/L)
Cations		
Ca^{2+}	5	1
K^+	4	159
Mg^{2+}	2	40
Na^+	142	10
Total	153	210
Anions		
Cl^-	103	3
HCO_3^- (bicarbonate)	25	7
Protein	17	45
Other	8	155
Total	153	210

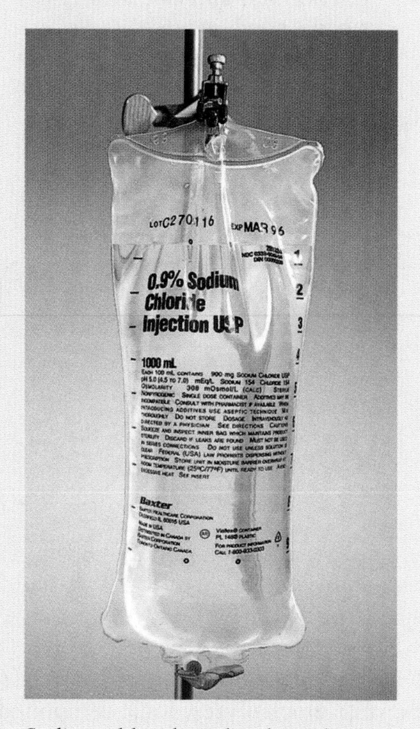

Sodium chloride is the electrolyte in this IV solution. *(Charles D. Winters)*

3.4 Covalent Bonding: Atoms Share Electrons in Molecules

Ionic compounds form when atoms of metals and nonmetals react with each other. This seems reasonable. Metals tend to lose electrons, and nonmetals can gain electrons to satisfy the octet rule. But what happens if two atoms with similar tendencies to gain or lose electrons come into contact with each other? When this happens, the atoms *share* electrons to satisfy the octet rule.

Let's consider two hydrogen atoms that come together. Each atom contains one electron in its outer shell (first shell). A filled first shell has two electrons; thus each hydrogen atom is one electron short of a filled outer shell. Each of the two hydrogen atoms shares its electron with the other atom. In this way, each H has a filled outer shell:

H· + ·H → H:H

Lewis structures of hydrogen atoms

Lewis structure of a hydrogen molecule

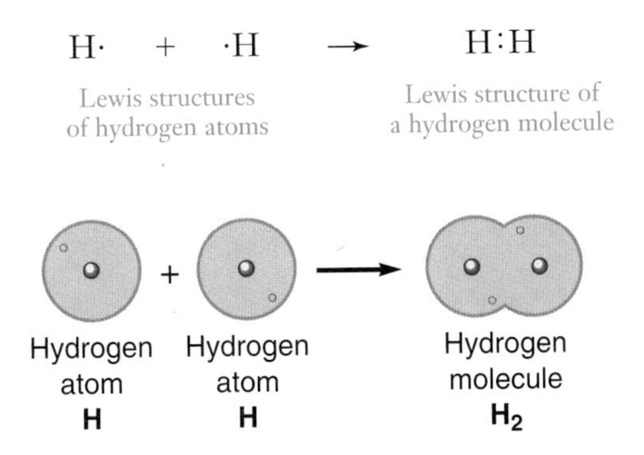

(a)

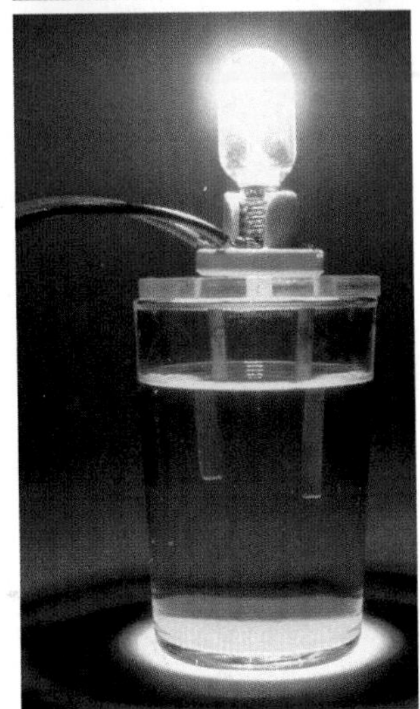
(b)

Figure 3.5 (a) When a nonelectrolyte, in this case table sugar, is dissolved in water it does not conduct electricity. (b) Ionic compounds are electrolytes; table salt conducts electricity when dissolved in water. *(Charles D. Winters)*

Each hydrogen atom has an equal attraction for the electrons, with the result that the two hydrogens share the two electrons *equally*. This sharing of electrons by two atoms results in a force of attraction called a **covalent bond.** The hydrogen atoms are no longer free atoms; they are joined together. A group of atoms joined together by covalent bonds is called a **molecule,** and a compound made up of molecules is called a **molecular compound.** Most molecular compounds do not contain ions and when tested as solutions or melts do not conduct electricity. Most molecular compounds are *nonelectrolytes* (Figure 3.5).

A covalent bond that involves only one pair of electrons is called a *single covalent bond* or, more simply, a **single bond.** A single bond can be represented as a pair of electrons between the two atoms—as in H:H—or more commonly, as a dash between the two atoms: H—H. Note that H:H and H—H are both valid Lewis structures for a hydrogen molecule because all of the electrons around the atoms are shown, either as dots or as the dash for the paired electrons. A molecule that contains only two atoms is a **diatomic molecule.**

A covalent bond can involve more than a single pair of electrons. Consider the bonds in a molecule of carbon dioxide. Each oxygen atom has six valence electrons and the carbon atom has four valence electrons. To get eight valence electrons, the atoms share electrons in this way:

:Ö· + ·Ċ· + ·Ö: → Ö::C::Ö

Lewis structures of two oxygen atoms and a carbon atom

Lewis structure of a carbon dioxide molecule

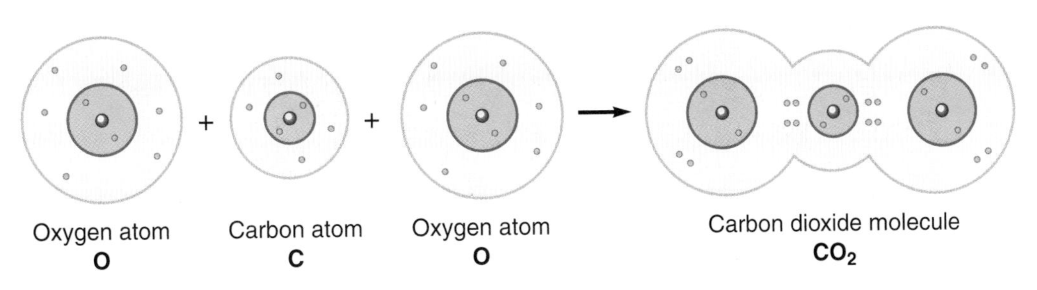

In the molecule the octet rule is satisfied for each atom. Each oxygen atom has four electrons it does not share and four electrons (shaded in blue) that it does share with the carbon atom. The carbon atom shares four electrons with each of the oxygen atoms. A covalent bond involving two pairs of shared electrons is called a **double bond.** Double bonds can be shown as two pairs of electrons between two atoms—as in the previous equation—but they are more often represented by two dashes. The two double bonds in carbon dioxide would be represented as O=C=O.

Nitrogen exists as a diatomic molecule held together by three pairs of shared electrons, a **triple bond:**

$$:\dot{\underset{\cdot}{N}}\cdot + \cdot\dot{\underset{\cdot}{N}}: \longrightarrow :N:::N: \qquad (N_2)$$

Lewis structures of nitrogen atoms — Lewis structure of a nitrogen molecule

Each atom has two unshared and six shared electrons. Molecular nitrogen is typically represented as :N≡N:, or simply N≡N.

Although you will quickly memorize the number of covalent bonds an atom can form, for now use Table 3.4 for this information.

The Polarity of Covalent Bonds

When two atoms of the same nonmetallic element react to form a covalent bond, they share electrons equally. However, covalent bonds also form between atoms of different nonmetals. The gas H—Cl (hydrogen chloride) is an example:

$$H\cdot + \cdot\ddot{\underset{\cdot\cdot}{Cl}}: \longrightarrow H:\ddot{\underset{\cdot\cdot}{Cl}}:$$

Lewis structures of a hydrogen and a chlorine atom — Lewis structure of a hydrogen chloride molecule

Table 3.4 Number of Covalent Bonds to a Nonmetal Atom

GROUP	ELEMENT(S)	NUMBER OF COVALENT BONDS	EXAMPLE
1A	Hydrogen	1	H—H (Molecular hydrogen, one single bond)
4A	Carbon, silicon	4	S=C=S (Carbon in carbon disulfide, two double bonds)
			CH_4 (H—C—H with H above and below) (Carbon in methane, four single bonds)
5A	Nitrogen	3	N≡N (Molecular nitrogen, one triple bond)
6A	Oxygen, sulfur	2	H—O—H (Oxygen in water, two single bonds)
7A	Fluorine, chlorine	1	Cl—Cl (Molecular chlorine, one single bond)

Is the covalent bond between the hydrogen atom and chlorine atom of H—Cl the same as the covalent bond between the two hydrogen atoms in H—H? The covalent bonds in these two molecules are similar because each is a shared pair of electrons, a single bond. The bonds are different because the electron pair is shared differently in the two bonds. The two hydrogen atoms in the hydrogen molecule (H—H) share the electrons equally, which means that the electrons spend the same amount of time near each hydrogen. In H—Cl, however, the pair of electrons is located closer to the chlorine atom than to the hydrogen atom. This is because a chlorine atom attracts shared pairs of electrons better than does a hydrogen atom.

The result of this unequal sharing of the electron pair is a partial separation of charge (Figure 3.6). The hydrogen end of the molecule has a *partial* positive charge, often shown as δ^+, where the lowercase delta means partial. This end is partially positive because the electrons of the covalent bond are pulled toward the chlorine atom. As a consequence, the positive charge of the hydrogen nucleus is no longer completely neutralized by the electrons. The chlorine end is partially negative (δ^-) because the pair of electrons is pulled more toward the chlorine atom. Now the positive charge of the chlorine nucleus is more than offset by electrons, giving the chlorine end of the molecule a slight excess of negative charge. This separation of charge results in a **dipole** (two poles). Just like the electric battery in a car, a hydrogen chloride molecule has two electric poles. A dipole in a molecule is shown by an arrow that has the arrowhead near the negative end of the dipole and a plus sign near the positive end (Figure 3.6). The covalent bond in HCl is called either a **polar covalent bond** or a *polar bond* because a dipole exists between the two atoms of the bond. The covalent bond in a hydrogen molecule is sometimes called a *nonpolar* covalent bond. This is because no dipole exists due to the equal sharing of electrons by the atoms. The bond lacks polarity.

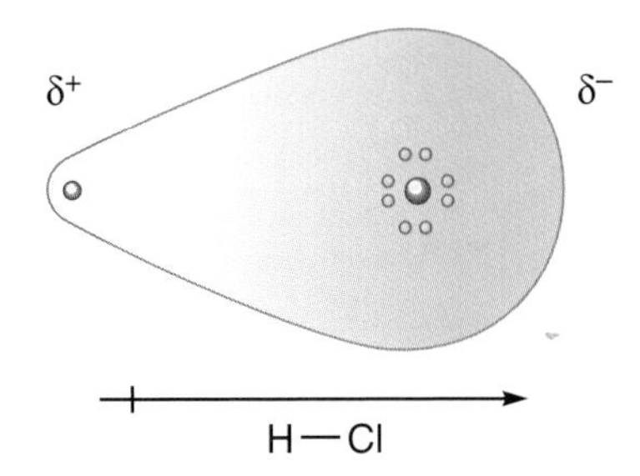

Figure 3.6 The shared electron pair in a hydrogen chloride molecule is attracted toward the chlorine atom. Thus the hydrogen end of the molecule is partially positive (δ^+), and the chlorine end is partially negative (δ^-). The bond is polar because a dipole exists.

3.5 Predicting Bond Type

A quick and fairly accurate rule for predicting whether a bond is covalent or ionic involves the metallic or nonmetallic properties of atoms. If one atom of the bond is a metal and the other a nonmetal, then the bond is usually ionic. If both atoms are nonmetallic, the bond is covalent.

Self-Test

Predicting Bond Type Using the Metallic Character of an Atom

Predict the bond type between an atom of each of these elements:

1. Nitrogen and fluorine
2. Calcium and oxygen
3. Iron and chlorine
4. Carbon and sulfur

ANSWERS

1. Covalent 2. Ionic 3. Ionic 4. Covalent

There is also a way to predict whether a covalent bond is polar or nonpolar, and this method also distinguishes between covalent and ionic bonds. The determining factor is the tendency of the bonding atoms to attract electrons to themselves, which is called **electronegativity.** If one of the bonding atoms attracts the electrons of a bond much more strongly than the other atom, then the electrons are transferred from the atom with weak attraction to the atom with the strong attraction. Ions are formed, and the bond is ionic.

$$\text{Na}\cdot + \cdot\ddot{\underset{\cdot\cdot}{\text{Cl}}}: \longrightarrow \text{Na}^+ + \left[:\ddot{\underset{\cdot\cdot}{\text{Cl}}}:\right]^- \qquad \text{Ionic bond}$$

Electron transferred to the more electronegative atom

An ionic bond forms between sodium and chlorine because chlorine is much more electronegative than sodium. A chlorine atom removes the electron from a sodium atom to form ions.

If each bonding atom has the same or nearly the same electronegativity, they will share the electrons equally. The bond between them is nonpolar covalent.

$$\text{H}\cdot + \cdot\text{H} \longrightarrow \text{H}:\text{H} \qquad (\text{H—H}) \qquad \text{Nonpolar covalent bond}$$

Electrons shared equally between atoms

A nonpolar covalent bond forms when two atoms share the electron pair equally.

If the attraction between two atoms for the shared electrons is different but not so great that ions form, then the bond between the atoms is polar covalent.

$$\text{H}\cdot + \cdot\ddot{\underset{\cdot\cdot}{\text{Cl}}}: \longrightarrow \overset{+\!\!\!\longrightarrow}{\text{H}:\ddot{\underset{\cdot\cdot}{\text{Cl}}}:} \qquad (\text{H—Cl}) \qquad \text{Polar covalent bond}$$

Unequal sharing of electrons

A polar covalent bond forms between a hydrogen atom and a chlorine atom because a chlorine atom attracts electrons to itself more strongly than a hydrogen atom does. A dipole exists in the molecule.

Elements are assigned electronegativity values based on their ability to attract shared electrons (Figure 3.7). The ability of an element to attract shared electrons is highest in the upper-right corner of the periodic table (fluorine's value of 4.0 is the largest) and lowest in the lower left. Electronegativity values are used to predict bond type. If the difference between the electronegativity of two atoms is large, say 1.8 or greater, then electron transfer occurs and an ionic bond forms (Table 3.5). Sodium has an electronegativity of 1.0, and chlorine has a value of 3.0. The difference, 2.0, is greater than 1.8; the bond that forms between sodium and chlorine is ionic. If the difference between the electronegativity of two atoms is small, less than 0.5, the electrons are shared equally and a nonpolar covalent bond forms. Carbon has an electronegativity of 2.5; hydrogen's value is 2.1. Because the difference, 0.4, is less than 0.5, the bond between carbon and hydrogen is nonpolar covalent. If the difference between the electronegativity of two atoms is in the range 0.5 to 1.7, the electrons are shared unequally. The difference between the electronegativities of hydrogen (2.1) and oxygen (3.5) is 1.4. A polar covalent bond exists between oxygen and hydrogen atoms.

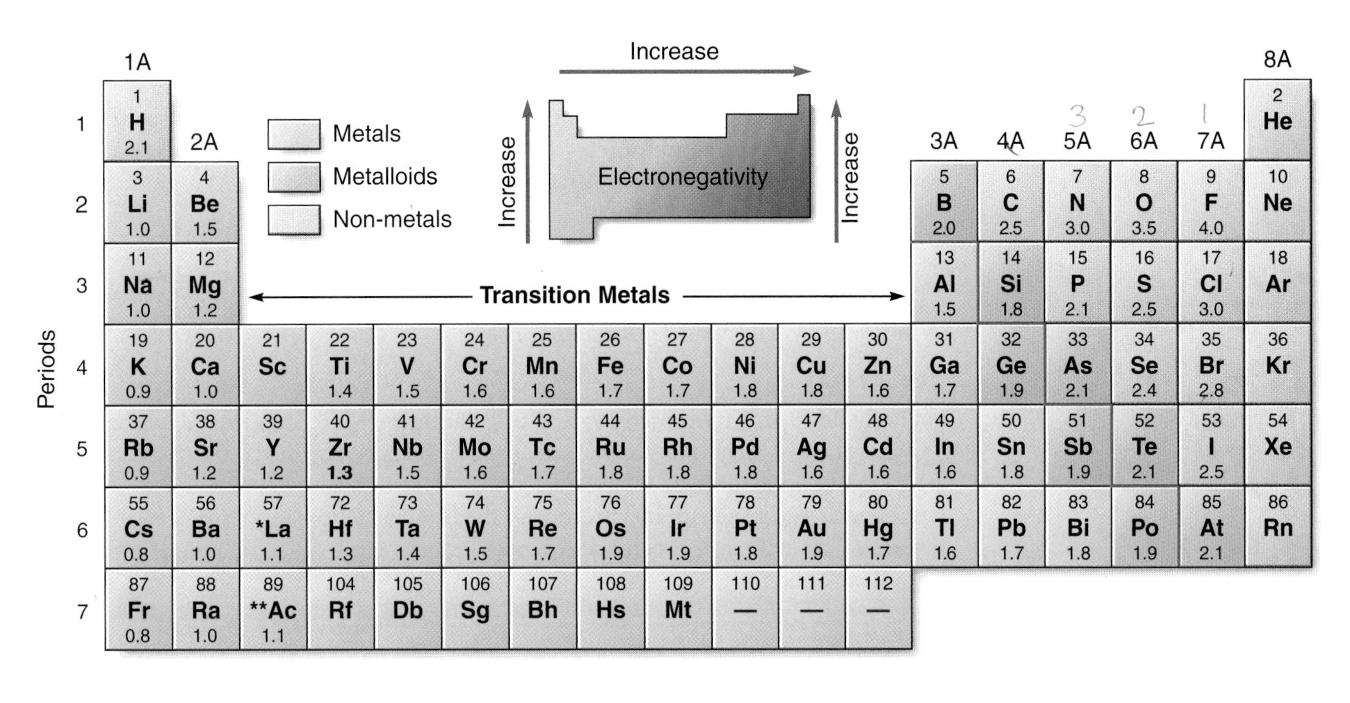

Figure 3.7 This periodic table contains the electronegativity values for some of the elements. These values range from 0.8 to 4.0, and, in general, nonmetals have larger electronegativity values than metals.

Table 3.5 Predicting Bond Type from Differences in Electronegativities

ELECTRONEGATIVITY DIFFERENCE	TYPE OF BOND	EXAMPLE
Less than 0.5	Nonpolar covalent	Bond between two identical atoms; bonds between nonmetals having similar electronegativity values; H—H, Br—Br
0.5 to 1.7	Polar covalent	Bond between nonmetals having somewhat different electronegativity values; $\overset{\delta^+}{\text{H}}$—$\overset{\delta^-}{\text{Cl}}$, $\overset{\delta^+}{\text{H}}$—$\overset{\delta^-}{\text{Br}}$
1.8 and larger	Ionic	Bond between metals and nonmetals; Na^+Cl^-, Li^+F^-

The three different bond types (ionic, nonpolar covalent, and polar covalent) have been discussed as though they are totally distinct from each other. In reality, the boundaries between the three types are not clear-cut. Some bonds seem to have some ionic properties and some covalent properties. We therefore must use differences in electronegativity only as a general guide to bond type.

Example 3.3 Predicting Bonds Between Atoms

Predict the type of bond that forms between (a) oxygen and sodium; (b) carbon and sulfur; (c) hydrogen and nitrogen.

SOLUTIONS

The bond type can be predicted by calculating the difference in the electronegativity values of the atoms. (a) $3.5 - 1.0 = 2.5$. Because the difference is greater than 1.8, the bond is ionic. (b) The electronegativity of both elements is 2.5. The difference in electronegativity is zero, and the bond is nonpolar covalent. (c) $3.0 - 2.1 = 0.9$. This electronegativity difference lies between 0.5 to 1.7, indicating a polar covalent bond.

Self-Test

Which type of bond forms between

1. Nitrogen and nitrogen
2. Potassium and oxygen
3. Chlorine and hydrogen
4. Carbon and phosphorus

ANSWERS

1. Nonpolar covalent 2. Ionic 3. Polar covalent 4. Nonpolar covalent

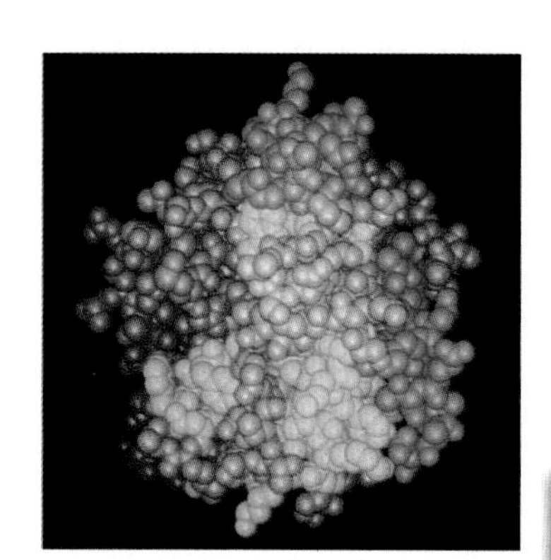

Large molecules like proteins and DNA have important roles in the body. These molecules perform their function because of their unique shapes. Shown here is the protein chymotrypsin. *(Charles Grisham)*

3.6 Shapes of Molecules

The shape of a molecule determines many of the properties of the compound. Proteins and DNA in the body, for instance, can carry out their specific function because they have their own unique shape. Smaller molecules also have specific properties because of their shape. Although the shapes of large molecules are hard to predict, the shapes of small molecules can be determined from the atoms and bonds they contain.

Diatomic molecules are always linear because the two atoms must lie in a straight line. Molecular hydrogen, H_2, nitrogen, N_2, and chlorine, Cl_2, are linear.

The shapes of molecules containing three or more atoms are predicted using Lewis structures to determine the arrangement of valence shell electrons. In general, *groups of electrons get as far away from each other as possible.* This makes sense because electrons have a negative charge and so repel each other. As an example, let's take an atom that has four pairs of electrons around it. The pairs are positioned as far apart as possible to minimize repulsions. The arrangement of the four electron

Just as opposite charges attract, like charges repel.

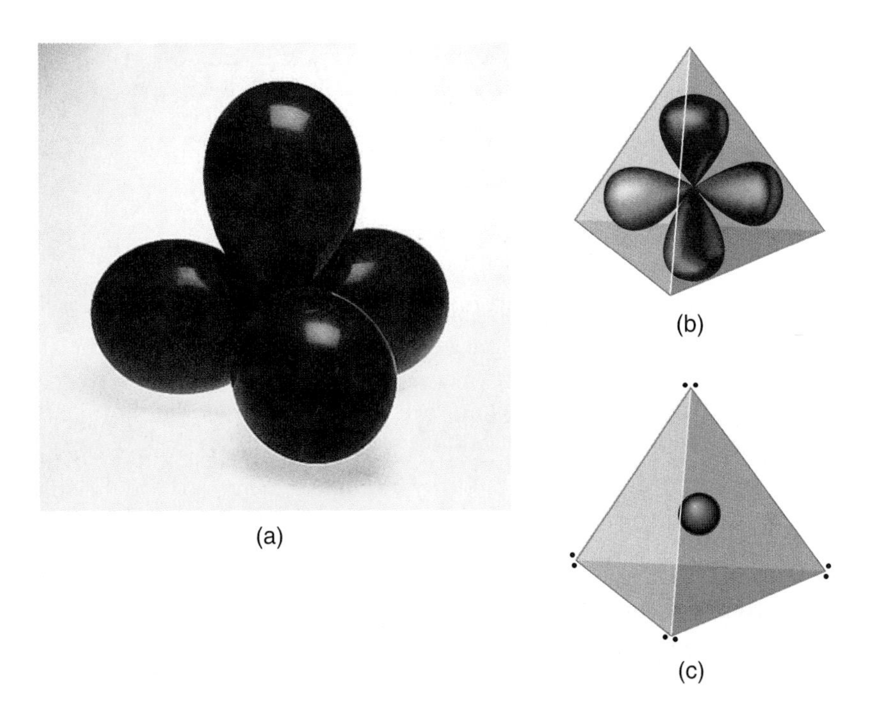

Figure 3.8 (a) When four balloons are tied together, they push against each other. (b) This mutual pushing (repulsion) forces the free end of each balloon toward one corner of a three-dimensional shape that is called a tetrahedron. (c) Four pairs of electrons (shown here around a carbon atom designated by a black sphere) also force each other to the corners of a tetrahedron. *(a, Charles D. Winters)*

pairs around an atom's nucleus can be compared to four balloons tied together (Figure 3.8). The pairs of electrons are forced apart just as the four balloons are forced apart. Each balloon, and each electron pair, points toward one corner of a tetrahedron.

Now consider a molecule of the compound known as methane, CH_4 (Figure 3.9). A methane molecule has a single carbon atom with four single bonds (the four pairs of electrons) to hydrogen atoms. The carbon atom lies in the middle of the molecule, and the four pairs of electrons force each other to point outward toward the corners of a tetrahedron. Each of the four pairs of electrons forms a covalent bond to a hydrogen atom. Because the four hydrogen atoms point to the four corners of a tetrahedron, the geometry around the carbon atom is said to be tetrahedral.

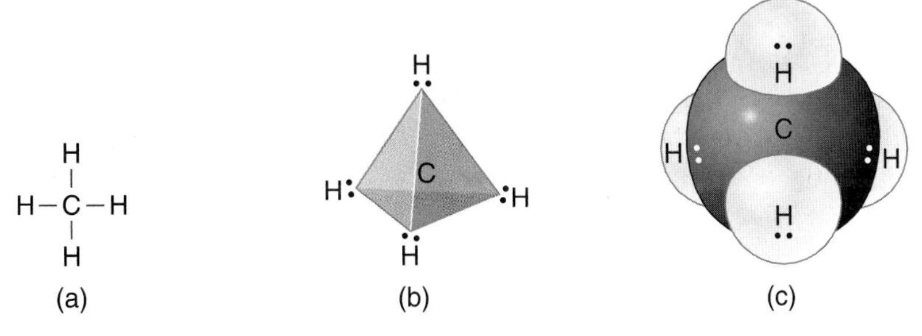

Figure 3.9 (a) This Lewis structure of methane shows methane's carbon atom bonded to four hydrogen atoms. (b) The carbon atom of methane lies in the middle of a tetrahedron, with the pairs of bonding electrons at each corner of the tetrahedron. There is a hydrogen atom at each of these corners. (c) This space-filling model of methane shows that methane is tetrahedral.

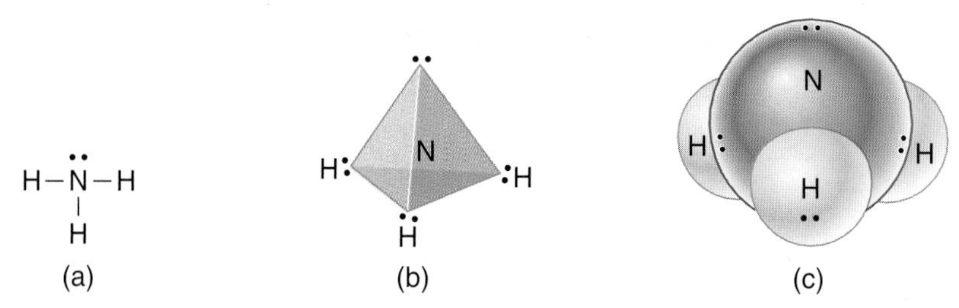

Figure 3.10 (a) This Lewis structure of ammonia shows ammonia's nitrogen atom bonded to three hydrogen atoms. The nitrogen atom has a fourth pair of electrons that is not involved in bonding. (b) The nitrogen atom of ammonia lies in the middle of a tetrahedron, with the electron pairs at the four corners of the tetrahedron. The three hydrogen atoms lie at three of the four corners of the tetrahedron. (c) The shape of a molecule of ammonia is pyramidal.

A molecule of the compound known as ammonia (NH_3) has a nitrogen atom covalently bonded to three hydrogen atoms (Figure 3.10). In this molecule, the nitrogen atom has four pairs of electrons around it, the three pairs that are the bonds to hydrogen and an *unshared* or *lone pair*. Like the electron pairs around the carbon atom in methane, the four pairs point toward the corners of a tetrahedron. In ammonia there are only three other atoms, however; thus one pair of electrons is not shared with another atom. The arrangement of electrons in ammonia and methane is the same, but the molecules are not represented in the same way because unshared electrons are not included in the description of molecular shape. The representation of the shape of any molecule is determined by the atoms making up the molecule. Therefore even though the four pairs of electrons in ammonia point toward the corners of a tetrahedron, the shape of ammonia is described as *pyramidal*. The three hydrogen atoms form the base of a pyramid, and the nitrogen atom forms the apex [Figure 3.10(c)].

Predicting the shape of water is similar to predicting the shapes of methane and ammonia. The Lewis structure of water shows two single bonds and two pairs of unshared electrons, thus the four pairs of electrons point toward the corners of a tetrahedron [Figure 3.11(a) and (b)]. But because only atoms (not unshared pairs of electrons) are considered in molecular shape, the three atoms in water are said to have a *bent* shape [Figure 3.11(c)].

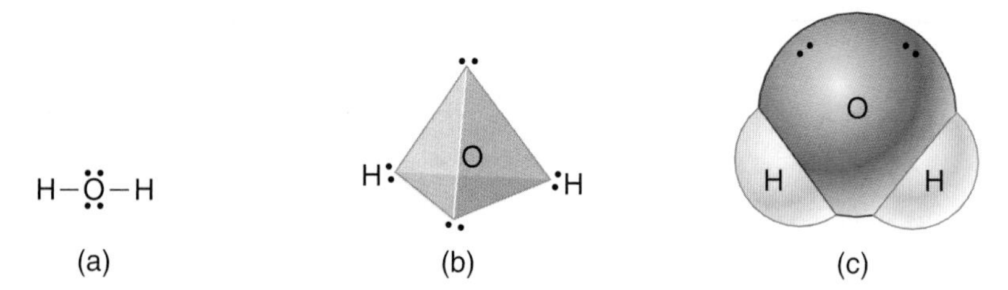

Figure 3.11 (a) The Lewis structure of water. (b) The arrangement of electron pairs and hydrogen atoms around the oxygen atom of water. (c) The shape of a water molecule is bent.

Example 3.4 Predicting Molecular Shapes

The Lewis structure of hydrogen sulfide (H_2S) is

$$H:\underset{\cdot\cdot}{\overset{\cdot\cdot}{S}}:H$$

1. What is the three-dimensional arrangement of electron pairs around the sulfur atom?
2. What is the shape of a hydrogen sulfide molecule?

SOLUTIONS

1. Because there are four pairs of electrons around the sulfur atom, the pairs must point to the corners of a tetrahedron (see figure in margin).
2. Each hydrogen atom of hydrogen sulfide goes to a corner of the tetrahedron. When the shape of a molecule is described, only the atoms are considered. The molecular shape of hydrogen sulfide is bent.

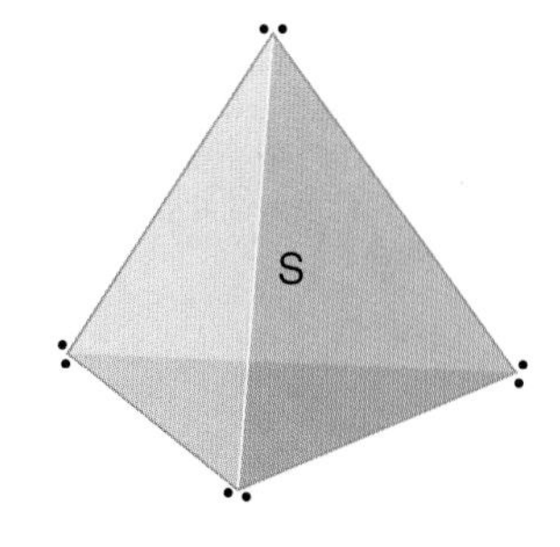

The four pairs of electrons around the sulfur atom point to the corners of a tetrahedron.

Self-Test

Use the Lewis structure of carbon tetrabromide (CH_4) to answer these questions.

$$\begin{array}{c} Br \\ Br:\underset{\cdot\cdot}{\overset{\cdot\cdot}{C}}:Br \\ Br \end{array}$$

1. What shape describes the arrangement of electron pairs around a carbon atom in carbon tetrabromide?
2. What is the shape of a carbon tetrabromide molecule?

ANSWERS

1. Tetrahedral 2. Tetrahedral

Shapes of Molecules

SHAPE	EXAMPLE
Linear	Hydrogen chloride (HCl)
Tetrahedral	Methane (CH_4)
Pyramidal	Ammonia (NH_3)
Bent	Water (H_2O)

There are other molecular shapes that you may see, but for now learn the ones discussed here: linear, tetrahedral, pyramidal, and bent.

3.7 Polarity of Molecules

Molecules of similar size can show large differences in properties. Consider methane and water, for example. Methane is a gas at room temperature. Water molecules have about the same mass as methane molecules, but water is a liquid at room temperature. What is the explanation for this? Similarly, when sugar is mixed with water, the sugar dissolves in the water, but if oil is mixed with water, the two do not dissolve into each other (Figure 3.12). Furthermore, sugar does not dissolve in oil but gasoline does. Why do some substances dissolve in some liquids and others do not? These questions will be answered in detail in later chapters, but for now let us look at the basis for these phenomena: molecular polarity.

What factors make a molecule polar? Earlier you learned that a covalent bond can be polar, and the bond in hydrogen chloride (H—Cl) was used as an example. Is the polarity in molecules related to bond polarity? Most definitely, but there is one additional consideration: molecular shape. To predict molecular polarity, we must know both the polarities of the covalent bonds in the molecule and the shape of the molecule.

If a molecule contains only nonpolar bonds, the electrons are shared equally throughout the molecule, and the molecule must be nonpolar regardless of its shape. Hydrogen, H—H, must be a nonpolar molecule because the electrons are shared equally, the bond is nonpolar. In methane each C—H bond is nonpolar, thus the electrons in the molecule are evenly distributed. A molecule of methane must be nonpolar.

Because the difference in electronegativity values between carbon and hydrogen is less than 0.5, the C—H bond is nonpolar.

If a molecule possesses one or more polar bonds, then the molecule may be either polar or nonpolar. Let's consider the molecules of several compounds to understand molecular polarity. We begin with hydrogen chloride with its polar bond. Because of the polar bond, the electrons spend more time around the chlorine atom, and thus a molecule of hydrogen chloride is polar. Next consider carbon dioxide, which is a linear molecule because the two pairs of electrons in the double bonds to carbon get as far apart as possible:

$$:\ddot{O}=C=\ddot{O}:$$

In carbon dioxide, the carbon atom is double bonded to two oxygen atoms and located between them. Each carbon-to-oxygen double bond is polar because the oxygen atoms are more electronegative than the carbon atom, and therefore the electrons of the bonds spend more time near the oxygen atoms. This results in two equal dipoles in a carbon dioxide molecule:

$$:\ddot{O}=C=\ddot{O}: \quad \text{(a linear molecule)}$$
$$\longleftarrow\!\!+ \quad +\!\!\longrightarrow$$

The dipoles point directly away from each other and so cancel each other. This results in no *net* dipole in the molecule. An analogy is a tug-of-war between two equal teams. If each team exerts an equal pull, the contest remains at a standstill; there is no net gain by either team. When equal dipoles oppose each other (pull in opposite directions), there is no *net* dipole in the molecule. A carbon dioxide molecule possesses two polar bonds, but the molecule is nonpolar because the bond dipoles cancel.

Figure 3.12 This oil and vinegar dressing shows that oil does not mix with water. (Vinegar is about 95% water.) The polarity of molecules determines which compounds dissolve in which other compounds. *(Charles D. Winters)*

Now consider the structure of water. Each hydrogen atom is single bonded to the oxygen atom. Because oxygen is more electronegative than hydrogen, the two bonds are polar covalent bonds, with the electrons attracted toward the oxygen atom. Two dipoles exist in the molecule:

$$\begin{array}{ccc} & :\ddot{O}: & \\ \nearrow & & \nwarrow \\ H & & H \end{array}$$

The negative ends of the dipoles are at the oxygen atom, and a positive end is at each hydrogen atom. Because water has a bent shape, the two dipoles do not cancel each other. Instead they *reinforce* each other—that is, they add up to yield a dipole that is characteristic of the molecule rather than of the individual bonds.

Table 3.6 Predicting the Polarity of a Molecule

THIS **BOND POLARITY**	WITH THIS **MOLECULAR SHAPE**	RESULTS IN THIS **POLARITY OF THE MOLECULE**	**EXAMPLE**
Nonpolar	Any molecular shape	Nonpolar	Methane CH_4 (H—C—H with H above and below C)
Polar	Shape that causes polar bonds to cancel	Nonpolar	Carbon dioxide O=C=O
Polar	Shape that does not cause polar bonds to cancel	Polar	Water H—O—H (bent)

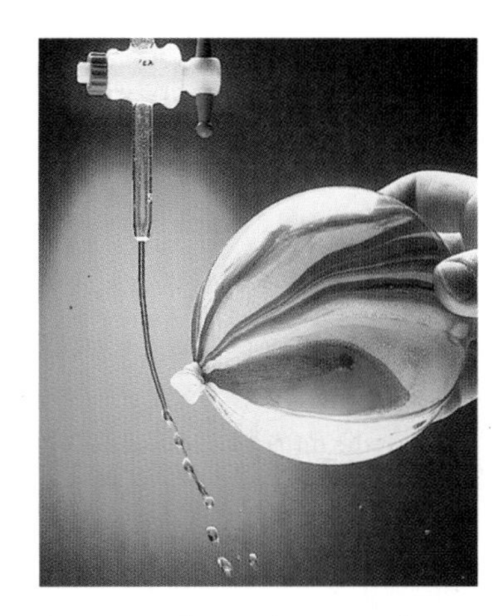

A stream of water is attracted toward this charged balloon because the water molecules are dipoles. You can try this at home by rubbing a balloon on your hair, clothes, or a cat to induce a static charge. *(Charles D. Winters)*

The oxygen end of the molecule is partially negative and the hydrogen end is partially positive. As a result, a water molecule is polar:

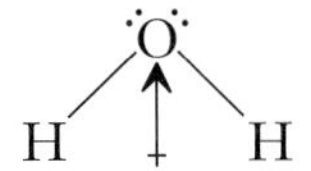

Table 3.6 summarizes the effects of bond polarity and molecular shape on molecular polarity.

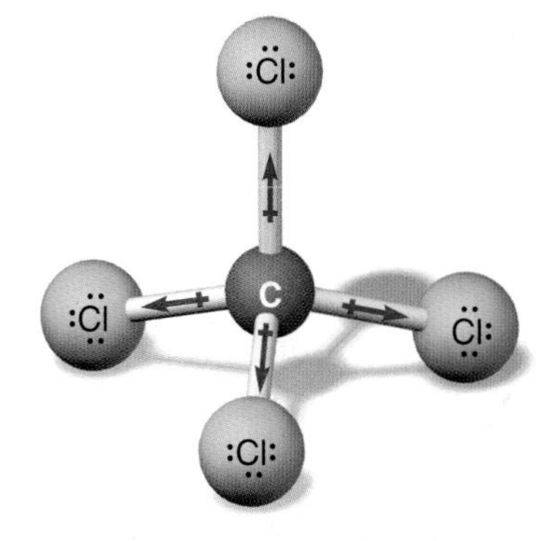

Carbon tetrachloride has a tetrahedral shape and four polar bonds, but it is a nonpolar molecule.

Example 3.5 Predicting Molecular Polarity

Carbon tetrachloride consists of a carbon atom single bonded to each of four chlorine atoms. Is a molecule of carbon tetrachloride polar or nonpolar?

SOLUTION

Carbon tetrachloride has four chlorine atoms, each single bonded to a carbon atom. The carbon-chlorine bond is polar, thus there are four polar bonds (and four dipoles) in the molecule. Carbon tetrachloride molecules are tetrahedral like methane molecules. Carbon tetrachloride molecules are nonpolar because the dipoles produced by the individual carbon-chlorine bonds cancel each other out. There is no *net* dipole in the molecule.

Self-Test

Is an ammonia molecule polar or nonpolar?

ANSWER

Polar; the molecule has three polar bonds that do not cancel each other out.

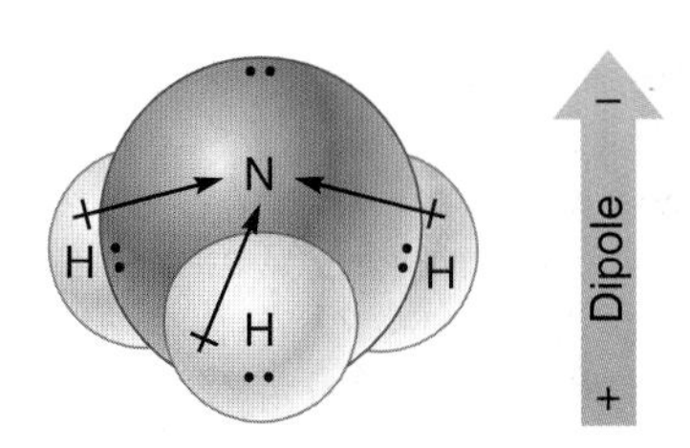

Ammonia is a polar molecule.

This section began with a comparison of the physical states of methane and water and with statements that point out that some substances dissolve in some liquids and others do not. These properties depend on molecular polarity.

Because water molecules are polar, the partially positive ends of water molecules are attracted to the partially negative ends of other water molecules. These attractions help hold water molecules together, and as a result water is a liquid at room temperature. In contrast, methane molecules are nonpolar; thus they are not attracted to each other in the same way as water molecules. Their attraction to each other is rather weak, and at room temperature methane is a gas.

Water molecules are also attracted to other polar molecules by attraction of opposite ends of their dipoles. As a result, polar molecules such as sugar dissolve into water. Similarly, many ionic compounds dissolve in water because the partially positive ends of water molecules are attracted to the anions and the partially negative ends of water molecules are attracted to the cations. In contrast, nonpolar substances do not form these same interactions with water molecules, and so they do not dissolve in water. Solubilities of compounds will be explained more fully in Chapter 6, but in general it helps to remember that like dissolves like. Polar and ionic substances dissolve in polar substances and nonpolar substances dissolve in nonpolar substances.

CHAPTER SUMMARY

The Group 8A elements, the noble gases, exist as free atoms because their valence shell contains eight electrons. The atoms of all other elements lack this feature, and these atoms lose, gain, or share electrons to obtain eight valence electrons (the **octet rule**). **Lewis structures** are used to display the arrangement of valence electrons in an atom, ion, or molecule. Atoms or groups of atoms that have gained or lost electrons are called **ions.** Ions bond to ions of opposite charge through **ionic bonds** in **ionic compounds.** If an aqueous solution of a compound conducts electricity, the compound is an **electrolyte.** If it does not, the compound is a nonelectrolyte. Atoms share pairs of electrons through **covalent bonds.** A **single bond, double bond,** and **triple bond** consist of one pair, two pairs, or three pairs of shared electrons, respectively. A collection of atoms joined by covalent bonds is a **molecule,** and a compound made up of molecules is a **molecular compound.** The electrons of a covalent bond are shared equally in a nonpolar bond but unequally in a **polar bond.** Unequal sharing of electrons results from differences in the **electronegativities** of the bonding atoms. The type of bond between atoms in a compound can be predicted using differences in electronegativity. A **dipole** is an unequal distribution of charge, and both bonds and molecules can exist as dipoles. The shapes of molecules can be predicted from the arrangement of electrons around the atoms in the molecule. Molecules can be polar, and molecular polarity can be predicted using the molecular shape and the polarity of the bonds in the molecule. Ultimately it is the structure that determines the properties of that molecule.

KEY TERMS

Section 3.1
octet rule
Lewis structure (electron dot structure)

Section 3.2
anion
cation
ion
monatomic ion
polyatomic ion

Section 3.3
electrolyte
ionic bond
ionic compound

Section 3.4
covalent bond
diatomic molecule
dipole
double bond
molecular compound
molecule
polar covalent bond
single bond
triple bond

Section 3.5
electronegativity

EXERCISES

Octet Rule and Lewis Structures (Section 3.1)

1. How many electrons are found in the valence shell of the atoms of most noble gases? Which noble gas is an exception?
2. Use valence shell electrons to explain why the noble gases are unreactive.
3. Describe the processes by which atoms of most elements achieve the noble-gas electron configuration.
4. In your own words define the octet rule.
5. Describe the location of electrons in shells for an atom of each of these elements. How many valence electrons are present in each atom?
   (a) Li **(b)** Beryllium **(c)** Potassium
6. Describe the location of electrons in shells of an atom of each of these elements. How many valence electrons are present in each atom?
   (a) Mg **(b)** Fluorine **(c)** Si
7. How many valence electrons are found in the atoms of these elements? Which elements satisfy the octet rule?
   (a) K **(b)** S **(c)** Chlorine **(d)** Argon **(e)** Ca **(f)** Boron
8. How many valence electrons are found in the atoms of these elements? Which elements satisfy the octet rule?
   (a) Nitrogen **(b)** O **(c)** Xenon **(d)** Mg **(e)** He **(f)** Sodium
9. In a Lewis structure, the element symbol represents what parts of an atom?
10. What do the dots represent in a Lewis structure?
11. Draw Lewis structures for atoms of these elements:
    (a) Li **(b)** Oxygen **(c)** Carbon **(d)** Al
12. Draw Lewis structures for atoms of these elements:
    (a) P **(b)** Neon **(c)** Ba **(d)** Hydrogen
13. Draw Lewis structures of the atoms in Exercise 7.
14. Draw Lewis structures of the atoms in Exercise 8.

Ions (Section 3.2)

15. What is the definition of an ion? How does a sodium atom differ from a sodium ion?
16. How does an atom become an ion? Describe how a lithium atom becomes a lithium ion and a fluorine atom becomes a fluoride ion.
17. Define an anion and give two examples.
18. Define a cation and give two examples.
19. How is the charge on an ion represented?
20. Write the correct representation for these monatomic ions, then identify the ion as a cation or anion.
    (a) Ion of potassium with a plus one charge
    (b) Ion of iodine with a minus one charge
    (c) Ion of iron that has a plus two charge
21. Write the correct representation for these monatomic ions, then identify the ion as a cation or anion:
    (a) Ion of sulfur with a minus two charge
    (b) Ion of silver with a plus one charge
    (c) Ion of chromium with a plus three charge
22. What two errors were made in the representation of this ion: Ba_{+2}?
23. What is the difference between a polyatomic and monatomic ion? Explain why the cyanide ion (CN^-) is monatomic or polyatomic.
24. What happens to a metal atom when it is involved in an electron-transfer reaction? Provide an example using a magnesium atom.
25. What happens to a nonmetal atom when it is involved in an electron-transfer reaction? Provide an example using a bromine atom.
26. Which of the following atoms would lose or gain electrons in an electron-transfer reaction? How many electrons would each atom gain or lose?
    (a) Li **(b)** S **(c)** Aluminum **(d)** Iodine
27. Which of the following atoms would lose or gain electrons in an electron-transfer reaction? How many electrons would each atom gain or lose?
    (a) O **(b)** Mg **(c)** Sodium **(d)** Te
28. After electron transfer, what charge would the monatomic ions in Exercise 26 have?
29. After electron transfer, what charge would the monatomic ions in Exercise 27 have?
30. Draw the Lewis structure of the monatomic ions formed in Exercise 26.
31. Draw the Lewis structure of the monatomic ions formed in Exercise 27.
32. Predict the ions that form when these atoms react. Draw their Lewis structures.
    (a) $K\cdot + :\dot{\underset{\cdot\cdot}{Br}}: \rightarrow$? **(b)** $2Na\cdot + :\dot{\underset{\cdot}{O}}: \rightarrow$?
    (c) $\cdot Mg\cdot + 2\cdot\ddot{\underset{\cdot\cdot}{F}}: \rightarrow$?

33. Predict the ions that form when these atoms react. Draw their Lewis structures.

(a) 2Li· + :S: → ?

(b) ·Ca· + 2:Br: → ?

(c) ·Al· + 3:Cl: → ?

34. How many atoms of bromine would react with an aluminum atom? What would be the charge on the ions that would be produced?

35. How many atoms of oxygen would react with a magnesium atom? What would be the charge on the ions that would be produced?

36. Why does this ion normally not exist? B^{3+}

Ionic Bonds and Ionic Compounds *(Section 3.3)*

37. What is an ionic bond?

38. Choose the pair(s) that would be bound by an ionic bond.
(a) Ca and F (b) Na and I (c) S and Te

39. Choose the pair(s) that would be bound by an ionic bond.
(a) H and I (b) Zn and Cl (c) P and F

40. What two factors affect the strength of an ionic bond?

41. Define an ionic compound. Explain how you can determine if a compound is ionic by looking at its composition.

42. A compound contains only sodium and fluorine. Is this compound ionic? Why or why not?

43. A compound contains only sulfur and oxygen. Is this compound ionic? Why or why not?

44. When an ionic compound is dissolved in water, the mixture conducts electricity. Why?

45. What is an electrolyte? Give an example of an electrolyte found in the body.

Covalent Bonding: Atoms Share Electrons in Molecules *(Section 3.4)*

46. What is a covalent bond? Provide an example of a covalent bond.

47. What is a molecule?

48. Compare the bonding and composition of a molecular compound and an ionic compound.

49. Compounds can be formed from the pairs of elements presented. Which of these compounds will be ionic and which will be covalent?
(a) Lead and oxygen (b) Sulfur and fluorine
(c) Phosphorus and hydrogen

50. Compounds can be formed from the pairs of elements presented. Which of these compounds will be ionic and which will be covalent?
(a) Chlorine and oxygen (b) Silver and iodine
(c) Carbon and hydrogen

51. A substance that, when dissolved in water, does not conduct electricity is called a(n) _____.

52. The compound chloroform contains only carbon, chlorine, and hydrogen. Is this compound ionic or molecular? Chloroform is neither acidic nor basic. Is chloroform an electrolyte or nonelectrolyte?

53. List the differences between a single, double, and triple covalent bond.

54. If an atom of carbon and an atom of oxygen were sharing six electrons, what type of bond is present between the atoms?

55. What is a diatomic molecule?

56. Give two examples of elements whose atoms normally exist as diatomic molecules.

57. What is a dipole?

58. What is a polar covalent bond? What would cause the bond to be polar?

59. An atom of fluorine attracts shared electrons better than an atom of phosphorus. Which end of a F—P bond would be partially negative? Partially positive? Draw a F—P bond and label it with the appropriate symbol to show the polarity of this bond.

Predicting Bond Type *(Section 3.5)*

60. Use the metallic character of elements to predict whether the bond between these pairs is ionic or covalent.
(a) Cr, Cl (b) Na, S (c) K, N (d) Se, F

61. Use the metallic character of elements to predict whether the bond between these pairs is ionic or covalent.
(a) Si, H (b) Li, Te (c) N, H (d) Cd, F

62. What is electronegativity?

63. The elements in which part of the periodic table would have the highest electronegativity values? Which would have the lowest?

64. From these diatomic molecules, select the one that would have a polar covalent bond and explain why the bond is polar: Br—Br H—Br

65. For the following pairs of atoms determine the difference in their electronegativities, then predict

whether the bond between them would be ionic, polar covalent, or nonpolar covalent.
(a) Na, Br **(b)** H, Br **(c)** Br, Br **(d)** C, O

66. For the following pairs of atoms determine the difference in their electronegativities, then predict whether the bond between them would be ionic, polar covalent, or nonpolar covalent.
(a) O, O **(b)** S, O **(c)** Ca, O **(d)** C, H

Shapes of Molecules (Section 3.6)

67. Electrons have negative charges. How are repulsions between the electrons minimized in an atom?

68. If an atom has four pairs of electrons, how in space will the pairs be arranged around the atom?

69. Ammonia, $\mathrm{H{-}\ddot{N}(H)_2}$ (N bonded to three H atoms, one lone pair), has a pyramidal shape.

What is the shape of $\mathrm{H{-}\ddot{P}(H)_2}$ (P bonded to three H atoms, one lone pair)?

Polarity of Molecules (Section 3.7)

70. A molecule contains only nonpolar bonds. Will it be polar or nonpolar?

71. A molecule is nonpolar. Does this mean it contains only nonpolar bonds? Why or why not?

72. Predict whether these molecules are polar or nonpolar.
(a) $H{-}Se{-}H$ (bent) **(b)** $F{-}C{-}H$ with F above and F below C **(c)** $C{\equiv}O$

73. Predict whether these molecules are polar or nonpolar.
(a) $S{=}C{=}S$ **(b)** $Cl{-}\ddot{P}$ with Cl above and Cl below P **(c)** $H{-}F$

74. Naphthalene is a nonpolar substance that is sometimes used in moth balls. Will it dissolve in water?

75. What does the expression *like dissolves like* mean? (Hint: Reread the end of Section 3.7.)

Challenge Exercises

76. **(a)** Determine the percent mass loss when a sodium atom becomes a sodium ion. **(b)** Does this mass loss seem significant? **(c)** In Section 4.5 you will learn that the mass of an atom and its corresponding monatomic ion are considered the same. Do your calculations support this consideration?

77. In one compound, a boron atom is single bonded to three fluorine atoms, but the boron atom has no lone pairs of electrons. What would be the arrangement of electron pairs around the boron atom?

The Language of Chemistry

4

CHEMISTS USE FORMULAS, NAMES, AND EQUATIONS TO DESCRIBE IONS, MOLECULES, AND CHEMICAL REACTIONS.

IN CHAPTER 3 you learned about ions and molecules. You also learned how electrons are transferred between atoms and how they are shared. When electrons are gained or lost, or when existing covalent bonds are broken and new ones formed, a chemical reaction occurs. Chemical reactions and substances can always be explained with language, but chemists have developed and now use a set of symbols, names, and conventions that represent substances and reactions more concisely. In this chapter you are introduced to these symbols, formulas, names, and equations. You will soon realize that these representations are not unique to chemistry. Many professions, including health care, incorporate chemical terminology into their language. As with any other language, you may struggle a bit with the language of chemistry at first, but you will see that the rules and terms are generally logical and follow a recognizable pattern. With a little practice, you will be speaking chemistry with some familiarity, and, more importantly, understanding it when you read or hear it.

Outline

4.1 Chemical Formulas and Names
Determine the composition of a compound from its formula.

4.2 Naming Ionic Compounds
Write the name of a binary ionic compound from its formula and determine its formula from its name.

4.3 Naming Covalent Compounds
Write the name of a binary covalent compound from its formula and determine its formula from its name.

4.4 Chemical Equations
Write simple chemical equations and interpret them.

4.5 The Mole
Define the mole and use moles to describe the reaction shown by a chemical equation.

Chemical reactions involve atoms, molecules, and ions and are represented by equations, formulas, and models. Here phosphorus (represented as four purple spheres) reacts with chlorine (two tan spheres) to produce phosphorus trichloride (one purple and three tan spheres). ***(Photo, Charles D. Winters; models, S. M. Young)***

4.1 Chemical Formulas and Names

There are literally millions of compounds, each with its own unique composition and structure. To deal with this huge number of compounds, chemists have developed a systematic way to both represent and name them.

Formulas for Compounds

Formulas (less commonly called *chemical formulas*) are used to show the composition or structure of compounds. The formula for a given compound always contains the symbol for each element in the compound; it also contains subscripted numbers if two or more ions or atoms of the element are present. For example, H_2O is the formula for water and $C_6H_{12}O_6$ is the formula for glucose (blood sugar).

The smallest whole-number ratio for the ions in sodium chloride is 1 : 1; thus its formula is NaCl. The formula Na_2Cl_2 would be incorrect.

Ionic Compounds

The formula for an ionic compound shows the composition of the compound as the smallest whole-number ratio of ions present in the compound. This group of ions is called the *formula unit* for that ionic compound. For sodium chloride, common table salt, the formula—NaCl—tells us that there is one sodium ion for each chloride ion and that NaCl is the formula unit for this compound (Figure 4.1). The ionic compound potassium oxide has this chemical formula: K_2O. This formula tells us (a) that both potassium and oxygen are present and (b) that there are two potassium ions (K^+) for each oxide ion (O^{2-}). The ionic compound aluminum oxide has this formula: Al_2O_3. What is the ratio of aluminum ions to oxide ions in aluminum oxide? The subscript 2 following Al means there are two aluminum ions (Al^{3+}). The subscript 3 following the O means there are three oxide ions (O^{2-}).

Aluminum oxide (Al_2O_3) is a hard, white solid used as the abrasive in some "sand" papers and is also the laboratory reagent known as alumina. Some gems—a ruby is shown as an example—are crystals of aluminum oxide that contain small amounts of colored ions. Small amounts of chromium(III) ions give rubies their characteristic color. *(Charles D. Winters)*

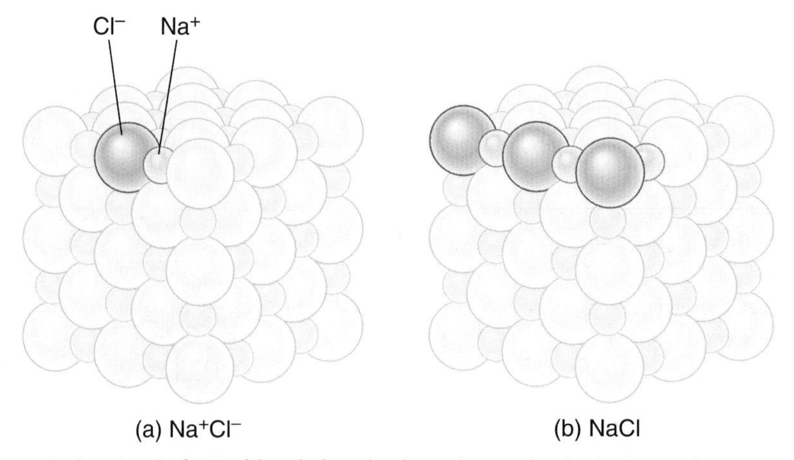

Figure 4.1 (a) Sodium chloride has the formula NaCl, which means there is one sodium ion per one chloride ion. (b) In a sample of sodium chloride, there is always one sodium ion per chloride ion.

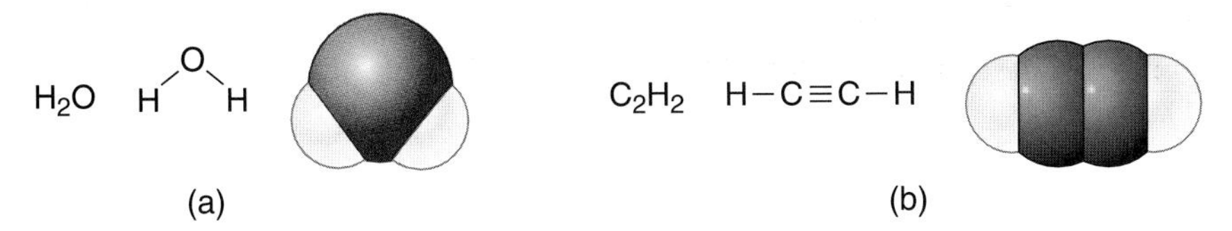

Figure 4.2 (a) The formula for water, a molecular compound, is H_2O. This means a water molecule has two hydrogen atoms and an oxygen atom. (b) The formula for acetylene is C_2H_2, which means there are two carbon atoms and two hydrogen atoms per acetylene molecule.

For ionic compounds, it is important to remember that the sum of the charges of the ions in the formula must equal zero. That is,

$$\text{The total charge on cations} + \text{the total charge on anions} = 0$$

For example, in aluminum oxide the two Al^{3+} ions have a total charge of 6+ and the three O^{2-} ions have a total charge of 6−:

$$\left(2 \text{ Aluminum cations} \times \frac{3^+}{\text{aluminum cation}}\right) + \left(3 \text{ oxide anions} \times \frac{2^-}{\text{oxide anion}} = 0\right)$$

$$\text{Total charge on aluminum cations } (6+) + \text{total charge on oxide anions } (6-) = 0$$

Thus the sum of the charges on the ions in aluminum oxide is zero. Check the formulas of NaCl and K_2O to see if the sums of the charges of the ions are zero in these formulas.

Note that although ionic charges are used to determine the correct formula, they do not appear in the formula.

Molecular Compounds

The formulas for molecular compounds **(molecular formulas)** show the number of atoms of each element in a molecule. Water consists of molecules that have two hydrogen atoms bound to one oxygen atom. The formula for water is H_2O [Figure 4.2(a)], which tells us (a) that both hydrogen and oxygen are present and (b) that there are two hydrogen atoms and one oxygen atom. Acetylene, a compound that is used in high-temperature torches, has this molecular formula: C_2H_2. How do we interpret this formula? Molecules of acetylene contain two carbon atoms and two hydrogen atoms [Figure 4.2(b)].

Example 4.1 Writing and Reading Chemical Formulas

1. How many ions of each element are present in these ionic compounds? (Table 3.3 provided guidelines for predicting the charge on monatomic ions).
   (a) Li_2S **(b)** $BaBr_2$ **(c)** $AlCl_3$
2. How many atoms of each element are present in these molecules?
   (a) HCl **(b)** HNO_3 **(c)** C_3H_8
3. When boron and fluorine react, they form a boron trifluoride molecule that has one atom of boron and three atoms of fluorine. What is the chemical formula for this compound?

SOLUTIONS

1. In a formula, each element symbol represents an atom or ion of the element. Subscripts are used if two or more ions or atoms are present. **(a)** There must be two lithium ions (Li^+) for each sulfide ion (S^{2-}). **(b)** There is one barium ion (Ba^{2+}) per two bromide ions (Br^-). **(c)** There is one aluminum ion (Al^{3+}) per three chloride ions (Cl^-).
2. **(a)** The elements hydrogen (H) and chlorine (Cl) are present in a molecule of this compound. Because no subscripts are present, each HCl molecule contains a single atom of each element. **(b)** An atom of hydrogen, an atom of nitrogen, and three atoms of oxygen are present. **(c)** Three carbon atoms and eight hydrogen atoms are present in one molecule of C_3H_8.
3. The formula is BF_3.

Self-Test

Determine the number of ions or atoms of each element in 1. NH_3 2. N_2O_4 3. CaI_2

ANSWERS

1. One nitrogen atom and three hydrogen atoms 2. Two nitrogen atoms and four oxygen atoms 3. One calcium ion (Ca^{2+}) and two iodide ions (I^-)

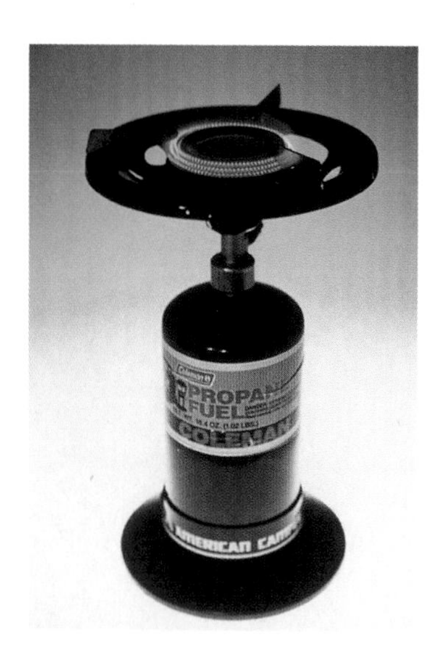

Propane stored in the tank fuels this backpacking stove. *(Charles D. Winters)*

The formulas discussed so far show the composition of a compound. A **structural formula** shows how the atoms are connected to each other in the molecule. In a structural formula, each atom and bond is shown. Water has this structural formula

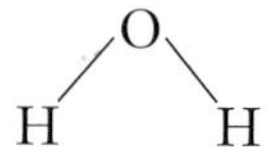

which tells us that in a water molecule the oxygen atom is connected to each of the hydrogen atoms by single bonds. Some structural formulas are simple, like this one for water, but other structural formulas, like this one for the fuel propane, are more complicated:

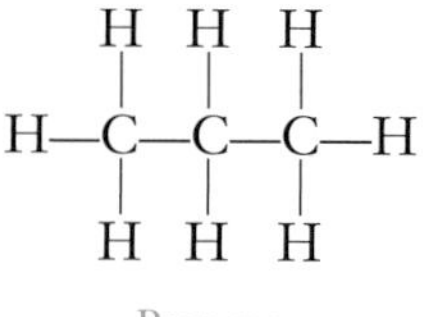

Propane

You will see many structural formulas in the organic chemistry and biochemistry sections of this textbook.

Chemical Names

Chemical formulas show the composition or structure of a compound, but compounds have names as well as formulas. Some older names are simply labels, like human names. For instance, the name *aspirin* is an old common name that provides no information about the composition or structure of this pain-killing compound. On the other hand, modern chemical names are generally more useful because the name includes information about the composition or structure of the compound. The International Union of Pure and Applied Chemistry (IUPAC) has developed rules that link chemical names to chemical composition and structure. There are many compounds, but by learning a few rules you will understand many of the names of the compounds frequently encountered in health care and chemistry.

4.2 Naming Ionic Compounds

Many ionic compounds contain ions of only two elements, and they are thus called **binary ionic compounds.** (*Bi-*, like *di-*, is a prefix that means two.) In binary ionic compounds, one element is a metal and the other is a nonmetal. All binary ionic compounds are named using one of two procedures. One procedure is used for naming compounds that contain a metal ion (cation) that always has the same charge. These cations form from atoms of the Groups 1A, 2A, and 3A metals and are called Type I cations. A different procedure is used for naming compounds that contain a metal that can form two or more cations (Type II cations). When naming binary ionic compounds, you must decide which procedure you should use. You do this by determining whether the cation can have only one charge (Type I) or can have more than one charge (Type II). With time, you may memorize this information, but in the meanwhile, use Tables 4.1 and 4.5 as guides.

The names for binary ionic compounds consist of two words, the first identifying the cation and the second identifying the anion. Let's consider binary ionic compounds containing Type I cations first. The name for a binary ionic compound containing a Type I cation consists of the metal name followed by a word derived from the nonmetal name and the suffix *-ide*. A compound containing monatomic ions of potassium and bromine, for example, would be called potassium bromide. Actually, you have already seen several examples of this naming, including sodium chloride (NaCl) and aluminum oxide (Al_2O_3). The procedure for naming these compounds is summarized in Table 4.2. You may also find Table 4.3 useful because it shows some common anion names.

Table 4.1 Some Common Type I Cations

METAL[a]	TYPE I CATION	METAL	TYPE I CATION
Hydrogen[b]	H^+	Magnesium	Mg^{2+}
Lithium	Li^+	Calcium	Ca^{2+}
Sodium	Na^+	Aluminum	Al^{3+}
Potassium	K^+		

[a]The metal cations of Group 1A always have a charge of 1+, Group 2A cations always have a charge of 2+, and the aluminum cation (Group 3A) always has a charge of 3+.

[b]Hydrogen is, of course, not a metal, but it forms a 1+ cation like the other Group 1A elements, and some hydrogen-containing compounds are named by these rules.

Table 4.2 Naming Binary Ionic Compounds Containing Type I Cations

1. Identify the cation first by writing down the element name. (For example, Ca^{2+} would be identified as calcium.)
2. Write the anion name second. The anion name is derived from the nonmetallic element's name by combining the element *stem* name, the first part of the element name, with the suffix *-ide*. (The anion for *sulf*ur, S^{2-}, would be *sulf*ide. Combining the two words yields calcium sulfide—CaS.)

Example 4.2 Naming Type I Binary Ionic Compounds

Name these compounds: 1. Li_2S 2. $MgBr_2$

SOLUTIONS

1. Lithium ions can have only one charge (1+), and so are Type I cations. This compound consists of lithium ions and sulfide ions. The name is *lithium sulfide*.
2. Magnesium ions have only one charge, and so Mg^{2+} is a Type I cation. Magnesium and bromide ions are present, and so the compound is *magnesium bromide*.

Self-Test

Name these Type I compounds: 1. $CaCl_2$ 2. $AlBr_3$ 3. K_2O

ANSWERS

1. Calcium chloride 2. Aluminum bromide 3. Potassium oxide

Iron(II) chloride (upper left) and iron(III) chloride. The properties of iron-containing compounds, including color, depend on the particular iron ion present in the compound. *(Richard Megna/ Fundamental Photographs)*

The names of binary ionic compounds containing Type II metal ions are very similar to those of Type I but take into account the charge on the ion in the compound. The charge on a Type II cation is shown with a Roman numeral in parentheses. Consider, for example, two compounds containing iron and chlorine. One of these compounds contains Fe^{2+} and the other contains Fe^{3+}. For the compound containing Fe^{2+}, the charge on the cation is indicated by adding (II) to the name; thus its name is iron(II) chloride. The other compound's name is iron(III) chloride. Naming binary ionic compounds containing Type II metal ions is summarized in Table 4.4. Look at Table 4.5 to see some common Type II metal ions.

Table 4.3 Some Common Anion Names

ELEMENT	ANION NAME	CHARGE ON ANION
Nitrogen	Nitride ion	N^{3-}
Oxygen	Oxide ion	O^{2-}
Sulfur	Sulfide ion	S^{2-}
Fluorine	Fluoride ion	F^-
Chlorine	Chloride ion	Cl^-
Bromine	Bromide ion	Br^-
Iodine	Iodide ion	I^-

Table 4.4 Assigning Stock Names to Type II Binary Ionic Compounds

1. Identify the cation first by writing the name of the metallic element. (For example, for iron, just write "iron.")
2. Indicate the charge on the cation by using Roman numerals in parentheses. [For Fe^{3+}, you would use (III).]
3. Name the anion second, using the *stem* of the nonmetallic element and the suffix *-ide* (Table 4.3). (For O^{2-} the anion name would be *oxide.*)
(The Stock name for Fe_2O_3 would be iron(III) oxide.)

The modern names for Type II binary ionic compounds are sometimes referred to as Stock names, in honor of Alfred Stock, who proposed this method of naming.

For Type II binary ionic compounds, Steps 1 and 3 are the same as Steps 1 and 2 for Type I. Using Step 2 for Type II compounds, however, requires a bit of thought. You must examine the formula of the compound to determine the charge on the metal ion. In Fe_2O_3, for example, you would find the charge on the iron ion by first determining the total charge on the oxide ions in the compound. Because each oxide ion has a 2− charge, three of them would have a total charge of 6−. Because an ionic compound is electrically neutral, the total charge of the iron ions must be 6+. There are two iron ions in the formula, therefore the total cation charge of 6+ must be divided by two to get the charge on a single cation: 3+. The iron ions in Fe_2O_3 are iron(III) ions. Example 4.3 provides some examples of naming Type II binary ionic compounds by these rules.

There is also an older way to name Type II ionic compounds. This older system does not use the element name to name the cation. That is, the cations of iron are called iron(II) and iron(III) in Stock names, but older names use the root *ferr-* to identify these ions as ferrous and ferric (Table 4.5). Furthermore, the older method uses the suffixes *-ous* and *-ic* to indicate the charge on the metal ion. The suffix *-ous* is used for the smaller charge that the metal ion can have. (For example, Fe^{2+} is called the *ferrous ion.*) The larger charge (Fe^{3+} for iron) is indicated by the suffix *-ic*. Thus Fe^{3+} is called the *ferric ion.* The compounds $FeCl_2$ and $FeCl_3$ are named ferrous chloride and ferric chloride, respectively. Because you will probably see examples of both Stock names and the older common names for Type II binary ionic compounds, you should become familiar with both.

Iron(III) oxide is a common component of the rust that forms when iron or steel is exposed to water and air. *(Darrell Gulin/Dembinsky Photo Associates)*

Table 4.5 Some Common Type II Cations

METAL[a]	STOCK TERMINOLOGY	OLDER TERMINOLOGY
Iron	Fe^{2+} Iron(II) ion	Ferrous
	Fe^{3+} Iron(III) ion	Ferric
Copper	Cu^{+} Copper(I) ion	Cuprous
	Cu^{2+} Copper(II) ion	Cupric
Tin	Sn^{2+} Tin(II) ion	Stannous
	Sn^{4+} Tin(IV) ion	Stannic
Mercury[b]	Hg_2^{2+} Mercury(I) ion	Mercurous
	Hg^{2+} Mercury(II) ion	Mercuric

[a]Many of the transition metals and the metals in the lower right portion of the periodic table can have two or more charges.

[b]Hg_2^{2+} is a diatomic ion in which each atom has a 1+ charge.

Example 4.3 Naming Type II Binary Ionic Compounds

What are the Stock and common names of 1. Cu_2O and 2. $SnCl_4$?

SOLUTIONS

1. Because copper ions can have more than one charge, rules for Type II cations must be used. An oxide ion has a charge of 2− (Table 4.3). The charge provided by the cation(s) must equal the charge of the anion(s), and so the total charge on the copper ions must be 2+. Because there are two copper ions, the charge on each must be 1+. To name the compound with the Stock name, write the cation name followed by its charge (Roman numeral, in parentheses), then write the anion name: copper(I) oxide.

   For common names, the cation carrying the smaller charge is given the *-ous* suffix (Table 4.5 shows some examples). This suffix is combined with the root for the element. For Cu^+ the name is *cuprous.* This is combined with the anion name to give the name *cuprous oxide.*
2. Tin requires Type II rules (Table 4.5). Chloride is the anion in the compound, and its charge is 1−. There are four anions, and so their total charge is 4−. The total charge on the cation must equal this, and so must be 4+. The name of this Type II binary ionic compound is *tin(IV) chloride.*

   Use Table 4.5 to derive the common name. *Stann-* is the root for tin. The 4+ charge on the tin ion is the larger of the two charges a tin ion can have; thus *-ic* is the appropriate suffix. *Stannic* is the proper word for tin in this compound. Combine this with *chloride* to get the older name: *Stannic chloride.*

Self-Test

What are the Stock and common names of 1. $FeBr_3$ 2. Hg_2S?

ANSWERS

1. Iron(III) bromide, ferric bromide 2. Mercury(I) sulfide, mercurous sulfide

Determining Chemical Formulas from Names of Ionic Compounds

Chemical names can be determined from formulas, and it also works the other way: Formulas can be figured out from chemical names. For Type I binary ionic compounds, the name tells you which ions are present in the compound. If you know or look up the charge on those ions, you can write the formula of the compound. Consider the name *magnesium bromide.* Magnesium is a metal, and bromine is a nonmetal, and so this is an ionic compound. Because magnesium is in Group 2A, a magnesium ion always has a 2+ charge (Table 4.1). Bromine is in Group 7A, and so the bromide ion has a 1− charge. In the formula, the sum of the positive and negative charges must be zero, thus there is one magnesium ion and two bromide ions. The formula for magnesium bromide is $MgBr_2$.

Writing the formula of Type II binary ionic compounds from their Stock names is straightforward. The charge on the cation is identified by the Roman numeral in the name. The charge on the anion can be looked up if it is not known. Consider the name chromium(III) chloride. In the formula, the sum of the positive and negative charges must be zero. Chloride ions have a 1− charge, and this

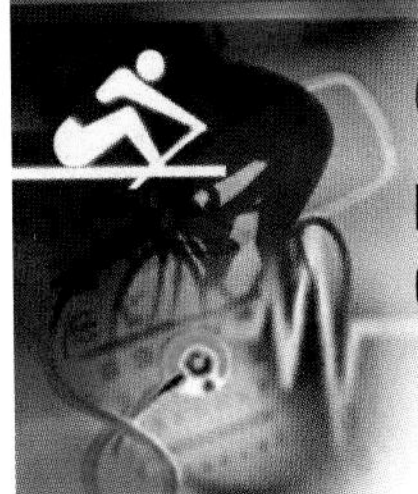

Connections

Hemoglobin's Iron Ions Help Transport Oxygen in the Blood

Hemoglobin is the substance in red blood cells that carries oxygen throughout the body. Hemoglobin is a macromolecule—a protein—containing four iron ions that actually bind to the oxygen molecules. These iron ions normally have a 2+ charge (the ions are ferrous ions), and in this form the binding of oxygen is reversible. As the blood flows through the lungs the iron ions bind oxygen, and then as the blood flows through the tissues of the body, the iron ions release the oxygen. Iron is one of the metals that form more than one cation, and so the iron(II) ions in hemoglobin can lose an electron to become iron(III) ions. If this happens, the iron ions can no longer bind oxygen reversibly, and the hemoglobin molecule loses its function as an oxygen carrier. This nonfunctional form of hemoglobin is called *methemoglobin* and has a brown color, rather than the characteristic red color of normal hemoglobin. Methemoglobin is responsible for the color of dried blood and old meat. Individuals whose blood contains methemoglobin are said to have methemoglobinemia, and their blood is chocolate brown.

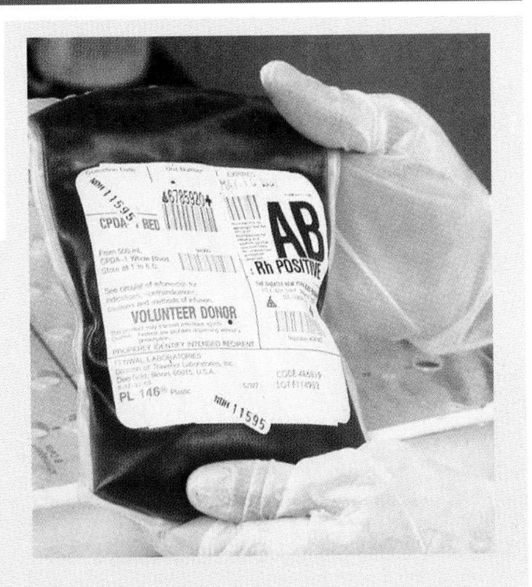

Hemoglobin is a protein in blood that contains iron ions that can bind oxygen only when their charge is 2+. The red color of blood is a result of heme, which is an iron-containing molecule that is found in hemoglobin. *(Larry Mulvehill/Photo Researchers, Inc.)*

chromium ion has a 3+ charge. There must, therefore, be one chromium ion and three chloride ions. The formula for chromium(III) chloride must be $CrCl_3$.

Example 4.4 Writing the Formulas of Type I and Type II Binary Ionic Compounds

1. What is the formula of sodium sulfide?
2. What is the formula of iron(II) oxide?

SOLUTIONS

1. Sodium is in Group 1A, and so sodium ions have a 1+ charge (the ions of all Group 1A elements always have a 1+ charge and are thus Type I cations). Sulfide ions (Group 6A) have a 2− charge. For the compound to be electrically neutral (to have a net charge of zero), there must be two sodium ions per one sulfide ion, and so the formula is Na_2S.
2. The Roman numeral clearly shows that the iron ion has a charge of 2+. Oxide ions have a charge of 2−. Because the net charge must be zero, there is one iron(II) ion per one oxide ion. The formula is FeO.

Self-Test

1. What is the formula of these Type I ionic compounds: **(a)** magnesium chloride **(b)** potassium telluride? (Hint: What is the stem of the word telluride?)
2. What is the formula of these Type II binary ionic compounds:
   (a) Copper(II) chloride **(b)** Iron(III) sulfide **(c)** Cuprous bromide?

ANSWERS

1. **(a)** $MgCl_2$ **(b)** K_2Te 2. **(a)** $CuCl_2$ **(b)** Fe_2S_3 **(c)** CuBr

This crystal of calcium carbonate contains calcium ions—Ca^{2+}—and carbonate ions—CO_3^{2-} (model in upper right). *(Photo, Charles D. Winters; model, S.M. Young)*

Table 4.6 Some Common Polyatomic Ions

NAME	FORMULA AND CHARGE
Cation	
Ammonium ion	NH_4^+
Anions	
Acetate ion	$C_2H_3O_2^-$
Bicarbonate (Hydrogen Carbonate) ion	HCO_3^-
Carbonate ion	CO_3^{2-}
Cyanide ion	CN^-
Hydroxide ion	OH^-
Nitrate ion	NO_3^-
Phosphate ion	PO_4^{3-}
Sulfate ion	SO_4^{2-}

Reminder: Polyatomic ions contain more than one atom; hydroxide ion (OH^-) and nitrate ion (NO_3^-) are examples.

Naming Ionic Compounds That Contain Polyatomic Ions

Ionic compounds containing polyatomic ions are named like binary ionic compounds except the polyatomic anion name replaces the monatomic anion name.

First decide whether the metal ion has only one charge (Type I) or can have more than one charge (Type II). If it is a Type I compound, just name the cation then the polyatomic ion. For instance, the formula Na_2SO_4 contains sodium ions (Na^+) and the polyatomic sulfate ion (SO_4^{2-}). The name of this compound is sodium sulfate. If the metal ion is a Type II cation, remember to include a Roman numeral to show the charge on the metal ion. The compound $FeSO_4$ is named iron(II) sulfate.

When polyatomic ions are in a formula, parentheses are put around the polyatomic ion if more than one polyatomic ion is present in the formula. A subscript then follows the closing parenthesis to indicate the number of the polyatomic ions that are present. Consider these examples. The formula unit of calcium nitrate has a single calcium ion (Ca^{2+}) and two nitrate ions (NO_3^-). The formula for calcium nitrate is written as $Ca(NO_3)_2$. Iron(III) sulfate contains two iron(III) ions and three sulfate ions. Its formula is $Fe_2(SO_4)_3$.

Table 4.6 shows the formula and charge of several common polyatomic ions. Note that ammonium ion is the only frequently encountered polyatomic cation. Treat it like a Type I cation with regard to naming.

Example 4.5 Naming Compounds Containing Polyatomic Ions

Name these ionic compounds: 1. $Fe(NO_3)_3$ 2. $(NH_4)_3PO_4$

SOLUTIONS

1. The ions of iron (Fe) can have more than one charge (Type II). Nitrate has a 1− charge, and the subscript 3 outside the parentheses tell us there are three nitrate ions. Therefore the total charge for the anions is 3−. The charge on the iron ion must therefore be 3+. The Stock name is iron(III) nitrate; the common name is ferric nitrate.

2. Ammonium ion (NH_4^+) always has a charge of 1+ and so is a Type I cation. The anion in the compound is PO_4^{3-}, which is the phosphate ion (Table 4.6). The compound name is ammonium phosphate.

Self-Test

Name these ionic compounds: 1. KOH 2. $Mg(NO_3)_2$ 3. Cu_2SO_4

ANSWERS

1. Potassium hydroxide 2. Magnesium nitrate 3. Copper(I) sulfate; cuprous sulfate

4.3 Naming Covalent Compounds

Some binary compounds contain atoms of two nonmetals connected by covalent bonds. These compounds are called **binary covalent compounds.** Carbon dioxide, CO_2, is an example of a binary covalent compound.

In the formulas for such a compound, the first element is the one that has the smaller electronegativity (Figure 3.7). In the periodic table, this element is located to the left or below the other element in the compound (for CO_2, carbon is to the left of oxygen in the periodic table). The second element in the formula is the more electronegative element, the one located farther to the right or farther up the periodic table (oxygen is located to the right of carbon in the periodic table).

When a binary covalent compound is named, the order of the elements in the name is the same as the order in the formula. For the compound with the formula NO, the nitrogen is named first, then the oxygen: nitrogen monoxide. Many binary covalent compounds contain the same elements, as carbon dioxide (CO_2) and carbon monoxide (CO) illustrate. Prefixes are used to indicate how many atoms of each element are present. For the first element in the name, a prefix is needed only if there is more than one atom of the element; thus nitrogen monoxide (NO) and dinitrogen monoxide (N_2O). Prefixes are used for the second element in the compound, as carbon dioxide (CO_2) and carbon monoxide (CO) show.

Numerical prefixes are needed in names for covalent compounds because two or more compounds may form from two nonmetals—carbon monoxide, CO, and carbon dioxide, CO_2, are examples.

Naming binary covalent compounds is summarized in Table 4.7. Table 4.8 contains the numerical prefixes that are used in covalent compound names. Although most binary covalent compounds are named by these rules, there are, of

Table 4.7 Naming Binary Covalent Compounds

1. Begin with the name of the less electronegative element in the formula. If there are two or more atoms of this element, indicate the number with the appropriate prefix.
2. Combine the stem name of the second element in the formula with the suffix *-ide* to make the second half of the name (Table 4.3). Indicate the number of atoms of this element with the appropriate prefix.

Table 4.8 Prefixes that Designate Numbers in Covalent Compound Names

NUMBER	PREFIX[a]	EXAMPLE OF COMPOUND NAME AND FORMULA
1	*mono-*	Carbon monoxide CO
2	*di-*	Carbon disulfide CS_2
3	*tri-*	Phosphorus tribromide PBr_3
4	*tetra-*	Carbon tetrachloride CCl_4
5	*penta-*	Diphosphorus pentoxide P_2O_5
6	*hexa-*	Sulfur hexafluoride SF_6
7	*hepta-*	Tetraphosphorus heptasulfide P_4S_7

[a]The final vowel of a prefix is sometimes dropped when the stem word begins with a vowel. Example: The correct name for CO is carbon monoxide, not carbon monooxide.

Connections

What's in a Name?

Many students in beginning chemistry courses are annoyed with having to learn chemical names. $Cr(NO_3)_3$ is chromium(III) nitrate and dinitrogen tetroxide is N_2O_4, and . . . who cares, anyway? Well, when you must keep track of hundreds of thousands of compounds, a systematic naming system linking name and structure is essential.

Consider the names of drugs (compounds with therapeutic use) as an example. Some drug names have a historical origin. *Curare,* for example, is a muscle relaxant obtained from the bark of several trees found in the Amazon. Its name was derived from the Indian name for the bark extracts that were used to poison arrowheads. Some drug names have a common, nonsystematic name. *Aspirin,* for example, is acetylsalicylic acid—*a-* from acetyl (a part of the compound's chemical name), *-spir-* from the German word "spirsaure", which means salicylic acid, and *-in,* which is a very commonly used medical suffix. Many drug names are trademarked names that were chosen primarily for marketing purposes, with no thought to linking structure and name. Worse yet, some drugs are known by several common and trademarked names. The antibacterial drug Bacitracin is also known as Altracin, Ayfivin, Fortracin, Penitracin, Topitracin, and Zutracin. The antibacterial drug ampicillin is known by 47 other nonchemical names. Fortunately, reference books such as *The Merck Index* provide cross-referencing for these names, and in these references the systematic chemical name is included to provide a link to the structure of the compound.

Aspirin is a synthetic drug that is similar in structure and pain-killing properties to the salicylates found in willow bark and roots. *(Charles D. Winters)*

course, exceptions. Many common substances, such as water and ammonia, have retained their nonsystematic names. Furthermore, the prefix *mono-* for the second element in the name is dropped for some compounds such as hydrogen chloride and hydrogen bromide.

Example 4.6 Naming Binary Covalent Compounds

Name these binary covalent compounds: 1. SO_3 2. N_2O_4

SOLUTIONS

1. A molecule of SO_3 contains a single sulfur atom and three oxygen atoms. The name is sulfur trioxide.
2. A molecule of N_2O_4 contains two nitrogen atoms and four oxygen atoms. The name is dinitrogen tetroxide.

Self-Test

Name these binary covalent compounds: 1. PCl_3 2. CS_2 3. N_2O_5

ANSWERS

1. Phosphorus trichloride 2. Carbon disulfide 3. Dinitrogen pentoxide

The name of a binary covalent compound can be used to determine the formula of the compound. The name *carbon tetrachloride* means one carbon atom and four chlorine atoms are present in the compound. The formula for carbon tetrachloride is CCl_4.

Example 4.7 Writing the Formulas of Binary Covalent Compounds

What is the formula for triphosphorus pentanitride?

SOLUTION

The name shows how many atoms of each element are present. In this compound there are three phosphorus atoms and five nitrogen atoms. The formula is P_3N_5.

Self-Test

What is the formula of 1. sulfur hexafluoride 2. carbon tetrabromide?

ANSWERS

1. SF_6 2. CBr_4

Systematic names are essential in chemistry, and with practice you will become comfortable with chemical names. As you learn, use Table 4.9 to review the naming of ionic and covalent compounds.

Table 4.9 Summary for Naming Ionic and Covalent Compounds

TYPE OF COMPOUND	PROCEDURE	EXAMPLE
Binary Ionic with Type I Cation	Name cation first with name of metal, then name anion second with nonmetal *stem* plus suffix *-ide*.	Sodium chloride (NaCl)
Binary Ionic with Type II Cation	Name cation first with name of metal and Roman numeral to show charge. Name anion second with nonmetal *stem* plus suffix *-ide*.	Iron(II) bromide ($FeBr_2$)
Ionic with Polyatomic Anion	Name cation first with either element name or element name and Roman numeral to show charge. Name polyatomic ion second.	Sodium hydroxide (NaOH) Copper(II) nitrate $Cu(NO_3)_2$
Covalent Compound	Name first element in formula with element name using the appropriate prefix if there is more than one atom of the element. Name the second element in the formula second using the stem of the element name, the suffix *-ide*, and the appropriate prefix.	Dinitrogen tetroxide (N_2O_4)

4.4 Chemical Equations

As you learned in Section 1.6, a chemical reaction occurs when substances are chemically changed into other substances. For example, we say sodium and chlorine react to yield sodium chloride, and methane (natural gas) and oxygen react to form carbon dioxide and water. When a bottle of soda or beer is opened, carbonic acid reacts (breaks down) to form water and carbon dioxide. The substances that

Two chemical reactions. (a) When this antacid tablet is dissolved in water, dissolved sodium bicarbonate reacts with dissolved citric acid to produce carbon dioxide (bubbles). (b) The reaction of nitric acid with copper produces brown nitrogen dioxide gas and a green copper(II) nitrate solution. *(Charles D. Winters)*

react are called **reactants,** and the substances that are produced are called **products.** The number of possible chemical reactions is very large.

Chemical reactions can be described with words, but more commonly they are represented by a concise shorthand method. A **chemical equation** (more commonly, *equation*) is a concise symbolic representation of a chemical reaction. Every equation is made up of two parts separated by an arrow. The formulas of all reactants are written to the left of the arrow, and the formulas of all products of the reaction are written to the right of the arrow. Often the physical state of a substance is also included using the symbols (g) for a gas, (l) for a liquid, and (s) for a solid. Sometimes other symbols or words are used to include additional information. If a substance is dissolved in water—if it is an aqueous solution—the symbol (aq) is used. Finally, whole numbers called **coefficients** are put in front of the formulas or symbols of the substances to indicate the relative number of molecules, atoms, or ions that participate in the reaction. This last step is necessary because the two sides of a chemical equation must have an equal number of atoms of each element present.

To see what all this means, look at this equation for the reaction of methane and oxygen:

$$CH_4(g) + 2O_2(g) \rightarrow CO_2(g) + 2H_2O(g)$$

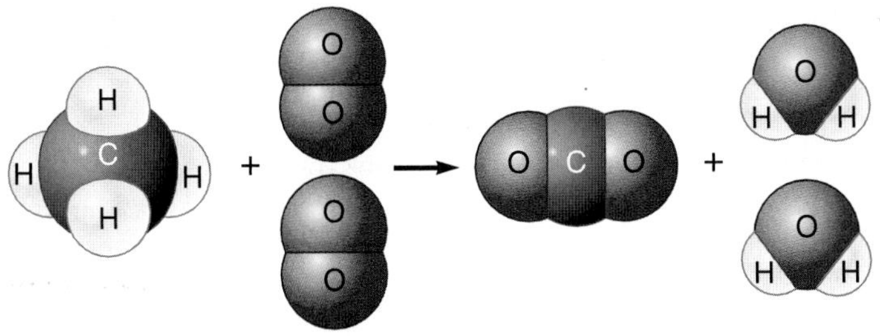

The reactants are methane, CH_4, and oxygen, O_2, both in the gaseous state. The products are CO_2 and H_2O, which are also gases. The language equivalent of this equation is "a molecule of methane gas reacts with two molecules of oxygen gas to yield a molecule of gaseous carbon dioxide and two molecules of steam (gaseous water)". The chemical equation states the same information more concisely.

Chemical Equations Are Balanced

In a chemical reaction substances change into other substances, but matter is neither created nor destroyed. This constancy of matter is called the **law of conservation of matter.** All of the atoms that exist before the reaction also exist after the reaction. The atoms are just rearranged into different combinations to yield products from reactants. To reflect the fact that the amount of matter does not change in a reaction, an equation must have the same number of atoms of each element on both sides of the equation. Because of this, it is necessary to *balance* an equation.

An equation is balanced when there are the same number of atoms of each element on both sides. The coefficients of the reactants and products are assigned to balance a chemical equation. Consider this unbalanced equation for the burning of propane (C_3H_8):

$$C_3H_8(g) + O_2(g) \rightarrow CO_2(g) + H_2O(g) \qquad \text{(Not balanced)}$$

Propane Oxygen Carbon dioxide Water

There are three carbon atoms in propane on the reactant side of the equation and one carbon atom in carbon dioxide on the product side. To balance carbon, put a 3 in front of the CO_2 to get three carbon atoms on the product side:

$$C_3H_8(g) + O_2(g) \rightarrow 3CO_2(g) + H_2O(g) \qquad \text{(Not balanced)}$$

3 carbon atoms 3 carbon atoms

The carbon atoms in propane are now balanced by the carbon atoms in carbon dioxide. Do hydrogen next. There are eight hydrogen atoms in propane and two in the water molecule. To balance the equation, eight hydrogen atoms are needed on each side of the equation. Put a 4 as the coefficient for water:

$$C_3H_8(g) + O_2(g) \rightarrow 3CO_2(g) + 4H_2O(g) \qquad \text{(Not balanced)}$$

8 hydrogen atoms 8 hydrogen atoms

Carbon and hydrogen are now balanced. Do oxygen next. There are two oxygen atoms in the oxygen molecule on the reactant side. There are now six oxygen atoms in the carbon dioxide molecules and four oxygen atoms in the water molecules. That is a total of ten oxygen atoms on the product side. To get ten oxygen atoms on the reactant side, use a 5 as the coefficient in front of the oxygen molecule.

$$C_3H_8(g) + 5O_2(g) \rightarrow 3CO_2(g) + 4H_2O(g) \qquad \text{(Balanced)}$$

10 oxygen atoms 6 oxygen atoms + 4 oxygen atoms = 10 oxygen atoms

Carbon, hydrogen, and oxygen are all balanced; thus the equation is balanced. Just as calculations in math should be checked, you should check the number of atoms of each element to make sure you have balanced the equation correctly.

$$C_3H_8(g) + 5O_2(g) \rightarrow 3CO_2(g) + 4H_2O(g) \qquad \text{(Balanced)}$$

3 C atoms + 8 H atoms + 10 O atoms = 3 C atoms + 8 H atoms + 10 O atoms

When you balance equations, *never* change the subscripts on a compound, because that would change the identity of the compound and depict a different reaction. You will also find that it is easier to balance equations if you begin with the elements that occur only once on each side of the equation. When the equation for the burning of propane was balanced, carbon and hydrogen were balanced first, then oxygen, which is present in both products, was balanced last.

Example 4.8 Balancing Chemical Equations

Balance these equations:

1. $Na_2SO_4(aq) + BaCl_2(aq) \rightarrow BaSO_4(s) + NaCl(aq)$
2. $C_6H_6(l) + O_2(g) \rightarrow CO_2(g) + H_2O(g)$

SOLUTIONS

1. There are two sodium ions on the reactant side. Use a coefficient of 2 for NaCl to get two sodium ions on the product side: $Na_2SO_4(aq) + BaCl_2(aq) \rightarrow$

$BaSO_4(s) + 2NaCl(aq)$. At this point the equation is balanced because there are the same number of atoms of each element on the two sides of the equation.

2. There are six carbon atoms on the left (reactant) side of the equation, so put a 6 in front of the CO_2 to balance carbon: $C_6H_6(l) + O_2(g) \rightarrow 6CO_2(g) + H_2O(g)$. Now balance hydrogen by using a coefficient of 3 in front of water: $C_6H_6(l) + O_2(g) \rightarrow 6CO_2(g) + 3H_2O(g)$. Carbon and hydrogen are balanced; now do oxygen. There are 15 oxygen atoms ($6 \times 2 = 12$ in the CO_2 plus 3 in the H_2O) on the right (product) side but only two on the left (reactant) side. If molecular oxygen is multiplied by $7\frac{1}{2}$, there will be 15 oxygens on each side: $C_6H_6(l) + 7\frac{1}{2}O_2(g) \rightarrow 6CO_2(g) + 3H_2O(g)$. The number of atoms of each element are now balanced, but a chemical equation should have *whole numbers* as coefficients. To get all whole numbers, multiply both sides of the equation by 2: $2C_6H_6(l) + 15O_2(g) \rightarrow 12CO_2(g) + 6H_2O(g)$.

Self-Test

Balance these equations:

1. $NH_3(aq) + H_3PO_4(aq) \rightarrow (NH_4)_3PO_4(aq)$
2. $C_2H_2(g) + O_2(g) \rightarrow CO_2(g) + H_2O(g)$

ANSWERS

1. $3NH_3(aq) + H_3PO_4(aq) \rightarrow (NH_4)_3PO_4(aq)$
2. $2C_2H_2(g) + 5O_2(g) \rightarrow 4CO_2(g) + 2H_2O(g)$

4.5 The Mole

A balanced chemical equation provides a description of a chemical reaction. It shows the reactants on the left, the chemical change with an arrow, and the products on the right. It also shows the *amounts* of reactants and products. *Stoichiometry* deals with the quantitative relationship between reactants and products in a chemical reaction.

The use of the descriptive information is, of course, helpful. If we want to make a compound, the equation shows what must be used to make it. Consider this equation for the synthesis of aspirin:

$$\underset{\text{Salicylic acid}}{C_7H_6O_3(s)} + \underset{\text{Acetic anhydride}}{C_4H_6O_3(l)} \rightarrow \underset{\text{Aspirin}}{C_9H_8O_4(s)} + \underset{\text{Acetic acid}}{HC_2H_3O_2(l)}$$

This equation shows that aspirin can be made from solid salicylic acid and liquid acetic anhydride. It also shows how much of each reactant is needed. One molecule of acetic anhydride and one molecule of salicylic acid react to yield one molecule of the desired product aspirin and one molecule of acetic acid as a by-product. Unfortunately, no one can measure out molecules directly because they are too small. Before we can use this chemical equation to make aspirin, we need to learn another chemical concept: the mole (abbreviated mol).

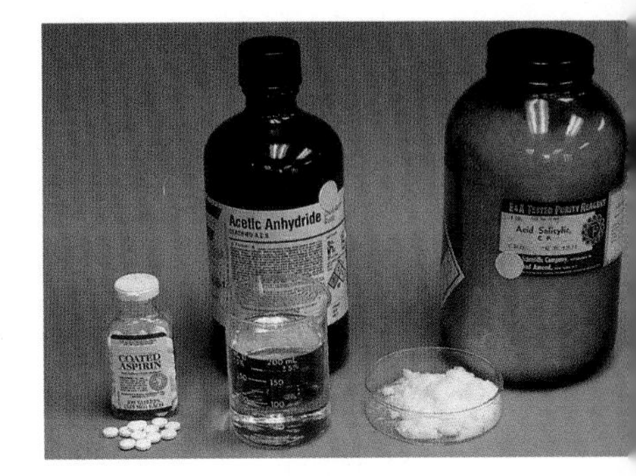

Aspirin is produced when acetic anhydride reacts with salicylic acid. *(Tom Pantages)*

Counting Particles with Moles

Atoms, molecules, and ions have masses and volumes, but these masses and volumes are very small. For example, a pinhead contains roughly 10^{18} iron atoms. We cannot observe, measure, or handle atoms, molecules, or ions directly. On the other hand, we can easily measure volumes and masses of larger samples of matter with accuracy and convenience. Is there some way we can link, or correlate, individual atoms and molecules to samples that are large enough to measure? The answer is yes. We can use the concept and definition of the *mole* to work with specific numbers of small particles.

The mole is a counting unit that represents a specific number of particles. One **mole** contains 6.022×10^{23} of anything. Just as the unit *dozen* signifies 12 of anything (eggs, ears of corn, donuts), the unit *mole* signifies 6.022×10^{23} of anything (sodium atoms, water molecules, marbles). The mole is a useful unit for small particles such as atoms, ions, and molecules, but it is too large for ordinary objects (Figure 4.3). This large number, 6.022×10^{23}, is called *Avogadro's number* in honor of the Italian scientist Amadeo Avogadro (1776–1856), whose work helped lead to our modern concept of the mole. When you work with moles, you are working with numbers of particles.

Look back at our equation for the synthesis of aspirin. The coefficients could be understood to mean molecules of reactants and products, but they could also mean moles. This is true because, as with a mathematical equation, you can multiply both sides of a chemical equation by the same number, and the equation is still valid. Multiply both sides of the aspirin equation by 6.022×10^{23} and it becomes 6.022×10^{23} molecules of salicylic acid react with 6.022×10^{23} molecules of acetic anhydride to yield 6.022×10^{23} molecules of aspirin and 6.022×10^{23} molecules of acetic acid. Because 1 mol = 6.022×10^{23} molecules, the equation also means one mole of salicylic acid reacts with one mole of acetic anhydride to yield one mole of aspirin and one mole of acetic acid. In other words, *the coefficients in a chemical equation can be interpreted as molecules of a substance or as moles of a substance.*

Dimensional analysis is a good way to convert moles of a substance into atoms, ions, molecules, or vice versa. How many atoms of aluminum would be in 0.35 mol of aluminum? Begin with the number of moles of aluminum, and multiply it by the appropriate conversion factor derived from the definition of a mole:

$$0.35 \text{ mol Al} \times \frac{6.022 \times 10^{23} \text{ atoms of Al}}{1 \text{ mol Al}} = 2.1 \times 10^{23} \text{ atoms of Al}$$

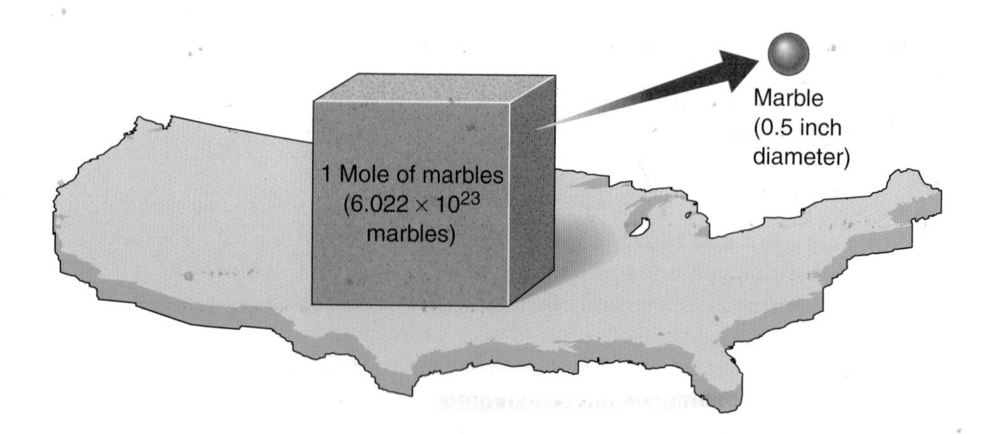

Figure 4.3 To get a sense of how large a mole is, picture a mole of half-inch marbles placed into a cube. The base of this cube would cover much of the central United States.

Going from particles of a substance to moles of the substance is equally straightforward. If you have 2.6×10^{24} ions of magnesium, how many moles of the ion do you have? Begin with the ions and multiply this term by the appropriate conversion factor derived from the definition of the mole:

$$2.6 \times 10^{24} \text{ Mg ions} \times \frac{1 \text{ mol Mg ions}}{6.022 \times 10^{23} \text{ Mg ions}} = 4.3 \text{ mol Mg ions}$$

Example 4.9 Using Moles in Stoichiometry

How many atoms are in 0.750 mol of lead?

SOLUTION

Use dimensional analysis (Section 1.5) and this conversion factor: 1 mol = 6.022×10^{23} particles (atoms in this case).

$$0.750 \text{ mol Pb} \times \frac{6.022 \times 10^{23} \text{ atoms Pb}}{1 \text{ mol Pb}} = 4.52 \times 10^{23} \text{ atoms Pb}$$

Self-Test

1. How many atoms are in 1.53 mol of sulfur?
2. 1.38×10^{19} molecules of water corresponds to how many moles of water?

ANSWERS

1. 9.21×10^{23} sulfur atoms 2. 2.29×10^{-5} mol of water

This 150-mL beaker contains 1.5×10^{24} atoms—2.5 mol—of lead. *(Charles D. Winters)*

Relationship Between Moles and Mass

Because individual molecules or atoms are much too small to handle individually, we cannot simply count them out to get the moles needed for a reaction. There is a relationship, however, between moles of matter and mass of matter. Because matter can be weighed easily and accurately with a balance, we can indirectly measure out moles of matter by measuring out mass.

Consider carbon as an example. From the periodic table, we can determine that one atom of carbon has a mass of 12.01 amu. (Put another way, as we learned in Section 2.2, 12.01 amu is the atomic mass of carbon.) One amu is equal to 1.661×10^{-24} g (Section 2.2); thus one carbon atom has this mass in grams:

$$\frac{12.01 \text{ amu}}{1 \text{ carbon atom}} \times \frac{1.661 \times 10^{-24} \text{ g}}{1 \text{ amu}} = \frac{1.995 \times 10^{-23} \text{ g}}{1 \text{ carbon atom}}$$

If we multiply this mass per carbon atom by 6.022×10^{23} carbon atoms per mole of carbon atoms we get this:

$$\frac{1.995 \times 10^{-23} \text{ g}}{1 \text{ carbon atom}} \times \frac{6.022 \times 10^{23} \text{ carbon atoms}}{1 \text{ mol carbon}} = \frac{12.01 \text{ g}}{1 \text{ mol carbon}}$$

We began this series of calculations with the atomic mass of carbon, which has a value of 12.01 amu/carbon atom. We ended with the value 12.01 g/mol of

Figure 4.4 One-mole samples of some common elements. Clockwise from top left: One mole (63.54 g) of copper beads; one mole (26.98 g) of aluminum foil; one mole (207.2 g) of lead shot; one mole (24.31 g) of magnesium chips; one mole (52.00 g) of chromium chunks; and one mole (32.06 g) of powdered sulfur. *(Charles D. Winters)*

carbon. The numerical value is the same in each case. Because the numerical value is the same, we can interpret the *atomic mass of an element* in two ways: (a) it is the mass of one atom expressed in amu, and (b), it is the mass of one mole of the element expressed in grams. The mass of one mole of atoms of an element is also known as the element's **molar mass,** and the molar mass is simply the atomic mass expressed in grams. If we need one mole of carbon (6.022×10^{23} C atoms) for a chemical reaction, we simply weigh out 12.01 g of carbon (Figure 4.4). If we need more or less than one mole of a substance, we can use dimensional analysis to determine the mass of the substance. For example, to determine the mass of 0.625 mol of sulfur, multiply the moles of sulfur by a conversion factor expressing the molar mass of the element:

$$0.625 \text{ mol S} \times \frac{32.07 \text{ g S}}{1 \text{ mol S}} = 20.0 \text{ g S}$$

A 20.0-g sample of sulfur corresponds to 0.625 mol of sulfur. Grams of a substance can be converted to moles in a similar manner. How many moles of potassium are present in 25.5 g of potassium?

$$25.5 \text{ g K} \times \frac{1 \text{ mol K}}{39.10 \text{ g K}} = 0.652 \text{ mol K}$$

Example 4.10 Determining Moles of an Element

1. How many grams of aluminum (Al) must be weighed out to get 1.00 mol of aluminum?
2. How many grams are in 0.500 mol of potassium (K)?

SOLUTIONS

1. Because we need 1.00 mol of aluminum, we need 1 molar mass of aluminum. From the periodic table, the atomic mass of aluminum is 26.98 amu; therefore, the molar mass is 26.98 g. To get 1.00 mol of aluminum, weigh out 27.0 g of aluminum.

2. We can answer this question by using dimensional analysis. We want to convert moles to grams, and so we begin with moles and multiply by a conversion factor (derived from the definition of molar mass) that contains both grams and moles:

$$0.500 \text{ mol K} \times \frac{39.10 \text{ g K}}{1 \text{ mol K}} = 19.6 \text{ g K}$$

0.500 mol of potassium has a mass of 19.6 g.

Self-Test

1. What is the mass of 1.00 mol of zinc (Zn)?
2. 120 g of iron (Fe) corresponds to how many moles of iron?

ANSWERS

1. 65.4 g zinc   2. 2.1 mol iron

Formula Mass, Molecular Mass, and Molar Mass

The formula unit is the simplest ratio of ions in an ionic compound; for example, there is one sodium ion for every chloride ion in NaCl.

It appears to be fairly simple to work with moles of an element, but what about compounds? Just as atomic mass is used to express the mass of an atom, the mass of the ions in an ionic compound (formula unit) is expressed by *formula mass*, and the mass of a molecule of a covalent compound (molecular formula) is expressed by *molecular mass*. The **formula mass** (formula weight) of an ionic compound is simply the total mass (in atomic mass units) of the ions in the formula of that compound. When you calculate formula masses just remember that the mass of a monatomic ion is the same as an atom, because the mass of the gained or lost electron(s) is negligible. Thus the formula mass of KCl is the atomic mass of potassium plus the atomic mass of chlorine: 39.10 amu + 35.45 amu = 74.55 amu. When you calculate the formula mass of a compound containing two or more of the polyatomic ions, remember to multiply the mass of the polyatomic ion by the number of polyatomic ions. The formula mass of $Mg(NO_3)_2$ (magnesium nitrate) is the atomic mass of magnesium plus two times the mass of the nitrate ion: 24.31 amu + 2 × (62.01 amu) = 148.33 amu. (The mass of the nitrogen atom and the three oxygen atoms in a nitrate ion is 62.01; for practice, calculate it yourself.)

The **molecular mass** (molecular weight) of a molecular compound is the total mass (in amu) of the atoms in a molecule of the compound. The molecular mass of methane (CH_4) is the mass of one carbon atom added to the mass of four hydrogen atoms: 12.01 amu + (4 × 1.01 amu) = 16.05 amu.

If we now use the same logic for compounds that we used previously for the element carbon, we can multiply the formula mass or molecular mass by the conversion factor 1.661×10^{-24} g/amu and Avogadro's number to get the *molar mass* of the compound. The molar mass of KCl is determined this way:

$$\frac{74.55 \text{ amu}}{1 \text{ formula unit}} \times \frac{1.661 \times 10^{-24} \text{ g}}{1 \text{ amu}} \times \frac{6.022 \times 10^{23} \text{ formula units}}{1 \text{ mol}} = \frac{74.55 \text{ g}}{1 \text{ mol}}$$

Because the molar mass of a compound has the same numerical value as the formula mass (or molecular mass) you do not need to use these calculations to determine the molar mass of a compound. Just change the formula mass units to g/mol. What is the molar mass of methane? Because the molecular mass is 16.05 amu/molecule, the molar mass is 16.05 g/mol.

Example 4.11 Determining the Molar Mass of a Compound

Calculate the molar masses of 1. $CaSO_4$ 2. aspirin ($C_9H_8O_4$)

SOLUTIONS

1. First determine the mass of the ions in the compound. The mass of a calcium ion is

$$1\ \text{Ca atom} \times \frac{40.08\ \text{amu}}{1\ \text{Ca atom}} = 40.08\ \text{amu}$$

Now determine the mass of the atoms in a sulfate ion:

$$1\ \text{S atom} \times \frac{32.06\ \text{amu}}{1\ \text{S atom}} = 32.06\ \text{amu}$$

$$4\ \text{O atoms} \times \frac{16.00\ \text{amu}}{1\ \text{O atom}} = 64.00\ \text{amu}$$

Sum these masses to get the mass of a sulfate ion: 32.06 amu + 64.00 amu = 96.06 amu.

Sum the masses of the calcium ion and sulfate ion to get the formula mass of calcium sulfate:

$$40.08\ \text{amu} + 96.06\ \text{amu} = 136.14\ \text{amu}.$$

The molar mass has the same numerical value as the formula mass, but has the units of g/mol. Thus the molar mass of calcium sulfate is 136.14 g/mol.

A calcium sulfate crystal and "dry wall" that is used in building construction. Calcium sulfate in these materials is complexed with water (hydrated). *(Charles D. Winters)*

2. First determine the mass of the atoms in the compound:

$$9 \text{ C atoms} \times \frac{12.01 \text{ amu}}{1 \text{ C atom}} = 108.1 \text{ amu}$$

$$8 \text{ H atoms} \times \frac{1.01 \text{ amu}}{1 \text{ H atom}} = 8.08 \text{ amu}$$

$$4 \text{ O atoms} \times \frac{16.00 \text{ amu}}{1 \text{ O atom}} = 64.00 \text{ amu}$$

Sum these values to get the molecular mass of aspirin:

108.1 amu + 8.08 amu + 64.00 amu = 180.2 amu per molecule of aspirin

Because the numerical value of molecular mass and molar mass are the same, the molar mass is 180.2 g/mol.

Self-Test

What is the molar mass of 1. sodium chloride (NaCl) 2. silver nitrate ($AgNO_3$)?

ANSWERS

1. 58.44 g 2. 169.88 g

We cannot work with individual atoms, but the concept of the mole allows us to work with balanced equations and determine how much of a product is formed or how much of a reactant is needed. Consider this reaction between phosphorus and oxygen that forms tetraphosphorus decaoxide:

$$4P(s) + 5O_2(g) \rightarrow P_4O_{10}(s)$$

Suppose we want to know how many moles of oxygen react with 10.0 mol of phosphorus? The balanced equation shows that four moles of P reacts with five moles of O_2. The coefficients of the balanced equation can be used to write *molar ratios:*

$$\frac{4 \text{ mol P}}{5 \text{ mol } O_2} \quad \text{or} \quad \frac{5 \text{ mol } O_2}{4 \text{ mol P}}$$

One of these molar ratios can then be used with dimensional analysis to determine the number of moles of oxygen that react with 10.0 mol of phosphorus:

$$10.0 \text{ mol P} \times \frac{5 \text{ mol } O_2}{4 \text{ mol P}} = 12.5 \text{ mol } O_2$$

Note that we chose the molar ratio that causes moles of phosphorus to cancel out, leaving the desired answer of moles of oxygen. When you use molar ratios, make sure you choose the correct ratio to obtain the correct units in your answer. Because molar ratios are exact numbers, you do not need to consider them when you are determining significant figures.

Example 4.12 Determining Moles of Reactants and Products in Reactions

1. Use the equation given previously to determine the number of moles of P_4O_{10} that form from 10.0 mol of phosphorus.
2. Use this equation to determine how many moles of dissolved chloride ion are needed to make 3.5 mol of solid lead(II) chloride:

$$Pb^{2+}(aq) + 2Cl^-(aq) \longrightarrow PbCl_2(s)$$

SOLUTIONS

1. The proper molar ratio for this calculation (the one that results in moles of P_4O_{10} and that results in the cancellation of moles of P) is 1 mol P_4O_{10}/4 mol P. The number of moles of product can be determined with this expression:

$$10.0 \text{ mol P} \times \frac{1 \text{ mol } P_4O_{10}}{4 \text{ mol P}} = 2.50 \text{ mol } P_4O_{10}$$

2. We are given moles of product and asked for moles of a reactant, so we work "backwards" from product to reactant. Begin with the moles of product, and multiply it by the proper molar ratio to get moles of reactant:

$$3.5 \text{ mol } PbCl_2 \times \frac{2 \text{ mol } Cl^-}{1 \text{ mol } PbCl_2} = 7.0 \text{ mol } Cl^-$$

The leaves of plants absorb sunlight and carbon dioxide, then use them and water to carry out photosynthesis. *(Charles D. Winters)*

Self-Test

Plants make glucose (the sugar found in blood) by the process of photosynthesis:

$$6CO_2(aq) + 6H_2O(l) \xrightarrow{\text{Light}} \underset{\text{Glucose}}{C_6H_{12}O_6(aq)} + 6O_2(g)$$

1. How many moles of glucose ($C_6H_{12}O_6$) are formed from 25 mol of carbon dioxide?
2. How many moles of oxygen are produced from 25 mol of CO_2?

ANSWERS

1. 4.2 mol glucose 2. 25 mol oxygen

Using a balanced equation and molar ratios, the number of moles of any reactant or product can be calculated from the number of moles of any other reactant or product. But how can we answer the following question: how many grams of P_4O_{10} can be made from 75.0 g of phosphorus, assuming there is enough oxygen for the reaction to occur? From the balanced equation, the relationship (molar ratio) between numbers of moles of these substances can be determined. We can also use the molar mass of each substance to convert between grams and moles. An outline of the approach to this question is:

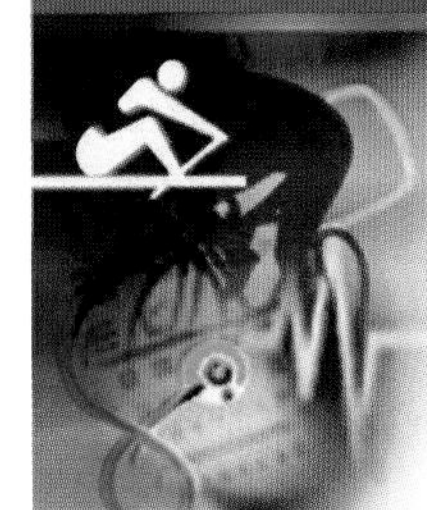

Connections

Isolation, Modification, and Synthesis of Drugs

Thousands of compounds have been or are currently used for medicinal purposes. Some of these compounds are obtained by isolating (purifying) them from a natural source. For example, the antimalarial drug quinine is isolated from the bark of the cinchona tree, and morphine, which is used to control pain, is obtained from the opium poppy. These purifications involve methods such as distillation or chromatography that allow the drug to be separated from the other substances found in the natural source.

Some of our modern drugs are prepared by using chemical reactions to modify natural substances. Several frequently used antibacterial drugs are modified forms of penicillin, which was originally isolated from a mold of the genus *Penicillium*. Drugs are modified to make them more stable or more effective than the natural product.

Many of the drugs in use today are synthesized entirely by chemical reactions. Some of these compounds, such as the antimicrobial sulfa drugs, do not occur naturally and must be produced. Others, such as the salicylates (aspirin is a salicylate), can be made by the pharmaceutical company at lower cost than they can be isolated from a natural source.

(a) 

(b) 

(a) Cinchona bark is the source of quinine. (b) A *penicillium* mold. *(a, © Walter H. Hodge/Peter Arnold, Inc.; b, Andrew McClenaghan/SPL/Photo Researchers)*

① Molar mass	② Molar ratio	③ Molar mass

Grams P ⟶ Moles P ⟶ Moles P_4O_{10} ⟶ Grams P_4O_{10}

The calculations to solve this question are:

$$\overset{}{75.0\ \text{g P}} \times \overset{①}{\frac{1\ \text{mol P}}{30.97\ \text{g P}}} \times \overset{②}{\frac{1\ \text{mol } P_4O_{10}}{4\ \text{mol P}}} \times \overset{③}{\frac{283.88\ \text{g } P_4O_{10}}{1\ \text{mol } P_4O_{10}}} = 172\ \text{g } P_4O_{10}$$

$4P + \overset{10}{O} \rightarrow P_4O_{10}$

Given enough oxygen, 172 g of P_4O_{10} can be made from 75.0 g of phosphorus.

Example 4.13 Determining Masses of Reactants and Products in Reactions

How much salicylic acid ($C_7H_6O_3$) is needed to make 175 g of aspirin ($C_9H_8O_4$)? The balanced equation for this reaction is

$$C_7H_6O_3(s) + C_4H_6O_3(l) \longrightarrow C_9H_8O_4(s) + C_2H_4O_2(l)$$

SOLUTION

(Because the two liquid reagents in the equation are not part of the question, they are ignored.) First use the molar mass of aspirin to convert grams of aspirin to moles of aspirin:

$$175\ \text{g aspirin} \times \frac{1\ \text{mol aspirin}}{180.17\ \text{g aspirin}} = 0.971\ \text{mol aspirin}$$

Next use the molar ratio from the balanced equation to convert moles of aspirin to moles of salicylic acid:

$$0.971 \text{ mol aspirin} \times \frac{1 \text{ mol salicylic acid}}{1 \text{ mol aspirin}} = 0.971 \text{ mol salicylic acid}$$

Finally, use the molar mass of salicylic acid to convert moles of salicylic acid to grams of salicylic acid:

$$0.971 \text{ mol salicylic acid} \times \frac{138.12 \text{ g salicylic acid}}{1 \text{ mol salicylic acid}} = 134 \text{ g salicylic acid}$$

To make 175 g of aspirin, 134 g of salicylic acid is needed.

Self-Test

Use this balanced equation to calculate the amount of bromine needed to make 4.75 g of aluminum bromide ($AlBr_3$):

$$2Al(s) + 3Br_2(l) \longrightarrow 2AlBr_3(s)$$

ANSWER

4.27 g bromine

CHAPTER SUMMARY

Chemical **formulas** show the composition or structure of compounds. The formula of an ionic compound shows the relative number of each ion; a **molecular formula** shows the number of atoms of each element that is present in the molecule. A **structural formula** shows all of the atoms in a molecule and the bonds that connect them. Common chemical names identify a compound but may provide little or no information about the compound. Systematic names always are related to the composition or structure of the compound. **Binary ionic** and **binary covalent compounds** are named using information contained in the formula of the compound. Conversely, the formulas of these compounds can be determined from their names. In chemical reactions, reactants are changed to products. **Chemical equations** are concise, symbolic representations of chemical reactions. When balanced with the proper **coefficients,** an equation provides quantitative information about **reactants** and **products.** The **mole** is a unit that indicates the number of particles in a sample of matter. The coefficients of an equation represent either the number of individual atoms, ions, or molecules in the reaction or the number of moles of each of them. The **molar mass** of an element or compound is the mass of one mole of that substance. Balanced equations, molar ratios, and molar masses can be used to determine the mass of reactants and products in a reaction.

KEY TERMS

Section 4.1
formula
molecular formula
structural formula

Section 4.2
binary ionic compound

Section 4.3
binary covalent compound

Section 4.4
chemical equation (equation)
coefficients
law of conservation of matter
products
reactants

Section 4.5
formula mass
molar mass
mole
molecular mass

EXERCISES

Chemical Formulas and Names (Section 4.1)

1. A chemical formula identifies and provides information about a compound. What specific information is contained in a chemical formula?
2. The chemical formula of the ionic compound magnesium chloride is $MgCl_2$. What does this formula tell you about the composition of magnesium chloride?
3. The formula of the molecular compound hydrogen peroxide is H_2O_2. What does this formula tell you about a molecule of hydrogen peroxide?
4. What is a structural formula? Interpret this structural formula for hydrogen peroxide: H–O–O–H.
5. How many ions of each element are indicated by these formulas?
   (a) MgO (b) Ag_2S (c) BaI_2 (d) CrF_3
6. How many ions of each element are indicated by these formulas?
   (a) NaBr (b) $NaNO_3$
   (c) K_2SO_4 (d) Fe_2O_3
7. How many atoms of each element are present in the molecules of these compounds?
   (a) PCl_5 (b) CS_2 (c) N_2H_4 (d) H_3PO_4
8. How many atoms of each element are present in the molecules of these compounds?
   (a) HCN (b) N_2O_5
   (c) H_2SO_4 (d) C_2H_6O

Naming Ionic Compounds (Section 4.2)

9. Before a compound can be named, it must be identified as ionic or covalent. How would you make this identification? Are these compounds ionic or covalent?
   (a) KI (b) PCl_5 (c) CBr_4
   (d) SrO (e) VF_3
10. How is the cation in a binary ionic compound named? How is the anion named?
11. What are the names of these ions?
   (a) Al^{3+} (b) Br^- (c) N^{3-}
   (d) Cs^+ (e) Fe^{3+}
12. What are the names of these ions?
   (a) S^{2-} (b) Mg^{2+} (c) P^{3-}
   (d) Cu^+ (e) Se^{2-}
13. Provide the names of these Type I binary ionic compounds.
   (a) NaI (b) MgO (c) K_3N (d) $CaBr_2$
14. Provide the names of these Type I binary ionic compounds.
   (a) Li_2Te (b) BaF_2 (c) $AlCl_3$ (d) Mg_3N_2
15. Give the Stock names for these Type II binary ionic compounds.
   (a) FeO (b) CuI
16. Give the Stock names for these Type II binary ionic compounds.
   (a) Fe_2O_3 (b) Hg_2Cl_2

Some common ionic compounds, clockwise from left: a calcium fluoride (CaF_2) crystal; a package of sodium chloride (NaCl); a crystal of calcite (calcium carbonate, ($CaCO_3$); and powdered and hydrated cobalt(II) chloride ($CoCl_2 \cdot 6H_2O$). *(Charles D. Winters)*

17. Use the appropriate (*-ous* and *-ic*) suffix to name these compounds:
(a) Cu_2S (b) $FeBr_2$ (c) $HgCl_2$

18. Use the appropriate suffix (*-ous* and *-ic*) to name these compounds:
(a) Cu_2O (b) Hg_2S (c) Fe_2O_3

19. Name these binary ionic compounds:
(a) Cs_2S (b) BeF_2 (c) $HgBr_2$

20. Name these binary ionic compounds:
(a) FeI_2 (b) Na_2S (c) KI

21. What are the formulas of these Type I binary ionic compounds?
(a) Potassium bromide (b) Magnesium iodide
(c) Lithium phosphide

22. What are the formulas of these Type I binary ionic compounds?
(a) Calcium chloride (b) Sodium selenide
(c) Aluminum fluoride

23. Give the formulas of these Type II binary ionic compounds.
(a) Copper(I) sulfide (b) Mercury(II) fluoride
(c) Iron(III) iodide

24. Give the formulas of these Type II binary ionic compounds.
(a) Tin(II) oxide (b) Iron(II) chloride
(c) Vanadium(V) oxide

25. What are the names of these ionic compounds?
(a) $Ca(OH)_2$ (b) Na_2SO_4 (c) $MgCO_3$

26. What are the names of these ionic compounds?
(a) NH_4Cl (b) $NaHCO_3$ (c) $Al(NO_3)_3$

27. Write the formulas of these ionic compounds.
(a) Sodium hydroxide (b) Stannous fluoride
(c) Iron(II) phosphate

28. Write the formulas of these ionic compounds.
(a) Potassium cyanide
(b) Copper(II) bicarbonate
(c) Ammonium sulfide

Naming Covalent Compounds (*Section 4.3*)

29. What is the difference between a covalent compound and an ionic compound? Identify two common covalent compounds.

30. Provide the names of these binary covalent compounds.
(a) CS_2 (b) N_2O (c) ClF_3
(d) Si_2F_6 (e) N_2O_4

31. Provide the names of these binary covalent compounds.
(a) P_2I_4 (b) S_2F_2 (c) BF_3
(d) ClF (e) Se_2Cl_2

32. What are the names of these binary covalent compounds?
(a) Br_2O (b) N_2O_5 (c) BrF_3
(d) S_4N_4 (e) PI_3

33. What are the formulas of these binary covalent compounds?
(a) Iodine monochloride
(b) Boron monophosphide
(c) Sulfur trioxide
(d) Selenium monosulfide
(e) Dinitrogen trioxide

34. Write the formulas of these binary covalent compounds.
(a) Nitrogen monoxide
(b) Oxygen difluoride
(c) Tetraphosphorus trisulfide
(d) Triphosphorus pentanitride
(e) Silicon tetraiodide

35. Give the formulas of these binary covalent compounds.
(a) Sulfur monoxide
(b) Iodine trichloride
(c) Iodine pentafluoride
(d) Selenium dioxide
(e) Disulfur heptoxide

Chemical Equations (*Section 4.4*)

36. What is a chemical reaction?

37. Which of these changes are chemical reactions?
(a) Burning wood
(b) Melting ice
(c) Dissolving sugar in coffee
(d) Rusting nail

38. Which of these changes are chemical reactions?
(a) Digestion of food
(b) Evaporating water from a lake
(c) Burning fuel in a cutting torch
(d) Cutting metal with a cutting torch

39. What is a chemical equation?

When answering Exercises 40–43, refer to this equation for the thermal decomposition of mercury(II) oxide:

$$2HgO(s) \longrightarrow 2Hg(l) + O_2(g)$$

40. Identify the product(s) of the reaction.

41. Identify the reactant(s) of the reaction.

42. What is the physical state of each of the substances in the equation?

43. List the coefficient for each of the substances in the equation.

When answering Exercises 44–47, refer to this equation for the complete combustion (oxidation) of glucose (blood sugar):

$$C_6H_{12}O_6(s) + 6O_2(g) \longrightarrow 6CO_2(g) + 6H_2O(g)$$

44. Identify the product(s) of this reaction.

45. Identify the reactant(s) of this reaction.

46. What is the physical state of each of the substances in the equation?

47. List the coefficient for each of the substances in the equation.

48. Why are coefficients necessary in a chemical equation?

49. Determine the number of atoms of each element on each side of these equations as they are written. Indicate whether the reaction is or is not balanced.

(a) $SO_2(g) + O_2(g) \longrightarrow SO_3(g)$
(b) $2Al(s) + 3Br_2(l) \longrightarrow 2AlBr_3(s)$
(c) $2CO(g) + O_2(g) \longrightarrow 2CO_2(g)$
(d) $Al(NO_3)_3(aq) + NaOH(aq) \longrightarrow Al(OH)_3(s) + NaNO_3(aq)$

50. Determine the number of atoms of each element on each side of these equations as they are written. Indicate whether the reaction is or is not balanced.

(a) $NaOH(aq) + HCN(aq) \longrightarrow H_2O(l) + NaCN(aq)$
(b) $Fe(s) + O_2(g) \longrightarrow Fe_2O_3(s)$
(c) $Zn(s) + HCl(aq) \longrightarrow ZnCl_2(aq) + H_2(g)$
(d) $(NH_4)_2SO_4(aq) + PbCl_2(aq) \longrightarrow 2NH_4Cl(aq) + PbSO_4(s)$

51. Examine these equations to determine if they are balanced. If an equation is not balanced, make changes in the coefficients to balance the equation.

(a) $Sn(s) + Cl_2(g) \longrightarrow SnCl_4(l)$
(b) $CH_4(g) + 2Cl_2(g) \longrightarrow CH_2Cl_2(l) + HCl(g)$
(c) $Pb(NO_3)_2(aq) + 2KI(aq) \longrightarrow KNO_3(aq) + PbI_2(s)$

52. Examine these equations to determine if they are balanced. If an equation is not balanced, make changes in the coefficients to balance the equation.

(a) $CaO(s) + H_2O(l) \longrightarrow Ca(OH)_2(s)$
(b) $2PbS(s) + O_2(g) \longrightarrow 2PbO(s) + 2SO_2(g)$
(c) $Fe_2O_3(s) + 3CO(s) \longrightarrow 2Fe(l) + 3CO_2(g)$

53. Balance these equations:

(a) $Na(s) + Cl_2(g) \longrightarrow NaCl(s)$
(b) $K(s) + H_2O(l) \longrightarrow KOH(aq) + H_2(g)$
(c) $LiOH(s) + CO_2(g) \longrightarrow Li_2CO_3(s) + H_2O(l)$

54. Balance these equations:

(a) $S(s) + O_2(g) \longrightarrow SO_3(g)$
(b) $C_2H_6(g) + O_2(g) \longrightarrow CO_2(g) + H_2O(g)$
(c) $HCl(aq) + Pb(NO_3)_2(aq) \longrightarrow PbCl_2(s) + HNO_3(aq)$

55. Balance these equations:

(a) $P(s) + Cl_2(g) \longrightarrow PCl_5(s)$
(b) $Ba(OH)_2(aq) + HCl(aq) \longrightarrow BaCl_2(aq) + H_2O(l)$
(c) $Ca(NO_3)_2(aq) + K_3PO_4(aq) \longrightarrow Ca_3(PO_4)_2(s) + KNO_3(aq)$

56. Balance these equations:

(a) $Cl_2O(g) + H_2O(g) \longrightarrow HClO(g)$
(b) $C_2H_2(g) + HCl(g) \longrightarrow C_2H_4Cl_2(g)$
(c) $Al(s) + HBr(aq) \longrightarrow AlBr_3(aq) + H_2(g)$

The Mole *(Section 4.5)*

57. What is a mole? How many atoms of zinc will be in 1 mol of zinc? How many apples are in a mole of apples?

58. Calculate the number of atoms in 0.316 mol of phosphorus.

59. Determine the number of atoms in 1.68 mol of tin.

60. How many moles of barium will contain 5.0×10^{20} atoms of barium?

61. Calculate the number of moles of sulfur that contain 1.35×10^{25} atoms of sulfur.

62. **(a)** Determine the number of molecules of HCl in 0.028 mol of HCl. **(b)** How many atoms of hydrogen and atoms of chlorine are present in 0.028 mol of HCl?

63. A 0.478-mol sample of water contains how many molecules of water? How many atoms of hydrogen? How many atoms of oxygen?

64. A sample containing 3.55×10^{24} molecules of N_2O_4 will contain how many moles of dinitrogen tetraoxide? How many moles of nitrogen atoms? How many moles of oxygen atoms?

65. How many moles of glucose, $C_6H_{12}O_6$, would contain 1.00×10^{23} molecules of glucose? How many moles of atoms of each element would be present in the sample?

66. What is the mass in grams of these samples of elements?

(a) 1.80 mol C **(b)** 27 mol He
(c) 0.40 mol of potassium

67. Give the mass (g) of these elemental samples.
(a) 2.35 mol uranium (b) 2.3×10^{-7} mol Ag
(c) 2.3 mol F_2
68. Convert these masses to moles.
(a) 750.0 g Si (b) 63.5 g of sulfur
(c) 3.0 mg Ar
69. Convert these masses to moles.
(a) 1.000 kg K (b) 25 μg Au
(c) 25.0 g of oxygen (Remember, oxygen is diatomic.)
70. Explain the difference between *molecular mass* and *formula mass*.
71. Define *molar mass* and provide a simple example of this concept.
72. What are the molar masses of these substances?
(a) calcium (b) Cl_2
(c) $NaClO_3$ (d) H_2SO_4
73. Provide the molar mass of these substances.
(a) Boron (b) I_2
(c) $(NH_4)_3PO_4$ (d) HCN
74. Convert these masses to moles.
(a) 25 g NaCl (b) 13.5 g PCl_3
(c) 2.5 kg $C_6H_{12}O_6$ (glucose)
75. Convert these masses to moles.
(a) 32 g HCN (b) 41.3 g $Cu(NO_3)_2$
(c) 23.5 mg KBr
76. Express each of these substances in grams.
(a) 1.50 mol LiI (b) 0.32 mol UF_6
(c) 3.0 mmol $CaCO_3$
77. Convert moles of these substances to grams.
(a) 0.83 mol $C_{12}H_{22}O_{11}$ (sucrose)
(b) 11.3 μmol PbS (c) 350 mol H_2

Ammonia is commonly used as a fertilizer and is also used to make other nitrogen-containing compounds. Use this equation for the production of ammonia to answer Exercises 78–81:

$$3H_2(g) + N_2(g) \longrightarrow 2NH_3(g)$$

78. How many moles of ammonia can be produced from 2.5 mol of nitrogen? Assume that there is enough hydrogen to react with all of the nitrogen.
79. Suppose you need to make 3.0 mol of ammonia. How many moles of nitrogen and moles of hydrogen must react to make 3.0 mol of ammonia?
80. Calculate the mass in grams of nitrogen that are needed to produce 100.0 g of ammonia. Assume there is enough hydrogen to react with the nitrogen.
81. How many grams of ammonia can be produced from 10.0 g of hydrogen? Assume there is enough nitrogen to react with the hydrogen.

Sodium bicarbonate—baking soda—decomposes when heated. Use this equation for the decomposition of sodium bicarbonate to answer Exercises 82–84:

$$2NaHCO_3(s) \longrightarrow Na_2CO_3(s) + H_2O(g) + CO_2(g)$$

82. How many moles of sodium bicarbonate must decompose to make 10.0 mol of sodium carbonate? During this decomposition, how many moles of water and carbon dioxide would also be produced?
83. Would 10 mol of sodium bicarbonate be sufficient to produce 400.0 g of sodium carbonate? Justify your answer.
84. A student heats a small quantity (25 g) of sodium bicarbonate. How many grams of sodium carbonate will be produced?

General Exercises

85. Name each of these compounds.
(a) NH_4F (b) FeS (c) SiC
(d) $Al(OH)_3$ (e) PBr_5 (f) $BeBr_2$
86. Name each of these compounds.
(a) IF_7 (b) $LiC_2H_3O_2$ (c) $MgCl_2$
(d) $CaSO_4$ (e) Hg_2O (f) SiS_2
87. Write the formulas for these compounds.
(a) Lead(II) oxide
(b) Diphosphorus tetraoxide
(c) Barium carbonate
(d) Mercuric sulfide
(e) Trisilicon tetranitride
(f) Sodium nitrate
88. Write the formulas for these compounds.
(a) Ferric sulfate
(b) Potassium phosphate
(c) Calcium phosphate
(d) Bromine dioxide
(e) Copper(II) acetate
(f) Diiodine tetraoxide

Challenge Exercises

89. The compound LiCl is used to treat depression. (a) What is the compound name? (b) Solid lithium can be reacted with aqueous hydrochloric acid, HCl, to produce the compound dissolved in water plus gaseous H_2 as a by-product. Write

and balance the equation for this reaction.
(c) What mass of the compound could be made from 125 g of lithium?

90. You have 1.000 ton of salicylic acid. If all of this compound were converted to aspirin, how many 325-mg tablets of aspirin could you make?

91. Magnesium hydroxide—$Mg(OH)_2$—is used as an antacid. How many grams of magnesium oxide are needed to make 175 g of magnesium hydroxide with this reaction:

$$MgO(s) + H_2O(l) \rightarrow Mg(OH)_2(aq)$$

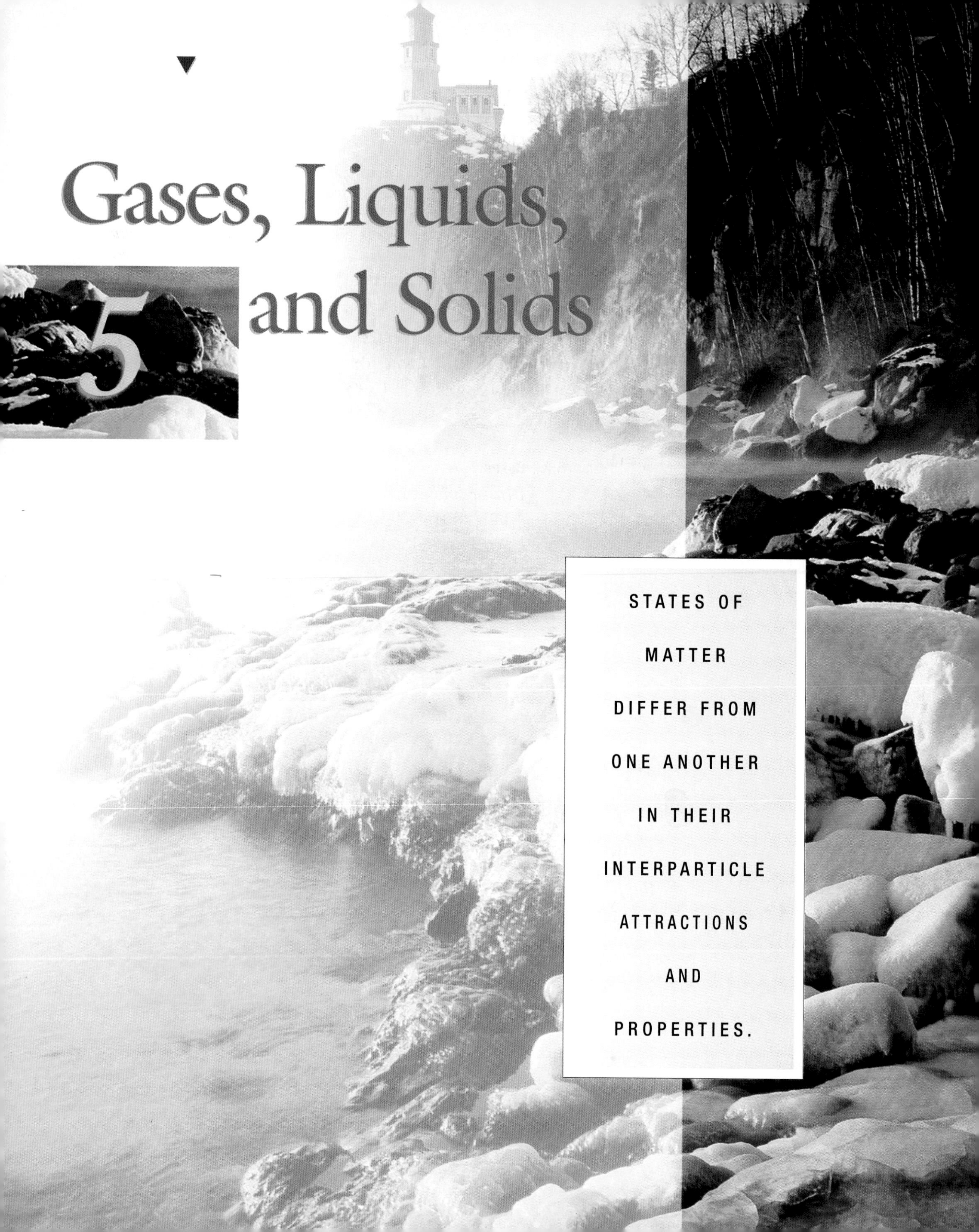

5 Gases, Liquids, and Solids

STATES OF MATTER DIFFER FROM ONE ANOTHER IN THEIR INTERPARTICLE ATTRACTIONS AND PROPERTIES.

MATTER CONSISTS of gases, liquids, and solids that exist either as pure substances or as mixtures. You are familiar with examples of all of these. Air is a mixture of gases, and water is a liquid that often has other substances dissolved in it. Many of the objects around you—your chair, desk, and house, to name just three—are solids. Matter can change from one form to another, and it can be mixed with or separated from other matter. This chapter helps you expand your understanding of matter and of the changes it can undergo.

Outline

5.1 Changes in State of Matter
List the changes that occur in the states of matter and recognize the role of energy in these changes.

5.2 Changes in Temperature
Understand the relationship between energy gain and loss and temperature changes in matter.

5.3 Gases
Define a gas, and explain why a gas exerts pressure.

5.4 Gas Laws
Use the gas laws to calculate volume, pressure, temperature, or moles of a gas.

5.5 Solids and Liquids
Define intermolecular forces and explain how they differ among gases, liquids, and solids.

Matter normally exists as either a gas, liquid, or solid. How can water molecules exist as a solid in ice, liquid in water, and a gas in air and clouds? How is water different in these three states? *(Tom Till/Tony Stone Images)*

5.1 Changes in State of Matter

Matter exists as solids, liquids, and gases. These three forms of matter are formally called *states of matter*, or more simply, **states.** For many common substances, we are familiar with one or more states. We usually think of iron as a solid, but we are familiar with water in all three states.

A sample of matter is not restricted to just one state because matter can change from one state to another. The melting of ice, freezing of water, and boiling of water are common examples. These physical processes are called **changes in state,** and all involve energy. If a solid such as ice is heated, it becomes liquid water. This change in state—the change of a solid to a liquid—is called **melting,** and the temperature at which it occurs is the **melting point** of that solid. When a liquid changes to a solid, it **freezes,** and the temperature at which this occurs is the **freezing point.** Freezing and melting occur at the same temperature. Ice melts and water freezes at 0°C.

When a liquid *boils* it becomes a gas; this liquid-to-gas change in state is called **vaporization.** For example, water is said to boil at 100°C. However, the temperature at which a liquid boils depends on the air pressure applied (pushing down) on the liquid surface, and that pressure decreases with increasing elevation (Table 5.1). Thus the **normal boiling point** of a liquid is defined as the temperature at which it boils at one atmosphere of pressure, typically the pressure at sea level. **Condensation** is the change from a gas to a liquid. For any substance, condensation begins when its gaseous state cools to its boiling point.

Three of the Group 7A elements are shown in their normal room-temperature states: gaseous chlorine, liquid bromine, and solid iodine. *(Charles D. Winters)*

Table 5.1 Effects of Altitude on the Boiling Point of Water

ELEVATION (m)	BOILING POINT (°C)	ATMOSPHERIC PRESSURE (TORR)	LOCATION
Sea level	100	760	Miami, FL
1600	95	630	Denver, CO
2300	93	585	Alamosa, CO
4400	87	467	Mt. Whitney (CA)
8800	73	270	Mt. Everest (S. Asia)

These freeze-dried coffee crystals were prepared by subliming water from frozen coffee. *(Charles D. Winters)*

Another change of state that occurs is **sublimation.** This is a transition from the solid state directly to the gaseous state. Dry ice (solid carbon dioxide) sublimes at room temperature. This solid does not melt; it goes straight from the solid state to the gaseous state. Freeze-dried coffee is prepared by freezing brewed coffee and then placing the frozen mixture in a cold vacuum. Under these conditions, the ice sublimes, leaving a residue of freeze-dried coffee. The reverse change, the change from the gaseous state directly to the solid state, is called **deposition.** Frost formation is a familiar example of deposition. Table 5.2 summarizes these changes in state.

Changes in state always involve energy. Water molecules gain energy as the water changes from ice to liquid water or from liquid water to steam. Water molecules lose energy as steam condenses to liquid water or liquid water freezes to ice. The amount of energy gained or lost in a change in state is characteristic of the substance undergoing the change. **Heat of fusion** is the formal term for heat of melting and is the amount of heat a substance gains when it melts or the amount of heat it loses when it freezes. For water, the heat of fusion is 79.7 cal/g. For every gram of ice that melts, 79.7 cal of energy is absorbed by the ice from the environment. This absorption of energy means that when ice melts, it cools its surroundings. Ice has a larger heat of fusion than other common substances (Table 5.3).

To calculate the amount of heat needed to melt a given amount of any solid, use this expression:

$$q = m \times \Delta H_{\text{fusion}}$$

In this expression q = heat in calories, m is the mass of the substance in grams, and ΔH_{fusion} is the heat of fusion in cal/g for the substance. This same equation is also used to calculate the amount of heat lost by a liquid when it freezes.

Solar energy melts ice and snow and evaporates water. The sun plays a vital role in the water cycle and our weather. *(Williard Clay/Dembinsky Photo Associates)*

Table 5.2 Changes in State

Melting	Solid to liquid	Freezing	Liquid to solid
Vaporization	Liquid to gas	Condensation	Gas to liquid
Sublimation	Solid to gas	Deposition	Gas to solid

Table 5.3 Heats of Fusion of Some Common Substances

SUBSTANCE	HEAT OF FUSION (cal/g)
Water	79.7
Rubbing alcohol (isopropyl alcohol)	21
Natural gas (methane)	15
Mercury	2.8

Example 5.1 Heat of Fusion

How much energy is needed to melt 454 g (one pound) of ice at 0°C?

SOLUTION

The mass of the water sample is 454 g, and the heat of fusion for water is 79.7 cal/g. Insert these values into the equation and solve:

$$q = m \times \Delta H_{\text{fusion}}$$

$$q = 454 \cancel{g} \times \frac{79.7 \text{ cal}}{1 \cancel{g}}$$

$$q = 3.62 \times 10^4 \text{ cal} = 36.2 \text{ kcal}$$

Self-Test

How much heat is lost by 55 g of water at 0°C when it freezes?

ANSWER

4.4 kcal

Energy is also gained when a liquid vaporizes and lost when a gas condenses. The amount of heat gained or lost in these changes at the normal boiling point is called the **heat of vaporization.** For water, 540 cal (actually 5.40×10^2, thus three significant figures) must be added to convert 1 g of water to steam. The heat of vaporization is different for each substance, and water has a larger heat of vaporization than other common substances (Table 5.4).

Table 5.4 Heats of Vaporization of Some Common Substances

SUBSTANCE	HEAT OF VAPORIZATION (cal/g)
Water	540
Rubbing alcohol (isopropyl alcohol)	159
Natural gas (methane)	138
Mercury	65

Calculations involving heats of vaporization are similar to those for heats of fusion, and use this equation:

$$q = m \times \Delta H_{vap}$$

The symbol q means heat in calories, m is the mass of the vaporizing (or condensing) substance in grams, and ΔH_{vap} is the heat of vaporization of the substance in cal/g.

Example 5.2 Heat of Vaporization

Steam burns are often serious because a large amount of energy is transferred to the victim when the steam at 100°C condenses to water. Calculate the amount of heat gained by the tissues of a person burned by steam if 5.0 g of steam condenses during contact.

SOLUTION

The amount of heat gained by the tissues equals the amount lost by the steam, which is determined by multiplying the mass of the condensed steam by the heat of vaporization for water:

$$q = m \times \Delta H_{vap}$$

$$q = 5.0 \cancel{g} \times \frac{540 \text{ cal}}{1 \cancel{g}}$$

$$q = 2.7 \times 10^3 \text{ cal} = 2.7 \text{ kcal}$$

Self-Test

1. How many kilocalories of heat are needed to vaporize 355 mL of 100°C water? Hint: Use the density of water (1.00 g/1.00 mL) to convert volume to mass.
2. A patient who has a fever is washed with 20.0 g (about 25 mL) of rubbing alcohol. How much energy (heat) is removed from this person by the evaporation of the alcohol?

ANSWERS

1. 192 kcal 2. 3.18 kcal

5.2 Changes in Temperature

Temperature change is another physical change, and it also involves heat. When cold water gains heat, its temperature increases. Hot water loses heat as it cools, and so its temperature decreases. The amount of energy needed to raise the temperature of a sample of matter depends on the mass of the sample, the temperature change, and a third factor called specific heat. The **specific heat** of any substance is the amount of heat needed to raise the temperature of 1 g of the substance by 1°C (Table 5.5). Water has the highest specific heat of any common substance. The

Table 5.5 Specific Heats of Some Common Substances

SUBSTANCE	SPECIFIC HEAT (cal/g °C)	SUBSTANCE	SPECIFIC HEAT (cal/g °C)
Aluminum	0.214	Lead	0.030
Air	0.240	Paraffin wax	0.694
Ethyl alcohol	0.581	Steam	0.48
Gold	0.0312	Water	1.00
Ice	0.500	Wood	0.42

expression for calculating heat gain or loss during a temperature change is:

$$\text{Heat} = \text{mass} \times \text{specific heat} \times \text{change in temperature}$$

This equation can be represented by symbols:

$$q = m \times SH \times \Delta T$$

Note that delta (Δ) means change, so ΔT means change in temperature.

Changes in temperature and changes in state can be summarized with heating curves or cooling curves (Figure 5.1). In these graphs, heat is added or removed at a constant rate along the horizontal axis. The temperature of the matter is plotted on the vertical axis. When the added heat increases the kinetic energy of the matter, it raises the temperature of the substance and the curve goes up. When the substance undergoes a change in state, the temperature does not change and the curve is flat. *Temperature remains constant when a change in state occurs.*

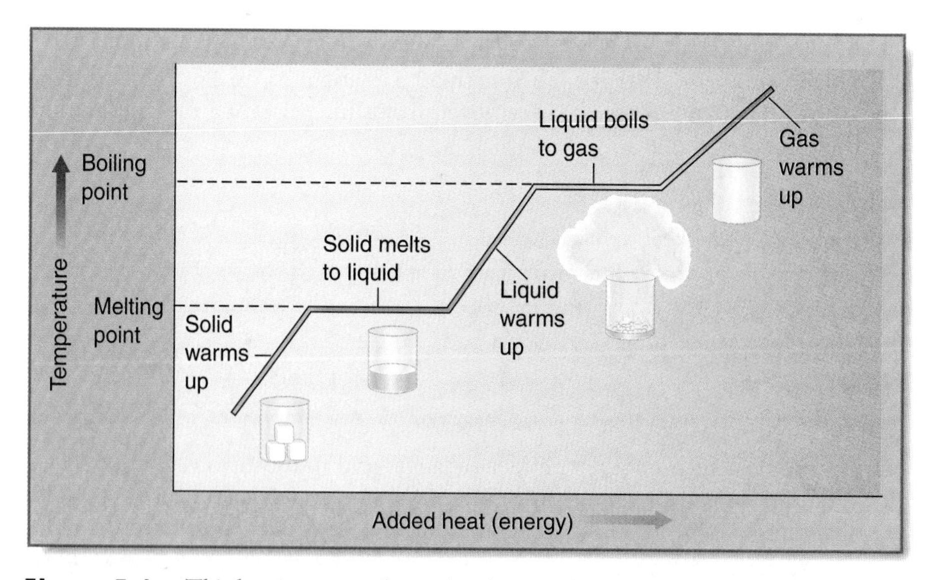

Figure 5.1 This heating curve shows the changes that occur in states of matter and temperature when heat is added to a substance.

Example 5.3 Specific Heats of Substances

1. How much heat does a 255-g aluminum pot gain as its temperature changes from 21°C to 93°C?
2. How much heat is required to change 14.5 g of ice at −22°C to liquid water at 0°C?

SOLUTIONS

1. The mass of the aluminum is given, and the specific heat of aluminum can be found in Table 5.5. The temperature change is the final temperature minus the initial temperature: $\Delta T = 93°C - 21°C = 72°C$. The expression and solution are

$$q = m \times SH \times \Delta T$$

$$q = 255\ \not{g} \times \frac{0.214\ \text{cal}}{1\ \not{g} \cdot 1°\not{C}} \times 72°\not{C} = 3900\ \text{cal} = 3.9\ \text{kcal}$$

2. This question has a two-part solution. In the first step, ice (solid water) is warmed from −22°C to 0°C. This is a change in temperature, but the state of the substance does not change. The second step is a change in state—the ice melts to liquid water—but the temperature *does not change.*

The first step uses the specific-heat expression:

$$q = m \times SH \times \Delta T$$

In the problem, the mass is 14.5 g, the specific heat of ice is 0.50 cal/g · °C, and the change in temperature is $0°C - (-22°C) = 22°C$:

$$q = 14.5\ \not{g} \times \frac{0.50\ \text{cal}}{1\ \not{g} \cdot 1°\not{C}} \times 22°\not{C} = 160\ \text{cal}$$

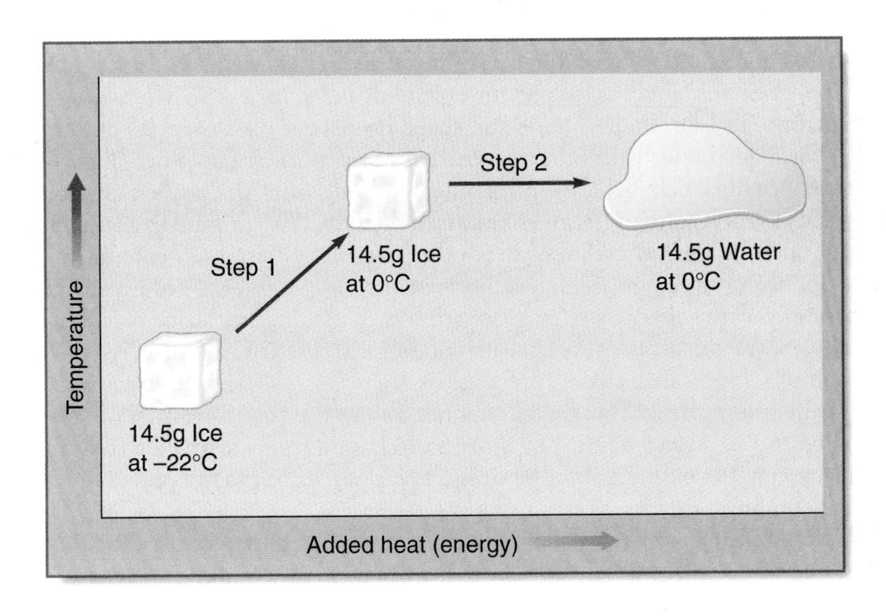

Figure 5.2 Gases have neither a definite shape nor a definite volume. Helium stored in a tank expanded as it filled this balloon. *(Mark C. Burnett/Photo Researchers)*

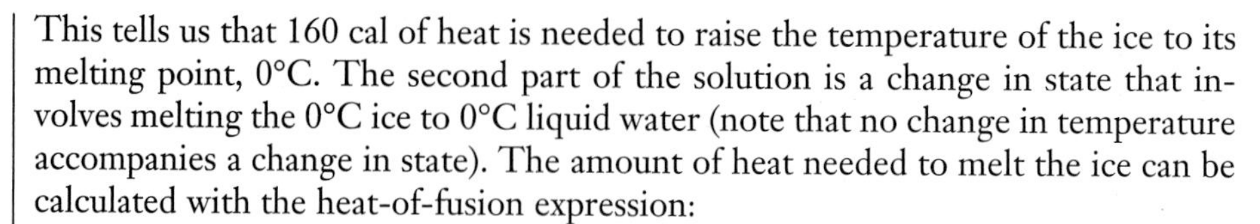

This tells us that 160 cal of heat is needed to raise the temperature of the ice to its melting point, 0°C. The second part of the solution is a change in state that involves melting the 0°C ice to 0°C liquid water (note that no change in temperature accompanies a change in state). The amount of heat needed to melt the ice can be calculated with the heat-of-fusion expression:

$$q = m \times \Delta H_{\text{fusion}}$$

The mass of the ice is still 14.5 g, and the heat of fusion for ice is 79.7 cal/g:

$$q = 14.5 \not{g} \times \frac{79.7 \text{ cal}}{1 \not{g}} = 1160 \text{ cal}$$

The total amount of heat needed to change 14.5 g of ice at −22°C to water at 0°C is

160 cal	Heating ice to 0°C
+ 1160 cal	Melting ice to form water
1320 cal	Total energy

Self-Test

How much heat is needed to warm a 6.0-g gold ring from 19°C to 37°C?

ANSWER

3.4 cal

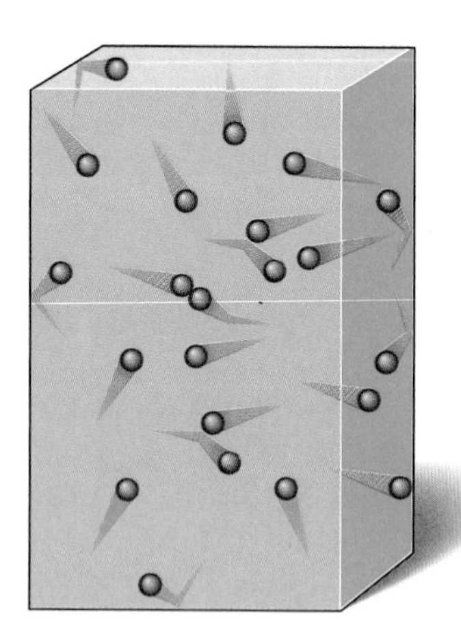

Figure 5.3 The particles in a gas are free and independent because they are too far apart to form bonds. The gas particles are moving and collide with each other and the container walls. These collisions with the walls produce pressure.

5.3 Gases

A **gas** has no definite shape and no definite volume (Figure 5.2). Gases assume the shape of their container, and they can expand or be compressed. There is a simple explanation for both their volume and shape properties. Gas particles do not normally form any bonds with other gas particles or with their container (Figure 5.3). In gases, each particle behaves independently of all other gas particles. Gas particles, such as molecules of oxygen or atoms of helium, "fly" around in a straight line at high speed until they collide with another particle or with their container. Then they bounce off in a new direction. Because the gas particles are not restricted by interactions with other particles, gases are free to assume any shape or volume.

Gases exert pressure on their surroundings. **Pressure** is defined as the force applied to a given area. The rapidly moving gas particles exert a force whenever they collide with the walls of their container, and so they exert pressure on the walls (see Figure 5.3). When you inflate a tire to 30 pounds per square inch (psi), for example, you put enough air into the tire so that the air exerts a pressure of 30 psi on its surroundings, the tire.

Our atmosphere exerts a continuous pressure on its surroundings, too. Normal atmospheric pressure is 14.7 psi, but other units of pressure are more common than

Connections

Measuring Blood Pressure

Liquids, like gases, exert pressure on their containers, and so blood exerts a pressure on the cardiovascular system. Blood pressure measurements have two values, with the higher value—*systolic pressure*—occurring when the heart contracts, and the lower value—*diastolic pressure*—occurring between heart contractions. Because high blood pressure is associated with a variety of health problems, the blood pressure of patients is routinely monitored.

Blood pressure is often measured with a *sphygmomanometer*, which consists of a bulb, a cuff, a manometer, and connecting tubing (a in figure). The cuff is placed around the arm and is inflated by squeezing the bulb. The cuff exerts pressure on the arm, which is measured by the manometer. When the cuff is inflated, the increasing pressure collapses the brachial artery, preventing blood flow. Next the cuff is slowly deflated, which results in decreasing pressure on the arm. When the pressure exerted by the cuff drops to the systolic pressure, blood begins to flow through the artery, generating a sound that can be heard with a stethoscope (b). At this moment, the pressure is read directly from the manometer and recorded as the systolic pressure. As the cuff pressure continues to decrease, more blood flows and the sound continues. When the sound stops, the pressure exerted by the cuff corresponds to the diastolic pressure (c), which is also read and recorded.

Measurements of blood pressure are reported as a ratio of systolic pressure over diastolic pressure. Although the manometer is marked in mm Hg (torr), the units are often not used. A blood pressure measurement of 120/80 means a systolic pressure of 120 mm Hg and a diastolic pressure of 80 mm Hg.

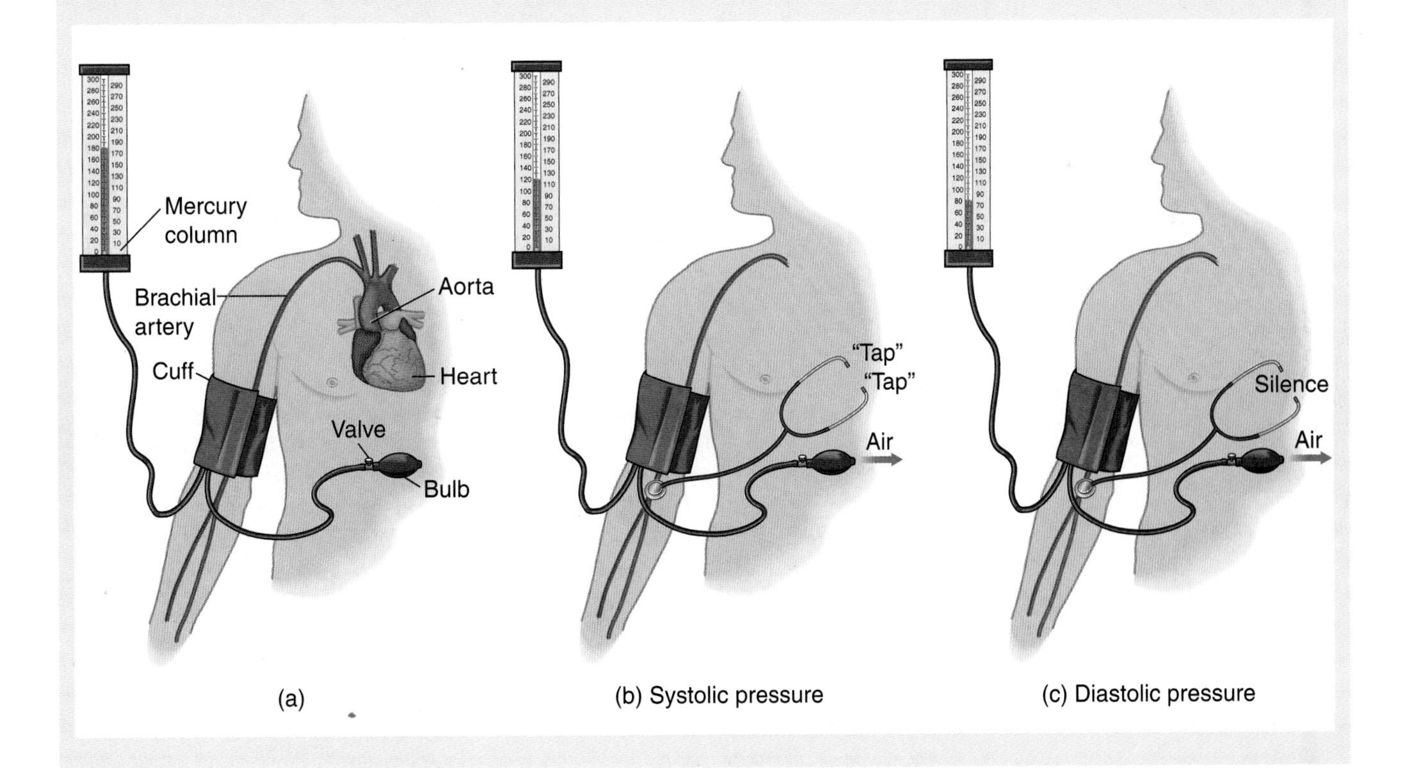

(a) (b) Systolic pressure (c) Diastolic pressure

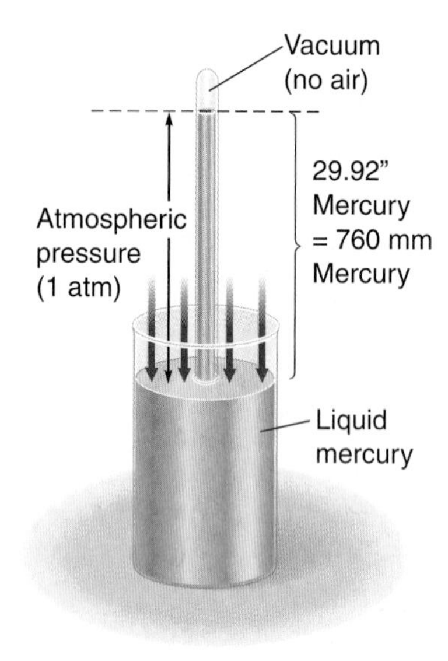

Figure 5.4 Normal atmospheric pressure is equal to the pressure exerted by a column of mercury that has a height of 29.92 in = 760 mm.

Hg is the symbol for mercury.

psi. Weather reports express atmospheric pressure in *inches of mercury.* An atmospheric pressure of 29.92 in of mercury means the atmosphere is exerting the same pressure as that exerted by a column of mercury 29.92 in high (Figure 5.4). The metric equivalent of inches of mercury is the *torr;* which is a unit named in honor of Evangelista Torricelli, whose work lead to our modern mercury barometers. One torr is the pressure exerted by 1 mm of mercury (Hg). Normal atmospheric pressure at sea level is called an atmosphere, and 1 atmosphere (atm) is defined as being equal to 760 torr.

Atmospheric pressure is not constant. It changes with weather patterns and altitude. Normal atmospheric pressure is standardized at 0°C and 760 torr (1 atm; Table 5.6). This temperature and pressure are referred to as **standard temperature and pressure (STP),** and these conditions are important in several common calculations involving gases. By the way, in the United States, the National Weather Service uses inches of mercury as their unit of pressure, and 29.92 in Hg is equivalent to 760 torr.

Pressure is normally measured with either a barometer or a manometer (Figure 5.5). In a Torricellian barometer, the pressure of the atmosphere presses on the pool of mercury in the barometer. The barometer has a glass tube that is evacuated of air. Because a vacuum exerts no pressure, the atmospheric pressure forces mercury up into the glass tube to a height where the downward pressure exerted by the mercury just equals the upward pressure of the atmosphere. The height of the mercury is measured in millimeters. Because 1 mm Hg = 1 torr, the height of the mercury in millimeters is the same as the pressure in torr. A manometer is used to measure the difference in pressure between two samples. If mercury is used in the manometer, the difference in the height, Δh (in millimeters), of the mercury in the two tubes is the pressure in mm Hg or torr.

In addition to pressure, temperature and the number of gas particles influence the behavior of gases.

Table 5.6 Standard Atmospheric Pressure at Sea Level and 0°C Expressed in Common Pressure Units

UNITS OF PRESSURE	ATMOSPHERIC PRESSURE
English System	
Pounds per square inch	14.7 psi
Inches of mercury	29.92 in Hg
Metric System	
Millimeters of mercury	760 mm Hg
Torr	760 torr
Atmospheres	1 atm

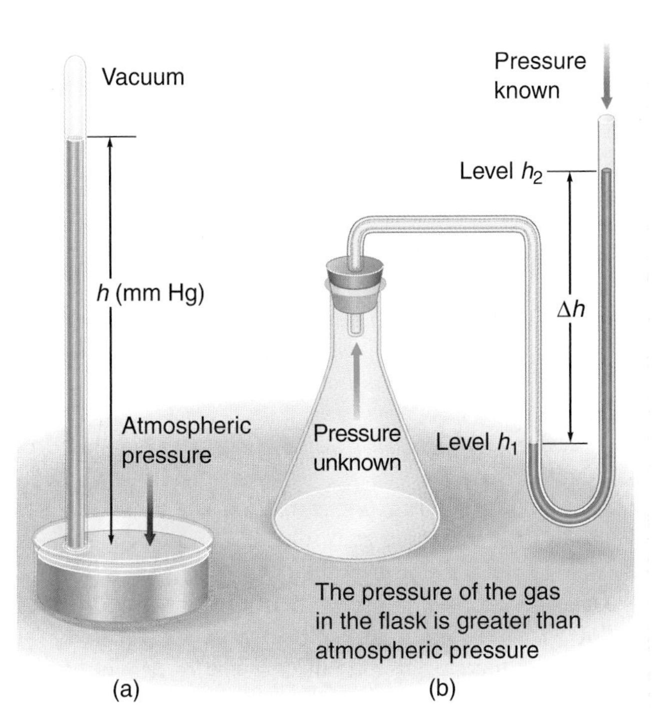

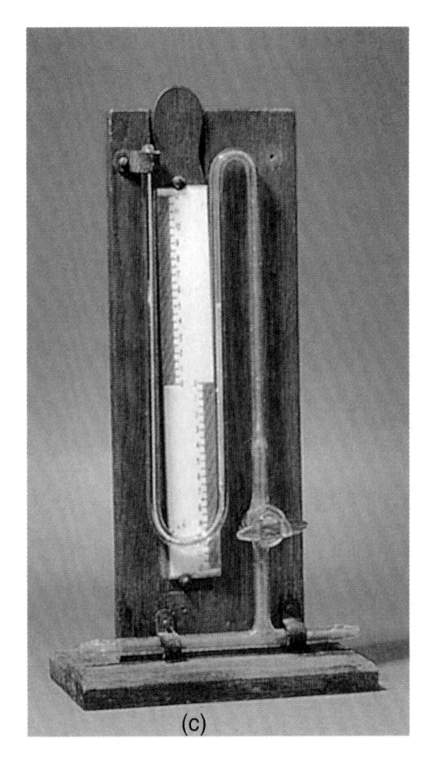

Figure 5.5 **(a)** The height, *h*, of a column of mercury in a Torricellian barometer is equal to the pressure of the atmosphere. The units used are millimeters of mercury (mm Hg), torr, or inches of mercury. **(b)** A manometer is used to measure a pressure difference between two gases. The difference in height of the liquid (Δh) in the two-armed tube is a measure of the pressure difference between the two gases. **(c)** A laboratory manometer. The glass tube at the bottom is connected to the gas sample and the difference in height between the levels of mercury in the two arms of the tube is read from a calibrated scale. *(Leon Lewandowski)*

5.4 Gas Laws

Remember, laws are summaries of observations, not explanations.

The fact that particles of a gas do not interact with each other results in some predictable properties for gases. These properties can be summarized by several simple mathematical laws known collectively as the gas laws. These laws express relationships among volume, pressure, temperature, and amounts of the gas.

Boyle's Law

The volume occupied by a gas depends on the pressure applied to the gas. This is easy to visualize. If gases consist of moving particles surrounded by lots of empty space, then applying pressure just moves the particles closer together, reducing the

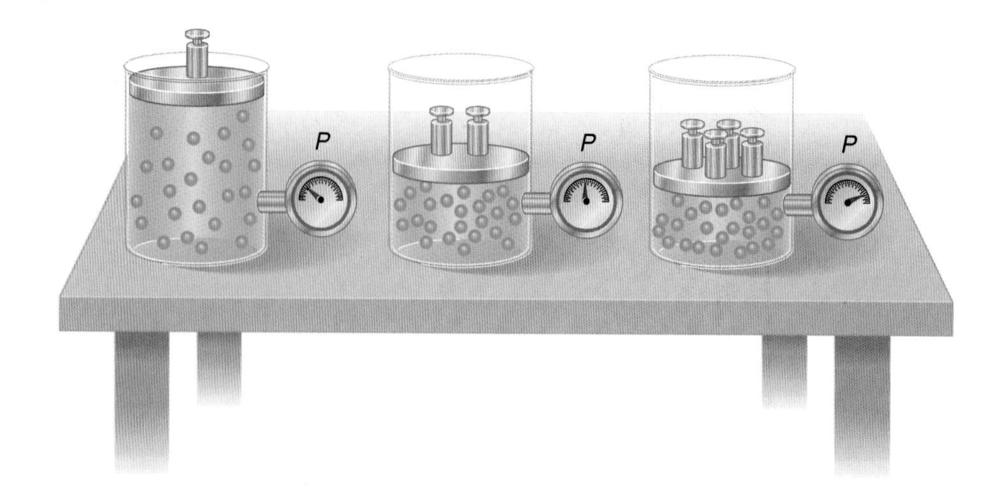

Figure 5.6 The volume occupied by a gas depends on the pressure applied to the gas, if other factors remain constant. The greater the pressure, the smaller the volume (Boyle's law).

empty space between them (Figure 5.6). The relationship between the pressure *(P)* and volume *(V)* of a gas is expressed as **Boyle's law:**

$$PV = \text{constant value}$$

This means that the product of pressure and volume equals a constant. If pressure increases, volume must decrease. If pressure goes down, volume goes up. If the pressure is doubled, the volume will be halved. Boyle's law is often written in this form:

$$P_1V_1 = P_2V_2$$

(These two terms are equal to each other because they both are equal to the same constant.) This means that pressure (P_1) times volume (V_1) under an initial set of conditions equals pressure (P_2) times volume (V_2) under a final set of conditions. Other factors besides pressure influence the volume of a gas. Because of this, it is important to remember that Boyle's law applies to changes in volume due to changes in pressure *only while temperature and amount of gas remain constant.*

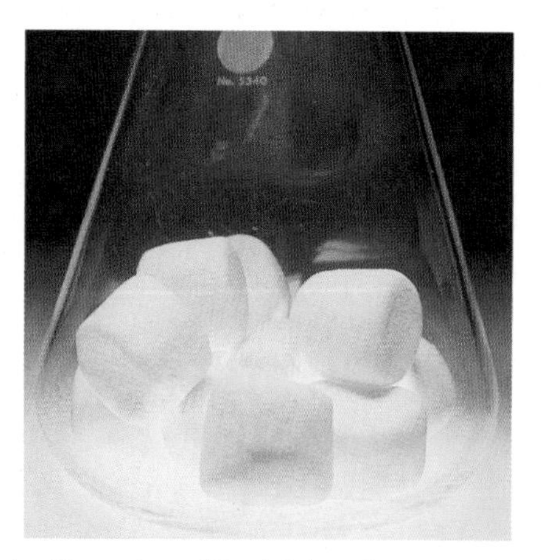

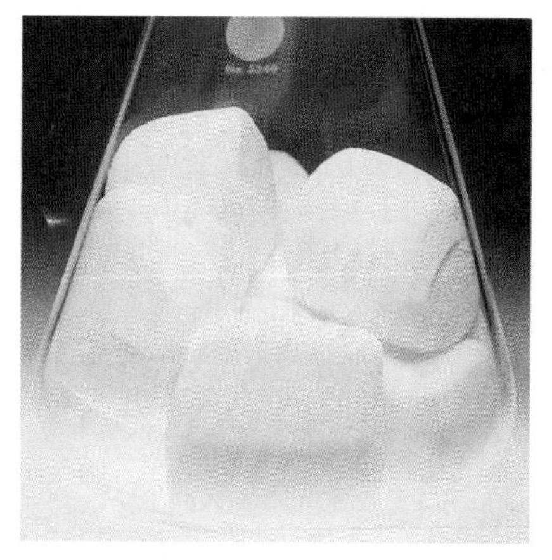

An illustration of Boyle's law. Some marshmallows are placed in a flask *(left)*. The flask is then evacuated of air by using a vacuum pump. The air in the marshmallows expands as the pressure is lowered, causing the marshmallows to expand *(right)*. *(Charles D. Winters)*

Example 5.4 Boyle's Law

A sample of helium has a pressure of 760.0 torr and occupies a volume of 125 mL. What is the volume if the pressure changes to 415.0 torr? (Temperature and amount of gas remain constant.)

SOLUTION

Because the pressure has decreased, the volume of the helium must increase. Use the expression for Boyle's law and rearrange to isolate the unknown, which in this case is the final volume:

$$P_1V_1 = P_2V_2$$

$$V_2 = \frac{P_1V_1}{P_2}$$

Substituting the appropriate values into the expression, and solving yields

$$V_2 = \frac{760.0\ \text{torr} \times 125\ \text{mL}}{415.0\ \text{torr}} = 229\ \text{mL}$$

It is always useful to check your answers. For the gas laws, it is helpful to ask if an answer is reasonable. For this problem, the pressure decreased, and so the volume must increase. The volume we calculated is larger than the original volume, thus the answer is reasonable.

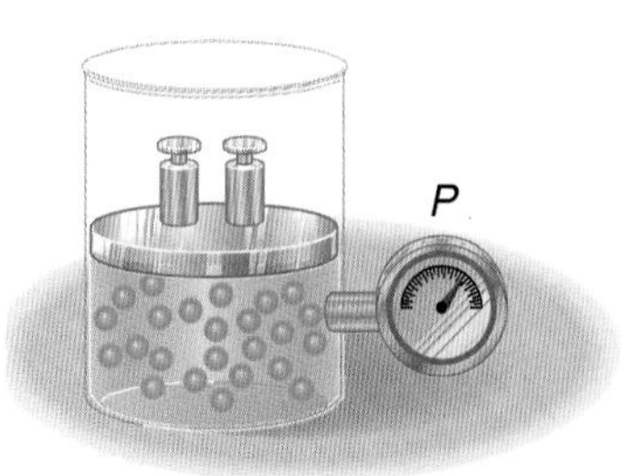

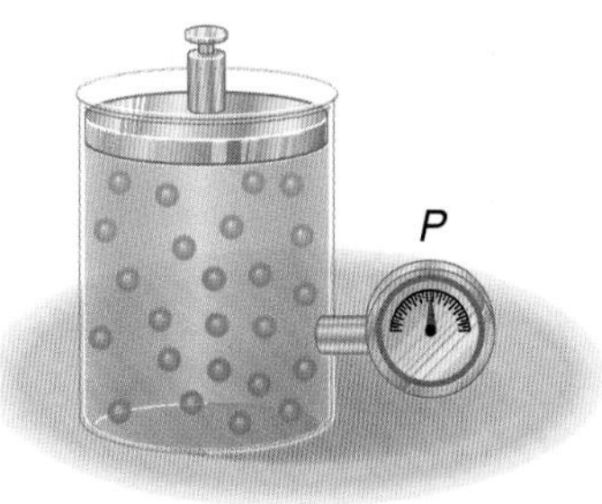

Self-Test

A hydrogen gas sample occupies 3.75 L at 743 torr. What is the pressure when it occupies 2.81 L? (All other factors remain constant.)

ANSWER

992 torr. Note that because the volume decreases, it is reasonable that the pressure increases.

Charles's Law

The volume of a gas is also affected by temperature. In general, volume increases when temperature increases, and it decreases when temperature decreases (Figure 5.7). This relationship between temperature (T) and volume (V) is summarized by **Charles's law:**

$$\frac{V}{T} = \text{constant}$$

If V divided by T is equal to a constant, then if either temperature or volume increases or decreases, the other must also increase or decrease. Like Boyle's law, Charles's law can be rewritten for initial and final conditions. Because V_1/T_1 and V_2/T_2 are equal to the same constant, they can be set equal to each other:

$$\frac{V_1}{T_1} = \frac{V_2}{T_2}$$

(a)

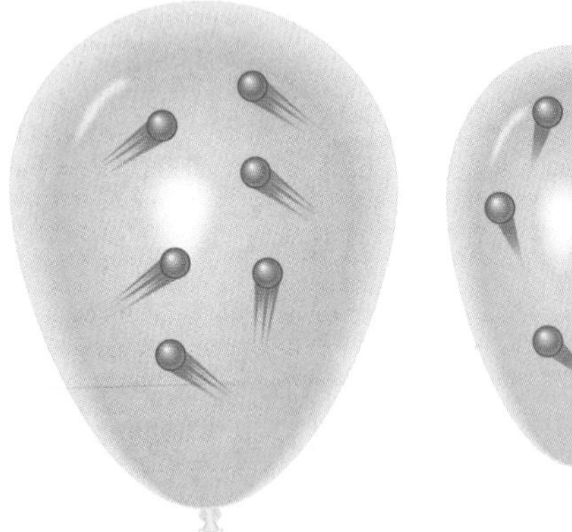

(b) Warm balloon Cold balloon

Figure 5.7 The volume occupied by a gas depends on the temperature of the gas, if other factors are constant. The greater the temperature, the greater the volume (Charles's law). **(a)** The volume of these balloons decreases greatly when they are placed into liquid nitrogen, which is very cold. The balloons reinflate to their original volume as they warm to room temperature. **(b)** Cold gas particles move more slowly and are closer together than gas particles in a warm sample. *(Charles D. Winters)*

Charles's law applies to conditions in which only temperature and volume change; other factors such as pressure and amount of gas must remain constant. When you use Charles's law, you must use kelvin temperature units (K) rather than degrees Celsius. To convert, remember to use this equation:

$$T_K = T_C + 273$$

Volume increases with temperature because as the temperature goes up, the kinetic energy of the gas particles goes up; they move faster. Because the gas particles move faster, they collide more forcefully with each other; thus they push harder against each other. The net effect is the particles are farther apart on the average; stated another way, there is more empty space between them. Look at Figure 5.7(b) again; notice the increased volume results from a larger average distance between the particles.

Example 5.5 Charles's Law

A balloon filled with helium has a volume of 35.0 L at 23°C. If the pressure and amount of gas remain constant, what is its volume at 11°C?

SOLUTION

To use the expression for Charles's law, the temperature must be in kelvins. Convert °C to K this way:

$$T_K = T_C + 273$$

$$T_1 = 23°C + 273 = 296 \text{ K}$$

$$T_2 = 11°C + 273 = 284 \text{ K}$$

The final volume can be determined by first rearranging the Charles's law expression to isolate the unknown:

$$\frac{V_1}{T_1} = \frac{V_2}{T_2}$$

$$V_2 = \frac{V_1 T_2}{T_1}$$

Now substitute the known values into the expression and solve:

$$V_2 = \frac{35.0 \text{ L} \times 284 \text{ K}}{296 \text{ K}} = 33.6 \text{ L}$$

The temperature decreases, and so the volume also decreases.

Self-Test

The balloon in Example 5.5 is taken from its initial conditions to a location at which its volume is 37.6 L. What is the temperature of this location? (The pressure remains constant.)

ANSWER

318 K = 45°C

This weather balloon is filled with helium. Will it expand or contract as it ascends? *(NASA/Science Source/Photo Researchers, Inc.)*

There are other relationships for gases that involve pressure, volume, and temperature. One such relationship—Gay-Lussac's law—involves pressure and temperature:

$$\frac{P_1}{T_1} = \frac{P_2}{T_2}$$

You may already be familiar with examples of this law; driving heats your tires, which increases tire pressure, and temperature increases when the piston of a diesel engine compresses the gases in the cylinder.

Combined Gas Law

Boyle's law and Charles's law apply to changes in volume that result from changes in pressure and changes in temperature, respectively. What happens if both pressure and temperature change? When both of these factors change (but the amount of gas remains constant), then the **combined gas law** is used:

$$\frac{P_1 V_1}{T_1} = \frac{P_2 V_2}{T_2}$$

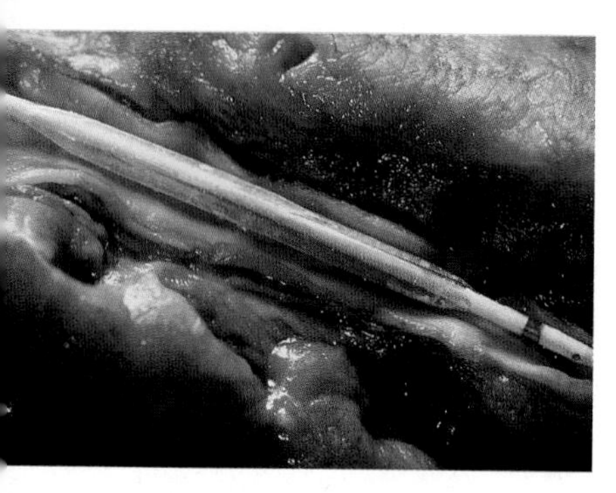

A balloon is inserted in a clogged coronary artery and inflated to open the artery and restore blood flow. *(SIU, Visuals Unlimited)*

Example 5.6 Using the Combined Gas Law

In angioplasty, a balloon is inflated in a blood vessel to increase the diameter of the vessel. Suppose a sample of gas is used to inflate an angioplasty balloon in a room in which the temperature is 21°C and the pressure is 751 torr. After inflation, the balloon has a volume of 0.150 mL. The same size gas sample is used to inflate the balloon when it is in an artery where the temperature is 37°C and the pressure is 875 torr. What is the volume of the balloon in the artery?

SOLUTION

Rearrange the combined gas law expression to solve for V_2:

$$V_2 = \frac{P_1 V_1 T_2}{T_1 P_2}$$

Substitute the known values and solve

$$V_2 = \frac{751 \text{ torr} \times 0.150 \text{ mL} \times 310 \text{ K}}{294 \text{ K} \times 875 \text{ torr}} = 0.136 \text{ mL}$$

The balloon is slightly smaller in the artery than it was under room conditions.

Self-Test

A 7.65-L balloon is released when the temperature is 27°C and the pressure is 754 torr. It rises until the temperature is −12°C and the pressure is 471 torr. What is the volume of the balloon at this point?

ANSWER

10.7 L

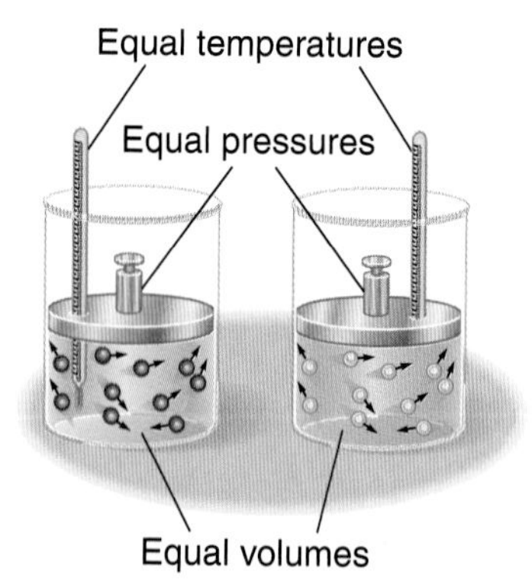

Figure 5.8 Two gas samples of equal volume at the same temperature and pressure contain the same number of particles (Avogadro's law).

Avogadro's Law

One additional factor affects the volume of a gas: the number of particles of the gas. Avogadro (Section 4.5) hypothesized that if two gas samples are at the same temperature, pressure, and volume, they must contain the same number of particles (Figure 5.8). From this work, a relationship **(Avogadro's law)** between volume and moles of gases can be expressed:

$$\frac{V}{n} = \text{constant}$$

In these expressions, V is volume and n is number of moles of the gas. The larger the number of particles, the greater the volume of the gas. Because V/n is equal to a constant, Avogadro's law can be rewritten as:

$$\frac{V_1}{n_1} = \frac{V_2}{n_2}$$

where the subscripts 1 and 2 mean initial and final conditions, respectively.

Example 5.7 Avogadro's Law

A 1.73-mol sample of oxygen occupies 38.8 L. Under the same conditions, what volume would 4.65 mol of oxygen occupy?

SOLUTION

We want to know volume, so begin with Avogadro's law, then rearrange to isolate the unknown:

$$\frac{V_1}{n_1} = \frac{V_2}{n_2}$$

$$V_2 = \frac{V_1 n_2}{n_1}$$

Substitute the known values into the expression and solve

$$V_2 = \frac{38.8 \text{ L} \times 4.65 \cancel{\text{mol}}}{1.73 \cancel{\text{mol}}} = 104 \text{ L}$$

Self-Test

At a certain temperature and pressure, 1.5 mol of a gas occupies 24 L. What volume would 2.0 mol of the gas occupy at the same temperature and pressure?

ANSWER

32 L

The Ideal Gas Law

Avogadro's law can be combined with the other gas laws to yield this expression:

$$\frac{PV}{nT} = \text{constant}$$

This is one form of what is called the **ideal gas law.** (An *ideal* gas would obey this law; *real* gases show slight variations from the behavior predicted by this law.) This law holds for any reasonable values of V, P, n, and T. For example, when 1 mol of a gas is held at STP (0°C, 1 atm), the gas occupies a volume of 22.4 L. This volume occupied by 1 mol of any gas at STP is called the **molar volume** of the gas. When the values for STP and molar volume are used to solve the previous equation, the constant has a specific value and is called the **universal gas constant (*R*):**

$$R = \frac{PV}{nT} = \frac{(22.4 \text{ L})(1 \text{ atm})}{(273 \text{ K})(1 \text{ mol})} = 0.0821 \frac{\text{L} \cdot \text{atm}}{\text{K} \cdot \text{mol}}$$

Note that when the value of 0.0821 L · atm/K · mol is used for R, pressure must be in atmospheres, volume must be in liters, and temperature must be in kelvins. The ideal gas law is usually written in this form:

$$PV = nRT$$

If all but one factor in the ideal gas law is known, the unknown factor can be calculated.

Example 5.8 The Ideal Gas Law

A gas cylinder containing 2.25 L of oxygen is held at 19°C. The pressure in the cylinder is 21.7 atm. How many moles of oxygen are in the cylinder?

SOLUTION

This problem can be solved with the ideal gas law rewritten to solve for n (number of moles):

$$n = \frac{PV}{RT}$$

Pressure and volume are given in the correct units, but temperature must be converted to kelvins:

$$T_K = T_C + 273 = 19°C + 273 = 292\ K$$

Insert the known values into the expression and solve

$$n = \frac{PV}{RT} = \frac{21.7\ atm \times 2.25\ L}{0.0821\ L \cdot atm/K \cdot mol \times 292\ K} = 2.04\ mol$$

Self-Test

If a 1.83-mol sample of a gas is placed in a 5.00-L cylinder and the temperature is 26°C, what is the pressure in the cylinder?

ANSWER

8.98 atm

Example 5.9 Molar Volume of a Gas

A sample of helium gas stored at 0°C and 1 atm of pressure occupies 17.6 L. How many moles of helium are in the sample?

SOLUTION

We could use the ideal gas law because temperature, pressure, and volume are known. However, because the temperature and pressure are at STP, we can also use Avogadro's law and our knowledge about molar volume of a gas:

$$\frac{V_1}{n_1} = \frac{V_2}{n_2}$$

If we let n_2 equal the moles of helium and rearrange the expression to isolate the unknown, we get

$$n_2 = \frac{n_1 V_2}{V_1}$$

Because 1 mol of a gas occupies 22.4 L at STP, we can let 1 mol equal n_1, and 22.4 L equal V_1. The helium volume, 17.6 L, is V_2. Substitute the known values into the expression and solve

$$n_2 = \frac{1.00 \text{ mol} \times 17.6 \cancel{\text{L}}}{22.4 \cancel{\text{L}}} = 0.786 \text{ mol}$$

Self-Test

How many moles of oxygen gas occupy 1.00×10^2 L at STP?

ANSWER

4.46 mol

Dalton's Law of Partial Pressures

The gas laws you have studied so far apply both to a pure gas and to a mixture of gases. An important question that arises, however, concerns the behavior of each gas in a mixture. Do they behave independently, or does the presence of the other gases in the mixture affect each gas in some way? With some exceptions, each gas behaves independently of the other gases, and because of this, the type of particle present does not matter. With respect to the gas laws, mixtures of gases behave like a pure gas.

Consider a balloon filled with a mixture of gases. Each gas in the mixture exerts its own pressure, its **partial pressure,** on the balloon. The total pressure in the balloon is expressed by **Dalton's law of partial pressures,** which states that the total pressure of a mixture of gases is equal to the sum of the pressures individually exerted by all the gases in the mixture (Figure 5.9):

$$P_{\text{total}} = P_1 + P_2 + \ldots P_n$$

Air provides an important example of Dalton's law of partial pressures. Air pressure at sea level and 0°C is 760 torr. Under these conditions, the nitrogen in air has a partial pressure of 593 torr (about 4/5 of the total air pressure), and oxygen has a partial pressure of 160 torr (about 1/5 of the total air pressure). All the other gases together exert a pressure of about 7 torr. Because all of the particles of the gas mixture behave independently of the others, each gas in the mixture can be considered independently. This proves to be a useful concept in health care. When the oxygen needs of a patient are considered, the partial pressure of oxygen in air or in a prepared mixture is the critical factor.

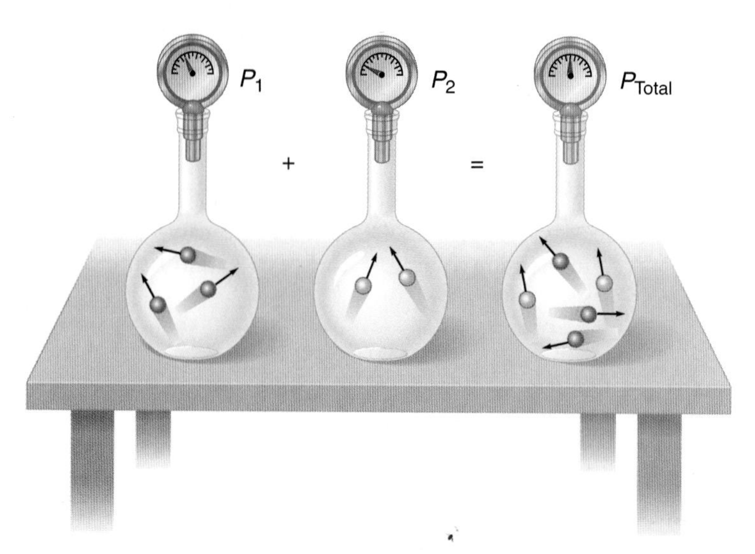

Figure 5.9 The total pressure of a mixture of gases is the sum of the pressures of all the gases in the mixture (Dalton's law of partial pressures).

Connections

Effects of Oxygen Partial Pressure on Oxygen Exchange in the Lungs

Breathing brings oxygen-rich air into the lungs and removes oxygen-poor air at the same time. The partial pressure of oxygen in air at sea level is about 160 torr, but as inhaled air mixes with residual gases in the lungs, the partial pressure of oxygen drops to about 100 torr. This partial pressure of oxygen is high enough to force oxygen from the alveoli of the lungs into the blood. For people with normal respiratory function, blood leaving the lungs is nearly saturated with oxygen.

This is not necessarily the case for patients with impaired respiratory function. Consider a patient with emphysema as an example. Emphysema is characterized by a decrease in both the number and elasticity of alveoli, which decreases gas exchange between the lungs and the blood. In many emphysema patients, a partial pressure of 100 torr is not enough to nearly saturate the blood, and as a result, the patient's activity is limited by this oxygen-poor blood. These patients may be given supplemental oxygen. When this oxygen is inhaled and mixed with the lungs' residual gases, the partial pressure of oxygen in the alveoli increases. At this higher partial pressure of oxygen, more oxygen enters the blood as it flows through the lungs, raising the amount of oxygen carried by the blood. This therapy increases both the comfort and potential for activity of the patient.

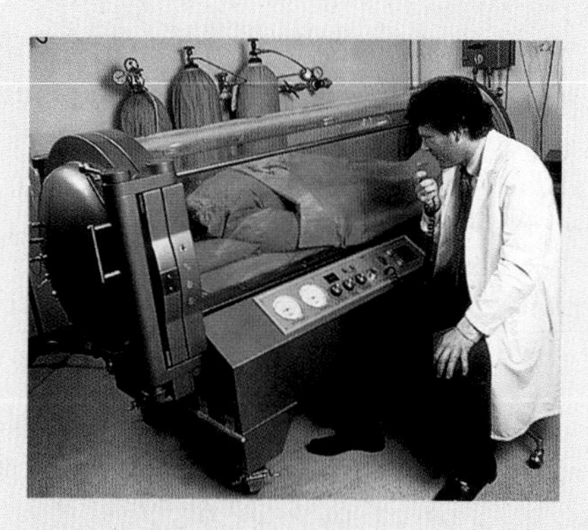

A hyperbaric chamber. The patient in the chamber is exposed to an oxygen-enriched atmosphere, which causes more oxygen to dissolve in the patient's blood. *(Peter Arnold, Inc.)*

Composition of Air

GAS	PARTIAL PRESSURE (mm Hg)	PERCENTAGE OF PRESSURE
Nitrogen	593	78.0
Oxygen	160	21.1
Other gases: including carbon dioxide and water vapor	7	0.9
Total	760	100.0

Example 5.10 Partial Pressure of Gases

A tank containing oxygen is prepared by first adding nitrogen to an empty tank. After adding the nitrogen, the tank pressure is 7.4 atm. Oxygen is then added to bring the total pressure in the tank to 12.0 atm. What is the partial pressure, in atmospheres, of oxygen in the tank?

SOLUTION

Dalton's law of partial pressures can be rearranged to solve for the pressure of oxygen.

$$P_T = P_{\text{nitrogen}} + P_{\text{oxygen}}$$

$$P_{\text{oxygen}} = P_T - P_{\text{nitrogen}}$$

$$P_{\text{oxygen}} = 12.0 \text{ atm} - 7.4 \text{ atm} = 4.6 \text{ atm}$$

Self-Test

In Alamosa, Colorado, at an elevation of 7544 feet, the air pressure is about 585 torr. If the partial pressure of nitrogen is 456 torr at that elevation and the pressure of all the other gases is 5 torr, what is the partial pressure of oxygen in the air?

ANSWER

124 torr. (The partial pressure of oxygen at sea level is 160 torr. The reduced partial pressure of oxygen at high altitudes has serious effects on patients with reduced respiratory capacity and also greatly influences aerobic athletic performance.)

5.5 Solids and Liquids

What are the differences among solids, liquids, and gases? Why do these differences exist? Let's compare gases to the other two states of matter to see if the comparison provides any clues. When pressure is applied to a gas, the volume of the gas decreases. Gases are compressible because a gas sample contains mostly empty space. When pressure is applied, the gas particles are forced closer together, which simply reduces the distance and empty space between them. When pressure is applied to liquids and solids, on the other hand, there is little or no change in their volume. This is because there is little empty space between the molecules or ions in liquids and solids. Because there is only a small distance between particles in solids and liquids, the particles are bonded to each other. It is this attraction that gives solids and liquids properties that differ from those of gases.

Intermolecular Forces

You are familiar with covalent bonds in molecules, bonds *within* a molecule, which can also be referred to as *intramolecular bonding*. Forces or bonds that exist *between* molecules in solids and liquids are called **intermolecular forces** or intermolecular bonds. Like all other forces at the atomic or molecular level, intermolecular forces involve attraction of opposite charges. How can this be? Cations and anions, of course, have opposite charges, and they are held together in a solid by ionic bonds, but how can molecules have opposite charges? The answer is they do not have opposite full charges, but they can have opposite *partial* charges. Think back to the discussion of polar molecules in Chapter 3. Polar molecules have dipoles, which means partial charges exist in the molecule. In a solid or liquid, polar molecules are close together and align themselves so that the oppositely charged ends of their dipoles are next to each other. The result is a **dipole–dipole interaction,** which is one of several kinds of intermolecular forces (Figure 5.10). Polar molecules bond to each other in liquids and solids through these weak electrical attractions. Dipole–dipole interactions are much weaker than covalent bonds, but they are significant in the solid and liquid states. In gases, these interactions are too weak to prevent the gas molecules from flying apart.

Hydrogen bonds, which are a special kind of dipole–dipole interaction, are a second type of intermolecular force. A **hydrogen bond** occurs between a very electronegative atom in one molecule and a hydrogen atom that is covalently bound to a very electronegative atom in another molecule (Figure 5.11). Highly electronegative atoms—oxygen, nitrogen, and fluorine—are necessary to form hydrogen bonds. Hydrogen bonds are somewhat stronger than other dipole–dipole interactions, and, as you will see, they play a very important role in biochemistry.

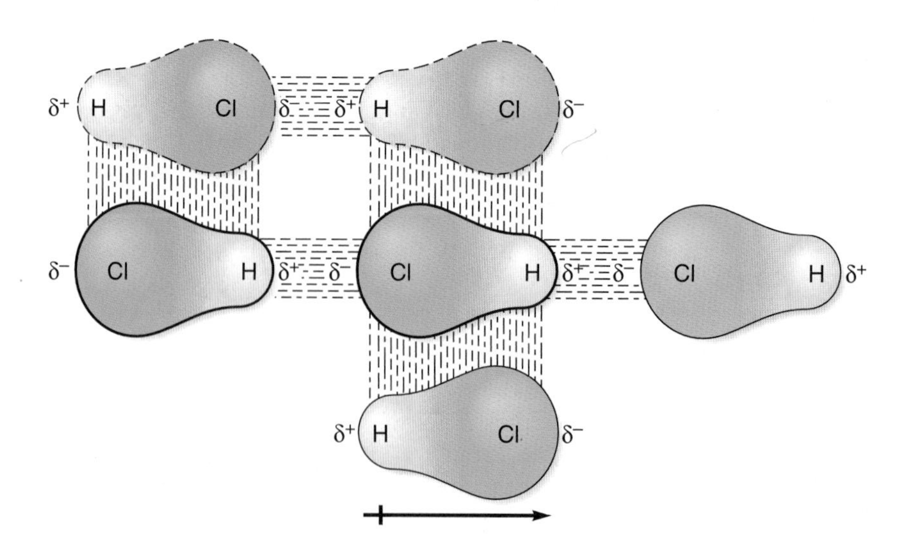

Figure 5.10 Dipole–dipole interactions form between oppositely charged ends of polar HCl molecules in the liquid or solid state.

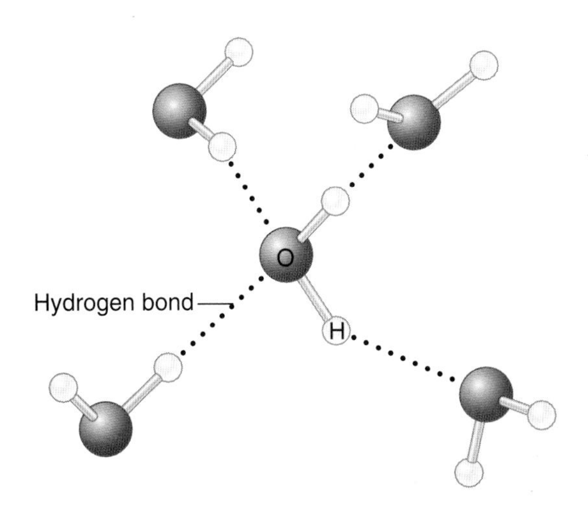

Figure 5.11 Hydrogen bonds form between the hydrogen atoms in these water molecules and an oxygen atom in adjacent water molecules. Hydrogen bonds are often shown as dashes or dots connecting the atoms of the hydrogen bond.

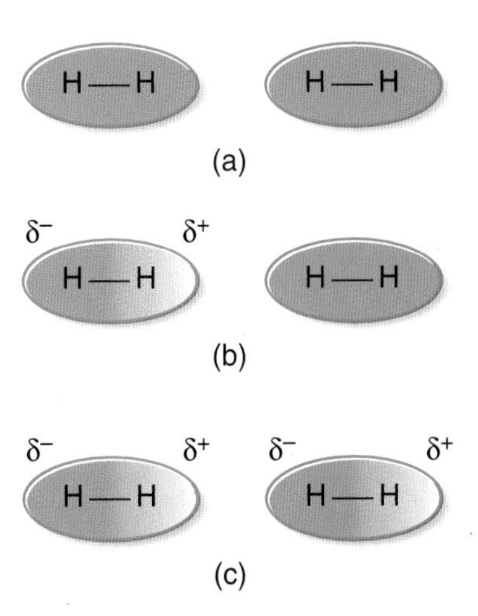

Figure 5.12 London dispersion forces between nonpolar molecules. Color intensity is used to show electron distribution. **(a)** On the average, the electrons are distributed equally around atoms in nonpolar molecules. **(b)** Sometimes electrons are temporarily distributed unevenly, as in the left-hand molecule, which results in a tiny dipole in the molecule. **(c)** When this happens, the dipole induces a tiny, temporary dipole in adjacent molecules.

Polar molecules bond to each other through dipole–dipole interactions or hydrogen bonds. How do nonpolar molecules interact in solids and liquids? They bond to each other through a third type of intermolecular interaction called **London dispersion forces.** These forces involve the temporary interaction between a *temporary dipole* and an *induced dipole.* (In this phrase, *induced* means one dipole *causes* another dipole.) These forces arise in this way. On average, the electrons in a nonpolar molecule are shared evenly throughout the molecule [Figure 5.12(a)]. However, at any brief moment in time, the electrons may be unevenly distributed, so that a *temporary dipole* exists in the molecule [Figure 5.12(b)]. This dipole can *induce* (cause) a temporary dipole in an adjacent molecule [Figure 5.12(c)]. A temporary dipole-induced dipole interaction then forms between the two molecules. To some extent, the strength of London dispersion forces depends on the number of electrons in the nonpolar molecules. The more electrons in the molecule, the greater the potential for interactions between temporary dipoles and induced dipoles. London dispersion forces range from very weak—like those between helium atoms or methane molecules—to quite strong—like those between paraffin or iodine molecules. Table 5.7 summarizes intermolecular forces.

Table 5.7 Intermolecular Forces

FORCE OR BOND	STRENGTH RELATIVE TO OTHER INTERMOLECULAR FORCES	BASIS FOR ATTRACTION	EXAMPLE
Dipole–dipole	Intermediate	Attraction between oppositely charged permanent dipoles in polar molecules	H–Cl · · · H–Cl
Hydrogen bond	Strong	Attraction between an atom of O, N, or F and a hydrogen atom covalently bonded to O, N, or F	H \\ O · · · H / \\ H O / H
London dispersion forces	Weak to strong	Attraction between temporary dipoles in nonpolar molecules	I–I · · · I–I

Figure 5.13 The atoms in these crystals of quartz are bonded strongly to their neighboring atoms. They are not free to move, and thus form a rigid, solid structure. *(Robert DeGuglieno/Science Photo Library/Photo Researchers, Inc.)*

Solids

A **solid** has a definite shape and a definite volume. The definite volume is consistent with the idea that there is little empty space between the atoms, molecules, or ions in solids. Applied or reduced pressure cannot significantly reduce or increase the volume of a solid.

Solids have a definite shape because the particles in solids are bound to the neighboring particles through a variety of attractions. The forces that hold these particles together prevent the particles from moving around. They are fixed in space relative to neighboring particles. Because they cannot move, the shape of the solid is fixed (Figure 5.13).

Ionic Solids

In ionic solids, the force holding the particles (ions) together is ionic bonding. Look at the structure of a crystal of sodium chloride in Figure 5.14. Each sodium ion is surrounded by six chloride ions. Similarly, each chloride ion is surrounded by six sodium ions. Because opposite charges attract, each sodium ion is attracted to the neighboring chloride ions and vice versa. There are also repulsive forces present in a crystal. The chloride ions repel each other, and the sodium ions repel each other. Given these repulsions, why do the ions remain in the crystal? Because the sum of all the attractive forces in the crystal is greater than the sum of all the repulsive ones. There is a *net attraction* that holds the ions together.

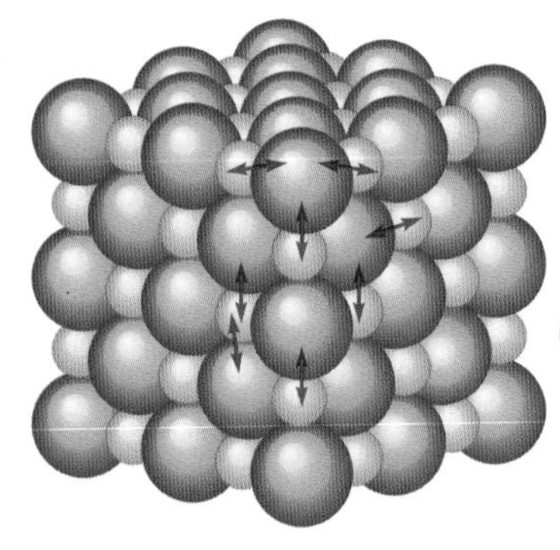

Figure 5.14 A model of a sodium chloride crystal. In the crystal each sodium ion (tan) is adjacent to six chloride ions (green), which in turn are surrounded by six sodium ions. Because oppositely charged ions attract, the ions are held rigidly in place.

Look again at the crystal in Figure 5.14. Note that most of the volume of the crystal is taken up by the ions; there is little empty space in the crystal. As pointed out at the beginning of this section, this lack of empty space is the reason solids cannot be compressed.

Molecular Solids

In molecular solids, the molecules bond to each other by some type of intermolecular force: dipole–dipole interactions, hydrogen bonds, or London dispersion forces. Consider ice as an example. Each water molecule is bound by four hydro-

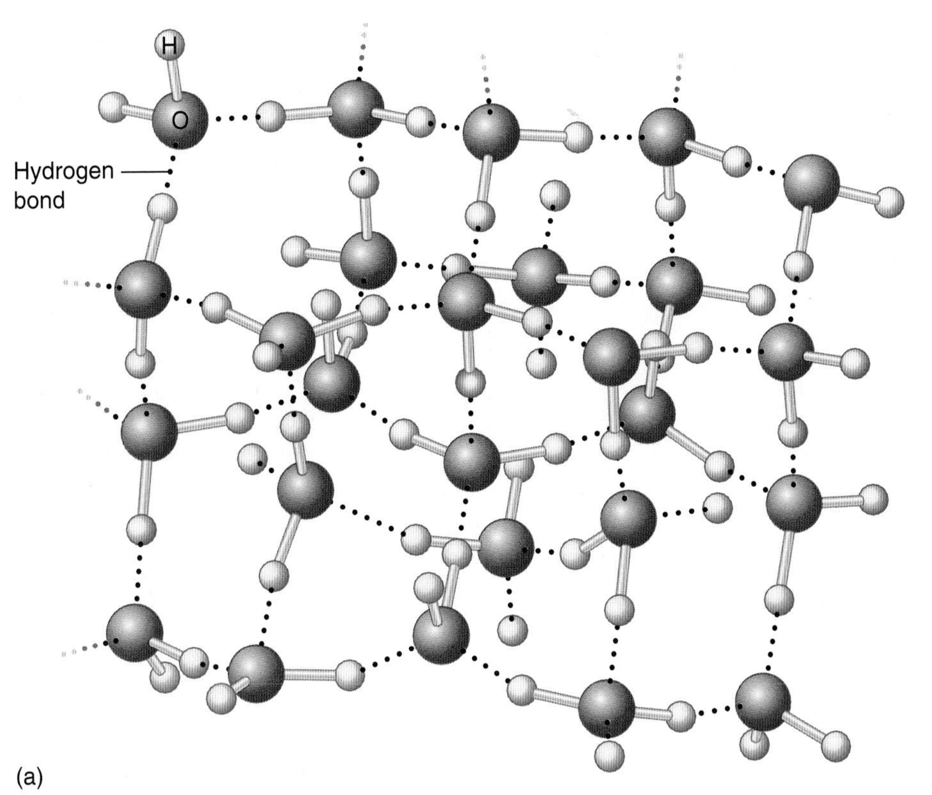

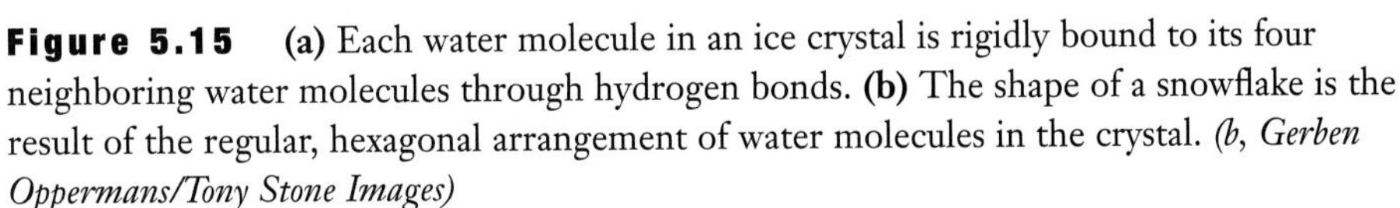
Figure 5.15 **(a)** Each water molecule in an ice crystal is rigidly bound to its four neighboring water molecules through hydrogen bonds. **(b)** The shape of a snowflake is the result of the regular, hexagonal arrangement of water molecules in the crystal. *(b, Gerben Oppermans/Tony Stone Images)*

gen bonds to four neighboring water molecules (Figure 5.15). Each water molecule is held in place by these bonds to its neighbors. Because the water molecules cannot move about, the ice has a definite shape and cannot expand. As with ionic solids, there is a limited amount of empty space in molecular solids; thus they cannot be compressed. Like other solids, they have a definite volume.

Network Solids

A third type of solid is the network solid. In these solids, each atom is covalently bonded to its neighboring atoms. These neighbors are in turn covalently bonded to their neighbors. In effect, each atom is bonded to all other atoms of the sample by a network of covalent bonds. Diamond (Figure 5.16) is an example of a network solid. As with the other two types of solids, the atoms in a network solid are held rigidly in place, and so the shape of the solid is fixed. Because there is little empty space, the volume does not change.

Metallic Solids

Metals are a fourth type of solid. Metallic solids consist of metal atoms packed closely together. The metal atoms are said to be *closest packed*, just like oranges stacked in a grocery store or marbles layered in a box. The valence electrons of an individual metal atom are held only loosely by that atom. In a metal, all of these

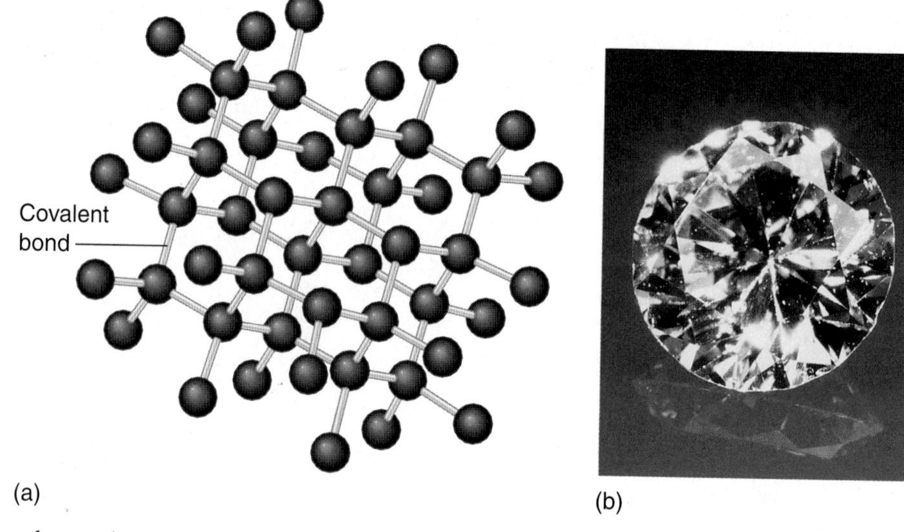

Figure 5.16 Diamond is an example of a network solid. **(a)** Each carbon atom in a diamond is covalently bound to its four neighboring carbon atoms. A network made up of all these covalent bonds effectively attaches all of the atoms in a diamond to each other. **(b)** As a result, diamond is the hardest known material. *(b, Charles D. Winters)*

valence electrons form a collection of electrons that are somewhat independent of the nuclei and core electrons of the atoms. The valence electrons form a "sea of electrons" that surround and bathe the atoms (Figure 5.17). Metals are good conductors of electricity because of these loosely held electrons. Because the atoms have in effect lost their outer electrons, they can be viewed as cations consisting of the nucleus and core electrons. The attraction between these cations and the sea of electrons effectively holds a metal sample together.

Each metal atom is close to its neighbors but not really bound to them. Thus metal atoms can move around each other more easily than the atoms, ions, or molecules in other types of solids. As a result, the shapes of metals can be changed more easily than the shapes of other solids. Metals are *ductile*, which means they can be pulled into wires, and *malleable*, which means they can be hammered into flat sheets (Figure 5.18).

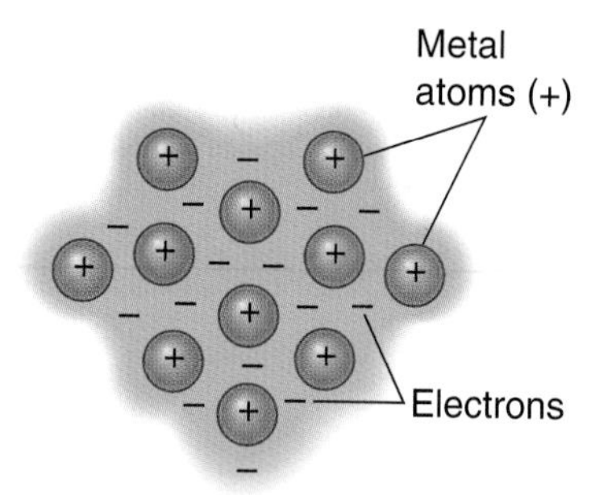

Figure 5.17 In metals, the outer electrons of the metal atoms form a sea of electrons that surrounds the nucleus and core electrons of the atoms. Bonding in metals is due to the attraction between the sea of electrons and the charged nucleus and core electrons.

Liquids

Liquids have a definite volume but an indefinite shape. In other words, their volume is fixed, but they can assume the shape of their container. What is it about the structure of a liquid that accounts for these properties? As in a solid, the particles of a liquid are close together, so that little empty space is between them. As a result, pressure changes do not cause significant changes in volume, which accounts for the fixed-volume property of liquids.

Why do liquids vary in shape while solids have a definite shape? This difference is due to the difference in intermolecular bonding found in liquids and solids. In solids, the particles are close together, and they are permanently bonded to each of their neighboring particles through the appropriate interaction—ionic bonds in ionic solids; dipole–dipole interactions, hydrogen bonds, and London forces in molecular solids; covalent bonds in network solids; and metallic bonds in metals. In liquids, the particles are also close together, but each particle is not bonded permanently to its neighbors. Instead, only some bonds are present to the neighboring particles, and these bonds can break while new ones form (Figure 5.19). As a consequence, with time each particle is free to move away from its neighboring particles and bond with new ones. Because each particle is somewhat free to move, the sample has no definite shape.

Figure 5.18 Metals are **(a)** malleable and **(b)** ductile. *(a, Stephen Saks/Photo Researchers; b, Peter Poulides/Tony Stone Worldwide)*

Let's use a sample of water to compare intermolecular forces in solids, liquids, and gases. At low temperatures, there is not enough energy to break any of the hydrogen bonds between adjacent water molecules. Therefore, each molecule is bound to four neighboring molecules, and the sample exists as solid ice. At 0°C, there is enough energy to break a few of these bonds. Now some water molecules are free to move about because they are no longer rigidly bound to their neighbors. The sample is now fluid rather than rigid, thus solid ice becomes liquid water at 0°C. At 100°C there is enough energy to break all of the hydrogen bonds between water molecules, and the molecules are now independent of each other. Liquid water changes to steam at 100°C.

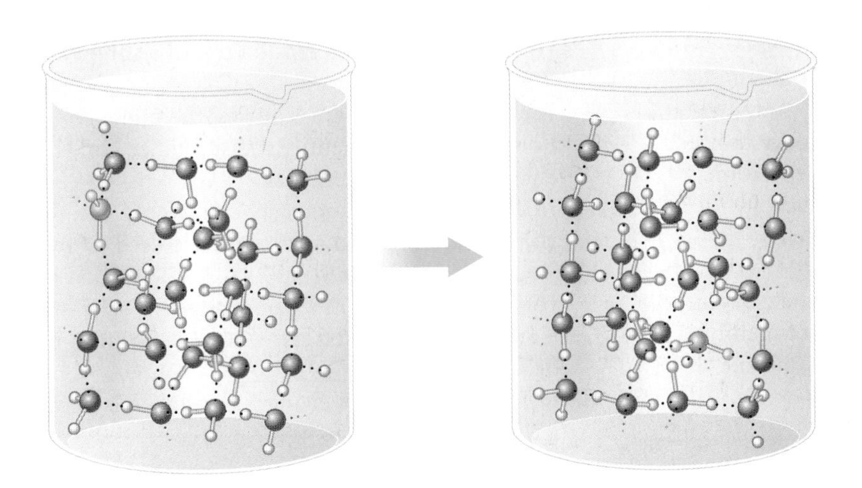

Figure 5.19 The molecules in water are bound to their neighboring molecules by temporary intermolecular forces. As a result, the water molecules can move around each other, and so the liquid has no fixed shape.

CHAPTER SUMMARY

Matter exists in either the gaseous, liquid, or solid **states.** Each state can change to any other state, and these **changes in state** involve energy. Energy is required to **melt** a solid or **vaporize** a liquid, and energy is released when a gas **condenses** or a liquid **freezes. Heat of fusion** is the energy involved in solid–liquid transitions, and **heat of vaporization** involves liquid–gas transitions. When matter in a state is heated or cooled it gains or loses energy, and its temperature increases or decreases. A **gas** has neither a definite shape nor volume because its particles are independent of each other. Gases exert **pressure,** and the volume and pressure of a gas are inversely proportional to each other **(Boyle's law).** The relationship between volume and temperature of a gas is expressed by **Charles's law,** and the volume of a gas and the number of particles in the gas sample are related by **Avogadro's law.** These three laws can be combined into the **ideal gas law:** $PV = nRT$. **Dalton's law of partial pressures** states that the sum of the **partial pressure** of each gas in a gas mixture is equal to the total pressure of the mixture. The particles in liquids and solids are bound to each other by **intermolecular forces** such as **dipole–dipole interactions, hydrogen bonds,** and **London dispersion forces.** Because the particles in solids are bonded continuously to their neighboring particles, a **solid** has a fixed shape and volume. The bonds between particles in a liquid are transient; thus the particles are free to move past neighboring particles. As a result, a **liquid** has a definite volume but no definite shape.

KEY TERMS

Section 5.1
changes in state
condensation
deposition
freeze
freezing point
heat of fusion
heat of vaporization
melting
melting point
normal boiling point
states
sublimation
vaporization

Section 5.2
specific heat

Section 5.3
gas
pressure
standard temperature and pressure (STP)

Section 5.4
Avogadro's law
Boyle's law
Charles's law
combined gas law
Dalton's law of partial pressures
ideal gas law
molar volume
partial pressure
universal gas constant (*R*)

Section 5.5
dipole–dipole interaction
hydrogen bond
intermolecular forces
liquid
London dispersion forces
solid

EXERCISES

Changes in State of Matter *(Section 5.1)*

1. Are changes in state physical changes or chemical changes? Justify your answer.
2. Describe the process of deposition. Identify the change in state that is the opposite of deposition.
3. In your own words, describe what is meant by the phrase *normal boiling point.* Name the change of state that is the opposite of boiling.
4. Explain what occurs when a substance melts. Name the change of state that is the opposite of melting.
5. For these events, identify the substance and the change in state that occurs.
   (a) Wet clothes dry on a clothes line.
   (b) A whitish layer forms on top of a soup cooling in the refrigerator.
   (c) Dew forms on grass during the night.
6. For these events, identify the substance and the change in state that occurs.
   (a) Ice forms on a moustache on a cold day.
   (b) A soldering iron and solder are used to attach a wire in a circuit.
   (c) The salt water in a pan changes to salt.
7. The energy gained or lost when a substance melts or freezes is called __________.
8. Describe the energy changes that take place when ice is placed in a beverage.

9. How much energy is required to melt 75 g of ice that is at 0°C?
10. A can of soda is put into a refrigerator that is too cold, and the contents freeze. Calculate the amount of energy given up by the 355 mL of water in the soda during the freezing process.
11. The name of the energy gained or lost when a substance vaporizes or condenses is ________.
12. Explain the energy changes that occur when alcohol applied to skin evaporates.
13. A 455-g sample of water at 100°C boils away to steam. How much energy is required for this change?
14. Calculate the amount of energy released when 1.00 kg of steam at 100°C condenses to water within a heating system.
15. Determine the energy change associated with these changes in state.
    - **(a)** 11.5 g of rubbing alcohol boils.
    - **(b)** 37.5 g of ice melts.
    - **(c)** 425 g of methane condenses.
    - **(d)** 183 mg of mercury freezes.
16. Determine the energy change associated with these changes in state.
    - **(a)** 138 g of steam condenses.
    - **(b)** 261 mg of solid mercury melts.
    - **(c)** 8.36 g of rubbing alcohol freezes.
    - **(d)** 52 g of liquid methane boils.

Changes in Temperature (Section 5.2)

17. Explain the meaning of the term *specific heat.*
18. Twenty-gram samples of water and alcohol are heated at the same rate. The temperature of the alcohol sample increases more rapidly than the temperature of the water sample. Why?
19. A 3.00-kg sample of water is heated from 14°C to 53°C. How much energy is absorbed by the water?
20. A hot piece of aluminum is cooled by placing it in water. Calculate the amount of energy lost by the 425-g piece of aluminum as it cools from 112°C to 53°C.
21. Calculate the heat gained or lost for these events:
    - **(a)** 24 g of water is heated from 11°C to 57°C.
    - **(b)** 225 g of ice is cooled from 0°C to −43°C.
    - **(c)** 385 g of paraffin wax is cooled from 79°C to 21°C.
22. Calculate the heat gained or lost for these events:
    - **(a)** 28 kg of air is heated from 17°C to 23°C.
    - **(b)** 83.4 g of water is cooled from 93.5°C to 24.2°C.
    - **(c)** 32 g of steam is heated from 110°C to 250°C.
23. Assess the change in heat that occurs during each of these events.
    - **(a)** 0.482 g of ethyl alcohol is heated from 19°C to 37°C.
    - **(b)** 1275 g of lead is cooled from 327°C to 165°C.
    - **(c)** 875 g of ice at −14°C is heated and melted to yield water at 0°C.
24. Assess the change in heat that occurs during each of these events.
    - **(a)** 21.7 g of 23°C water is heated and vaporized to steam at 100°C.
    - **(b)** 47.1 g of ice at −37°C is heated and melted to yield water at 0°C.
    - **(c)** A 6.9-g sample of steam at 150°C is cooled and condensed to water at 100°C.
25. Draw and label the heating curve representing the changes that occur as −40°C ice is converted to 150°C steam.
26. Draw and label the cooling curve representing the changes that occur as steam at 115°C is cooled to ice at −10°C.

Gases (Section 5.3)

27. Matter that has no definite shape or volume is in what physical state? List three substances or mixtures that have this physical state.
28. What is present in liquids and solids but lacking in gases that allows gas molecules to be independent?
29. Use the motion of gas particles to explain why gases exert pressure. Why would the pressure double if the amount of a gas in a closed container doubled?
30. Name two units that are commonly used for pressure in the metric system.
31. "STP" is an abbreviation used for a set of conditions. State the names and values for the conditions summarized by STP.
32. Explain how you would read the pressure from a Torricellian barometer. If the mercury were halfway between the 762 and 763 marks, what is the atmospheric pressure?
33. An atmospheric pressure of 746 mm Hg is measured with a barometer. Express this pressure in:
    - **(a)** Torr **(b)** Atmospheres
    - **(c)** Inches of mercury
34. Describe how a manometer is used to measure pressure.

Gas Laws (*Section 5.4*)

35. State Boyle's law using your own words. What symbols are used in Boyle's law? Would a Boyle's law calculation be valid if the temperature changed during the measurements? Why or why not?
36. If the pressure on a gas sample increases, what will happen to the volume of the sample? If the volume of a gas sample increases, what will happen to the pressure of the sample?
37. If the volume of a gas is reduced to a third of the original volume, what will happen to the pressure? If the pressure of a gas is decreased by 50%, what will happen to its volume?
38. Weather balloons carry instruments into the upper atmosphere to gather information. **(a)** Assuming no other conditions change, what will happen to the volume of the balloon as it rises? **(b)** What can you say about the atmospheric pressure where the weather balloon is twice the size as when it was released?
39. Use Boyle's law to calculate the value of the unknown pressure or volume.
    (a) $V_1 = 231$ mL, $P_1 = 757$ torr; $V_2 = 286$ mL, $P_2 = ?$
    (b) $P_1 = 1.13$ atm, $V_1 = 841$ mL; $P_2 = 1.18$ atm, $V_2 = ?$
    (c) $V_1 = 3.7$ L, $P_1 = 2.6$ atm; $V_2 = 2.1$ L, $P_2 = ?$
    (d) $P_1 = 212$ torr, $V_1 = 1.7$ L; $P_2 = 459$ torr, $V_2 = ?$
40. Use Boyle's law to calculate the value of the unknown pressure or volume.
    (a) $P_1 = 19.2$ atm, $V_1 = 18$ L; $P_2 = 7.1$ atm, $V_2 = ?$
    (b) $V_1 = 25.7$ L, $P_1 = 783$ torr; $V_2 = 8.43$ L, $P_2 = ?$
    (c) $V_1 = 125$ mL, $P_1 = 1.05$ atm; $V_2 = 137$ mL, $P_2 = ?$
    (d) $P_1 = 87.4$ torr, $V_1 = 26.3$ mL; $P_2 = 113$ torr, $V_2 = ?$
41. A gas occupies 11.9 L at 745 torr. What volume will it occupy when the pressure is 816 torr?
42. A gas occupies 413 mL at 1.1 atm. What pressure would yield a volume of 655 mL?
43. State Charles's law in your own words. What effect would a change in pressure have on the measurements of a Charles's law experiment?
44. What will happen to a balloon's volume if it is moved from a hot room to the outside on a very cold day?
45. What would happen to a balloon if it were held in the hot air coming out of a heating register?
46. Use Charles's law to calculate the value of the unknown temperature or volume.
    (a) $T_1 = 23.4$°C, $V_1 = 476$ mL; $T_2 = 38.4$°C, $V_2 = ?$
    (b) $V_1 = 22.4$ L, $T_1 = 13$°C; $V_2 = 23.8$ L, $T_2 = ?$
    (c) $T_1 = 303$ K, $V_1 = 2.74$ L; $T_2 = 251$ K, $V_2 = ?$
    (d) $V_1 = 31.9$ mL, $T_1 = 278$ K; $V_2 = 18.3$ mL, $T_2 = ?$
47. Use Charles's law to calculate the value of the unknown temperature or volume.
    (a) $V_1 = 2575$ L, $T_1 = 18$°C; $V_2 = 2611$ L, $T_2 = ?$
    (b) $T_1 = 467$ K, $V_1 = 83.26$ mL; $T_2 = 525$ K, $V_2 = ?$
    (c) $T_1 = 203$°C, $V_1 = 81.7$ L; $T_2 = 92.6$°C, $V_2 = ?$
    (d) $V_1 = 316$ mL, $T_1 = 311$ K; $V_2 = 277$ mL, $T_2 = ?$
48. A gas at 22°C occupies 1.83 L. What will be its volume at −43°C?
49. You are going to have a party at your house and you tie a balloon in front to help people locate your house. The balloon had a volume of 27 L when you tied it up in the 14°C morning air. Later in the day, the temperature rises to 31°C. What is the volume of the balloon at this temperature?
50. In words or symbols, state the combined gas law.
51. A balloon is transferred to a location where the temperature is higher and the pressure lower than the original location. Will the balloon's volume increase or decrease? Justify your answer.
52. A balloon is transferred to a location where the temperature is lower and the pressure lower than the original location. Will the balloon's volume increase or decrease? Justify your answer.
53. A 7.2-L gas sample is held at 283 K and 1.6 atm. What will be the sample's volume at 238 K and 2.2 atm?
54. A 1150-L balloon is launched where the pressure is 762 torr and the temperature is 29°C. The balloon rises until the pressure is 289 torr and the temperature is −36°C. What is its volume under these conditions?
55. A gas sample occupies 61.9 mL at 755 torr and 21°C. Conditions are changed to give a temperature of 28°C and a volume of 52.7 mL. What is the pressure of the gas now?

56. A balloon has a volume of 8.2 L at 17°C and 764 torr. When released into the air, the balloon has a volume of 11.6 L, where the pressure is 513 torr. What is the temperature of the balloon?

57. A 432-L balloon is launched when the temperature is 27°C and the pressure is 753 torr. It rises to an elevation where the temperature and pressure are now −8°C and 516 torr. What is the volume of the balloon?

58. A 2.80-L balloon is released when the temperature is 34°C and 748 torr. When it stops rising, the balloon has a volume of 3.10 L and the temperature is 13°C. What is the pressure at this altitude?

59. Avogadro's law is another of the gas laws. How does this law differ from Boyle's and Charles's laws?

60. Explain why increasing the amount of gas in a balloon causes the balloon to expand, assuming temperature and pressure remain constant. (Hint: Read the beginning of Section 5.3 again.)

61. At a fixed temperature and pressure, 2.6 mol of a gas occupies 51 L. What volume would 3.3 mol occupy?

62. A 0.47-mol gas sample occupies 10.6 L. How many moles would be in a gas sample that occupies a volume of 17.9 L? The pressure and temperature remain constant.

63. In words or symbols, state the ideal gas law.

64. At STP, what is the molar volume of a gas? Use the ideal gas law to calculate the molar volume of a gas.

65. A 0.63-mol sample of a gas is put into a 10.0-L tank and held at 22°C. What will be the pressure in the tank?

66. A 0.17-mol sample of a gas is put into a balloon when the temperature is 26°C and the pressure is 757 torr. What is the volume of the balloon?

67. How many mol of gas are in a sample that occupies 345 mL at a pressure of 772 torr and 18°C?

68. What is the temperature of a 1.02-mol gas sample that occupies 17.3 L at 753 torr?

69. What is meant by the term *partial pressure* of a gas? Use air as an example in your explanation.

70. A tank contains 2.6 mol of nitrogen and 1.3 mol of oxygen. If 0.65 mol of oxygen were added to the tank, what effect would this have on the pressure in the tank?

71. Oxygen is added to an empty tank until the pressure in the tank is 11 atm. Nitrogen is then added to the tank until the total pressure is 27 atm. What is the partial pressure of each gas?

72. For a particular air sample that has a pressure of 756 torr, the partial pressure of nitrogen is 597 torr and the partial pressure of oxygen is 151 torr. Determine the partial pressure exerted by all the other gases in the sample.

Solids and Liquids (Section 5.5)

73. Name the general term that is used to describe the forces that occur between molecules in solids and liquids.

74. Describe dipole–dipole interaction using a common substance as an example.

75. What is a hydrogen bond? What common substance has hydrogen bonding between its molecules in the liquid and solid state?

76. Describe how London dispersion forces arise at the atomic or molecular level. What type of compound would have London dispersion forces between molecules in the liquid and solid states?

77. A sample of matter that has a definite shape and definite volume is a __________.

78. What forces are present in an ionic solid?

79. List the forces that may be present in a molecular solid.

80. Explain the bonding in a network solid.

81. Metals are good conductors of electricity. Why?

82. Why are metals malleable and ductile?

83. Identify the state of matter that has a definite volume and indefinite shape.

84. Use the differences between intermolecular bonding in solids and liquids to explain the difference in shapes of solids and liquids.

Challenge Exercises

85. Determine the amount of energy needed to change 25.0 g of ice at −35°C to steam at 145°C. 180°C

86. A balloon has a volume of 4.75 gal at 72°F and 29.92 in of mercury. What volume in liters will the balloon occupy at 13°C and 768 torr?

87. How many oxygen molecules and oxygen atoms will be in 45.0 L of oxygen at 24°C and 753 torr?

6 Solutions, Dispersions, and Suspensions

MIXTURES DIFFER IN COMPOSITION, CONCEN-TRATIONS, AND PROPERTIES.

IN CHAPTER 5 we studied mostly pure gases, liquids, and solids. However, most of Earth's matter exists as mixtures rather than pure substances. In this chapter we will examine mixtures and their properties, as well as the intermolecular forces that are present in the mixture. You will find that solutions, dispersions, and suspensions have important roles in health care.

Outline

The water in the oceans is a solution, a dispersion, and a suspension. (*Sharon Cummings/Dembinsky Photo Associates*)

6.1 Solutions and Solubility

Solutions

Solutions are the most important type of mixture. A **solution** is a homogeneous mixture of two or more pure substances. (A *homogeneous mixture* is one that is the same throughout, whereas in a *heterogeneous mixture* the composition varies from one part of the mixture to another.) The most abundant component of a solution is called the **solvent.** The solvent in saltwater or sugar water is water. Water is a very common solvent, and solutions prepared with water are called *aqueous solutions.* The other component(s) is called the **solute** (Figure 6.1). We say that a solute is dissolved in the solvent. Sugar and salt are the solutes in sugar water and saltwater. Blood plasma is a solution of amino acids, sugar, salt, and many other solutes dissolved in water.

Figure 6.1 A solution consists of one or more solutes dissolved in a solvent. *(Charles D. Winters)*

Each of the three states of matter can be either the solute or the solvent of a solution. Carbonated water consists of the gas carbon dioxide (solute) dissolved in liquid water (solvent). Tincture of iodine consists of solid iodine (solute) dissolved in alcohol (solvent). Alloys such as brass and bronze consist of solid metals (solute) dissolved into other solid metals (solvent). Alloys are formed by first melting the individual metals, which are mixed while they are liquids. The mixture is then cooled to yield the solid alloy.

Solubility

Not all possible solutes dissolve in all possible solvents to yield all possible solutions. Stated another way, substances differ in solubility. The **solubility** of a substance is the tendency of that substance to dissolve into another substance. What factors determine solubility? The solubility of a solute in a solvent is determined by the intermolecular forces that occur between particles of solute and solvent.

Let's consider saltwater (sodium chloride in water) as an example of a solution. First consider a crystal of salt [Figure 6.2(a)] with its sodium and chloride ions attracted to each other by ionic bonds. Then consider water with its water molecules hydrogen bonded to neighboring water molecules [Figure 6.2(a)]. What happens when these pure substances are mixed? New forces called *ion–dipole interactions* occur between the water molecules (dipoles) and the ions [Figure 6.2(b)]. These interactions are strong enough to overcome the ion–ion interactions in the crystal and the hydrogen bonding between water molecules. The ions leave the crystal, are surrounded by water molecules, and are thus in solution [Figure 6.2(c)]. These solutions have particle interactions that were not present in the pure solvent or pure solute.

Polar solutes dissolve in polar solvents. In both pure solute and pure solvent the molecules are held together by dipole–dipole interactions or hydrogen bonding. These same interactions form between polar solute molecules and polar solvent molecules [Figure 6.3(a)]. Because the interactions between molecules in the solution are roughly the same strength as those between molecules in the pure substances, the solute dissolves in the solvent.

In general, a solute will not dissolve in a solvent unless the attractions between the molecules of the solute and solvent are roughly equal to the attractions between solute molecules and the attraction between solvent molecules.

When a nonpolar solute dissolves in a nonpolar solvent, the interactions between the solute and solvent particles are London dispersion forces [Figure 6.3(b)]. These forces are about the same strength as the London dispersion forces between pure solute molecules or pure solvent molecules, and thus a solution forms. Gasoline is an example of a solution formed from nonpolar substances.

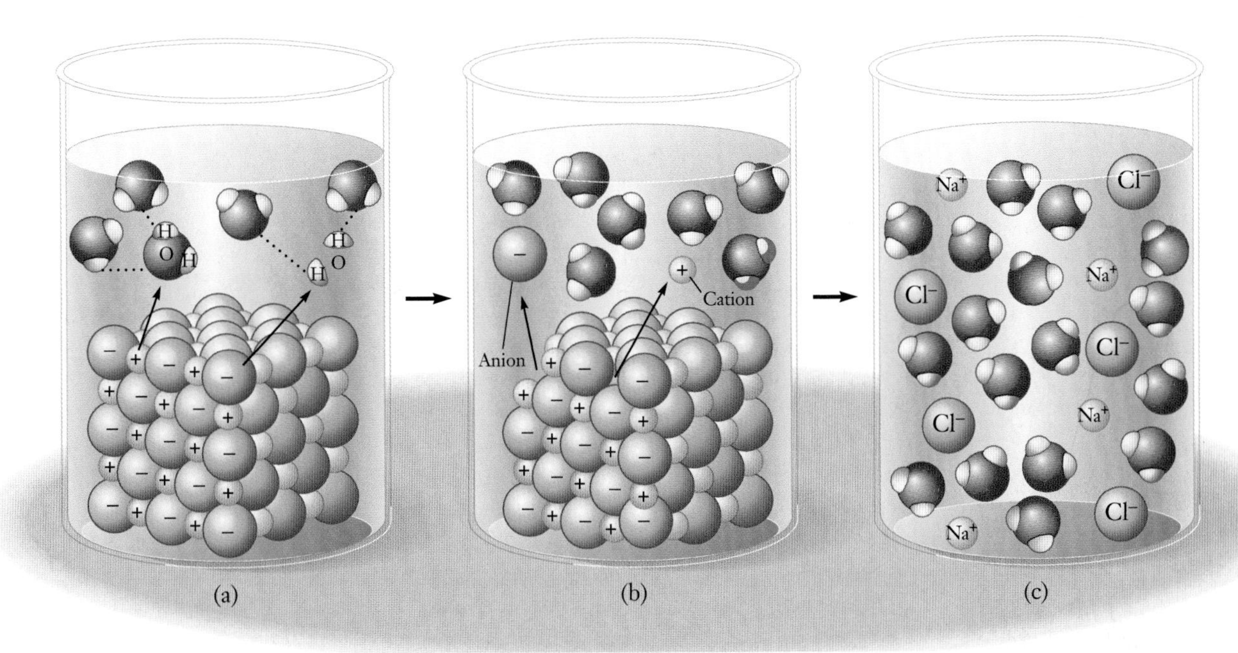

Figure 6.2 **(a)** Ionic bonds hold the ions together in a sodium chloride (NaCl) crystal and water molecules are held together by hydrogen bonds. **(b)** When sodium chloride dissolves in water, each charged ion is surrounded by and bonded to polar water molecules. **(c)** The attraction between the water molecules and the ions is sufficient to lift the ions from the crystal forming a solution.

Now that we understand why solutions form we can make this generalization about solubility: *Like dissolves like.* Many polar and ionic substances dissolve in polar solvents, and many nonpolar substances dissolve in nonpolar solvents.

Example 6.1 Solubility

1. Will these substances dissolve in water? KBr; paraffin
2. Will these substances dissolve in the nonpolar solvent hexane? table salt; cooking oil

SOLUTIONS

1. Potassium bromide (KBr) is an ionic substance that, like sodium chloride, will dissolve in water because of the ion-dipole interactions between the water molecules and the ions. Paraffin is a nonpolar substance, and will not dissolve in the polar solvent water.
2. The ionic substance table salt (NaCl) will not dissolve in a nonpolar solvent such as hexane. Cooking oil is nonpolar and will dissolve in this nonpolar solvent (like dissolves like).

Self-Test

1. Will these substances dissolve in water? $CaCl_2$; motor oil
2. Will these substances dissolve in the nonpolar solvent hexane? gasoline; corn syrup

ANSWERS

1. $CaCl_2$ dissolves in water but motor oil does not.
2. Gasoline dissolves in hexane but corn syrup does not.

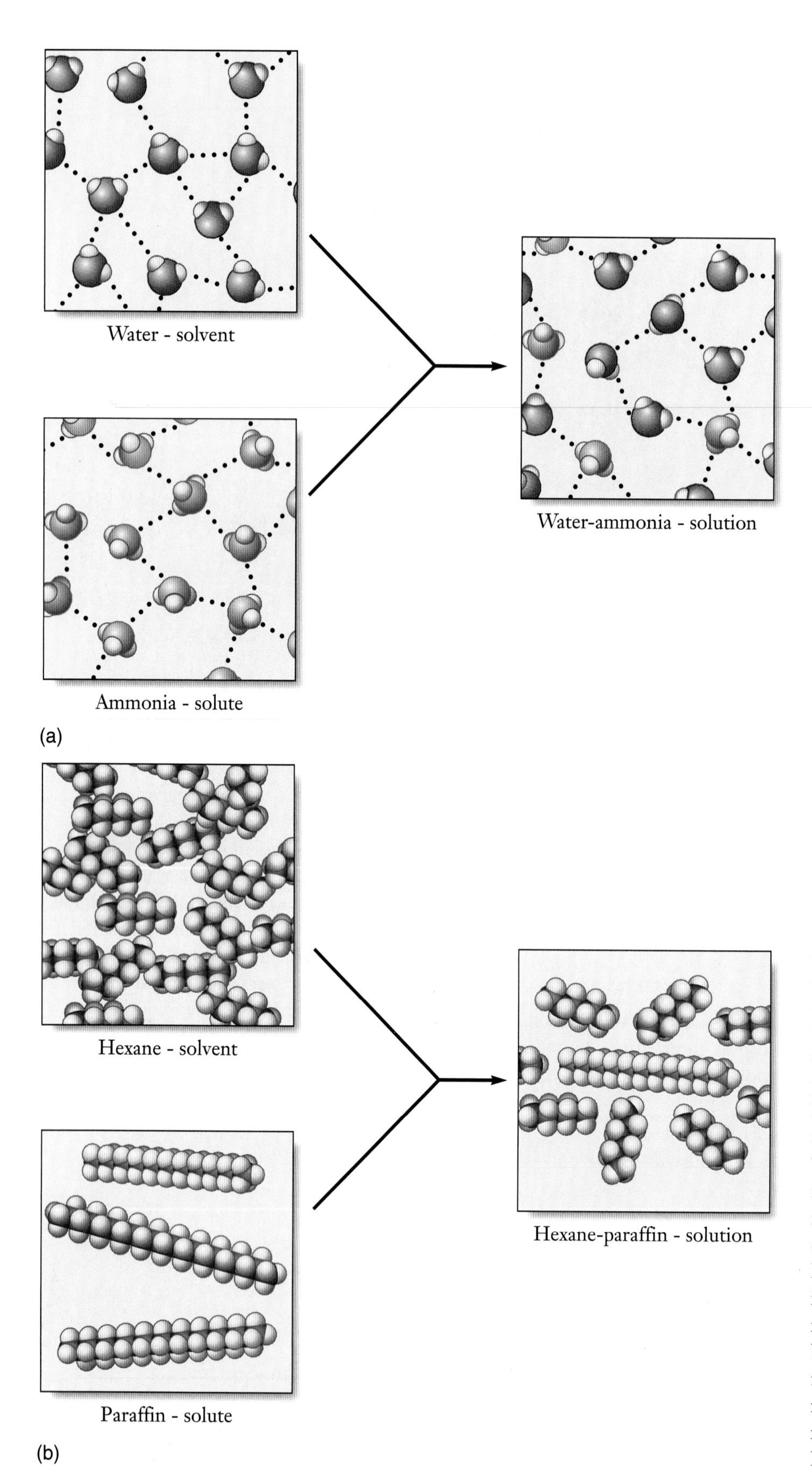

Figure 6.3 A solution can form if the interactions between the molecules of solvent and solute are similar in strength to the interactions found in the pure solvent or pure solute. **(a)** The molecules in either polar solutes or polar solvents are held together by dipole–dipole interactions or hydrogen bonds. The forces between polar solute molecules and polar solvent molecules in solution are also dipole–dipole interactions or hydrogen bonds. **(b)** The molecules in either nonpolar solutes or nonpolar solvents are held together by London dispersion forces. The forces between nonpolar solute molecules and nonpolar solvent molecules in solution are also London dispersion forces.

Effects of Temperature and Pressure on Solubility

The solubility of many solid or liquid solutes in a solvent increases with temperature. For example, 180 g of sucrose dissolve in 100 g of water at 0°C, but 490 g dissolve in 100g of water at 100°C. Unfortunately, there are exceptions to this rule. The solubility of sodium sulfate in water decreases as temperature increases. Although solubility increases with temperature for many solids and liquid solutes, the solubility of gases in a solvent always decreases with increasing temperature. This has environmental consequences, because as water warms, the amount of dissolved oxygen decreases, which may adversely affect aquatic organisms in the water.

Pressure has no affect on the solubility of a solid or a liquid in a solvent. However, increases in pressure cause more gas to dissolve in a solvent. This idea will be discussed more fully after we learn about concentration.

6.2 Concentrations

If a solution is made by adding more and more solute to a solvent until no more solute will dissolve, then a **saturated solution** has been prepared. Any solution containing less solute than that found in a saturated solution is an **unsaturated solution.** The word *unsaturated* provides no numerical data, and thus we have no way of knowing how much solute is present in the solution. It is useful and necessary to express the amount of dissolved substance in a quantitative way. The **concentration** of a solution expresses how much solute is present in a given quantity of a solution.

Molarity

The most common way of expressing concentration in chemistry is moles of solute per liter of solution. This measure of concentration is called **molarity** **(*M*).** A one-molar solution (usually written 1 *M*) contains 1 mol of solute per 1 L of solution.

$$1\ M = \frac{1\text{ mole of solute}}{1\text{ liter of solution}} = \frac{1\text{ mol}}{1\text{ L}}$$

Chemists routinely need to know how many moles of a substance are present. The number of moles can be easily calculated from molarity.

Preparing solutions of known molarity is a common laboratory activity. Here are the steps that are used.

1. Weigh out the amount of solute that provides the number of moles (Section 4.5) needed for the solution [Figure 6.4(a)].
2. Place the solute in a volumetric flask of the desired final volume, and add some solvent to dissolve the solute [Figure 6.4(b)].
3. Now add enough solvent to bring the final volume to the mark on the neck of the flask [Figure 6.4(c)].

(a)

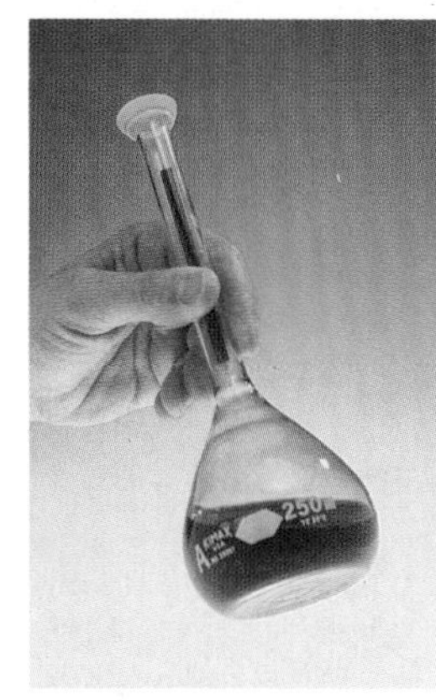

(b)

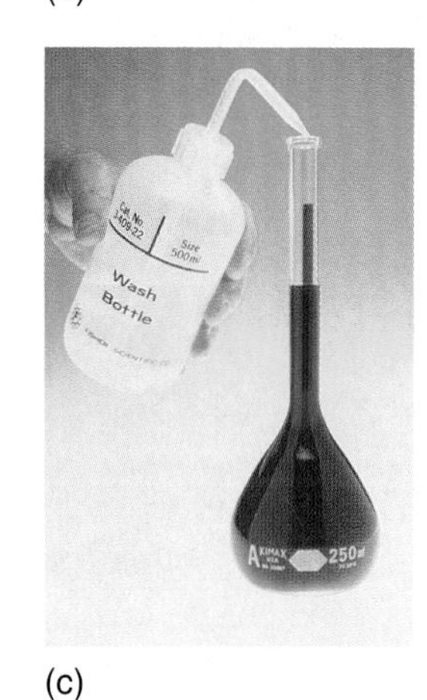

(c)

Figure 6.4 Preparation of a solution of known molarity. The text describes the specific steps. *(Charles D. Winters)*

Example 6.2 Molarity

1. If 0.38 mol of sodium chloride (NaCl) is dissolved in water to yield 0.85 L of solution, what is the molarity of the solution?
2. How many moles of lead (II) nitrate, $Pb(NO_3)_2$, are in 1.50 L of a 0.250 *M* solution?

SOLUTIONS

1. Molarity is moles of solute divided by liters of solution. Begin with the expression for molarity, and make the appropriate substitutions:

$$M = \frac{\text{moles of solute}}{\text{L of solution}}$$

$$M = \frac{0.38 \text{ mol}}{0.85 \text{ L}} = 0.45 \text{ mol/L} = 0.45\ M$$

The solution has a concentration of 0.45 *M*.

2. The volume of the solution is known, and the molarity is known. The expression for molarity is

$$M = \frac{\text{moles of solute}}{\text{L of solution}}$$

This can be rearranged to yield

$$\text{moles of solute} = M \times \text{L of solution}$$

Substituting molarity and volume into this expression yields

$$\text{Moles of solute} = \frac{0.250 \text{ mol}}{1 \not{\text{L}}} \times 1.50 \not{\text{L}} = 0.375 \text{ mol}$$

There is 0.375 mol of $Pb(NO_3)_2$ in the solution.

Self-Test

1. What is the molarity of **(a)** 0.15 mol of glucose dissolved in water to yield 0.75 L of solution? **(b)** 0.0381 mol of NaCl dissolved in water to yield 11.43 mL of solution?
2. How many moles of solute are in **(a)** 0.400 L of 0.25 *M* HCl? **(b)** 125 mL of 0.11 *M* CaI_2?

ANSWERS

1. **(a)** 0.20 mol/L = 0.20 *M* **(b)** 3.33 mol/L = 3.33 *M* (Did you remember to change mL to L?)
2. **(a)** 0.10 mol **(b)** 1.4×10^{-2} mol

Mass and Volume Percent

The amount of solute in a solution can also be expressed as a percentage. The percentage of solute may be expressed on a mass basis—that is, as *mass/mass (m/m)*, *percent.* If a saline solution is prepared from 5.0 g of salt and 95 g of water, then the solution is a 5% (m/m) solution. Mass/mass percent can be determined with this expression:

$$\%\ (\text{mass/mass}) = \frac{\text{grams of solute}}{\text{grams of solution}} \times 100$$

A solution can also be expressed as a *mass/volume (m/v) percent.* A 5% (m/v) saline solution is prepared by adding enough water to 5 g of sodium chloride to yield 100 mL of solution. These solutions are easily prepared and are commonly used in science and health care.

$$\%\ (\text{mass/volume}) = \frac{\text{grams of solute}}{\text{milliliters of solution}} \times 100$$

Solutions can also be prepared by *volume/volume (v/v) percent.* A 25% (v/v) aqueous solution of alcohol is prepared by adding enough water to 25 mL of alcohol to yield 100 mL of solution. Always use the same units of volume for the solute and solvent. Volume percent can be determined with this expression:

$$\%\ (\text{volume/volume}) = \frac{\text{volume of solute}}{\text{volume of solution}} \times 100$$

Solutions whose concentrations are expressed as a percentage are a bit easier to prepare than molar solutions, because it is not necessary to calculate the number of moles of solute in the solution. Many laboratory and medicinal solutions are prepared and labeled as percent concentration.

Example 6.3 Mass Percent and Volume Percent

1. You need 2.50×10^2 g of a 2.00% (m/m) aqueous solution of sucrose (table sugar). How will you prepare it?
2. How many grams of epsom salt do you need to make 500.0 mL of a 3.0% (m/v) aqueous solution?
3. How many milliliters of alcohol are needed to make 50.0 mL of a 12.5% (v/v) aqueous solution of alcohol?

SOLUTIONS

1. Use the expression for mass percent to calculate how much sucrose you need.

$$\%\ (\text{mass/mass}) = \frac{\text{grams of solute}}{\text{grams of solution}} \times 100$$

Rearrange the expression to isolate mass of solute:

$$\text{Grams of solute} = \frac{\text{grams of solution} \times \text{mass }\%}{100}$$

Make the appropriate substitution and solve

$$\text{Grams of solute} = \frac{2.50 \times 10^2 \text{ grams} \times 2.00\%}{100} = 5.00 \text{ g}$$

A mass of 5.00 g of sucrose is needed. The amount of water is 250. g − 5.00 g = 245 g. Dissolve the 5.00 g of sucrose in 245 g of water.

2. To determine the number of grams of epsom salt that are needed, use this expression:

$$\% \text{ (mass/volume)} = \frac{\text{grams of solute}}{\text{milliliters of solution}} \times 100$$

Rearrange to solve for grams of solute:

$$\text{Grams of solute} = \frac{\text{milliliters of solution} \times \% \text{ (m/v)}}{100}$$

Make the appropriate substitutions and solve

$$\text{Grams of solute} = \frac{500.0 \text{ mL solution} \times 3.0\%}{100} = 15 \text{ g solute}$$

A 15-g sample of epsom salt must be dissolved in enough water to yield 500 mL of solution.

3. The expression for volume percent is needed here.

$$\% \text{ (volume/volume)} = \frac{\text{volume of solute}}{\text{volume of solution}} \times 100$$

Rearrange to isolate volume of solute, and make appropriate substitutions:

$$\text{Volume of solute} = \frac{12.5\% \times 50.0 \text{ mL}}{100} = 6.25 \text{ mL}$$

A 6.25-mL sample of alcohol is needed.

Self-Test

1. What mass of glucose is needed to prepare 1.5 kg of an aqueous solution that is 2.2% (m/m)?
2. What volume of acetone (a somewhat polar liquid) is needed to prepare 1.50×10^2 mL of an aqueous solution that is 25.0% (v/v) acetone?
3. What mass of iodine is needed to prepare 35.0 mL of a 2.50% (m/v) tincture of iodine?

ANSWERS

1. 33 g 2. 37.5 mL 3. 0.875 g

Dilution

Sometimes it is useful to dilute a solution of known molarity that you have rather than to prepare a new one. To determine how much of the available solution to use in the dilution, use this expression:

$$V_{\text{available}} \times C_{\text{available}} = V_{\text{needed}} \times C_{\text{needed}}$$

In this expression, V is volume and C is concentration. Usually you know the volume and concentration of the solution that is needed, and you know the concentration of the solution that is available. You can rearrange this expression to solve for the volume of the available solution that will be diluted:

$$\text{Volume}_{\text{available}} = \frac{\text{volume}_{\text{needed}} \times \text{concentration}_{\text{needed}}}{\text{concentration}_{\text{available}}}$$

If any three of the four terms in this expression are known, the fourth can be determined.

Example 6.4 Preparing Solutions by Dilution

You have an aqueous solution that is 1.00 M sodium chloride. Use it to prepare 75.0 mL of a 0.350 M NaCl solution.

SOLUTION

Use the expression for dilution:

$$\text{Volume}_{\text{available}} = \frac{\text{volume}_{\text{needed}} \times \text{concentration}_{\text{needed}}}{\text{concentration}_{\text{available}}}$$

Make the appropriate substitutions and solve:

$$\text{Volume}_{\text{available}} = \frac{75.0\text{ mL} \times 0.350\ \cancel{M}}{1.00\ \cancel{M}} = 26.3\text{ mL}$$

You must take 26.3 mL of the 1.00 M solution and add enough water to bring the final volume to 75.0 mL.

Self-Test

You have a 12.0 M solution of HCl. How much of this solution do you need to prepare 1.50×10^2 mL of a 3.00 M solution?

ANSWER

37.5 mL

Henry's Law: Pressure Affects the Concentration of Gases in Liquids

There are no general rules for predicting how much solid or liquid solute will dissolve in a solvent, but for gases we can apply **Henry's law.** If a gas is in contact with a liquid, the concentration of the gas in the liquid is directly proportional to the pressure of the gas:

$$\text{Concentration of gas in liquid} = \text{constant} \times \text{pressure of the gas}$$

The higher the gas pressure, the more gas that dissolves in the solvent. At lower pressures, less gas dissolves (Figure 6.5).

When a bottle of soda is opened, the partial pressure of carbon dioxide over the water decreases, and carbon dioxide gas bubbles escape. *(Charles D. Winters)*

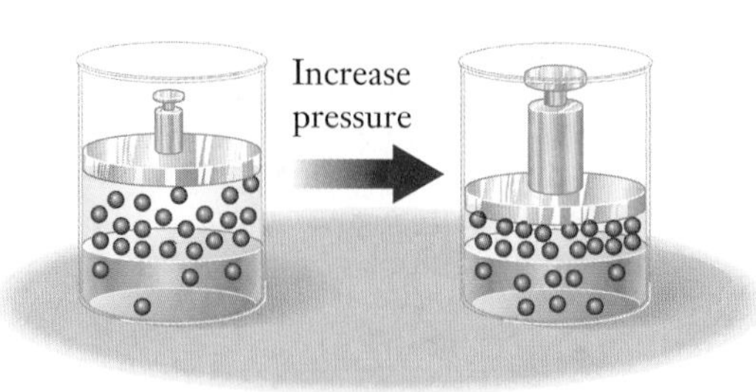

Figure 6.5 Doubling the pressure of a gas above a liquid doubles the amount of the gas dissolved in the liquid.

Carbonated water provides a common example of Henry's law. Carbonated water is prepared by exposing water to gaseous carbon dioxide under high pressure. According to Henry's law, carbon dioxide dissolves in the water, and the amount that dissolves is determined by the partial pressure of the carbon dioxide. After soda is bottled, the space above the water has a large partial pressure of carbon dioxide. Under these conditions, the gas remains dissolved in the water. When the cap is removed, the partial pressure of carbon dioxide above the water drops greatly. There is no longer enough pressure of carbon dioxide to keep the carbon dioxide dissolved in the water, and gaseous carbon dioxide bubbles from the soda.

Connections

The Effects of Nitrogen on Divers

Henry's law states that the amount of a gas that dissolves in a solvent is directly proportional to the gas's partial pressure. Henry's law applies to all gases, including nitrogen in air and blood. On land and during shallow dives, the amount of inhaled nitrogen that dissolves in blood is small, but during deep dives significant amounts may dissolve.

During a dive, the pressure on the lungs increases with depth. As the pressure increases, more nitrogen dissolves in the blood. During shallow dives, this increased amount of blood nitrogen is too small to cause any adverse effects; however, the situation is rather different in deep dives where relatively large amounts of nitrogen dissolve in the blood. These larger amounts of dissolved nitrogen can produce *nitrogen narcosis,* which is a state of altered consciousness. Furthermore, if a diver returns rapidly to the surface, the reduced pressure results in decreased nitrogen solubility, which results in some of the nitrogen forming small bubbles in the blood, which tend to accumulate in joints. This leads to excruciating pain that causes the diver to fold up into a shape that is called *the bends.*

To avoid the bends, divers return slowly to the surface from deep dives. This allows sufficient time for the nitrogen to be exhaled before it can form bubbles in the blood. The bends can also be avoided by using a mixture of helium and oxygen instead of compressed air. Helium is less soluble in blood than nitrogen, which means much less helium will dissolve at any pressure. This reduces the chance of bubbles forming in blood. When divers use a helium–oxygen mixture they gain an additional benefit because helium does not cause narcosis.

(Susan Blanchet/Dembinsky Photo Associates)

A second example of Henry's law involves the blood gases, oxygen and carbon dioxide. Inhaled air in the alveoli of lungs has an oxygen partial pressure of around 100 torr. The blood entering the lungs is oxygen-poor. As a result, oxygen dissolves in the blood, nearly saturating it with oxygen. The blood leaving the lungs has an oxygen partial pressure of around 100 torr. Body tissues contain much less oxygen—less than 40 torr—than oxygenated blood. When oxygen-rich blood enters body tissues, oxygen diffuses from the blood to the tissues. The blood is now oxygen-poor, having an O_2 partial pressure less than 40 torr. It returns to the lungs and is reoxygenated (Figure 6.6).

Gas concentrations in blood are expressed in torr. A blood-oxygen partial pressure of 100 torr means the amount of dissolved oxygen in the blood corresponds to the amount that would dissolve from a gas having an oxygen partial pressure of 100 torr.

Carbon dioxide produced by cells leaves the tissues raising the blood's carbon dioxide concentration (the concentration of carbon dioxide varies with body activity). The partial pressure of carbon dioxide in blood entering the lungs—greater than 45 torr—is larger than the carbon dioxide partial pressure in the lungs—40 torr. Thus carbon dioxide leaves the blood and enters the lungs. This cyclic flow of

The amount of oxygen carried by blood is greatly influenced by the oxygen-carrying protein hemoglobin.

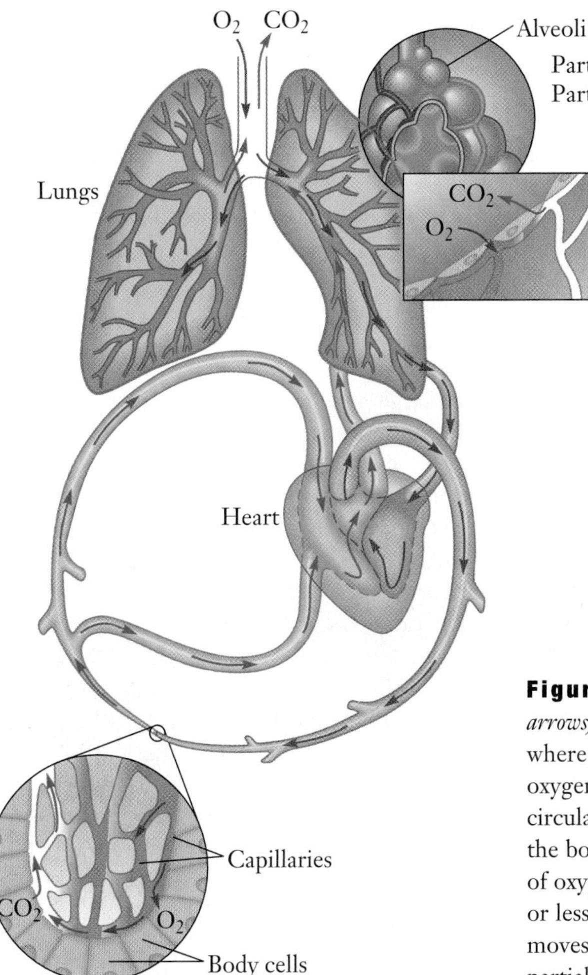

Figure 6.6 Oxygen *(red arrows)* moves from the lungs—where the partial pressure of oxygen is 100 torr—through the circulatory system to the tissues of the body, where the partial pressure of oxygen is much lower—40 torr or less. Carbon dioxide *(blue arrows)* moves from tissues—where the partial pressure of carbon dioxide is 45 torr—to the lungs where its partial pressure is 40 torr.

gas-rich and gas-poor blood between lungs and tissues provides for delivery and removal of blood gases by the body (Figure 6.6).

Respiratory therapy takes advantage of Henry's law. If a patient is unable to get enough oxygen from air, an oxygen-enriched mixture of gases can be substituted for air. The larger oxygen partial pressure in the gas mixture favors oxygen uptake in the patient's lungs.

6.3 Dispersions and Suspensions

Solutions are homogeneous mixtures in which a solute is dissolved in a solvent. The key word is *dissolved.* The solvent and solute are not just mixed together; instead the solute is dissolved in the solvent. The solute and solvent particles are bound to each other, and they are roughly the same size—that is, they are in the nanometer size range. A solution is stable, which means it remains a homogeneous mixture over time. *Dispersions* and *suspensions* are mixtures that resemble solutions but differ in some important ways.

Let's examine suspensions first, using this mental exercise. Imagine taking some very fine sand and adding it to water. Stir it well, then sit back and watch. With time, the particles of sand settle out of the water. This happens because the particles are much larger (micrometers or larger) than water molecules and are not truly dissolved in the water. They are large enough to be pulled down slowly by gravity. The forces of attraction between the sand particles and water molecules are too weak to overcome gravity. We say that the fine sand–water mixture is a **suspension,** which is defined as an unstable mixture of larger particles suspended (hanging) in a liquid. Without an outside influence such as stirring, the larger particles settle out. In rivers, turbulence keeps mixing the mud and water; thus the mud does not settle out. Because of the large size difference between suspended particles and solvent molecules, the particles in a suspension can be removed from the solvent by *filtration* [Figure 6.7(a)].

The phrase *shake well before using* is found on the label of many medicines and cosmetics that are suspensions.

Blood is another example of a suspension. The red and white blood cells are too large to form a true solution. They are suspended in the aqueous plasma. In the cardiovascular system, blood is continuously mixed as it passes through the vessels of the bloodstream. There is sufficient turbulence to maintain the suspension. In the laboratory, blood cells are easily separated from the plasma by *centrifugation* [Figure 6.7(b)].

Dispersions, which are also known as **colloids,** fall between solutions and suspensions in their properties. In a dispersion, the dispersed particles in the solvent are larger than the solute particles in a solution but smaller than the particles in a suspension (size ranges roughly from tens of nanometers up to a micrometer). In a dispersion, the particles are too big to form a true solution but small enough to be unaffected by gravity. They are stable over time.

Blood plasma is an example of a dispersion. Protein molecules are too large to form a true solution, but their small size keeps them from settling out of the water part of the plasma. Dispersions are sufficiently similar to solutions that the two are often considered the same. However, dispersions differ from solutions in two discernible ways. One, the particles in a dispersion can often be separated from the fluid by centrifugation and filtration. Proteins in blood can be separated from wa-

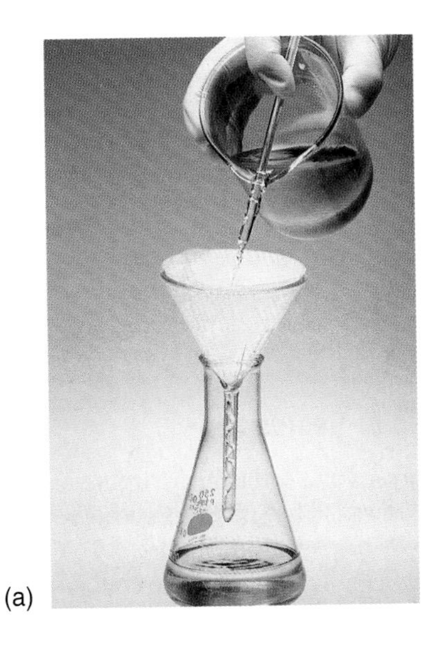

(a)

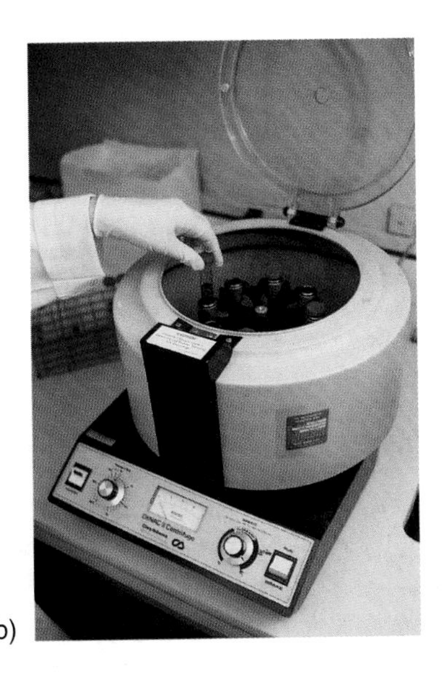

(b)

Figure 6.7 **(a)** Suspended particles can be removed from solvents and solutes by filtration. The suspended particles are too large to pass through the paper or membrane, but the smaller molecules of the solvent and solute readily pass through. **(b)** Blood is placed into tubes prior to centrifugation. In a centrifuge, the sample is subjected to a centrifugal force that forces the larger particles down the tube. *(Charles D. Winters)*

ter and smaller solutes by filtering the blood or by high-speed centrifugation. Solutions cannot be separated by either of these techniques.

A second difference between dispersions and solutions involves light. In normal lighting, both solutions and dispersions appear transparent (Figure 6.8). Light passes through them. But in intense light, they appear different. If a beam of intense light is passed through a solution, the solution appears transparent. None of the particles in a solution are large enough to deflect light. But if the beam is passed through a dispersion, the larger particles in the dispersion deflect the light as it passes through. This deflected light travels to the observer's eye, giving the appearance of a beam passing through the liquid. This is called the **Tyndall effect.** The sunbeam you see passing through a room is caused by this same effect. Tiny dust particles in air form a dispersion. Solutions, dispersions, and suspensions are compared in Table 6.1.

Figure 6.8 The Tyndall effect. The bottle on the left contains a solution and the one on the right contains a dispersion. When intense light (a laser beam) passes through the solution, the light is not scattered, but the larger particles in the dispersion deflect some of the light, which appears as a "light beam." *(Charles D. Winters)*

Table 6.1 **Some Properties of Solutions, Dispersions, and Suspensions**

MIXTURE	APPEARANCE	SIZE OF PARTICLES	SEPARATION	EXAMPLES
Solution	Clear	Solute and solvent particles are small, subnanometer to nanometer	Distillation	Saltwater, sugar water
Dispersion	Generally clear	Solute particles larger than solvent particles, nanometers to submicrometer, and can scatter light; the Tyndall effect	Usually by filtration or centrifugation	Blood plasma, colloidal gold
Suspension	Cloudy or murky	Solute particles larger than a micrometer and affected by gravity	Filtration, centrifugation	Whole blood, muddy water

6.4 Colligative Properties

Colligative properties of solutions are those that depend only on the amount of solute in the solution, not on the nature of the solute. Three common colligative properties are boiling-point elevation, freezing-point depression, and osmotic pressure.

Boiling-Point Elevation and Freezing-Point Depression

The boiling point and freezing point of a solvent in a solution are influenced by the amount of solute. The higher the concentration of the solute, the lower the freezing point and the higher the boiling point of the solution (Figure 6.9). Commercial automotive "antifreeze" is a very practical application of these colligative properties. In cold weather, the water in engines and radiators would freeze. Because water expands when it freezes, the engine and radiator could be damaged. In addition, water is somewhat limited as a coolant because it turns to steam at 100°C (pressurized radiators reduce this problem). When ethylene glycol, the primary ingredient in commercial antifreeze, is dissolved in water, the solution freezes at a lower temperature, which reduces the risk of engine damage by freezing. The solution also boils at a higher temperature, which increases the cooling efficiency of the cooling system.

Ethylene glycol is the primary ingredient in commercial antifreeze. These jars were filled with water *(left)* and a mixture of water and antifreeze *(right)*, then placed in a freezer overnight. The water froze, but the solution containing the antifreeze did not. *(Charles D. Winters)*

Osmotic Pressure/Dialysis

Substances tend to diffuse from regions in which they are more concentrated to regions in which they are less concentrated. For example, if pure water were layered over saltwater, the ions of the salt diffuse into the pure water and water molecules from the pure water diffuse into the saltwater. These diffusions continue until the concentration of the ions and water molecules are the same throughout the mixture. However, movement of some substances in solutions can be restricted by *semipermeable membranes* that allow only some substances to pass through them (Figure 6.10). Consider a semipermeable membrane that allows passage of water but not solutes. If pure water were placed on one side of this semipermeable membrane and an aqueous solution were placed on the other, water molecules could

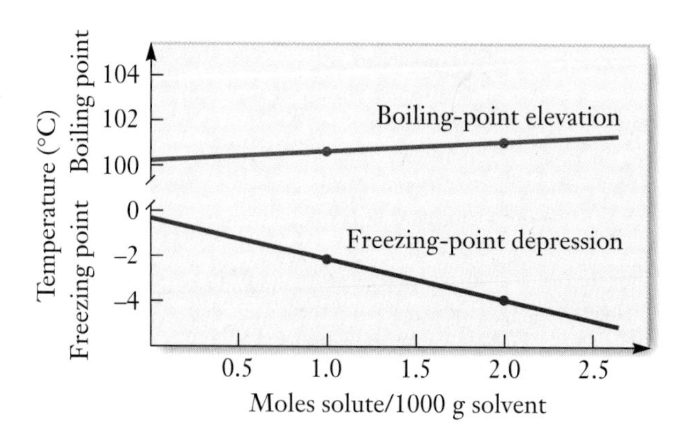

Figure 6.9 Freezing-point depression and boiling-point elevation are colligative properties of solutions. These properties depend on the number of solute particles in solution.

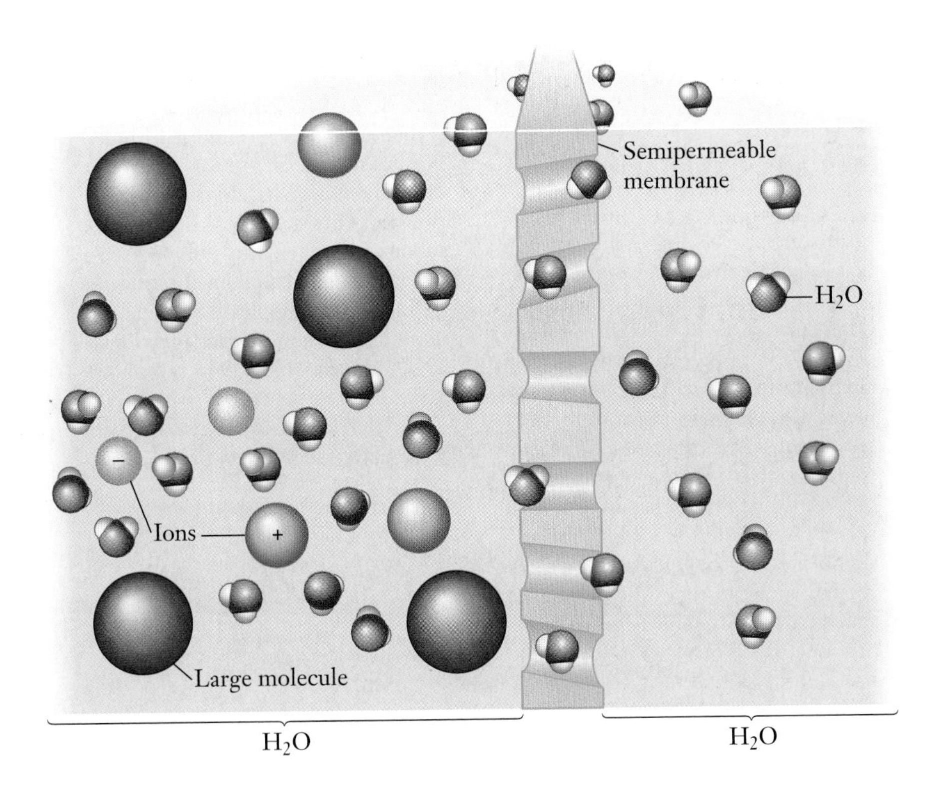

Figure 6.10 This semipermeable membrane allows water molecules to pass through but prevents larger molecules and ions from passing.

move across the membrane, but the solute particles could not. With time *osmosis* occurs, which is the movement of the water molecules from the water side of the membrane, across the membrane, and into the solution (Figure 6.11). The movement of water across a semipermeable membrane into a solution can be prevented by applying pressure to the solution side of the membrane. **Osmotic pressure** is the pressure that must be applied to the solution to prevent movement of water across the membrane. Osmotic pressure is also equal to the difference in height of the liquids on either side of the semipermeable membrane after water has moved across the membrane [Figure 6.11(b)]. Osmotic pressure depends only on the number of solute particles in the solution; thus it is a colligative property. The greater the number of particles dissolved in solution, the higher its osmotic pressure.

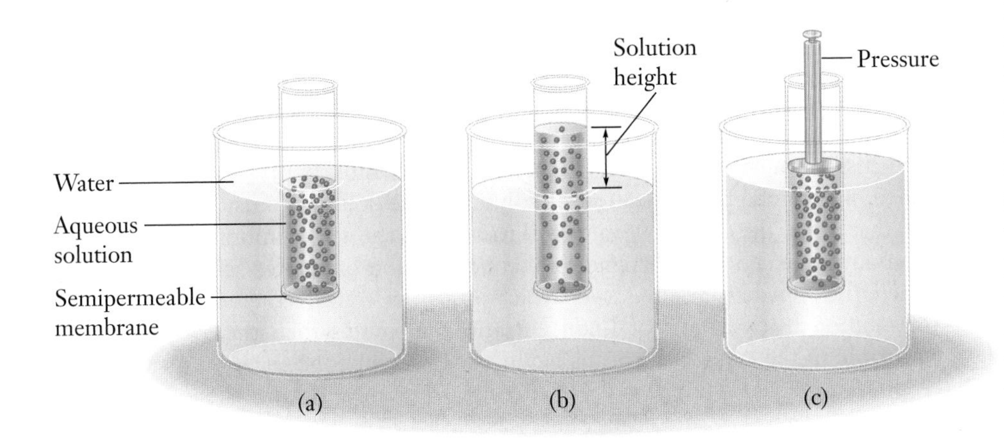

Figure 6.11 **(a)** A tube fitted with a semipermeable membrane is partially filled with an aqueous solution and placed in pure water. **(b)** Water can pass through the membrane but solutes cannot, thus water moves across the membrane into the tube. After some time net water movement stops, and the height of the solution above the surface of the water corresponds to the osmotic pressure of the original solution. **(c)** If a pressure equal to the osmotic pressure is applied to the top of the tube, water will be forced from the solution back into the pure water.

Connections

Intravenous Solutions and Osmolarity

Because blood cells undergo hemolysis or crenation in solutions with an osmotic pressure significantly different than blood, osmotic pressure must be considered carefully when intravenous (IV) solutions are prepared. To avoid blood cell damage, IV solutions are usually prepared with osmotic pressures similar to that of blood and are therefore isotonic solutions. Proper preparation of IV solutions requires an understanding of *osmolarity*, which is a measure of concentration that considers all solute particles in a solution.

When molecular substances dissolve, they yield one mole of solute particles per mole of the solute. When an ionic substance dissolves, the ions dissociate, yielding more than a mole of solute particles. For example, salt—NaCl—yields two moles of ions (Na^+, Cl^-) per mole of salt, and calcium nitrate—$Ca(NO_3)_2$—yields three moles of ions (Ca^{2+}, $2NO_3^-$) per mole of calcium nitrate. Molarity is expressed in moles of solute per liter of solution, but osmolarity is expressed as moles of solute particles per liter of solution. A one-osmolar solution has one mole of solute particles per liter of solution. Osmolarity can be calculated by multiplying molarity times the number of solute particles obtained from a mole of solute (i):

$$\text{Osmolarity} = \text{molarity} \times i$$

For a molecular substance, i is 1; thus molarity and osmolarity are the same for a solution of a molecular substance like glucose (blood sugar). Because ionic substances dissociate into ions, i is the number of ions in the formula. For example, a one-molar solution of sodium chloride has an osmolarity of two because $i = 2$; there is a mole of sodium ions and a mole of chloride ions per mole of sodium chloride.

Blood is a complex mixture containing many solutes that contribute to blood's osmolarity, which is about 0.30.

The table shows the osmolarity of some common IV solutions.

Osmolarity of Some Intravenous Solutions

SOLUTION	OSMOLARITY	COMPARISON TO BLOOD
0.9% sodium chloride	0.308	Isotonic[a]
5% dextrose	0.278	Somewhat hypotonic
0.33% sodium chloride	0.113	Hypotonic

[a]Exerts same osmotic pressure as blood.

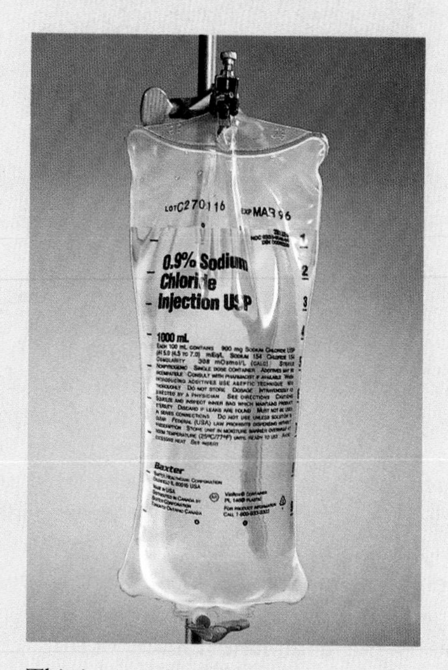

This intravenous (IV) solution is isotonic to blood. *(Charles D. Winters)*

Osmotic pressure is a colligative property that has great importance in health care. The membranes of cells are semipermeable. The solutions in the body—blood, tissue fluids, lymph, and plasma—all exert osmotic pressure. As a consequence, if the total concentration of solutes in the solution outside a cell differs from that inside the cell, then water will pass across the cell membrane. When a solution outside a cell has the same total concentration of solutes as the solution in the cell, the outside solution is said to be **isotonic** (*iso-* means the same). If the outside solution has a lower total concentration of solutes than the cell, then it is **hypotonic** to the cell (*hypo-* means below). If the total concentration of solutes in the solution outside the cell is greater than that in the cell, then the outside solution is **hypertonic** (*hyper-* means above).

Connections

Hemodialysis

Kidneys perform the vital function of clearing wastes and excess water and salts from the blood. If the kidneys fail, these wastes accumulate, proper salt and water balance is lost, and life-threatening conditions may result. *Hemodialysis* (renal or kidney dialysis) is a treatment for kidney failure that removes wastes and restores salt and water balance.

During hemodialysis, the patient's blood is passed through a dialyzer (artificial kidney), which contains tiny, semipermeable tubules that allow passage of the small molecules and ions found in blood but that keep cells and blood proteins in the blood. In the dialyzer these tubules are bathed in an aqueous solution—the dialysate—containing normal concentrations of needed substances such as glucose and electrolytes. Because both the blood and the dialysate contain about the same concentration of these needed substances, there is no net movement of the substances between these fluids. On the other hand, because the dialysate contains no waste substances, the concentration of the wastes is higher in the blood, and there is a net movement of the wastes from the blood to the dialysate. In addition, any excess electrolytes in the blood will also move into the dialysate. With time, the composition of the blood returns to normal. For a patient with chronic kidney failure, treatment normally takes a few hours and occurs three times a week.

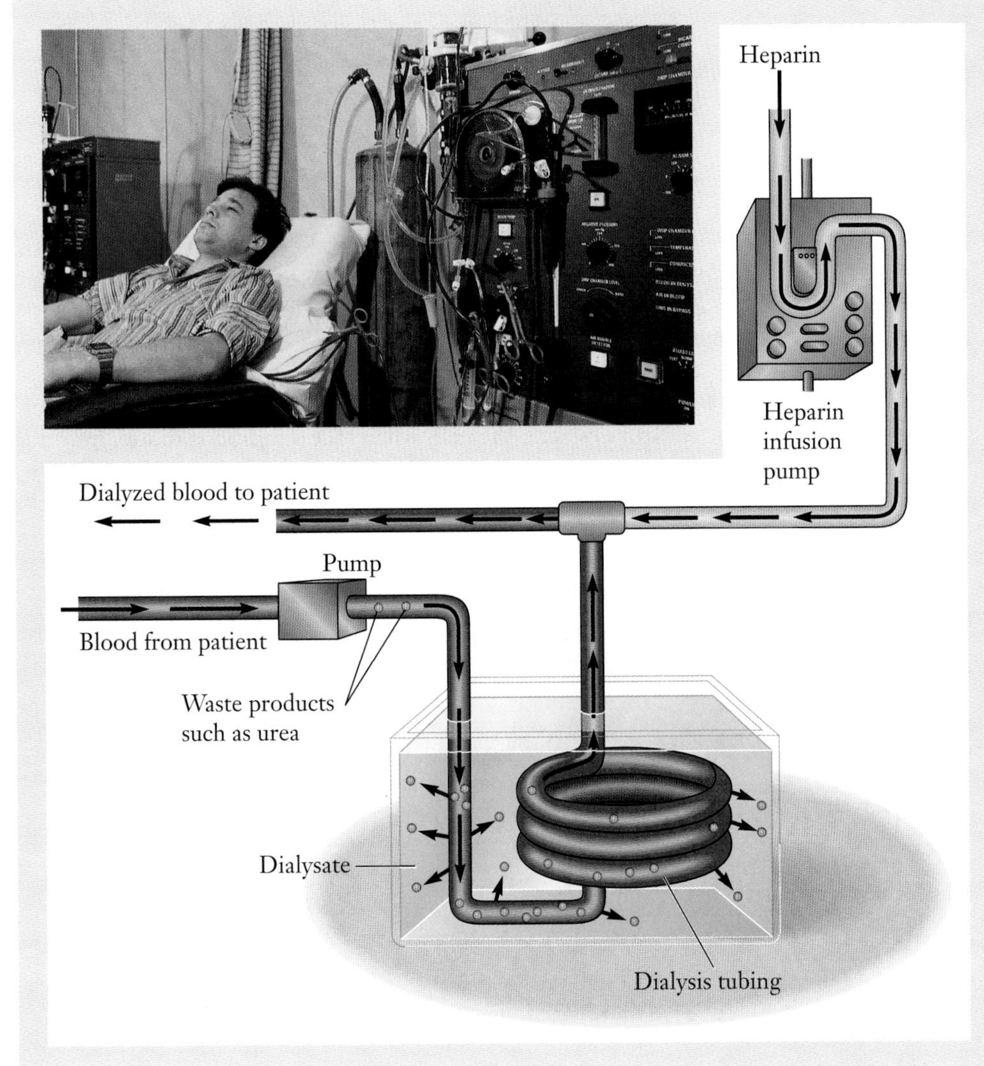

This patient is undergoing hemodialysis. As blood flows through a dialyzing coil, waste products diffuse out of the coil and into the dialysate liquid. (© *Richard Hutchings/ Photo Researchers, Inc.*)

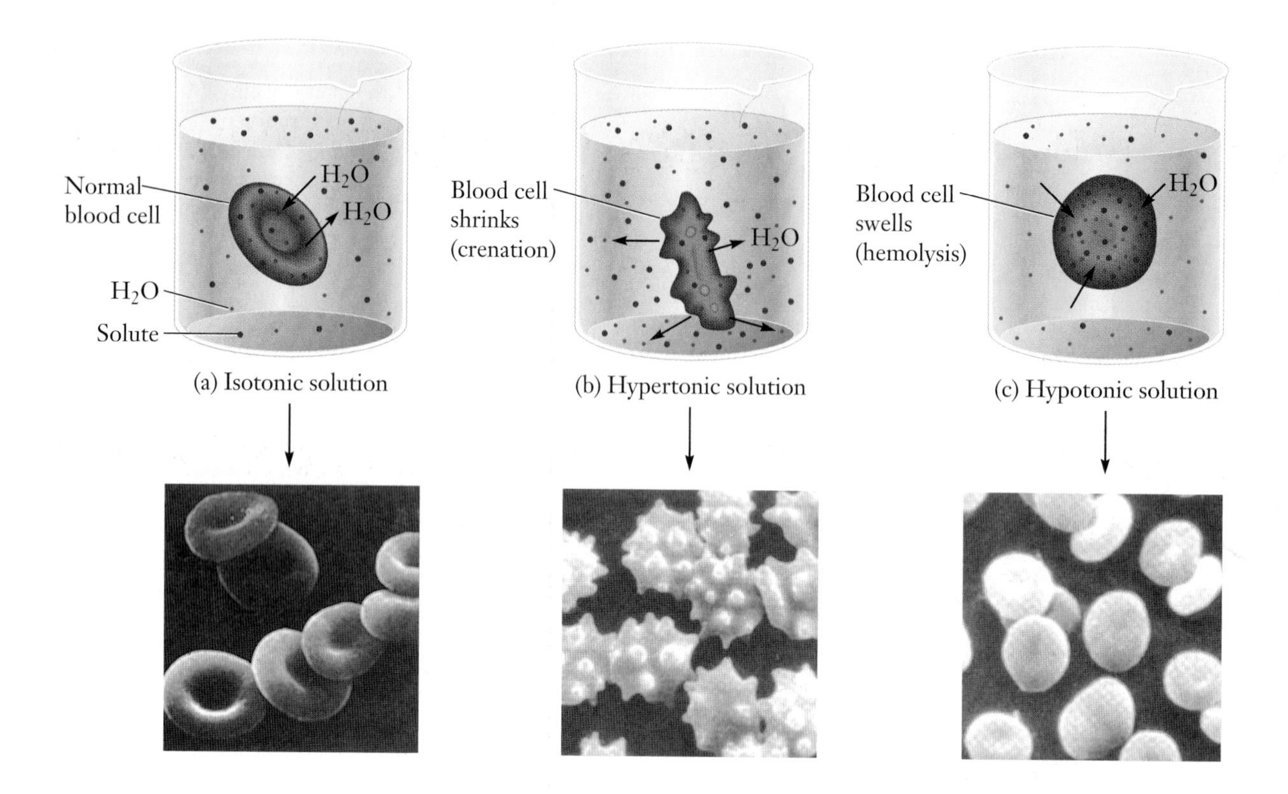

Figure 6.12 **(a)** Red blood cells have a biconcave shape in blood and in isotonic solutions. **(b)** In hypertonic solutions, red blood cells shrink or crenate. **(c)** In hypotonic solutions red blood cells expand, and may rupture. *(R. F. Baker/University of Southern California Medical School)*

When cells are immersed in an isotonic solution, there is no *net* movement of water into or out of the cells. Body fluids like blood are isotonic solutions, and blood cells in these fluids have a normal, characteristic shape [Figure 6.12(a)]. When cells are placed in a hypertonic solution, water passes through the cell membrane from the inside of the cells to the outside, and as a result, the cells shrink up, they *crenate* [Figure 6.12(b)]. When cells are placed into either water or hypotonic solutions, water moves from outside the cell, where the solution is dilute, into the cell, where the solution is more concentrated [Figure 6.12(c)]. As water enters the cell, the internal pressure increases and the cell may burst *(hemolysis).*

Dialysis is a separation technique that uses semipermeable membranes and concentration differences to separate mixtures. In the laboratory, dialysis is used to separate large molecules such as proteins and nucleic acids from smaller molecules by using a semipermeable membrane that separates by size. Dialysis is also used to supplement or replace kidney function in patients with kidney failure.

CHAPTER SUMMARY

Solutions are homogeneous mixtures of a **solvent** and one or more **solutes.** Substances differ in their **solubility** in a solvent, with polar and ionic solutes generally dissolving in polar solvents and nonpolar solutes generally dissolving in nonpolar solvents. Increasing temperature generally increases the solubility of a solid or liquid but always decreases the solubility of a gas. Increased pressure increases the solubility of a gas but has no effect on a solid or liquid. **Concentration** is an expression of the amount of solvent and solute in a solution. **Molarity,** mass percent, and volume percent are common measures of concentration. **Henry's law** states that the amount of gas that dissolves in a liquid depends on the partial pressure of the gas. Unfortunately, similar quantitative expressions for dissolving solids and liquids in a solvent are not known. Homogeneous mixtures containing a solvent and large molecules are called **dispersions** or colloids, and the unstable mixtures containing a liquid and large particles are called **suspensions. Colligative properties** of solutions, which include **osmotic pressure,** freezing-point depression, and boiling-point elevation, depend on the number of solute particles in the solution.

KEY TERMS

Section 6.1
solubility
solute
solution
solvent

Section 6.2
concentration
Henry's law
molarity *(M)*
saturated solution
unsaturated solution

Section 6.3
dispersion (colloid)
suspension
Tyndall effect

Section 6.4
colligative properties
hypertonic solution
hypotonic solution
isotonic solution
osmotic pressure

EXERCISES

Solutions and Solubility *(Section 6.1)*

1. What is a solution? Give an example of a solution with which you are familiar.

2. Explain the meaning of the word solvent. Identify the solvent in saltwater.

3. Explain the meaning of the word solute. Identify the solute in saltwater.

Use the following information for Exercises 4 and 5. Tincture of iodine is an antiseptic that is prepared by dissolving elemental iodine in alcohol.

4. What is the solvent in tincture of iodine?

5. What is the solute in tincture of iodine?

6. When an ionic substance dissolves in water, what forces form between the ions and the water molecules in the solution?

7. What forces can occur between polar solvent molecules and polar solute molecules?

8. Describe the forces that occur between nonpolar solute molecules and nonpolar solvent molecules.

9. Describe or sketch the interaction between a sodium ion (Na^+) and a water molecule in a solution.

10. Describe or sketch the interaction between a chloride ion (Cl^-) and a water molecule in a solution.

11. Water molecules are polar; benzene molecules are nonpolar. Predict the solubility of these substances in water and benzene:
(a) Gasoline, nonpolar **(b)** Glucose (blood sugar), polar **(c)** KI, ionic **(d)** Methane (natural gas), nonpolar

Concentrations *(Section 6.2)*

12. Compare a saturated solution to an unsaturated solution.

13. A solution is prepared by adding 0.3 g of KBr to 500 mL of water. All of the solid KBr disappears and the resulting mixture is clear. Is this solution saturated or unsaturated?

14. A solution is prepared by adding 100 g of salt to 100 mL of water. After heating and stirring, a clear solution overlies a layer of salt. Is this solution saturated or unsaturated?

15. A sample of saltwater is left by a window for several weeks. When observed, the volume of water is reduced and crystals of salt line the bottom of the container. Is the remaining saltwater saturated or unsaturated?

16. What word is used to describe the amount of solute that is present in a given amount of a solution?

17. Define molarity in your own words, and illustrate your definition with a mathematical formula.

18. Calculate the molarity of these solutions:
(a) 0.45 mol of H_2SO_4 in 1.25 L of solution
(b) 0.180 mol of $Ca(NO_3)_2$ in 5.00 L of solution
(c) 0.25 mol of KBr in 575 mL of solution

19. Calculate the molarity of these solutions:
(a) 1.4 mol of NaCl in 2.50 L of solution
(b) 0.025 mol of I_2 in 125 mL of solution
(c) 0.185 mol of HCl in 4.5 L of solution

20. A 0.12-mol sample of lactose (milk sugar) is dissolved in water to yield 0.107 L of solution. Calculate the molarity of this milk–sugar solution.

21. Calculate the molarity of an aqueous caffeine solution prepared by dissolving 0.014 mol of caffeine in water to get 225 mL of solution.

22. 2.7 mol of ammonia are dissolved in water yielding 1.45 L of solution. What is the molar concentration of the ammonia?

23. Determine the number of moles of solute needed to prepare these solutions:
(a) 2.0 L of 0.15 *M* NaCl
(b) 275 mL of 2.5 *M* HNO_3
(c) 41.6 mL of 2.75 *M* KCl
(d) 450 L of 1.1 *M* $(NH_4)_2SO_4$

24. Determine the number of moles of solute needed to prepare these solutions:

(a) 41.5 L of 0.10 *M* LiI
(b) 12.5 mL of 0.0055 *M* $LiNO_3$
(c) 1200 mL of 0.050 *M* $CaCl_2$
(d) 1.50 L of 11.5 *M* NH_3

25. Solutions of silver nitrate have been used as topical antiseptics to treat burns. Calculate the number of moles of $AgNO_3$ that are needed to prepare 0.100 L of a 0.10 *M* solution of silver nitrate.
26. You have 238 mL of a 0.103 *M* solution of $NaHCO_3$. How many moles of $NaHCO_3$ are present in the solution?
27. Describe how you would prepare 0.250 L of a solution that is 0.10 *M* for KBr and 0.25 *M* for NaCl.
28. Calculate the final volume of a 0.50 *M* solution of potassium iodide that contains 0.13 mol of KI.
29. Determine the mass/mass percent for the solute in these solutions:
    (a) 2.6 g of $AgNO_3$ in 125 g of aqueous solution
    (b) 114 g sucrose dissolved in 1.000 kg of water
    (c) Solder prepared with 28.5 g tin and 66.5 g lead
30. Determine the mass/mass percent for these solutions:
    (a) 2.5 g of wood alcohol dissolved in enough grain alcohol to yield 35 g of solution.
    (b) 212 mg of caffeine dissolved in alcohol to yield 15.0 g of solution.
    (c) 14.0 g of melted beeswax mixed with enough melted paraffin to yield a 125-g candle. (The mass of the wick is negligible.)
31. You want to prepare a 1.5% (m/m) zinc oxide ointment for use as a sun screen. How many grams of zinc oxide are needed to prepare 35 g of the ointment?
32. Calculate the mass/volume percent for potassium nitrate in these aqueous solutions:
    (a) 2.5 g KNO_3 in 55 mL (b) 0.24 g in 15 mL
    (c) 453 g in 2.000 L
33. Calculate the mass/volume percent for the solute in these aqueous solutions:
    (a) 1.00 g ascorbic acid in 150.0 mL (b) 0.50 g creatine in 200.0 mL (c) 3.00 g NH_3 in 235 mL
34. The mass of silver nitrate needed to prepare 65 mL of a 0.20% (m/v) aqueous solution is __________ grams.
35. Determine the volume/volume percent for these aqueous solutions:
    (a) 35 mL acetone in 150 mL
    (b) 18.5 mL alcohol in 0.125 L
36. Determine the volume/volume percent for these aqueous solutions:
    (a) 175 mL alcohol in 0.500 L
    (b) 0.125 L acetone in 1.75 L
37. Calculate the volume of alcohol needed to prepare 125 mL of an aqueous solution that is 22.5% (v/v) alcohol.
38. Rubbing alcohol is isopropyl alcohol. How many mL of isopropyl alcohol would be needed to prepare 125 mL of a 45% (v/v) aqueous solution of isopropyl alcohol?
39. Concentrated hydrochloric acid, HCl, is 12 *M*. Calculate the volume of concentrated HCl needed to prepare 2.5 L of a 1.0 *M* solution of HCl.
40. Concentrated acetic acid, $HC_2H_3O_2$, is 17.4 *M*. How many mL of concentrated acetic acid do you need to prepare 125 mL of a 0.150 *M* solution of acetic acid?
41. Your laboratory has a 6.0 *M* solution of nitric acid, but you need 2.0 *M* nitric acid. What volume of the 6.0 *M* nitric acid solution do you need to prepare 85 mL of 2.0 *M* nitric acid?
42. Explain Henry's law in your own words and give an example.
43. A patient is suffering from emphysema, a condition that reduces the transfer of gases from the lungs to the blood. This patient is given a gas mixture that contains more oxygen than air. Explain why this mixture improves oxygen transfer to the patient's blood.
44. If you open a bottle of soda and do not drink all of it, why does the soda go "flat" over time?

Dispersions and Suspensions *(Section 6.3)*

45. What is the difference between a solution and a dispersion? Provide an example of a dispersion.
46. Explain the difference between a solution and a suspension. Give an example of a suspension.
47. What is the difference between a dispersion and a suspension?
48. Is milk a solution, dispersion, or suspension?
49. Is gasoline a solution, dispersion, or suspension?
50. Starch consists of large molecules (macromolecules). When starch is added to water and heated, a mixture is obtained that has an opalescent appearance that is most obvious in brighter light. Is this mixture a solution, dispersion, or suspension?
51. A mixture containing water has an orange color, but appears clear. Is the mixture a solution, disper-

sion, or a suspension? If needed, what additional test could you do to decide?

52. Why do particles settle out of suspensions?

53. Some medicines are suspensions containing a finely divided solid medication in an aqueous solution. If you did not shake these medicines before taking them, would you be getting the correct dose, an inadequate dose, or an overdose of the medication? Explain.

54. Dispersions differ from solutions in what two properties?

55. What is the Tyndall effect? Give one or more common examples of the Tyndall effect.

56. In this chapter blood plasma was described as both a solution and a dispersion. How can plasma be both?

Colligative Properties (Section 6.4)

57. Define a colligative property. What common characteristic distinguishes colligative properties?

58. Name the three colligative properties of solutions described in this textbook.

59. When a solute is present in a solution, what happens to the freezing point of the solution compared to the freezing point of the pure solvent? What happens to the freezing point if additional solute is added?

60. When a solute is present in a solution, what happens to the boiling point of the solution compared to the boiling point of the pure solvent?

61. What is osmotic pressure? What would happen to the osmotic pressure of a solution if the concentration of solute increased?

62. Pure water and a 2.5% solution of sugar are separated by a semipermeable membrane.

(a) Which liquid, the water or the solution, has the higher osmotic pressure?

(b) In what direction will water flow across the membrane?

63. Animal cells, unlike plant cells and bacteria, lack cell walls. What happens when an animal cell is placed in a hypotonic solution? A hypertonic solution? An isotonic solution?

64. Explain how a renal (kidney) dialysis machine removes wastes from blood.

65. Compare the boiling-point elevation, freezing-point depression, and osmotic pressure of a 0.50 M aqueous solution of sucrose (table sugar) and a 0.50 M aqueous solution of glucose (blood sugar).

66. Why are the colligative properties of the solutions in Exercise 65 the same or different?

67. A can of soda, a can of diet soda, and a can of beer are cooled to a degree or two below freezing. What will happen to the contents of each can? What is the explanation for these observations?

68. Biological membranes are examples of semipermeable membranes. Explain what is meant by *semipermeable.*

69. A solution is said to be isotonic to blood. What does this mean?

Challenge Exercises

70. The alcohol ethanol (grain alcohol) is less polar than water but more polar than benzene. Which of the compounds in Exercise 11 might dissolve in ethanol? Justify your answers.

71. Calculate the molarity of a solution prepared by dissolving 38.5 g of calcium chloride in enough water to yield 1.00 L. What is the molarity of both the calcium ion and chloride ion in this solution?

72. Blood normally contains 60 to 100 mg of glucose per 100 mL of blood. Express this range of glucose in molarity. (Assume three significant figures for your calculations; the molar mass of glucose, 180 g/mol.)

73. How many grams of sucrose (molar mass = 342 g/mol) do you need to prepare 1.00 L of a 0.100 M aqueous solution of sucrose?

Paper is made from visible and microscopic fibers of bleached wood (cellulose) that are pressed and matted together. Consider a filter paper with interfiber spaces between 0.1 mm and 10 μm. Use this information to answer the next three questions.

74. Will filtration using this filter paper separate the components of a solution? Why or why not?

75. Will filtration using this filter paper separate the components of a dispersion? Why or why not?

76. Will filtration using this filter paper separate the components of a suspension? Why or why not?

7 Chemical Reactions

REACTIONS DIFFER BY TYPE, AND VARY IN ENERGY, RATES, AND POINTS OF EQUILIBRIUM.

THERE ARE many chemical reactions, and for convenience they are divided into three main types: acid–base reactions, precipitation reactions, and oxidation–reduction (redox) reactions. In this chapter you will be introduced to these three types of reactions and will learn about some of the properties that are common to all chemical reactions.

Outline

Volcanoes spew tons of hydrochloric acid and other acidic compounds into the earth's atmosphere and waters, where they react with naturally occurring bases, such as bicarbonate ion, HCO_3^-, and carbonate ion, CO_3^{2-}. *(Greg Vaughn/Tony Stone Images)*

7.1 Acids, Bases, and Salts

A very important group of chemical reactions are the acid–base reactions. The term *acid* may be familiar to you. You may have heard of battery acid (which is sulfuric acid), citric acid in citrus fruits, lactic acid in blood, and ascorbic acid (vitamin C). Acids in foods typically have a sour taste. You may also be familiar with some bases, such as ammonia in household cleaning agents and sodium hydroxide, which is present in some drain cleaners. Solutions of bases have a *slippery* feel to them, and bases in foods often have a bitter taste. **Salts** are the ionic compounds that form when acids and bases react with each other.

An **acid** is defined as any compound that can donate (give up) a hydrogen ion (H^+) (Figure 7.1). (When a hydrogen atom loses an electron it becomes a proton. As a result, some chemists use the word *proton* as a synonym for hydrogen ion.) Consider hydrochloric acid as an example:

An acid is a hydrogen ion (H^+) donor.

$$HCl(aq) \rightarrow H^+(aq) + Cl^-(aq)$$

Because the molecule of HCl gives up a hydrogen ion, it is an acid. The process that converts a molecule such as HCl into ions such as H^+ and Cl^- is called *ionization.*

Many common acids can be recognized as such because hydrogen is the first element in their formula: HNO_3 = nitric acid, H_2SO_4 = sulfuric acid.

A **base** is a hydrogen ion *acceptor* (Figure 7.2). In the reaction shown, hydroxide ion accepts a hydrogen ion to become a water molecule:

A base is a hydrogen ion (H^+) acceptor.

$$OH^-(aq) + H^+(aq) \rightarrow H_2O(l)$$

Because hydroxide ion accepts a hydrogen ion, it is a base; however, more commonly we think of compounds that contain hydroxide ion as bases. Sodium hydroxide (NaOH) and potassium hydroxide (KOH) are common bases because they yield hydroxide ion when they dissolve in water:

When an ionic compound dissolves yielding ions in solution, we say the compound *dissociated.* Bases containing hydroxide ion are ionic compounds, thus they dissociate yielding ions.

$$NaOH(aq) \rightarrow Na^+(aq) + OH^-(aq)$$

The base sodium hxydroxide yields hydroxide ion (hydrogen-ion acceptor)

Ammonia (NH_3) and bicarbonate ion (HCO_3^-) are additional examples of bases. These equations show the reactions of ammonia and bicarbonate ion with hydrogen ion:

$$NH_3(aq) + H^+(aq) \rightarrow NH_4^+(aq)$$

$$HCO_3^-(aq) + H^+(aq) \rightarrow H_2CO_3(aq)$$

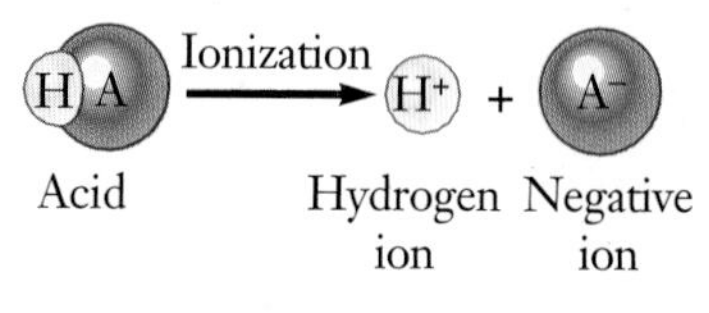

Figure 7.1 An acid is a compound that can liberate a hydrogen ion.

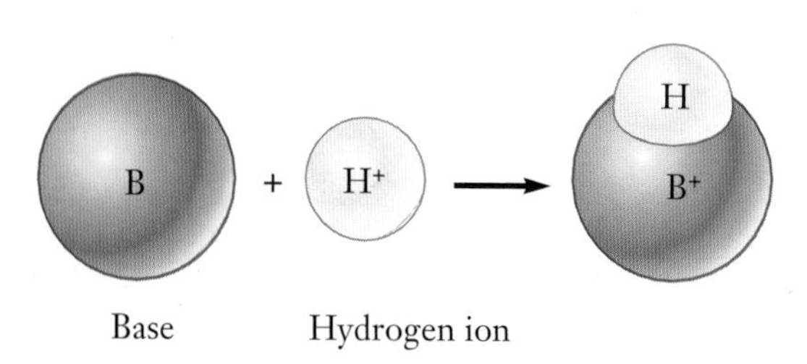

Figure 7.2 A base is a compound that accepts a hydrogen ion.

A hydrogen ion cannot be donated to empty space; there must be something there to accept it. In an acid–base reaction, an acid donates its hydrogen ion to a base, and a base accepts a hydrogen ion from the acid. Consider again hydrochloric acid, HCl. This compound acts as an acid in water, which means it donates a hydrogen ion. Water accepts the hydrogen ion; thus water is acting as a base in this reaction (Figure 7.3):

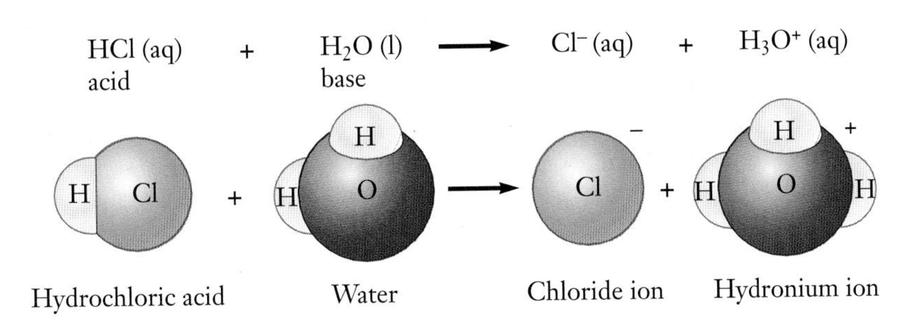

Figure 7.3 In this reaction, water accepts a hydrogen ion from hydrochloric acid. Hydrochloric acid is acting as an acid (hydrogen ion donor) and water is acting as a base (hydrogen ion acceptor).

The H_3O^+ ion is called the **hydronium ion.** It forms when water accepts a hydrogen ion. In water, acid molecules donate their hydrogen ions to water to form hydronium ions, but often water is left out of the equation for the reaction. The acid is then simply shown reacting to yield ions:

In aqueous solutions, the symbols H_3O^+(aq) and H^+(aq) are often used interchangeably.

$$HCl(aq) \rightarrow H^+(aq) + Cl^-(aq)$$

Acids differ in their tendency to donate hydrogen ions. With some acids, the **strong acids,** all the acid molecules donate their hydrogen ions in water. The acid is 100% ionized. Hydrochloric acid, nitric acid (HNO_3), and sulfuric acid (H_2SO_4) are the most common strong acids. With other acids, the **weak acids,** only some of the acid molecules in water donate their hydrogen ions at any given time. Many common substances are weak acids. Weak acids are less than 100% ionized when they dissolve in water. In an aqueous solution of a weak acid, only a few percent of the acid molecules have donated their hydrogen ions. The equation for the reaction of acetic acid, the weak acid in vinegar, with water is

$$HC_2H_3O_2(aq) \rightleftharpoons H^+(aq) + C_2H_3O_2^-(aq)$$

Acetic acid (Most of the molecules in the original solution remain as molecules.)

Acetate ion (Only a fraction of the molecules in the original sample form ions.)

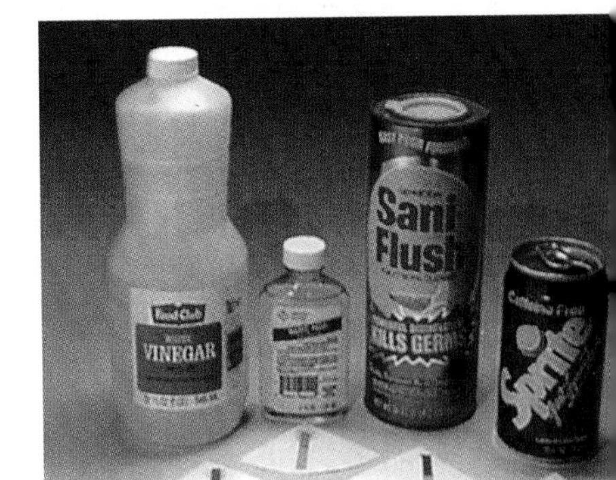

Weak acids are found in many common materials. In the foreground are strips of litmus paper that have been dipped into solutions of these materials. Acids turn litmus paper red. *(Charles D. Winters)*

Note the double arrow in this equation. Double arrows in chemical equations mean the reaction is a reversible one (you will learn more about reversible reactions later in this chapter). Note that the equations used for weak acids use double arrows ($\rightleftharpoons$) to imply less than 100% reaction, and those for strong acids use single arrows ($\rightarrow$) to imply complete, 100%, reaction.

Bases are also classified as strong or weak. A **strong base** is a hydroxide-containing compound that dissociates in water to yield a high concentration of hydroxide ion (OH^-). There are only a few strong bases, including the Group-1A hydroxides—sodium hydroxide and potassium hydroxide are examples—and the

Weak bases are also common in many household products. These cleaning agents all contain weak bases. Bases turn litmus paper blue. *(Charles D. Winters)*

larger Group-2A hydroxides—barium hydroxide and strontium hydroxide. Most other hydroxide-containing compounds are either insoluble or only slightly soluble in water.

Solutions of **weak bases** contain only small quantities of hydroxide ions because most of the base molecules do not yield hydroxide ion in the solution. Ammonia is an example of a weak base. Ammonia is quite water-soluble, but when ammonia is dissolved in water, only a small amount of it reacts with water to produce a small amount of hydroxide ion:

$$\underset{\text{Base}}{NH_3(aq)} + \underset{\text{Acid}}{H_2O(l)} \rightleftharpoons NH_4^+(aq) + OH^-(aq)$$

Most of the ammonia remains as ammonia and only a little hydroxide ion is formed. In Chapter 10 you will learn about amines, which are organic compounds that are also weak bases.

Use Table 7.1 as a guide to learning the definitions and properties of acids and bases.

Table 7.1 A Comparison of Acids and Bases

SUBSTANCE	PROPERTIES	COMMON EXAMPLES
Acid	H^+ donor; tastes sour	
Strong acid	Ionizes completely in water	Hydrochloric acid (HCl), nitric acid (HNO_3), and sulfuric acid (H_2SO_4)
Weak acid	Ionizes slightly in water	Acetic acid ($HC_2H_3O_2$), and citric acid ($H_3C_6H_5O_7$)
Base	H^+ acceptor; tastes bitter, feels slippery	
Strong base	Dissolves completely in water, yielding significant amounts of OH^-	Sodium hydroxide (NaOH), and potassium hydroxide (KOH)
Weak base	Compound that reacts slightly with water, yielding small amounts of hydroxide ion	Ammonia (NH_3)

7.2 Ionization of Water

When water reacts with ammonia, the water acts as an acid. Water also acts as a base when it accepts a hydrogen ion from an acid:

$$\underset{\text{Acid}}{HCl(aq)} + \underset{\text{Base}}{H_2O(l)} \rightarrow Cl^-(aq) + H_3O^+(aq)$$

In a pure water sample, a tiny number of the water molecules act as acids and donate their hydrogen ions to other water molecules (Figure 7.4):

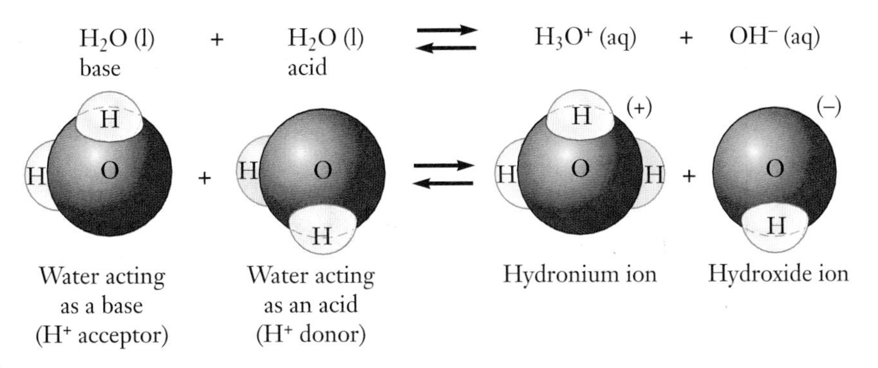

Figure 7.4 Water can act as either an acid or a base. Although all the hydrogen atoms in water are equivalent, one of them in this illustration has been colored yellow to help you see the hydrogen ion transfer.

In this reaction, one water molecule acts as an acid and donates a hydrogen ion, and another water molecule acts as a base and accepts the hydrogen ion. A hydronium ion and a hydroxide ion form. This ionization of a water molecule means that pure water contains very small and equal amounts of hydrogen ion (as hydronium ion) and hydroxide ion.

When water or an aqueous solution contains equal numbers of hydrogen ions and hydroxide ions, the solution is said to be *neutral.* If there is a larger amount of hydrogen ion than hydroxide ion, then the solution is *acidic.* If there are more hydroxide ions than hydrogen ions, then the solution is *basic* (another word for basic is *alkaline;* Table 7.2).

At room temperature (25°C) the concentration of hydrogen ions and hydroxide ions in pure water or neutral solutions is 1.0×10^{-7} M for both ions. In acidic solutions, the concentration of hydrogen ion will be higher than that of hydroxide ion, and in basic solutions, the concentration of hydroxide ion will be higher than that of hydrogen ion. We can now look at acidic and basic solutions in a more numerical way.

There is an expression that allows us to calculate the concentration of one of these ions if the concentration of the other is known:

$$1.0 \times 10^{-14} = [H^+][OH^-]$$

Brackets around a chemical formula mean molar concentration.

The number 1.0×10^{-14} is known as the **ion-product constant of water** and is symbolized by K_w:

$$K_w = 1.0 \times 10^{-14} = [H^+][OH^-]$$

Table 7.2 Neutral, Acidic, and Basic Solutions

Neutral solutions and pure water . . .	contain equal amounts of hydrogen ions and hydroxide ions.
Acidic solutions . . .	always contain more hydrogen ions than hydroxide ions.
Basic solutions . . .	always contain more hydroxide ions than hydrogen ions.

Consider pure water, in which, by definition, $[H^+] = 1.0 \times 10^{-7}$ *M* and $[OH^-] = 1.0 \times 10^{-7}$ *M*. If we substitute these numbers into the ion-product constant expression, we get

$$K_w = [H^+][OH^-]$$

$$1.0 \times 10^{-14} = [1.0 \times 10^{-7}] \times [1.0 \times 10^{-7}]$$

$$1.0 \times 10^{-14} = 1.0 \times 10^{-14}$$

Because the two sides are equal to each other, the expression must be correct for neutral water.

What if the concentration of hydrogen ion were 5.0×10^{-4} *M*? What would be the $[OH^-]$? Simply rearrange the expression for the ion-product constant of water to isolate the unknown, then solve

$$K_w = [H^+][OH^-]$$

$$[OH^-] = \frac{K_w}{[H^+]} = \frac{1.0 \times 10^{-14}}{5.0 \times 10^{-4}} = 2.0 \times 10^{-11}\ M$$

Note that the product of $[H^+]$ and $[OH^-]$ equals 1.0×10^{-14} for pure water *and* for aqueous solutions.

We said that an acidic solution has more hydrogen ions than hydroxide ions. That means that in an acidic solution, the molar hydrogen-ion concentration, which is symbolized as $[H^+]$, must be greater than 1.0×10^{-7} *M*. It also follows from the ion-product expression that in acidic solutions, the $[OH^-]$ must be less than 1.0×10^{-7} *M*. The same argument holds true for hydrogen-ion and hydroxide-ion concentrations in basic solution. A basic solution always has a hydrogen-ion concentration less than 1.0×10^{-7} *M* and a hydroxide-ion concentration greater than 1.0×10^{-7} *M*.

Example 7.1 Working with the Ion-Product Constant of Water

A solution has a hydroxide-ion concentration of 4.6×10^{-6} *M*. Is the solution neutral, basic, or acidic? What is the molar concentration of hydrogen ion in the solution?

SOLUTION

Because the hydroxide-ion concentration is greater than 1.0×10^{-7} *M*, the solution is basic. The hydrogen-ion concentration can be determined by first rearranging the ion-product constant expression for water to isolate $[H^+]$, then substituting and solving:

$$K_w = [H^+][OH^-]$$

$$[H^+] = \frac{K_w}{[OH^-]} = \frac{1.0 \times 10^{-14}}{4.6 \times 10^{-6}} = 2.2 \times 10^{-9}\ M$$

Self-Test

Determine the molar concentration of 1. H^+ in a solution that is 1.7×10^{-8} *M* OH^- and 2. OH^- in a solution that is 5.3×10^{-5} *M* H^+. State whether each solution is acidic, neutral, or basic.

ANSWERS

1. $[H^+] = 5.9 \times 10^{-7}$ *M*, acidic 2. $[OH^-] = 1.9 \times 10^{-10}$ *M*, acidic

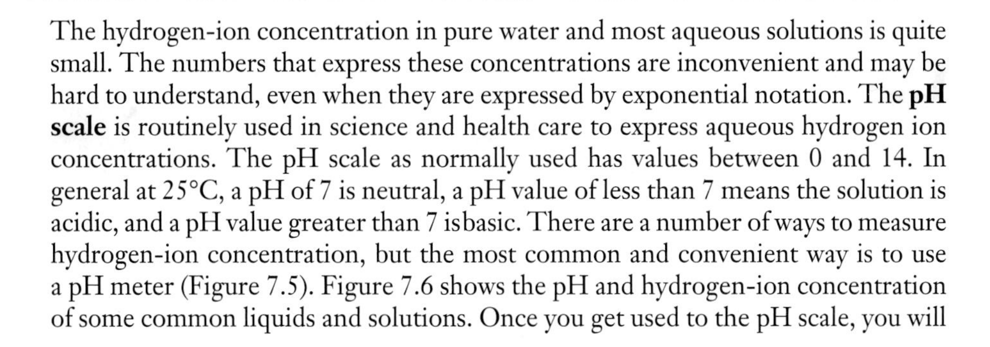

7.3 pH

The hydrogen-ion concentration in pure water and most aqueous solutions is quite small. The numbers that express these concentrations are inconvenient and may be hard to understand, even when they are expressed by exponential notation. The **pH scale** is routinely used in science and health care to express aqueous hydrogen ion concentrations. The pH scale as normally used has values between 0 and 14. In general at 25°C, a pH of 7 is neutral, a pH value of less than 7 means the solution is acidic, and a pH value greater than 7 isbasic. There are a number of ways to measure hydrogen-ion concentration, but the most common and convenient way is to use a pH meter (Figure 7.5). Figure 7.6 shows the pH and hydrogen-ion concentration of some common liquids and solutions. Once you get used to the pH scale, you will

The term pH is from the French *pouvoir hydrogène*, meaning hydrogen power.

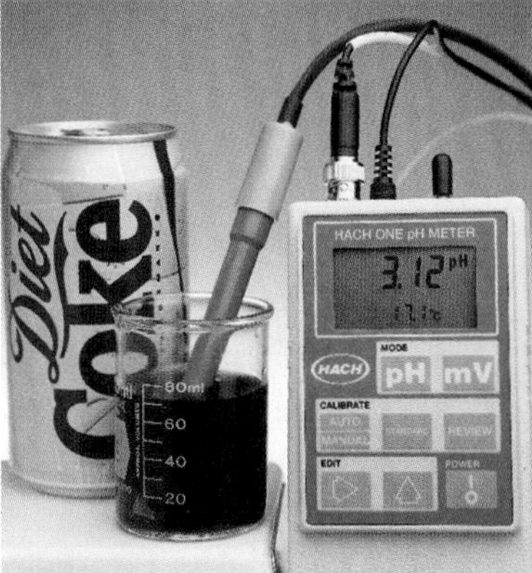

Figure 7.5 The pH (a measure of hydrogen ion concentration) of a solution is most easily measured with a pH meter. These devices are common laboratory instruments. Here the pH of a soft drink is measured as 3.12. Soft drinks are often quite acidic. *(Charles D. Winters)*

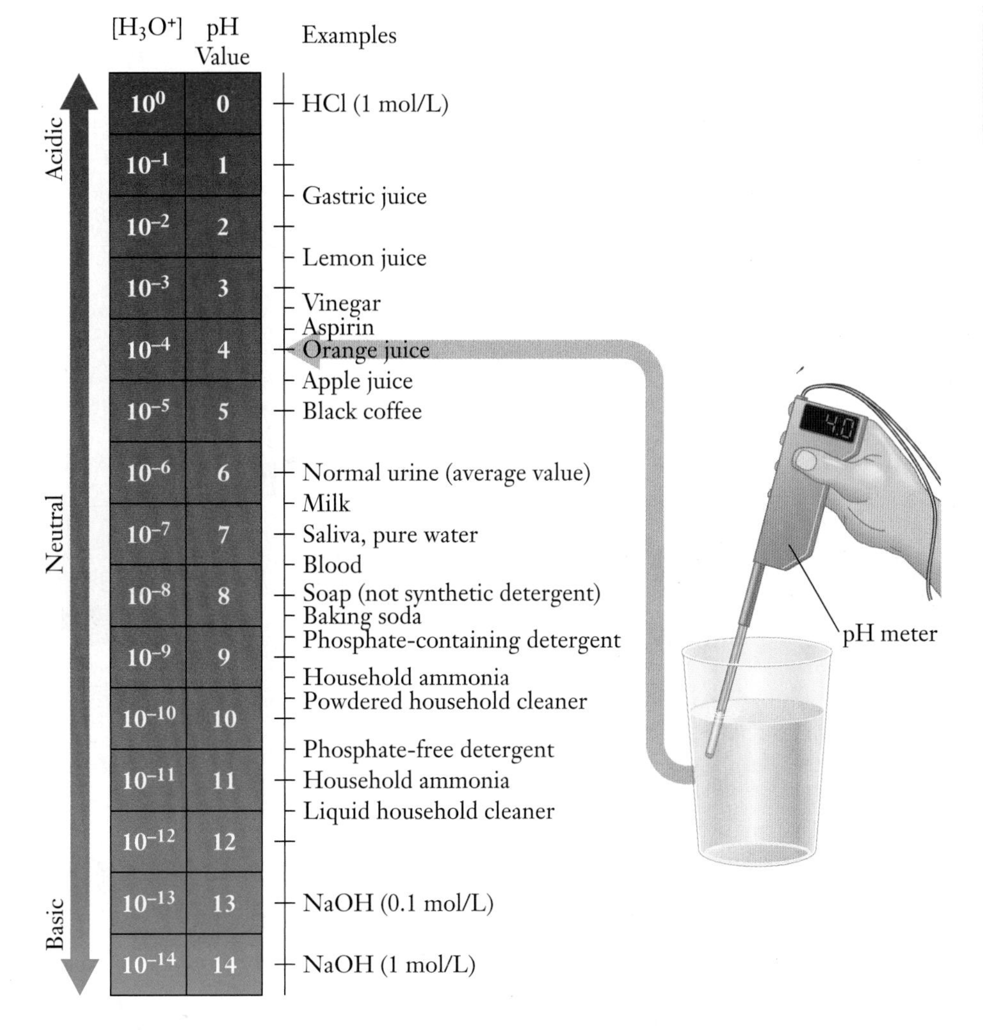

Figure 7.6 The pH values and hydrogen ion concentrations of some common solutions. Are you surprised by the acidity or basicity of some of them?

find it more useful and convenient than hydrogen-ion concentrations.

Example 7.2 The Acidity or Basicity of Solutions

1. Determine if these solutions are acidic, basic, or neutral: coffee, vinegar, baking soda, saliva.
2. Compare the acidity or basicity of any solutions that are acidic or basic.

SOLUTIONS

1. We could measure the pH of these solutions and state that any solution that has a pH less than 7 is acidic, any with a pH of 7 is neutral, and any with a pH greater than 7 is neutral. For convenience, Figure 7.6 lists the pH of several common solutions, including these examples, so a glance at the figure shows that coffee and vinegar are acidic, a baking soda solution is basic, and saliva is neutral.
2. The further a pH value is from neutral, the more acidic or basic the solution. Thus vinegar is more acidic than coffee.

Self-Test

1. Determine if these solutions are acidic or basic: (a) phosphate-free detergent (b) gastric juice (c) normal urine.
2. Which of these solutions is the most acidic?

ANSWERS

1. (a) Basic (b) Acidic (c) Acidic 2. Gastric juice

The pH scale is based on the logarithm (base 10) of hydrogen-ion concentration. Neutral water has a hydrogen-ion concentration of 1.0×10^{-7} *M*, and the pH of this solution is 7.00. By definition, $pH = -\log[H^+]$, thus $7.00 = -\log[1.00 \times 10^{-7}]$. Take a moment to review logarithms if you are not comfortable with them.

A Quick Review of Logarithms

We have already learned that any number can be expressed in scientific notation, which is a number times ten raised to a power. For example:

$$229 = 2.29 \times 10^2$$

These numbers can also be written another way, as ten raised to a power; 229 expressed this way is $10^{2.360}$. Thus

$$229 = 2.29 \times 10^2 = 10^{2.360}$$

Do this calculation on your calculator. Enter the number 10, press the y^x key, then enter 2.360. Now press the equal key. You should get 229, to three significant figures.

The logarithm (log) of a number is the exponent that 10 must be raised to to provide the number:

$$10^{2.360} = 229$$

Stated another way:

$$\log 229 = 2.360$$

Calculations involving logarithms are easily solved with a calculator using a few key strokes. Do this calculation on your calculator. Enter 229 then press the **log** key. You should get 2.360 to three significant figures.

Calculating pH is straightforward with a calculator. Enter 1.0×10^{-7} in your calculator and push the **log** key. The display should read -7. Substituting this into the pH expression yields pH $= -(-7)$. Minus a negative value is positive (press the **+/–** key to change sign), so the pH is 7.00. Thus in pure water or a solution in which $[H^+]$ is 1.0×10^{-7} *M*, the pH is 7.00, and is neutral.

When you calculate pH, be aware of two things. First, be careful to take the logarithm of the hydrogen-ion concentration correctly. Your calculator probably has a log key, but be sure you know what it is and how to use it (check with your instructor if you need help). Second, for our purposes, remember that the number of significant figures to the right of the decimal point in a pH value represents the number of significant figures in the molarity value. The number 1.0×10^{-7} *M* has two significant figures. A pH of 7.00 also has two significant figures. The 7 in the pH value has the same purpose as the 10^{-7} in the exponential notation; it shows size rather than the number of significant figures.

Example 7.3 Calculating pH

(a) A sample of muscle tissue has a hydrogen-ion concentration of 6.13×10^{-7} *M*. What is the pH of this sample?
(b) Is the sample acidic, neutral, or basic?

SOLUTION

(a) Use the expression for pH: pH $= -\log [H^+]$. Substituting the molarity into the expression yields

$$\text{pH} = -\log(6.13 \times 10^{-7})$$

To obtain the answer, enter 6.13×10^{-7} into your calculator, and press the **log** key. You should see $-6.2125\ldots$, which is the log of 6.13×10^{-7}. Substituting this value into the pH expression yields

$$\text{pH} = -(-6.2125\ldots)$$

Changing the sign **+/–** for the right side of the expression yields pH $= 6.213$. The 6 in 6.213 is related to the size of the pH value, and the 213 to the right of the decimal point shows there are three significant figures, just as the hydrogen-ion molarity, 6.13×10^{-7} *M* has three significant figures.
(b) The sample is acidic because the pH is less than 7 (and the $[H^+]$ is greater than 1.00×10^{-7} *M*).

To enter 6.13×10^{-7} on your calculator, press 6.13 **EE** or **Exp** 7 **+/–**.

Self-Test

Convert these hydrogen-ion concentrations to pH values: 1. 4.31×10^{-4} *M* 2. 7.2×10^{-8} *M* 3. 2.8×10^{-10} *M*

ANSWERS

1. 3.366 (Remember: Three digits to the right of the decimal point because there are three significant figures in 4.31×10^{-4}.) 2. 7.14 3. 9.55

Values expressed in pH units can be converted to hydrogen-ion concentrations with the expression

$$[H^+] = \text{inverse log}(-\text{pH})$$

First change the sign on the pH value (make it negative). Then take the inverse log by pressing the 10^x key on your calculator. (Your instructor can show you your corresponding key if your calculator uses a different label for this key.) For example, the hydrogen-ion concentration of a solution of pH 4.76 would be calculated this way:

$$[H^+] = \text{inverse log}(-\text{pH})$$

$$[H^+] = \text{inverse log}(-4.76)$$

$$[H^+] = 1.7 \times 10^{-5}\ M$$

Example 7.4 Converting pH to Hydrogen-Ion Concentration

What is the hydrogen-ion concentration in a blood sample that has a pH of 7.30?

SOLUTION

Use the expression $[H^+] = \text{inverse log}(-\text{pH})$. Enter 7.30 then press +/− key to give −7.30. With −7.30 displayed, press the 10^x key. The hydrogen-ion concentration is $5.0 \times 10^{-8}\ M$. Note that the pH value and the answer both have two significant figures.

Self-Test

The contents of the stomach can be rather acidic. What is the hydrogen-ion concentration in a sample of stomach fluid that has a pH of 2.13?

ANSWER

$7.4 \times 10^{-3}\ M$

7.4 Acid–Base Reactions

The reaction of a strong acid with a strong base is called **neutralization.** We say the base neutralizes the acid or the acid neutralizes the base. Consider the neutralization of hydrochloric acid with sodium hydroxide:

$$\underset{\text{Acid}}{HCl(aq)} + \underset{\text{Base}}{NaOH(aq)} \longrightarrow \underset{\text{Water}}{H_2O(l)} + \underset{\text{Salt}}{NaCl(aq)}$$

In the neutralization of a strong acid like HCl with a strong base like NaOH, two products are always formed, water and a salt. The water forms when the hydrogen ion from the acid combines with the hydroxide ion from the base. The salt consists of the cation from the base (Na^+ from NaOH in this case) and the anion from the acid (Cl^- from HCl). Water and a salt form whenever a strong acid reacts with a strong base.

Although many acid–base reactions involve solutions, a solid or liquid acid or base can also react. Consider the reaction of magnesium hydroxide, $Mg(OH)_2$, with aqueous sulfuric acid. Magnesium hydroxide is not very soluble in water, but the hydrogen ions from the acid can react with the hydroxide ions on the solid to form a salt and water:

$$\underset{\text{Acid}}{H_2SO_4(aq)} + \underset{\text{Base}}{Mg(OH)_2(s)} \longrightarrow \underset{\text{Water}}{2H_2O(l)} + \underset{\text{Salt}}{MgSO_4(aq)}$$

Many antacids are solid bases that react with hydrochloric acid in the stomach.

Weak bases do not contain hydroxide ions. When these bases react with an acid, water is not formed because the base has no hydroxide ion that can accept the hydrogen ion. Consider the reaction of ammonia with hydrogen chloride:

$$\underset{\text{Base}}{NH_3(g)} + \underset{\text{Acid}}{HCl(g)} \longrightarrow \underset{\text{Salt}}{NH_4Cl(s)}$$

When these two gases react, the NH_3 accepts a hydrogen ion from the HCl to become an ammonium ion. After HCl has lost the hydrogen ion, it is a chloride ion. Ammonium chloride is the product of the reaction. In general, reactions of acids with strong bases produce water and a salt and reactions of acids with weak bases produce a salt.

Example 7.5 Predicting the Products of Acid-Base Reactions

What are the products of these reactions?

1. $HNO_3(aq) + KOH(aq) \longrightarrow ?$
2. $HF(aq) + NaOH(aq) \longrightarrow ?$

SOLUTIONS

1. The products of the reaction between an acid and a base that contains an hydroxide ion are always water and a salt. The cation of the salt comes from the base, and the anion comes from the acid. Therefore, the products are water and potassium nitrate (KNO_3):

$$HNO_3(aq) + KOH(aq) \longrightarrow H_2O(l) + KNO_3(aq)$$

2. The argument in the first example applies in this case, too. The products are water and sodium fluoride (NaF):

$$HF(aq) + NaOH(aq) \longrightarrow H_2O(l) + NaF(aq)$$

Self-Test

Determine the products and write the balanced equation for this reaction:

$$HBr(aq) + Ba(OH)_2(s) \longrightarrow ?$$

ANSWER

Water and barium bromide ($BaBr_2$); $2HBr(aq) + Ba(OH)_2(s) \longrightarrow 2H_2O(l) + BaBr_2(aq)$

Acid–base reactions are used to determine the amount (concentration) of acid or base in a sample such as stomach acid, blood, or rainwater. Acidic or basic solutions of unknown concentration are often analyzed by **titration.** In a typical analysis for an acid, a known amount of the acidic solution, for example HCl, of unknown concentration is placed in a flask [Figure 7.7(a)]. Small amounts of a base, such as NaOH, of known concentration (the titrant) are then added with a buret until the acid is just neutralized. A pH meter can be used to measure the pH of the solution, or as an alternative, a *pH indicator*, a compound that has different colors at different pH values, may be used [Figure 7.7(b) and (c)]. The volume of added NaOH (base) that reacted with the HCl (acid) is read from the buret. The number of moles of NaOH can be calculated from the volume and molarity of the base. (Example 7.6 shows the calculations of a typical titration.) The moles of HCl can then be determined by using the balanced equation for the reaction. Finally, the moles of acid and the volume of the acid solution can be used to determine the mo-

Figure 7.7 **(a)** A known volume of a solution containing an unknown amount of acid has been placed in this flask for analysis by titration. The titrant, which is a basic solution of known concentration, is in the buret. **(b)** Titrant is being slowly added to the unknown acidic solution. A pH indicator that is colorless in acidic solutions has also been added. **(c)** When all of the acid is neutralized by the base in the titrant, the pH of the solution changes. This indicator is red at the pH where all the acid has reacted. *(Charles D. Winters)*

Connections

Antacids

Heartburn is caused by an excess amount of acid in the stomach and is treated by many people with over-the-counter antacids. The composition of these antacids varies from brand to brand, but many contain one or more of the following: calcium carbonate, $CaCO_3$, aluminum hydroxide, $Al(OH)_3$, magnesium hydroxide, $Mg(OH)_2$, or sodium or potassium bicarbonate, $NaHCO_3$ and $KHCO_3$.

If a strong base such as sodium hydroxide (NaOH) or potassium hydroxide (KOH) were used as an antacid, it would dissolve in the mouth, throat, and esophagus and react with the lining of these structures, causing serious tissue destruction. The makers of antacids get around this problem by using either a base that has low solubility such as calcium carbonate or magnesium hydroxide, or a base that is too weak to cause immediate tissue damage such as sodium or potassium bicarbonate. When these bases enter the stomach, they react with the hydrogen ion from the stomach acid, reducing or eliminating heartburn.

Perhaps the most common base currently used in antacids is calcium carbonate. The reaction of this base with hydrochloric acid (stomach acid) is shown in this equation:

$$2HCl(aq) + CaCO_3(s) \longrightarrow CaCl_2(aq) + H_2CO_3(aq)$$

The carbonic acid (H_2CO_3) that forms then breaks down in this reaction:

$$H_2CO_3(aq) \longrightarrow H_2O(l) + CO_2(g)$$

The use of calcium carbonate as an antacid has the additional advantage of providing calcium, which is a mineral that is present in inadequate amounts in many diets.

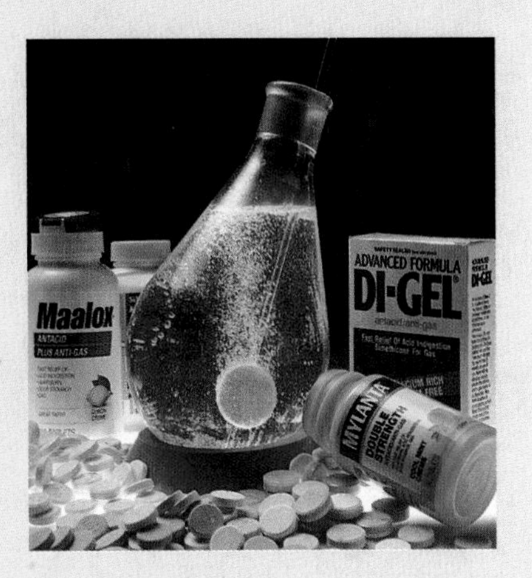

Some common antacids, all of which contain various weakly basic compounds. *(Charles D. Winters)*

larity of the acid. This equation and flowchart summarize the calculations used for the reaction of sodium hydroxide and hydrochloric acid:

$$NaOH(aq) + HCl(aq) \longrightarrow H_2O(l) + NaCl(aq)$$

Volume and molarity of NaOH → Moles of NaOH → Moles of HCl → Molarity of HCl

The concentration of a basic solution can be determined by titrating it with an acid.

Example 7.6 Titration of an Acidic Solution of Unknown Concentration

A 20.0-mL sample of clear fluid is removed from a patient's stomach and analyzed for hydrochloric acid by titration. A 0.100 *M* sodium hydroxide solution is used as the titrant. In the titration, 2.37 mL of the base is needed to neutralize all the acid in the sample. What is the molarity of the hydrochloric acid in the stomach fluid?

SOLUTION

First determine the number of moles of base needed to neutralize the acid. Begin with the expression for molarity, then rearrange to isolate moles:

$$M = \frac{\text{moles}}{\text{L}}$$

$$\text{Moles} = M \times \text{L}$$

Substitute in the appropriate values (remember to convert milliliters into liters), and solve to determine the number of moles of NaOH needed to neutralize the acid:

$$\text{Moles} = \frac{0.100 \text{ mol NaOH}}{\cancel{\text{L}}} \times 0.00237 \cancel{\text{L}} = 0.000237 \text{ mol NaOH}$$

Now write the balanced equation for the reaction of hydrochloric acid with sodium hydroxide:

$$HCl(aq) + NaOH(aq) \longrightarrow H_2O(l) + NaCl(aq)$$

Look at the balanced equation to determine the molar ratio of HCl to NaOH, and use this molar ratio to calculate the number of moles of HCl neutralized. Here the molar ratio is 1 (one mole of the base reacts with one mole of the acid), and so we have

$$0.000237 \cancel{\text{mol NaOH}} \times \frac{1 \text{ mol HCl}}{1 \cancel{\text{mol NaOH}}} = 0.000237 \text{ mol HCl}$$

Determine the molarity of the HCl by dividing the number of moles of HCl by the volume of the sample in liters:

$$M = \frac{0.000237 \text{ mol HCl}}{0.0200 \text{ L}} = 0.0119\ M \text{ HCl}$$

The molarity of the hydrochloric acid in the sample is 0.0119 M. For practice, calculate the pH of this sample. (It is 1.924.)

Self-Test

A 15.0-mL sample of a solution of nitric acid (HNO_3) requires 41.7 mL of a 0.250 M solution of sodium hydroxide (NaOH) to neutralize it. What is the molarity of the nitric acid solution?

ANSWER

0.695 M

Equivalents

The mole is the most commonly used measure of amount, but there is another term that is frequently used in health care. For acids, an **equivalent** is the amount of an acid that produces one mole of hydrogen ions. For bases, it is the amount of a base that produces one mole of hydroxide ions. The concept of equivalence is useful because in many reactions the most important thing is the amount of hydrogen ion or hydroxide ion present. It does not matter which acid or base provides it.

To help understand the idea of equivalents, let's look at some examples. For many acids and bases, one mole of the acid or base produces one mole of the H^+ or OH^- ion:

$$HCl(aq) \rightarrow H^+(aq) + Cl^-(aq)$$

$$NaOH(aq) \rightarrow Na^+(aq) + OH^-(aq)$$

For these substances, the equivalent is the same as the mole. One mole of HCl (36.5 g) produces one mole of hydrogen ion, and one mole of NaOH (40.0 g) produces one mole of hydroxide ion.

The mass of an acid or base that yields one mole of hydrogen ion or hydroxide ion is called the **equivalent mass** of the acid or base.

$$\text{Equivalent mass} = \frac{\text{molar mass (g)}}{\text{number of equivalents}}$$

For the acids and bases that yield one mole of hydrogen ion or hydroxide ion per mole of the acid or base, the equivalent mass is the same as the molar mass. For HCl, for instance, the equivalent mass is the same as the molar mass (36.5 g/mol) because an HCl molecule yields one hydrogen ion (Table 7.3). Likewise, the equivalent mass of NaOH is the same as the molar mass (40.0 g/mol) because one NaOH formula unit yields one hydroxide ion.

Some acids and bases yield more than one mole of hydrogen ion or hydroxide ion per mole of acid or base. For example, one mole of sulfuric acid yields two moles of hydrogen ion (it is a diprotic acid):

$$H_2SO_4(aq) \rightarrow 2H^+(aq) + SO_4^{2-}(aq)$$

One mole of sulfuric acid (98.08 g) yields two moles of hydrogen ions, which means one mole of sulfuric acid is two equivalents of hydrogen ion. Said another way, 0.5 mol of sulfuric acid is 1 equivalent. Because half a mole of sulfuric acid is one equivalent of hydrogen ion, the equivalent mass of sulfuric acid is 49.04 g, which is 0.5 mol (98.08 g/2) of sulfuric acid:

$$\text{Equivalent mass of sulfuric acid} = \frac{98.08 \text{ g}}{2 \text{ equivalents}} = \frac{49.04 \text{ g}}{\text{equivalent}}$$

When one mole of calcium hydroxide dissolves completely in water, it yields two moles of hydroxide ion; we say that it is *dibasic:*

$$Ca(OH)_2(s) \rightarrow Ca^{2+}(aq) + 2OH^-(aq)$$

One mole of calcium hydroxide is two equivalents of hydroxide ion—that is, one equivalent of hydroxide ion is provided by 0.5 mol of calcium hydroxide. The

Table 7.3 Equivalents and Equivalent Mass of Some Common Acids and Bases

ACID OR BASE	EQUIVALENTS/MOL	MOLAR MASS (g/mol)	EQUIVALENT MASS (g/eq)
HCl	1	36.5	36.5
HNO_3	1	63.0	63.0
H_2SO_4	2	98.1	49.1
H_3PO_4	3	98.0	32.7
NaOH	1	40.0	40.0
$Ca(OH)_2$	2	74.1	37.1

molar mass of calcium hydroxide is 74.1 g/mol, but its equivalent mass is half that, or 37.1 g/eq.

Example 7.7 Equivalents

What mass of sulfuric acid (H_2SO_4) yields 1.00 mol of H^+? What mass of phosphoric acid (H_3PO_4) yields 6.00 mol of H^+?

SOLUTION

One equivalent mass of an acid or base yields one mole of H^+ or OH^-. From Table 7.3, the equivalent mass of sulfuric acid is 49.1 g, thus 49.1 g of sulfuric acid provides 1.00 mol of hydrogen ion. For phosphoric acid, the equivalent mass is 32.7 g, thus this mass provides one mole of hydrogen ion. The mass of phosphoric acid needed to yield 6.00 mol of hydrogen ion is 32.7 g phosphoric acid/mol $H^+ \times$ 6.00 mol $H^+ = 196$ g phosphoric acid.

Self-Test

What mass of calcium hydroxide, $Ca(OH)_2$, is needed to provide 1.50 mol of hydroxide ion?

ANSWER

55.6 g

Equivalents are sometimes used to express the amounts of ions in solutions. One equivalent of an ion is the amount of that ion that yields one mole of charge. One mole of Na^+ has one equivalent of charge; one mole of Ca^{2+} has two equivalents of charge. Look at the Connections on Electrolytes in Body Fluids in Section 3.3 again. Does *milliequivalents* now make sense? (Remember *milli-* simply means 1/1000.)

Buffers

Acids react with bases and bases react with acids. Is there anything that can react with both acids and bases? Yes. **Buffered solutions,** or **buffers,** are solutions that resist changes in pH when acid or base is added to them. Buffers work because they contain both an acid to react with any added base and a base to react with any added acid. The only trick to making buffers is to choose the acid and base so they do not react with each other. This is accomplished by using either a weak acid and a salt of the same weak acid or a weak base and one of its salts.

It is essential to use *weak* acids or *weak* bases to prepare buffers. This is because a weak acid or base is only partially ionized in water, thus much of the acid or base remains in a solution and can react with any added base or acid. A second important point is that the salts of weak acids act as weak bases and the salts of weak bases act as weak acids. Thus a mixture of a weak acid and one of its salts is really a mixture of a weak acid and a weak base, and similarly, a mixture of a weak base and one of its salts is really a mixture of a weak base and a weak acid.

Let's consider the acetate buffer to see how a buffer works. We could use an acetate buffer whenever we need a buffered solution with a pH in the range of 4.5

Connections

The Buffering of Blood

Buffers in our blood maintain the pH near a value of 7.4. If blood were not buffered, blood pH would vary greatly because both breathing and reactions in the body affect the amount of acids and bases in the blood. At high or low pH values, many proteins in the body would lose their normal function, resulting in serious medical conditions or death.

There are two major buffering systems that contribute to stable blood pH: (1) the bicarbonate system and (2) the blood proteins. The bicarbonate system depends on carbon dioxide that is produced by the cells of the body. This carbon dioxide diffuses into blood and reacts with water to form carbonic acid. This in turn ionizes to form bicarbonate ion:

$$CO_2(aq) + H_2O(l) \rightleftharpoons \underset{\text{Carbonic acid}}{H_2CO_3(aq)} \rightleftharpoons \underset{\text{Bicarbonate ion}}{HCO_3^-(aq)} + H^+(aq)$$

The weak acid carbonic acid and the anion of its salt, bicarbonate, are the components of the buffering system. Carbonic acid can react with any added base, and bicarbonate ion reacts with any added acid. The components of this system are maintained at appropriate concentrations because the formation of carbon dioxide in cells is balanced by the exhalation of carbon dioxide in the lungs.

The proteins found in blood, especially the most abundant blood protein, hemoglobin, make up the second blood buffering system. Proteins are macromolecules (really BIG molecules) that contain groups that are either weak acids or weak bases. Thus collectively these groups can react with either added acid or base.

Hyperventilation is rapid breathing, causing inappropriately low amounts of $CO_2(aq)$ in the blood. As $CO_2(aq)$ levels drop in the blood, there is a decline of H^+ and HCO_3^- amounts in the blood plasma, causing blood pH to rise above its normal pH of 7.4. Breathing in a paper bag helps to restore CO_2 levels in the blood. *(Charles D. Winters)*

to 5.0. This buffer is prepared by mixing an appropriate amount of acetic acid ($HC_2H_3O_2$) with an appropriate amount of a salt that contains the acetate ion ($C_2H_3O_2^-$). When prepared, this mixture will have some acetic acid (a weak acid) and some acetate ion (a weak base).

When a base that contains hydroxide ion is added to an acetate buffer, the acetic acid in the buffer reacts with the added hydroxide ion in this reaction:

$$\underset{\substack{\text{Acetic acid}\\ \text{weak acid}}}{HC_2H_3O_2(aq)} + \underset{\substack{\text{Hydroxide ion}\\ \text{from base}}}{OH^-(aq)} \rightarrow H_2O(l) + C_2H_3O_2^-(aq)$$

Because the added hydroxide ion reacts with the acetic acid rather than with hydrogen ion, the concentration of hydrogen ion in the buffer remains nearly the same, and so the pH changes very little (Figure 7.8).

When an acid is added to the buffer, the acetate ions in the buffer react with the added hydrogen ions in this reaction:

$$\underset{\substack{\text{Acetate ion}\\ \text{weak base}}}{C_2H_3O_2^-(aq)} + \underset{\substack{\text{Hydrogen ion}\\ \text{from acid}}}{H^+(aq)} \rightarrow HC_2H_3O_2(aq)$$

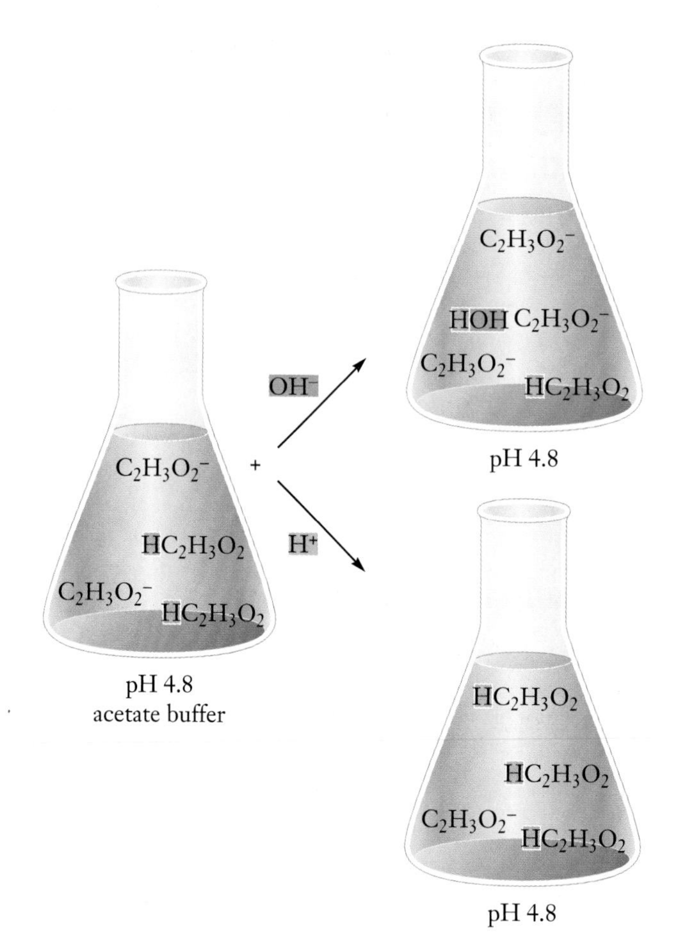

Figure 7.8 A pH 4.8 acetate buffer contains acetic acid ($HC_2H_3O_2$) and acetate ions ($C_2H_3O_2^-$). When base (OH^-) is added to the buffer, some of the acetic acid in the buffer reacts with the added base. The pH of the buffer remains essentially the same. When acid (H^+) is added, some of the acetate ion in the buffer reacts with the added acid. Again, the pH remains essentially the same.

Because the acetate ion reacted with the added H^+ the concentration of H^+ in the solution does not change much, and the pH of the solution remains nearly the same (Figure 7.8).

Buffers play an extremely important role in chemistry and health care.

Example 7.8 Choosing the Components of a Buffer

Which of the following solutions would be buffered? Why would they work or not work as a buffer?

A solution containing 1. H_2CO_3 and $NaHCO_3$ 2. HCl and NaCl

SOLUTIONS

1. Carbonic acid, H_2CO_3, is a weak acid, and sodium bicarbonate, $NaHCO_3$, is a salt of this acid. This solution is buffered because the carbonic acid can react with any added base, and the bicarbonate ion can react with any added acid.

2. HCl is a strong acid, and NaCl is a salt of this acid. This is not a buffered solution. HCl ionizes completely to H^+ and Cl^-. There is no HCl left in the solution to react with added base. In addition, the salts of strong acids cannot act as bases, so NaCl would not react with any added acid.

Self-Test

Which of these solutions would be buffered? If they are not buffered, explain why. A solution containing 1. HNO_3 and KNO_3 2. NH_4Cl and NH_3

ANSWERS

1. Not buffered; nitric acid is a strong acid. 2. Buffered

Connections

Acid Rain and the Environment

Acid rain is simply rain that is more acidic than normal. It is formed primarily from the reaction of atmospheric water with some emissions formed during the combustion of fossil fuels. Many fossil fuels, such as coal, gasoline, and diesel fuel, contain sulfur-containing compounds. When these fuels are burned in coal-fired electrical generation plants and vehicles, the sulfur is converted into sulfur dioxide:

$$S \text{ (in fuel molecules)} + O_2(g) \rightarrow SO_2(g)$$

If the sulfur dioxide is not removed from the exhaust or emissions, it is converted into sulfur trioxide in the atmosphere:

$$2SO_2(g) + O_2(g) \rightarrow 2SO_3(g)$$

Sulfur trioxide in the atmosphere reacts with water to form sulfuric acid:

$$SO_3(g) + H_2O(g) \rightarrow H_2SO_4(g)$$

This sulfuric acid dissolves in the water in clouds making the water acidic. When this water falls it is *acid rain.*

Just as our blood is buffered, much of the water on Earth is also buffered and is relatively unaffected by acid rain. However, some natural waters, including many lakes in the northeastern United States and adjacent parts of Canada, are relatively unbuffered. Runoff from acid rain enters these lakes and lowers their pH. Because many aquatic organisms cannot live in these acidic conditions, they die, and the normal aquatic community disappears. Acid rain also kills many trees and other plants, and the acid in the rain may react with structures made of limestone or marble.

The production of acid rain can be reduced by using low-sulfur fuels or by removing sulfur dioxide from emissions. The effects of acid rain can be reduced by treating the affected area with a base such as calcium carbonate, which is both inexpensive and, because of its low solubility, effective over a relatively long period of time.

The effect of acid rain on stone is readily seen in the pitting and corrosion of this statue. Acid rain is also responsible for the devastation of many forests and trees. *(statue, Adam Hart-Davis, Science Photo Library/ Photo Researchers, Inc.; trees, Bobbie Kingsley/Photo Researchers, Inc.)*

7.5 Oxidation–Reduction and Precipitation Reactions

A second major group of chemical reactions—those involving loss or gain of electrons by the reactants—are the **oxidation–reduction (redox) reactions.** These are common chemical reactions and are responsible for producing electrical energy in batteries and corroding metals. Our interest in these reactions, however, is prompted by the many redox reactions that occur in the body. Some of our oxidation–reduction reactions break down fuel molecules such as glucose to produce energy and others are involved in the building of molecules such as fats and cholesterol.

Oxidation is formally a loss of electrons. In the reaction of magnesium with oxygen, the magnesium atoms are oxidized to magnesium ions:

$$2Mg(s) + O_2(g) \longrightarrow 2MgO(s)$$

Magnesium is in Group 2A, and forms stable 2+ cations. We can illustrate the loss of electrons by the magnesium atoms in this way:

$$2Mg \longrightarrow 2Mg^{2+} + 4e^-$$

The mnemonic *oilrig* can be used to remember that *o*xidation *i*s *l*oss and *r*eduction *i*s *g*ain.

Reduction is formally a gain of electrons. Oxygen forms stable 2− anions, thus during the reaction of magnesium and oxygen the oxygen atoms each gain two electrons to become oxide ions:

$$O_2 + 4e^- \longrightarrow 2O^{2-}$$

In the reaction between magnesium and oxygen, the electrons lost by the magnesium atoms are gained by the oxygen atoms. Just as an acid and a base must be present for an acid–base reaction to occur, both an oxidation and a reduction must occur in a redox reaction. In general, during reactions between metals and nonmetals the metal atoms become oxidized and the nonmetal atoms become reduced.

Example 7.9 Predicting the Products of an Oxidation–Reduction Reaction

Decide which reactants are oxidized and which are reduced, then predict the products of this oxidation–reduction reaction.

$Mg(s) + Cl_2(g) \longrightarrow ?$

SOLUTION

Because magnesium is a metal, it will be oxidized, which means it must lose electron(s). A magnesium atom must lose two electrons to become a stable ion because magnesium is in Group 2A of the periodic table. $Mg \longrightarrow Mg^{2+} + 2e^-$. Nonmetals tend to gain electrons, thus the chlorine atoms will each gain one electron to become reduced: $Cl_2 + 2e^- \longrightarrow 2Cl^-$. The products are magnesium ion (Mg^{2+}) and two chloride ions (Cl^-).

Self-Test

What are the products of the oxidation–reduction reaction $Li(s) + Br_2(l) \longrightarrow ?$ (Remember from Section 3.3 that metal atoms tend to lose electrons and nonmetal atoms tend to gain them.)

ANSWER

$2Li(s) + Br_2(g) \rightarrow 2LiBr(s)\ (2Li^+ + 2Br^-)$

In the previous examples involving a metal and a nonmetal it was easy to see which atoms were oxidized—the ones that lost electrons to become cations—and which atoms were reduced—the ones that gained electrons to become anions. In some other redox reactions, however, oxidation and reduction are less obvious. For many of these reactions, oxidation can be recognized by the addition of oxygen atoms to an atom or a loss of hydrogen atoms from an atom. Reduction is recognized by a gain of hydrogen atoms or a loss of oxygen atoms. Consider the reaction of methane (CH_4) with oxygen (O_2):

$$CH_4(g) + 2O_2(g) \rightarrow CO_2(g) + 2H_2O(g)$$

The carbon atom in the reactant methane molecule has hydrogen bound to it, but the carbon atom in the product carbon dioxide molecule has oxygen bound to it. Both the loss of hydrogen and the gain of oxygen indicate that the carbon atom has been oxidized during this reaction. During the reaction, the oxygen atoms of molecular oxygen that gained hydrogen have been reduced.

Example 7.10 Recognizing Oxidation and Reduction by Loss of Hydrogen or Gain of Oxygen

1. Hydrogen sulfide reacts with oxygen to produce sulfur dioxide and water: $2H_2S(g) + 3O_2(g) \rightarrow 2SO_2(g) + 2H_2O(g)$. Which atoms have been oxidized and which have been reduced?
2. Sulfur dioxide reacts with oxygen to produce sulfur trioxide: $2SO_2(g) + O_2(g) \rightarrow 2SO_3(g)$. During the reaction, is the sulfur atom oxidized or reduced?

SOLUTION

1. Because the sulfur atom in hydrogen sulfide has lost hydrogen during the reaction and gained oxygen, the sulfur atom has been oxidized. The oxygen atoms in molecular oxygen that gained hydrogen have been reduced.
2. The sulfur atom in the reactant sulfur dioxide molecule has two oxygen atoms, and the sulfur atom in the product sulfur trioxide molecule has three oxygen atoms. The sulfur atom has gained oxygen, thus it has been oxidized.

Self-Test

Methanol is a fuel used by race cars and is often added to gasoline to reduce emissions. When methanol is burned, which atoms are oxidized and which are reduced?

$$CH_3OH(l) + O_2(g) \rightarrow CO_2(g) + H_2O(l) \quad \text{(Unbalanced)}$$

ANSWER

The carbon atoms of methanol are oxidized (they lost hydrogen and gained oxygen). The oxygen atoms of molecular oxygen have been reduced (gained hydrogen).

Connections
Gallstones and Kidney Stones

The concentrations of molecules and ions in body fluids are normally kept within a narrow range. As a result, dissolved compounds normally remain dissolved. However, if conditions change, some of these compounds can precipitate to form stones or deposits.

Gallstones form from bile, which is a thick fluid produced by the liver and stored in the gallbladder for later use in the small intestine. Bile contains cholesterol, and the most common type of gallstone forms if cholesterol becomes too concentrated in the bile. These stones are usually found in either the bile ducts or the gallbladder.

Kidney stones may form in the kidney pelvis during the formation of urine. These stones are usually composed of ionic compounds containing calcium ion as the cation, and either urate, oxalate, phosphate, or carbonate as the anions (urate and oxalate are organic anions that are topics of organic chemistry). From a chemical standpoint we can describe kidney stone formation in terms of precipitation reactions, but medical researchers do not yet know the specific physiological conditions that lead to the formation of kidney stones.

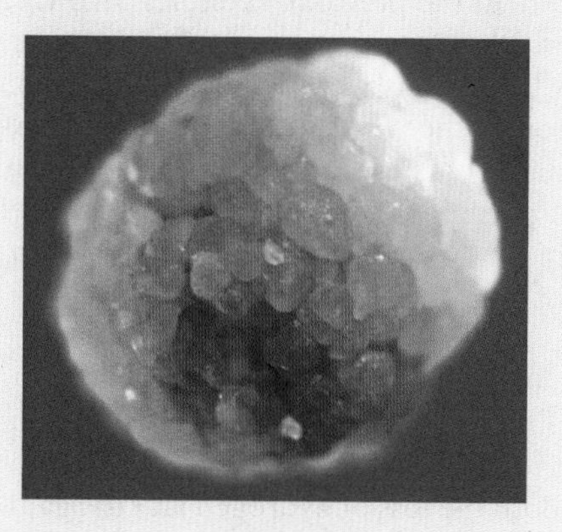

This gall stone formed by precipitation of cholesterol in the gallbladder. *(© Carolina Biological Supply Company, Phototake, NYC)*

A third group of chemical reactions—**precipitation reactions**—occurs whenever an insoluble product is formed during a chemical reaction. Once formed, the insoluble product then precipitates (falls out) of solution. These reactions typically occur when two ions come in contact to form an insoluble salt (Figure 7.9). Although these reactions are of limited importance in health care, formation of deposits or "stones" from low-solubility substances does occur. Gout, kidney stones, and gallstones are examples (see Connections on Gallstones and Kidney Stones).

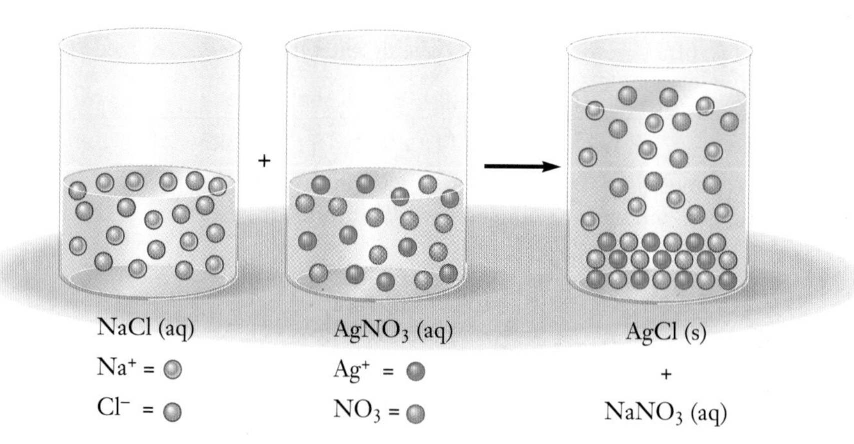

Figure 7.9 Silver ions and chloride ions in solution collide and form insoluble silver chloride (AgCl) as a precipitate.

7.6 Properties of Chemical Reactions

Thermochemistry

Thermochemistry is the part of chemistry that deals with heat in chemical reactions. When reactants are converted to products in chemical reactions, energy as heat is either produced or consumed. Reactions that produce heat are called **exothermic reactions** [Figure 7.10(a)]. The thermal energy produced by any exothermic reaction is given off to the surroundings and can be considered a product of these reactions:

$$\text{Reactants} \longrightarrow \text{products} + \text{energy} \qquad \text{(exothermic)}$$

If reactants take up heat as they react to form products, then the reaction is an **endothermic reaction** [Figure 7.10(b)]. The surroundings cool as the reaction occurs. Energy can be considered a reactant in these reactions:

$$\text{Reactants} + \text{energy} \longrightarrow \text{products} \qquad \text{(endothermic)}$$

Energy cannot be created or destroyed, so where is this energy released by exothermic reactions coming from? It comes from potential energy stored in the bonds of the reactant molecules. When the reaction occurs, this potential energy is converted into kinetic energy (usually heat). In endothermic reactions the opposite is true: Some kinetic energy (heat) from the surroundings is converted to potential energy stored in the bonds of the product molecules. Figure 7.11 shows the relative potential energies of reactant and product molecules for exothermic and endothermic reactions. Note that the reactants of exothermic reactions have more potential energy than the products. The reverse is true for endothermic reactions.

(a)

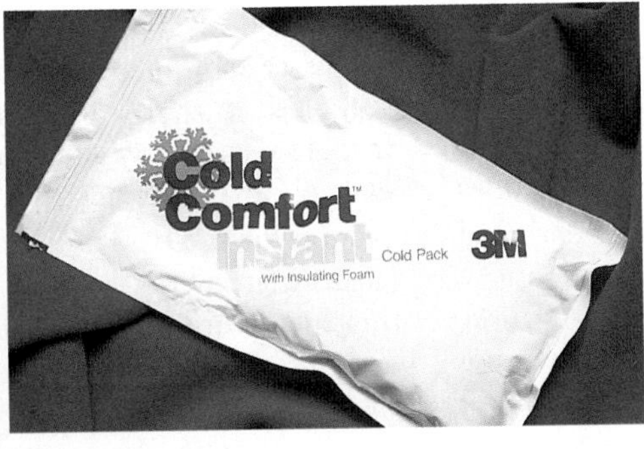

(b)

Figure 7.10 Chemical reactions either yield or consume energy. **(a)** The combustion of natural gas in the burner flame and the oxidation of iron filings are both *exothermic* reactions; they *yield* energy. **(b)** A cold pack uses an *endothermic* reaction to remove heat from the surroundings to produce the sensation of cold. The reaction *consumes* energy. The reactants are in separate compartments, and when cold is needed, the barrier between them is broken, allowing the reaction to take place. *(Charles D. Winters)*

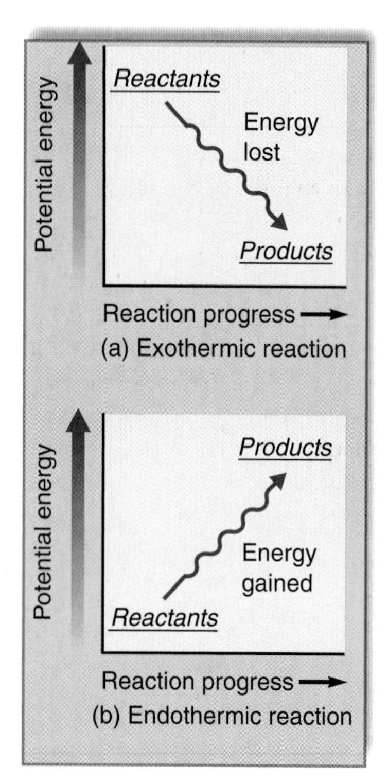

Figure 7.11 **(a)** The reactants of exothermic reactions have more potential energy than their products. **(b)** The products of endothermic reactions have more potential energy than the reactants.

Example 7.11 Exothermic and Endothermic Reactions

1. Which of these reactions or physical changes is exothermic or endothermic?
   (a) Burning of methane **(b)** Use of a chemical cold pack.
2. The heat of fusion of ice (Chapter 5) is 80 cal/g. Is the melting of ice endothermic or exothermic?

SOLUTIONS

1. **(a)** We burn methane to produce heat, thus this reaction must be exothermic. **(b)** When a chemical cold pack is activated, two solutions mix, and an endothermic reaction takes place, absorbing heat from the surroundings such as an injured knee. 2. Because the ice gains the heat, melting is an endothermic process.

Self-Test

Are these reactions/processes endothermic or exothermic?

1. When ammonium nitrate is dissolved in water, the container becomes colder.
2. When a sodium hydroxide solution is mixed with a solution of hydrochloric acid, the container becomes warm.

ANSWERS

1. Endothermic 2. Exothermic

The amount of energy gained or lost varies from one reaction to another. The amount of energy given up or taken in by a reaction is called the **enthalpy of reaction.** The term *heat of reaction* is a synonym. Glucose (blood sugar) is a carbohydrate and is a principal fuel molecule for living organisms. When one mole of glucose is oxidized 686 kcal of energy is released:

$$C_6H_{12}O_6(aq) + 6O_2(aq) \longrightarrow 6CO_2(aq) + 6H_2O(l) + \text{energy as heat}$$

The enthalpy of reaction for the oxidation of glucose is therefore −686 kcal/mol. (The minus sign means that energy is given off—that is, the reaction is exothermic.) Portions of this energy are needed to maintain the highly organized state of a living system. Fats, oils, and amino acids are also major fuel molecules for the body. When carbohydrates are oxidized in the body, about 4 kcal of energy is released per gram of carbohydrate. The same amount of energy is obtained from amino acids from dietary protein. Fats and oils yield about 9 kcal/g.

Kinetics

Chemical reactions vary greatly in how fast they occur. The reaction in a high explosive or the reaction of gasoline with oxygen in your car engine occurs in tiny fractions of a second. The rusting of iron may take days, weeks, or years. The weathering of rock, which is a chemical process, may take hundreds or thousands of years. The branch of chemistry that deals with how fast reactions occur—the rates of reactions—is called **kinetics.** Before we look at specific factors that affect rates of reactions, let's review the process that occurs when a reaction takes place, beginning with collisions between molecules.

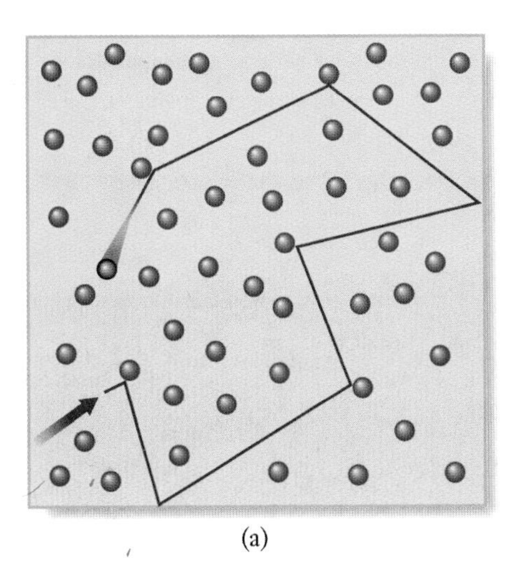

(a)

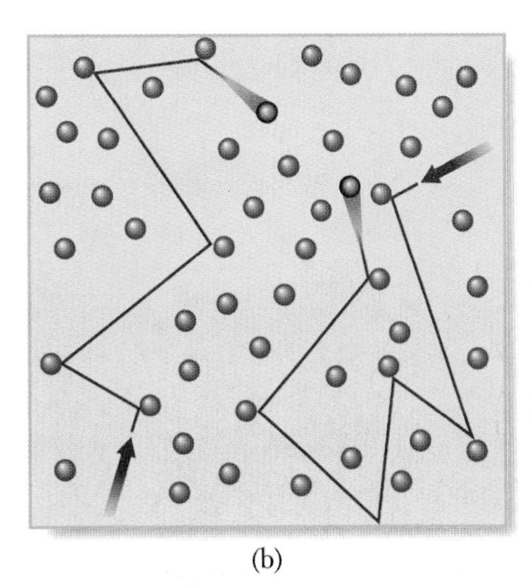

(b)

Figure 7.12 The concentration of a reactant affects the rate of a chemical reaction because the increased concentration increases the number of collisions between reactant molecules. **(a)** In a given period of time, a red molecule collides with five blue reactant molecules. **(b)** If the number of red molecules doubles (if the concentration is doubled) this doubles the number of collisions between reactant molecules.

Chemical reactions involve collisions between molecules of reactants. These collisions possess energy. If the colliding molecules are moving rapidly, the collision has lots of energy. If they are moving slowly, there is little energy in the collision. At low temperatures, on the average, most molecules move slowly; thus most collisions that take place at low temperatures have little energy. As temperature increases, the particles on average move faster, and the collisions have more average energy. It turns out that collisions affect the rates of chemical reactions in two important ways: (1) the larger the number of collisions that occur between reactant molecules, the faster the reaction, and (2) the greater the average energy of the collisions, the faster the reaction.

Let's first consider point (1): The larger the number of collisions, the faster the reaction. For most reactions, when more of a reactant is present, the rate of the reaction is faster. Why does reactant concentration affect the rate of the reaction? If the concentration of a reactant is increased, there will be more collisions between the reactant molecules (Figure 7.12). If more collisions occur in a period of time, more reactants form product in that period of time. The rate of the reaction is faster at higher concentrations of reactant.

Now let's consider point (2): The greater the average energy of the collisions, the faster the reaction. When temperature increases, the rate of a reaction increases. Again, let's think of this in terms of collisions. If the temperature increases, then on the average, collisions will have more energy. More of the collisions have enough energy to break the bonds in the reactant molecules, so more of them will react to form product molecules (Figure 7.13).

Energy of Activation

Consider for a moment the reaction between natural gas and air that occurs in a gas stove. If no pilot light is present, when you open a gas valve to a stove burner and leave it open, gas enters the burner and escapes into the room. A flame or explosion does not automatically occur. Apparently methane (the main component of natural gas) and oxygen (in the air) can mix without reacting. Something appears to be missing that is needed to cause the reaction between them. Now imagine striking a match or causing a spark in the gas-filled room. The methane and oxygen

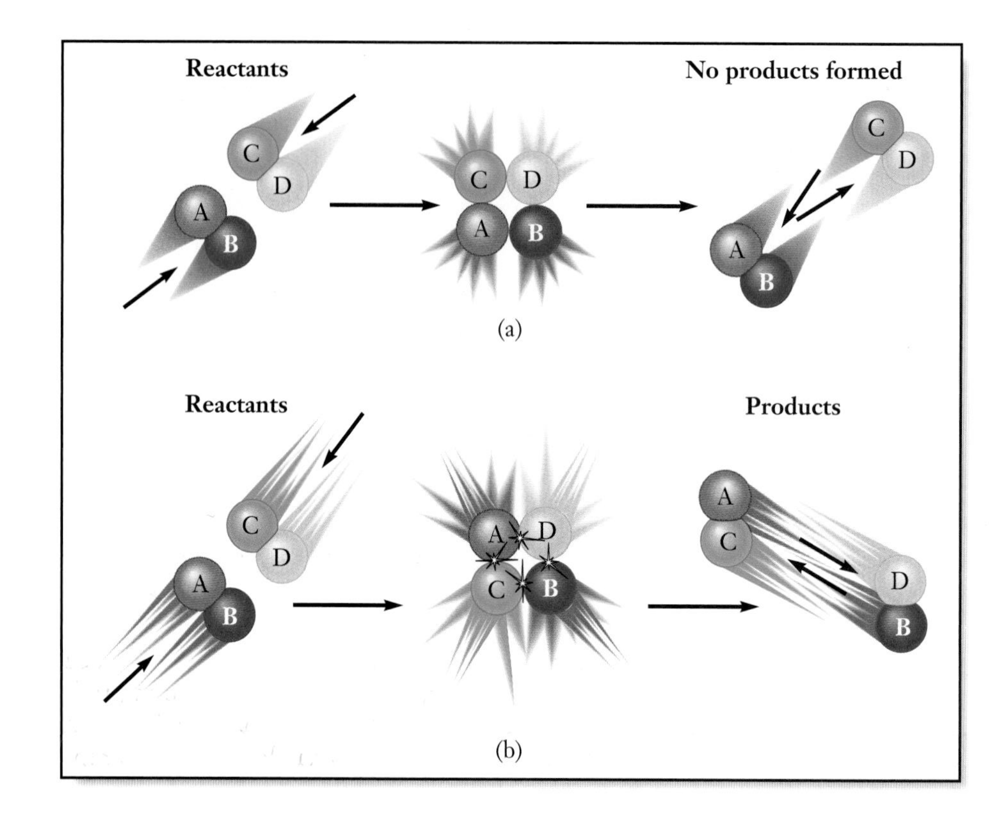

Figure 7.13 **(a)** This collision does not have enough energy to break the bonds in the reactant molecules, thus products do not form. **(b)** This collision possesses enough energy to break bonds in the reactant molecules, which allows formation of new bonds in the product molecules.

would react explosively. Energy from the match or spark is necessary for oxygen and methane to react.

For a reaction to occur, bonds between atoms in reactant molecules must be broken, and new bonds between atoms in product molecules must form. Sometimes the energy from collisions between reactant molecules is not enough to break these bonds, and an outside energy source such as a spark is needed to heat the molecules so that their collisions have more energy. We can view this need for energy to break bonds as an *energy barrier* that must be overcome.

Let's look at this energy barrier in another way. Figure 7.14 shows the potential energy of the reactants and products of a reaction. This diagram is similar to Figure 7.11, but now there is an energy barrier drawn between the reactants and products. This energy barrier is called the **energy of activation,** and it represents the energy that must be provided by collisions before reactants can become products. If no collisions provide this amount of energy, then the rate of the reaction is zero; no product can form. If a small fraction of the collisions have this amount of energy, the reaction rate is slow. The larger the fraction of collisions that have this amount of energy or more, the faster the rate. For very fast reactions, many of the collisions possess sufficient energy to overcome the energy barrier.

Catalysts are common in the chemical industry. For example, a mixture of iron oxide and small amounts of potassium oxide and aluminum oxide are used to catalyze the formation of ammonia. Furthermore, the reactions of the body are catalyzed by proteins called enzymes.

The rates of chemical reactions are affected by temperature and concentrations. The rate can also be affected by adding a catalyst. A **catalyst** speeds up the rate of a chemical reaction but is unchanged by the reaction. Catalysts allow reactants to become products by lowering the energy of activation between them (Figure 7.15). If the energy of activation of a reaction is lower, then at any given temperature, more collisions have enough energy to meet or exceed the energy of activation. The result is that the reaction occurs more rapidly.

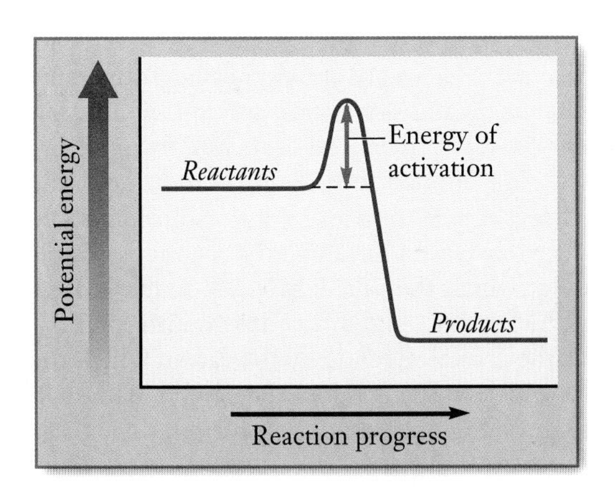

Figure 7.14 The energy of activation is an energy barrier that exists between the reactants and products of a reaction.

Chemical Equilibrium

Most of the chemical reactions you have seen so far have been written in the form

$$\text{Reactants} \rightarrow \text{products}$$

This equation says that the reaction goes in one direction (left to right, called the *forward* direction). In the reaction all that happens is the reactants form products. In reality, most chemical reactions are *reversible:* Reactants form products, but products can also react to form reactants. This idea of reversibility can be best represented in this way:

$$\text{Reactants} \rightleftharpoons \text{products}$$

The double arrows mean the reaction can go in both directions, forward or backward. This means that, under some conditions, the reactants form products, and under other conditions, the products react to yield reactants.

You have seen this double arrow before when you were introduced to weak acids and weak bases.

Figure 7.16 shows the effects of temperature on the reversible reaction between dinitrogen tetraoxide, N_2O_4, and nitrogen dioxide, NO_2:

$$N_2O_4(g) \rightleftharpoons 2NO_2(g)$$

At high temperatures, the reaction that converts N_2O_4 to NO_2 occurs readily. At low temperatures, the reverse reaction—$2NO_2(g) \rightarrow N_2O_4(g)$—occurs more readily.

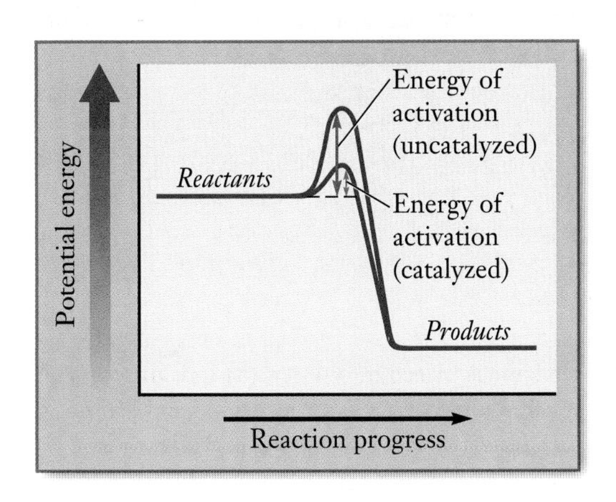

Figure 7.15 A catalyst lowers the energy of activation for a reaction. As a result, reaction rate increases because more collisions have enough energy to meet or exceed this lowered energy barrier.

Figure 7.16 Gaseous N_2O_4 is colorless and gaseous NO_2 is brown. The tube on the left is in hot water and appears dark brown because significant amounts of $NO_2(g)$ are present. The tube on the right is in ice water and appears light brown because less $NO_2(g)$ is present and more $N_2O_4(g)$ is present. *(Charles D. Winters)*

Chemical reactions can go forward or backward—that is, they are reversible. What happens when both the forward and the backward reaction occur at the same time? It turns out that this is usually the case. Consider a large amount of table salt added to water. Initially, at the moment the salt is added, all of it is solid NaCl, and all of the water is pure water [Figure 7.17(a)]. Immediately after the addition, some sodium and chloride ions begin to dissolve. Once there are ions in solution, some of the ions collide with salt crystals and go back to the solid—that is, they precipitate [Figure 7.17(b)]. Initially the rate at which the salt dissolves is much greater than the rate at which ions precipitate, but as more and more ions go into solution, more and more of them combine with the crystal. Eventually the rate at which the salt dissolves is the same as the rate at which the ions rejoin the crystal. Although at any given moment some ions are dissolving and some are precipitating, these processes are occurring *at the same rate*, with the result that no *net* change occurs [Figure 7.17(c)]. The salt and solution are now at **chemical equilibrium.**

Look again at Figure 7.16. The contents of the tube on the right are at chemical equilibrium, with most of the molecules being N_2O_4 molecules, which are colorless, and a few of them being NO_2 molecules, which are brown. The rates at which N_2O_4 forms and breaks down are the same, and so chemical equilibrium exists. Chemical equilibrium also exists for the gases in the tube on the left, but at this temperature more $NO_2(g)$ exists than in the tube on the right, and thus the tube appears darker brown.

In the previous section you learned that catalysts speed up chemical reactions. Catalysts *do not* affect equilibrium. When a catalyst is present, the reaction will reach equilibrium faster, but the equilibrium is the same in the presence or absence of a catalyst.

Le Chatelier's Principle

When a reversible reaction is not at equilibrium, the reaction proceeds in the appropriate direction until equilibrium is reached. Once equilibrium is reached, will

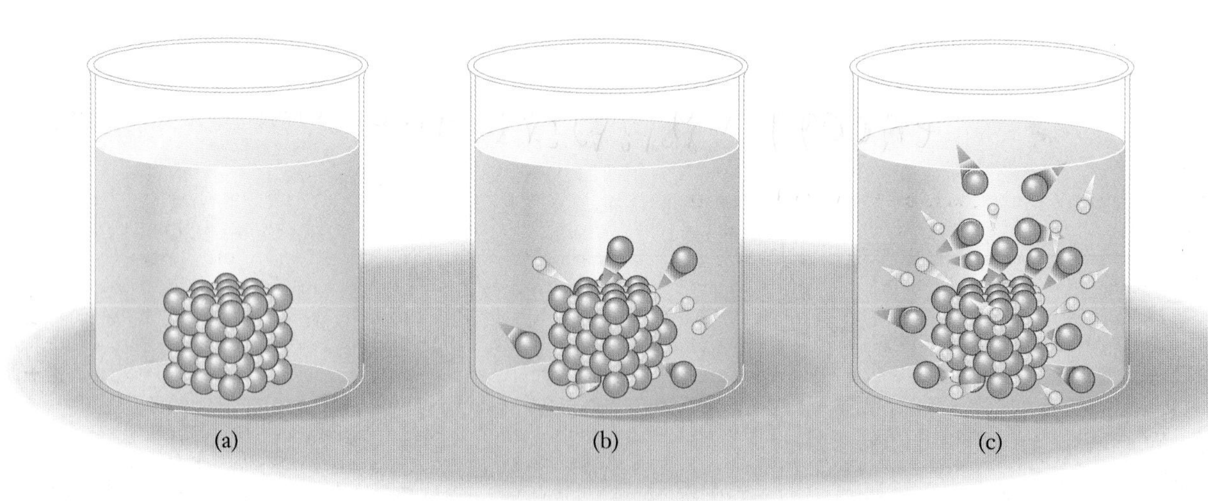

Figure 7.17 **(a)** A crystal of sodium chloride is placed into water. **(b)** Ions of sodium and chloride begin to dissolve, but some of these ions collide with the crystal and leave solution. **(c)** At equilibrium, the rate at which ions dissolve and the rate at which they return to the crystal are equal.

equilibrium always remain? Stated another way, can an equilibrium be disturbed in such a way that either the forward or backward reaction occurs more rapidly? The answer is yes: Equilibria can be disturbed or changed. Look at the tubes containing N_2O_4 and NO_2 in Figure 7.16. If the temperature is lowered, a net reaction occurs until equilibrium is reestablished. As a result, NO_2 was consumed and N_2O_4 produced until a new equilibrium is established. This is one example of **Le Chatelier's principle:** If a chemical equilibrium is disturbed, changes will occur to reestablish equilibrium.

Why Buffers Work

To illustrate Le Chatelier's principle, consider buffered solutions once again. If acetic acid and acetate are used to make a buffer, the buffer can be written as a chemical equilibrium between the acetic acid and the acetate ion:

$$HC_2H_3O_2(aq) \rightleftharpoons H^+(aq) + C_2H_3O_2^-(aq)$$

Because the rates of the forward and reverse reactions are the same, equilibrium exists. What happens if the concentration of H^+ is increased because we add acid? The increase in $[H^+]$ disturbs the equilibrium. There is too much H^+ now; the equilibrium is disturbed (stressed). As a result, a change occurs to reduce the stress. The rate of the backward reaction (right to left) increases to lower the $[H^+]$. This means that some acetate ions accept H^+ ions to become acetic acid molecules. This reduces the $[H^+]$. This backward reaction continues until equilibrium is again established. At this new equilibrium, the $[H^+]$ is almost the same as it was before the acid was added, but the amount of acetate ion is less and the amount of acetic acid is greater than before the acid was added. Because there is no overall increase in $[H^+]$, however, the pH does not change.

What happens if we add base, OH^-? Hydroxide would react with some H^+ to form water, which would reduce the $[H^+]$. This disturbs the equilibrium. Acetic acid would now ionize, a step that releases hydrogen ion. As a result, the $[H^+]$ does not change; instead some acetic acid is consumed because of the added base.

CHAPTER SUMMARY

Acids are hydrogen-ion donors, and **bases** are hydrogen-ion acceptors. A **salt** is an ionic compound that forms in an acid–base reaction. In pure water and neutral solutions, a small number of water molecules ionize to yield an equal concentration of **hydronium ion** and hydroxide ion. In acidic solutions, the concentration of hydronium ion exceeds that of hydroxide ion, but in basic solutions the hydroxide-ion concentration is higher. The **pH scale** is a logarithmic expression for hydrogen (hydronium) ion concentration. The amount of an acid or base is often determined by **titration.** The amount of hydrogen ion or hydroxide ion in acids or bases can be expressed as **equivalents. Buffers** are solutions that resist changes in pH when an acid or base is added to them, and are prepared with a **weak acid** (base) and a salt of the weak acid (base). In **oxidation–reduction (redox) reactions** electrons are transferred between reactants. **Oxidation** is a loss of electrons and **reduction** is a gain of electrons. Oxidation can also be recognized by loss of hydrogen atoms or gain of oxygen atoms. Reduction can be recognized by gain of hydrogen atoms or loss of oxygen atoms. In **precipitation reactions,** an insoluble product forms from soluble reactants. Reactions that yield energy as heat are **exothermic;** those that require heat are **endothermic.** The amount of energy yielded or taken up by a reaction is expressed as the **enthalpy of reaction. Kinetics** is the study of reaction rates. Reactants require energy **(energy of activation)** to change to products. **Catalysts** speed up reactions by reducing the energy of activation. Chemical reactions are reversible, and **chemical equilibrium** exists whenever the rates of formation of reactants and products are equal. When a reaction at equilibrium is disturbed, the forward or reverse reaction increases until equilibrium again exists **(Le Chatelier's principle).**

KEY TERMS

Section 7.1
acid
base
hydronium ion
salt
strong acid
strong base
weak acid
weak base

Section 7.2
ion-product constant of water (K_w)

Section 7.3
pH scale

Section 7.4
buffered solution (buffer)
equivalent
equivalent mass
neutralization
titration

Section 7.5
oxidation
oxidation–reduction (redox) reaction
precipitation reaction
reduction

Section 7.6
catalyst
chemical equilibrium
endothermic reaction
energy of activation
enthalpy of reaction
exothermic reaction
kinetics
Le Chatelier's principle

EXERCISES

Acids, Bases, and Salts *(Section 7.1)*

1. What is an acid? List two or more common examples of acids.

2. A synonym for hydrogen ion, commonly used by chemists, is __________.

3. What is a salt? In what type of reaction are salts produced? List two salts commonly found in the home.

4. Define a base. What ion is found in the formula of all strong bases? List two or more common bases.

5. What information in a chemical formula identifies a compound as an acid? Use this information to determine which of these compounds are acids.
(a) H_3PO_4 (b) LiOH (c) KCl
(d) $HC_2Cl_3O_2$ (e) HBr

6. Which of the following compounds are acids?
(a) HCN (b) NH_3 (c) PH_3
(d) $HClO_4$ (e) Li_2CO_3

7. A solution of sodium carbonate, Na_2CO_3, feels slippery. What does this tell you about sodium carbonate?

8. To most people, grapefruit tastes very sour. From this trait, what type of compound is present in grapefruit?

9. What word is used to describe the formation of ions from a molecule?

10. Provide a definition of *hydronium ion*. Write a chemical equation showing the formation of a hydronium ion.

11. Write the equation for the reaction of HCl(aq) with water.

12. What is a strong acid?

13. List the common strong acids that were introduced to you in this textbook.

14. What is a weak acid? List at least two common examples.

15. A solution is 0.50 *M* for an acid, but the concentration of H^+ in the solution is 0.0041 *M*. Is this a strong or weak acid?

16. A solution is 0.025 *M* for hydroiodic acid, HI, and 0.025 *M* for H^+. Is HI a strong or weak acid?

17. Formic acid, $HCHO_2$, is a weak acid that is a component of many ant venoms. Write an equation to show its ionization.

18. What is a strong base? Give an example.

19. What is a weak base? Give an example.

20. Ammonia contains no hydroxide ions, but aqueous solutions of ammonia are basic. Write an equation that explains these observations.

Ionization of Water *(Section 7.2)*

21. Identify the ions present in a sample of pure water.

22. Why are the concentrations of H_3O^+ and OH^- the same in pure water?

23. A solution containing a higher concentration of hydrogen ions than hydroxide ions is __________.

24. A solution containing a higher concentration of hydroxide ions than hydrogen ions is __________.

25. Listed are the concentrations of an ion in a solution. Indicate whether the solution is acidic, neutral, or basic.
(a) $[OH^-] = 4.3 \times 10^{-6}\ M$
(b) $[H^+] = 2.91 \times 10^{-3}\ M$
(c) $[H_3O^+] = 2.07 \times 10^{-10}\ M$
(d) $[OH^-] = 1.0 \times 10^{-7}\ M$

26. Listed are the concentrations of an ion in a solution. Indicate whether the solution is acidic, neutral, or basic.

(a) $[H_3O^+] = 1.8 \times 10^{-6}\ M$
(b) $[H^+] = 4.43 \times 10^{-9}\ M$
(c) $[OH^-] = 4.55 \times 10^{-9}\ M$
(d) $[OH^-] = 2.7 \times 10^{-3}\ M$

27. K_w is the symbol for what constant? What is the numerical value of K_w at 25°C?

28. You have some solutions that have the following $[H^+]$ in them. Calculate the concentration of hydroxide ion in each of them.
(a) $4.11 \times 10^{-5}\ M$
(b) $2.09 \times 10^{-3}\ M$
(c) $5.72 \times 10^{-6}\ M$
(d) $8.22 \times 10^{-8}\ M$

29. You have some solutions that have the following $[H^+]$ in them. Determine the concentration of hydroxide ion in each of them.
(a) $5.26 \times 10^{-12}\ M$
(b) $1.78 \times 10^{-9}\ M$
(c) $3.92 \times 10^{-5}\ M$
(d) $6.94 \times 10^{-4}\ M$

30. You have some solutions that have the following $[OH^-]$ in them. What is the concentration of hydrogen ion in each of them?
(a) $7.82 \times 10^{-5}\ M$
(b) $1.96 \times 10^{-3}\ M$
(c) $8.05 \times 10^{-11}\ M$
(d) $4.92 \times 10^{-8}\ M$

31. You have some solutions that have the following $[OH^-]$ in them. Calculate the concentration of hydrogen ion in each of them.
(a) $9.04 \times 10^{-4}\ M$
(b) $2.48 \times 10^{-9}\ M$
(c) $1.32 \times 10^{-12}\ M$
(d) $5.38 \times 10^{-6}\ M$

pH (Section 7.3)

32. A logarithmic scale that is used to express hydrogen-ion concentration is called the __________.

33. Determine the pH of these solutions of known hydrogen-ion concentration.
(a) $4.04 \times 10^{-5}\ M$
(b) $7.418 \times 10^{-10}\ M$
(c) $5.02 \times 10^{-2}\ M$
(d) $5.838 \times 10^{-9}\ M$

34. What is the pH of a solution containing these concentrations of hydrogen ion?
(a) $9.804 \times 10^{-3}\ M$
(b) $4.47 \times 10^{-4}\ M$
(c) $1.69 \times 10^{-13}\ M$
(d) $3.06 \times 10^{-7}\ M$

35. Given these pH values, provide the hydrogen-ion concentration in these solutions.
(a) 4.83
(b) 6.07
(c) 12.41
(d) 8.273

36. What is the hydrogen-ion concentration in these solutions that have this pH?
(a) 5.334
(b) 10.72
(c) 2.93
(d) 9.441

37. Which of these solutions are acidic? Which are basic?
(a) Lemon juice, pH = 2.5
(b) Urine, pH = 7.7
(c) Milk, pH = 6.8
(d) Blood, pH = 7.4

38. Which of these solutions are acidic? Which are basic?
(a) Soap solution, pH = 8.5
(b) Vinegar, pH = 2.8
(c) Tomato juice, pH = 4.8
(d) Seawater, pH = 8.3

39. Arrange the solutions in Exercise 37 from most acidic to most basic.

40. Arrange the solutions in Exercise 38 from most basic to most acidic.

41. The pH value, $[H^+]$, $[OH^-]$, and the acid-base characteristics of a solution are interrelated. Use these relationships to complete this table:

pH	$[H^+]$	$[OH^-]$	ACIDIC, NEUTRAL, OR BASIC?
6.41	—	—	—
—	4.82×10^{-9}	—	—
—	—	2.77×10^{-7}	—

Acid–Base Reactions (Section 7.4)

42. What is a neutralization reaction? Write a chemical equation illustrating a neutralization reaction.

43. Predict the product(s) and balance these acid–base reactions:
(a) $HBr(aq) + NaOH(aq) \longrightarrow$?
(b) $Ba(OH)_2(aq) + HNO_3(aq) \longrightarrow$?
(c) $H_3PO_4(aq) + KOH(aq) \longrightarrow$?

44. Predict the product(s) and balance these acid–base reactions:

(a) $H_2SO_4(aq) + KOH(aq) \longrightarrow$?
(b) $NH_3(g) + HI(g) \longrightarrow$?
(c) $NaOH(aq) + HC_2H_3O_2(aq) \longrightarrow$?

45. $Mg(OH)_2$ is the active ingredient of some antacids. Write the equation for the reaction of magnesium hydroxide with stomach acid (HCl).

46. State the meaning of the word *titration.*

47. Provide a description or definition of a pH indicator. How can a pH indicator be used in a titration?

48. A 15.0-mL sample of HNO_3 of unknown concentration was titrated with 0.109 *M* NaOH. A volume of 18.6 mL of the base was needed to neutralize the acid sample. What is the molarity of the HNO_3 solution?

49. A 25.0-mL sample of H_2SO_4 of unknown concentration was titrated with 0.102 *M* NaOH. A volume of 41.7 mL of the base was needed to neutralize the acid sample. What is the molarity of the sulfuric acid solution? (Sulfuric acid is diprotic; it donates 2 H^+.)

50. A 50.0-mL solution of caffeine (monobasic) is titrated with 0.0050 *M* HCl, and 12.7 mL of the titrant is required to complete the titration. What is the molarity of the caffeine? How many moles of caffeine were in the solution?

51. How many equivalents of hydrogen ion or charge are in a mole of these compounds or ions?
(a) $HClO_4$
(b) H_2SO_4
(c) Na^+
(d) Ca^{2+}

52. What is the equivalent mass of these compounds?
(a) H_3PO_4 (b) $Ba(OH)_2$

53. Provide a definition of a buffer. From that definition, what is *buffered aspirin?*

54. How does a buffer resist changes in pH?

55. These solutions contain the following substances. Which of them would be buffered?
(a) HF + KF (b) $HNO_3 + NaNO_3$

56. Use equations to show how a lactate buffer resists changes in pH. The components of the lactate buffer are lactic acid, $HC_3H_5O_3$, and lactate ion, $C_3H_5O_3^-$, often provided as the sodium salt.

Oxidation–Reduction and Precipitation Reactions (Section 7.5)

57. What is an oxidation–reduction reaction? Give one or more examples.

58. What is oxidation? How is an oxidation recognized when a metal and nonmetal react?

59. What is reduction? How is a reduction recognized when a metal and nonmetal react?

60. Predict the products of these reactions and indicate which atoms are oxidized and which are reduced.
(a) $Na(s) + Cl_2(g) \longrightarrow$?
(b) $Ag(s) + S(l) \longrightarrow$? (sulfur is heated to melt it)
(c) $Ca(s) + O_2(g) \longrightarrow$?

61. Predict the products of these reactions and indicate which atoms are oxidized and which are reduced.
(a) $Ba(s) + O_2(g) \longrightarrow$?
(b) $Al(s) + Br_2(l) \longrightarrow$?
(c) $Mg(s) + I_2(g) \longrightarrow$?

62. Explain how gain and loss of hydrogen or oxygen can be recognized as oxidation or reduction in a redox reaction.

63. Indicate which atoms in the reactants become oxidized and which become reduced.
(a) $2C_2H_2(g) + 5O_2(g) \longrightarrow 4CO_2(g) + 2H_2O(g)$
(b) $CH_2O(g) + H_2(g) \longrightarrow CH_4O(g)$
(c) $ZnO(s) + C(s) \longrightarrow Zn(s) + CO(g)$

64. Indicate which atoms in the reactants become oxidized and which become reduced.
(a) $WO_3(s) + 3H_2(g) \longrightarrow W(l) + 3H_2O(g)$
(b) $C_2H_4(g) + H_2(g) \longrightarrow C_2H_6(g)$
(c) $FeO(l) + CO(g) \longrightarrow Fe(l) + CO_2(g)$

65. You have left an open container of a salt solution—NaCl(aq)—out in the sun. What will happen to the solution with time, and why will the change happen?

Properties of Chemical Reactions (Section 7.6)

66. What does *exothermic* mean? What is the meaning of *endothermic?*

67. What is meant by *enthalpy of reaction?*

68. What is kinetics?

69. List two ways that collisions influence the rates of chemical reactions.

70. What is *energy of activation?*

71. What is a catalyst? How do catalysts work?

72. What is meant by the term *reversible reaction?*

73. What is chemical equilibrium? What is true about the rates of a reversible reaction at equilibrium?

74. What is Le Chatelier's principle?

75. Use Le Chatelier's principle to explain how a buffer would resist a change in pH if acid were added to it.

Challenge Exercises

76. A solution is 0.000000000046 M in hydroxide ion.
  (a) Express this value in scientific notation.
  (b) Determine the concentration of hydrogen ion in the solution.
  (c) What is the pH of the solution?

77. Hyperventilation (fast and deep breathing) can cause alkalosis (blood pH more alkaline than normal). Hypoventilation (slow and shallow breathing) can cause acidosis (blood pH more acidic than normal). Use your knowledge of blood buffering to explain these observations.

78. How many kcal and kJ will be produced when 125 g of glucose ($C_6H_{12}O_6$) are oxidized in the body?

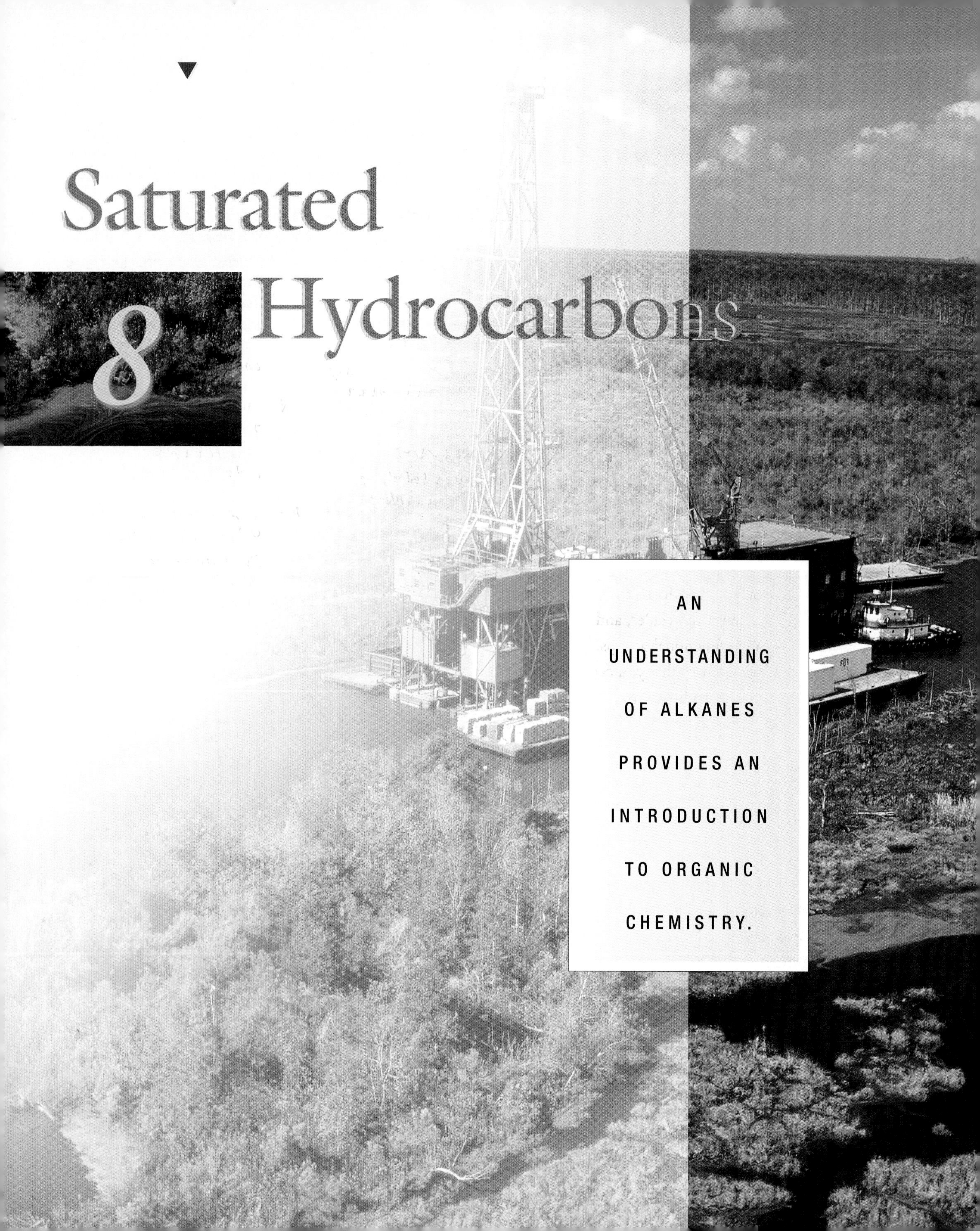

8 Saturated Hydrocarbons

AN UNDERSTANDING OF ALKANES PROVIDES AN INTRODUCTION TO ORGANIC CHEMISTRY.

SO FAR in this course you have learned about some fundamentals of chemistry, collectively called *general chemistry*. The topics in Chapters 1 through 7 deal with concepts that apply to all of chemistry. Now you begin your study of a major subdivision of chemistry—*organic chemistry*, which is the chemistry of carbon-containing compounds. Organic chemistry is an important area of science and technology that has contributed much to modern life and health care. In addition, a knowledge of organic chemistry is needed before you can study biochemistry, which is another branch of chemistry that has a central role in medicine.

In this chapter you are introduced to alkanes, the simplest organic molecules, and to groups of atoms called functional groups. These two topics provide a base that you can then use to build an understanding of the rest of organic chemistry.

Outline

8.1 Organic Chemistry

Distinguish inorganic compounds from organic compounds.

8.2 Alkanes—The Simplest Organic Compounds

Name and draw structures of the first ten straight-chain alkanes.

8.3 Branched Alkanes

Identify branched alkanes and use IUPAC rules to name them.

8.4 Cycloalkanes

Identify and name simple cycloalkanes.

8.5 Shapes of Organic Molecules

Differentiate among positional isomers, geometric isomers, and chiral compounds, and explain the difference between configuration and conformation.

8.6 Properties and Reactions of Alkanes

Describe the physical properties and reactions of alkanes.

8.7 Common Functional Groups of Organic Chemistry

Recognize the common functional groups of organic chemistry.

An oil exploration platform. When petroleum is pumped out of the ground, it is a complex mixture of mostly alkanes of varying sizes, whose composition varies greatly with the source. *(Keith Wood/Tony Stone Images)*

8.1 Organic Chemistry

Today we define **organic chemistry** as the chemistry of carbon compounds. Two centuries ago, chemistry was divided into two branches: inorganic chemistry and organic chemistry. *Inorganic chemistry* deals with compounds found primarily in the nonliving parts of the Earth: rocks, water, and air. Collectively, inorganic compounds contain atoms of all of the naturally occurring elements. In contrast, the original organic chemistry of 200 years ago dealt with compounds that came from living organisms. These compounds always contain carbon atoms, and they usually contain atoms of one or more other elements. Out of this early organic chemistry has developed two modern branches of chemistry: *biochemistry*, which deals with the chemistry of living systems, and modern organic chemistry.

Organic chemistry studies carbon-containing compounds, and inorganic chemistry studies all other compounds (Table 8.1). It may seem odd that a single element receives so much attention, but there are important reasons for this division. First, many extremely important compounds, including most biological compounds, are organic. A second and perhaps more important reason is the unique chemistry of carbon. Each element has its own chemistry, true, but *the chemistry of carbon is more diverse than that of any other element.*

Consider carbon's electronegativity, which is 2.5, about in the middle of the electronegativity range for elements (refer back to Figure 3.7). In Chapter 3 we learned that the difference between the electronegativities of two bonding atoms

Inorganic chemistry deals with substances that lack carbon, such as substances found in air and many rocks and the metals in the piton and carabiner of the climber. Organic chemistry deals with carbon-containing compounds like those found in the woman, her rope, and her clothing. Biochemistry deals with the biomolecules and chemical processes that support life. *(Horst Schafer/Peter Arnold, Inc.)*

Table 8.1 A Comparison of Some Properties of Organic and Inorganic Compounds[a]

ORGANIC COMPOUNDS	INORGANIC COMPOUNDS
Always contain carbon	Rarely contain carbon
Contain mostly covalent bonds (Although carbon forms only covalent bonds, other atoms in the organic compound may form ionic bonds.)	Often contain ionic bonds
Are generally nonelectrolytes	Are often electrolytes
Are usually soluble in nonpolar solvents and insoluble in water	Are generally soluble in water and insoluble in nonpolar solvents
Usually have low melting points	Usually have high melting points
Usually have low boiling points	Often have high boiling points
Are usually combustible	Are often noncombustible
Often exist in many similar forms (isomers)	Usually exist as a single form
Often contain numerous atoms	Usually contain only a few atoms
Examples: methane (natural gas), sugar, aspirin	Examples: sodium chloride (table salt), sulfuric acid (battery acid), sodium hydroxide, water

[a] Although these differences are generally true, numerous exceptions exist.

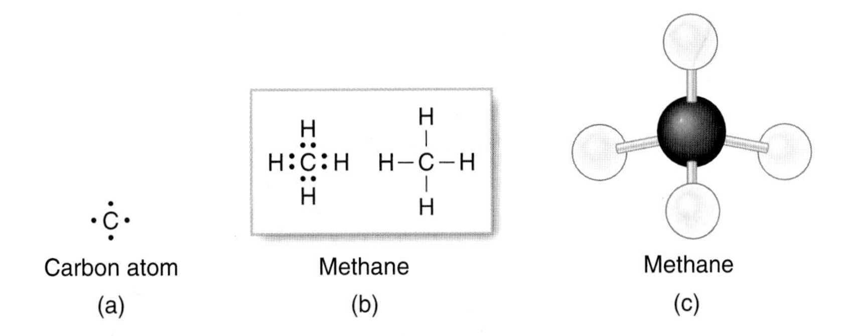

The Lewis structures of **(a)** a carbon atom and **(b)** a molecule of methane. In molecules, carbon atoms always form four covalent bonds. The ball-and-stick model of methane is shown in **(c).**

determines the nature of the bond between them. If the difference is 2.0 or greater, the bond is ionic. If the difference is less than 2.0, the bond is covalent. Because of its electronegativity, carbon forms mostly covalent bonds. (In a few inorganic compounds, such as some metal carbides, carbon does form ionic bonds.) Now look at the periodic table and note carbon's location in Group 4A. The atoms of Group 4A elements have four electrons in their outer shells. To gain eight electrons in the outer shell, a carbon atom must therefore form four covalent bonds. In addition, carbon is versatile because its atoms can form covalent bonds to atoms of many other elements, including other carbon atoms. Furthermore, there is no theoretical limit to how many carbon atoms can be linked through covalent bonds. Carbon thus forms four covalent bonds, it can bond to many different kinds of atoms, and there are no restrictions on how many atoms can be linked together. These properties allow carbon to form many different organic compounds.

Now consider another factor—the formation of double and triple bonds between a carbon atom and another atom. This possibility of double and triple bonds increases both the number of possible organic compounds and the number of different kinds of reactions in which organic compounds can participate. As a consequence of carbon's bonding properties, huge numbers of organic compounds either exist in nature or can be made in the laboratory.

8.2 Alkanes: The Simplest Organic Compounds

Hydrocarbons are organic compounds that contain only carbon and hydrogen atoms. The **alkanes** are hydrocarbons that contain only carbon–carbon single bonds and carbon–hydrogen bonds. Alkanes are sometimes called *saturated hydrocarbons* because no more hydrogen atoms can be added to them; that is, they are *saturated* with hydrogen. Alkanes serve a variety of roles in modern society, with their use as fuels being the largest. Gasoline, diesel, heating oil, and natural gas are mixtures that contain mostly alkanes. Alkanes also have a variety of other commercial and industrial uses, including use as lubricants, propellants in spray cans, and solvents.

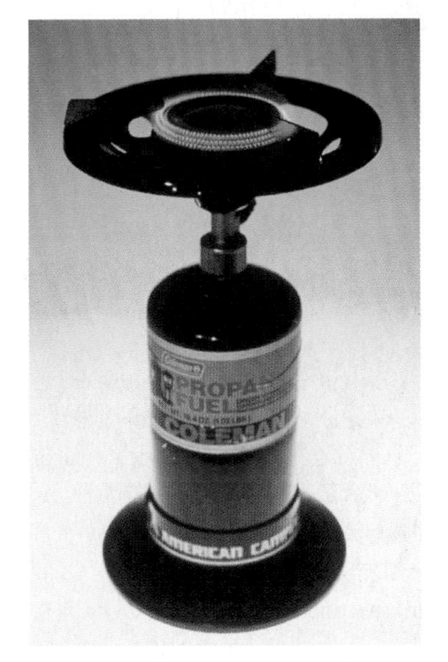

Alkanes have many common uses. Propane is used as a fuel for this camping stove. *(Charles D. Winters)*

The simplest alkanes are called **straight-chain alkanes,** or *unbranched alkanes,* or by an older term *n-alkanes* (the "n" stands for normal). In the straight-chain

```
   H   H   H   H   H
   |   |   |   |   |
H— C — C — C — C — C —H
   |   |   |   |   |
   H   H   H   H   H
```

(a) Pentane

```
   H   H   H   H
   |   |   |   |
H— C — C — C — C —H
   |   |   |   |
   H   H   |   H
         H—C—H
           |
           H
```

(b) 2-Methylbutane

Figure 8.1 **(a)** In straight-chain alkanes, all the carbon atoms are linked to each other in one straight, unbranched chain. **(b)** If an alkane contains one or more carbon atoms branching from the main chain, it is a branched alkane, not a straight-chain alkane.

alkanes, the carbon atoms are linked by single bonds in one continuous chain; there are no branches present (Figure 8.1).

The names of the straight-chain alkanes consist of two parts: (1) a *parent* that tells us the number of carbon atoms in the molecule and (2) the suffix *-ane*, which tells us the compound is an alk*ane*. The smallest alkane is methane (note the parent, *meth*, which means one, and the suffix *-ane*). Methane is the main component of natural gas. A molecule of methane consists of a single carbon atom bonded by single bonds to four hydrogen atoms; thus the molecular formula for methane is CH_4. The carbon atom sits in the middle of a tetrahedron, and the four hydrogen atoms sit at the corners (Figure 8.2). Methane can also be represented by its structural formula:

```
   H
   |
H— C —H
   |
   H
```

Methane

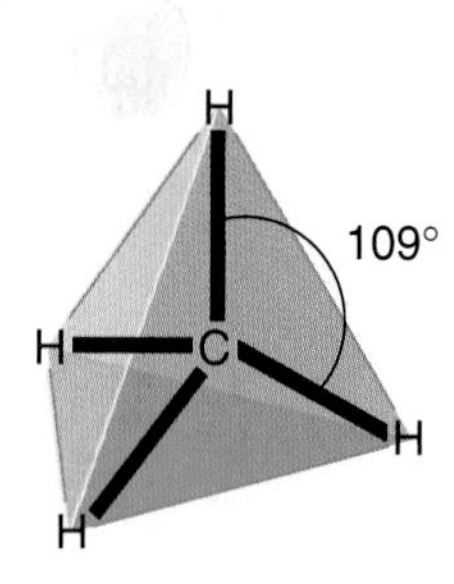

Figure 8.2 A methane molecule has the shape of a tetrahedron. The carbon atom is in the middle, and there is a hydrogen atom at each corner. This configuration yields 109° bond angles, which is the typical bond angle found in a carbon atom that has four single bonds.

The next unbranched alkane is ethane (*eth* means two):

```
   H   H
   |   |
H— C — C —H
   |   |
   H   H
```

Ethane

Note that each carbon atom in ethane forms four covalent bonds, three to hydrogen atoms and the fourth to the other carbon atom.

The next straight-chain alkane, *propane*, has a chain of three (*prop* means three) carbon atoms. The carbon atom on each end of propane has three hydrogen atoms attached to it, and the middle carbon atom has only two hydrogen atoms:

```
   H   H   H
   |   |   |
H—C—C—C—H
   |   |   |
   H   H   H
```

Propane

The next unbranched alkane is butane (what would you guess is the meaning of *but?*):

```
   H   H   H   H
   |   |   |   |
H—C—C—C—C—H
   |   |   |   |
   H   H   H   H
```

Butane

The first ten straight-chain alkanes are listed in Table 8.2.

Before we continue with these alkanes, let's consider some additional ways of showing a molecule. Structural formulas are very useful, but they are larger than molecular formulas and, for larger molecules, time-consuming to draw. **Condensed structural formulas** are another way to represent the structure of molecules. In a condensed structural formula, some or all of the lines that represent bonds are not shown. Although the bonds are left out, they are *understood to be there.* The hydrogen atoms bonded to a carbon atom are drawn right next to the carbon atom, and often no lines are used between carbon atoms.

Let's use ethane to illustrate the interpretation of condensed structural formulas. As these formulas are explained, compare them to the structural formula of ethane:

```
   H   H
   |   |
H—C—C—H
   |   |
   H   H
```

Ethane

Table 8.2 The First Ten Straight-Chain Alkanes

ALKANE	MOLECULAR FORMULA	CONDENSED STRUCTURAL FORMULA
Methane	CH_4	CH_4
Ethane	C_2H_6	CH_3CH_3
Propane	C_3H_8	$CH_3CH_2CH_3$
Butane	C_4H_{10}	$CH_3CH_2CH_2CH_3$
Pentane	C_5H_{12}	$CH_3CH_2CH_2CH_2CH_3$
Hexane	C_6H_{14}	$CH_3CH_2CH_2CH_2CH_2CH_3$
Heptane	C_7H_{16}	$CH_3CH_2CH_2CH_2CH_2CH_2CH_3$
Octane	C_8H_{18}	$CH_3CH_2CH_2CH_2CH_2CH_2CH_2CH_3$
Nonane	C_9H_{20}	$CH_3CH_2CH_2CH_2CH_2CH_2CH_2CH_2CH_3$
Decane	$C_{10}H_{22}$	$CH_3CH_2CH_2CH_2CH_2CH_2CH_2CH_2CH_2CH_3$

Note: When a molecule contains several copies of the same structural feature (such as $—CH_2—$), the structure can be condensed further by placing that unit in parentheses and using a subscript to show how many units are present. Thus the condensed structural formula for decane can be written $CH_3(CH_2)_8CH_3$.

Three condensed structural formulas of ethane are

CH_3CH_3	CH_3-CH_3	H_3CCH_3
(Preferred)	(Common)	(Less common)

These formulas are interpreted this way: The first carbon atom is bonded to three hydrogen atoms. The fourth bond to this carbon is to the next carbon atom. The three remaining bonds to this second carbon atom are to the three hydrogen atoms written next to it. Once you get used to condensed structural formulas, they are usually as informative as structural formulas.

Look at the condensed structural formulas of the straight-chain alkanes in Table 8.2. In each of them (except methane), the carbon atoms at the two ends are bonded to three hydrogen atoms and one carbon atom, and all the other carbon atoms are bonded to two hydrogen atoms and two carbon atoms each.

Example 8.1 Using and Understanding Condensed Structural Formulas

Draw the structural formula of butane using the information in its condensed structural formula: $CH_3CH_2CH_2CH_3$

SOLUTION

In condensed structural formulas, the hydrogen atoms that are bonded to a carbon atom are written next to the carbon atom. Therefore, this structure tells us that the first carbon atom of butane is bonded to three hydrogen atoms

$$\begin{array}{ccc} & H & \\ & | & \\ H- & C & - \\ & | & \\ & H & \end{array}$$

The second and third carbon atoms are bonded to two hydrogen atoms each:

$$\begin{array}{ccccccc} & H & & & & H & \\ & | & & & & | & \\ - & C & - & \quad & - & C & - \\ & | & & & & | & \\ & H & & & & H & \end{array}$$

The fourth carbon atom is bonded to three hydrogen atoms, just as the first one is:

$$\begin{array}{ccc} & H & \\ & | & \\ - & C & -H \\ & | & \\ & H & \end{array}$$

The four carbon atoms are bonded to each other to form a chain, and so the complete structural formula is

$$\begin{array}{cccccccccc} & H & & H & & H & & H & \\ & | & & | & & | & & | & \\ H- & C & - & C & - & C & - & C & -H \\ & | & & | & & | & & | & \\ & H & & H & & H & & H & \end{array}$$

Butane

Note that each carbon atom of butane is bonded to four other atoms through single covalent bonds.

Self-Test

1. What is the *condensed structural formula* for propane?
2. What is the *structural formula* of CH_3CH_3?

ANSWERS:

1. $CH_3CH_2CH_3$
2.
```
   H  H
   |  |
H—C—C—H
   |  |
   H  H
```

Because there is no theoretical limit to the number of carbon atoms that can be linked into a chain, there are many possible unbranched alkanes. In nature, there are alkanes (those found in paraffin) containing dozens of carbon atoms. Polyethylene and polypropylene are synthetic compounds that contain hundreds or thousands of carbon atoms in chains.

Structural formulas and condensed structural formulas are useful representations of molecules, but they provide only two-dimensional images. Because three-dimensional images are sometimes needed for better understanding, **molecular models** are frequently used to show organic molecules in three dimensions. Figure 8.3 shows two types. A *ball-and-stick model* shows the atoms present in the molecule, and it also shows the bond lengths and bond angles very well. A *space-filling model* gives a better impression of the overall shape of a molecule, but individual atoms and bonds are often obscured.

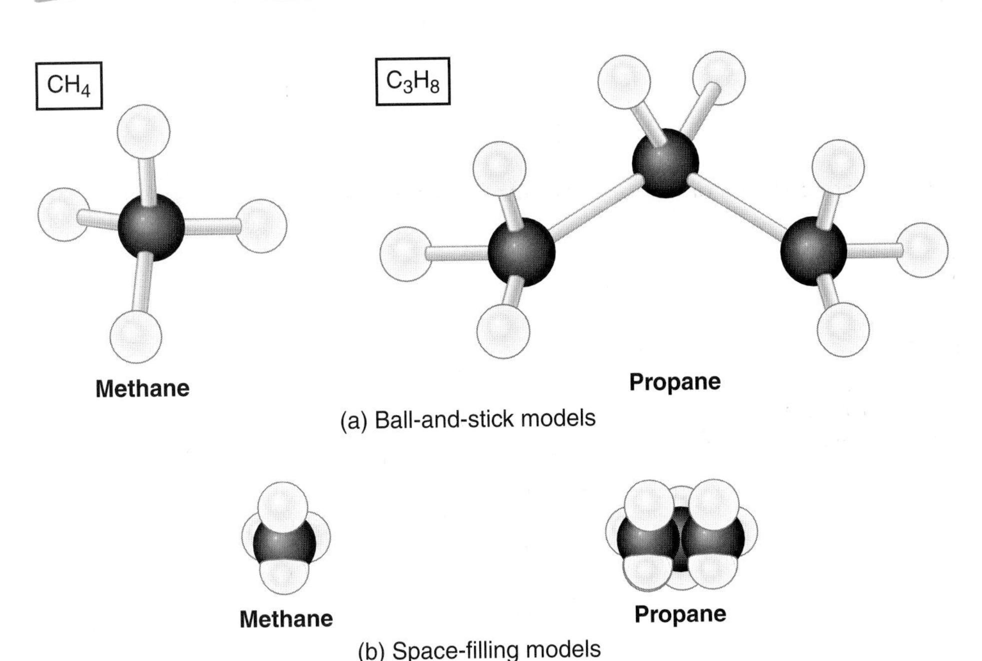

Figure 8.3 (a) Ball-and-stick models of methane and propane show the atoms and bonding in the molecules as well as their three dimensional shapes. (b) Space-filling models of the same molecules show the overall shapes of the molecules well, but some atoms and some bonding may be obscured.

Still another common way to represent molecules is with a *perspective formula*. In these formulas, solid lines represent bonds that lie in the plane of the paper. Dashed or dotted lines represent bonds that project below (behind) the plane of the paper, and wedges represent bonds that project up and out of (in front of) the plane of the paper. Here are the structural and perspective formulas of methane:

```
     H
     |
  H—C—H
     |
     H
```

This hydrogen atom drawn in yellow lies below the plane of the paper; it is projecting away from you.

This hydrogen atom drawn in green lies above the plane of the paper; it is projecting toward you.

Methane structural formula

Methane perspective formula

In the perspective formula, the dashed bond to the yellow hydrogen tells you that this hydrogen lies below the plane of the paper; the wedged bond tells you the green hydrogen is located above the plane of this page.

8.3 Branched Alkanes

Isomers

These are the structural and condensed structural formulas of the straight-chain alkane butane, C_4H_{10}:

```
     H   H   H   H
     |   |   |   |
  H—C—C—C—C—H
     |   |   |   |
     H   H   H   H
```

$CH_3CH_2CH_2CH_3$

Now look at these formulas:

```
            H
            |
    H    H—C—H    H
    |       |       |
 H—C———C———C—H
    |       |       |
    H       H       H
```

```
     CH3
      |
  CH3CHCH3
```

What is the composition of the molecule shown in the second structure? It has the same molecular formula as butane, C_4H_{10}. Is this just another way to represent a butane molecule? No; the structures are different because the carbon atoms are not connected in the same way in the two molecules. In butane, the carbon atoms form a chain of four carbon atoms. In the second molecule, there is a three carbon atom chain with a fourth carbon atom attached to the middle carbon atom. This second

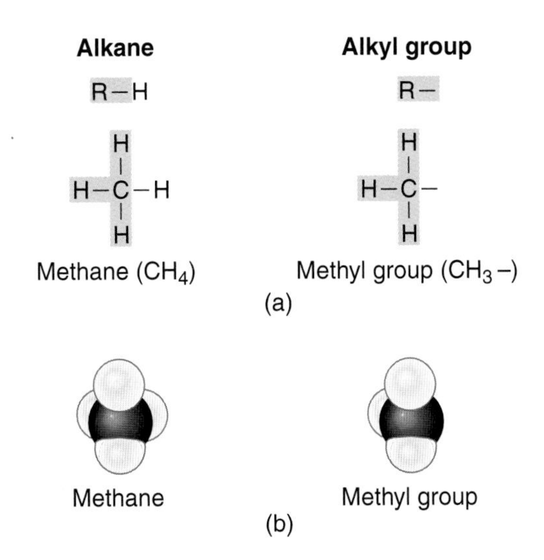

Figure 8.4 Alkyl groups are made by removing a hydrogen atom from an alkane. **(a)** Formula representations and **(b)** space-filling models of methane and the methyl group.

structure represents a *branched alkane,* which is any alkane that has one or more carbon atoms branching from the main chain of carbon atoms.

There are many organic compounds that have the same composition but different structures, and such compounds are called **isomers.** (The number of isomers increases greatly as the size of a formula increases; for example, there are two alkanes of molecular formula C_4H_{10}, 75 alkanes with a formula of $C_{10}H_{22}$, and more than 300,000 with a formula of $C_{20}H_{42}$.) Because isomers have *different structures,* they have *different properties.* In chemistry it is the structure of a molecule that determines its properties, not the composition of the molecule. The two alkanes shown earlier are isomers.

Alkyl Groups

We have a name for the straight-chain alkane represented by the first structure shown previously—butane—but what shall we call the other isomer, the branched C_4H_{10} compound? The names of branched alkanes are derived from the longest chain of carbon atoms in the molecule and from the groups *(substituents)* attached to the carbon atoms of the longest chain. Before we can name branched alkanes, we need to learn how to name these substituents. An **alkyl group** is a substituent that is formed from an alkane by removing one hydrogen atom (Figure 8.4). The name of the alkyl group is the parent part of the alkane name combined with the suffix *-yl.* Methane (CH_4) is the smallest alkane, and the *methyl group* (—CH_3) is the alkyl group that is formed from methane. Table 8.3 shows the common alkyl groups used in this textbook.

Alkyl groups never exist as free substances; they are parts of larger molecules and are identified in the names of the compounds. Often we show each alkyl group in a compound, but sometimes it is useful to represent alkyl groups in a more general way. The letter "R" is used as a general symbol for alkyl groups, as noted in Figure 8.4.

Table 8.3 The Alkyl Groups Formed from the Smallest Alkanes

ALKANE	(H—R)	ALKYL GROUP	(—R)
Methane	CH_4	Methyl	$—CH_3$
Ethane	CH_3CH_3	Ethyl	$—CH_2CH_3$
Propane[a]	$CH_3CH_2CH_3$	Propyl	$—CH_2CH_2CH_3$
		Isopropyl	CH_3CHCH_3 (attached at middle carbon)

[a]Two alkyl groups can be formed from propane. The propyl group, like the ethyl group, has the hydrogen atom removed from the end of the chain. The other alkyl group is made from propane by removing a hydrogen atom from the middle carbon atom. This alkyl group is an isomer of the propyl group (why?) and is called the isopropyl group.

Naming Branched Alkanes

The names of branched alkanes and many other organic compounds consist of three parts—(1) a prefix, (2) a parent, and (3) a suffix. The *parent* is derived from the name of the longest continuous chain of carbon atoms in the molecule. The suffix *-ane* is added to show that the compound is an alkane. The prefix indicates which alkyl groups are present in the molecule and where on the chain these groups are located.

Consider again the isomer of butane we saw at the beginning of this section:

$$\begin{array}{c} CH_3 \\ | \\ CH_3CHCH_3 \end{array}$$

The correct name for this compound is 2-methylpropane; prefix = *2-methyl*, parent = *prop*, suffix = *ane*. The parent *prop*, means there are three carbon atoms in the chain, the suffix *ane* means the compound is an alkane, and the prefix *2-methyl* means there is a methyl group attached to carbon 2 of the chain.

The rules for naming branched alkanes are shown in Table 8.4. Let us now use them to name this branched alkane:

$$\begin{array}{l} CH_3CH_2CHCH_2CH_2CH_3 \\ \qquad\quad\;\; | \\ \qquad\quad\; CH_3 \end{array}$$

Applying Rule 1: Find the longest chain of carbon atoms in the structure. It does not matter whether you count left, right, up, down, or in two or more directions. Just start at any end and count to get the longest chain. Starting at the left in our example and going straight across yields six carbon atoms. Starting at the left and going down yields four. Starting at the right and going across yields six. Starting at the right and going down yields five. Beginning at the bottom, you get four if you go left and five if you go right. Six is the longest chain of carbon atoms, and so *-hex-* is the parent for this branched alkane (Table 8.2). Combine the parent with suffix *-ane* to get *-hexane*. Now number the parent chain from the left because the substituent is closest to that end.

Table 8.4 Rules for Naming Branched Alkanes

Rule 1 Find the longest chain of carbon atoms. The length of this chain is used to determine the parent portion of the branched alkane name. Combine the parent with the suffix *-ane*. Now *number the carbon atoms of the chain beginning with the end closest to a substituent.*

Rule 2 Name the alkyl group or groups attached to the longest (parent) chain. Assign to each group the number of the carbon atom to which it is attached. (If you correctly numbered the carbon atoms of the longest chain using Rule 1, the substituents will have the smallest possible numbers for their positions.) The alkyl groups and their numbers make up the prefix of the name for the branched alkane.

Rule 3 Assemble the name. The prefix consists of the *position* (number) of the substituent first, then the *name* of the substituent. Add to this prefix the parent portion of the name, and finish with the suffix *-ane*. When applying Rule 3, follow these conventions:

(a) If the same substituent occurs more than once, indicate the number of each substituent using the appropriate prefix: *di-* is 2, *tri-* is 3, *tetra-* is 4. (There are prefixes for larger numbers, but you will not need them.)

(b) Always use commas between numbers.

(c) Always use a hyphen between numbers and letters.

(d) If two or more different substituents are present, arrange them alphabetically based on the alkyl name, not on the prefix or number.

```
1     2     3    4     5     6
CH3—CH2—CH—CH2—CH2—CH3
             |
            CH3
```

Applying Rule 2: Identify the alkyl group or groups attached to the longest chain. In our example, there is only one, and it is a methyl group (—CH_3). Because the methyl group is on carbon 3 of the chain, 3-*methyl-* is the prefix that identifies the substituent and its location.

Applying Rule 3: Begin the name with the prefix, then write the parent and suffix. By convention, a hyphen is used to separate numbers from letters, and so the name is 3-*methylhexane.*

Example 8.2 Naming Branched Alkanes

Name these compounds:

1.
```
CH3—CH—CH2—CH3
     |
    CH2
     |
    CH3
```
3-methylpentane

2.
```
            CH3
             |
CH3—CH2—CH—CH—CH2—CH2—CH3
                 |
                CH3
```
3,4 dimethylheptane

3.

$$
\begin{array}{c}
CH_2CH_3 \\
| \\
CH_3-CH_2-CH_2-CH_2-CH-CH-CH_2-CH_2-CH_3 \\
| \\
CH_3
\end{array}
$$

SOLUTIONS

1. Applying Rule 1 to this alkane yields a longest chain of five carbon atoms. The combined parent and suffix is therefore *pentane.* The numbering can be from either end of the five-carbon chain, because the substituent is on carbon 3 in both cases:

$$
\begin{array}{l}
3\quad\quad 4\quad\quad\quad 5 \\
CH_3-CH-CH_2-CH_3 \\
| \\
2\ CH_2 \\
| \\
1\ CH_3
\end{array}
$$

When Rule 2 is applied, there is one methyl group that is attached to carbon 3 of the chain. The group and its position yield the prefix 3-*methyl-*. Using Rule 3 gives us the name of this compound: *3-methylpentane.*

2. Begin by counting the carbon atoms in the longest chain: seven. The parent is therefore *-hept* which combines with the suffix *-ane* to yield *-heptane.* The numbering must be from the left to give the substituents the lowest numbers (try numbering from the right; what do you get?):

$$
\begin{array}{c}
CH_3 \\
1\quad\quad\quad 2\quad\quad\quad 3|\quad\quad 4\quad\quad\quad 5\quad\quad\quad 6\quad\quad\quad 7 \\
CH_3-CH_2-CH-CH-CH_2-CH_2-CH_3 \\
| \\
CH_3
\end{array}
$$

Now identify the alkyl groups making up the branches. There are two methyl groups, one attached to carbon 3 of the chain, the other to carbon 4. This information goes into the prefix, with a comma separating the two digits and a hyphen between the final digit and the first letter: *3,4-dimethyl-*. Now combine the parts of the name to get *3,4-dimethylheptane.*

3. The longest chain of carbon atoms contains nine atoms. The parent and suffix, therefore, are *nonane.* The numbering must be from the right end to give the substituents the lowest numbers (what do you get if you number from the left?):

$$
\begin{array}{c}
CH_2CH_3 \\
9\quad\quad\quad 8\quad\quad\quad 7\quad\quad\quad 6\quad\quad\quad 5|\quad\quad 4\quad\quad\quad 3\quad\quad\quad 2\quad\quad\quad 1 \\
CH_3-CH_2-CH_2-CH_2-CH-CH-CH_2-CH_2-CH_3 \\
| \\
CH_3
\end{array}
$$

There are two substituents; one is a methyl group, the other is an ethyl group. The methyl group is attached to carbon 4 of the longest chain, the ethyl group is on carbon number 5. In the prefix, the substituents are arranged alphabetically, thus the prefix is *5-ethyl-4-methyl-*. Combining the prefix, parent, and suffix yields *5-ethyl-4-methylnonane.*

Self-Test

Name these alkanes:

```
                  CH3          CH3
                  |            |
1.  CH3—CH—CH2—C—CH3
                               |
                               CH3

              CH2CH3
              |
2.  CH3—CH2—C—CH2—CH—CH2—CH2—CH3
              |          |
              CH3        CH3
```

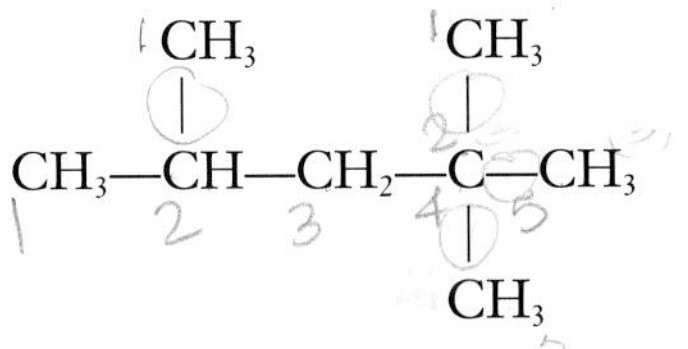

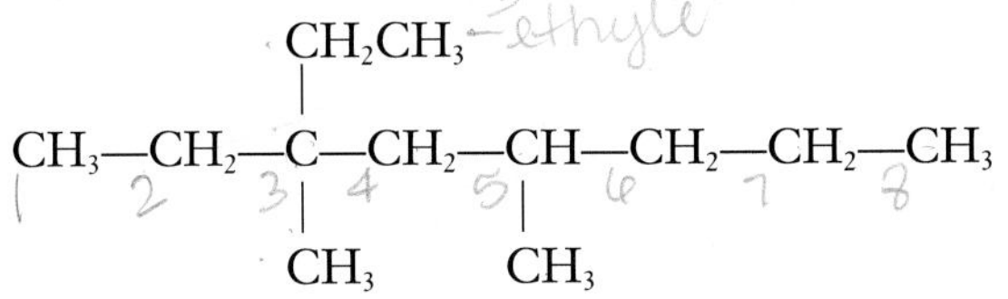

ANSWERS

1. 2,2,4-trimethylpentane 2. 3-ethyl-3,5-dimethyloctane

The names of compounds are based on compound structure. Thus if you know the name of a compound, you can write its structure. Example 8.3 provides some tips for writing structures from names.

Example 8.3 Writing the Structures of Branched Alkanes

What is the structure of 4-ethyl-2-methylheptane?

SOLUTION

Because the parent is *-hept-*, draw the seven carbon atoms of the chain first (leave room to insert hydrogen atoms and subscripts):

```
C—C—C—C—C—C—C
```

Now attach the substituents at the appropriate places—the methyl group goes on carbon 2, the ethyl group on carbon 4:

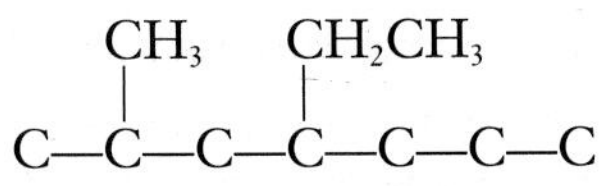

```
   CH3   CH2CH3
   |     |
C—C—C—C—C—C—C
```

Add hydrogen atoms to the carbon atoms of the chain to give each carbon atom a total of four bonds:

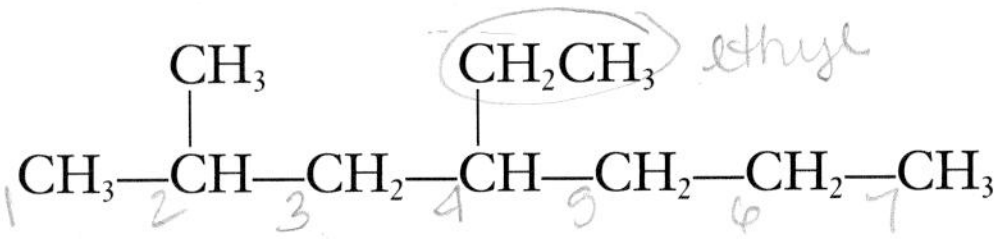

```
      CH3          CH2CH3
      |            |
CH3—CH—CH2—CH—CH2—CH2—CH3
```

You could also have numbered the chain from the right and then placed the groups on the appropriate carbon atoms.

Self-Test

Draw the structure of 3,3,5-trimethyloctane.

ANSWER

$$
\begin{array}{c}
\qquad\qquad\quad CH_3 \qquad\quad CH_3 \\
\qquad\qquad\quad | \qquad\qquad\quad | \\
CH_3\text{—}CH_2\text{—}C\text{—}CH_2\text{—}CH\text{—}CH_2\text{—}CH_2\text{—}CH_3 \\
\qquad\qquad\quad | \qquad\qquad\qquad\qquad\qquad\qquad\qquad \\
\qquad\qquad\quad CH_3 \qquad\qquad\qquad\qquad\qquad\qquad\qquad
\end{array}
$$

The International Union of Pure and Applied Chemistry (IUPAC) has devised a systematic nomenclature for organic compounds. The rules you have just learned yield IUPAC names for alkanes. Alkanes are usually identified by IUPAC names, but common names are sometimes used. *Isobutane* is a common name for the compound you now know as 2-methylpropane. *Isooctane* is the common name for 2,2,4-trimethylpentane. IUPAC nomenclature always relates name to structure. Common names do not necessarily give enough information to draw structures.

The octane rating on a fuel pump is a measure of the burning properties of the gasoline. The original octane ratings were based on heptane, rating = 0, which causes engines to knock badly, and isooctane, rating = 100, which burns well without knocking. Modern fuels, which are blends of many hydrocarbons and additives, can be prepared in a wide range of octane ratings. *(Diane Schiumo/Fundamental Photographs)*

8.4 Cycloalkanes

The alkanes discussed so far contain only open chains of carbon atoms that end with methyl groups. The carbon atoms of some other alkanes form a ring. **Cycloalkanes** have carbon atoms bonded together to form a ring. Cycloalkanes can be represented as structural formulas, but they are more commonly represented by geometric figures in which each corner of the figure represents a carbon atom and its attached hydrogen atoms (Figure 8.5).

To name cycloalkanes, count the number of carbon atoms in the ring to determine the parent, then combine this parent with the suffix *-ane* and add the prefix *cyclo-*. If there is an alkyl group attached to the ring, add the alkyl group name to the cycloalkane name:

CH_2CH_3 (on a hexagon ring)

Ethylcyclohexane

CH_3 (on a pentagon ring)

Methylcyclopentane

There is no need to use a number if there is only one alkyl group on the ring.

Figure 8.5 (a) Structural formulas of some cycloalkanes. (b) Geometric representations of these cycloalkanes.

C_3H_6

Cyclopropane

C_4H_8

Cyclobutane

C_5H_{10}

Cyclopentane

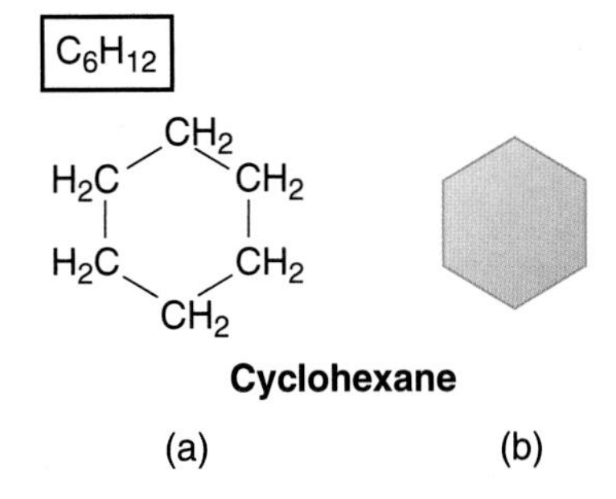

If two groups are bonded to the ring, you must add numbers to indicate where on the ring the groups are located. Choose one of the ring carbons carrying a group as carbon 1, then number around the ring to give the other substituent the smallest possible number:

1,2-dimethylcyclohexane

(Note that numbering the carbon atoms in the reverse direction would yield a numerical prefix of *1,6-*, which would be incorrect.) If the two groups are not identical, assign the number 1 to the group that comes first in the alphabet:

1-ethyl-3-methylcyclopentane

Because *ethyl* precedes *methyl* in an alphabetical list, 1-methyl-3-ethylcyclopentane would be an incorrect name for this compound.

Example 8.4 Names and Structures of Cycloalkanes

1. What is the name of this cycloalkane?

2. Draw the structure of 1-isopropyl-3-methylcyclohexane.

SOLUTIONS

1. This substituted ring contains five carbon atoms, and so this molecule is a cyclopentane. There are two ethyl groups on adjacent carbon atoms. The ring should be numbered to give the lowest possible numbers to these two substituents. The correct name is 1,2-diethylcyclopentane.
2. Because it is a substituted cyclohexane, a hexagon is used to represent the ring of this cycloalkane:

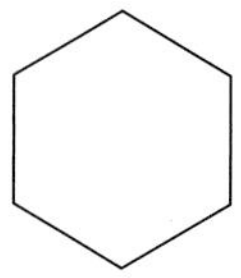

There are two groups, an isopropyl group on carbon 1 of the ring and a methyl group on carbon 3 of the ring. Put the isopropyl group on any ring carbon atom and call this carbon 1. (Remember, because isopropyl precedes methyl in an alphabetical list, it gets the smaller number.) Now attach the methyl group to carbon 3 of the ring (the second carbon atom away from carbon 1 in either direction):

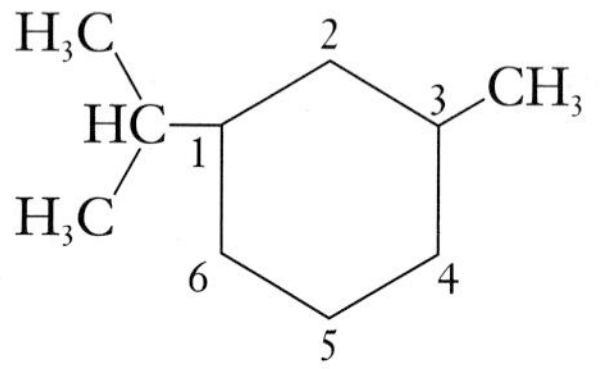

1-isopropyl-3-methylcyclohexane

Check back to Table 8.3 if necessary to review the structure of the isopropyl group.

Self-Test

1. What is the name of this compound:

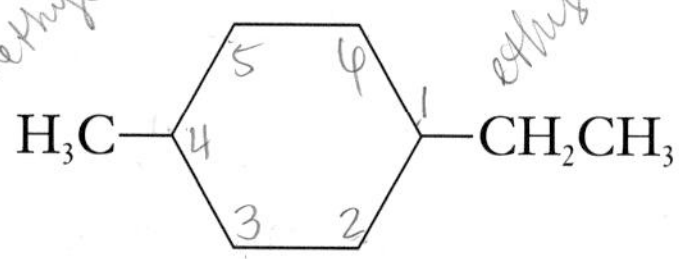

2. What is the structure of 1,1-dimethylcyclobutane?

ANSWERS

1. 1-ethyl-4-methylcyclohexane (the alphabetical rule makes 1-methyl-4-ethylcyclohexane invalid)
2.

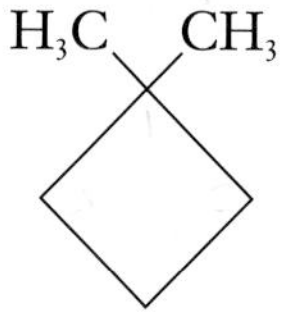

Geometric Isomerism in Cycloalkanes

Just as branched alkanes can exist as isomers, so too can some substituted cycloalkanes. 1,1-Dimethylcyclohexane and 1,2-dimethylcyclohexane are examples. These isomers are sometimes called positional isomers, because they differ in the position of the substituent: 1,1- versus 1,2-.

1,1-dimethylcyclohexane 1,2-dimethylcyclohexane

Two positional isomers of dimethylcyclohexane

In addition to positional isomers, disubstituted cycloalkanes can show *geometric isomerism.* A pair of **geometric isomers** have the same composition, and they also have the same atoms bonded to the same atoms through the same bonds, but they *differ* from each other in how some of the atoms are *arranged in space.* Look at these molecules:

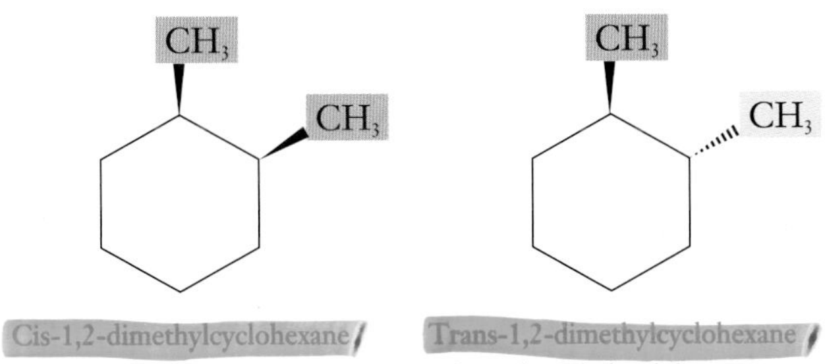

Cis-1,2-dimethylcyclohexane Trans-1,2-dimethylcyclohexane

The geometric (cis-trans) isomers of 1,2-dimethylcyclohexane

Just as in the perspective drawings of methane at the end of Section 8.2, a wedge indicates a group pointing up out of the paper toward you and a dashed line indicates a group that projects below (behind) the plane of the paper. Both of these molecules are 1,2-dimethylcyclohexane, but they are not identical. In the one on the left, both methyl groups are on the same side of the ring. In the one on the right, the methyl groups are on opposite sides of the ring. Because these molecules have the same composition, and the atoms are bonded to the same atoms but with different arrangements in space, these molecules are geometric isomers of each other.

In ring compounds, one geometric isomer is distinguished from the other by a prefix on the compound name: *cis-* if both groups lie on the same side of the ring, and *trans-* if the groups are on opposite sides of the ring.

Example 8.5 Geometric Isomerism in Rings

1. Draw the structure of *trans*-1-ethyl-3-methylcyclohexane.
2. What is the name of this cycloalkane?

SOLUTIONS

1. Cyclohexane means the compound has a six-carbon ring, and an ethyl group will go on carbon 1 of the ring and a methyl group on carbon 3. Because the groups are *trans*-, a wedge must be used for one of them to indicate that group is above the ring, and dashed lines used for the other group to show it is below the plane of the ring:

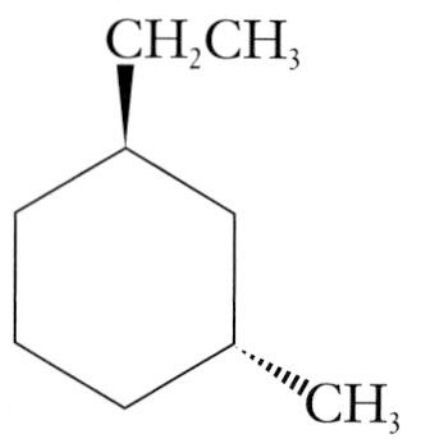

CH_2CH_3

CH_3

2. A five-membered ring of carbon atoms means this is cyclopentane and it is dimethyl, with the proper numbering being 1,2-. The methyl groups are on the same side of the ring, so this is the *cis*-isomer. The name is *cis*-1,2-dimethylcyclopentane.

Self-Test

1. Draw the structure of *cis*-1,4-dimethylcyclohexane.
2. What is the name of this cycloalkane?

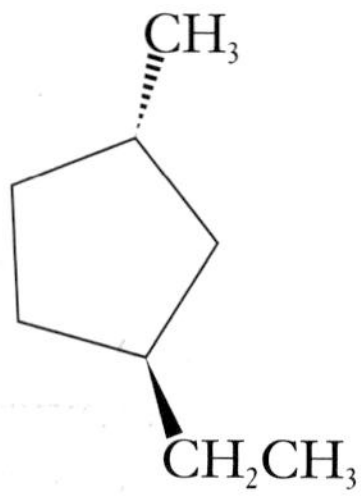

CH_3

CH_2CH_3

ANSWERS

1. CH_3

CH_3

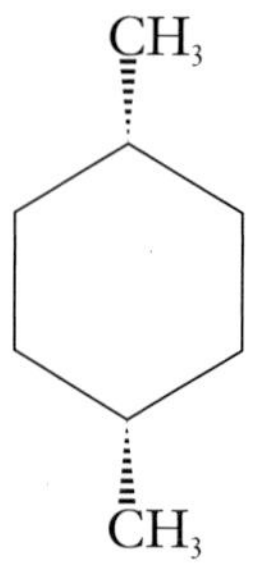

2. *trans*-1-ethyl-3-methylcyclopentane

8.5 Shapes of Organic Molecules

Chiral Molecules

As we learned at the beginning of Section 8.3, two (or more) molecules can have the same composition but different structures, and these molecules are positional isomers of each other. In our discussion of cycloalkanes, we learned about another kind of isomerism, geometric isomerism, involving the arrangement of groups attached to a ring. Geometric isomerism is a part of a broader category of isomerism that is known as stereoisomerism, which depends on the three-dimensional shape of molecules. Like other isomers, **stereoisomers** have the same composition. Unlike other isomers, though, stereoisomers have the same atoms attached to the same atoms. Stereoisomers differ from each other only in the arrangement of atoms in three-dimensional space.

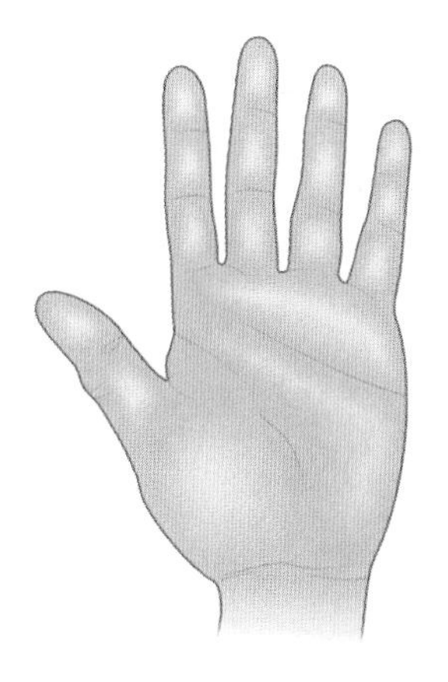

You are already familiar with the basis for one kind of stereoisomerism. Consider a pair of hands. Each hand has a back, a palm, four different fingers, and a thumb. Each hand of the pair has these same components. The components are connected to the same parts in the two hands. Are the two hands identical? Look at the hand in the margin. Is there any doubt in your mind that it is a left hand? The fact that you immediately identified it as a left hand tells you that your hands cannot be identical. If they were, you could not distinguish one from the other; you could only guess which hand is shown in the margin. Your hands are an example of a pair of **chiral** objects, which are nonsuperimposable mirror images. A pair of chiral molecules are stereoisomers of each other.

Let's do two simple tests on your hands. Put one hand over the other and try to merge them (Figure 8.6). No matter how you try, you can never arrange your hands so all of the parts of one are exactly aligned with the parts of the other. This first test shows your hands are *nonsuperimposable*. (If your hands were superimposable, they would be identical.) For the second test stand in front of a mirror. Hold your hands up in front of you, with your left palm facing the mirror and your right palm facing you (Figure 8.7). Compare the image of your left palm with your real right palm and you will find the two are identical. The results of this test show that your hands are mirror images of each other. Because your hands are both nonsuperimposable (first test), and mirror images (second test), they are chiral. In fact, the root of the word *chiral* comes from the Greek word for *hand*.

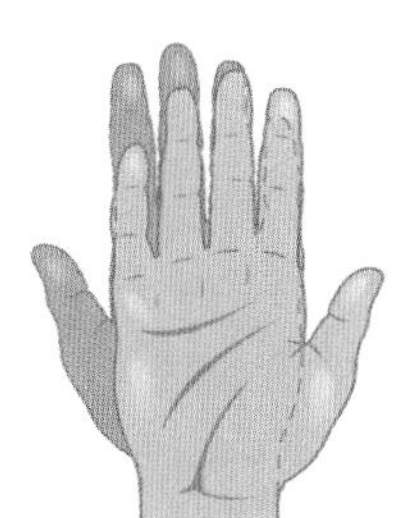

Figure 8.6 A test for superimposability: Hold your hands out in front of you, side by side, palm side up. Move your right hand until it is above your left hand (keep those palms up). Lower your right hand until it touches your left hand, and imagine that your right keeps moving downward and merges into your left hand. Now wiggle your thumbs. Are your thumbs exactly aligned with each other; do you see one thumb? Because your hands cannot be merged into a single image, they are nonsuperimposable.

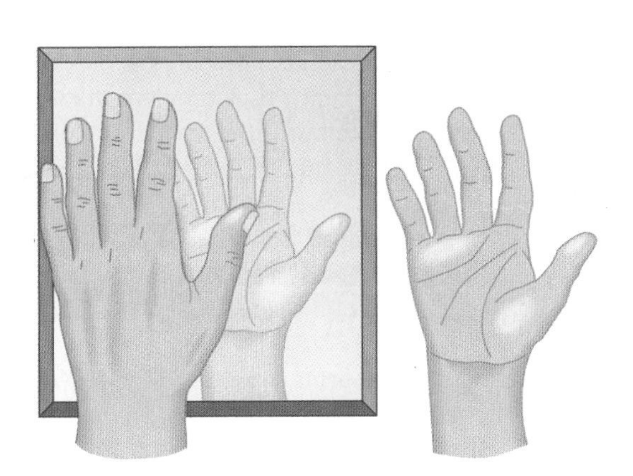

Figure 8.7 Your hands are mirror images of each other.

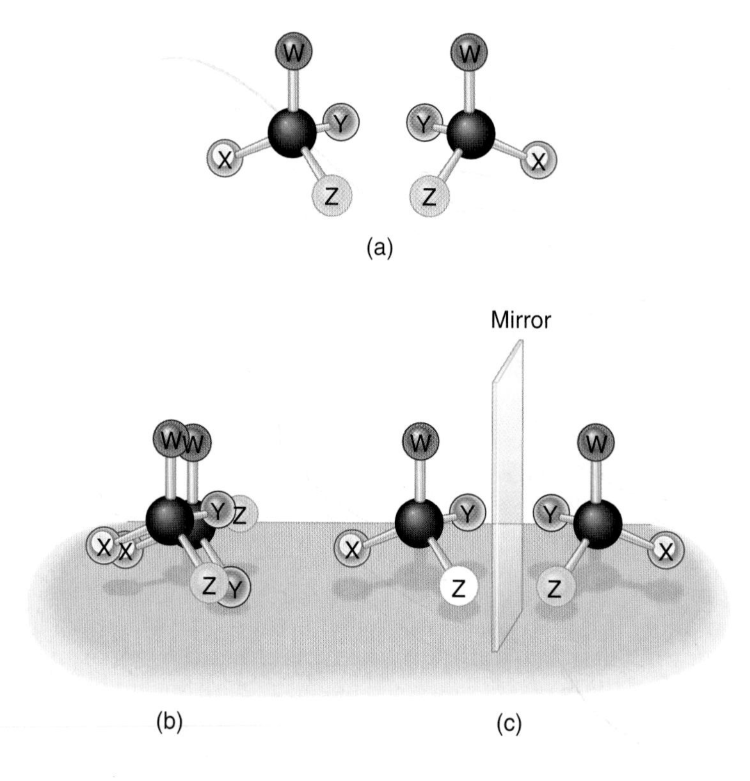

Figure 8.8 **(a)** These two molecules are a set of chiral objects; they are a pair of enantiomers. **(b)** These molecules are not superimposable. Because these molecules are nonsuperimposable mirror images of each other, they are enantiomers. **(c)** Each molecule is the mirror image of the other.

Some molecules are also chiral. Look at the molecules represented by the two ball-and-stick models shown in Figure 8.8(a). Both of them have the same composition: a central black atom and a single copy of atoms W, X, Y, and Z. Both have the same atoms attached to the central black atom. Are the molecules identical? In Figure 8.8(b), one of the molecules has been placed on the other to see if they are superimposable. But no matter how the molecules are arranged, only the central atom and two other atoms (in this case W and X) are superimposed, atoms Y and Z are not. Because the two molecules are not superimposable, they cannot be identical. For two molecules to be chiral, they must be both nonsuperimposable and mirror images of each other. In Figure 8.8(c) the molecule on the right is in front of a mirror, and the one on the left is behind the mirror. The molecule on the left is identical to the mirror image of the molecule on the right, thus they are mirror images of each other. If a pair of objects or molecules are both nonsuperimposable and mirror images, they must be chiral. A pair of chiral molecules that are nonsuperimposable mirror images are called **enantiomers.**

There is an easy way to tell whether a molecule might be chiral. Look for an atom in the molecule (usually a carbon atom) that has four *different* substituents attached to it; such an atom is called a **chiral center.** If a chiral center is present in a molecule, the molecule is usually chiral.

So some molecules are chiral. Is this of any real importance? Consider caraway and spearmint. Do they smell at all alike to you (think rye bread and chewing gum)? Figure 8.9 shows two molecules. The molecule on the left is responsible for the smell of spearmint, the one on the right for the smell of caraway. The shaded

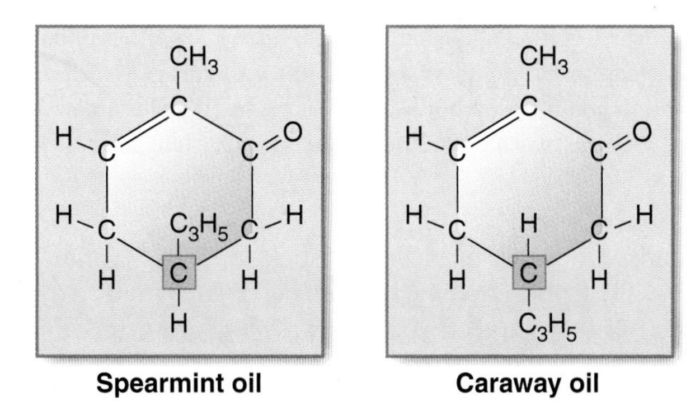

Figure 8.9 These molecules are responsible for the scent of spearmint and caraway. In each of these enantiomers, the chiral center is highlighted.

carbon atom in each molecule is a chiral center, thus the two molecules are enantiomers (they are chiral). They smell different because the sense of smell depends on very large molecules (proteins) called scent receptors. Receptor molecules bind specific atoms or groups of atoms in a molecule (Figure 8.10). Most molecules cannot bind because they lack the correct groups to bind. In fact, only one of a pair of enantiomers can bind to a receptor. You smell caraway when one enantiomer binds to a particular set of scent receptors in your nose, and you smell spearmint when the other enantiomer binds to a different set of scent receptors.

More important than scent receptors are enzymes, which are proteins that catalyze chemical reactions in the body. Enzymes can bind only one or a few molecules in the same way that a scent receptor can bind only with certain scent molecules. Usually an enzyme will bind only one of a pair of enantiomers. In Chapter 15 you will learn how enzymes work and why stereoisomerism is so important to proper enzyme function.

Conformation

The *configuration* of a molecule is the molecular shape that is the result of the covalent bonding that is present between atoms in the molecule. A water molecule is bent and a methane molecule is tetrahedral because of the configurations of the water and methane molecules. There is another aspect of molecular shape. The **conformation** of a molecule is that part of its shape that depends on free rotation

Molecular conformations are three-dimensional shapes that result from rotation about single bonds.

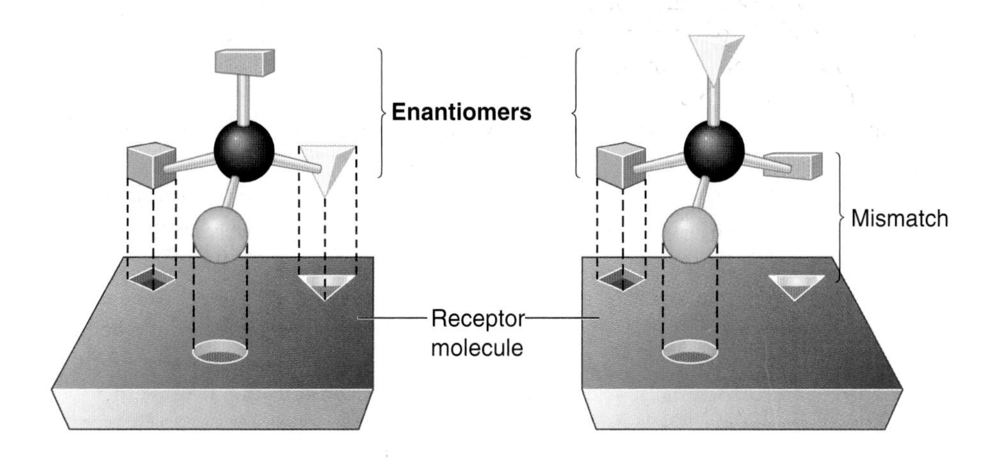

Figure 8.10 Only one of these enantiomers matches the receptor molecule. The other enantiomer, and other molecules in general, cannot effectively bind to the receptor.

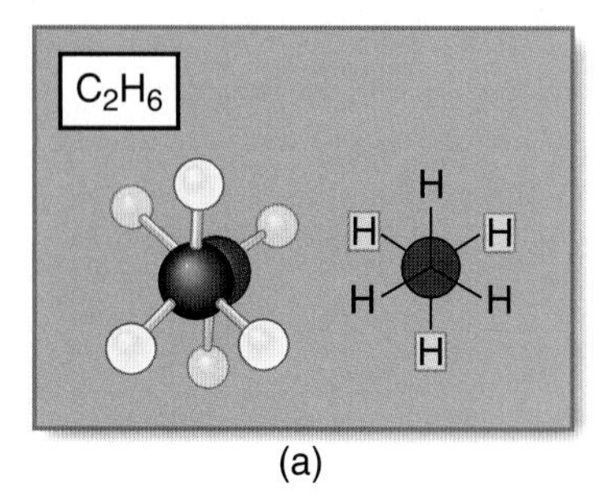

(a)

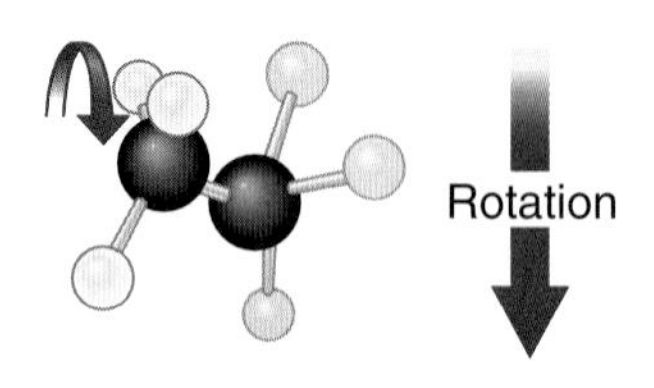

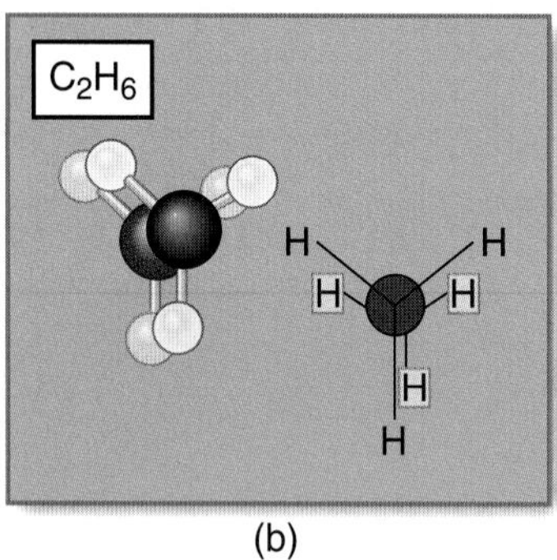

(b)

Figure 8.11 Free rotation about single bonds allows for different conformations. **(a)** The most stable conformation of ethane, with the six hydrogens as far apart as possible. Free rotation about the carbon–carbon single bond yields a new and less stable conformation **(b)**, with the hydrogens closer to each other. (The back carbon atom is shown in red.)

about single bonds. Consider the ball-and-stick model of a molecule of ethane shown in Figure 8.11. Single covalent bonds have free rotation (although those between ring atoms have restricted free rotation). This means that in ethane the hydrogens on one carbon can be aligned to alternate with the hydrogens of the other carbon [Figure 8.11(a)], or the model can be twisted so the hydrogens on both carbon atoms overlap [Figure 8.11(b)]. There are also all possibilities in between. Each arrangement is a unique *conformation*, a unique shape of the molecule. The same atoms are still attached to the same atoms, but the overall shape of the molecule is different. Although there are an infinite number of possible conformations for a molecule, only one or a few are stable, and the molecule often exists in this most stable conformation much of the time.

Cycloalkanes contain only carbon-to-carbon single bonds and thus they provide a useful example of conformations. Although we have learned that single bonds allow free rotation in molecules, the ring structure of cycloalkanes greatly restricts rotation about single bonds in the ring. This reduces the number of conformations that can reasonably exist. Figure 8.12 shows some conformations of cyclohexane. Of these three conformations, the chair conformation is the most stable [Figure 8.12(b)]. Other rings that contain six atoms have this same chair shape. You will see this shape again and again in biochemistry as we talk about sugars such as glucose, and lipids such as cholesterol.

You will study large molecules like DNA and proteins in the biochemistry section. Each of these large molecules has a unique shape because of its composition, configuration, and conformation. Its shape (structure) gives it the properties that allow it to carry out its role in the maintenance of life.

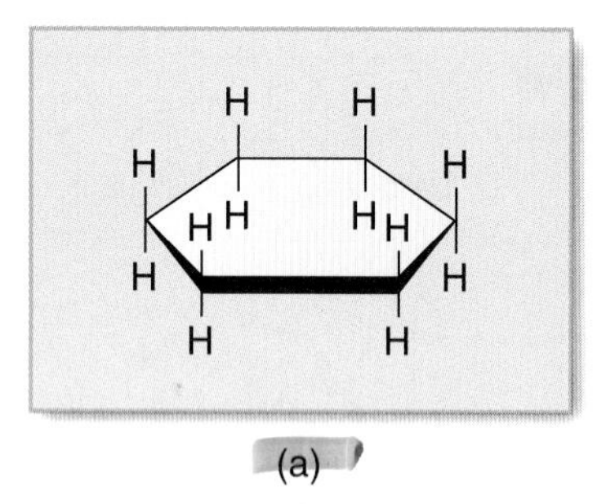

(a)

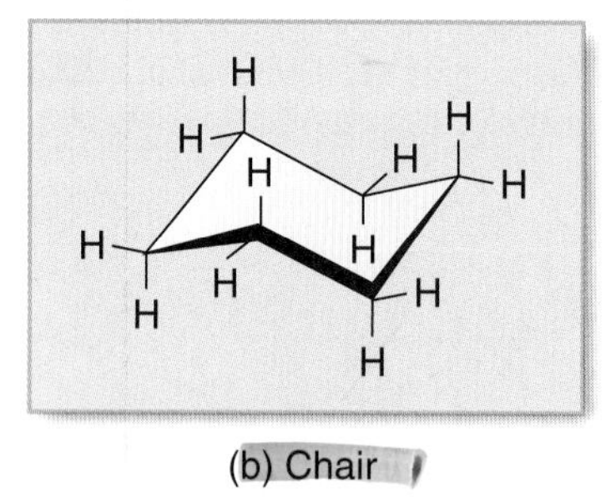

(b) Chair

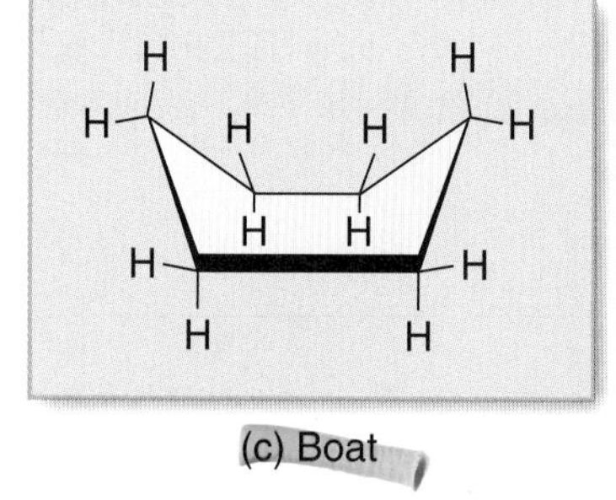

(c) Boat

Figure 8.12 Each carbon atom in cyclohexane, like those in other alkanes, forms four single bonds. This means the bond angles of these carbon atoms are 109°. **(a)** The carbon atoms of cyclohexane cannot lie in a plane because that shape would result in bond angles of 120°. **(b)** Instead, cyclohexane assumes the *chair conformation*, in which the carbon atoms have 109° bond angles. This is the most stable conformation for cyclohexane and other rings containing six carbon atoms. **(c)** This conformation of cyclohexane (the boat form) also has 109° bond angles, but it is less stable than the chair conformation because the ends of the boat are too close to each other.

8.6 Properties and Reactions of Alkanes

Physical Properties

Alkanes are nonpolar molecules (see Section 3.7) because carbon and hydrogen have very similar electronegativity values. This means these atoms share electrons more or less equally, and as a result their bonds are nonpolar. There are no significant dipoles in alkanes, and these molecules bond to each other only through London forces (see Section 5.4).

The amount of attraction between alkane molecules depends on molecular size. Longer chains of carbon atoms have more surface area that can contact adjacent molecules. This increased contact results in more London forces between the molecules (Figure 8.13), which means more energy is needed to break these interactions. As a consequence, larger alkanes have higher melting points and boiling points than smaller alkanes. Depending on size, alkanes may be gases, liquids, or solids at room temperature (Figure 8.14). The same trends are seen in other classes of organic compounds because their surface area also increases as their molecules get larger.

Branching in alkanes lowers the boiling point and often lowers the melting point relative to those of the straight-chain isomer (Figure 8.15). Because branched alkanes are more compact, they possess a smaller surface area than the corresponding straight-chain alkane, they form fewer London forces with neighboring molecules, and less energy is required to change the state of the compound.

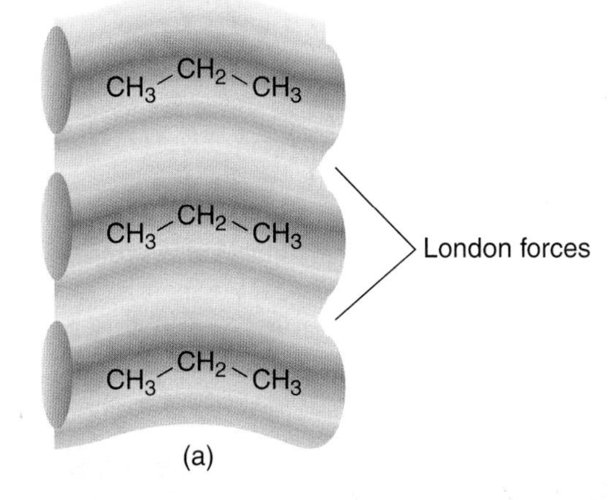

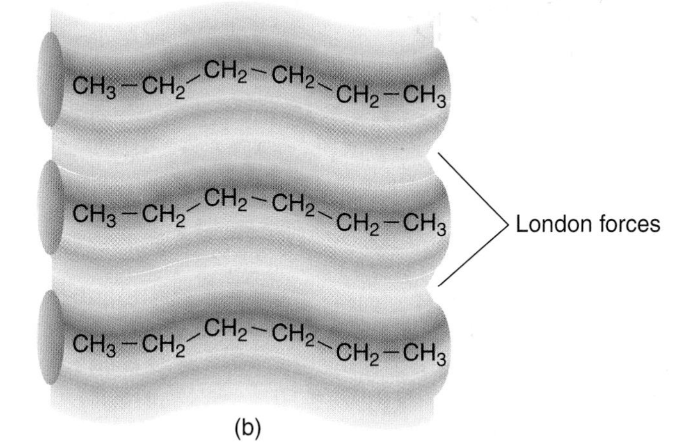

Figure 8.13 The size of a nonpolar molecule determines the surface area of the molecule, which in turn affects the amount of contact with adjacent molecules. London dispersion forces depend on contact between two molecules. Note that propane **(a)** has less surface area and less contact with adjacent molecules than hexane **(b).**

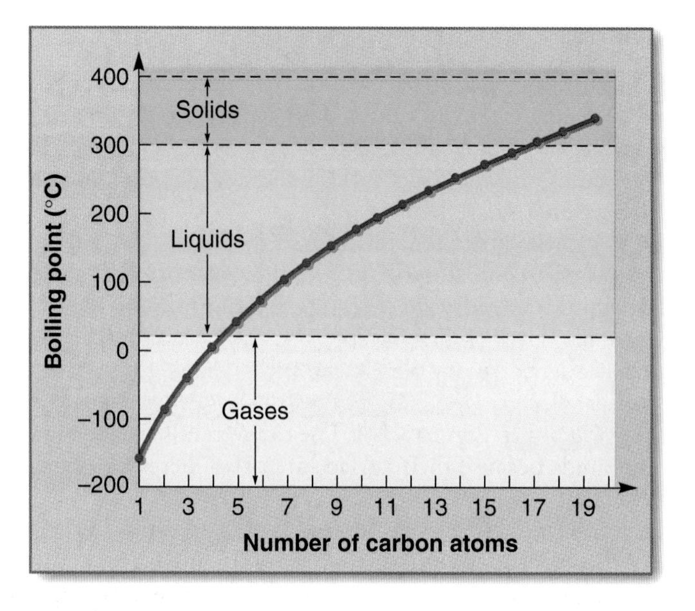

Figure 8.14 The boiling points of straight-chain alkanes increase with molecular size. This trend is a result of increased London dispersion forces, which in turn result from increased contact between molecules. The room-temperature physical state of these straight-chain alkanes is also shown.

Alkanes are insoluble in water [Figure 8.16(a)]. If alkane molecules were to dissolve in water, the strong hydrogen bonds between the polar water molecules would be broken and replaced by weaker bonding between the water molecules and the alkane molecules. This does not happen. Instead, the alkane molecules float on the water because they are less dense. Alkanes dissolve in most nonpolar substances [Figure 8.16(b)], but perhaps it is better to say the reverse because some alkanes are commonly used as solvents. Most nonpolar organic compounds dissolve in the liquid alkanes. Hexane, C_6H_{14}, is a common laboratory solvent, and some people use gasoline (which is primarily a mixture of alkanes) as a solvent. Although gasoline is a great nonpolar solvent, its flammability makes it unsafe for normal use. Many less flammable commercial solvents contain turpentine, diesel, kerosene, or petroleum spirits—all of which are mixtures that are primarily or entirely alkanes.

Chemical Reactivity

Compared to other organic compounds, alkanes show little chemical reactivity. This is because the carbon-to-carbon single bond (C—C) and the carbon-to-

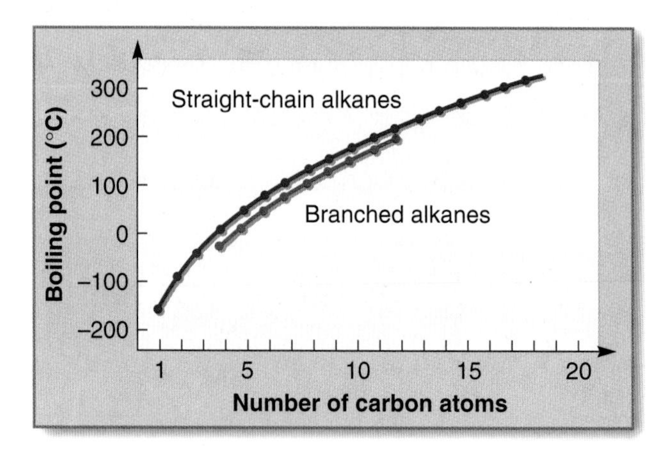

Figure 8.15 Any branched alkane boils at a lower temperature than the straight-chain alkane containing the same number of carbon atoms.

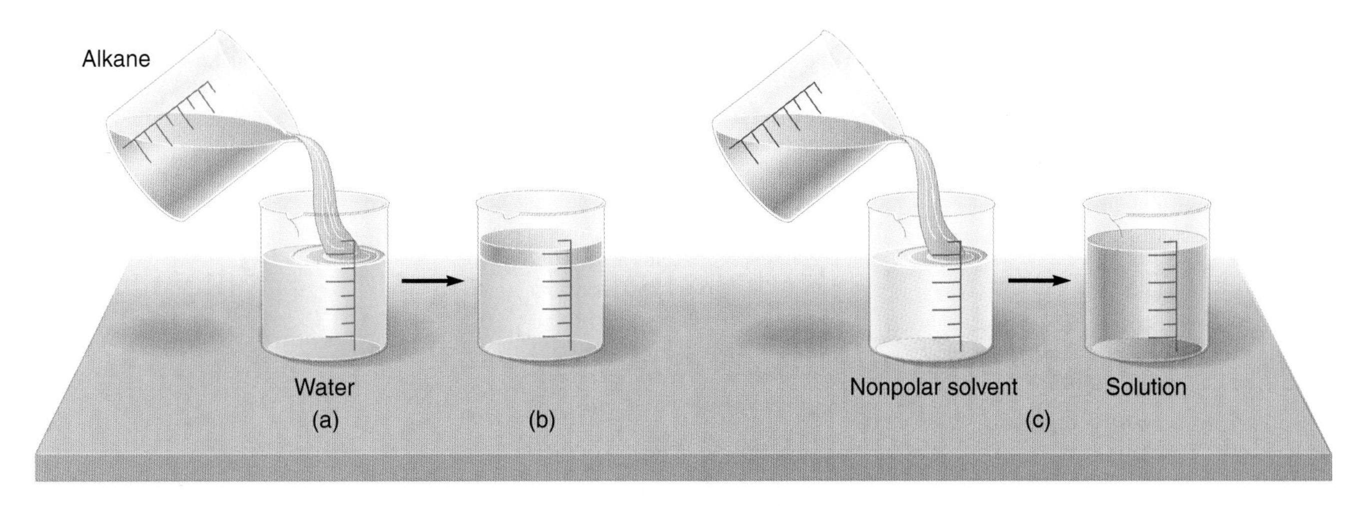

Figure 8.16 **(a)** Alkanes are nonpolar, and are therefore insoluble in water, which is polar. **(b)** Alkanes float on water because they are less dense than water. **(c)** Alkanes dissolve in nonpolar solvents because both are nonpolar (like dissolves like).

hydrogen bond (C—H) are relatively stable. There are two reactions of alkanes of importance to us. The first is a reaction that affects virtually all organic compounds—combustion. *Combustion* is an oxidation-reduction reaction between a combustible material (one that burns) and oxygen. When an alkane is heated in the presence of oxygen, the compound burns. Because alkanes contain only carbon and hydrogen, normally the only products of alkane combustion are carbon dioxide and water. Consider the combustion of methane (natural gas) and propane (a common home and RV fuel):

Connections

The Products of Incomplete Combustion

Although complete combustion of hydrocarbons produces only carbon dioxide and water, limited oxygen results in incomplete combustion that yields other products. One of these products is the toxic, odorless, and colorless gas carbon monoxide (CO). This gas is produced in engines and heating devices when oxygen is limited or if the device is not functioning properly. When inhaled, carbon monoxide diffuses into red blood cells and binds tightly to the iron ions in the *heme* of hemoglobin. While a carbon monoxide is bound to the heme of hemoglobin, an oxygen molecule cannot bind; thus the amount of oxygen carried by the blood from the lungs to the body is reduced. If much of the hemoglobin has carbon monoxide bound to it, the lack of oxygen in the body can lead to death or permanent injury. Fortunately, the binding of carbon monoxide to hemoglobin is not permanent, and if a victim is removed from the presence of carbon monoxide quickly enough, fresh air or treatment with oxygen may yield recovery.

Soot is another product of incomplete combustion. Diesel and gasoline engines may produce soot, which contains particles that give smog its visible properties. In addition, soot may contain some polycyclic aromatic ring compounds that cause cancer (Chapter 9).

The products of incomplete combustion are in auto emissions that contribute to smog. *(Ed Pritchard/Tony Stone Images)*

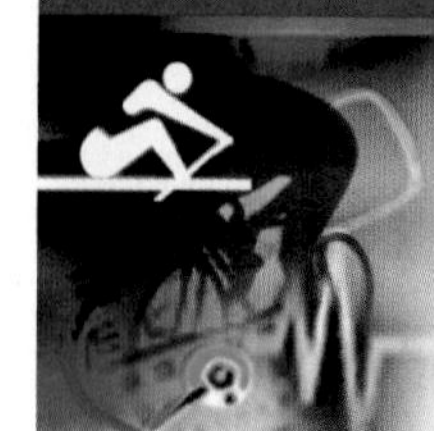

Connections

Ozone, CFCs, and Skin Cancer

The oxygen we breathe is molecular oxygen, O_2, but oxygen also exists as ozone, O_3. This highly reactive form of oxygen is produced from molecular oxygen when it absorbs ultraviolet (uv) light in the upper atmosphere in the first half of the ozone cycle:

$$3O_2(g) + \text{uv light} \rightarrow 2O_3(g)$$

The ozone thus formed then absorbs uv light and forms molecular oxygen in the second half of the ozone cycle:

$$2O_3(g) + \text{uv light} \rightarrow 3O_2(g)$$

This formation and destruction of ozone by uv light in the upper atmosphere removes much of the uv light that comes from the sun before it reaches the surface of the Earth. Because uv light causes skin cancer, the ozone cycle effectively reduces the risk of uv light–induced skin cancer.

Unfortunately, the normal removal of ultraviolet light by the ozone cycle is threatened by the presence of chlorofluorocarbons (CFCs) in the atmosphere. These compounds are among the many halogenated hydrocarbons that are or have been in common use. CFCs were commonly used as refrigerants in older air conditioners, refrigerators, and freezers. Since their introduction years ago, significant amounts of these gases have been released and diffused throughout the atmosphere.

In the upper atmosphere, CFCs are broken down by uv light to yield chlorine atoms that act as catalysts to break ozone down to oxygen. This effectively reduces the amount of ozone, which means less uv light is absorbed in the atmosphere and more uv light thus strikes the Earth's surface. This increased uv radiation increases the risk of skin cancer and may also adversely affect the environment and agriculture through its impact on plants and animals.

The new commercial use of CFCs is banned, and with time the amounts of CFCs in the atmosphere should decline.

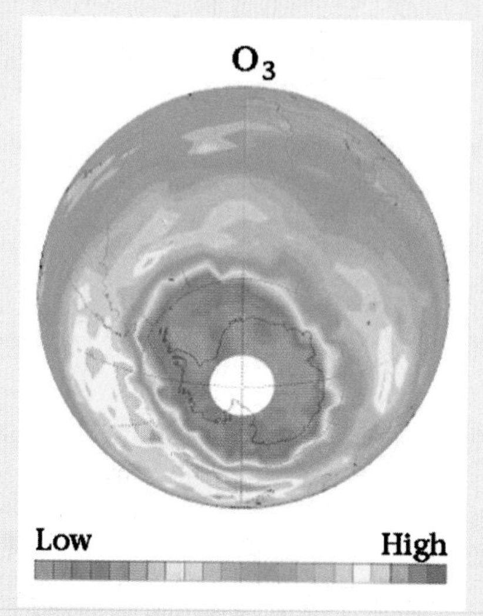

This photo taken in August 1996 shows the depletion of ozone that occurs over the Antarctic each winter. *(Courtesy of NASA)*

$$CH_4(g) + 2O_2(g) \rightarrow CO_2(g) + 2H_2O(g) + \text{energy}$$

$$C_3H_8(g) + 5O_2(g) \rightarrow 3CO_2(g) + 4H_2O(g) + \text{energy}$$

These reactions yield energy as heat. Natural gas, propane, butane, gasoline, home heating fuel, and diesel are commonly used fuels that contain only alkanes or primarily alkanes.

The second reaction of alkanes is *halogenation*. In this reaction, a halogen atom, usually Cl or Br, replaces (substitutes for) a hydrogen atom on the alkane, and the hydrogen atom combines with the other atom of the halogen molecule. Light or heat catalyzes (speeds up) the reaction. The reaction of ethane with chlorine is

You will learn to name chloroethane and other organic compounds in the chapters that follow. They are included in these equations to help you become familiar with them.

$$\underset{\text{Ethane}}{CH_3CH_3(g)} + Cl_2(g) \xrightarrow{\text{light}} \underset{\text{Chloroethane}}{CH_3CH_2Cl(g)} + HCl(g)$$

The reaction is not limited to the replacement of a single hydrogen atom. If enough halogen is present, more or all of the hydrogen atoms are replaced. Complete bromination of methane, for instance, yields

These laboratory, pet, and garden products on the left contain a variety of organohalogens. The bottle on the right contains methylene chloride, which is a common name for dichloromethane, CH_2Cl_2. *(Charles D. Winters)*

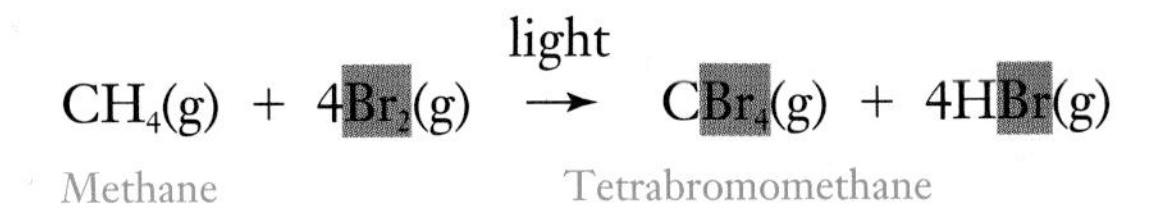

$$CH_4(g) + 4Br_2(g) \xrightarrow{\text{light}} CBr_4(g) + 4HBr(g)$$

Methane Tetrabromomethane

The products of these halogenation reactions are called *alkyl halides*, which are one group of *organohalogens*—organic compounds that contain halogen atoms. Many of these compounds serve important and, as discussed in the Connections, controversial roles in our society.

8.7 Common Functional Groups of Organic Chemistry

There are millions of known organic compounds. Each of these unique compounds has its own set of properties. Is there any way to make sense of this large number of compounds short of memorizing the properties of each? The **structural theory** of chemistry states that the *properties* of a compound depend on the *structure* of the compound.

Each alkane has a unique structure made up of carbon and hydrogen atoms and single bonds. This unique structure results in a unique set of properties for that compound. Like the alkanes, nearly all organic compounds contain carbon and hydrogen atoms linked by single bonds. These other organic compounds also contain additional atoms or bonds that contribute to the unique structure and properties of these compounds. Atoms or groups of atoms that give a molecule its specific chemical and physical properties are called **functional groups.**

All organic compounds contain carbon atoms and most contain hydrogen atoms, but many organic compounds contain one or more atoms of other elements as well. In organic compounds, these atoms are referred to as *hetero-atoms*, atoms other than carbon or hydrogen atoms. A hetero-atom is a functional group or it makes up part of a functional group. Table 8.5 summarizes the common functional groups and shows an organic compound that contains each of them. (These functional groups are covered in the next three chapters.)

Table 8.5 Common Organic Functional Groups

CLASS OF ORGANIC COMPOUND	FUNCTIONAL GROUP	EXAMPLE
Alkenes	$>C=C<$ (double bond)	$CH_2=CH_2$
Alkynes	$-C\equiv C-$ (triple bond)	$HC\equiv CH$
Aromatic compounds[a]	(benzene ring)	(benzene ring with CH_3)
Alcohols and phenols	$-OH$	CH_3CH_2OH
Ethers	$-O-$	CH_3OCH_3
Thiols	$-SH$	CH_3SH
Amines	$-N<$	CH_3NH_2
Organohalogens	$-X$ (X = halogen atom)	CH_3CH_2Br
Aldehydes	$-\overset{O}{\overset{\Vert}{C}}-H$ or $-CHO$	CH_3CHO
Ketones	$-\overset{O}{\overset{\Vert}{C}}-$	$CH_3\overset{O}{\overset{\Vert}{C}}CH_3$
Carboxylic Acids	$-\overset{O}{\overset{\Vert}{C}}OH$ or $-COOH$ or $-CO_2H$	CH_3COOH
Esters	$-\overset{O}{\overset{\Vert}{C}}OR$ or $-COOR$	$CH_3COOCH_2CH_3$
Amides	$-\overset{O}{\overset{\Vert}{C}}N<$	$CH_3\overset{O}{\overset{\Vert}{C}}NH_2$
Anhydrides	$-\overset{O}{\overset{\Vert}{C}}O\overset{O}{\overset{\Vert}{C}}-$	$CH_3\overset{O}{\overset{\Vert}{C}}O\overset{O}{\overset{\Vert}{C}}CH_3$

[a]The representation of the aromatic ring is discussed in Section 9.5.

CHAPTER SUMMARY

Organic chemistry is the study of carbon-containing compounds. Carbon atoms have four valence electrons and can form single, double, and triple covalent bonds with atoms of many elements. As a consequence of this bonding, the number of carbon-containing compounds is huge. **Hydrocarbons** are organic compounds containing only carbon and hydrogen atoms. **Alkanes** are hydrocarbons containing only single covalent bonds. Because no more hydrogen atoms can be added to alkane molecules, they are said to be saturated. Alkanes may have a single chain of carbon atoms **(straight-chain alkanes)** or they may have branched chains. Compounds with the same composition but different structures are called **isomers.** Condensed structural formulas and molecular models are often used to represent organic compounds. **Alkyl groups** are groups of carbon and hydrogen atoms that are used in some organic compound names. **Cycloalkanes** contain three or more carbon atoms bonded to form a ring. Because of this ring, some cycloalkanes can exist as *cis-* or ***trans-geometric isomers.*** A **chiral center** is an atom that has four different substituents attached to it. Molecules with a chiral center are **chiral** and exist as a pair of **stereoisomers** called **enantiomers.** Free rotation occurs around single bonds, and as a result a molecule can exist in one or more shapes called **conformations.** Alkanes are nonpolar molecules, thus they are water insoluble. The smallest alkanes (methane through butane) are gases, pentane through alkanes containing about 18 to 20 carbon atoms are liquids, and the larger ones are solids. Alkanes burn and are major components of natural gas and petroleum products. Alkanes also react with halogens in a substitution reaction to produce alkyl halides. The **structural theory** of chemistry says structure determines the properties of compounds. **Functional groups** are atoms or groups of atoms that give a molecule specific properties.

Summary of Reactions of Alkanes

Combustion

$$\text{Alkane} + O_2 \longrightarrow CO_2 + H_2O + \text{energy}$$

$$CH_4(g) + 2O_2(g) \longrightarrow CO_2(g) + 2H_2O(g) + \text{energy}$$

Halogenation

$$\text{Alkane} + X_2(\text{halogen}) \xrightarrow{\text{light}} \text{alkyl halide} + HX$$

$$CH_4(g) + Cl_2(g) \xrightarrow{\text{light}} CH_3Cl + HCl$$

KEY TERMS

Section 8.1
organic chemistry
hydrocarbons
molecular models
straight-chain alkane

Section 8.2
alkanes
condensed structural formulas

Section 8.3
alkyl group
isomer

Section 8.4
cycloalkane
geometric isomers
conformation
enantiomers
stereoisomers

Section 8.5
chiral
chiral center

Section 8.7
functional groups
structural theory

EXERCISES

Organic Chemistry *(Section 8.1)*

1. Organic chemistry is a very large branch of chemistry. What do organic chemists study?

2. Why is organic chemistry included in many introductory chemistry textbooks? (Stated another way, how come you gotta learn it?)

3. Which of these compounds are organic and which are inorganic?
(a) HNO_3 (b) C_2H_6
(c) C_2H_5OH (d) H_2O

4. Which of these compounds are organic and which are inorganic?
(a) $CaSO_4$ (b) CH_3NH_2
(c) C_6H_6 (d) N_2O_4

5. A compound dissolves in water and is an electrolyte. Is it more likely to be an inorganic or organic compound? Support your decision.

6. In general, which would be more volatile, organic or inorganic compounds?

7. Would you expect organic or inorganic compounds to have lower melting points? What is the basis for your choice?

8. A compound dissolves in water, its aqueous solution conducts electricity, and it melts at 628° C. Is this compound likely to be organic or inorganic?

9. A compound is a gas at room temperature, dissolves in hexane (nonpolar solvent) but not in water, and burns readily. Is this compound likely to be organic or inorganic?

10. Why does carbon normally form only covalent bonds?

11. Carbon normally forms four bonds. Why?

12. What is the maximum theoretical size for a carbon-containing compound?

Alkanes—The Simplest Organic Compounds *(Section 8.2)*

13. What is the definition of a hydrocarbon? Provide three examples of common hydrocarbons.

14. The word *saturated* is commonly applied to some organic compounds. What is meant when a compound is said to be saturated?

15. List the kinds of bonds found in alkanes.

16. What is a straight-chain alkane?

17. What is the shape of methane?

18. Sketch the shape of a methane molecule, and label the bond angles in your sketch. Compare the bond angles in methane to the bond angles for any carbon atom that has four single bonds.

19. What is the name of the alkane that contains three carbon atoms?

20. What is the name of the straight-chain alkane that has six carbon atoms?

21. Draw the structural formula of propane and pentane.

22. Draw the structural formula of hexane and octane.

23. Add hydrogen atoms and appropriate carbon-to-hydrogen bonds to complete the structural formulas of these alkanes.

(a) C—C—C

(b) C—C—C—C (with a C branch below the third carbon)

24. Add hydrogen atoms and appropriate carbon-to-hydrogen bonds to complete the structural formulas of these alkanes.

(a) C—C—C—C

(b) C—C—C—C (with a C branch above the second carbon and a C branch below the third carbon)

25. Write the condensed structural formula for both ethane and nonane.

26. Write the condensed structural formula for both butane and octane.

27. Use this structural formula of propane to draw its condensed structural formula.

```
   H   H   H
   |   |   |
H—C—C—C—H
   |   |   |
   H   H   H
```

28. Use this condensed structural formula of pentane to draw its structural formula.

$CH_3CH_2CH_2CH_2CH_3$

29. What advantage do molecular models have over structural formulas?

30. Compare the advantages of ball-and-stick models to those of space-filling models.

31. A perspective formula of an organic compound includes wedges and dashes. What does a wedge represent in these formulas? Explain the meaning of a dashed line in these formulas.

Branched Alkanes *(Section 8.3)*

32. Compare a straight-chain alkane to a branched alkane. Feel free to sketch an example of each to support your comparison.

33. Determine which of the following compounds are straight-chain alkanes and which are branched alkanes.

(a) $CH_3CH_2CH_2CH_2CH_3$

(b) $CH_3CH_2CHCH_2CH_2CH_3$ (with CH_3 below the CH)

(c) H_3C—CH_2—CH_2—CH_3 (zigzag: H_3C, CH_2, CH_2, CH_3)

34. Determine which of the following compounds are straight-chain alkanes or branched alkanes.

(a) CH_3CH_2 with $CH_2CH_2CH_3$ below the CH_2

(b) $CH_3CH_2CCH_3$ with CH_3 above and CH_3 below the C

(c) H_3C—CH_2—CH_2—CH—CH_3 (zigzag, with CH_3 above the CH)

35. Isomerism is a common feature of organic compounds. What are isomers?

36. Identify each pair of structures as representing the same compound, isomers, or different compounds.

(a) $CH_3CH_2CH_2CH_2CH_3$ and $CH_3CHCH_2CH_3$ (with CH_3 below the CH)

(b) CH_3—CH_2—CH_2—CH_3 (zigzag) and $CH_3CH_2CH_2CH_3$

(c) $CH_3CHCH_2CH_2CH_2CH_3$ (with CH_3 below the CH) and $CH_3CHCH_2CH_2CH_2CH_2CH_3$ (with CH_3 below the CH)

37. Identify each pair of structures as representing the same compound, isomers, or different compounds.

(a) $CH_3CH_2CHCH_3$ (with CH_3 on C3) and $CH_3CHCH_2CH_3$ (with CH_3 on C2)

(b) $CH_3CHCH_2CH_2CH_3$ (with CH_3 on C2) and $CH_3CH_2CHCH_2CH_3$ (with CH_3 on C3)

(c) $CH_3CHCH_2CH_2CH_3$ (with CH_3 on C2) and $CH_3CH_2CH_2CH_2CHCH_3$ (with CH_3 on C5)

38. Draw the two isomers for the molecular formula C_4H_{10}.

39. Draw the isomers that have the formula C_6H_{14}.

40. Provide a definition of an alkyl group including a comparison of the methyl group to the compound methane.

41. How is the name of an alkyl group derived? Use propane and the propyl group as a specific example.

42. Look at the structures of the ethyl and propyl groups. Now draw the structure of a butyl group. (I know you haven't seen one yet, but I bet you can do it.)

43. Systematic nomenclature includes specific rules for naming organic compounds. Identify the three parts of the name for a branched alkane. Explain what information is contained in each part of the name.

44. Provide the names of these alkanes.

(a) $CH_3CH_2CHCH_2CH_2CH_3$ (with CH_2CH_3 on C3)

(b) $CH_3CHCHCH_2CH_3$ (with CH_3 on C2 and CH_3 on C3)

(c) $CH_3CH_2CHCHCH_2CH_2CH_3$ (with CH_2CH_3 on C3 and CH_3 on C4)

(d) $CH_3CHCHCH_2CH_2CH_3$ (with CH_2CH_3 on C2 and $CH(CH_3)_2$, drawn as CH bonded to H_3C and CH_3, on C3)

45. Provide the names of these alkanes.

(a) $CH_3CH_2CHCH_2CHCH_3$ (with $CH_2CH_2CH_3$ on C3 and CH_3 on C5)

(b) $CH_3CCH_2CH_2CHCH_3$ (with two CH_3 groups on C2 and CH_3 on C5)

(c) $CH_3CHCH_2CH_2CHCH_3$ (with CH_2CH_3 on C2 and $CH_2CH_2CH_3$ on C5)

(d) $CH_3CCH_2CH_2CCH_3$ (with two CH_3 groups on C2 and two CH_2CH_3 groups on C5)

46. Draw the condensed structural formula of these alkanes:

(a) 4-propylnonane
(b) 5-ethyl-3-methyldecane
(c) 2,3,4-trimethylhexane
(d) 5-ethyl-2,2-dimethyloctane

47. Draw the condensed structural formula of these alkanes:

(a) 3-ethyl-2-methyl-4-propylnonane
(b) 2,2,3,3-tetramethylheptane
(c) 2-methylbutane
(d) 4-isopropyl-3,3-dimethyloctane

48. The name of this compound is incorrect: 2-propylhexane. Draw the structure implied by this name, then name the compound correctly. What rule was broken?

Cycloalkanes (Section 8.4)

49. What is a cycloalkane?

50. Draw the geometric figures that would be used to represent cyclopentane and cyclobutane.

51. Name these cycloalkanes:

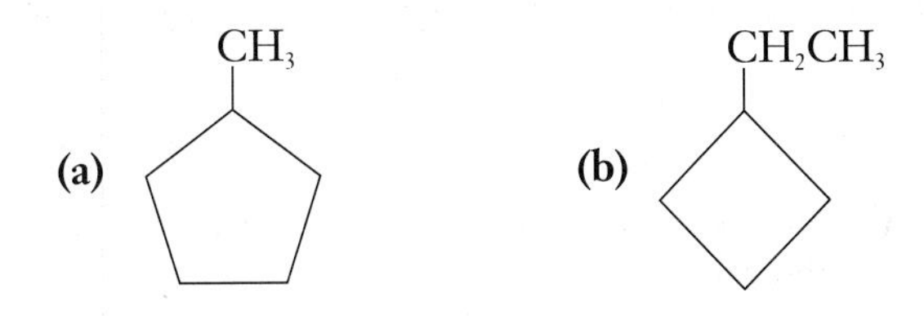

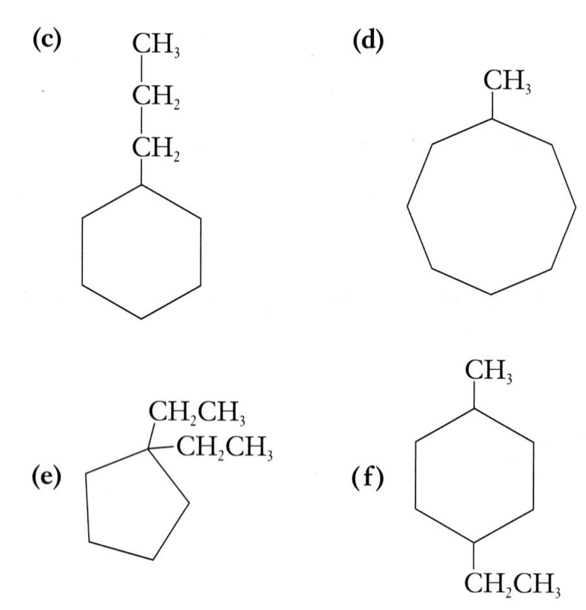

52. Name these cycloalkanes:

(a) CH_2CH_3 (b) CH_3 CH CH_3

(c) CH_3 CH_3 (d) CH_3 CH_2CH_3

(e) CH_3 CH_3 CH_3 (f) CH_3 CH_3 CH_3

53. Draw the structures of these cycloalkanes.
   (a) Methylcyclopropane
   (b) Ethylcycloheptane
   (c) 1,3-dimethylcyclopentane
   (d) 1,1,2-trimethylcyclohexane
   (e) 1-ethyl-3,4-dimethylcyclohexane

54. Draw the structures of these cycloalkanes.
   (a) Methylcyclopentane
   (b) Propylcyclohexane
   (c) 1,2-dimethylcyclobutane
   (d) 1,2,3-trimethylcyclohexane
   (e) 1-ethyl-3,4-dimethylcyclooctane

55. Why is this name wrong?
   2-ethyl-3-methylcyclobutane

56. Draw the structure of *trans*-1,3-dimethylcyclopentane.

57. Name and draw the structure of the geometic isomer of the compound in Exercise 56.

58. Draw the chair conformation of cyclohexane. Use Figure 8.12 as a guide if needed.

59. Why is cyclohexane not planar? (It may be useful to include a sketch in your explanation.)

Shapes of Organic Molecules *(Section 8.5)*

60. Describe how a pair of stereoisomers differ from each other.

61. Two objects that are nonsuperimposable mirror images are examples of ________ objects.

62. Which of the following items are chiral objects?
   (a) A glove (b) A baseball (c) A fishhook

63. Which of the following items are chiral objects?
   (a) A coffee cup (b) An ear (c) A foot

64. Provide a definition for enantiomers. How many enantiomers are in a set of enantiomers?

65. The central carbon atom in these molecules has four substituents. (By the way, these are *not* hydrocarbons. Do you know why?) Which of these carbon atoms are chiral centers?

   (a) Cl—C—H with H above and Br below (b) Cl—C—Br with H above and I below

66. The central carbon atom in these molecules has four substituents. Which of these carbon atoms are chiral centers?

   (a) CH_3—C—Cl with H above and Br below (b) H—C—CH_2CH_3 with CH_3 above and Cl below

67. Draw the structure of 3-methylheptane. Which carbon atom in this molecule is chiral?

68. What is the configuration of a compound?

69. Free rotation is found around what type of covalent bond?

70. The infinite number of shapes that result from rotation about single covalent bonds are called ________.

Properties and Reactions of Alkanes (Section 8.6)

71. Compare the London dispersion forces between two methane molecules to those between two butane molecules. Now provide a summarizing statement for the interactions between nonpolar molecules of increasing size.
72. Which boils at a higher temperature, hexane or 2-methylpentane? Why?
73. Are alkanes and other nonpolar organic compounds water soluble?
74. Gasoline and small alkanes are good solvents for grease and other nonpolar substances, but when possible, kerosene or diesel would be a better choice. Why?
75. List the products of complete combustion of an alkane.
76. Write balanced equations for the combustion of:
    (a) Methane (b) Pentane (c) 2-Methylbutane
77. Write balanced equations for the combustion of:
    (a) Propane (b) Hexane (c) Cyclopentane
78. Assuming one mole of bromine reacts with methane, write the balanced equation for the bromination of methane.
79. Complete and balance these equations. Assume each product molecule contains one halogen atom:

    (a) $CH_3CH_3 + Br_2 \xrightarrow{\text{light}} ?$

    (b) $CH_4 + Cl_2 \xrightarrow{\text{light}} ?$
80. Write a balanced equation for chlorination of methane, assuming that each mole of methane reacts with four moles of chlorine.

Common Functional Groups of Organic Chemistry (Section 8.7)

81. According to the structural theory, what determines the properties of a compound?
82. Define a functional group and give three examples.
83. Alkanes are hydrocarbons, but not all hydrocarbons are alkanes. What is the functional group of the alkynes? The alkenes? The aromatic hydrocarbons?
84. An organic compound is classified as an alcohol if the molecule contains a(n) ________ .
85. Provide the structure of the functional group found in amines.
86. What are the functional groups in each of these molecules found in living organisms?

    (a) 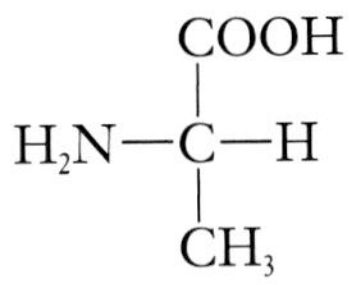 $H_2N{-}C(COOH)(CH_3){-}H$ (COOH above C, CH_3 below C)

    (b) 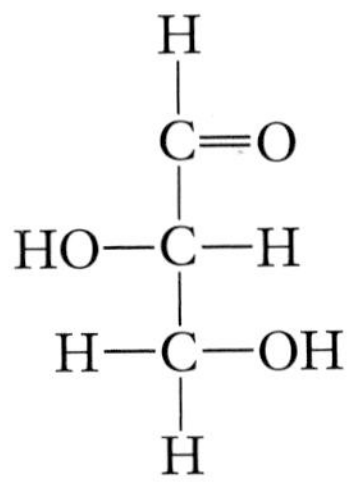 H–C(=O)–C(H)(OH)–C(H)(OH)–H (drawn vertically: H, C=O, HO–C–H, H–C–OH, H)

87. What are the functional groups in each of these molecules?
    (a) Aspirin

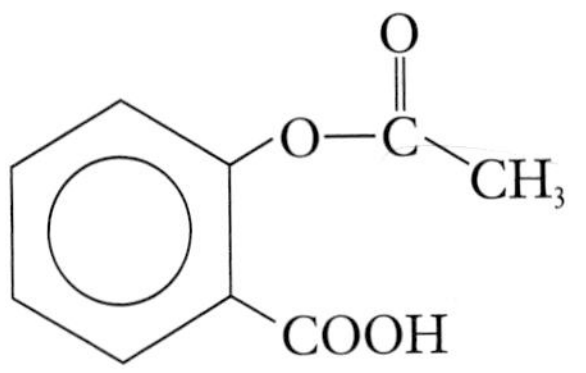

    benzene ring bearing $-O-C(=O)-CH_3$ and, on the adjacent carbon, $-COOH$

    (b) Tyrosine

    $HO-C_6H_4-CH_2-CH(NH_2)-C(=O)-OH$

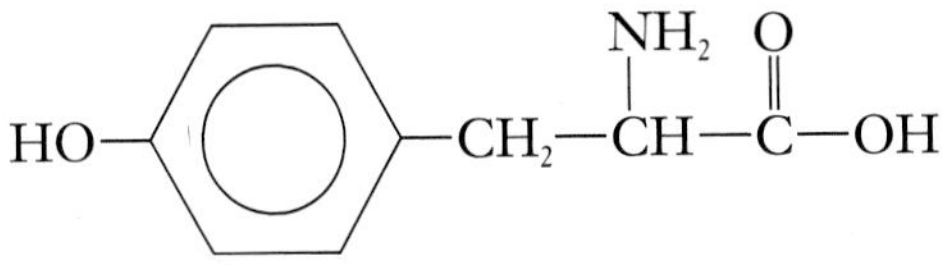

Challenge Exercises

88. Identify the chiral center in these molecules. (Hint: It may be helpful to convert the condensed structural formula to a structural formula.)
    (a) $HOCHClCH_3$
    (b) $CH_3CHBrCH_2CH_3$
89. Propane is a common fuel sold in cylinders of various sizes.
    (a) Determine the amount of heat in both kcal and kJ that would be produced from a 20.0 lb (net weight) cylinder of propane. One mole of propane releases 526.3 kcal when burned completely.
    (b) Write the balanced equation for the complete oxidation of propane.
    (c) Determine the number of pounds of carbon dioxide produced when 20.0 lb of propane burns.
90. (a) Write an equation for the monobromination of butane.
    (b) Name and draw the structure of all products that could form.
    (c) Identify any chiral centers in the product molecules.

9 Unsaturated Hydrocarbons

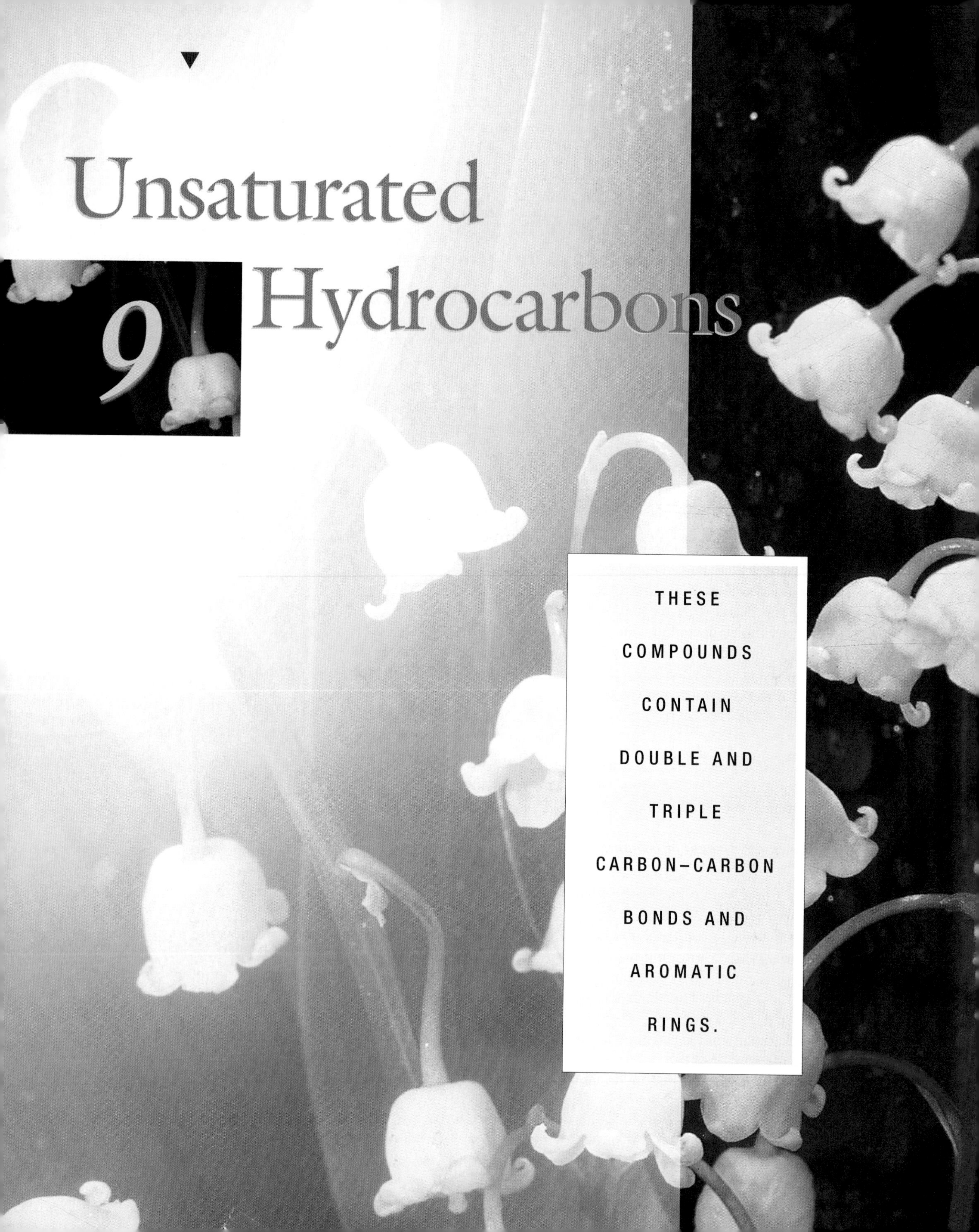

THESE COMPOUNDS CONTAIN DOUBLE AND TRIPLE CARBON–CARBON BONDS AND AROMATIC RINGS.

As we learned in Chapter 8, alkanes are saturated hydrocarbons, which means they have the maximum number of hydrogen atoms on each carbon atom. *Unsaturated hydrocarbons* are those that contain fewer than the maximum number of hydrogen atoms. There are three classes of unsaturated hydrocarbons: the alkenes, which contain one or more carbon–carbon double bonds; the alkynes, which possess one or more carbon–carbon triple bonds; and the aromatic hydrocarbons, which contain one or more rings in which electrons are shared by the carbon atoms forming the ring(s). Compounds of all three classes contribute to our modern society, and many of these compounds play an important role in the processes of life.

Outline

Farnesol, from lily-of-the-valley, contains three double bonds and is unsaturated. Farnesol is used in the perfume industry to emphasize sweet floral scents. *(Barry L. Runk/Grant Heilman Photography, Inc.)*

9.1 Alkenes

The **alkenes** are unsaturated hydrocarbons that contain one or more carbon–carbon double bonds, and the carbon–carbon double bond is the functional group of alkenes:

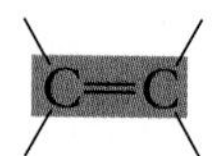

Most alkenes are produced during petroleum refining, which yields billions of pounds of alkenes annually. Alkenes have numerous industrial uses, and the two smallest ones, ethene (common name, ethylene) and propene (propylene), are two of the most common industrial organic compounds. Alkenes are used to make plastics and fibers such as polyethylene and polypropylene, ethanol (alcohol), ethylene glycol (antifreeze), and acetic acid (vinegar).

The structural formula of ethene, C_2H_4, the smallest alkene, is

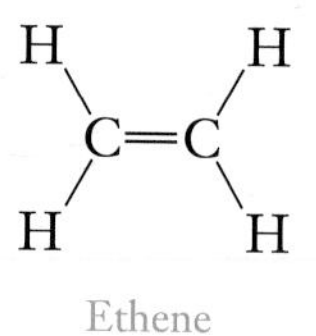

Ethene

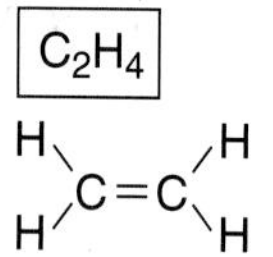

Ethene

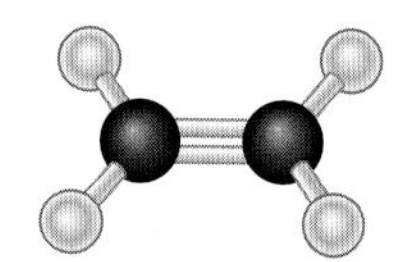

Ball-and-stick model

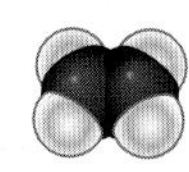

Space-filling model

The most common condensed structural formula for ethene is $CH_2{=}CH_2$. Note that in the formula for ethene, there are two hydrogen atoms for each carbon atom—C_2H_4. The formula for alkenes in general is C_nH_{2n}, where n is any positive integer. Propene, C_3H_6, is the next alkene:

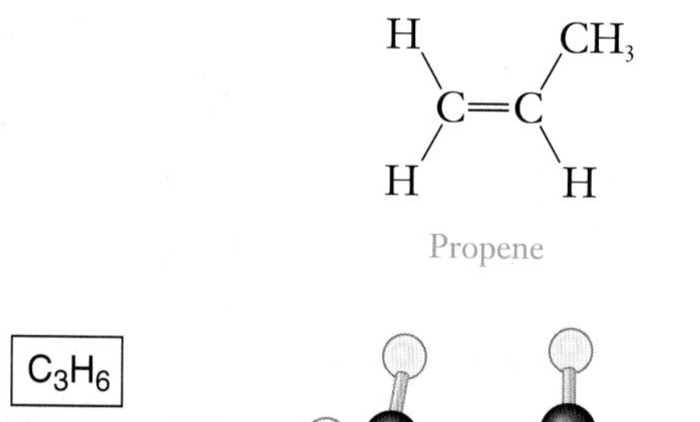

Propene

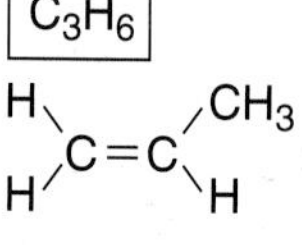

Propene

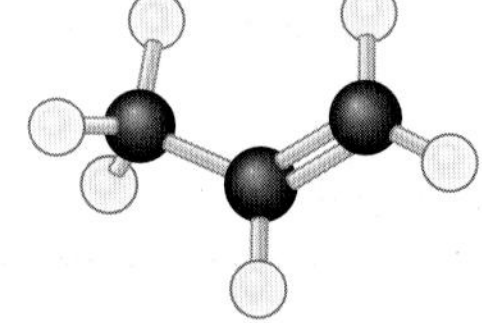

Ball-and-stick model

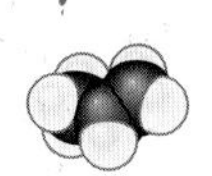

Space-filling model

Propene is often shown as $CH_2{=}CHCH_3$ ($CH_3CH{=}CH_2$ is an equivalent formula). Do you notice a parallel between alkane names and alkene names—ethane, propane and ethene, propene? What do you guess is the name of the next alkene?

Naming Alkenes

Just as the *-ane* suffix tells us that a compound is an alkane, the *-ene* suffix identifies a compound as an alkene. The rules for naming alkenes are similar to the rules for naming alkanes. The major difference is in the numbering of the parent chain. For alkenes, the chain must be numbered to give the carbon atoms of the double bond the smallest possible numbers. The other rules are the same as for alkanes, as Table 9.1 shows.

The smaller alkenes are sometimes identified by common names. Ethene and propene, for example, are sometimes called ethylene and propylene, respectively. Some larger alkenes do have common names that also end in *-ylene*, but these names are not frequently used.

Here are some examples of alkene structures and names:

$$CH_2{=}CHCH_2CH_3 \qquad \text{1-butene}$$

$$CH_3CH_2CH{=}CHCH_2CH_3 \qquad \text{3-hexene}$$

$$\underset{\displaystyle CH_3}{CH_2{=}CH\overset{}{C}HCH_2CH_3} \qquad \text{3-methyl-1-pentene}$$

Two or more double bonds can be present in an alkene. 1,3-Butadiene is an example:

$$CH_2{=}CH{-}CH{=}CH_2$$

Note two things in the name of this compound: (1) that *di-* is added to the suffix *-ene* to indicate that two double bonds are present, and (2) that the position of each double bond is identified by a number.

Double bonds may be present in a ring, which makes these compounds *cycloalkenes*. The rules for naming cycloalkenes are very similar to those for cycloalkanes, but in a cycloalkene the double bond is assumed to be between carbon atoms 1 and 2. If only one double bond is present, there is no need to state the location of the double bond in the name. Cyclopentene, for instance, is a complete

Table 9.1 Rules for Naming Alkenes

Rule One	Find the longest continuous chain of carbon atoms that includes the double bond. The length of this chain determines the parent part of the alkene name. Combine the parent with the suffix *-ene*.
Rule Two	Number the chain beginning with the end closer to the double bond. Use a prefix corresponding to the smaller of the two carbon atom numbers in the double bond to designate the location of the bond. Place this prefix just before the parent part of the name.
Rule Three	The rules for substituents are the same as those for alkanes.
Rule Four	The rules for numerical prefixes, commas, and hyphens are the same as those for alkanes.

and correct name. If one or more groups are present on the ring, each is assigned the lowest possible number (remember, the double bond is between carbon atoms 1 and 2):

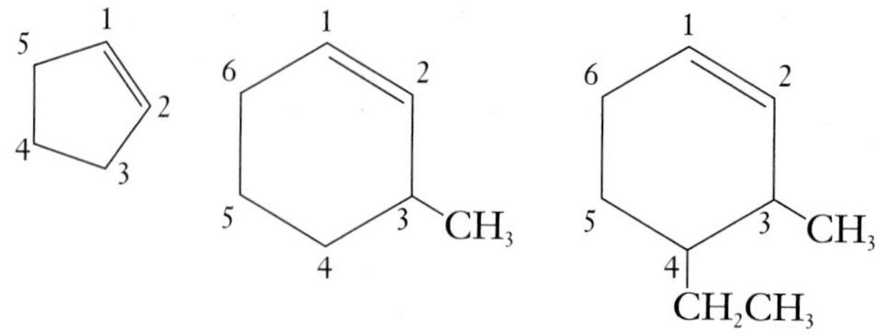

Cyclopentene 3-methylcyclohexene 4-ethyl-3-methylcyclohexene

When there are two or more groups on the ring, remember to place these groups in alphabetical order in the name—it is 4-ethyl-3-methylcyclohexene, not 3-methyl-4-ethylcyclohexene.

Example 9.1 Naming Alkenes

1. Name the alkene $CH_3—CH=CH—CH_2—CH_3$
2. What is the name of $CH_3—CH_2—C(CH_3)=CH_2$?
3. What is the structure of 2-methyl-3-hexene?
4. What is the structure of 4,4-dimethylcyclohexene?

SOLUTIONS

1. There are five carbon atoms in the chain, and so *-pent-* is the parent. There is a double bond, and so the suffix is *-ene.* These combine to form *-pentene.* The chain must be numbered in such a way that the double bond has the lowest possible number, and so the double bond is between carbon atoms 2 and 3. Use the smaller of these numbers in the prefix to indicate the location of the bond. The correct name is 2-pentene.
2. The longest continuous chain that includes the double bond has four carbon atoms. The parent is combined with *-ene* to yield *-butene.* The double bond is between carbon atoms 1 and 2, and so the correct prefix gives us 1-butene. Because the substituent is a methyl group on carbon 2, the complete name is 2-methyl-1-butene.
3. The *-3-hexene* part of the name tells us the longest continuous chain contains six carbon atoms and a double bond between carbons 3 and 4:

$$\underset{1}{C}—\underset{2}{C}—\underset{3}{C}=\underset{4}{C}—\underset{5}{C}—\underset{6}{C}$$

The *2-methyl-* tells us that a methyl group is on carbon 2:

$$C—\overset{CH_3}{\overset{|}{C}}—C=C—C—C$$

Add the appropriate number of hydrogen atoms to get the final structure:

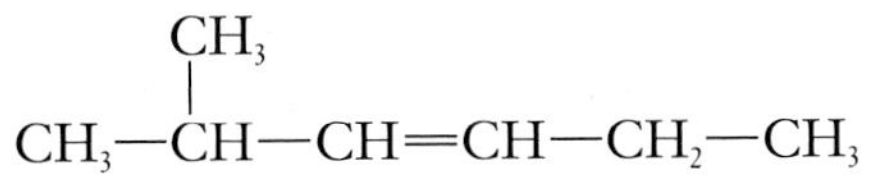

$$\begin{array}{l} \quad\quad\;\; CH_3 \\ \quad\quad\quad | \\ CH_3{-}CH{-}CH{=}CH{-}CH_2{-}CH_3 \end{array}$$

4. *Cyclohex-* tells us that the compound contains a ring of six carbon atoms. The *-ene* means a carbon–carbon double bond is present in the ring, and we give the carbon atoms of this bond the lowest possible number:

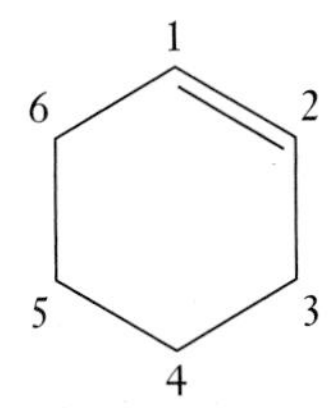

Begin counting with the two carbon atoms of the double bond and count to carbon number 4, then put two methyl groups on this carbon. The structure is

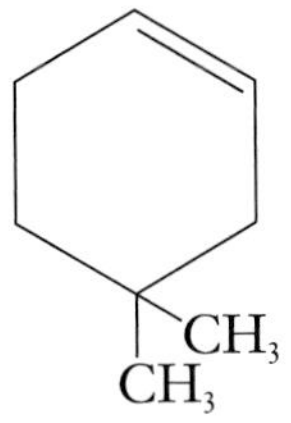

CH_3 CH_3

Self-Test

1. Name this alkene:

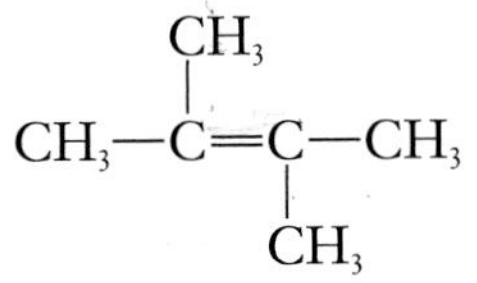

$$\begin{array}{l} \quad\quad CH_3 \\ \quad\quad\; | \\ CH_3{-}C{=}C{-}CH_3 \\ \quad\quad\quad\;\; | \\ \quad\quad\quad CH_3 \end{array}$$

2. What is the structure of 5-ethyl-3-heptene?
3. What is the structure of 1-methylcyclopentene?

ANSWERS

1. 2,3-dimethyl-2-butene
2. 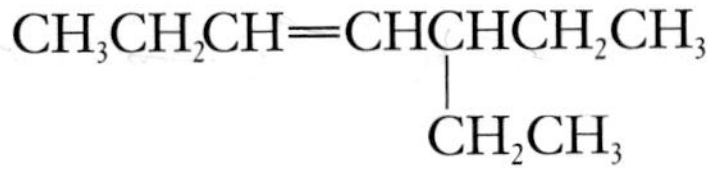 $$\begin{array}{l} CH_3CH_2CH{=}CHCHCH_2CH_3 \\ \quad\quad\quad\quad\quad\quad\;\; | \\ \quad\quad\quad\quad\quad\quad CH_2CH_3 \end{array}$$
3. CH_3 or CH_3

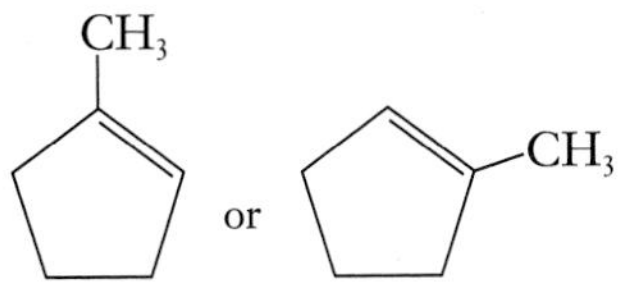

These are two ways of representing the same molecule; in the first structure the numbering of the ring carbons was clockwise, in the second it was counterclockwise.

9.2 Alkene Structure

The carbon–carbon double bond has two significant effects on alkene structure. First, there is no free rotation about a double bond because the energy required for such a rotation would break the bond. Thus the presence of a double bond prevents any conformations at the double bond. (Of course, there are still all the possible conformations around any carbon–carbon single bonds present in the structure.)

The second result of the double bond is the possibility of geometric isomerism (see Section 8.4). Some alkenes exist as a pair of geometric isomers. Consider these two 2-butenes:

```
   H       H        H3C       H
    \     /            \     /
     C=C                C=C
    /     \            /     \
 H3C       CH3        H       CH3
```

These 2-butenes are different compounds because they have different physical properties (for example, the one on the left boils at 3.7°C and the one on the right at 0.9°C). Look closely at the two structures. Both have the same composition (C_4H_8), and both have the same atoms attached to the same atoms. *The molecules differ only in the spatial arrangement of the atoms.* As you saw in Section 8.4, some cycloalkanes show geometric isomerism, with the prefixes *cis*- and *trans*- used to designate geometry. The same prefixes are used in alkenes. The 2-butene on the left in the previous example has two hydrogen atoms *on the same side of the double bond*, and so is named *cis*-2-butene.

```
   H       H        H3C       H
    \     /            \     /
     C=C                C=C
    /     \            /     \
 H3C       CH3        H       CH3
```

Cis-2-butene *Trans*-2-butene

The 2-butene on the right is *trans*-2-butene because the hydrogen atoms are on opposite sides of the double bond. Normally the *cis*- or *trans*- prefix is at the beginning of the name.

When deciding on *cis*- and *trans*- for designating geometry in an alkene, you can use any two similar groups; they do not need to be hydrogen atoms. Go back and name the two 2-butenes using the methyl groups instead of the hydrogen atoms. Did you get the same result?

Please note that an alkene with a *terminal* double bond (one at the end of a chain) cannot exist as geometric isomers. This is because carbon 1 has two hydrogen atoms attached to it; because they are the same, geometric isomers cannot exist. There is only one 1-butene, and it does not exist as *cis*- and *trans*-isomers.

```
 H       H
  \     /
   C=C
  /     \
 H       CH2CH3
```

1-butene
(No geometric isomers exist)

Similarly, *any* alkene with two copies of the same group on a carbon atom of a double bond cannot have *cis*- or *trans*-isomers. For example, carbon 2 in 2-methyl-2-butene has two methyl groups. As a consequence, there is only one possible structure:

$$(H_3C)_2C{=}CH(CH_3)$$

2-methyl-2-butene

Connections

Alkene Chemistry in the Vision Cycle

Did you know that chemistry was involved in vision? What do you think happens after light is focused on the retina? This light is absorbed by the substance *retinal*, which is made from vitamin A. Before it is exposed to light, retinal exists as 11-*cis* retinal. (Don't worry about the carbon numbering in retinal because it does not follow IUPAC rules.)

11-*cis* retinal

When light strikes 11-*cis* retinal, the energy from the light causes the *cis* carbon–carbon double bond to change to the trans form yielding all-*trans* retinal:

All-*trans* retinal

Vision involves the light-induced isomerization of a double bond in retinal. *(VCG, 1998/FGP International)*

This isomerization to all-*trans* retinal triggers a series of events that leads to signals traveling from the eye to the brain, resulting in the perception of light. The retinal does not absorb more light in the all-*trans* form, and so an enzyme in the retina catalyzes the isomerization of all-*trans* retinal back to 11-*cis* retinal, which sets the stage for the next vision cycle.

Example 9.2 Geometric Isomerism in Alkenes

1. Name the alkene

$$\begin{array}{ccc} H & & H \\ & C{=}C & \\ H_3CH_2C & & CH_2CH_2\underset{\displaystyle CH_3}{\underset{|}{C}}HCH_3 \end{array}$$

2. Draw the structure of *trans*-2-pentene.

SOLUTIONS

1. The longest continuous chain that contains the double bond has eight carbon atoms, and the double bond is between carbon atoms 3 and 4. The name thus must include *-3-octene*. A methyl group is on carbon 7, and so *7-methyl-3-octene* must be in the name. The hydrogen atoms on the carbon atoms of the double bond are on the same side of the double bond. The prefix *cis-* is used to designate this; thus the complete name is *cis*-7-methyl-3-octene.
2. The double bond is between carbons 2 and 3 of a five-carbon chain. *Trans-* means similar groups are on opposite sides of the double bond. The structure of this compound is therefore

$$\begin{array}{ccc} H_3C & & H \\ & C{=}C & \\ H & & CH_2CH_3 \end{array}$$

Self-Test

1. Name the alkene

$$\begin{array}{ccc} H_3CH_2C & & CH_2CH_2CH_3 \\ & C{=}C & \\ H & & H \end{array}$$

2. What is the structure of *trans*-5-methyl-2-hexene?

ANSWERS

1. cis-3-heptene
2.

$$\begin{array}{ccc} H & & CH_2\underset{\displaystyle CH_3}{\underset{|}{C}}HCH_3 \\ & C{=}C & \\ H_3C & & H \end{array}$$

9.3 Alkynes

The **alkynes** are unsaturated hydrocarbons containing the carbon–carbon triple bond:

$$-C{\equiv}C-$$

The carbon–carbon triple bond is the functional group of the alkynes. The smallest alkyne is ethyne, C_2H_2:

$$H{-}C{\equiv}C{-}H$$

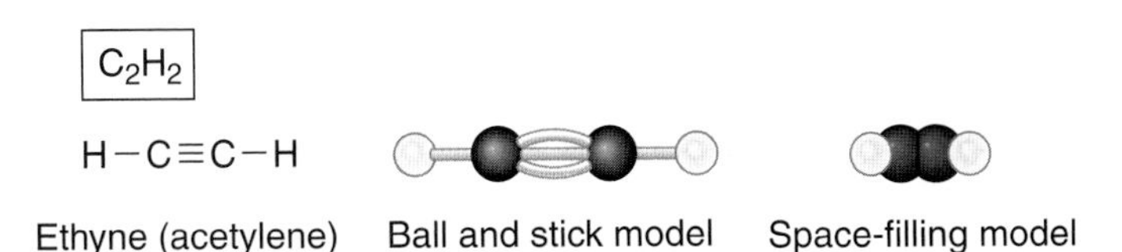

The general formula for an alkyne is C_nH_{2n-2}, where n is any positive integer.

Although alkynes can contain more than one triple bond, this is uncommon. Cycloalkynes are also uncommon. Large rings can have triple bonds in them, but small ones cannot because the 180° bond angles at a triple bond prevent a small ring from forming.

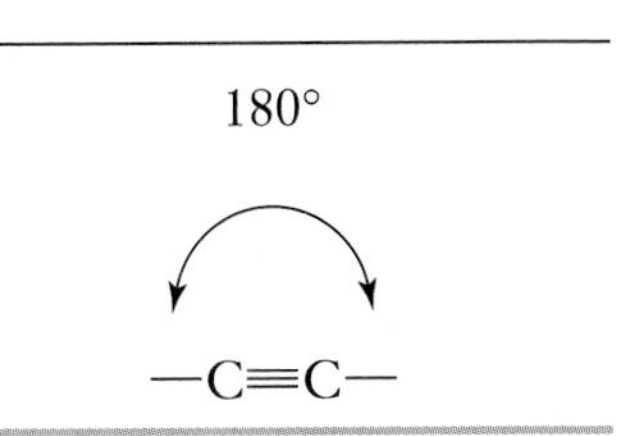

Alkynes are the least common of the hydrocarbons. Some quantities are used in industry, with the largest single use being that of acetylene (ethyne) in oxyacetylene torches.

Alkynes are usually represented by condensed structural formulas. Ethyne, for instance, is represented as HC≡CH. The next alkyne is propyne (did you anticipate the name?): $HC{\equiv}CCH_3$. Alkyne names are regular and systematic, and they are closely related to the names of alkanes and alkenes.

Naming Alkynes

The suffix *-yne* indicates a compound that contains a carbon–carbon triple bond. The rules for naming alkynes, shown in Table 9.2, are essentially the same as those for naming alkenes. Note that the designations *cis-* and *trans-* are not needed for alkynes because the presence of a carbon–carbon triple bond does not yield geometric isomers.

A few alkynes are identified by common names. Ethyne, the smallest alkyne, is often called acetylene, as noted previously.

Table 9.2 Rules for Naming Alkynes

Rule One	Find the longest continuous chain of carbon atoms that includes the triple bond. The length of this chain determines the parent part of an alkyne name. Combine the parent with the suffix *-yne*.
Rule Two	Number the chain beginning with the end closer to the triple bond. Use a prefix corresponding to the smaller of the two carbon atom numbers in the triple bond to designate the location of the bond. Place this prefix just before the parent part of the name.
Rule Three	The rules for substituents are the same as those for alkanes and alkenes.
Rule Four	The rules for numerical prefixes, commas, and hyphens are the same as for alkanes and alkenes.

Example 9.3 Names and Structures of Alkynes

1. Name the alkyne $HC{\equiv}C{-}\underset{\displaystyle CH_3}{\underset{|}{C}}HCH_3$.
2. What is the structure of 4-methyl-2-pentyne?

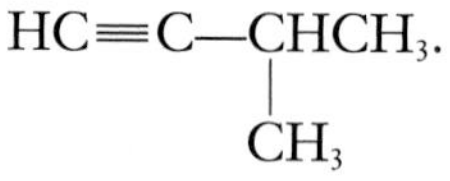

SOLUTIONS

1. Because the longest continuous chain that contains the triple bond consists of four carbon atoms, the name includes *-butyne.* The triple bond is between carbon atoms 1 and 2, and so the name so far is *-1-butyne.* The substituent is a methyl group on carbon 3, and so the complete name is 3-methyl-1-butyne.
2. The *2-pentyne* means the longest continuous chain containing the triple bond consists of five carbon atoms, and the triple bond is between carbons 2 and 3. There is a methyl group on carbon 4, and so the structure is

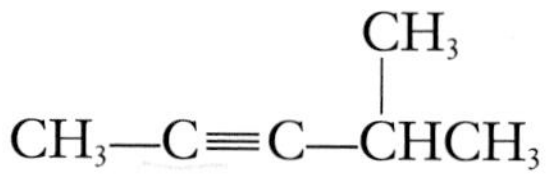

$$CH_3{-}C{\equiv}C{-}\overset{\displaystyle CH_3}{\overset{|}{C}}HCH_3$$

Self-Test

1. Draw the structure of 5-methyl-3-heptyne.
2. Why must 5-methyl-4-octyne be an incorrect name?

ANSWERS

1.

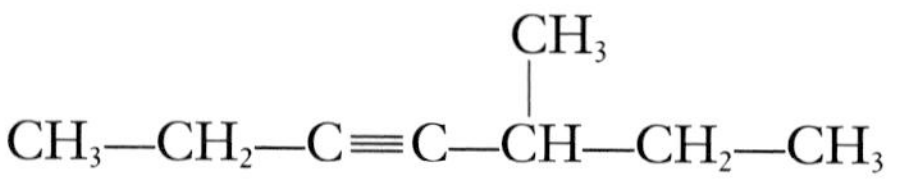

$$CH_3{-}CH_2{-}C{\equiv}C{-}\overset{\displaystyle CH_3}{\overset{|}{C}}H{-}CH_2{-}CH_3$$

2. Because carbon 5 is next to carbon 4, it must be part of the triple bond, and it is also part of the longest chain. This bonding uses all four bonds to carbon 5. There cannot be a methyl group attached to carbon 5, therefore, because that would require this carbon atom to form five bonds.

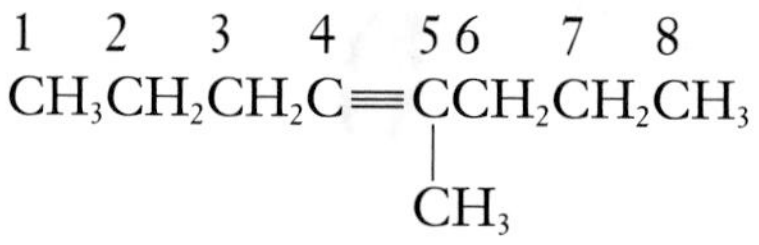

$$\overset{1}{C}H_3\overset{2}{C}H_2\overset{3}{C}H_2\overset{4}{C}{\equiv}\underset{\displaystyle CH_3}{\underset{|}{\overset{5}{C}}}\overset{6}{C}H_2\overset{7}{C}H_2\overset{8}{C}H_3$$

This is the structure derived from the name 5-methyl-4-octyne; this molecule cannot exist.

9.4 Properties and Reactions of Alkenes and Alkynes

Physical Properties

Like alkanes, alkenes and alkynes are nonpolar compounds. This means that the interactions between their molecules are London forces, which are weak. As a con-

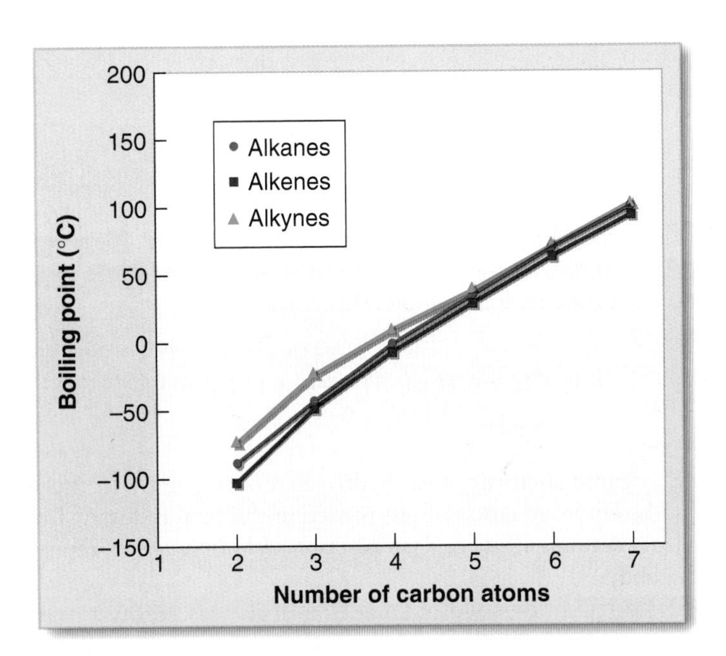

Figure 9.1 The boiling points of alkanes, alkenes, and alkynes. The nonpolar molecules of these compounds bond through London dispersion forces. For this reason, these hydrocarbons have similar boiling points.

sequence, the boiling points and melting points of alkanes, alkenes, and alkynes are similar (Figure 9.1). Like alkanes, alkenes and alkynes are soluble in nonpolar solvents and insoluble in water.

Reactions of Alkenes

Alkenes undergo three major kinds of reactions: addition reactions, oxidation reactions, and polymerization reactions. Before we discuss any specific alkene reaction, let's introduce a convention that is used in organic chemistry, which you will see in the rest of the organic chemistry chapters. Organic chemical equations often show only the functional group that is involved in the reaction, rather than the entire molecule. An alkene, for instance, is often shown this way,

$$\gt C{=}C \lt$$

with the reader expected to understand that there are hydrogen atoms or alkyl groups located at the ends of the four bonds.

Addition Reactions

In addition reactions, parts of the reacting molecule *add* to each of the carbon atoms of a double bond, and the double bond becomes a single bond. Alkenes can undergo many different addition reactions, but four important ones will give you an idea of what addition reactions are all about.

Hydrogen can be added to alkenes in a reaction called **hydrogenation**, in which each carbon atom of the double bond gets one of the hydrogen atoms of a molecule of hydrogen:

$$\underset{\text{Alkene}}{>C{=}C<} + H_2 \xrightarrow{Pt} \underset{\text{Alkane}}{-\overset{H}{\overset{|}{\underset{|}{C}}}-\overset{H}{\overset{|}{\underset{|}{C}}}-}$$

Finely divided metals like platinum (Pt) catalyze the reaction. Note that an unsaturated hydrocarbon, the alkene, is changed to a saturated one, the alkane. For the hydrogenation of propene, for example, this reaction is

$$\underset{\text{Propene}}{CH_3CH{=}CH_2} + H_2 \xrightarrow{Pt} \underset{\text{Propane}}{CH_3CH_2CH_3}$$

Synthetic organic chemistry uses hydrogenation regularly in the synthesis of new compounds. Hydrogenation of oils is used in the food industry. Living organisms use hydrogenation and its reverse reaction, *dehydrogenation*, repeatedly in reactions of the body.

Halogenation is another addition reaction of alkenes. In this reaction, the two atoms of a halogen molecule (bromine is used in this example) add to the carbon atoms of the double bond:

$$\underset{\text{Alkene}}{>C{=}C<} + Br_2 \rightarrow \underset{\text{Alkyl halide}}{-\overset{Br}{\overset{|}{\underset{|}{C}}}-\overset{Br}{\overset{|}{\underset{|}{C}}}-}$$

The chlorination of 1-butene is represented this way:

$$\underset{\text{1-butene}}{CH_3CH_2CH{=}CH_2} + Cl_2 \rightarrow \underset{\text{1,2-dichlorobutane}}{CH_3CH_2CHClCH_2Cl}$$

The products of alkene halogenation are organohalogens (alkyl halides). The uses of some of these compounds were discussed in Section 8.6. In addition, this reaction can be used to measure the unsaturation of fats and oils.

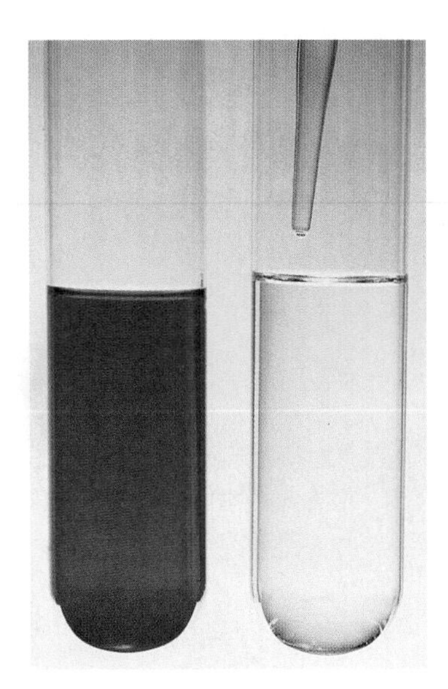

The bromine in this solution is red. When a few drops of an alkene are added, the color disappears as the bromine reacts with the alkene. *(Charles D. Winters)*

Example 9.4 Predicting the Products of Hydrogenation and Halogenation

What are the products of these reactions:

1. $CH_2{=}CH_2 + H_2 \xrightarrow{Pt}$
2. $CH_3CH{=}CHCH_3 + Cl_2 \rightarrow$

SOLUTIONS

1. In hydrogenation, a hydrogen atom adds to each carbon of the double bond of ethene to yield an alkane. The alkane produced is ethane: CH_3CH_3.
2. In halogenation, each carbon atom of the double bond receives a halogen atom.

The product is

$$\begin{array}{c} \quad\ \ Cl\ \ Cl \\ \quad\ \ |\ \ \ \ | \\ CH_3CHCHCH_3 \end{array} \quad \text{(2,3-dichlorobutane)}$$

Self-Test

1. What is the structure of the product of this reaction?

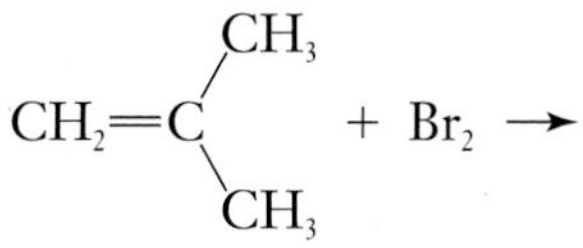

$$CH_2{=}C\begin{array}{l} \diagup CH_3 \\ \diagdown CH_3 \end{array} + Br_2 \rightarrow$$

2. If hydrogen were used instead of bromine, what would be the product?

ANSWERS

1. 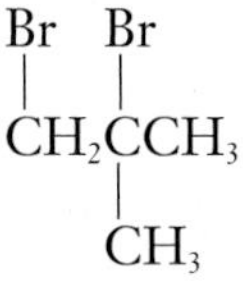 $$\begin{array}{l} Br\ \ \ Br \\ |\ \ \ \ \ \ | \\ CH_2CCH_3 \\ \qquad | \\ \quad\ CH_3 \end{array}$$ (1,2-dibromo-2-methylpropane)

2. 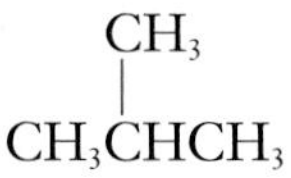 $$\begin{array}{l} \qquad CH_3 \\ \qquad\ | \\ CH_3CHCH_3 \end{array}$$ (2-methylpropane)

A third important addition reaction of alkenes is **hydration**. In this addition reaction, water adds across the double bond in the presence of hydrogen ion, which acts as a catalyst:

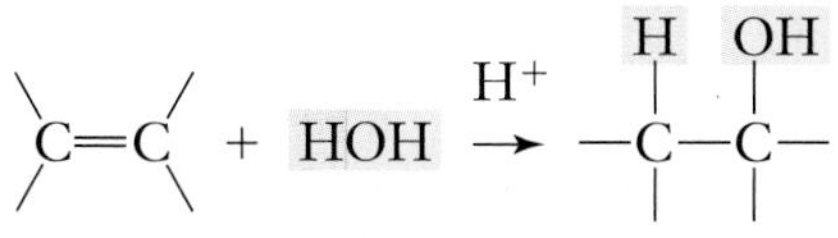

$$\begin{array}{c} \diagdown\ \ \ \diagup \\ C{=}C \\ \diagup\ \ \ \diagdown \end{array} + HOH \xrightarrow{H^+} \begin{array}{c} H\ \ \ OH \\ |\ \ \ \ \ | \\ -C{-}C- \\ |\ \ \ \ \ | \end{array}$$

Alkene Alcohol

This equation represents the hydration of ethene:

$$CH_2{=}CH_2 + HOH \xrightarrow{H^+} HCH_2CH_2OH$$

Ethene Ethanol

In a hydration reaction, one carbon atom of the double bond gets a hydrogen atom from the water molecule, and the other gets the $-OH$ (hydroxyl group) from the water molecule. The product molecule is an alcohol (Chapter 10). Hydration, like hydrogenation, is an important reaction in the body.

The hydration of alkenes differs from hydrogenation and halogenation in one very important way. In hydrogenation and halogenation, an atom of the same element adds to both carbon atoms of the double bond. The adding molecule is

Connections
Hydrogenation of Fats and Oils

Fat and oils—high-energy substances found in many foods—are solids and liquids at room temperature, respectively. The molecules of fats and oils contain fatty acids (Chapter 13) that may contain *cis*-carbon–carbon double bonds, and the number of these bonds changes the melting point of these compounds. Molecules containing few or no *cis*- double bonds have a higher melting point and are thus fats, whereas those containing larger numbers of these bonds have lower melting points and are thus oils (see the table).

Cis- double bonds in oils present two potential problems to the food industry: (1) they can be cleaved by oxidization—a reaction you will learn shortly—which yields smaller, smelly compounds that are responsible for foods becoming rancid, and (2) they make the substance a liquid that adversely affects the consistency and texture of some foods. Both of these problems can be reduced or eliminated by partial or total hydrogenation of oils. Hydrogenation of an oil reduces the number of double bonds, which reduces the chance of the food going rancid, and if the hydrogenation is extensive or complete, the oils become semisolid or solid fats.

Unfortunately, hydrogenation of oils has two potentially adverse effects on health. Hydrogenation makes the oils more saturated (they contain fewer double bonds) and this increased saturation increases the risk of higher blood cholesterol (see Chapter 13 for more details). In addition, during hydrogenation of *cis*- double bonds, some of them are simply isomerized to *trans*- double bonds, which may also increase blood cholesterol.

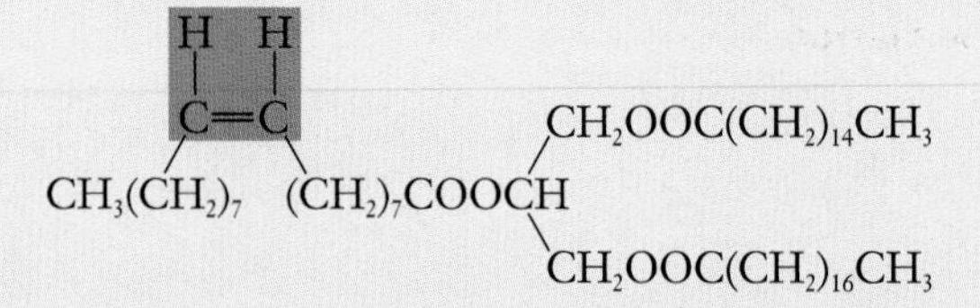

A fat molecule containing a single *cis*-carbon–carbon double bond

Percentage of Saturation in Fats and Oils

FAT OR OIL	PERCENTAGE SATURATED FATTY ACIDS
Safflower oil	9
Olive oil	14
Peanut oil	18
Lard	41
Butter	68

The hydrogenation of an oil to form a fat is called hardening because the liquid oil is converted into a solid or semisolid fat. *(Charles D. Winters)*

symmetrical—that is, the two adding groups are the same. In hydration, each carbon atom of the double bond gets a different group, because the molecule is not symmetrical. Because the adding groups are different, there are two possible products for a hydration reaction. Consider the acid-catalyzed hydration of propene as an example. If the hydroxyl group goes to carbon atom 1 of propene, the reaction is

$$\underset{\text{Propene}}{CH_3CH{=}CH_2} + HOH \xrightarrow{H^+} \underset{\text{1-propanol}}{CH_3\overset{\overset{\displaystyle H}{|}}{C}HCH_2{-}OH}$$

If the hydroxyl group goes to carbon 2, the reaction is

$$CH_3CH{=}CH_2 + HOH \xrightarrow{H^+} CH_3\underset{\displaystyle OH}{\underset{|}{C}}HCH_2{-}H$$

Propene 2-propanol

(You will learn how to name alcohols such as 1-propanol and 2-propanol in Chapter 10.) Clearly the products are different, but which one is produced in a hydration reaction?

In the nineteenth century, the Russian chemist Vladimir Markovnikov studied hydration of alkenes as well as other similar reactions. He made observations that are today summarized as **Markovnikov's rule:** When an asymmetrical, hydrogen-containing reagent adds to a carbon–carbon double bond, the carbon atom of the double bond that has more hydrogens gets the hydrogen atom of the adding molecule ("Them that has, gits"). Look again at the hydration of propene. According to Markovnikov's rule, 2-propanol should be the product, because carbon atom 1 gets the hydrogen atom from water, because this carbon atom has more hydrogen atoms (two) than carbon atom 2, which has one hydrogen atom. Carbon 1 gets the hydrogen atom, carbon atom 2 gets the hydroxyl group from the water molecule.

A fourth addition reaction of alkenes is **hydrohalogenation:**

$$\gt C{=}C\lt + HX \rightarrow H{-}\overset{|}{\underset{|}{C}}{-}\overset{|}{\underset{|}{C}}{-}X$$

Alkene Alkyl halide

In this reaction, a molecule of hydrogen halide, such as HCl or HBr, is added to the carbon atoms of the double bond. Because hydrogen halide molecules are asymmetrical, Markovnikov's rule is used to predict the product. One carbon atom of the double bond gets the hydrogen atom and the other gets the halogen atom. The hydrochlorination of propene is

$$CH_3CH{=}CH_2 + HCl \rightarrow CH_3CHClCH_2{-}H$$

Propene 2-chloropropane

Example 9.5 Predicting the Products of Hydration and Hydrohalogenation

1. What is the product when 2-methylpropene reacts with water in this acid-catalyzed reaction?

$$CH_3{-}\overset{\displaystyle CH_3}{\overset{|}{C}}{=}CH_2 + H_2O \xrightarrow{H^+} ?$$

2. What is the product when HCl reacts with 2-methyl-2-butene:

$$CH_3{-}\overset{\displaystyle CH_3}{\overset{|}{C}}{=}CH{-}CH_3 + HCl \rightarrow ?$$

SOLUTIONS

1. Water is an asymmetrical reagent, and its addition to a double bond follows Markovnikov's rule. Carbon 1 of 2-methylpropene has two hydrogen atoms on it; carbon 2 has no hydrogen atoms. Therefore the hydrogen atom from water goes to carbon 1, with the product being 2-methyl-2-propanol (you will learn to name alcohols in Chapter 10):

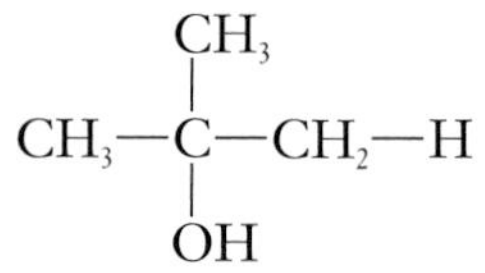

$$CH_3-\overset{\overset{\displaystyle CH_3}{|}}{\underset{\underset{\displaystyle OH}{|}}{C}}-CH_2-H$$

2. Hydrogen chloride is also an asymmetrical molecule, and its addition to a double bond also obeys Markovnikov's rule. Carbon 2 of the reactant has no hydrogen atoms attached to it; carbon 3 has one hydrogen. The hydrogen of HCl therefore goes to carbon number 3, producing 2-chloro-2-methylbutane:

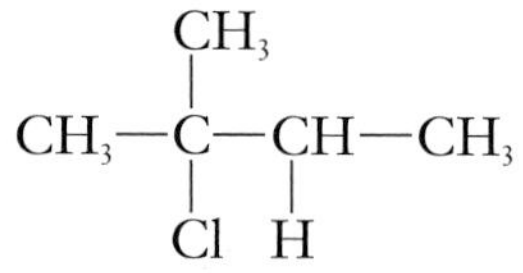

$$CH_3-\overset{\overset{\displaystyle CH_3}{|}}{\underset{\underset{\displaystyle Cl}{|}}{C}}-\underset{\underset{\displaystyle H}{|}}{CH}-CH_3$$

Self-Test

Predict the product of the reaction

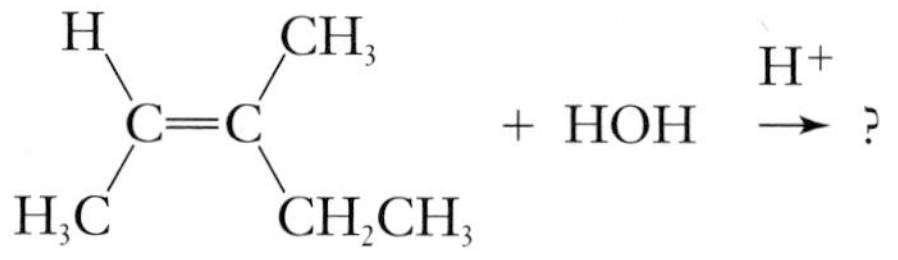

$$\begin{matrix} H & & CH_3 \\ & C{=}C & \\ H_3C & & CH_2CH_3 \end{matrix} + HOH \xrightarrow{H^+} ?$$

ANSWER

$$CH_3\overset{\overset{\displaystyle H}{|}}{\underset{\underset{\displaystyle H}{|}}{C}}-\overset{\overset{\displaystyle CH_3}{|}}{\underset{\underset{\displaystyle OH}{|}}{C}}CH_2CH_3$$

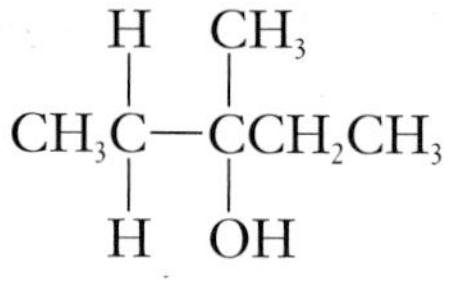

Oxidation Reactions

Combustion is an oxidation reaction, and alkenes undergo combustion just as most other organic compounds do. (Pure alkenes are too expensive to use as fuel, however.) Two important oxidative reactions of alkenes are hydroxylation and oxidative cleavage. **Hydroxylation** is an oxidation reaction that yields a product having a hydroxyl group on each of the carbons of the double bond:

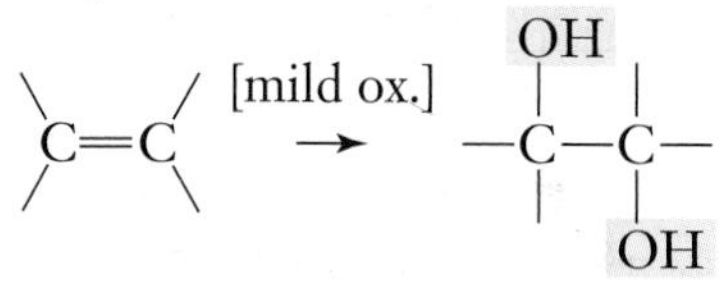

$$>C{=}C< \xrightarrow{\text{[mild ox.]}} -\overset{\overset{\displaystyle OH}{|}}{\underset{|}{C}}-\overset{|}{\underset{\underset{\displaystyle OH}{|}}{C}}-$$

There are several reactants and conditions that can be used for the mild oxidation necessary for this reaction. The symbol [mild ox.] will be used here rather than any specific compound. Ethylene glycol, the major ingredient of antifreeze, is formed by hydroxylation of ethylene (ethene):

$$\underset{\text{Ethene}}{CH_2{=}CH_2} \xrightarrow{\text{[mild ox.]}} \underset{\text{1,2-ethanediol (ethylene glycol)}}{HOCH_2CH_2OH}$$

Oxidative cleavage of a carbon–carbon double bond involves stronger oxidation of the double bond. Under strong oxidizing conditions that can be produced in several ways [strong ox.], the double bond is broken. The molecule is cleaved at the double bond to yield two new molecules. The product molecules of these cleavage reactions are small organic compounds called aldehydes, ketones, and carboxylic acids (more about these compounds in Chapter 11). In the example shown, 2-methylpropene yields two molecules, one of which is an aldehyde and the other a ketone:

$$\underset{\text{Alkene}}{(H)_2C{=}C(CH_3)_2} \xrightarrow{\text{[strong ox.]}} \underset{\text{Aldehyde}}{(H)_2C{=}O} + \underset{\text{Ketone}}{O{=}C(CH_3)_2}$$

The small organic products from oxidative cleavage have an unpleasant odor. Many foods—vegetables oils are one example—contain compounds that have carbon–carbon double bonds. These foods can go *rancid.* This means that some of their carbon–carbon double bonds have undergone oxidative cleavage. The bad smell in rancid foods is due to the aldehydes and carboxylic acids that are formed.

Polymerization

In **polymerization** reactions, small molecules called *monomers* are joined to form very large molecules called *polymers.* Alkenes can be joined together in polymerization reactions to yield a variety of polymers; ethylene (ethene) is the monomer used to make polyethylene, and propylene (propene) is used to make polypropylene:

$$\text{Many } \underset{\text{Ethylene (monomer)}}{CH_2{=}CH_2} \rightarrow$$

$$\underset{\text{Polyethylene (polymer)}}{\ldots{-}CH_2{-}CH_2{-}CH_2{-}CH_2{-}CH_2{-}CH_2{-}CH_2{-}CH_2{-}CH_2{-}CH_2{-}\ldots}$$

The ellipses at each end of the polymer shows that the molecule continues indefinitely. Although the product is like an alkane, it is much larger than the molecules found in petroleum or synthesized in the laboratory by other reactions. Polymers like polyethylene, polypropylene, and polystyrene possess physical properties that make them ideal for use in materials like fibers, films, and containers.

(a)

(b)

These consumer products are made from polymers: **(a)** packaging materials made from polyethylene (polyethene) and **(b)** from polystyrene. *(Charles D. Winters)*

Reactions of Alkynes

The combustion of acetylene (ethyne) provides the hot flame of this oxyacetylene torch. *(Charles D. Winters)*

Alkynes burn, but they are generally too expensive to use as fuels. An exception is the use of acetylene (ethyne) in an oxyacetylene torch:

$$2C_2H_2(g) + 5O_2(g) \longrightarrow 4CO_2(g) + 2H_2O(g) + \text{heat}$$

This combustion yields energy that produces the very high temperatures characteristic of these torches.

Alkynes undergo some of the same addition reactions as alkenes. The main difference is that two moles of the reactant can add to the triple bond, whereas only one mole adds to a double bond. Our first example is the hydrobromination of 1-butyne:

$$\underset{\text{1-butyne}}{HC\equiv C-CH_2-CH_3} + 2HBr \longrightarrow \underset{\text{2,2-dibromobutane}}{H-\overset{\overset{H}{|}}{\underset{\underset{H}{|}}{C}}-\overset{\overset{Br}{|}}{\underset{\underset{Br}{|}}{C}}-CH_2-CH_3}$$

(Naming will be covered in Chapter 10.) Do you see where the two bromine atoms and the hydrogen atoms are added? Does hydrohalogenation of an alkyne obey Markovnikov's rule?

A second example of addition to alkynes is the chlorination of 2-butyne:

$$\underset{\text{2-butyne}}{CH_3C\equiv CCH_3} + 2Cl_2 \longrightarrow \underset{\text{2,2,3,3-tetrachlorobutane}}{CH_3\overset{\overset{Cl}{|}}{\underset{\underset{Cl}{|}}{C}}-\overset{\overset{Cl}{|}}{\underset{\underset{Cl}{|}}{C}}CH_3}$$

Example 9.6 Addition Reactions of Alkynes

Predict the product of the reaction $HC\equiv CCH_3 + 2HCl \longrightarrow ?$

SOLUTION

The addition of asymmetric reagents to alkynes obeys Markovnikov's rule. Carbon 1 of propyne will get both hydrogen atoms, and carbon 2 will get both chlorine atoms. The product is

$$CH_3\overset{\overset{Cl}{|}}{\underset{\underset{Cl}{|}}{C}}CH_3 \quad \text{(2,2-dichloropropane)}$$

Self-Test

What is the product of the reaction $HC\equiv CCH_3 + 2Br_2 \longrightarrow ?$

ANSWER

```
    Br Br H
    |  |  |
H—C—C—C—H
    |  |  |
    Br Br H
```

or $CHBr_2CBr_2CH_3$ (1,1,2,2-tetrabromopropane)

9.5 Aromatic Compounds

The **aromatic hydrocarbons,** the third group of unsaturated hydrocarbons, contain one or more aromatic rings rather than double or triple bonds. The *aromatic ring* is represented this way (the origin of this representation is explained next):

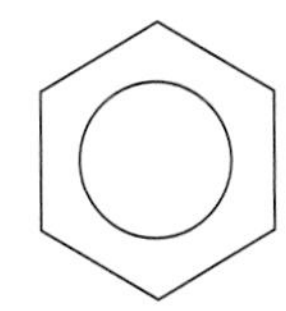

The aromatic ring is the functional group of the aromatic hydrocarbons and all other aromatic compounds. The term *aromatic* is historical rather than an accurate description of these compounds. Many fragrant compounds such as vanilla and cinnamon are indeed aromatic compounds, but others are not. Pine and banana smell good, but their scents are due to nonaromatic compounds. The modern chemical meaning of *aromatic* refers to the presence of an aromatic ring in the compound rather than fragrance.

Although aromatic hydrocarbons are unsaturated, their chemical reactivities differ greatly from those of alkenes and alkynes. Under conditions that result in addition to or cleavage of double bonds in alkenes, for instance, aromatic compounds do not react. Why do these compounds differ in chemical behavior from the other unsaturated hydrocarbons? To answer this question, let's look at benzene, the simplest aromatic hydrocarbon. Benzene, C_6H_6, is a six-carbon cyclic hydrocarbon that has a hydrogen attached to each carbon. Based on this information alone, we can represent benzene with this incomplete structure:

```
        H
        |
  H    / \    H
    \ /   \ /
     |     |
     |     |
    / \   / \
  H    \ /    H
        |
        H
```

This structure has three single bonds to each carbon atom, but each carbon should have four bonds. To get bonds to each carbon atom, we need to add three double bonds (three pairs of electrons), which alternate with single bonds around the ring. When we do this, however, we can draw two different structures:

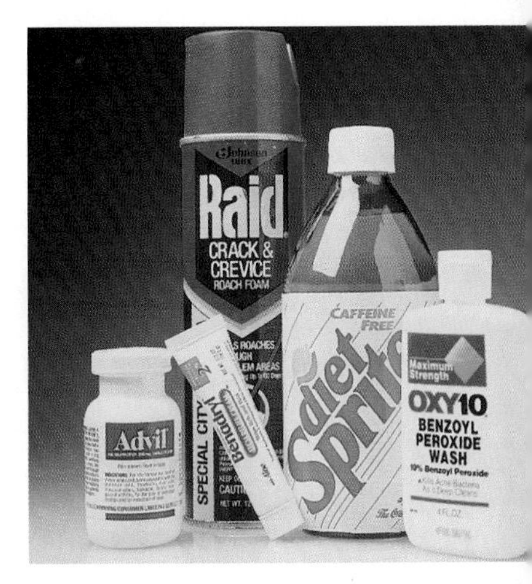

Many common products use compounds containing aromatic rings. Examples include sodium benzoate in soft drinks, ibuprofen in Advil®, and benzoyl peroxide in Oxy-10®. *(Charles D. Winters)*

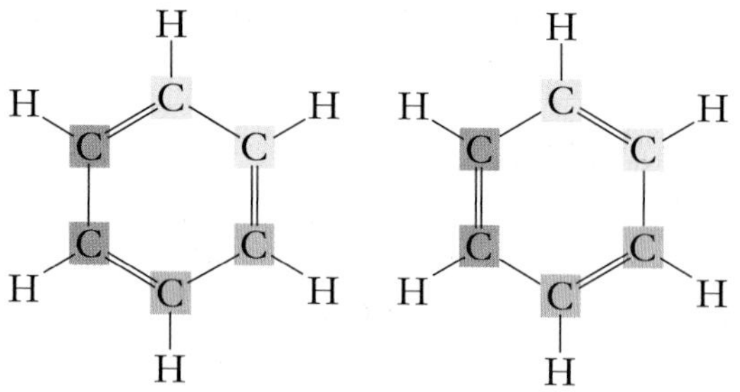

So which is the structure of benzene, the one on the left or the one on the right? Actually neither one is correct, because each implies that carbon–carbon double bonds are present. Because benzene does not undergo addition reactions the way alkenes do, we must conclude that double bonds are absent.

We get around this problem by looking again at the two structures of benzene. Look at the top carbon atom and the carbon to its right (the yellow carbons) in each structure. In one structure, there is a single bond between these atoms, and in the other one there is a double bond. This is also true for each other pair (marked in gray and green) of carbon atoms around the ring. Let's mentally merge the two structures into one. The average of a single and double bond is a bond and a half, so let's put a bond and a half between each pair of carbon atoms. We represent a *half-bond* by a dashed line:

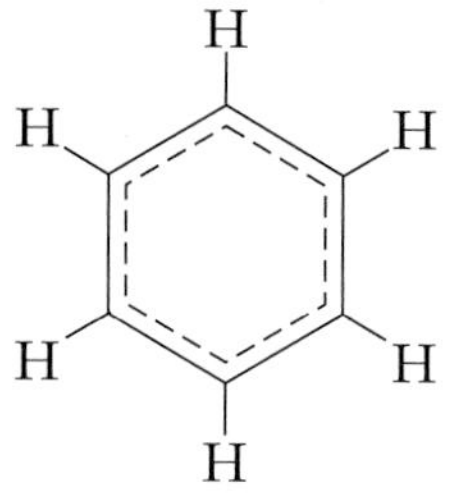

This structure shows that a partial double bond exists between each pair of carbon atoms. This means there are no carbon–carbon double bonds. Instead, the three pairs of electrons that appeared to be double bonds in our original drawings are actually shared by all six ring atoms. A chemist would say that these electrons are *delocalized* throughout the ring rather than being located between a specific pair of carbon atoms. If the electrons are distributed in this way (and they in fact are), then there are no double bonds and as a consequence no reactions typical of double bonds. To emphasize the delocalization of electrons, the structure of benzene is often represented as

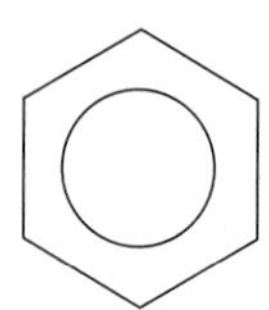

The circular ring means the six electrons are spread out—delocalized—around the ring. Note that the hydrogen atom on each carbon atom of benzene is understood to be there, and is not normally shown in these benzene structures. In the beginning of this section, we used this same symbol to represent the aromatic ring. Benzene is the simplest aromatic hydrocarbon.

Naming Aromatic Compounds

Aromatic compounds contain the aromatic ring of benzene plus one or more additional groups (substituents) attached to the ring. Many of these compounds are named by combining a prefix that identifies the substituent with the parent benzene. Ethylbenzene and chlorobenzene are the IUPAC names for the two compounds shown here:

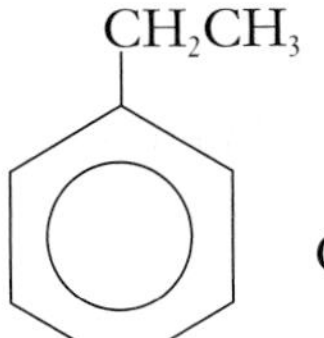

CH_2CH_3 Cl—

Ethylbenzene 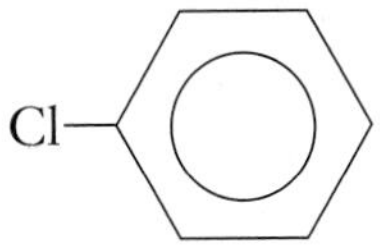   Chlorobenzene

When a compound contains a single substituent (as here), the substituent can be shown connected to any carbon atom.

Some aromatic compounds do not follow this simple rule of combining a prefix with the word *benzene*. Instead, their historical (common) names are retained. *Toluene* is used instead of *methylbenzene* and *aniline* instead of *aminobenzene:*

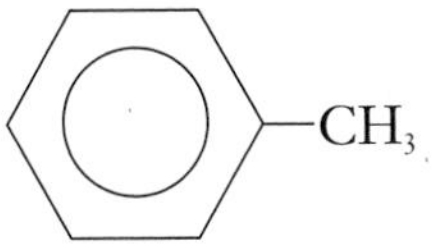

—CH_3 H_2N—

Toluene 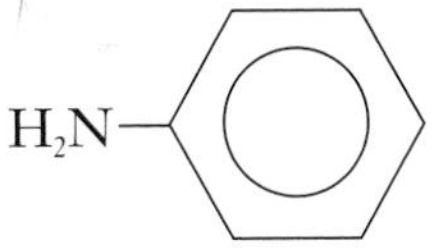   Aniline

The names of some larger aromatic compounds identify the aromatic ring as a substituent using the word *phenyl* to identify the aromatic ring. Just as an alkyl group is an alkane minus a hydrogen atom, the phenyl group is an aromatic ring minus a hydrogen atom. The phenyl group is represented as

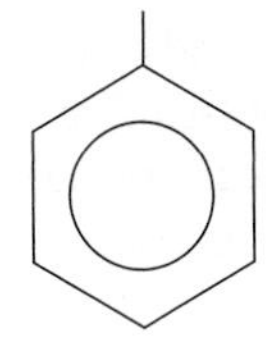

2-Phenyldecane and 2-methyl-3-phenyloctane are examples of organic names identifying the aromatic ring (the phenyl group) as a substituent.

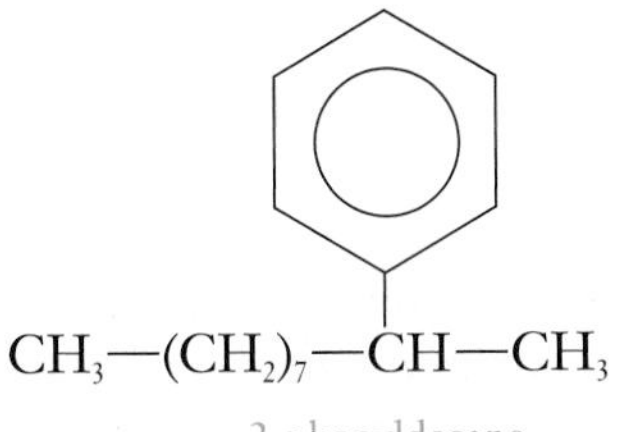

$CH_3—(CH_2)_7—CH—CH_3$ $CH_3—(CH_2)_4—CH—CH(CH_3)—CH_3$

2-phenyldecane 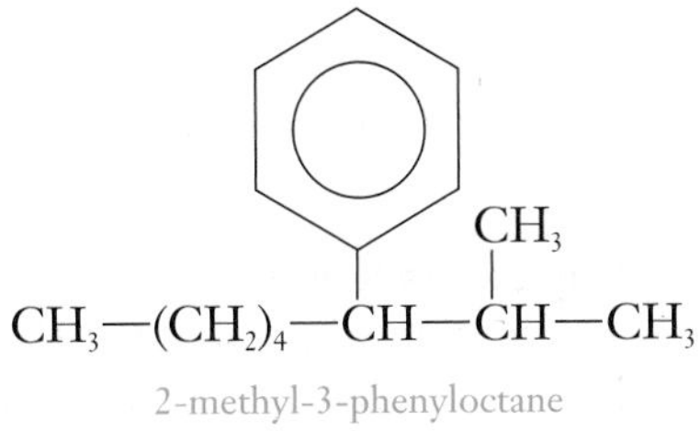   2-methyl-3-phenyloctane

Example 9.7 Naming Aromatic Compounds that Have One Substituent

What is the name of the compound

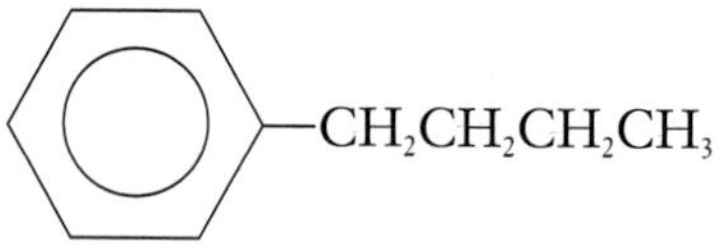

C_6H_5—$CH_2CH_2CH_2CH_3$

SOLUTION

The compound consists of an aromatic ring (benzene) attached to a butyl group. The prefix is made from the substituent name, which is combined with the parent benzene: The answer is butylbenzene.

Self-Test

Name the compound

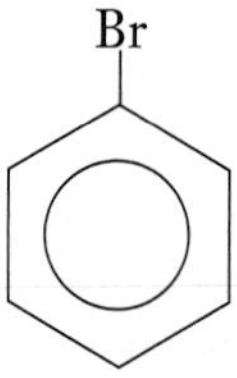

Br

ANSWER

bromobenzene

If two groups are present on an aromatic ring, IUPAC rules use numbers to indicate ring position, just as with the cycloalkanes. The groups are numbered to give the smallest possible numbers, then the groups are arranged in alphabetical order in the prefix. If the compound contains a group that is governed by a historical name, the historical name is retained and that group has precedence, which means it is assumed to be attached to carbon 1. The rules you have learned for commas, hyphens, and numerical prefixes remain unchanged.

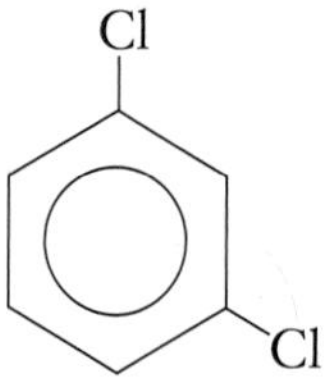

Cl Cl Br Cl CH_3 Br

1,3-dichlorobenzene (m-dichlorobenzene) 1-bromo-4-chlorobenzene (p-bromochlorobenzene) 2-bromotoluene (o-bromotoluene)

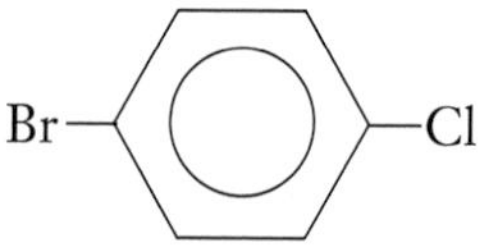

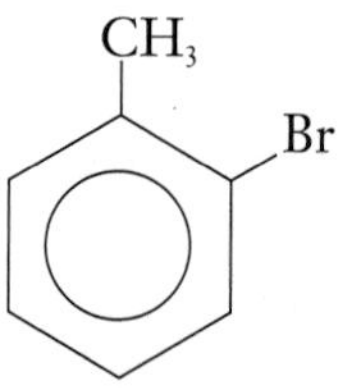

Note that because toluene is known by its historical name, the methyl group has precedence in 2-bromotoluene; 1-bromotoluene or 2-bromo-1-methylbenzene are incorrect names.

Connections

Antioxidants

Earlier the oxidation of carbon–carbon double bonds was discussed. These reactions are often undesirable, and the oxidation of these bonds in oils in foods is an example. One way to prevent the rancidity that these oxidations cause is to add some antioxidative food preservatives. These compounds prevent oxidation of double bonds by interfering with the oxidation reaction (they react with intermediates formed during the reaction).

Two common antioxidants are butylated hydroxytoluene (BHT) and butylated hydroxyanisole (BHA).

Oxidation of double bonds can also occur in cells, and to prevent this cells use antioxidants called tocopherols, which collectively are known as vitamin E.

Note that α-tocopherol, BHA, and BHT all contain the aromatic ring, and as you will learn in Chapter 10, they are substituted phenols.

OH, $(CH_3)_3C$, $C(CH_3)_3$, CH_3

BHT

OH, $C(CH_3)_3$, OCH_3

BHA
(One isomer shown)

Many foods, including crackers and cereals, contain BHT to inhibit spoilage. *(Charles D. Winters)*

OH, H_3C, CH_3, H_3C, O, H_3C, CH_3, CH_3, CH_3, CH_3

Vitamin E
(α-tocopherol)

As an alternative, the terms *ortho-*, *meta-*, and *para-* can be used to designate substituent position. *Ortho-* (o-) means the two groups are on adjacent carbon atoms; in the examples given previously, 2-bromotoluene would be called o-bromotoluene. *Meta-* (m-) indicates the groups are separated by a ring carbon atom that has no substituent group attached to it; 1,3-dichlorobenzene can be called m-dichlorobenzene. *Para-* (p-) means the groups are on carbon atoms opposite each other in the ring; the compound 1-bromo-4-chlorobenzene could be

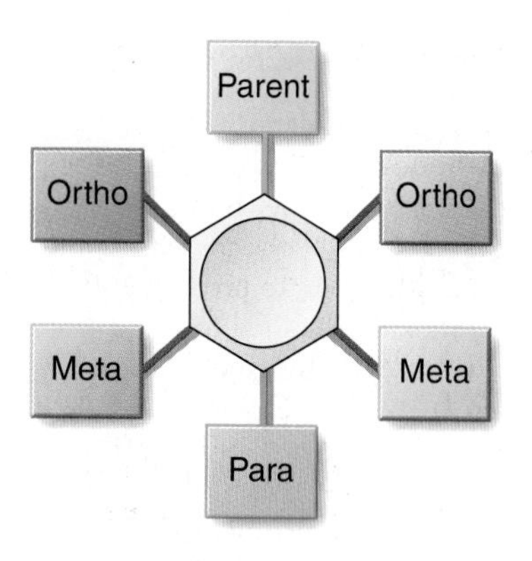

Figure 9.2 The prefixes *ortho-* (o-), *meta-* (m-), and *para-* (p-) are used to designate the location of a second substituent on an aromatic ring.

called p-bromochlorobenzene. Figure 9.2 shows the proper prefix to use for each position in the benzene ring.

If three or more groups are present on an aromatic ring, numbers are used to designate all positions. Choose the smallest possible numbers to designate the positions—for instance, 1,2,4-trichlorobenzene rather than 1,3,4-trichlorobenzene.

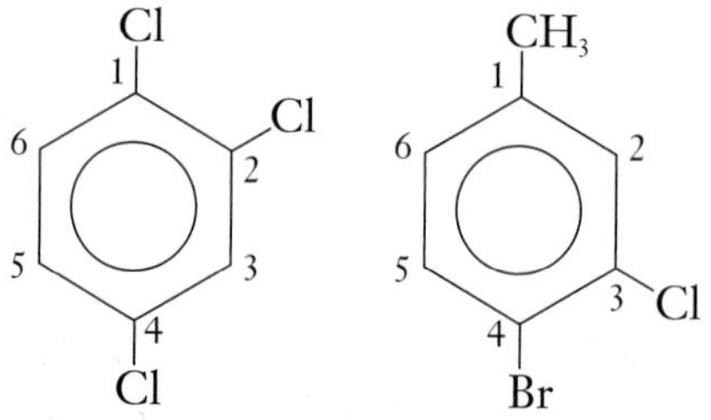

1,2,4-trichlorobenzene 4-bromo-3-chlorotoluene

Note that in the right-hand compound the ring carbons are numbered to give the smallest numbers to the substituents, and the prefixes are then arranged in alphabetical order in the name.

Example 9.8 Naming Aromatic Compounds Having Two or More Substituents

Use the two ways you have learned to provide two names for this compound:

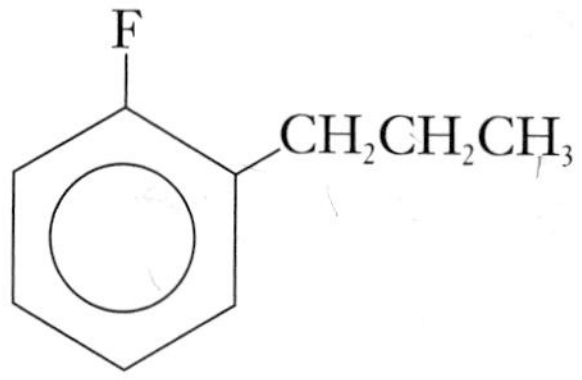

SOLUTION

There are two substituents on a benzene ring, a fluorine atom and a propyl group. Arrange the groups alphabetically, and give them the smallest possible numbers: 1-fluoro-2-propylbenzene. An alternative name would use *ortho-* instead of numbers: o-fluoropropylbenzene.

Self-Test

Draw the structure of m-bromotoluene (3-bromotoluene).

ANSWER

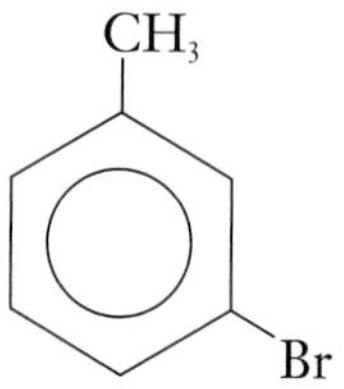

Some aromatic compounds have two or more aromatic rings fused together. These compounds are called *polycyclic* aromatic compounds. Naphthalene is the simplest polycyclic hydrocarbon, and it is sometimes used as a repellent in mothballs. Anthracene is another common polycyclic aromatic hydrocarbon.

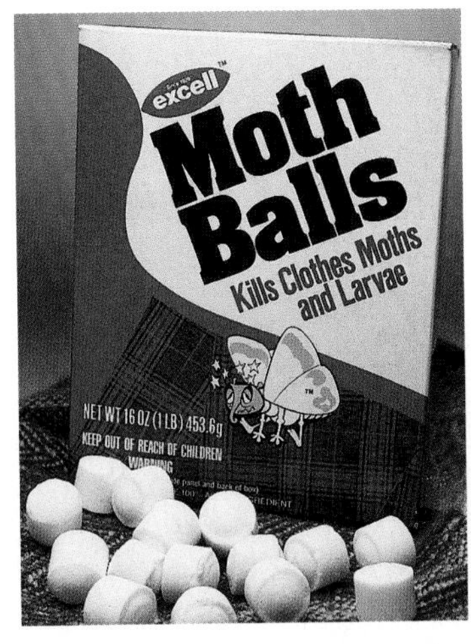

Naphthalene is the major ingredient in some brands of mothballs. *(Charles D. Winters)*

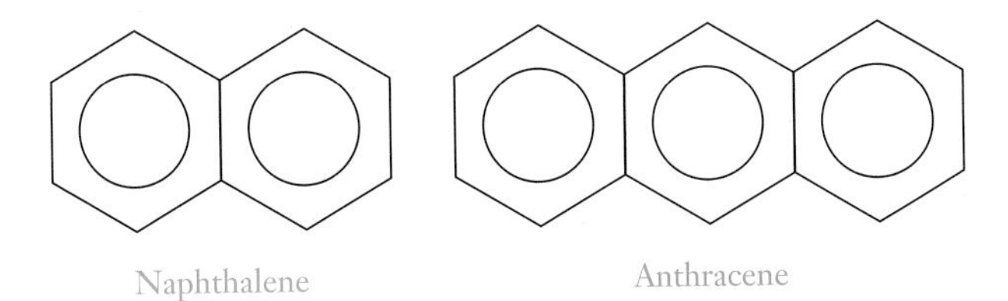

Benzo[a]pyrene is a polycyclic compound that is produced during incomplete combustion of organic compounds. It is a carcinogen (cancer-causing agent), as are benzene and a number of other aromatic compounds.

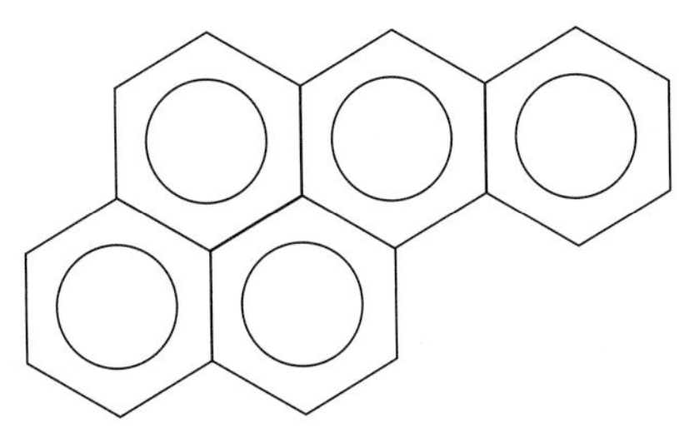

9.6 Properties and Reactions of Aromatic Hydrocarbons

Properties

Because aromatic hydrocarbons are nonpolar compounds, their molecules bond to each other through London forces. Even though these forces of attraction are weak, however, these compounds are liquids or solids at room temperature because the smallest of them contains at least six carbon atoms. Aromatic hydrocarbons and many other aromatic compounds are water-insoluble, but they are generally soluble in organic solvents. Some of the liquid aromatic hydrocarbons like toluene are good solvents themselves. Benzene was a commonly used nonpolar solvent, but it has been replaced by toluene because benzene is both toxic and a suspected carcinogen. Some smaller aromatic hydrocarbons are used in motor fuels to improve octane rating.

Automobile exhaust and cigarette smoke both contain measurable amounts of the carcinogen benzo[a]pyrene. *(©Martin Bough, 1992, Fundamental Photographs)*

Aromatic Substitution Reactions

Even though they are unsaturated, aromatic compounds do not undergo addition reactions because they do not have double bonds (Figure 9.3). Instead, the typical reaction of an aromatic ring is a *substitution* reaction, in which part of a reactant

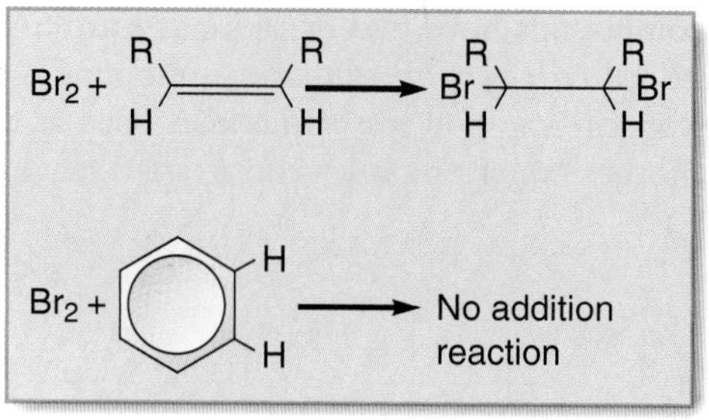

Br_2 + R(H)C=C(R)H → Br(R)(H)C–C(R)(H)Br

Br_2 + benzene → No addition reaction

Figure 9.3 Alkenes undergo addition reactions. Under the same conditions, aromatic hydrocarbons do not react to form addition products.

molecule substitutes for a hydrogen atom on the aromatic ring. (Remember, each carbon atom of the benzene ring has a hydrogen atom attached to it even though the ring representation does not show the hydrogens.) More than one substituent can be put onto a ring, and substituents can be put onto specific positions of the ring by using a variety of reactants.

In the substitution reaction known as *alkylation*, an alkyl group replaces a hydrogen atom on the ring. Because a variety of alkyl groups can be added, many different aromatic compounds can be made through this reaction.

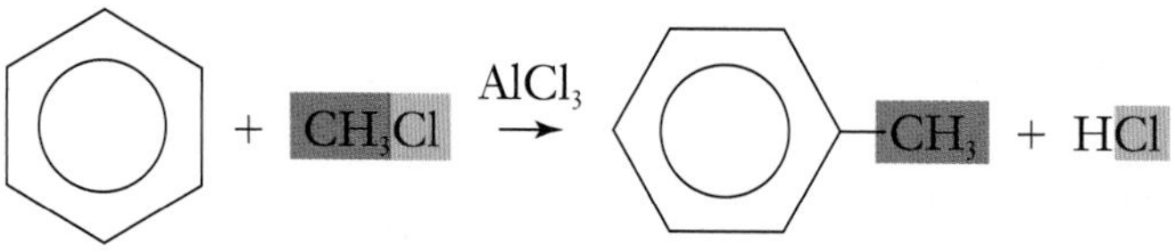

$$\text{Benzene} + CH_3Cl \xrightarrow{AlCl_3} \text{Toluene } (C_6H_5\text{–}CH_3) + HCl$$

In a second type of substitution reaction, *halogenation*, a halogenated aromatic ring is formed.

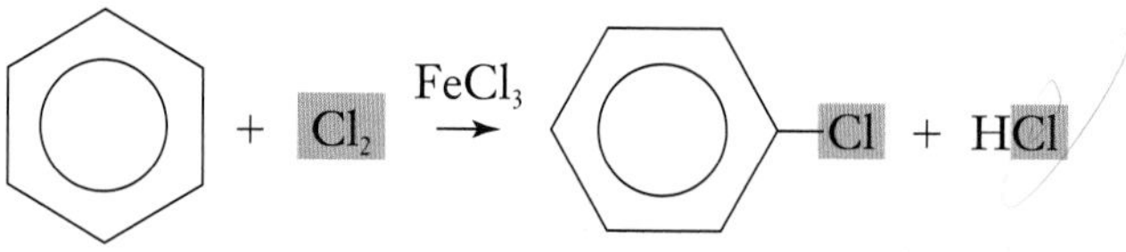

$$\text{Benzene} + Cl_2 \xrightarrow{FeCl_3} \text{Chlorobenzene } (C_6H_5\text{–}Cl) + HCl$$

A third substitution reaction, *nitration*, introduces a nitro group (—NO_2) onto the aromatic ring.

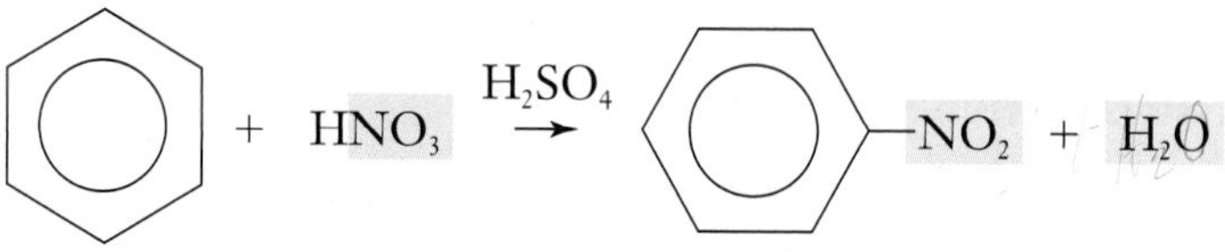

$$\text{Benzene} + HNO_3 \xrightarrow{H_2SO_4} \text{Nitrobenzene } (C_6H_5\text{–}NO_2) + H_2O$$

In *sulfonation,* a sulfonic acid group is introduced onto the aromatic ring. Sulfonic acids are important for the synthesis of some dyes and some pharmaceuticals, including sulfa drugs.

$$\text{Benzene} + SO_3 \xrightarrow{H_2SO_4} \text{Benzenesulfonic acid } (C_6H_5{-}SO_3H)$$

Benzene Benzenesulfonic acid

These reactions are only a few of many possible aromatic-ring substitution reactions. Through these reactions, vast numbers of aromatic compounds are synthesized for use in many aspects of modern life.

CHAPTER SUMMARY

Alkenes, alkynes, and **aromatic hydrocarbons** are unsaturated hydrocarbons containing one or more carbon–carbon double bonds, triple bonds, or aromatic rings, respectively. Alkenes are identified by the *-ene* or *-ylene* suffix in their names. Because there is no free rotation about double bonds, some alkenes can exist as *cis-* or *trans-*geometric isomers. Alkynes are identified by the *-yne* suffix and do not exist as geometric isomers. Alkenes and alkynes are nonpolar compounds and are thus insoluble in water but soluble in nonpolar organic solvents. The smallest alkenes and alkynes are gases, larger ones are liquids or solids. Alkenes undergo addition reactions such as **hydrogenation** and **hydration,** and can be oxidized or polymerized. Alkynes have chemical properties somewhat similar to alkenes. Aromatic hydrocarbons and all other aromatic compounds contain the aromatic ring, which consists of a ring of six carbon atoms linked by six single bonds and six additional electrons that are shared equally (delocalized) throughout the ring. Benzene is the smallest aromatic hydrocarbon, and many other aromatic compounds are named as derivatives of benzene. Polycyclic aromatic hydrocarbons have two or more rings fused together. Aromatic hydrocarbons and many other aromatic compounds are nonpolar. Thus they are soluble in organic solvents but are water-insoluble. These compounds are either liquids or solids. The typical reaction of an aromatic ring is a substitution reaction in which a part of a reactant substitutes for a hydrogen atom on the ring.

SUMMARY OF REACTIONS

Alkenes

Addition Reaction: Hydrogenation

$$\text{Alkene} + \text{hydrogen} \xrightarrow{\text{catalyst}} \text{alkane}$$

$$CH_2{=}CH_2 + H_2 \xrightarrow{Pt} CH_3CH_3$$

Addition Reaction: Halogenation

$$\text{Alkene} + \text{halogen} \rightarrow \text{organohalogen (alkyl halide)}$$

$$CH_2{=}CH_2 + Cl_2 \rightarrow CH_2ClCH_2Cl$$

Addition Reaction: Hydration

$$\text{Alkene} + \text{water} \xrightarrow{H^+} \text{alcohol}$$

$$CH_2{=}CHCH_3 + H_2O \xrightarrow{H^+} CH_3CH(OH)CH_3$$

Addition Reaction: Hydrohalogenation

$$\text{Alkene} + \text{hydrogen halide} \rightarrow \text{organohalogen (alkyl halide)}$$

$$CH_2{=}CHCH_3 + HBr \rightarrow CH_3CHBrCH_3$$

Oxidation Reaction: Hydroxylation

$$\text{Alkene} \xrightarrow{\text{[mild ox.]}} \text{diol (glycol)}$$

$$CH_2{=}CH_2 \xrightarrow{\text{[mild ox.]}} HOCH_2CH_2OH$$

Oxidation Reaction: Oxidative Cleavage

$$\text{Alkene} \xrightarrow{\text{[strong ox.]}} \text{aldehydes or ketones}$$

$$CH_2{=}C(CH_3)_2 \xrightarrow{\text{[strong ox.]}} H_2C{=}O + O{=}C(CH_3)_2$$

Polymerization

$$\text{Alkene} \xrightarrow{\text{[cat.]}} \text{polymer}$$

$$CH_2{=}CH_2 \xrightarrow{\text{[cat.]}} \ldots CH_2CH_2CH_2CH_2CH_2CH_2 \ldots$$

Alkynes

$$\text{Alkyne} + 2 \text{ hydrogen halide} \rightarrow \text{organohalogen (alkyl halide)}$$

$$HC{\equiv}CCH_3 + 2HBr \rightarrow CH_3CBr_2CH_3$$

Aromatic Hydrocarbons (Aromatic Compounds)

Aromatic Substitution (Alkylation)

$$\text{Aromatic ring} + \text{alkyl halide} \xrightarrow{AlCl_3} \text{substituted aromatic ring} + HX$$

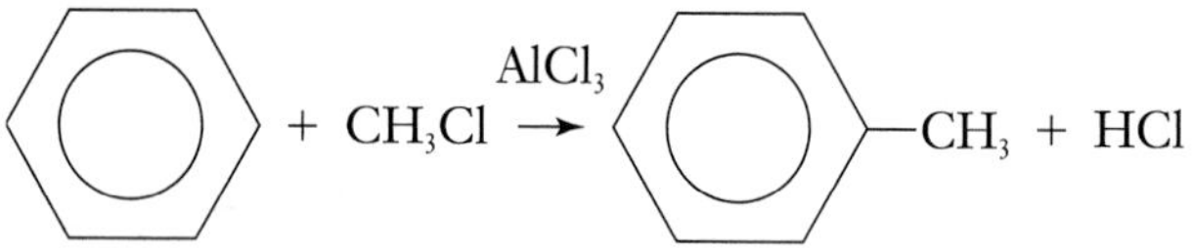

$$C_6H_6 + CH_3Cl \xrightarrow{AlCl_3} C_6H_5{-}CH_3 + HCl$$

Aromatic Substitution (Halogenation)

$$\text{Aromatic ring} + \text{halogen} \xrightarrow{FeBr_3} \text{halogenated aromatic ring} + HX$$

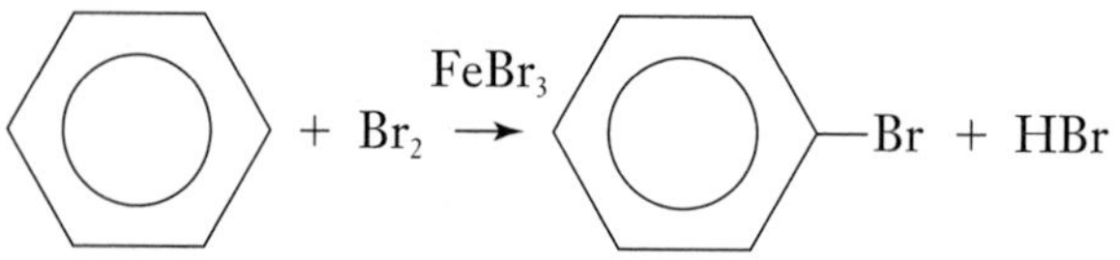

$$C_6H_6 + Br_2 \xrightarrow{FeBr_3} C_6H_5{-}Br + HBr$$

Aromatic Substitution (Nitration)

$$\text{Aromatic ring} + \text{nitric acid} \xrightarrow{H_2SO_4} \text{nitrated aromatic ring} + H_2O$$

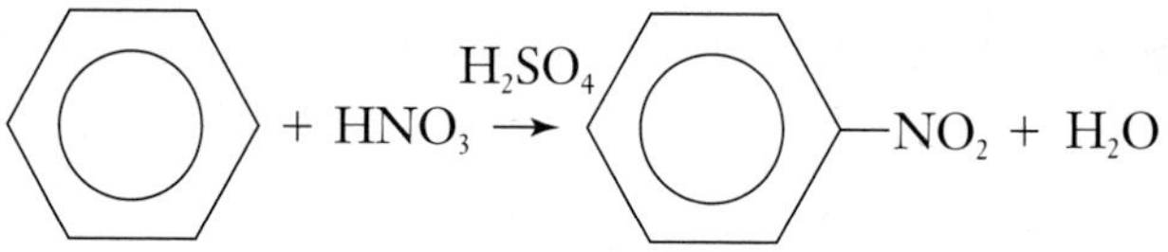

$$C_6H_6 + HNO_3 \xrightarrow{H_2SO_4} C_6H_5{-}NO_2 + H_2O$$

Aromatic Substitution (Sulfonation)

$$\text{Aromatic ring} + SO_3 \xrightarrow{H_2SO_4} \text{sulfonated aromatic ring}$$

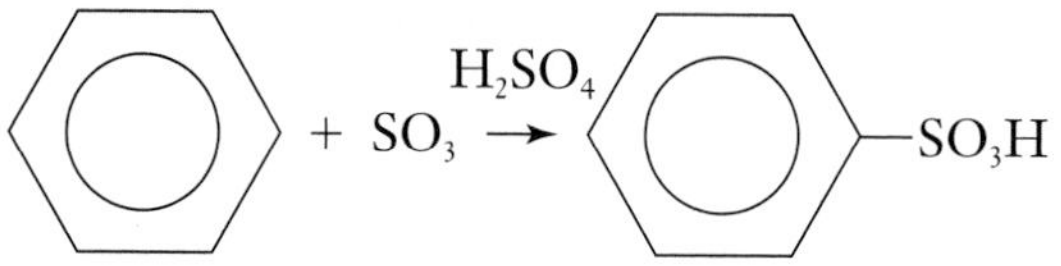

$$C_6H_6 + SO_3 \xrightarrow{H_2SO_4} C_6H_5{-}SO_3H$$

KEY TERMS

Section 9.1
alkene

Section 9.3
alkyne

Section 9.4
halogenation
hydration
hydrogenation
hydrohalogenation
hydroxylation
Markovnikov's rule
oxidative cleavage
polymerization

Section 9.5
aromatic hydrocarbon

Alkenes (*Section 9.1*)

1. What is an unsaturated hydrocarbon?
2. Three structural features can cause a hydrocarbon to be unsaturated. What are they?
3. List the three classes of hydrocarbons that are unsaturated.
4. What is the functional group of an alkene?
5. (a) Identify the principal source of alkenes.
   (b) List some major industrial uses of alkenes.
6. Draw the structure of the smallest alkene and name it.
7. Write the IUPAC name of any alkene. In the name, draw a circle around or highlight the part that indicates the presence of a carbon–carbon double bond.
8. Draw the structures of these alkenes.
   (a) Propene (b) 1-pentene
   (c) 2-hexene (d) 4-octene
9. Sketch the structure of each of these alkenes.
   (a) 1,4-hexadiene (b) Cycloheptene
   (c) Propylene (d) 3-heptene
10. What is the structure of these alkenes?
    (a) 4-methyl-1-hexene
    (b) 2-methyl-2-heptene
    (c) 3,4-dimethyl-2-pentene
    (d) 1,2,4-trimethylcyclopentene
11. Provide the name for each of these compounds.
    (a) $CH_3CH_2CH{=}CHCH_2CH_2CH_3$
    (b) $CH_2{=}CHCH_2CH(CH_3)CHCH_3$ with a CH_3 on the fifth carbon ($CH_2{=}CHCH_2CH(CH_3)CH(CH_3)CH_3$)
    (c) [cyclohexene ring bearing CH_3 and $CH_2CH_2CH_3$]
12. Provide the name of each of these compounds.
    (a) $CH_2{=}CHCH_2CH_3$
    (b) $CH_3CH{=}CHCHCH_3$ with $CH_2CH_2CH_3$ on the fourth carbon
    (c) H_3C [cyclopentene ring] CH_3
13. Draw the structure of these compounds:
    (a) Cyclohexene (b) 3-methylcyclopentene
14. Name these alkenes:
    (a) 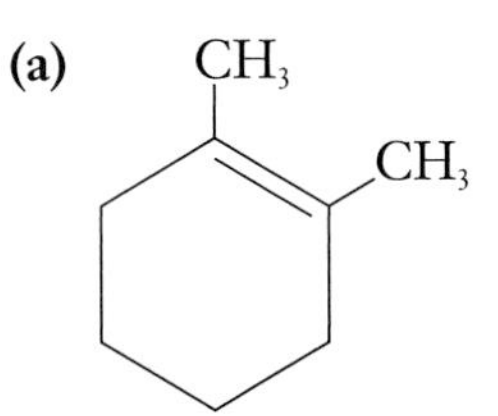

    (b) H_3C [cyclopentene ring] H_3C CH_2CH_3

Alkene Structure (*Section 9.2*)

15. Free rotation can occur around what type of covalent bond?
16. Why is there no free rotation about a carbon–carbon double bond?
17. Two kinds of geometric isomers can exist because of carbon–carbon double bonds. What is the difference between these two isomers, and what words are used to identify them?
18. Draw the two geometric isomers of 2-pentene. Correctly label each isomer.
19. Are there geometric isomers of 1-pentene? Why or why not?
20. Draw the structure of 2,3-dimethyl-2-octene. Can you draw geometric isomers of this compound? Why or why not?
21. Draw the structure of these compounds.
    (a) *Trans*-3-heptene
    (b) *Cis*-3-octene
    (c) *Cis*-4-methyl-2-pentene
    (d) *Trans*-2,2-dimethyl-4-nonene
22. What is the structure of these alkenes?
    (a) *Cis*-3-methyl-4-nonene
    (b) *Trans*-5-methyl-2-hexene
    (c) *Trans*-2,3-dimethyl-4-octene
    (d) *Cis*-3-methyl-2-pentene
23. The name 3-butene is incorrect. What is wrong with this name?
24. Provide the name of these alkenes.
    (a) H_3C, H on one carbon; H, $CH_2CH(CH_3)CH_3$ on the other ($C{=}C$)

(b) 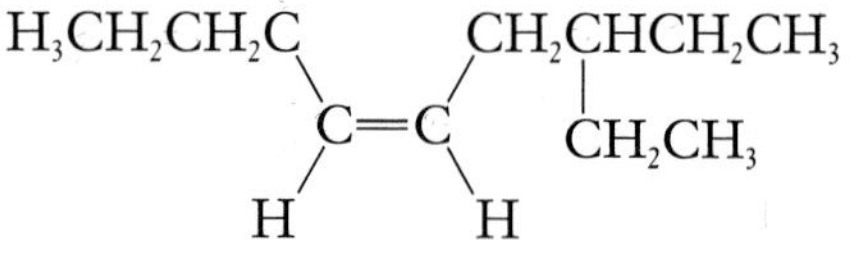
$$\begin{array}{ccc} H_3CH_2CH_2C & & CH_2CHCH_2CH_3 \\ & C{=}C & \quad | \\ & & \quad CH_2CH_3 \\ H & & H \end{array}$$

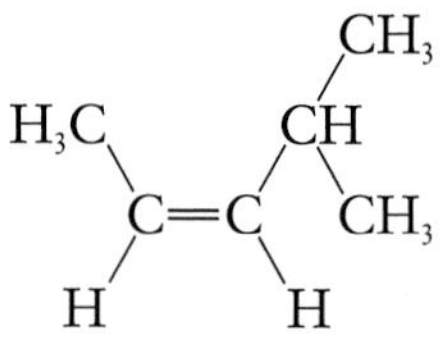

(c)
$$\begin{array}{ccc} CH_3 & & H \\ & C{=}C & \\ H_3CH_2CH_2C & & CH_3 \end{array}$$

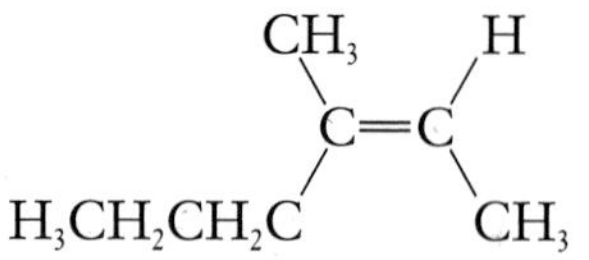

25. Provide the name of these alkenes.

(a) 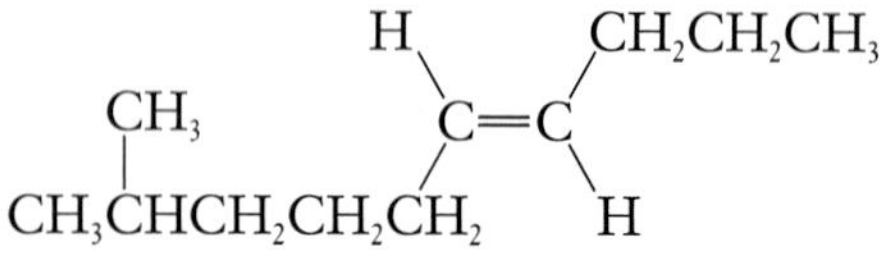
$$\begin{array}{ccc} & H & CH_2CH_2CH_3 \\ CH_3 & C{=}C & \\ | & & \\ CH_3CHCH_2CH_2CH_2 & & H \end{array}$$

(b)
$$\begin{array}{ccc} & & CH_3 \\ H_3C & & CH \\ & C{=}C & \quad CH_3 \\ H & & H \end{array}$$

(c)
$$\begin{array}{ccc} H_3C & & CH_2CH_2CH_3 \\ & C{=}C & \\ H_3CH_2C & & CH_3 \end{array}$$

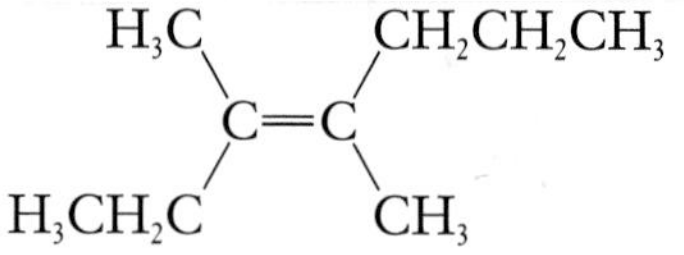

Alkynes (Section 9.3)

26. What is the functional group of the alkynes?

27. Name and draw the structure of the most common alkyne. What is its principal use?

28. The suffix that is used to identify a compound as an alkyne is ____________.

29. Could you draw this geometric isomer: *cis*-2-butyne? Why or why not?

30. You were not told how to name cyclic alkynes, but based on what you have learned about cycloalkanes and cycloalkenes, what is the structure of cyclononyne?

31. Write the IUPAC name of each of these compounds.

(a) $HC{\equiv}CH$ (IUPAC and common name)

(b) $CH_3CH_2CH_2C{\equiv}CCH_3$

(c) 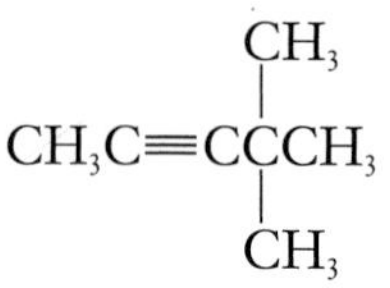
$$\begin{array}{c} CH_3 \\ | \\ CH_3C{\equiv}CCCH_3 \\ | \\ CH_3 \end{array}$$

(d) 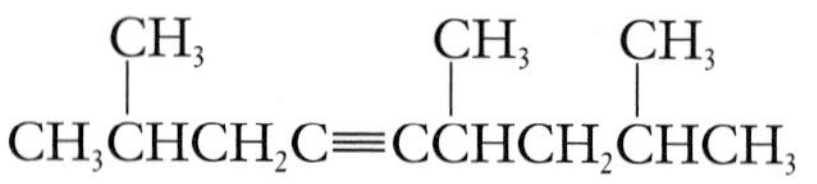
$$\begin{array}{c} CH_3 \qquad\qquad CH_3 \quad CH_3 \\ | \qquad\qquad\quad | \qquad | \\ CH_3CHCH_2C{\equiv}CCHCH_2CHCH_3 \end{array}$$

32. Provide the IUPAC name for these compounds.

(a) $HC{\equiv}CCH_2CH_3$

(b) $CH_3CH_2C{\equiv}CCH_3$

(c) 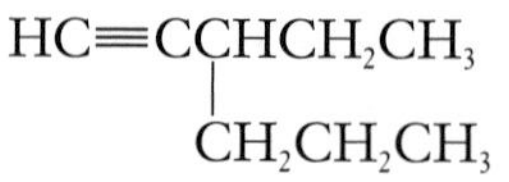
$$\begin{array}{c} HC{\equiv}CCHCH_2CH_3 \\ | \\ CH_2CH_2CH_3 \end{array}$$

(d) 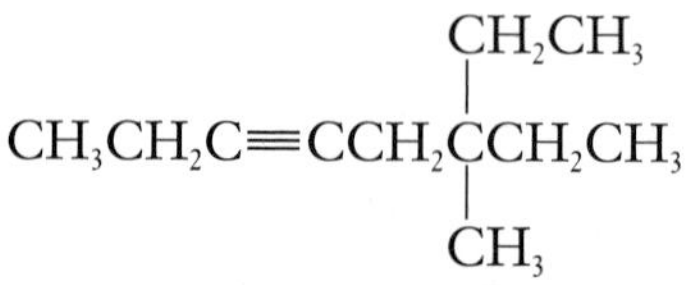
$$\begin{array}{c} CH_2CH_3 \\ | \\ CH_3CH_2C{\equiv}CCH_2CCH_2CH_3 \\ | \\ CH_3 \end{array}$$

33. What is the structure of these alkynes?

(a) 3-heptyne

(b) 2-methyl-4-nonyne

(c) 5-ethyl-3-decyne

(d) 2,2,5,5-tetramethyl-3-hexyne

34. Draw the structure of these alkynes.

(a) Propyne

(b) 2-pentyne

(c) 6,6-diethyl-2-octyne

(d) 3-propyl-1-heptyne

Properties and Reactions of Alkenes and Alkynes *(Section 9.4)*

35. Identify the intermolecular forces that hold alkene and alkyne molecules together in the liquid and solid state. How does the strength of these forces change with the size of the molecules?

36. Describe the solubility properties of the alkenes and alkynes.

37. Name the three types of chemical reactions that commonly occur in the alkenes.

38. The product in the reaction between an alkene and hydrogen belongs to what class of organic compounds?

39. Complete the following reactions by drawing the structure of the product.

(a) $CH_2{=}CHCH_2CH_3 + H_2 \xrightarrow{Pt} ?$

(b) 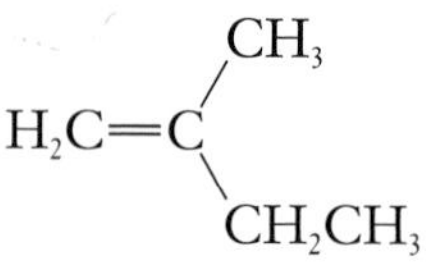
$$\begin{array}{ccc} & CH_3 & \\ H_2C{=}C & & + H_2 \xrightarrow{Pt} ? \\ & CH_2CH_3 & \end{array}$$

(c) $CH_2{=}CH{-}CH{=}CH_2 + 2H_2 \xrightarrow{Pt} ?$

40. Complete the following reactions by drawing the structure of the product.

(a) $CH_3C{\equiv}CCH_2CH_3 + 2H_2 \xrightarrow{Pt}$?

(b) 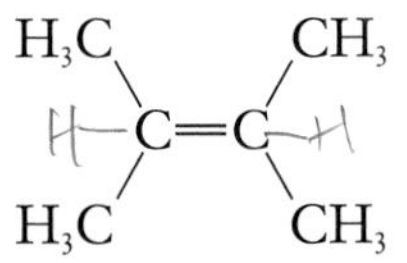 $(H_3C)_2C{=}C(CH_3)_2 + H_2 \xrightarrow{Pt}$?

(c) 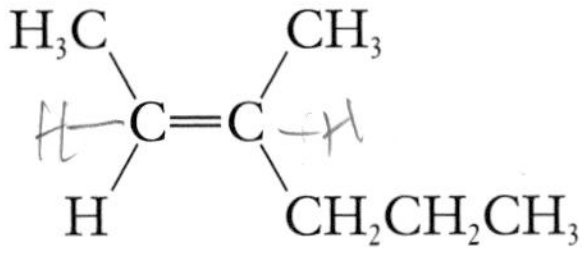 $(H_3C)(H)C{=}C(CH_3)(CH_2CH_2CH_3) + H_2 \xrightarrow{Pt}$?

41. When alkenes are halogenated, to which carbon atoms do the halogen atoms become bonded?

42. Draw the structure of the product formed in these reactions:

(a) $CH_3CH{=}CH_2 + Br_2 \rightarrow$?

(b) 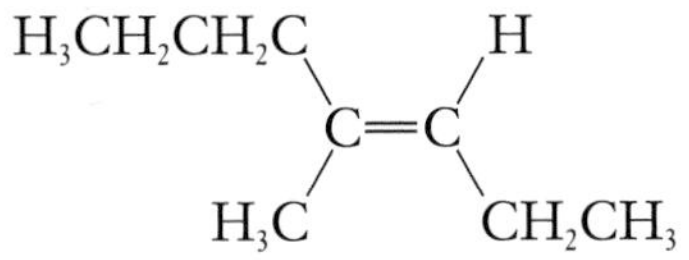 $(H_3CH_2CH_2C)(H_3C)C{=}C(H)(CH_2CH_3) + Cl_2 \rightarrow$?

(c) 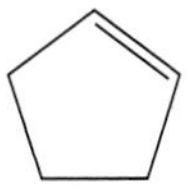 cyclopentene $+ Cl_2 \rightarrow$?

43. Draw the structure of the product formed in these reactions:

(a) $CH_3CH{=}CHCH_2CH_3 + Cl_2 \rightarrow$?

(b) 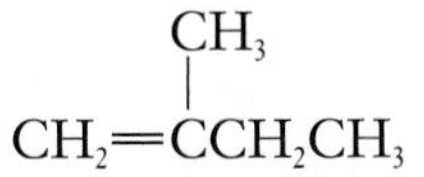 $CH_2{=}C(CH_3)CH_2CH_3 + Br_2 \rightarrow$?

(c) $CH_2{=}CHCH{=}CH_2 + 2Cl_2 \rightarrow$?

44. How do water and hydrogen chloride (HCl) differ from hydrogen (H_2) and chlorine (Cl_2) as adding reagents to alkenes?

45. State Markovnikov's rule. Would Markovnikov's rule apply to the hydration of ethene? Why or why not?

46. Predict the product of these reactions and balance them.

(a) $CH_2{=}CH_2 + H_2O \xrightarrow{H^+}$?

(b)  $CH_3C(CH_3){=}CHCH_3 + H_2O \xrightarrow{H^+}$?

(c) 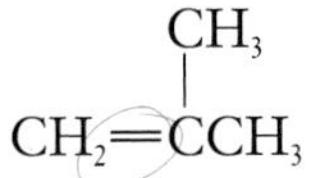 $CH_2{=}C(CH_3)CH_3 + HBr \rightarrow$?

47. Predict the product of these reactions and balance them.

(a) $CH_3CH_2CH{=}CH_2 + HCl \rightarrow$?

(b) 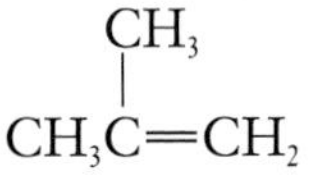 $CH_3C(CH_3){=}CH_2 + HBr \rightarrow$?

(c) 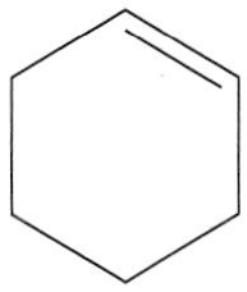 cyclohexene $+ H_2O \xrightarrow{H^+}$?

48. Hydration of what alkene would give you this product: $CH_3CH_2C(OH)(CH_3)CH_3$

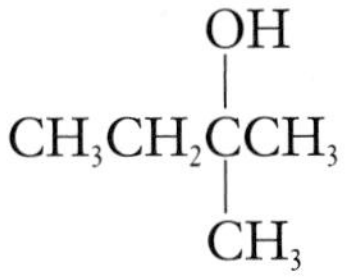

49. Complete and balance these reactions.

(a) $CH_2{=}CHCH_3 \xrightarrow{[mild\ ox.]}$?

(b) $CH_3CH{=}CH_2 \xrightarrow{[strong\ ox.]}$?

(c) 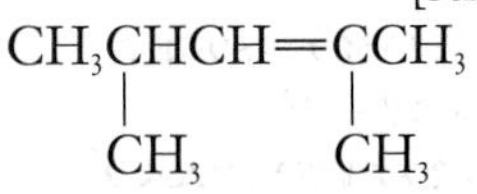 $CH_3CH(CH_3)CH{=}C(CH_3)CH_3 \xrightarrow{[strong\ ox.]}$?

50. Complete and balance these reactions.

(a) $CH_2{=}CH_2 \xrightarrow{[strong\ ox.]}$?

(b) $CH_3CH{=}CHCH_3 \xrightarrow{[mild\ ox.]}$?

(c) 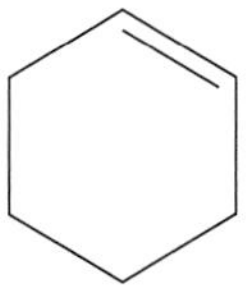 cyclohexene $\xrightarrow{[mild\ ox.]}$?

51. Write an equation that illustrates the polymerization of ethene (ethylene).
52. Write an equation that illustrates the polymerization of propene (propylene). (Hint: Think of propene as being a methyl-substituted ethylene molecule.)

Aromatic Compounds *(Section 9.5)*

53. A compound is considered to be aromatic if what structural feature is present?
54. A circle (ring) within a hexagon is often used to represent an aromatic ring. What does the circle represent?
55. *Delocalized electrons* are found in a number of organic and inorganic compounds. What are delocalized electrons, and what class of hydrocarbons always contain them?
56. Draw the structures for these aromatic hydrocarbons:
    (a) Benzene
    (b) Toluene
    (c) Ethylbenzene
57. Draw the structures for these aromatic hydrocarbons:
    (a) o-ethyltoluene
    (b) m-propyltoluene
    (c) 1,3-diethylbenzene
58. Xylenes are aromatic hydrocarbons that could (incorrectly) be called dimethylbenzenes. Draw the structures of the three known xylenes.
59. Name each of these aromatic hydrocarbons. When it is appropriate, name them by both IUPAC (numbers) and common (o-, m-, p-) rules.

(a) 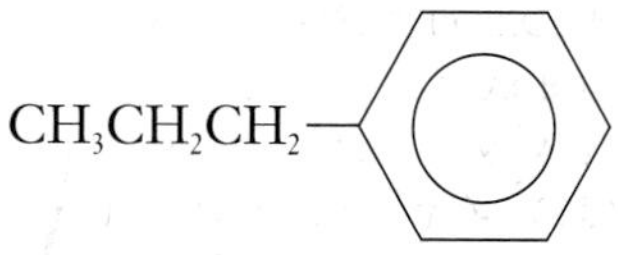 $CH_3CH_2CH_2$–

(b) 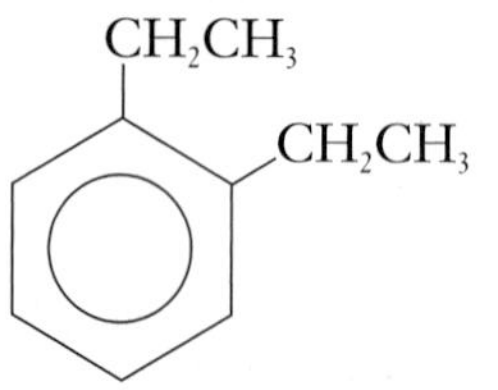 CH_2CH_3 CH_2CH_3

(c) 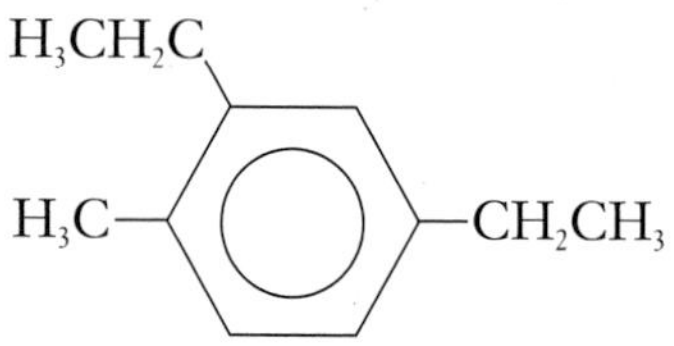 H_3CH_2C H_3C– –CH_2CH_3

60. Name each of these aromatic hydrocarbons. When it is appropriate, name them by both IUPAC (numbers) and common (o-, m-, p-) rules.

(a) 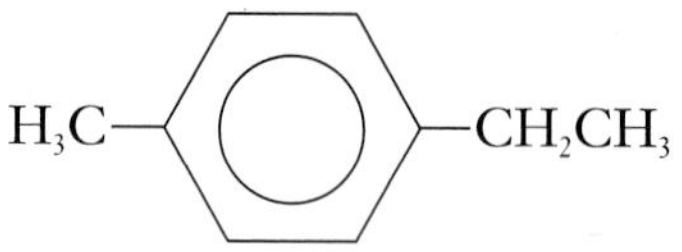 H_3C– –CH_2CH_3

(b) 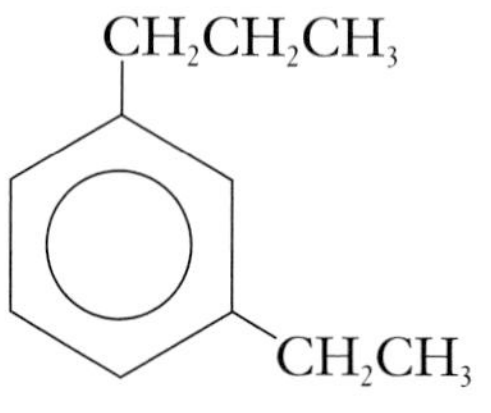 $CH_2CH_2CH_3$ CH_2CH_3

(c) 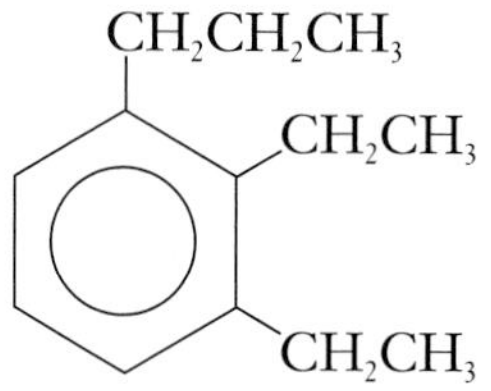 $CH_2CH_2CH_3$ CH_2CH_3 CH_2CH_3

61. Draw the structure of these aromatic compounds:
    (a) Chlorobenzene **(b)** Bromobenzene
    (c) p-bromotoluene **(d)** 1,2-difluorobenzene
    (e) 1-bromo-2,4-dichlorobenzene
62. What is the structure of each of these compounds?
    (a) Nitrobenzene
    (b) m-dichlorobenzene
    (c) o-bromochlorobenzene
    (d) 3-chlorotoluene
    (e) 2,4,6-trinitrotoluene
63. Name each of these aromatic compounds, giving two names where appropriate:

(a) 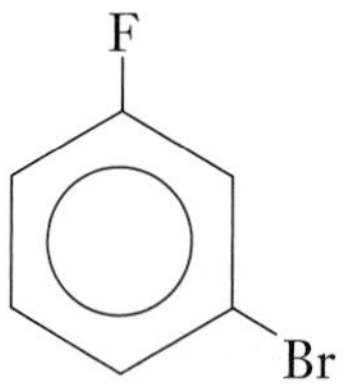 F Br

(b) 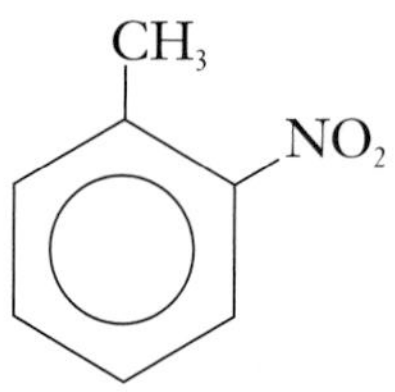 CH_3 NO_2

(c) 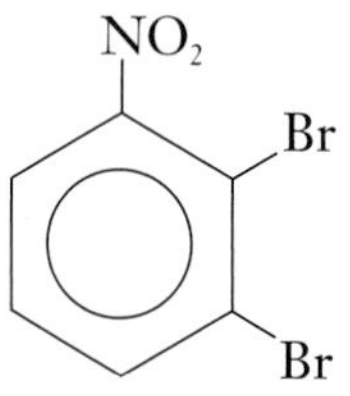 NO_2 Br Br

64. Name each of these aromatic compounds, giving two names where appropriate.

(a) 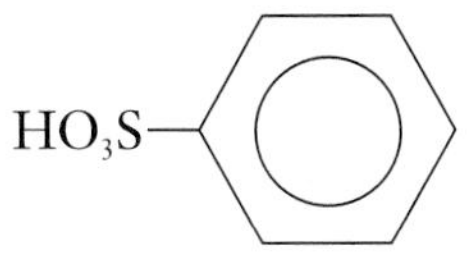 HO_3S–(benzene ring)

(b) 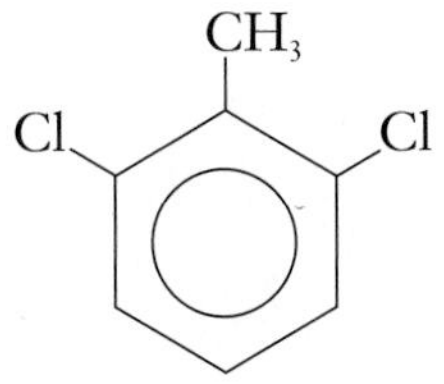 benzene ring with CH_3, Cl, Cl

(c) 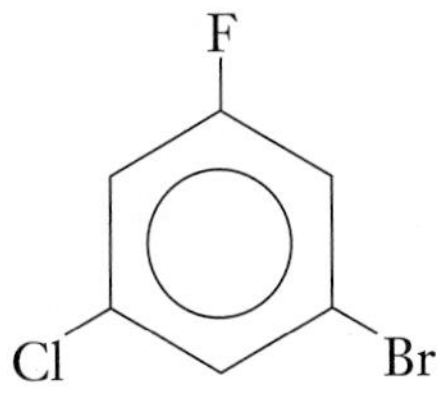 benzene ring with F, Cl, Br

65. What structural feature characterizes a polycyclic compound? Name two common polycyclic aromatic hydrocarbons.

Properties and Reactions of Aromatic Hydrocarbons *(Section 9.6)*

66. The aromatic hydrocarbons are nonpolar. Why?

67. Describe the solubility properties of the aromatic hydrocarbons.

68. Complete and balance these reactions.

(a) 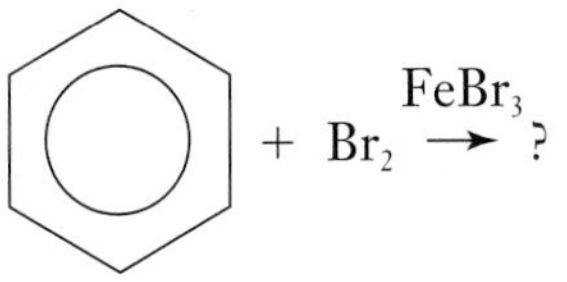 benzene $+ Br_2 \xrightarrow{FeBr_3}$?

(b) 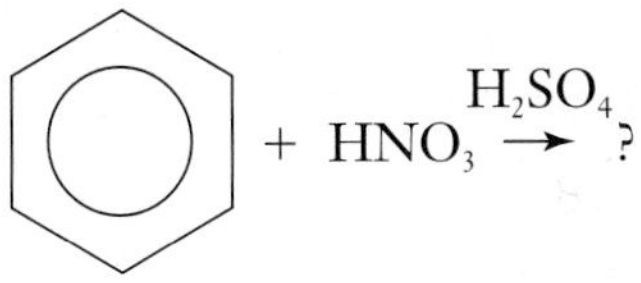 benzene $+ HNO_3 \xrightarrow{H_2SO_4}$?

(c) 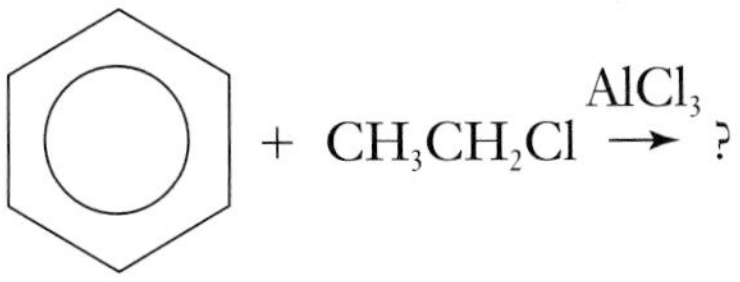 benzene $+ CH_3CH_2Cl \xrightarrow{AlCl_3}$?

69. Draw the structure of the organic product of each of these reactions and balance the equation.

(a) 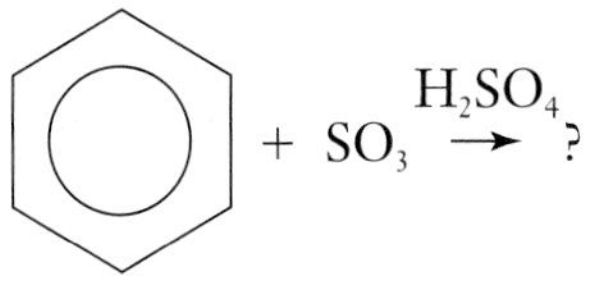 benzene $+ SO_3 \xrightarrow{H_2SO_4}$?

(b) 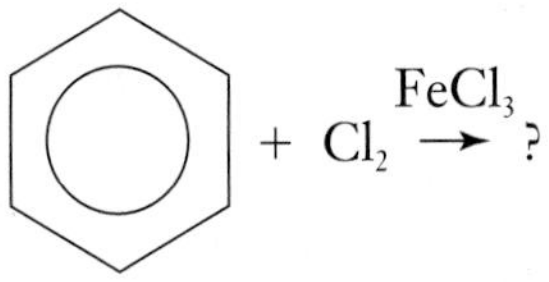 benzene $+ Cl_2 \xrightarrow{FeCl_3}$?

(c) 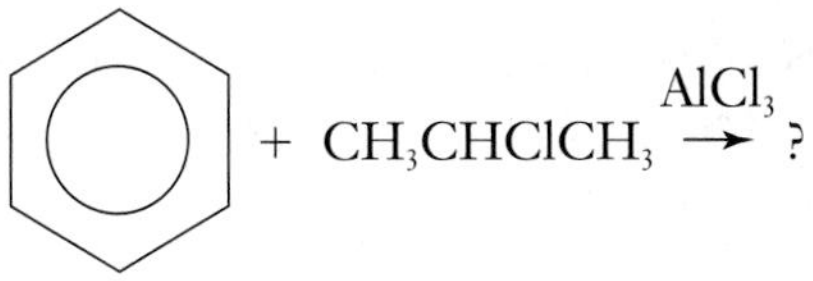 benzene $+ CH_3CHClCH_3 \xrightarrow{AlCl_3}$?

Challenge Exercises

70. **(a)** A C_5H_{10} hydrocarbon is mixed with hydrogen and a catalyst yet no reaction occurs. Assign a structure to the hydrocarbon that is consistent with this data. **(b)** A C_5H_8 hydrocarbon, when hydrogenated, yields the hydrocarbon in part (a) as the product. Assign a structure to the hydrocarbon that is consistent with this data. **(c)** Look up the structure of isopropyl alcohol (rubbing alcohol), then write a balanced equation for its one-step synthesis from a hydrocarbon.

71. Acetylene (ethyne) is the fuel in many high-temperature torches. **(a)** Write the balanced equation for the combustion of acetylene. **(b)** Determine the energy released, in both kcal and kJ, when 100.0 g of acetylene burns completely. When one mole of acetylene burns, 312 kcal of energy are released. **(c)** Determine the volume of oxygen in liters that is needed to react completely with the 100.0 g of acetylene.

72. Draw the structures and name five or more drugs or medications that contain carbon–carbon double or triple bonds or an aromatic ring. Good sources for this information include textbooks from some of your other science courses, the *Merck Index*, and the Internet.

10 Alcohols, Ethers, Thiols, Amines, and Organohalogens

THESE ORGANIC COMPOUNDS CONTAIN A HETERO-ATOM SINGLE-BONDED TO A CARBON ATOM.

HYDROCARBONS ARE organic compounds that contain only atoms of carbon and hydrogen. The compounds of other classes of organic compounds also contain one or more atoms of some other element. For these compounds, this different atom, known as a **hetero-atom,** is part of a functional group. Alcohols and ethers contain an oxygen atom, thiols a sulfur atom, and amines a nitrogen atom. Organohalogens contain a halogen atom bonded to an alkyl group or aromatic ring. In this chapter you will examine these new functional groups and some of the compounds that contain them.

Outline

Anethole, an ether, is a chief component of anise and fennel oils, which are used as flavoring agents in cooking. *(Charles D. Winters)*

10.1 Alcohols and Phenols

Many useful everyday compounds are alcohols or phenols, including many biomolecules, drugs, and disinfectants. These compounds contain an oxygen atom in a **hydroxyl group:** —OH. An **alcohol** has the hydroxyl group attached to a carbon atom of an alkyl group, and a **phenol** has the hydroxyl group attached to a carbon atom of an aromatic ring.

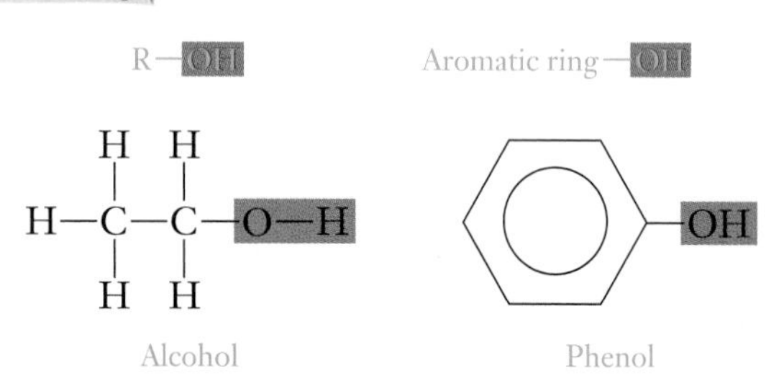

Some of the useful compounds in these products are alcohols and phenols. The —OH alcohol group is found in the menthol of mint and the —OH phenol group is found in the oil of clove. *(Charles D. Winters)*

Naming Alcohols and Phenols

The IUPAC rules for naming alcohols and phenols are simply extensions of the rules used to name hydrocarbons (Table 10.1). The parent and prefix of the corresponding hydrocarbon are retained, but the letter "e" of the suffix for the hydrocarbon (*-e*) is replaced by the suffix *-ol*. Methan*ol* and cyclohexan*ol*, for instance, are the names of the alcohols derived from the names methan*e* and cyclohexan*e*.

CH_4	CH_3OH	(cyclohexane ring)	HO—(cyclohexane ring)
Methan*e*	Methan*ol*	Cyclohexan*e*	Cyclohexan*ol*

Example 10.1 Names and Structures of Alcohols

1. What is the IUPAC name of the alcohol $CH_3CHCH_2CH_3$?

2. What is the structure of 2-methyl-3-pentanol?

SOLUTIONS

1. The longest continuous chain of carbon atoms that includes the hydroxyl-bearing carbon atom contains four carbon atoms. The parent and suffix must therefore be *-butanol.* The hydroxyl group is on carbon 2, and so the name is 2-butanol.
2. The parent, *-pentan-*, tells us that the longest continuous chain contains five carbon atoms. The *-ol* and *-3-* mean there is a hydroxyl group on carbon 3:

```
        OH
        |
C—C—C—C—C
```

The *2-methyl-* tells us that there is a methyl group on carbon 2:

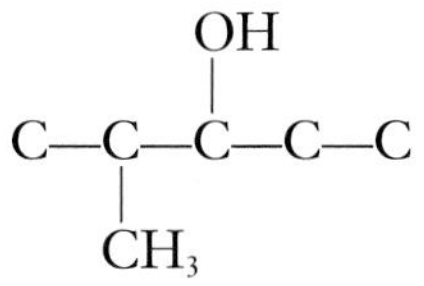

```
         OH
         |
C—C—C—C—C
    |
    CH3
```

Now add the appropriate number of hydrogen atoms to the carbon atoms to get the final structure:

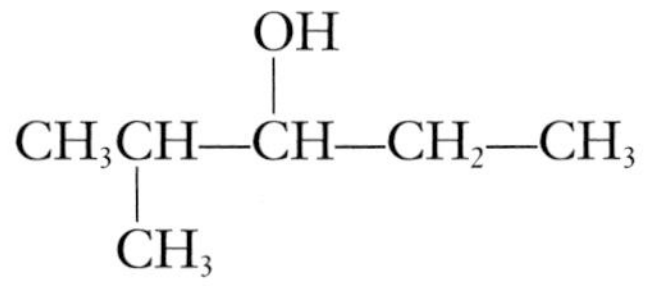

```
            OH
            |
CH3CH—CH—CH2—CH3
     |
     CH3
```

Self-Test

1. Name the compound

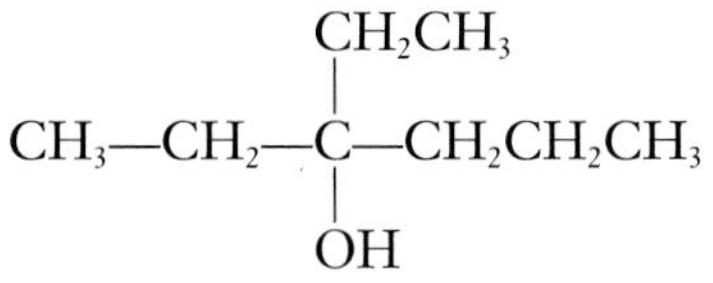

```
                CH2CH3
                |
CH3—CH2—C—CH2CH2CH3
                |
                OH
```

2. Draw the structure of 5-ethyl-6-methyl-1-heptanol.

ANSWERS

1. 3-ethyl-3-hexanol

2.

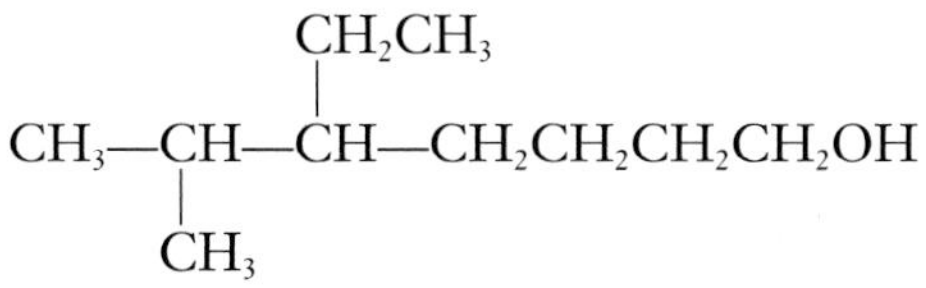

```
              CH2CH3
              |
CH3—CH—CH—CH2CH2CH2CH2OH
       |
       CH3
```

Two or more hydroxyl groups can be in a molecule. The IUPAC names for these compounds clearly identify both the number of hydroxyl groups and their location. For example, the IUPAC name for $HOCH_2CH_2OH$ is 1,2-ethanediol. The *di-* in *-diol* means there are two hydroxyl groups, and the *1,2-* prefix indicates they are attached to carbons 1 and 2. Alcohols containing two hydroxyl groups are commonly called *glycols.* Thus 1,2-ethanediol is also known as ethylene glycol, which is the primary ingredient in automotive antifreeze.

Table 10.1 IUPAC Rules for Naming Alcohols

Rule One	Find the longest continuous chain of carbon atoms that includes the hydroxyl-bearing carbon atom. This chain determines the parent. Combine the parent and *an* with the suffix *-ol.*
Rule Two	Number the chain from the end that gives the hydroxyl-bearing carbon atom the lowest possible number. Use this number as a prefix in front of the parent name.
Rule Three	The rules for substituents are the same as those for alkanes.
Rule Four	The rules for numerical prefixes, commas, and hyphens are the same as for alkanes.

Many of the smaller alcohols are also identified by common names. These names consist of two words. The first word identifies the alkyl group in the alcohol, and the second word is *alcohol.* For example, methyl alcohol is the common name of methanol, CH_3OH, and ethyl alcohol is the common name of ethanol, CH_3CH_2OH.

Example 10.2 Common Names of Alcohols

What are the common names of these alcohols:

1. $CH_3CH_2CH_2CH_2OH$
2. 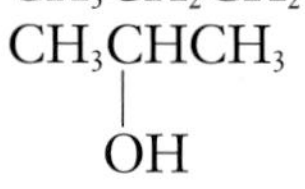

SOLUTIONS

The common name of an alcohol is derived from the alkyl group in the alcohol and the word *alcohol.*

1. The alkyl group is the butyl group (derived from butane), and so the common name is butyl alcohol.
2. The alkyl group is the isopropyl group, and so this is isopropyl alcohol. This alcohol is also known as rubbing alcohol because it is used in "rub downs".

Self-Test

1. What is the common name of $CH_3CH_2CH_2CH_2CH_2OH$?
2. What is the structure of propyl alcohol?

ANSWERS

1. Pentyl alcohol 2. $CH_3CH_2CH_2OH$

Alcohols are separated into three classes based on the number of carbon atoms attached to the hydroxyl-bearing carbon atom. A *primary alcohol* has one carbon atom attached to the hydroxyl-bearing carbon atom (Table 10.2). A *secondary alcohol* has two carbon atoms attached to the hydroxyl-bearing carbon atom, and a *tertiary alcohol* has three carbon atoms attached to the hydroxyl-bearing carbon atom.

Example 10.3 Classifying Alcohols

Classify these alcohols as primary, secondary, or tertiary.

1. 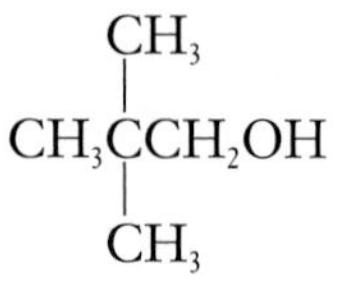
2. 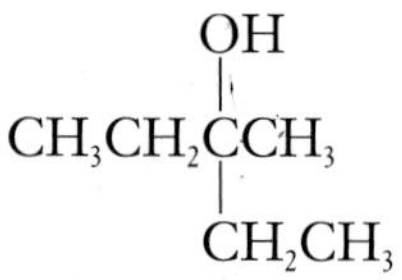

3. HO—(cyclohexane ring)

SOLUTION

To classify an alcohol, count the number of carbon atoms bonded to the hydroxyl-bearing carbon atom.

1. There is just one carbon atom on the hydroxyl-bearing carbon atom, thus this alcohol, 2,2-dimethyl-1-propanol, is a primary alcohol.
2. The hydroxyl-bearing carbon atom of 3-methyl-3-pentanol has three carbon atoms bound to it, thus this is a tertiary alcohol.
3. The hydroxyl-bearing carbon atom of cyclohexanol has two carbon atoms bound to it, and so it is a secondary alcohol.

Self-Test

Classify these alcohols as primary, secondary, or tertiary:

1. 3-octanol, $CH_3CH_2CH_2CH_2CH_2\underset{}{\overset{OH}{\overset{|}{C}}}HCH_2CH_3$
2. 1-methylcyclopentanol, (cyclopentane ring bearing CH_3 and OH on the same carbon)
3. 4-methyl-1-heptanol, $HOCH_2CH_2CH_2\underset{\underset{CH_3}{|}}{C}HCH_2CH_2CH_3$

ANSWERS

1. Secondary 2. Tertiary 3. Primary

Table 10.2 Classification of Alcohols

CLASS OF ALCOHOL	DEFINITION	EXAMPLE	
Primary alcohol (1°)	Hydroxyl-bearing carbon atom is bonded to one carbon atom	Ethanol	CH_3CH_2OH
Secondary alcohol (2°)	Hydroxyl-bearing carbon atom is bonded to two carbon atoms	2-Propanol	$CH_3\underset{\underset{OH}{|}}{C}HCH_3$
Tertiary alcohol (3°)	Hydroxyl-bearing carbon atom is bonded to three carbon atoms	2-Methyl-2-propanol	$CH_3-\underset{\underset{OH}{|}}{\overset{\overset{CH_3}{|}}{C}}-CH_3$

The nomenclature of phenols is similar to the nomenclature of other aromatic compounds (Chapter 9). The smallest and simplest phenol is given the name phenol:

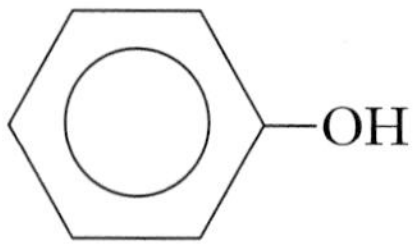

OH

Phenol

Other phenols are named as derivatives of phenol. The substituent name is placed before the word *-phenol,* and numbers (or *ortho-*, *meta-*, *para-* prefixes for disubstituted molecules) are used to designate the position of the substituent groups. The hydroxyl group is assumed to be on carbon 1, and a number indicating this position is not included in the name.

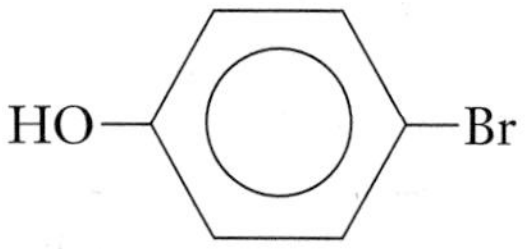

OH Cl HO Br

4-bromophenol (*p*-bromophenol)

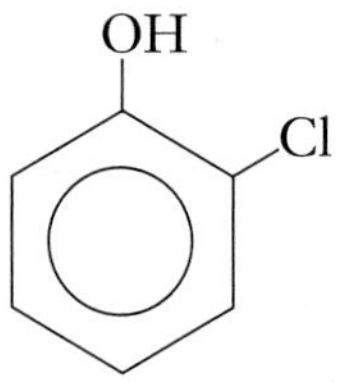

2-chlorophenol (*o*-chlorophenol)

Some phenols are identified by non-IUPAC names. Salicylic acid, the active ingredient in aspirin, is an example. This name, like many other common names, provides no information about the structure of this phenol.

HO O HO

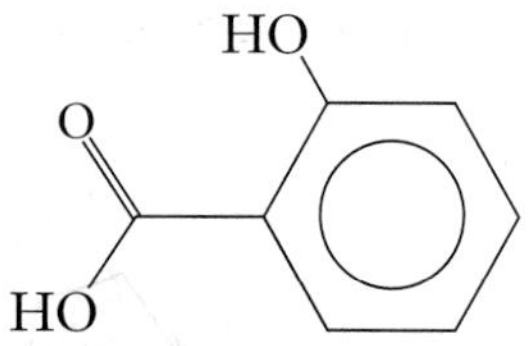

Salicylic acid

Example 10.4 Names and Structures of Phenols

What is the structure of *m*-ethylphenol (3-ethylphenol)?

SOLUTION

A phenol has a hydroxyl group attached to an aromatic ring. The hydroxyl group is assumed to be on carbon 1, and so the ethyl group is located two carbon atoms from the hydroxyl group.

The structure of this phenol is

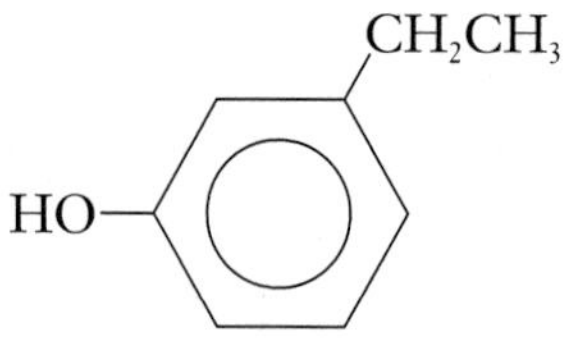

CH_2CH_3

HO

Self-Test

What is the name of

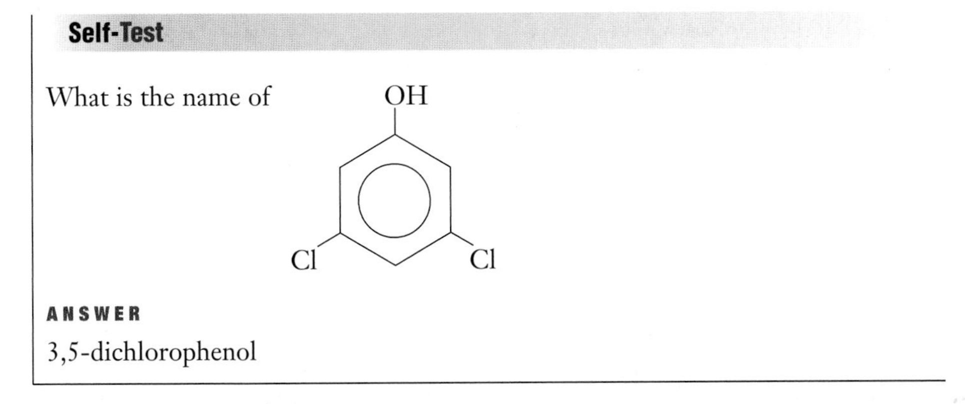

ANSWER

3,5-dichlorophenol

Properties of Alcohols and Phenols

Alcohol and phenol molecules contain the hydroxyl group, which is both polar (Figure 10.1) and capable of forming hydrogen bonds. These molecules also contain a nonpolar alkyl group or aromatic ring, and as a result, they have both polar and nonpolar characteristics. This dual polarity–nonpolarity has significant influence on the physical properties of these molecules.

In the smaller alcohols (three carbons or less) the hydroxyl group forms significant hydrogen bonding and dipole–dipole interactions with water, and as a result these smaller alcohols are completely soluble in water. They are also good solvents for other polar organic compounds. In larger alcohols these hydrogen bonds and dipole–dipole interactions also exist, but the larger alkyl group results in significant London dispersion forces between the alcohol molecules, which results in less of the alcohol dissolving in the water. Water solubility decreases from 1-butanol through 1-hexanol, and alcohols larger than 1-hexanol are essentially water-insoluble (Figure 10.2). Phenol and other small phenols are somewhat

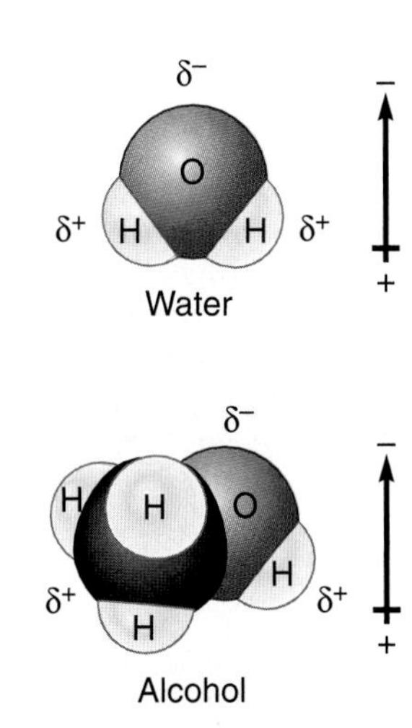

Figure 10.1 As we saw in Section 3.5, oxygen is more electronegative than hydrogen, and so oxygen pulls shared electrons toward itself. As a result, water molecules are polar. Oxygen is more electronegative than either hydrogen or carbon. As a result, alcohol molecules are polar.

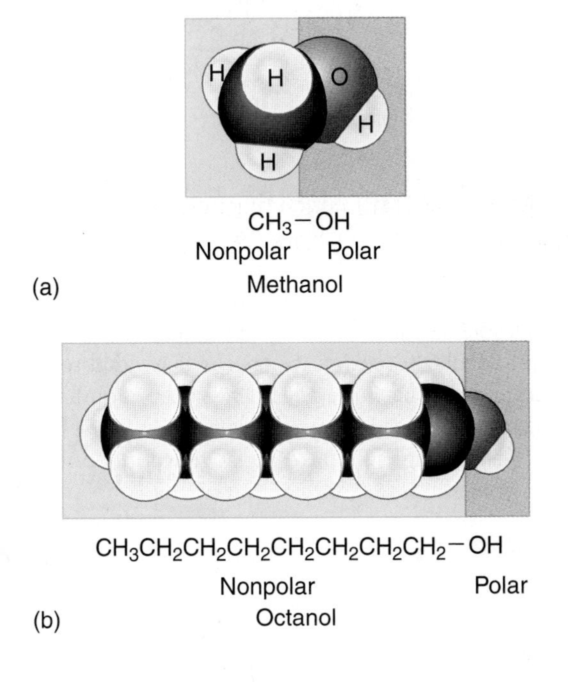

Figure 10.2 **(a)** The alkyl group of small alcohols is nonpolar, but it is too small to overcome the polarity of the hydroxyl group. Thus the molecule is polar, and small alcohols are water-soluble. **(b)** Larger alcohols have larger alkyl groups. The nonpolarity of these larger groups outweighs the polarity of the hydroxyl group. Large alcohols are therefore primarily nonpolar and they are water-insoluble.

Connections

Phenols as Antiseptics and Disinfectants

Open wounds and surgery were very dangerous in the past because of infections. In the 1860s, Joseph Lister began applying solutions of phenol—then called carbolic acid—to wounds and surgical cuts to remove the "bad air" that was thought to cause infections. (Of course the cause of infections was later shown to be bacteria.) With time, the use of phenolic solutions as antiseptics (used to kill or inhibit bacteria on living tissues) and as disinfectants (used to kill bacteria on inanimate objects) became common, and many patients' lives were saved as a result. Indeed, in 1897 a patient of Lister's, Queen Victoria, made him a baron.

Unfortunately, phenol has some dangerous and undesirable side effects. It is a strong irritant that can burn tissue, and it is absorbed by the body, where it produces several adverse effects. Phenol also has a quite disagreeable smell. Researchers began searching for both an understanding of how phenol worked, and for a replacement for phenol.

We now know that many phenols disrupt bacterial cell membranes and damage cell proteins. In addition, many substituted phenols are better antibacterial agents than phenol itself, and have much less adverse effects on human tissues. (They don't smell as bad either.) Today phenol is rarely used as an antiseptic or disinfectant. Hexylresorcinol is a phenol that is commonly used in dilute solutions as an antiseptic, and several phenols, including *o*-phenylphenol, are among the many agents used in modern disinfectants.

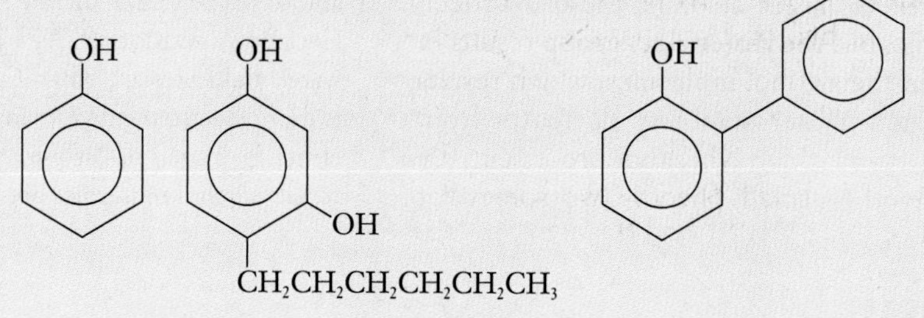

These disinfectants and insecticide contain phenol. *(Charles D. Winters)*

water-soluble, but larger phenols containing one or more alkyl groups or some other nonpolar groups are generally water-insoluble.

The physical state of any substance depends on the strength of the intermolecular interactions that exist in the compound. Hydrogen bonding occurs between alcohol molecules and phenol molecules, and because hydrogen bonds are relatively strong intermolecular forces, more energy is needed to separate these molecules than to separate nonpolar molecules held by London forces (Figure 10.3). (It may be useful for you to review intermolecular bonding, explained in Section 5.5.) The temperature must be higher before hydrogen-bonded molecules boil, and so the boiling points of alcohols are always higher than those of alkanes or alkenes of similar molecular weight (Figure 10.4). Larger alcohols and many phenols are solids at room temperature, because the attractions between the molecules involve both hydrogen-bonding and significant London forces between the large molecules.

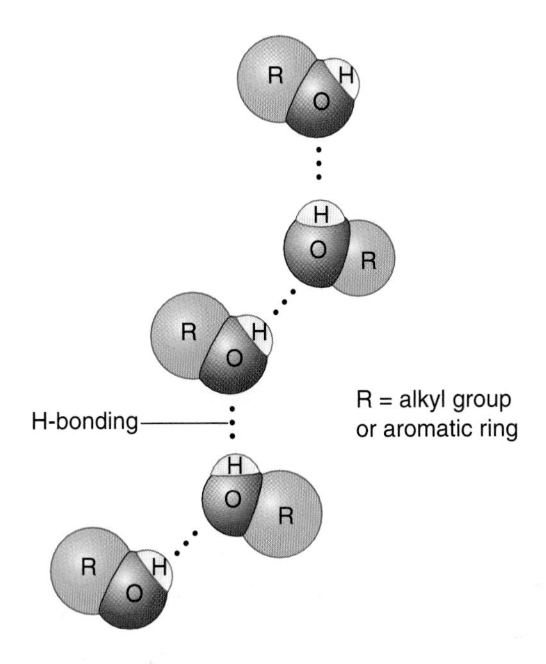

Figure 10.3 The hydroxyl group of alcohols and phenols allows hydrogen bonding between these molecules. These intermolecular forces are relatively strong. (In this figure, R- represents the rest of the molecule—that is, an alkyl group or an aromatic ring.)

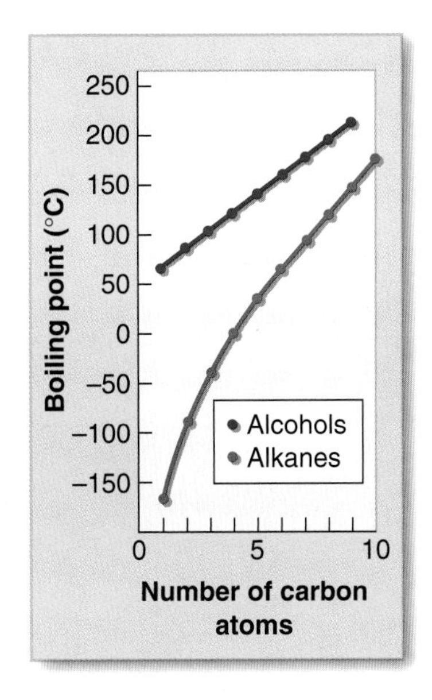

Figure 10.4 Because more energy is needed to overcome the hydrogen bonding in alcohols and phenols, these compounds boil at a higher temperature than nonpolar organic molecules of similar size.

Reactions of Alcohols and Phenols

The functional group in a compound determines the reactivity of that compound. Given this, it should be no surprise to find that the hydroxyl group is the key to understanding the reactions of alcohols and phenols.

Dehydration

An important reaction of alcohols is the reversal of a reaction of alkenes. By adding water, alkenes can be hydrated to form alcohols (Chapter 9), and through the loss of water, alcohols can be *dehydrated* to form alkenes:

```
  H  H  OH          H       H
  |  |  |     H+    |      /
H—C——C——C—H  ──→ H—C——C=C      +  H2O
  |  |  |           |  |   \
  H  H  H           H  H    H

  1-propanol          Propene
```

In this acid-catalyzed *elimination* reaction that often requires heat, the hydroxyl group of the alcohol and a hydrogen atom on an adjacent carbon atom are *eliminated* as water. In larger alcohols, there may be as many as three carbon atoms adjacent to the hydroxyl-bearing carbon, but they are not all equally likely to take part in the elimination reaction. As a general rule, the adjacent carbon atom bonded to the *fewest* hydrogen atoms donates the hydrogen atom:

```
  H  H  OH  H              H           H
 4|  3|  2|  1/     H+     |           |
H—C——C——C——C—H   ──→  H—C——C=C——C—H  +  H2O
  |  |  |   |              |  |  |  |
  H  H  H   H              H  H  H  H

    2-butanol                2-butene
```

The dehydration of an alcohol is an elimination reaction in which the eliminated molecule is water.

In this reaction, the hydroxyl group is on carbon atom 2, and so carbons 1 and 3 are the adjacent carbon atoms. Carbon atom 3 is bonded to two hydrogen atoms, and carbon atom 1 is bonded to three hydrogen atoms. The hydrogen given up in the elimination comes from carbon number 3 to yield the product shown.

Note that *phenols cannot be dehydrated.*

Example 10.5 Dehydration of Alcohols

Predict the product of this dehydration:

$$\mathrm{H{-}\underset{H}{\overset{H}{C}}{-}\underset{H}{\overset{OH}{C}}{-}\underset{H}{\overset{H}{C}}{-}\underset{H}{\overset{H}{C}}{-}\underset{H}{\overset{H}{C}}{-}H \xrightarrow{H^+} ? + H_2O}$$

SOLUTION

In dehydration reactions, the hydroxyl group of the alcohol is lost along with a hydrogen atom from an adjacent carbon atom. The hydrogen comes from the adjacent carbon atom bonded to the fewest hydrogen atoms, which is carbon 3 in this case:

$$\mathrm{H{-}\underset{H}{\overset{H}{C}}{-}\underset{H}{\overset{OH}{C}}{-}\underset{H}{\overset{H}{C}}{-}\underset{H}{\overset{H}{C}}{-}\underset{H}{\overset{H}{C}}{-}H}$$

Loss of the highlighted atoms yields 2-pentene:

$$\mathrm{H{-}\underset{H}{\overset{H}{C}}{-}\underset{H}{C}{=}\underset{H}{C}{-}\underset{H}{\overset{H}{C}}{-}\underset{H}{\overset{H}{C}}{-}H}$$

Self-Test

What is the product of this reaction:

$$\mathrm{CH_3{-}\underset{CH_3}{\overset{OH}{C}}{-}CH_2CH_3 \xrightarrow{H^+} ? + H_2O}$$

ANSWER

2-methyl-2-butene $\mathrm{CH_3{-}\underset{CH_3}{C}{=}CHCH_3}$

Oxidation

Some alcohols can be oxidized. Oxidation is formally defined as a loss of electrons (Chapter 7), but oxidation of alcohols can be recognized by the loss of two hydrogen atoms. During the oxidation, a hydrogen atom is lost from the hydroxyl group, and another is lost from the carbon atom that bears the hydroxyl group. The product of alcohol oxidation is either an aldehyde or a ketone (these compounds are discussed in Chapter 11), depending on the class of alcohol that is oxidized.

In general, $R{-}\overset{\overset{\displaystyle O}{\|}}{C}{-}H$ is an aldehyde and $R{-}\overset{\overset{\displaystyle O}{\|}}{C}{-}R'$ is a ketone, where R and R′ are alkyl groups or aromatic rings.

Oxidation of a primary alcohol yields an aldehyde:

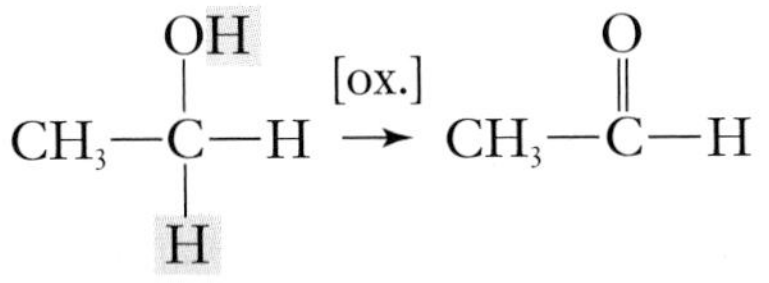

$$\underset{\text{1° alcohol}}{CH_3{-}\underset{\underset{\displaystyle H}{|}}{\overset{\overset{\displaystyle OH}{|}}{C}}{-}H} \xrightarrow{[ox.]} \underset{\text{Aldehyde}}{CH_3{-}\overset{\overset{\displaystyle O}{\|}}{C}{-}H}$$

Although this equation is correct, aldehydes are very easily oxidized to carboxylic acids, RCOOH, and oxidation of a primary alcohol often does not stop at the aldehyde; thus the final product may be the carboxylic acid (Chapter 11).

When a secondary alcohol is oxidized, the product is a ketone:

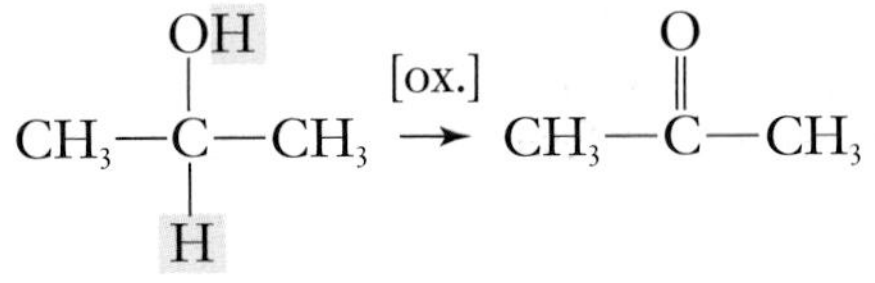

$$\underset{\text{2° alcohol}}{CH_3{-}\underset{\underset{\displaystyle H}{|}}{\overset{\overset{\displaystyle OH}{|}}{C}}{-}CH_3} \xrightarrow{[ox.]} \underset{\text{Ketone}}{CH_3{-}\overset{\overset{\displaystyle O}{\|}}{C}{-}CH_3}$$

Tertiary alcohols and phenols do not react when exposed to conditions that would oxidize a primary or secondary alcohol.

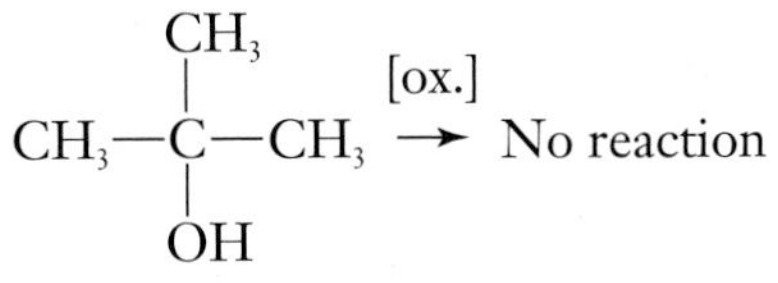

$$\underset{\text{3° alcohol}}{CH_3{-}\underset{\underset{\displaystyle OH}{|}}{\overset{\overset{\displaystyle CH_3}{|}}{C}}{-}CH_3} \xrightarrow{[ox.]} \text{No reaction}$$

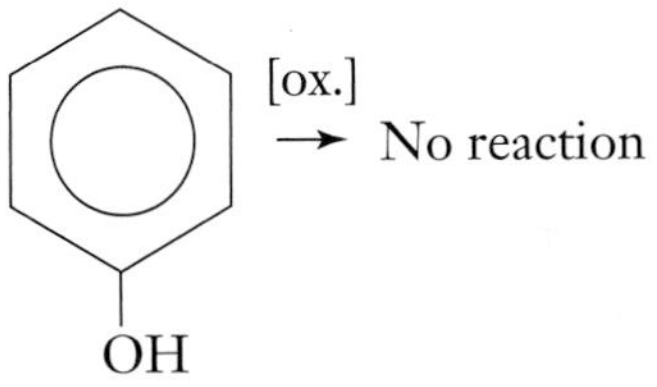

$$\underset{\text{Phenol}}{C_6H_5{-}OH} \xrightarrow{[ox.]} \text{No reaction}$$

Example 10.6 Oxidation of Alcohols

What is the structure of the product of these reactions?

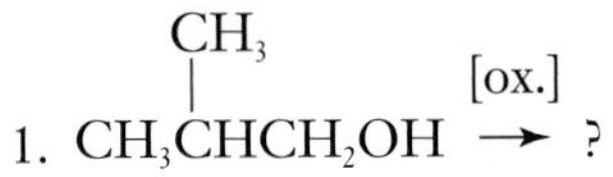

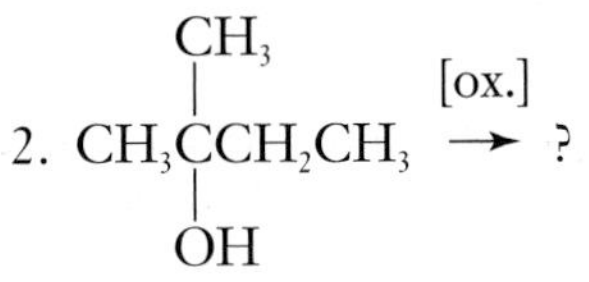

1. $CH_3\overset{\overset{\displaystyle CH_3}{|}}{C}HCH_2OH \xrightarrow{[ox.]} ?$

2. $CH_3\underset{\underset{\displaystyle OH}{|}}{\overset{\overset{\displaystyle CH_3}{|}}{C}}CH_2CH_3 \xrightarrow{[ox.]} ?$

SOLUTION

1. This is a primary alcohol because the carbon bearing the hydroxyl group is bonded to only one carbon atom. Primary alcohols yield aldehydes, and so the product is

$$CH_3-\overset{\overset{\displaystyle CH_3}{|}}{C}H-\overset{\overset{\displaystyle O}{\|}}{C}-H$$

2. This is a tertiary alcohol because it has three carbon atoms attached to the carbon bearing the hydroxyl group. Tertiary alcohols are not easily oxidized, and so no reaction will occur.

Self-Test

What alcohol, when oxidized, yields

$$CH_3-\overset{\overset{\displaystyle CH_3}{|}}{C}H-\overset{\overset{\displaystyle O}{\|}}{C}-CH_3 \quad ?$$

ANSWER

$$CH_3-\overset{\overset{\displaystyle CH_3}{|}}{C}H-\overset{\overset{\displaystyle OH}{|}}{C}H-CH_3 \quad \text{(3-methyl-2-butanol)}$$

Sterno® is a canned solid fuel that contains methanol and an ingredient that makes the methanol a solid gel. *(Charles D. Winters)*

Acid–Base Reactions

Although alcohols are neither acidic nor basic, phenols are weakly acidic. When a phenol is mixed with sodium hydroxide, the hydroxyl group of the phenol loses its hydrogen atom, yielding a salt containing sodium cation and phenoxide anion:

$$C_6H_5OH + NaOH(aq) \longrightarrow C_6H_5O^-Na^+ + H_2O$$

Phenol — Sodium phenoxide

Combustion

Alcohols and phenols burn. For this reason, care must be taken to avoid fires when small, volatile alcohols are used as solvents. Methanol is used as a fuel in some racing cars and in the solid-fuel product sold as Sterno. Most fuels contain hydrocarbons, but oxygenated fuels also contain some molecules that have one or more oxygen atoms. Oxygenated fuels tend to burn more cleanly and reduce car emissions, thus ethanol is often added to gasoline to reduce emissions.

In general, an ester is

$$R-\overset{\overset{\displaystyle O}{\|}}{C}-O-R'$$

, where R and R′ are alkyl groups or aromatic rings.

Esterification

Another important reaction of alcohols and phenols is the formation of esters, compounds important in both organic and biological chemistry. We postpone discussion of ester formation until Chapter 11.

10.2 Ethers

An **ether** is an organic compound that contains an oxygen atom bonded to two carbon atoms. We are looking at an ether whenever two alkyl groups or rings are connected by an oxygen atom:

R—O—R   R—O—ring   ring—O—ring

CH_3—O—CH_3 CH_3CH_2—O—(cyclopentyl)   (phenyl)—O—(phenyl)

Ethers are not as common as alcohols, but they have a variety of uses. Some ethers are good solvents for organic compounds, and some have general anesthetic properties. Many of these compounds are volatile and flammable, and so they must be used with caution.

Naming Ethers

Although there are IUPAC names for ethers, we will not introduce this naming to you because common names are still widely used. Many common names are derived from a systematic but non-IUPAC naming procedure. These common names list (in alphabetical order) each alkyl group or ring that is present, followed by the word *ether:*

Connections

Ether as an Anesthetic

Can you imagine having a tooth pulled or undergoing surgery without an anesthetic? Surgery without an anesthetic is both highly traumatic and dangerous. Local anesthetics reduce or eliminate pain where they are administered, and general anesthetics cause loss of sensation and loss of consciousness. In 1842 diethyl ether (ether) became the first effective general anesthetic, and its use continued until the middle of the twentieth century. Ether was replaced because it is both explosively flammable and produces undesirable side effects in many patients.

Today several inhalation general anesthetics have replaced ether. These gases or volatile liquids are nonflammable and generally yield less serious side effects. The agents are typically halogenated ethers such as enflurane, $CF_2H—O—CF_2CHFCl$, or halogenated hydrocarbons such as halothane $CHBrClCF_3$. Anesthesiologists often use combinations of anesthetics, sedatives, and relaxants to obtain a safe, desired effect in their patients.

The dentist W. T. G. Morton discovered the anesthetic properties of ether. This painting by Robert Hinckley illustrates the first use of ether as a general anesthetic in surgery. *(Boston Medical Library in the Francis A. Countway Library of Medicine)*

$CH_3—O—CH_2—CH_2—CH_3$

Methyl propyl ether

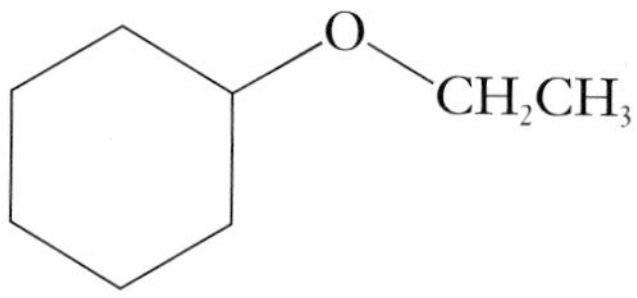

Cyclohexyl ethyl ether

If both groups of the ether are the same, add the prefix *di-* to the group name:

$CH_3—CH_2—O—CH_2—CH_3$

Diethyl ether

If a benzene ring is attached to the oxygen atom, use the word *phenyl:*

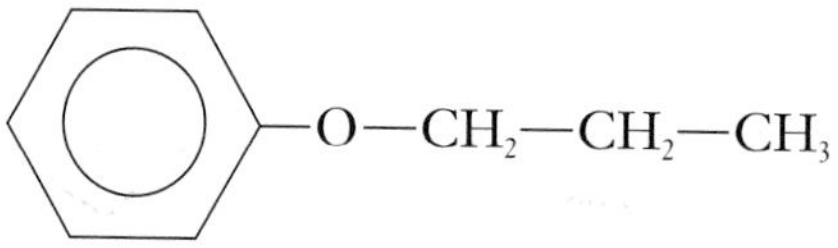

Phenyl propyl ether

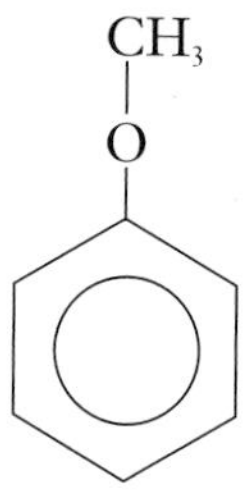

Anisole
(methyl phenyl ether)

Note that methyl phenyl ether is normally called anisole. Some ether names provide little or no structural information; anisole and the cyclic ether tetrahydrofuran are examples:

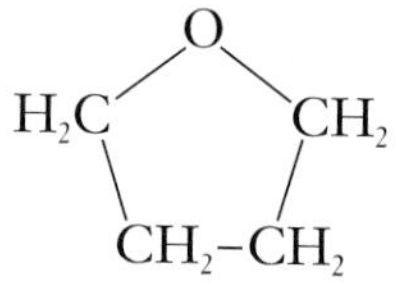

Tetrahydrofuran

Several ethers that are used as general anesthetics are identified by common names: divinyl ether (vinethene), $CH_2{=}CH—O—CH{=}CH_2$, and ethrane (enflurane), $F_2HC—O—CF_2CHFCl$, are examples.

Example 10.7 Names and Structures of Ethers

1. What is the structure of dimethyl ether?
2. What is the name of the ether with the structure $CH_3CH_2—O—CH_2CH_2CH_3$?

SOLUTIONS

1. There are two methyl groups, one on either side of an oxygen atom. The structure is $CH_3—O—CH_3$.
2. There is an ethyl group and a propyl group. List the groups in alphabetical order, then add the word ether: ethyl propyl ether.

Self-Test

1. What is the name of $CH_3—\underset{\displaystyle CH_3}{\underset{|}{CH}}—O—CH_3$

2. What is the structure of ethyl phenyl ether?

ANSWERS

1. Isopropyl methyl ether

2. $CH_3—CH_2—O—$(benzene ring)

Properties and Reactions of Ethers

Properties

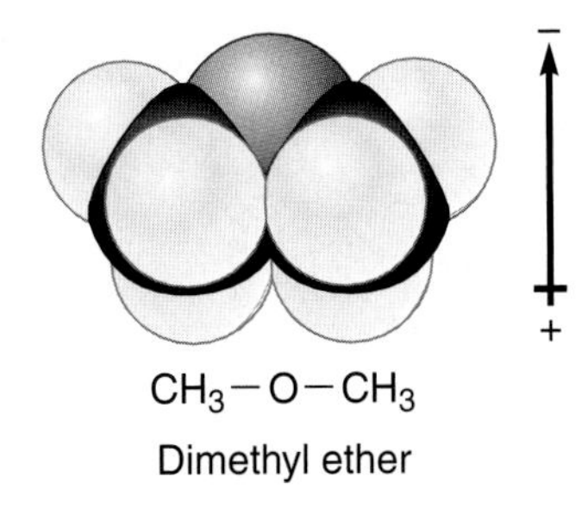

Oxygen makes an ether molecule slightly polar.

Because oxygen is more electronegative than carbon, there is a weak dipole where the oxygen atom is located and weak dipole–dipole interactions occur between ether molecules. Because there are no hydrogen atoms bound to the oxygen atom, hydrogen bonding cannot occur between ether molecules. As a result, the boiling points of ethers are much lower than alcohols of similar size, and in fact the boiling points of ethers are similar to those of hydrocarbons of similar size (Figure 10.5).

Because oxygen is present, ether molecules can form hydrogen bonds to some other molecules such as water. Thus small ethers are water-soluble because they

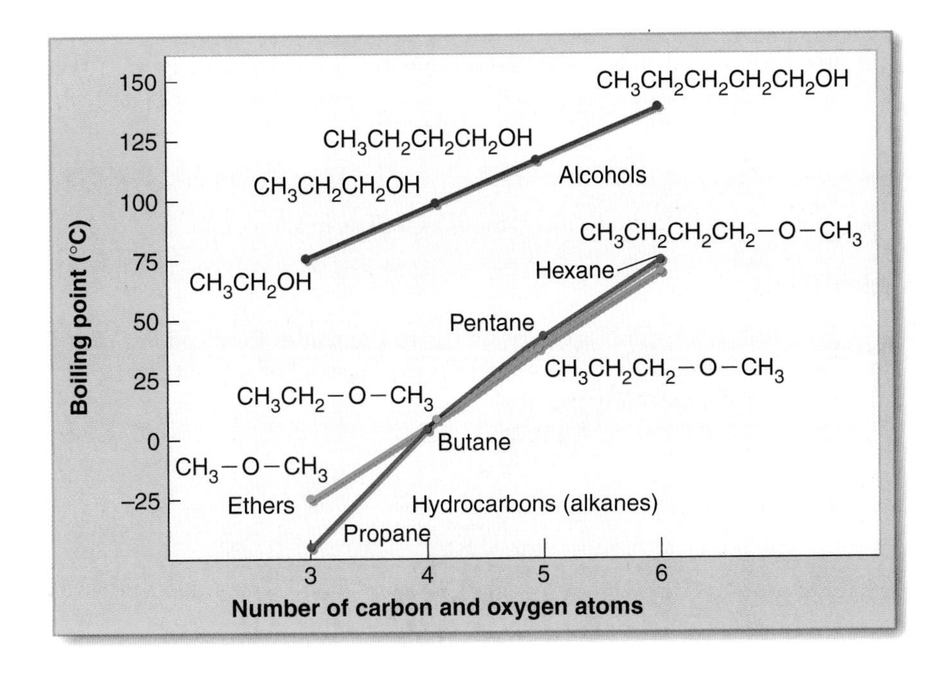

Figure 10.5 Ethers boil at temperatures near those of hydrocarbons of similar size, and they boil at much lower temperatures than alcohols of similar size.

hydrogen bond with water. Larger ethers are water-insoluble because of the nonpolarity of their larger alkyl groups or rings. Diethyl ether is an excellent solvent but must be used carefully because it is highly flammable. It dissolves many organic compounds that range from nonpolar to fairly polar.

Reactions

Ethers, like alkanes, undergo a limited number of chemical reactions, one of which is combustion. The highly volatile diethyl ether, when used carefully, can be used to start engines in cold weather. Methyl tert-butyl ether (MTBE) is mixed with hydrocarbons to make oxygenated gasoline blends that burn more cleanly than those that contain only hydrocarbons. However, the use of MTBE may be restricted or eliminated because it has been found repeatedly in ground water.

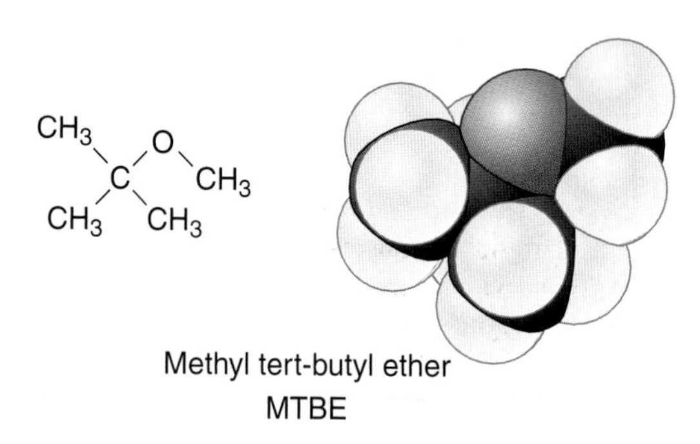

Methyl tert-butyl ether
MTBE

Methyl tert-butyl ether (MTBE) is added to gasoline to decrease some emissions from cars.

COOH

HS SH
Asparagus

CH2SH
O
Coffee

CH3 S S CH3
Onion

S—S
S
S S
Mushroom

CH2 S S CH2
Garlic

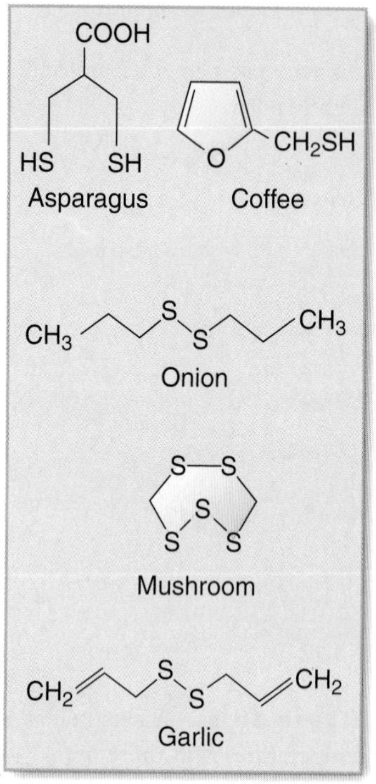

Thiols and disulfides contribute to the distinctive smell of several common foods and beverages. *(Charles D. Winters)*

10.3 Thiols

Sulfur-containing organic compounds are not particularly common, but some of them have important roles in biochemistry. You have learned that the —OH group is the hydroxyl group and that it is found in alcohols. The —SH group is called the *sulfhydryl group*, and it is the functional group of the **thiols.** The sulfhydryl group can be attached to an alkyl group or to an aromatic or nonaromatic ring:

R—SH ring—SH

Thiols

These compounds are found in many foods as well as in skunk odor and proteins.

Naming Thiols

The IUPAC names for thiols are very similar to the names for alcohols. The two main differences are that (1) the suffix *-thiol* is used instead of *-ol*, and (2) the entire parent name is used, including the final *-e:*

$CH_3—SH$ Methanethiol

$CH_3—CH(CH_3)—CH_2—CH_2—CH_2—SH$ 4-methyl-1-pentanethiol

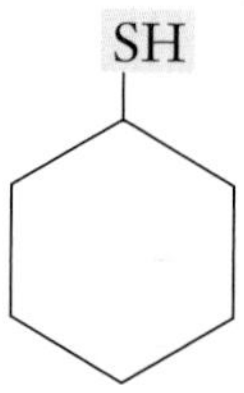

Cyclohexane ring with SH: Cyclohexanethiol

$CH_3CH_2CH_2SH$ 1-propanethiol

CH_3CHCH_3 with SH on C2 — 2-propanethiol

An older nomenclature identifies these compounds as mercaptans. In this system, the alkyl group is named first, and then the word *mercaptan* is added. Methyl mercaptan is thus a synonym for methanethiol.

Example 10.8 Names and Structures of Thiols

What is the IUPAC name of the thiol

$$CH_3-CH_2-\underset{\displaystyle |}{\overset{\displaystyle CH_2CH_3}{\overset{|}{C}H}}-CH_2-SH?$$

SOLUTION

The longest chain that contains the sulfhydryl group is four carbon atoms. The parent name is therefore butanethiol. The sulfhydryl group is on carbon 1, and this is designated in this way: 1-butanethiol. The substituent is an ethyl group on carbon 2. The full name is therefore 2-ethyl-1-butanethiol.

Self-Test

What is the structure of 2-hexanethiol?

ANSWER

$$CH_3-\underset{\displaystyle SH}{\underset{|}{C}H}-CH_2-CH_2-CH_2-CH_3$$

Properties and Reactions of Thiols

Thiols are less polar than alcohols because sulfur is less electronegative than oxygen. Because thiols are relatively nonpolar, they are generally soluble in nonpolar solvents and insoluble in polar solvents. Even the smallest ones have little solubility in water. The boiling points of thiols are lower than those of alcohols of similar size, and the smallest thiols are gases or volatile liquids (Figure 10.6).

The only reactions of thiols that we study in this chapter involve oxidation and reduction. Under oxidative conditions, two thiols combine to form a *disulfide:*

$$2CH_3-SH \xrightarrow{[ox.]} CH_3-S-S-CH_3$$

Thiols → Disulfide

The bond between the sulfur atoms is called a *disulfide bond* or disulfide bridge. It can be broken by reduction to the thiols:

$$CH_3-S-S-CH_3 \xrightarrow{[red.]} 2CH_3-SH$$

Disulfide → Thiols

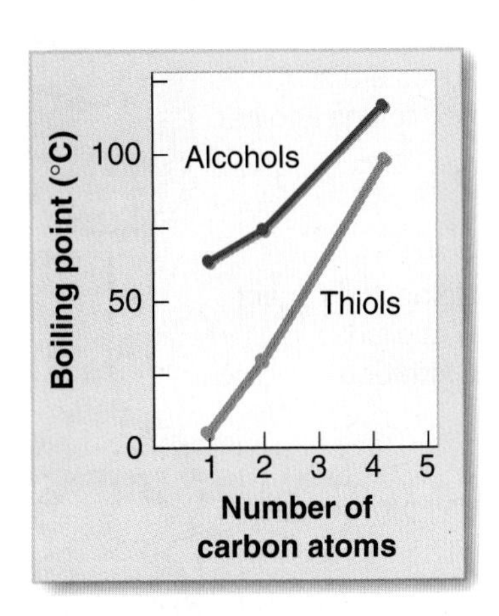

Figure 10.6 Thiols boil at lower temperatures than alcohols of similar size.

Hair protein whose shape is fixed by disulfide bonds

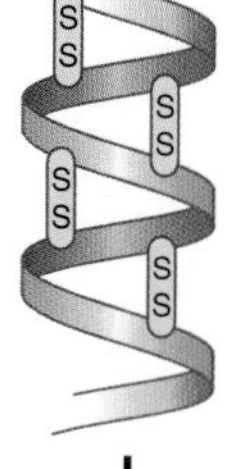

Hair is rolled into rods and hair-releasing agent is applied: **Reduction**

The protein now is not fixed by the disulfide bonds

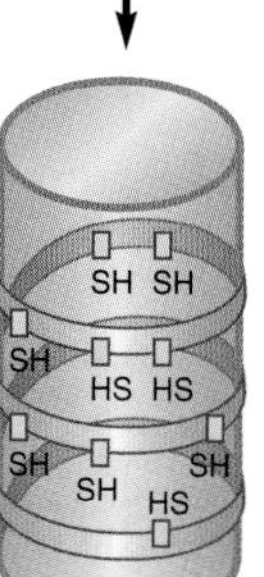

Protein is flexible and different S—H bonds are now next to one another

Hair-setting agent is applied: **Oxidation**

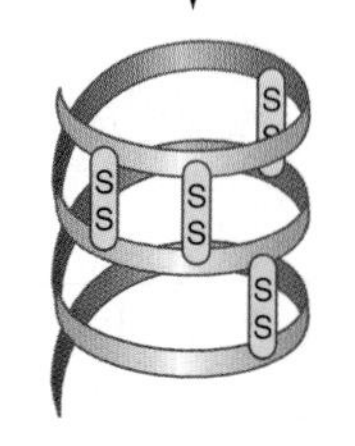

Hair protein now has a new shape

(b)

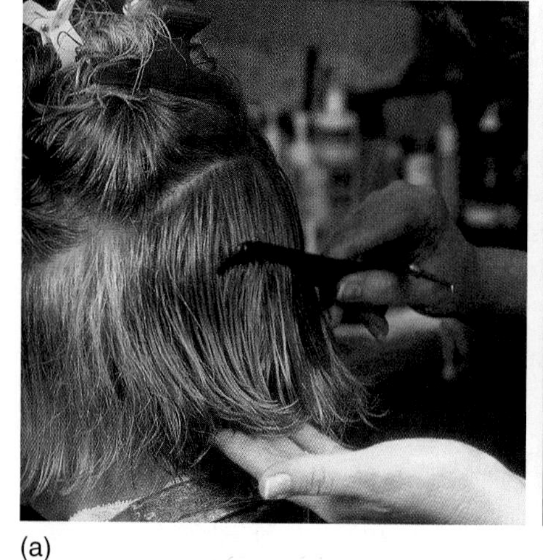

(a)

Figure 10.7 **(a)** A "permanent" wave alters the shape of hair through reduction and oxidation of disulfides and thiols. **(b)** The specific steps of a "perm". *(Charles D. Winters)*

As we shall see, disulfides and thiols have a special role in protein chemistry. Many proteins possess sulfhydryl groups. In some proteins, these sulfhydryl groups form disulfides that help stabilize the protein. Proteins in hair have disulfide bonds, and the amount of "curl" in hair can be changed by using this thiol chemistry (Figure 10.7).

10.4 Amines

Amines are organic compounds that contain a nitrogen atom bonded to a carbon atom. It is helpful to think of amine molecules forming when a carbon chain or ring replaces a hydrogen atom of ammonia:

$$NH_3 \qquad R{-}NH_2 \qquad \text{ring}{-}NH_2$$

Ammonia Amines

In amines, the nitrogen atom and its hydrogens are called the **amino group.**

There are several classes of amines, depending on the number of alkyl groups or rings that substitute for the hydrogens in ammonia (Figure 10.8). If a single group is attached to the nitrogen atom, the compound is a **primary amine.** If two groups are attached, it is a **secondary amine,** and if three are attached it is a **tertiary amine.** A related group of compounds, **quaternary ammonium salts,** have four groups attached to the nitrogen atom. Some quaternary ammonium salts have important roles in the body. In physiology, you have learned or will learn that acetylcholine is a neurotransmitter that contains the quaternary ammonium group. Can you pick out this group in acetylcholine?

$$CH_3\overset{\overset{\large O}{\|}}{C}{-}O{-}CH_2CH_2N(CH_3)_3{}^+$$

Acetylcholine
(Quaternary ammonium salt)

Among the many important amines found in nature are the *alkaloids*, the nitrogenous bases found in plants. Caffeine and morphine are two common examples of alkaloids:

Caffeine

Morphine

Caffeine and morphine are also examples of *heterocyclic* compounds. This means they contain one or more rings that contain a noncarbon atom.

Naming Amines

When amines are given IUPAC names, the word *amino* is used as a prefix and is combined with the hydrocarbon name. Aminomethane, for instance, has this con-

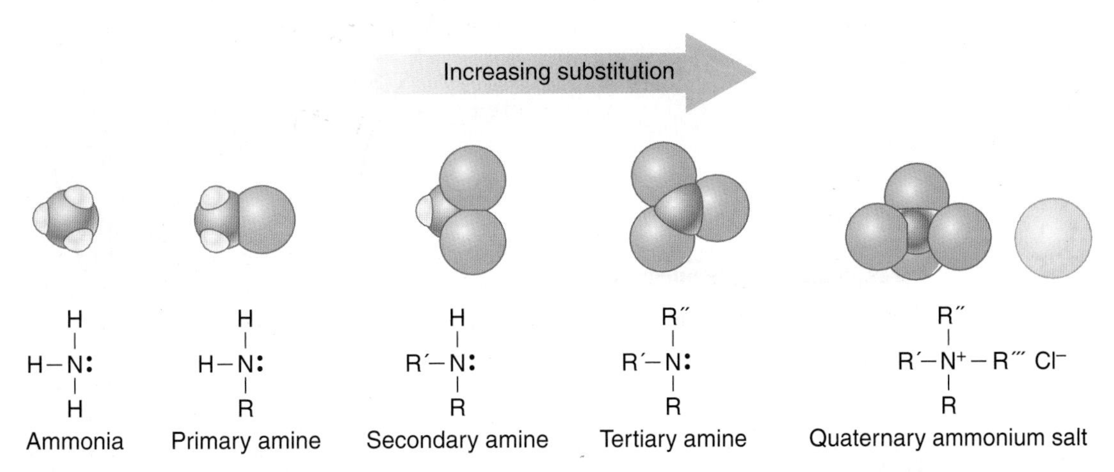

Figure 10.8 Amines are classified according to the number of groups attached to the nitrogen atom. A quaternary ammonium salt has four groups attached to a nitrogen atom.

densed structure CH_3NH_2, and 2-aminobutane is

$$\begin{array}{c} NH_2 \\ | \\ CH_3CHCH_2CH_3 \end{array}$$

2-aminobutane
(Primary amine)

Often smaller amines are named by a common naming system. These common names combine the alkyl group name of the amine with the suffix *-amine*. Methylamine is the common name of aminomethane. If more than one alkyl group is present, each is named in alphabetical order. If two or three copies of an alkyl group are present, the prefixes *di-* and *tri-* are used.

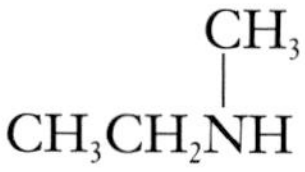

$$CH_3CH_2\overset{CH_3}{\overset{|}{N}}H \qquad CH_3\overset{CH_3}{\overset{|}{N}}H$$

Ethylmethylamine (Secondary amine) — Dimethylamine (Secondary amine)

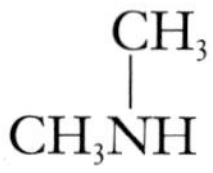

In addition, the common names of some amines do not follow this pattern (caffeine and morphine are examples), and they provide no structural information.

The names of aromatic amines are based on the naming rules for aromatic compounds you learned in Chapter 9. For naming purposes, aniline is the parent aromatic amine, and it is used to name other aromatic amines.

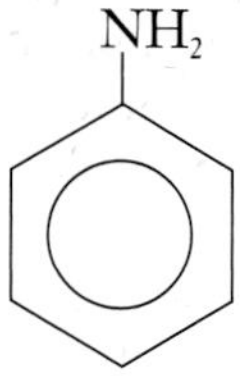

[Structures: benzene ring with NH_2; benzene ring with NH_2 and Br on adjacent carbon]

Aniline 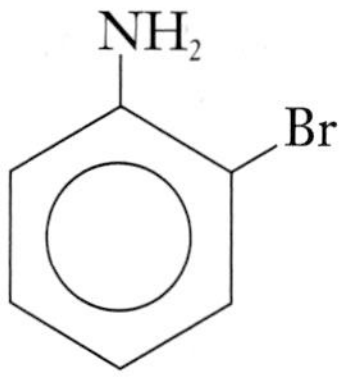 — 2-bromoaniline (*o*-bromoaniline)

For secondary and tertiary aromatic amines, a capital *N*- is included to show that a group is bonded to the nitrogen atom of the amine:

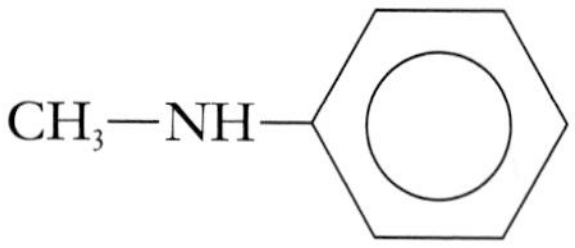

$CH_3—NH—C_6H_5$

N-methylaniline

Example 10.9 Naming Amines

Name these compounds:

1. $NH_2CH_2CH_2CH_3$
2. $CH_3CH_2NHCH_2CH_3$

SOLUTION

1. This compound has nitrogen bonded to a carbon atom, and so it is an amine. The alkyl group of this amine is the propyl group. The name is formed by combining the alkyl group name and the suffix *-amine* to form propylamine.

2. Again we recognize this as an amine because a nitrogen atom is bound to a carbon atom. There are two ethyl groups attached to the nitrogen, and we designate this with the prefix *di-* before the alkyl name: diethylamine.

Self-Test

1. Name the tertiary amine CH_3NCH_3 (with CH_3 bonded to N, drawn below it).

2. What is the structure of *N*-propylaniline?

ANSWER

1. Trimethylamine

2. 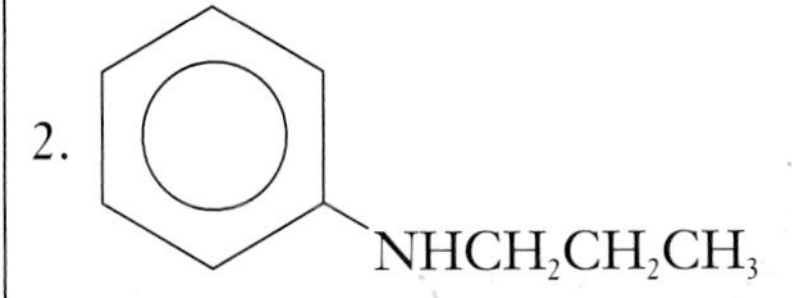

Properties of Amines

Amines are polar compounds because nitrogen is significantly more electronegative than carbon. Nitrogen is also one of the elements that can form hydrogen bonds. For these reasons, smaller amines are water-soluble. Larger amines are water-insoluble, however, again because the nonpolarity of the larger alkyl group or groups becomes more important than the polarity of the amino group. Amines are generally soluble in both polar and nonpolar organic solvents.

The boiling points of amines are higher than those of similar alkanes but lower than those of similar alcohols. The hydrogen bonds between amine molecules are weaker than those between alcohol molecules because nitrogen is less electronegative than oxygen. As a result, the dipole in amines is smaller than the dipole in alcohols and the attraction between molecules is smaller (Figure 10.9).

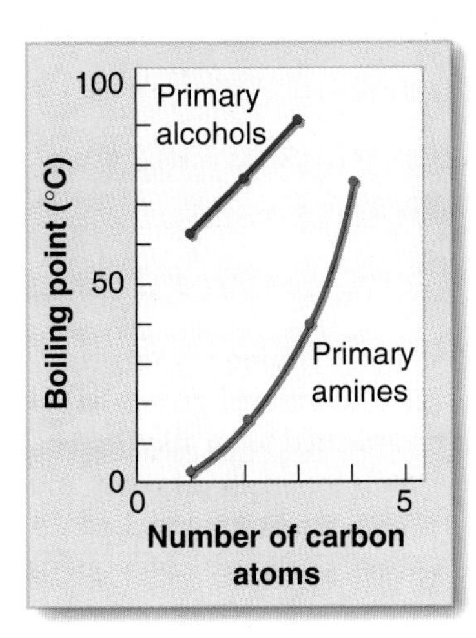

Figure 10.9 Amines boil at lower temperatures than alcohols because hydrogen bonding between molecules is weaker in amines than in alcohols.

Reactions of Amines

Acid–Base Reactions

Amines are similar to ammonia in acid–base properties. Ammonia is basic. It can accept a hydrogen ion because it has an unshared pair of electrons. Amines are also bases for the same reason. Like all other bases, amines react with acids:

$$\underset{\substack{\text{Amine}\\\text{(Base)}}}{R-\overset{\overset{\large H}{|}}{\underset{\underset{\large H}{|}}{N}}:} + \underset{\text{Acid}}{HCl} \rightarrow \underset{\text{Ammonium salt}}{R-\overset{\overset{\large H}{|}}{\underset{\underset{\large H}{|}}{N}}-H^{+}Cl^{-}}$$

The product of this reaction is called an **ammonium salt.** The names of ammonium salts come from the name of the amine. The suffix *-ammonium* replaces the suffix-*amine*, and the anion name follows. Methylamine forms methylammonium chloride when it reacts with hydrochloric acid:

$$\underset{\text{Methylamine}}{CH_3NH_2} + HCl \rightarrow \underset{\text{Methylammonium chloride}}{CH_3NH_3{}^{+}Cl^{-}}$$

Ammonium salts are weak acids; thus they can donate a hydrogen ion. You will see this important reaction when you study protein chemistry (Chapter 14).

Formation of Amides

Amides are another class of organic compounds that contain a nitrogen atom. However, these compounds are not classified as amines because the nitrogen atom is adjacent to a carbon–oxygen double bond (this is a functional group called the carbonyl group), which gives these compounds a different set of properties:

$$\underset{\text{Amide group}}{-\overset{\overset{\large O}{\|}}{C}NH_2}$$

Amides are formed from amines and include such important compounds as proteins and nylon. Amides are described in Chapter 11.

10.5 Organohalogens—Alkyl Halides and Aryl Halides

As you learned in Chapter 8, organohalogens are halogen-containing organic compounds that have a halogen atom—F, Cl, Br, or I—as their functional group. **Alkyl halides** are organohalogens that have the halogen atom attached to an alkyl group. They are often represented as R—X. The R, as usual, stands for an alkyl group, and the X stands for a halogen atom. **Aryl halides** have a halogen atom attached directly to an aromatic ring. The nomenclature and synthesis of these compounds were discussed in Chapter 9.

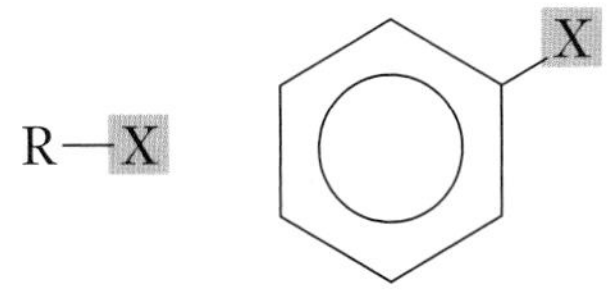

R—X

Alkyl halide

Aryl halide

CH_3CH_2—Br

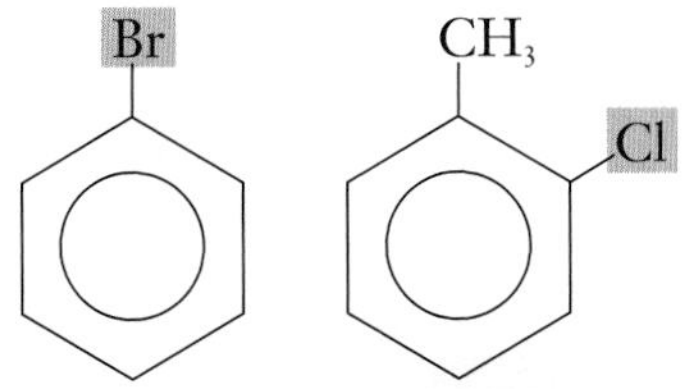

Bromoethane or ethyl bromide
(Alkyl halide)

Bromobenzene
(Aryl halide)

2-chlorotoluene
(Aryl halide)

Naming Alkyl Halides

The IUPAC rules for the names of alkyl halides are similar to the rules for alkanes. The halogen atom is considered a substitute and is represented by the prefixes *fluoro-*, *chloro-*, *bromo-*, or *iodo-*. The compound CH_3Cl is called chloromethane, and 2-bromopropane is

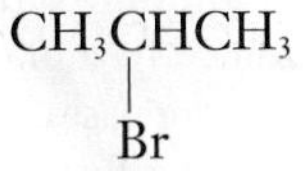

$$CH_3\underset{\displaystyle \mathrm{Br}}{\underset{|}{C}}HCH_3$$

In addition, these compounds have common names that generally consist of the alkyl name (Chapter 8) followed by the ion name for the halogen. The common name for chloromethane is methyl chloride, and 2-bromopropane is also known as isopropyl bromide. Three common alkyl halides are dichloromethane (methylene chloride), CH_2Cl_2; trichloromethane (chloroform), $CHCl_3$; and tetrachloromethane (carbon tetrachloride), CCl_4. The common names of these compounds do not obey the general rule given previously.

Example 10.10 Names and Structures of Alkyl Halides

Draw the structure of the alkyl halide 2,4-dichloropentane.

SOLUTION

There are five carbon atoms in the parent chain. There is a chlorine atom attached to the second carbon atom and another one to the fourth carbon atom. The structure is

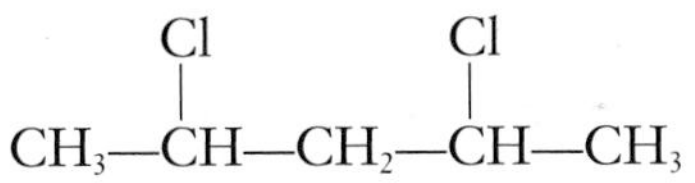

$$CH_3\text{—}\overset{\displaystyle \mathrm{Cl}}{\overset{|}{C}}H\text{—}CH_2\text{—}\overset{\displaystyle \mathrm{Cl}}{\overset{|}{C}}H\text{—}CH_3$$

Self-Test

What is the name of CBr_4?

ANSWER

Tetrabromomethane (carbon tetrabromide)

Properties and Reactions of Organohalogens

The organohalogens are nonpolar compounds. Because they are nonpolar, they have relatively low boiling points for their molar mass. They are generally insoluble in water and soluble in organic solvents. Several organohalogens are very good nonpolar solvents, but they are used less as solvents now than in the past. This is because many organohalogens are toxic.

Organohalogens are used in some reactions that involve the synthesis (building) of organic compounds, but they are relatively uncommon in living organisms. Because of this, we do not discuss any specific reactions.

CHAPTER SUMMARY

In an organic compound, any atom that is not a carbon or hydrogen is a **hetero-atom.** The **hydroxyl group** contains an oxygen atom and a hydrogen atom. An **alcohol** has a hydroxyl group attached to a carbon atom in an alkyl group, and a **phenol** has a hydroxyl group attached to a carbon atom of an aromatic ring. Alcohols and phenols are recognized by the *-ol* suffix in their names. Primary, secondary, and tertiary alcohols have one, two, or three carbon atoms bonded to the hydroxyl-bearing carbon, respectively. The hydroxyl group is polar, thus small alcohols are polar compounds, but as the size of the alkyl group or ring increases, the polarity of the molecule decreases. Alcohols participate in several common reactions, including dehydration, oxidation, combustion, and esterification. An **ether** molecule has an oxygen atom single-bonded to two carbon atoms. Common ether names end with the word *ether.* Ethers are relatively nonpolar compounds. The **sulfhydryl group** contains a sulfur atom and a hydrogen atom, and is the functional group of the relatively nonpolar compounds known as thiols. **Thiols** are recognized by either the suffix *-thiol* in their name or the word *mercaptan* in the name. The **amino group** is a nitrogen atom single-bonded to a carbon atom, and is the functional group of amines. The words *amine* or *amino-* in a name identifies the compound as an **amine. Primary, secondary,** and **tertiary amines** have one, two, or three carbon atoms single-bonded to the nitrogen atom, respectively. A **quaternary ammonium salt** has four carbon atoms bound to the nitrogen atom. The amino group is polar, thus small amines are polar. Amines are bases, and when they react with acids in reactions, the product formed is an **ammonium salt.** Organohalogens are organic compounds containing one or more halogen atoms. **Alkyl halides** have the halogen atom attached to a carbon atom in an alkyl group, and in **aryl halides,** the halogen is attached to a carbon atom in an aromatic ring. Alkyl halides are named as substituted alkanes, or by the alkyl group and halogen they contain. Organohalogens are nonpolar compounds.

SUMMARY OF REACTIONS

Alcohols and Phenols

Dehydration of Alcohols

$$\text{alcohol} \xrightarrow{H^+} \text{Alkene} + \text{water}$$

$$CH_3CH_2OH \xrightarrow{H^+} CH_2{=}CH_2 + H_2O$$

Oxidation of Alcohols

$$\text{Primary alcohol} \xrightarrow{[ox.]} \text{aldehyde (which oxidizes to a carboxylic acid)}$$

$$CH_3CH_2OH \xrightarrow{[ox.]} CH_3CHO$$

$$\text{Secondary alcohol} \xrightarrow{[ox.]} \text{ketone}$$

$$CH_3CH(OH)CH_3 \xrightarrow{[ox.]} CH_3C(=O)CH_3$$

Acidity of Phenols

$$\text{Phenol} + \text{NaOH(aq)} \longrightarrow \text{sodium phenoxide} + \text{water}$$

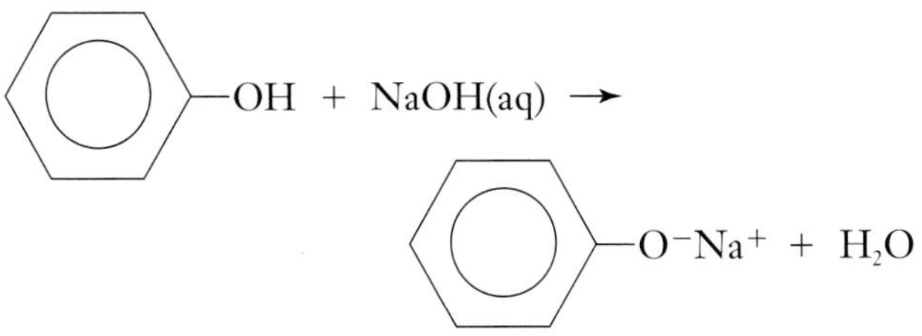

$C_6H_5{-}OH + NaOH(aq) \rightarrow C_6H_5{-}O^-Na^+ + H_2O$

Combustion

Alcohol + oxygen → carbon dioxide + water

$2CH_3OH + 3O_2 \rightarrow 2CO_2 + 4H_2O$

Thiols

Oxidation

$$2 \text{ thiols} \xrightarrow{[ox.]} \text{disulfide}$$

$$2CH_3SH \xrightarrow{[ox.]} CH_3{-}S{-}S{-}CH_3$$

Reduction

$$\text{disulfide} \xrightarrow{[red.]} 2 \text{ thiols}$$

$$CH_3CH_2{-}S{-}S{-}CH_2CH_3 \xrightarrow{[red.]} 2CH_3CH_2SH$$

Amines

Acid–Base Reactions

amine + acid → ammonium salt

$CH_3CH_2NH_2 + HCl \rightarrow CH_3CH_2NH_3^+Cl^-$

KEY TERMS

Introduction
hetero-atom

Section 10.1
alcohol
hydroxyl group
phenol

Section 10.2
ether

Section 10.3
sulfhydryl group
thiol

Section 10.4
amine
amino group
ammonium salt
primary amine
quaternary ammonium salt
secondary amine
tertiary amine

Section 10.5
alkyl halide
aryl halide

EXERCISES

Introduction

1. What distinguishes all other classes of organic compounds from the hydrocarbons?

Alcohols and Phenols (Section 10.1)

2. Identify the functional group that distinguishes the alcohols and phenols from all other classes of organic compounds.

3. What distinguishes a phenol from an alcohol? Use a sketch of any phenol and any alcohol to support your answer.

4. Draw the structure of these alcohols. (Hint: You may want to look at Section 10.5 before naming the alcohol that contains a halogen atom.)
(a) 1-butanol
(b) 3-methyl-2-heptanol
(c) Isopropyl alcohol
(d) Ethylene glycol
(e) 2-chlorocyclopentanol

5. Draw the structures of these alcohols.
(a) 4-methylcyclohexanol
(b) 2,4-dimethyl-3-octanol
(c) 1,3-propanediol
(d) 2-methyl-2-butanol
(e) Hexyl alcohol

6. Name each of these alcohols. (Hint: You may want to look at Section 10.5 before naming the alcohol that contains a halogen atom.)

(a)

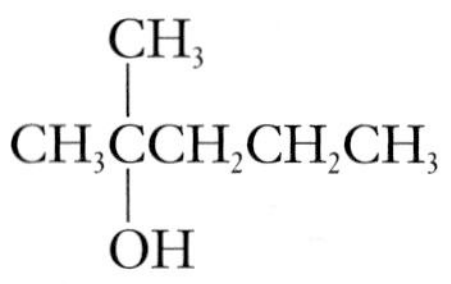

CH_3
$CH_3CCH_2CH_2CH_3$
OH

(b)

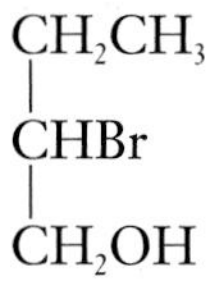

CH_2CH_3
$CHBr$
CH_2OH

(c)

OH
(cyclopentane ring) — CH_3

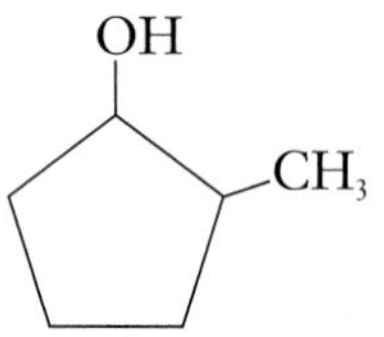

(d) $HOCH_2CH_2CH_2CH_2CH_3$ (IUPAC and common)

(e) 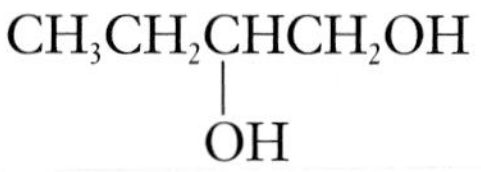 $CH_3CH_2CHCH_2OH$
OH

7. Name each of these alcohols.

(a)

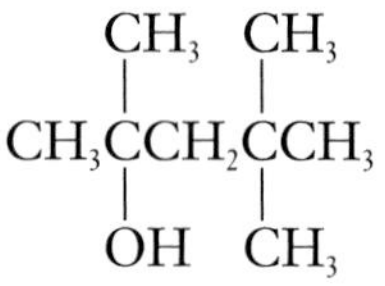

CH_3 CH_3
$CH_3CCH_2CCH_3$
OH CH_3

(b)

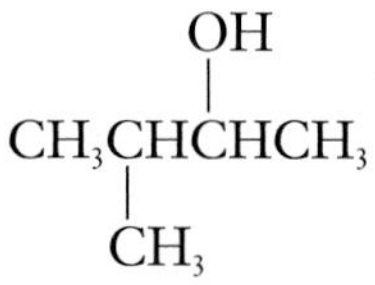

OH
$CH_3CHCHCH_3$
CH_3

(c)

OH, CH_3, CH_3
(cyclohexane ring)
H_3C

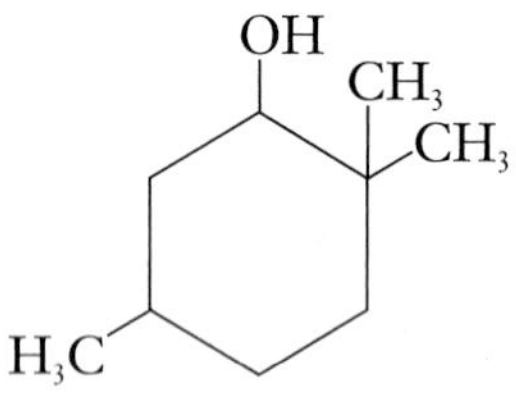

(d) $HOCH_2CH_2CH_2CH_2CH_2CH_2OH$

(e)

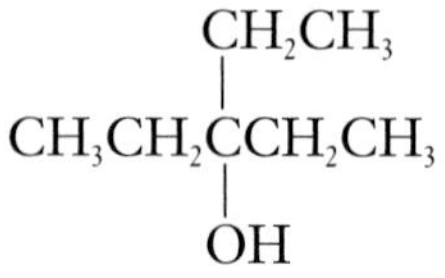

CH_2CH_3
$CH_3CH_2CCH_2CH_3$
OH

8. Classify each of these alcohols as primary, secondary, or tertiary:

(a)

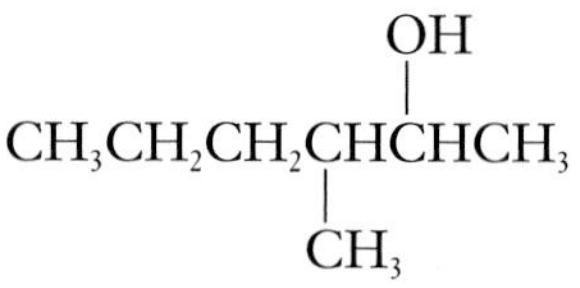

OH
$CH_3CH_2CH_2CHCHCH_3$
CH_3

(b)

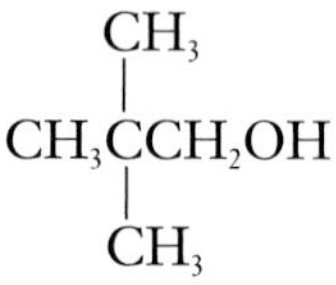

CH_3
CH_3CCH_2OH
CH_3

(c)

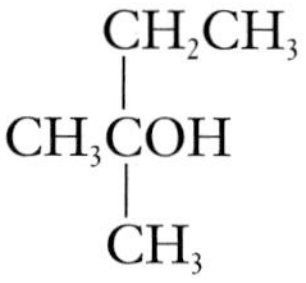

CH_2CH_3
CH_3COH
CH_3

(d)

OH
(cyclopentane ring)
CH_3

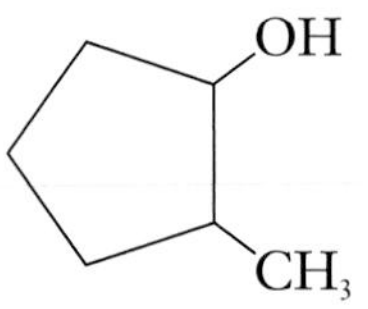

9. Classify each of these alcohols as primary, secondary, or tertiary:

(a)

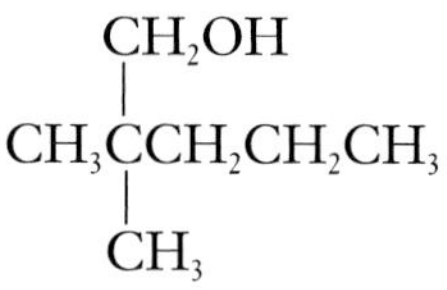

CH_2OH
$CH_3CCH_2CH_2CH_3$
CH_3

(b)

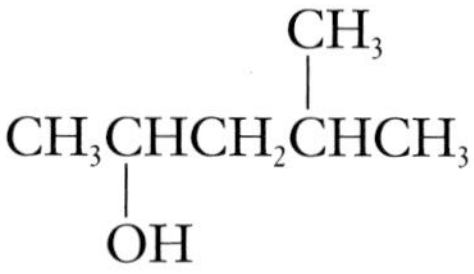

CH_3
$CH_3CHCH_2CHCH_3$
OH

(c)

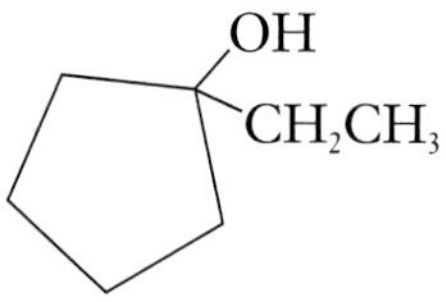

OH
(cyclopentane ring) — CH_2CH_3

(d)

CH_2CH_2OH
(cyclopentane ring) — CH_3

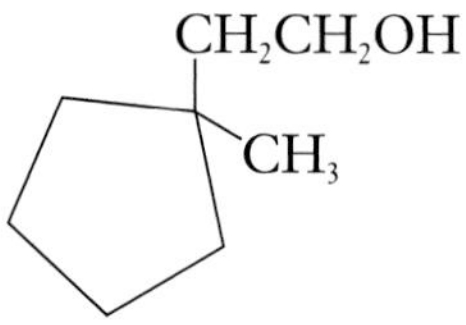

10. Name each of these phenols.

(a) 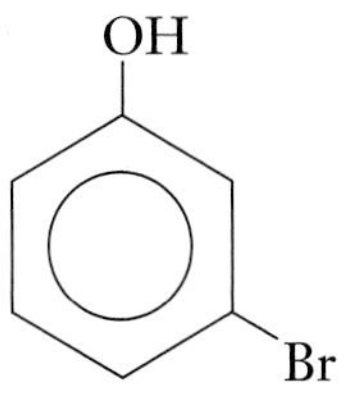 OH, Br

(b) 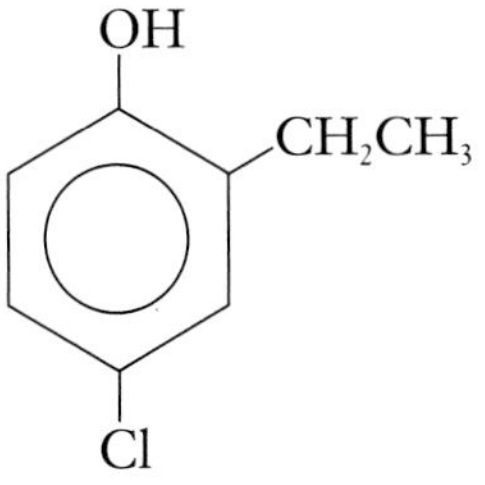 OH, CH_2CH_3, Cl

(c) 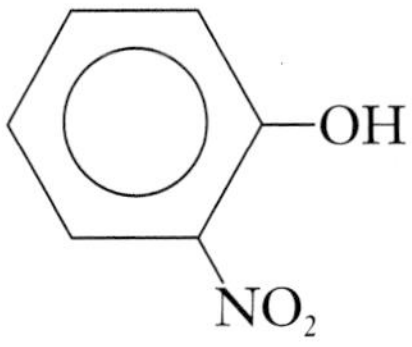 OH, NO_2

11. What are the names of these phenols?

(a) 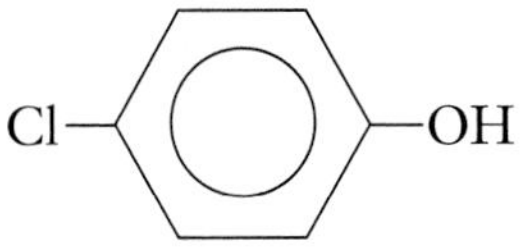 Cl—, —OH

(b) 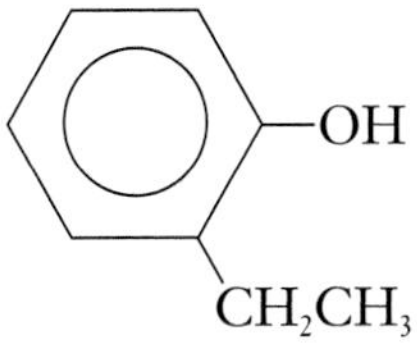 —OH, CH_2CH_3

(c) 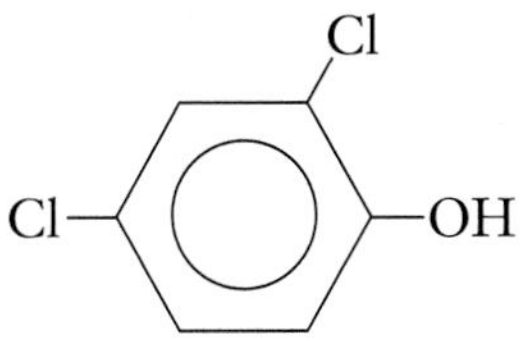 Cl, Cl—, —OH

12. Draw the structure of these phenols.

(a) 2-bromophenol
(b) *m*-phenylphenol
(c) 3,5-diisopropylphenol
(d) 4-fluorophenol

13. Use the information in these names to sketch the structures of these phenols.

(a) *p*-nitrophenol
(b) *o*-propylphenol
(c) 2,4-dinitrophenol
(d) 2,4,6-trichlorophenol

14. Why is 1-bromo-2-phenol an incorrect name?

15. Methanol and ethanol are water-soluble, but methane and ethane are not. Why?

16. Why does the water-solubility of alcohols decrease as the size of the alcohol increases?

17. Describe the water-solubility of phenols.

18. In what physical state do we expect to find the smaller alcohols at room temperature? What intermolecular interaction is largely responsible for these alcohols existing in this state?

19. Provide both the original and current name of the first disinfectant. This disinfectant is no longer used. Why?

20. Complete these equations and balance them.

(a) 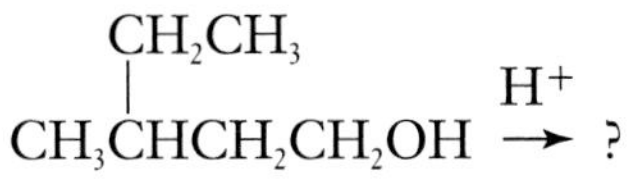 $CH_3CH(CH_2CH_3)CH_2CH_2OH \xrightarrow{H^+}$?

(b) 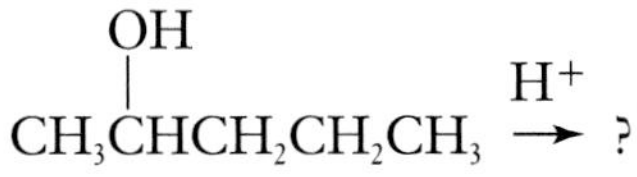 $CH_3CH(OH)CH_2CH_2CH_3 \xrightarrow{H^+}$?

(c) 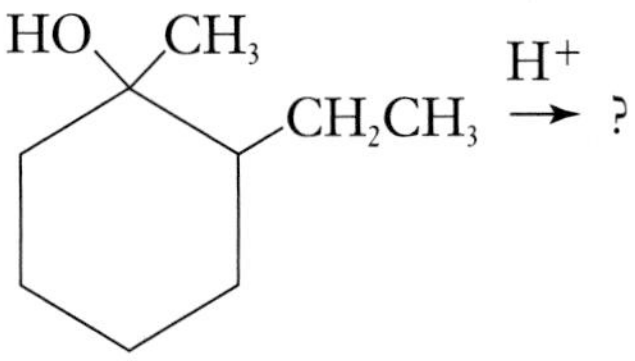 HO, CH_3, CH_2CH_3 $\xrightarrow{H^+}$?

21. Complete these reactions and balance them.

(a) 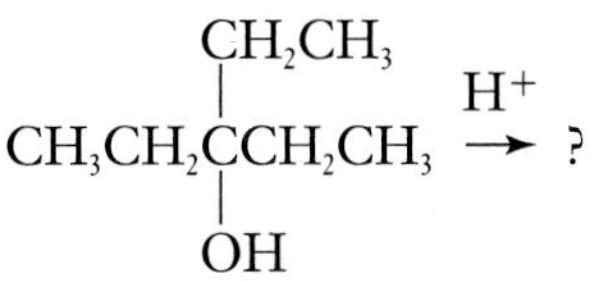 $CH_3CH_2C(CH_2CH_3)(OH)CH_2CH_3 \xrightarrow{H^+}$?

(b) 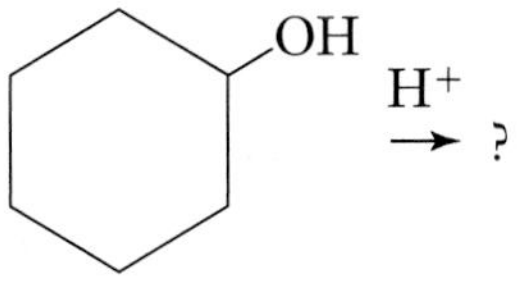 OH $\xrightarrow{H^+}$?

(c) 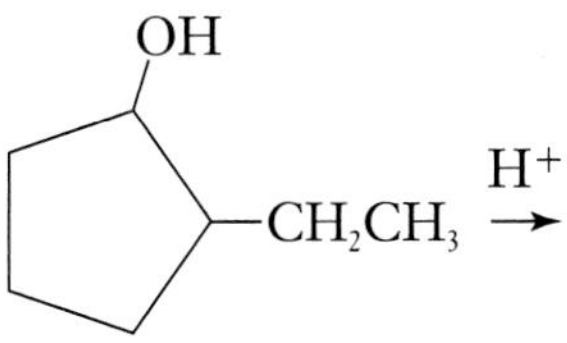 OH, CH_2CH_3 $\xrightarrow{H^+}$

22. What is the product of each of these reactions?

(a) 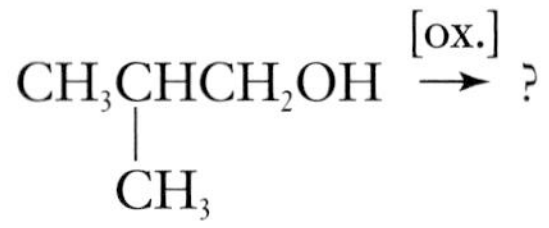 $CH_3CHCH_2OH \xrightarrow{[ox.]} ?$ (with CH_3 on the second carbon)

(b)  $(CH_3)_2CHOH \xrightarrow{[ox.]} ?$

(c) 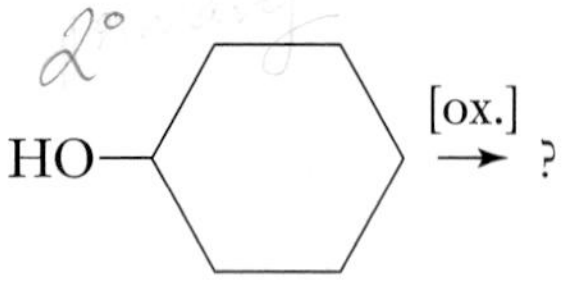 HO–(cyclohexane ring) $\xrightarrow{[ox.]} ?$

(d) 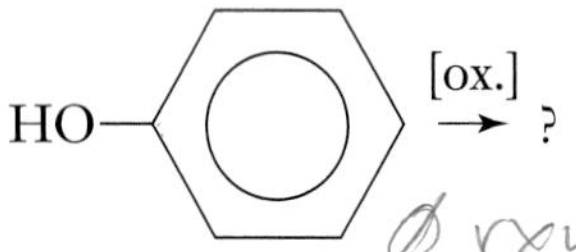 HO–(benzene ring) $\xrightarrow{[ox.]} ?$

23. Complete each of these equations.

(a) 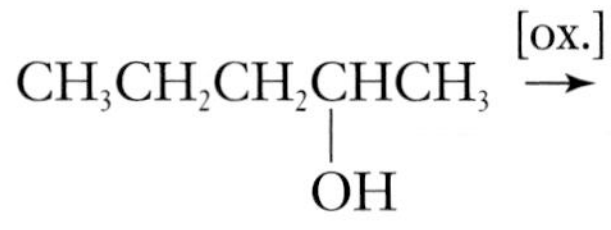 $CH_3CH_2CH_2CHCH_3 \xrightarrow{[ox.]}$ (with OH on the fourth carbon)

(b) 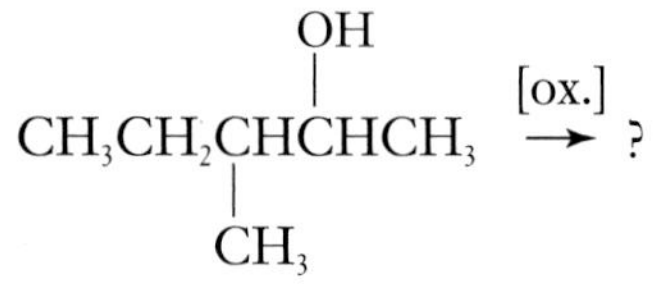 $CH_3CH_2CHCHCH_3 \xrightarrow{[ox.]} ?$ (with OH on the third carbon and CH_3 on the fourth carbon)

(c) 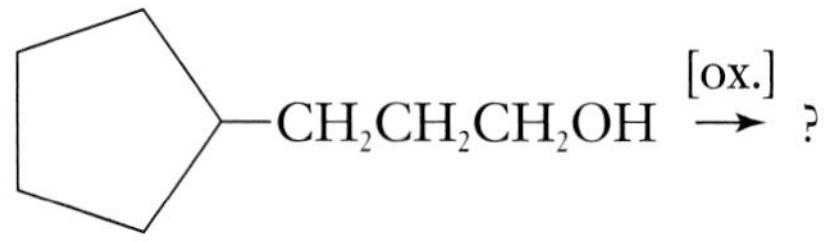 (cyclopentane ring)–$CH_2CH_2CH_2OH \xrightarrow{[ox.]} ?$

(d) 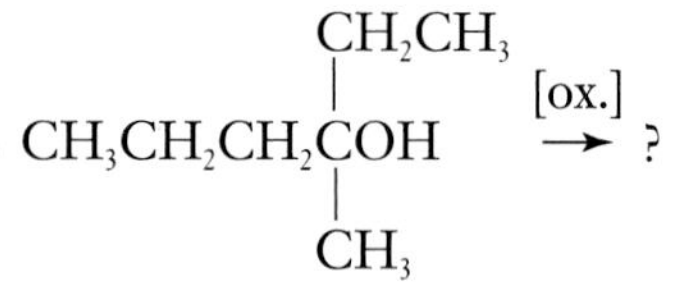 $CH_3CH_2CH_2COH \xrightarrow{[ox.]} ?$ (with CH_2CH_3 and CH_3 on the carbon bearing OH)

24. Name and draw the structure of the reactant needed to make this product.

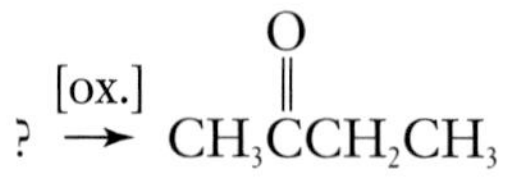

$$? \xrightarrow{[ox.]} CH_3\overset{\overset{\displaystyle O}{\|}}{C}CH_2CH_3$$

25. Draw the structure and name the product of this reaction:

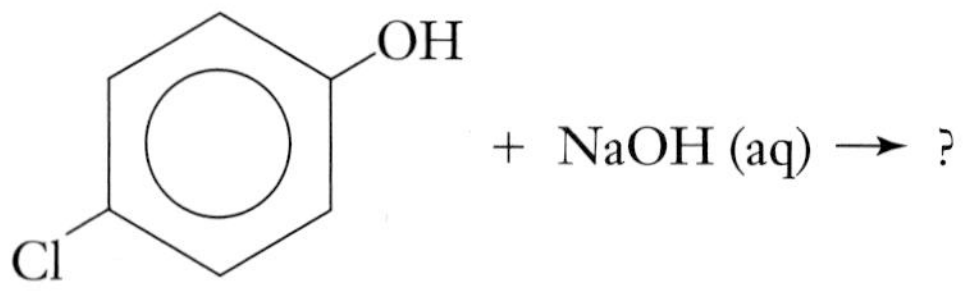

(benzene ring with OH and Cl para) $+ \text{NaOH (aq)} \rightarrow ?$

26. Would a chemical reaction occur if 1-propanol were mixed with a dilute aqueous solution of sodium hydroxide? Why or why not?

Ethers (Section 10.2)

27. Identify the functional group that is associated with ethers.

28. Provide the names of these ethers.
(a) $CH_3OCH_2CH_3$
(b) $CH_3CH_2OCH_2CH_3$

(c) 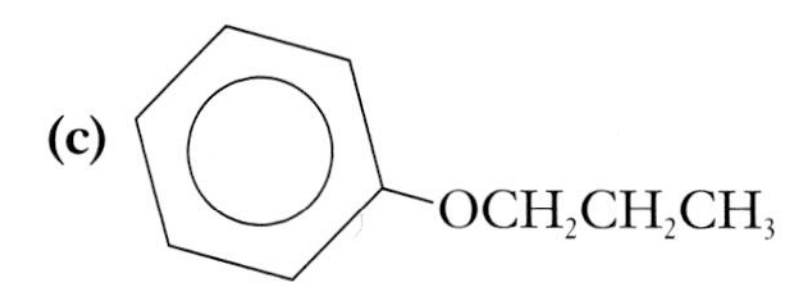

(d) 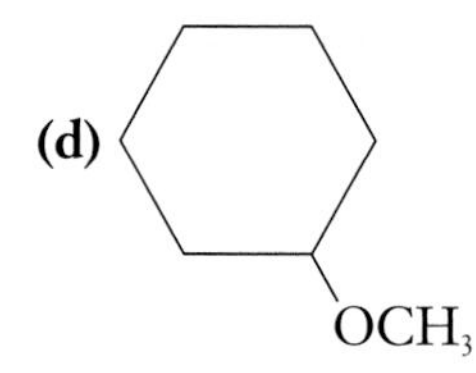

29. What are the names of these ethers?
(a) $CH_3OCH_2CH_2CH_2CH_3$
(b) $CH_3CH_2CH_2OCH_2CH_3$
(c) 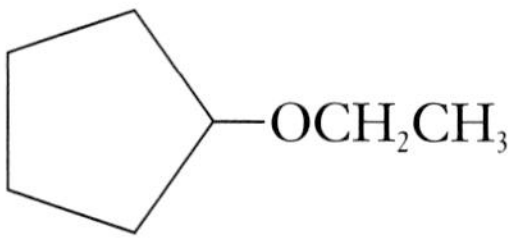 (cyclopentane ring)–OCH_2CH_3
(d) $CH_3CH_2CH_2OCH_2CH_2CH_3$

30. Draw the structures of these ethers.
(a) Anisole
(b) Dimethyl ether
(c) Isopropyl methyl ether
(d) Cyclopentyl methyl ether

31. Use the information contained in these names to draw the structures of these ethers.
(a) Isopropyl phenyl ether
(b) Cyclohexyl propyl ether
(c) Diphenyl ether
(d) Butyl isopropyl ether

32. Dimethyl ether and ethanol have the same formula, C_2H_6O, but the ether has a much lower boiling point than the alcohol. Why?

33. Although diethyl ether and other small ethers are nonpolar, they have a higher water-solubility than many other small, nonpolar molecules. Provide an explanation for this increased water solubility.

34. **(a)** MTBE is an abbreviation for ___________.
(b) What is the primary use for this compound?

Thiols (Section 10.3)

35. **(a)** Identify the functional group of a thiol.
(b) Compare this functional group to that of an alcohol. From this comparison, what can you predict about the properties of an alcohol and a thiol?

36. Name these compounds:
(a) CH_3CH_2SH
(b) $CH_3CH(CH_3)CH_2CH_2SH$ (CH$_3$ on the second carbon)
(c) cyclopentane—SH

37. Name these compounds:
(a) $HSCH_3$
(b) $CH_3CH_2CH(SH)CH_2CH_3$
(c) $CH_3CH_2CH_2CH(CH_2CH_3)CH_2CH_2CH_2SH$

38. Draw the structures of these thiols.
(a) 2-methyl-1-hexanethiol
(b) 3-ethylcyclopentanethiol
(c) Ethyl mercaptan

39. Use the information contained in these names to draw the structures of these thiols.
(a) 3-methyl-3-pentanethiol
(b) Isopropyl mercaptan
(c) Benzenethiol

40. What is the product when a thiol is oxidized? Write an equation to illustrate this reaction.

41. The reaction in Exercise 40 can be reversed. Write the reverse reaction.

Amines (Section 10.4)

42. Identify the functional group of the amines.

43. Classify the following compounds as primary, secondary, or tertiary amines:
(a) $CH_3CH_2CH_2CH_2CH_2NH_2$
(b) $CH_3N(CH_2CH_3)CH_2CH_3$
(c) cyclohexane—NH—CH$_3$

44. Classify the following compounds as primary, secondary, or tertiary amines:
(a) $CH_3—CH_2—NH—$(benzene ring)
(b) $CH_3CH_2CH_2CH_2CH_2N(CH_3)CH_3$
(c) $NH_2CH(CH_3)_2$

45. Distinguish a quaternary ammonium salt from an amine, or sketch an example of each and highlight the structural differences.

46. What is an alkaloid?

47. What is a heterocycle?

48. Name these amines:
(a) $CH_3CH_2NH_2$
(b) $HN(CH_3)CH_2CH_3$
(c) cyclopentane—N(CH$_3$)$_2$

49. Name these amines:

(a) $CH_3CH_2CH_2NH_2$

(b) H–N bonded to two cyclohexane rings

(c) CH_3–CH_2–N(–CH_2CH_3)–C_6H_5

50. Name the amines shown in Exercise 43.
51. Name the amines shown in Exercise 44.
52. Provide the structural formula of each of these amines.
(a) *N*-methyl-1-aminopropane
(b) Triethylamine
(c) Cyclopentylamine
53. Use the information in these names to draw the structures of these amines.
(a) Dimethylpropylamine
(b) *N*-ethylaniline
(c) Ethylisopropylamine
54. Would you expect amines to be polar or nonpolar? Explain your choice.
55. What strong interaction occurs between amine molecules?
56. Compare the water solubility of the smaller amines to larger amines. Why are they different?
57. How would you distinguish an organic ammonium salt from other salts?
58. Complete these equations for the reaction of an amine with an acid.
(a) $CH_3CH_2CH_2NH_2 + HCl \rightarrow$?
(b) $CH_3NHCH_3 + HNO_3 \rightarrow$?
59. Complete these equations for the reaction of an amine with an acid.

(a) cyclopentyl–NH–CH_3 + HCl → ?

(b) $2(CH_3)_3N + H_2SO_4 \rightarrow$?

60. Name the products formed in Exercise 58.
61. Name the products formed in Exercise 59.

Organohalogens—Alkyl Halides and Aryl Halides (Section 10.5)

62. By which functional group do we recognize organohalogens?
63. Identify the structural difference between an aryl halide and an alkyl halide.
64. Name these alkyl halides:
(a) $CH_3CH_2CH_2CH_2CH_2Br$

(b) $CH_3CH(CH_3)CH(Cl)CH_3$

(c) 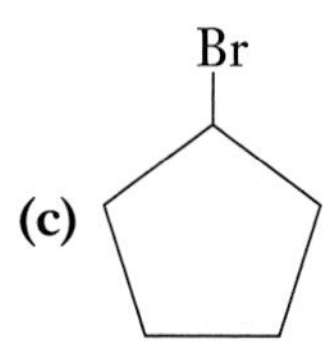

65. Name these alkyl halides:
(a) $ClCH_2CH_2Cl$
(b) $CHCl_3$

(c) cyclohexane ring with adjacent CH_3 and Br substituents

66. Illustrate the structures of these alkyl halides.
(a) 1,2-dichloropropane
(b) 2-iodo-2,3-dimethylbutane
(c) Ethyl bromide
67. Illustrate the structures of these compounds.
(a) Cyclohexyl chloride
(b) Carbon tetraiodide
(c) 3-fluoro-2,4-dimethylhexane
68. Draw the structures of these aryl halides.
(a) 2-fluorotoluene
(b) *m*-bromochlorobenzene
69. What are the structures of these aryl halides?
(a) *p*-dichlorobenzene
(b) 1,2,4-tribromobenzene
70. Are the organohalogens polar or nonpolar?
71. Describe the solubility properties of the organohalogens.

Challenge Exercises

72. An oxygen-containing organic compound has the formula C_6H_6O. The compound is a solid at room temperature, is somewhat water-soluble, cannot be dehydrated, and reacts with dilute solutions of sodium hydroxide. Is the compound an ether, alcohol, or a phenol? What is the name of this compound?

73. Amino acids are the building blocks of proteins and are discussed in Chapter 14. Why are these compounds called amino acids? Refer to Chapter 14, then identify any hydroxyl or thiol containing amino acids.

11 Aldehydes, Ketones, Carboxylic Acids, Esters, and Amides

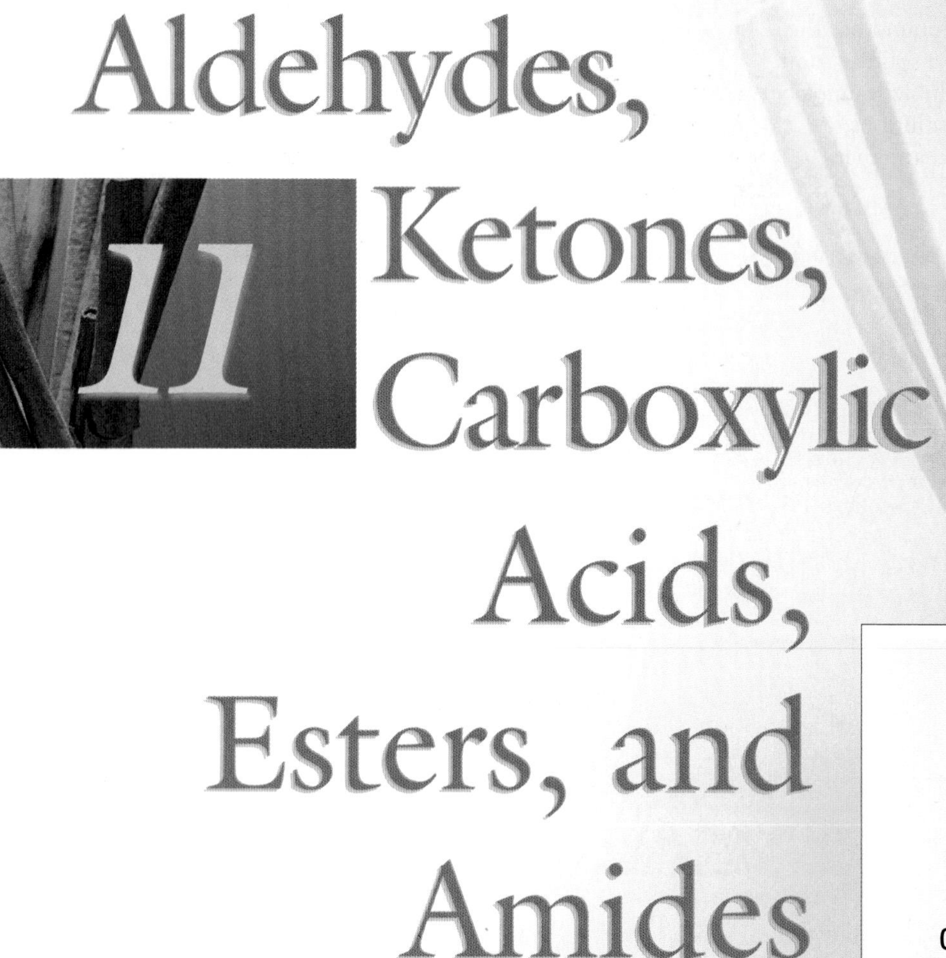

THESE ORGANIC COMPOUNDS CONTAIN THE CARBON-TO-OXYGEN DOUBLE BOND.

THE ORGANIC compounds examined so far include the hydrocarbons and the compounds that contain a hetero-atom single-bonded to a carbon atom. We now move on to several classes of organic compounds that share a common feature, the *carbon–oxygen double bond.* This structural feature is the *carbonyl group,* which is found in many common biomolecules, including sugars, amino acids, fats, and oils.

Outline

11.1 The Carbonyl Group

Distinguish among the classes of organic compounds that contain the carbonyl group.

11.2 Aldehydes

Name and draw structures of simple aldehydes and recognize some common reactions of them.

11.3 Ketones

Name and draw structures of simple ketones and recognize some common reactions of them.

11.4 Carboxylic Acids

Recognize carboxylic acids and write an equation that illustrates their acidity.

11.5 Esters, Amides, and Other Derivatives of Carboxylic Acids

Distinguish among the carboxylic acid derivatives and name simple examples of them.

The principal components of the fragrances of almond and cinnamon are benzaldehyde and cinnamaldehyde, respectively. ***(Charles D. Winters)***

11.1 The Carbonyl Group

In organic chemistry, the **carbonyl group** consists of a carbon atom double-bonded to an oxygen atom:

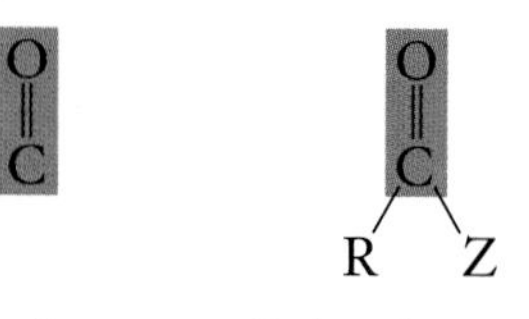

Carbonyl group Carbonyl compound

In all carbonyl compounds, R represents an alkyl group, aromatic ring, or hydrogen atom. Z represents any of several different atoms or groups of atoms. It is Z and the carbonyl group that collectively make up the functional group of the compound and thus determine the properties of the compound. In *aldehydes* Z is a hydrogen atom; in *ketones* Z is an alkyl group or aromatic ring; in *carboxylic acids* Z is a hydroxyl group; in *esters* Z is an oxygen atom that is bonded to an alkyl group or aromatic ring; and in *amides* Z is a nitrogen atom. Table 11.1 summarizes the classes of carbonyl-containing compounds and their functional groups.

Table 11.1 The Functional Groups of Carbonyl-Containing Compounds

CLASS OF COMPOUND	FUNCTIONAL GROUP[a]	EXAMPLE
Aldehyde	$R-\overset{O}{\overset{\|}{C}}-H$	$CH_3\overset{O}{\overset{\|}{C}}-H$ Ethanal (acetaldehyde)
Ketone	$R-\overset{O}{\overset{\|}{C}}-C$	$CH_3\overset{O}{\overset{\|}{C}}-CH_3$ Propanone (acetone)
Carboxylic Acid	$R-\overset{O}{\overset{\|}{C}}-OH$	$CH_3\overset{O}{\overset{\|}{C}}-OH$ Ethanoic acid (acetic acid)
Ester	$R-\overset{O}{\overset{\|}{C}}-OC$	$CH_3\overset{O}{\overset{\|}{C}}-OCH_3$ Methyl ethanoate (methyl acetate)
Amide	$R-\overset{O}{\overset{\|}{C}}-N$	$CH_3\overset{O}{\overset{\|}{C}}-NH_2$ Ethanamide (acetamide)

[a]R stands for an alkyl group, aromatic ring, or hydrogen atom.

11.2 Aldehydes

What do formaldehyde, vanilla, and glucose (blood sugar) have in common? All of these very different substances are aldehydes. An **aldehyde** is a compound that contains a carbonyl group that has a hydrogen atom attached to the carbon atom. The aldehyde group can be attached to either a hydrogen atom, an alkyl group, or an aromatic ring:

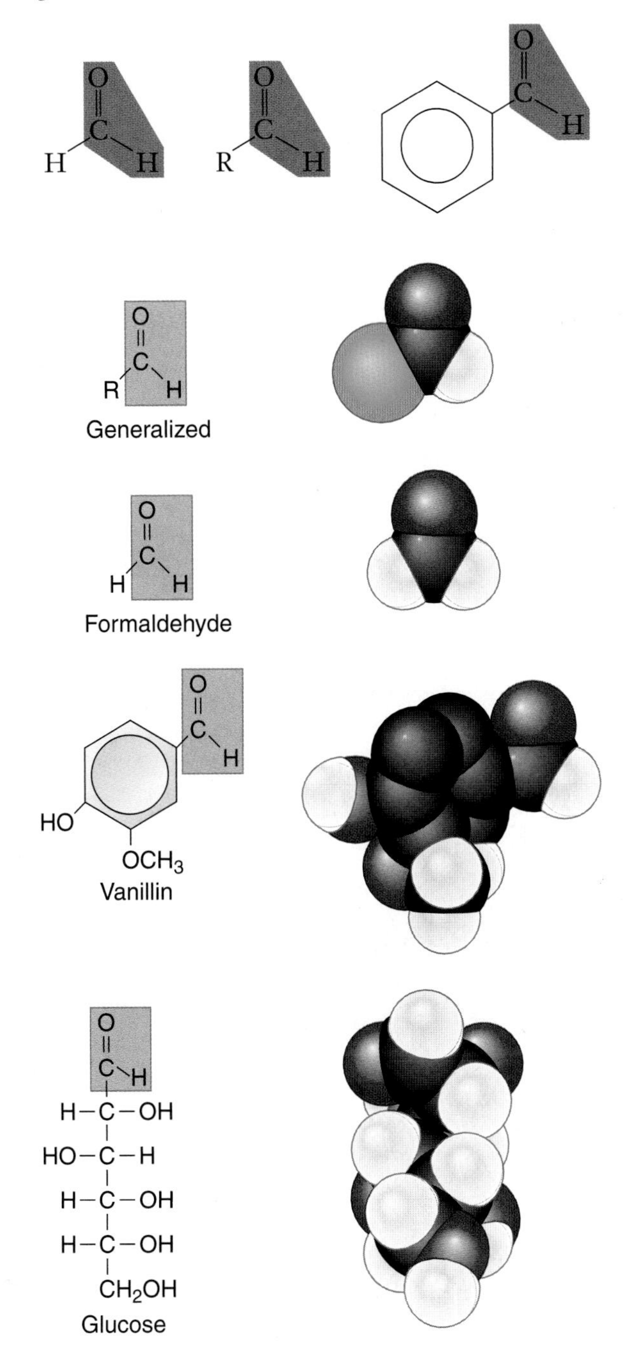

Aldehydes are common and widespread. Formaldehyde was used as a preservative of biological specimens, and is currently used to make a variety of structural materials. Vanillin is the principal flavoring agent in vanilla. Many sugars contain an aldehyde group.

Table 11.2 IUPAC Rules for Naming Aldehydes

Rule One	Find the longest continuous chain of carbon atoms that includes the carbon atom of the carbonyl group. This is the parent. Combine the parent and *an* with the suffix *-al.*
Rule Two	The carbon atom of the carbonyl group is carbon 1 of the chain.
Rule Three	The rules for substituents are the same as those for alkanes.
Rule Four	The rules for prefixes, commas, and hyphens are the same as for alkanes.

The aldehyde group is often represented as —CHO, where the carbon–oxygen double bond is implied.

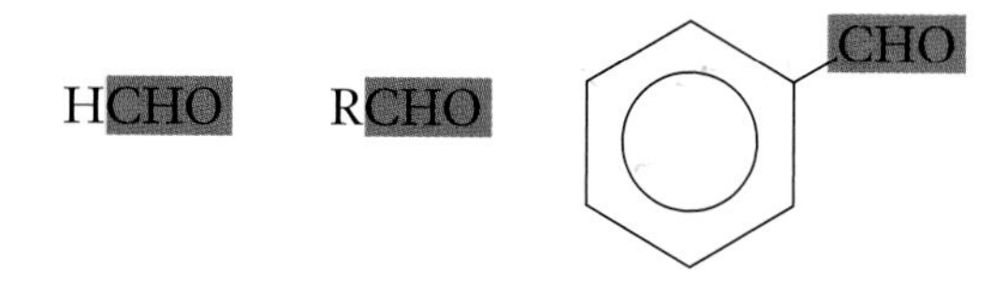

In this textbook, the aldehyde group will be shown by either this condensed representation or a structural formula.

Naming Aldehydes

The IUPAC nomenclature for aldehydes uses the corresponding alkane name as the parent. The suffix *-e* is replaced with the suffix *-al.* The carbonyl carbon is carbon 1 in the chain, and the rules for designating substituents are the same as for other organic compounds (Table 11.2).

In IUPAC nomenclature, the aldehyde group has priority over all of the functional groups you have studied so far. For example, if an —OH group is present in a molecule that also contains the aldehyde group, the compound is named as an aldehyde rather than an alcohol. The hydroxyl group is designated by the word *hydroxy:*

$$CH_3CH(OH)CH_2CH{=}O$$

3-hydroxybutanal

Example 11.1 Names and Structures of Aldehydes

1. What are the names of these compounds?

(a) $CH_3CH_2CH(CH_3)CH_2CHO$

(b) $CH_3CH(Cl)CH(OH)CHO$

2. What is the structure of 3,3-dimethylhexanal?

SOLUTIONS

1. (a) The longest continuous chain containing the carbonyl carbon has five carbon atoms, and the compound is an aldehyde. This yields *-pentanal.* The carbon atom of the carbonyl group is numbered 1; thus the methyl group is attached to carbon 3. The name is 3-methylpentanal.

   (b) A chain of four carbon atoms yields a parent of *-butan-*. When this is combined with the aldehyde suffix, the result is *-butanal.* This compound has two substituents, a hydroxyl group on carbon 2 and a chloro group on carbon 3. The name is 3-chloro-2-hydroxybutanal (remember, *chloro-* comes before *hydroxy-* because of the alphabetical rule).

2. *Hexan-* means there are six carbon atoms in the chain, and the *-al* ending means the functional group is an aldehyde group:

$$\text{C—C—C—C—C—CHO}$$

There are two methyl groups on carbon 3:

$$\begin{array}{c} \qquad\quad CH_3 \\ \qquad\quad | \\ \text{C—C—C—C—C—CHO} \\ \qquad\quad | \\ \qquad\quad CH_3 \end{array}$$

Adding hydrogen atoms to those carbon atoms that do not already have four bonds yields

$$\begin{array}{c} \qquad\qquad CH_3 \\ \qquad\qquad | \\ CH_3\text{—}CH_2\text{—}CH_2\text{—}C\text{—}CH_2\text{—}CHO \\ \qquad\qquad | \\ \qquad\qquad CH_3 \end{array}$$

Self-Test

1. Draw the structure of 2-chloro-3-methylbutanal.
2. What is the name of $BrCH_2CHCHO$ (with CH_3 attached to the CH)?

ANSWERS

1. $$\begin{array}{l} \quad\;\; CH_3 \;\; O \\ \quad\;\; | \quad\;\; \| \\ CH_3CHCHCH \\ \qquad\;\; | \\ \qquad\;\; Cl \end{array}$$

2. 3-bromo-2-methylpropanal

Some aldehydes are identified by common names. These names use the suffix *-aldehyde,* with a root (parent) for the chain length. These roots are identified in Table 11.3. They are worth memorizing. The two smallest aldehydes,

Table 11.3 Roots That Identify Chain Length in Some Common Aldehyde Names

NUMBER OF CARBON ATOMS IN CHAIN	ROOT	ALDEHYDE
1	*Form-*	Formaldehyde
2	*Acet-*	Acetaldehyde
3	*Propion-*	Propionaldehyde
4	*Butyr-*	Butyraldehyde

methanal and ethanal, have the common names formaldehyde and acetaldehyde, respectively:

$HCHO$ — Formaldehyde (Methanal)

CH_3CHO — Acetaldehyde (Ethanal)

Some aldehydes have common names that provide no structural information. Glucose and vanillin are examples:

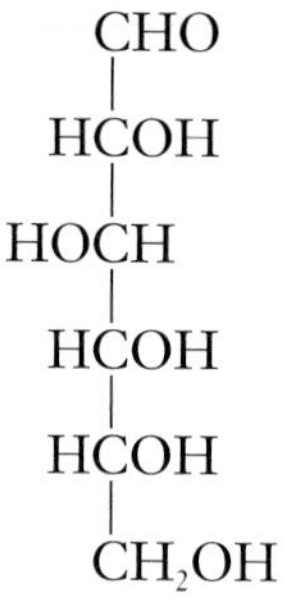

CHO, HCOH, HOCH, HCOH, HCOH, CH_2OH

Glucose

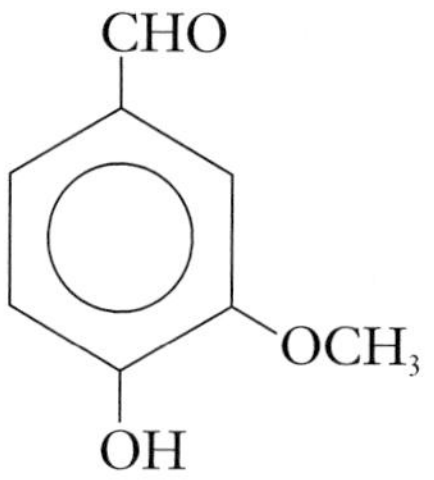

Benzene ring with CHO, OCH_3, OH

Vanillin

The simplest aromatic aldehyde is called benzaldehyde:

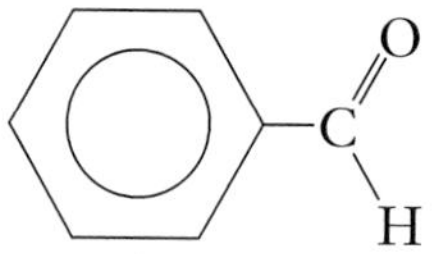

Benzene ring–C(=O)H

This name was originally a common name, but it is now the parent name for all aromatic aldehydes. Other aromatic aldehydes are named by combining their substituent names with the word benzaldehyde:

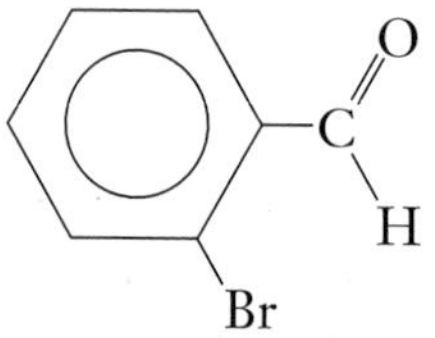

Benzene ring (Br)–C(=O)H

2-bromobenzaldehyde or *o*-bromobenzaldehyde

Remember, the location of the substituent(s) can be identified by either numbers (IUPAC) or by the prefixes *o*-, *m*-, or *p*- (common).

Example 11.2 Names and Structures of Aromatic Aldehydes

Draw the structure of 3-ethylbenzaldehyde:

SOLUTION

The word *benzaldehyde* means the compound contains an aromatic ring with an aldehyde group on carbon 1, and there is an ethyl group on carbon number 3:

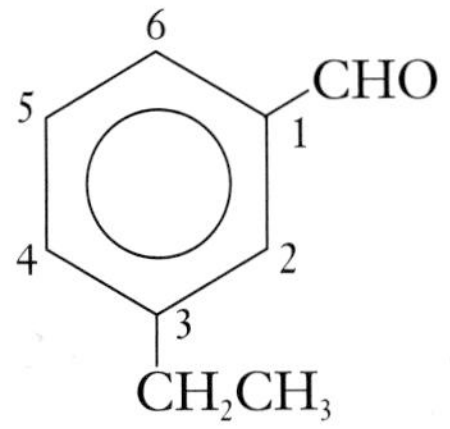

6
5 CHO
1
4 2
3
CH_2CH_3

Self-Test

What is the name of this compound?

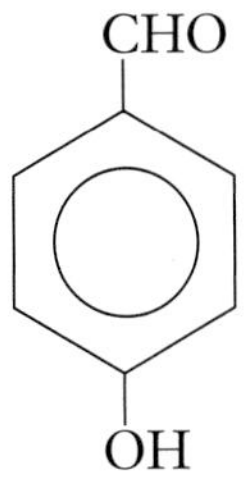

CHO

OH

ANSWER

4-hydroxybenzaldehyde (*p*-hydroxybenzaldehyde)

Physical Properties of Aldehydes

The carbonyl group gives aldehydes their physical and chemical properties. This group is polar because oxygen is more electronegative than carbon. Thus the oxygen atom pulls the electrons of the carbon–oxygen double bond toward itself:

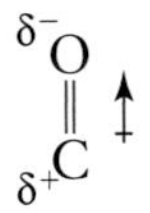

$$\overset{\delta-}{O} = \overset{\delta+}{C}$$

A permanent dipole exists. Because of the polarity of the carbonyl group, the smallest aldehydes such as methanal and ethanal are completely soluble in water. Water-solubility decreases in order from propanal to pentanal, and hexanal and larger aldehydes—including aromatic aldehydes—are water-insoluble because the nonpolar alkyl group or aromatic ring has a larger effect on solubility than the polar carbonyl group. These larger aldehydes are soluble in nonpolar solvents.

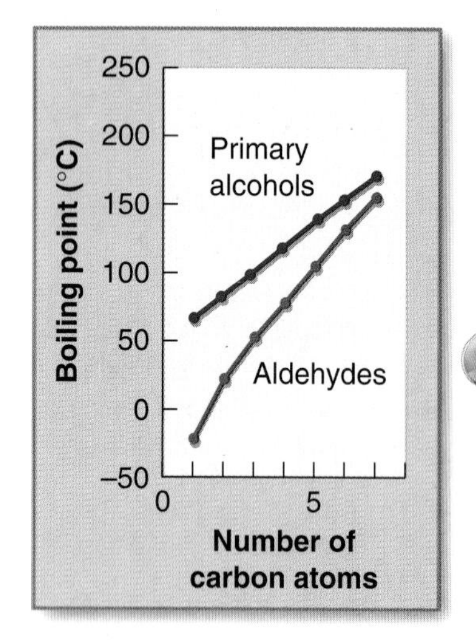

Figure 11.1 The boiling points of aldehydes are lower than the corresponding alcohols. The dipole–dipole bonding between the carbonyl groups of the aldehyde molecules is not as strong as the hydrogen bonding between the alcohol molecules.

Aldehyde molecules cannot hydrogen bond to each other like alcohol molecules do, thus the boiling points of aldehydes are lower than the corresponding alcohols (Figure 11.1). The smallest aldehyde, formaldehyde (methanal), is a gas at room temperature, although it is also available as a 37% aqueous solution called formalin. Many other common aldehydes are liquids. Although the smallest aromatic aldehyde—benzaldehyde—is a liquid, many aromatic aldehydes are solids.

Reactions of Aldehydes

Aldehydes undergo a variety of reactions. In this section, we concentrate on a few that are of particular importance to organic chemistry and biochemistry.

Reduction

Aldehydes can be reduced to primary alcohols by a variety of reducing agents, and this reduction of aldehydes is one way to make primary alcohols:

$$\underset{\text{Aldehyde}}{R{-}\overset{\overset{\displaystyle O}{\|}}{C}H} \xrightarrow{[\text{red.}]} \underset{1^\circ\text{ alcohol}}{R{-}\overset{\overset{\displaystyle H}{|}}{\underset{\underset{\displaystyle H}{|}}{C}}{-}O{-}H}$$

In the reduction, a hydrogen atom adds to each atom of the carbonyl group, and the double bond becomes a single bond. In the catalytic reduction of propanal to 1-propanol by hydrogen, for example, we have

$$\underset{\text{Propanal}}{CH_3CH_2\overset{\overset{\displaystyle O}{\|}}{C}{-}H} + H_2 \xrightarrow{Pt} \underset{\text{1-propanol}}{CH_3CH_2\overset{\overset{\displaystyle H}{|}}{\underset{\underset{\displaystyle H}{|}}{C}}{-}O{-}H}$$

In the reverse reaction, aldehydes are made by oxidation of primary alcohols (Section 10.1).

Oxidation

Aldehydes are readily oxidized to carboxylic acids:

$$\underset{\text{Aldehyde}}{R{-}\overset{\overset{\displaystyle H}{|}}{C}{=}O} \xrightarrow{[\text{ox.}]} \underset{\text{Carboxylic acid}}{R{-}\overset{\overset{\displaystyle OH}{|}}{C}{=}O}$$

The oxidation of ethanal to ethanoic acid (acetic acid) is

$$\underset{\substack{\text{Ethanal}\\(\text{acetaldehyde})}}{CH_3CHO} \xrightarrow{[\text{ox.}]} \underset{\substack{\text{Ethanoic acid}\\(\text{Acetic acid})}}{CH_3COOH}$$

(a)

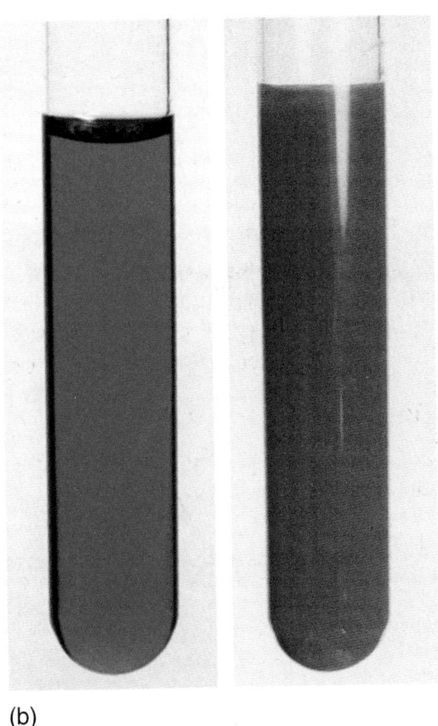

(b)

Figure 11.2 **(a)** Metallic silver produced in a Tollen's test for aldehydes has plated the inside of this flask. **(b)** When the soluble blue copper(II) ions found in Benedict's solution react with an aldehyde, they are reduced to red insoluble copper(I) ions. These tests are used to detect aldehydes or reducing sugars or to measure the amount of these compounds in a solution.

The ease with which aldehydes can be oxidized provides several tests for aldehydes and the many sugars that contain the aldehyde group. The *Tollen's test* [Figure 11.2(a)] uses silver ion as the oxidizing agent. If an aldehyde or a sugar containing an aldehyde group (a reducing sugar) is present in the tested sample, silver metal plates out onto the surface of the container because the metal ion becomes reduced as it oxidizes the aldehyde to a carboxylic acid:

$$\text{R—CHO} + \text{Ag}^+ \longrightarrow \text{R—COOH} + \text{Ag}^0 \quad \text{(Not balanced)}$$

Two other solutions used to test for aldehydes or reducing sugars—Benedict's and Fehling's—contain copper(II) ions. If an aldehyde is present, the characteristic blue color of soluble Cu^{2+} changes to the brick-red color of Cu^+ in copper(I) oxide [Figure 11.2(b)]. Test papers are sometimes used to test for glucose in body fluids. Some test papers contain copper(II) ion, and if glucose is present in the fluid, the metal ion is reduced (remember, glucose contains an aldehyde group). The color of the paper changes, and the amount of glucose in the fluid sample can be estimated.

Test papers can be used to measure the amount of glucose in body fluids, such as blood or urine. *(Charles D. Winters)*

Formation of Hemiacetals and Acetals

Aldehydes can form compounds called hemiacetals and acetals. A **hemiacetal** is an organic compound formed from an aldehyde or ketone and an alcohol. An **acetal** is formed from a hemiacetal and an alcohol. Hemiacetals and acetals are important to us because many sugars show this chemistry. This equation shows the reaction of aldehydes with alcohols to yield hemiacetals:

$$\underset{\text{Aldehyde}}{R{-}\underset{\underset{O}{\|}}{C}{-}H} + \underset{\text{Alcohol}}{HO{-}R'} \overset{H^+}{\rightleftharpoons} \underset{\text{Hemiacetal}}{R{-}\overset{\overset{OR'}{|}}{\underset{\underset{OH}{|}}{C}}{-}H}$$

In this reaction a molecule of an aldehyde and a molecule of an alcohol condense into a single molecule of a hemiacetal. The reaction of acetaldehyde with ethanol provides a specific example:

$$\underset{\text{Acetaldehyde}}{CH_3{-}\underset{\underset{O}{\|}}{C}{-}H} + \underset{\text{Ethanol}}{HOCH_2CH_3} \overset{H^+}{\rightleftharpoons} \underset{\text{Hemiacetal}}{CH_3{-}\overset{\overset{OCH_2CH_3}{|}}{\underset{\underset{OH}{|}}{C}}{-}H}$$

Hemiacetals react with an alcohol to yield an acetal:

$$\underset{\text{Hemiacetal}}{R{-}\overset{\overset{OR'}{|}}{\underset{\underset{OH}{|}}{C}}{-}H} + \underset{\text{Alcohol}}{HO{-}R''} \overset{H^+}{\rightleftharpoons} \underset{\text{Acetal}}{R{-}\overset{\overset{OR'}{|}}{\underset{\underset{OR''}{|}}{C}}{-}H} + H{-}O{-}H$$

When a molecule of an acetal forms, a molecule of water forms as a by-product. The hemiacetal formed from acetaldehyde and ethanol reacts with a second ethanol molecule to form this acetal:

$$\underset{\text{Hemiacetal}}{CH_3{-}\overset{\overset{OCH_2CH_3}{|}}{\underset{\underset{OH}{|}}{C}}{-}H} + \underset{\text{Ethanol}}{HOCH_2CH_3} \overset{H^+}{\rightleftharpoons} \underset{\text{Acetal}}{CH_3{-}\overset{\overset{OCH_2CH_3}{|}}{\underset{\underset{OCH_2CH_3}{|}}{C}}{-}H} + H_2O$$

We shall see hemiacetals and acetals involving aldehydes again in Chapter 12, when we begin our study of biochemistry.

Example 11.3 Reactions of Aldehydes

1. Which aldehyde, when reduced, would yield 3-methyl-1-butanol?
2. What product will result from the oxidation of 3-ethylpentanal?
3. Draw the structure of the hemiacetal that would form from ethanal, CH_3CHO, and 2-propanol, $CH_3CHOHCH_3$.

SOLUTIONS

1. The *but-* in the product name tells us there are four carbon atoms in the parent chain. Carbon one of the product has a hydroxyl group bound to it that had to come from the carbonyl group of the aldehyde. Carbon three has a methyl

group on it. Combining these three structural features and adding hydrogen atoms to complete the structure yields 3-methylbutanal:

$$\begin{array}{c} \text{O} \quad \text{CH}_3 \\ \| \qquad | \\ \text{HCCH}_2\text{CHCH}_3 \end{array}$$

2. Aldehydes are oxidized to carboxylic acids. All of the structure of the aldehyde remains unchanged except for the hydrogen on the carbonyl carbon, which becomes a hydroxyl group. The product is 3-ethylpentanoic acid:

$$\begin{array}{c} \text{CH}_2\text{CH}_3 \\ | \\ \text{CH}_3\text{CH}_2\text{CHCH}_2\text{COOH} \end{array}$$

3. One molecule of ethanal and one molecule of 2-propanol would combine to form a hemiacetal. The methyl group and H attached to the carbonyl carbon of the aldehyde would remain unchanged, but the oxygen atom of the carbonyl group becomes a hydroxyl group. The alcohol becomes bonded to the carbonyl carbon through its oxygen atom:

$$\begin{array}{c} \text{OH} \\ | \\ \text{CH}_3\text{—C—H} \\ | \\ \text{O} \\ | \\ \text{CH}_3\text{CHCH}_3 \end{array}$$

Self-Test

1. Reducing what aldehyde would yield 1-heptanol?
2. Draw the structure of the compound formed when benzaldehyde is oxidized.
3. What compounds are needed to make this hemiacetal?

$$\begin{array}{c} \text{OH} \\ | \\ \text{CH}_3\text{—C—H} \\ | \\ \text{OCH}_3 \end{array}$$

ANSWERS

1. Heptanal
2. $C_6H_5\text{—}\overset{\overset{\text{O}}{\|}}{\text{C}}\text{—O—H}$
3. Ethanal and methanol

11.3 Ketones

Ketones are organic compounds that have two carbon atoms bonded to a carbonyl group. The carbon atoms can be in alkyl groups or aromatic rings:

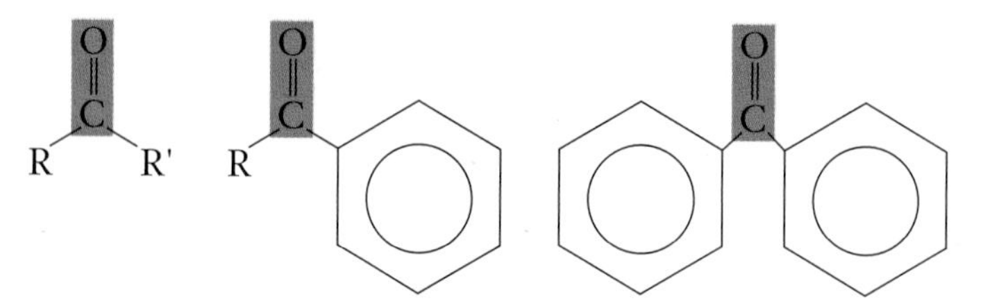

(The prime symbol in R' indicates an alkyl group that may be different from the group represented by plain R.)

The smallest ketone is acetone (common name), which is a very common solvent you may have used (it is the primary ingredient of some nail polish removers):

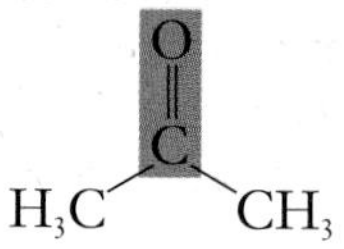

$$H_3C-\overset{O}{\overset{\|}{C}}-CH_3$$

Acetone

Ketones are sometimes represented by a condensed structural formula that places the carbon and oxygen atoms of the carbonyl group together as in the condensed structural formula of acetone: CH_3COCH_3 .

Ketones have a variety of uses, and some are important biological molecules. Examples include pyruvic acid, which is an important molecule in the metabolic pathway called glycolysis, and the sugar fructose.

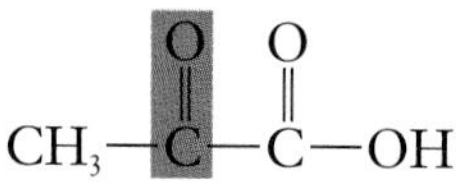

$$CH_3-\overset{O}{\overset{\|}{C}}-\overset{O}{\overset{\|}{C}}-OH$$

Pyruvic acid

$$\begin{array}{c} CH_2-OH \\ | \\ C=O \\ | \\ HO-C-H \\ | \\ H-C-OH \\ | \\ H-C-OH \\ | \\ CH_2OH \end{array}$$

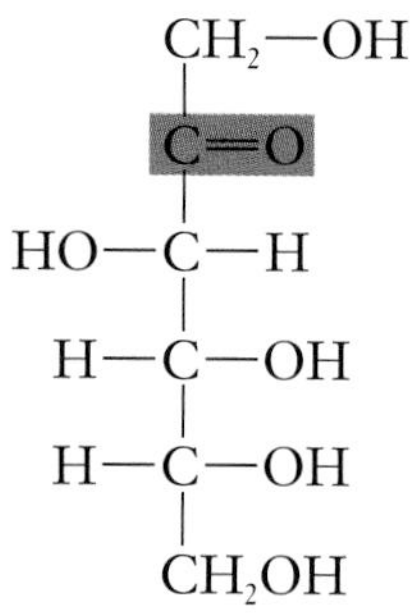

Fructose

Naming Ketones

The IUPAC rules for naming ketones are similar to the rules you have already learned (Table 11.4). The suffix indicating ketones is *-one*. A ketone group has priority over a hydroxyl group, which is named as a substituent using the word *hydroxy*. 4-Hydroxy-2-butanone illustrates this priority:

Table 11.4 IUPAC Rules for Naming Ketones

Rule One	The longest continuous chain of carbon atoms that includes the carbonyl carbon atom is the parent. Combine the parent and *an* with the suffix *-one.*
Rule Two	Number the chain to give the carbon atom of the carbonyl group the smallest possible number.
Rule Three	The rules for substituents are the same as those you have learned for other compounds.
Rule Four	The rules for prefixes, commas, and hyphens are the same as those for other compounds.

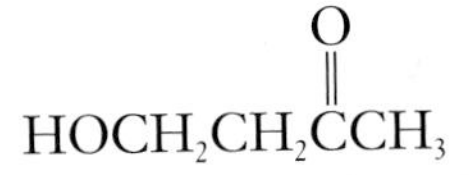

Several ketones are often identified by their common name. These include the smallest, most common ketone, which is acetone (its IUPAC name is propanone). Other small ketones are sometimes identified by a "systematic common" nomenclature. They are named by first identifying, in alphabetical order, the alkyl groups attached to the carbonyl group. These alkyl group names are then followed by the word ketone:

$$CH_3CH_2\overset{\overset{\displaystyle O}{\|}}{C}CH_3$$

Ethyl methyl ketone

The two most common aromatic ketones are acetophenone and benzophenone. Other aromatic ketones are usually named as derivatives of these two compounds.

$$C_6H_5\overset{\overset{\displaystyle O}{\|}}{C}CH_3 \qquad C_6H_5\overset{\overset{\displaystyle O}{\|}}{C}C_6H_5$$

Acetophenone — Benzophenone

Ethyl methyl ketone (2-butanone), sometimes abbreviated MEK, is the solvent in some cements and polishes. *(Charles D. Winters)*

Example 11.4 The Structures and Names of Ketones

1. What is the IUPAC name of the ketone whose common name is ethyl methyl ketone?
2. What is the structure of 5-chloro-2-pentanone?

SOLUTIONS

1. This ketone has an ethyl group, a carbonyl group, and a methyl group. Combined, these groups contain four carbon atoms, and so the name is derived from butane by dropping the *-e* suffix and adding the suffix *-one:* butanone. Use the prefix 2- to indicate the location of the carbonyl group: 2-butanone.

2. *Pentan-* indicates a five-carbon skeleton:

$$\mathrm{C{-}C{-}C{-}C{-}C}$$

Put a carbonyl group on carbon 2, then put a chlorine atom on carbon 5:

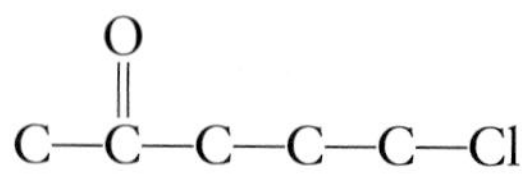

$$\mathrm{C{-}\overset{\overset{\displaystyle O}{\|}}{C}{-}C{-}C{-}C{-}Cl}$$

Add appropriate hydrogen atoms to get the correct structure:

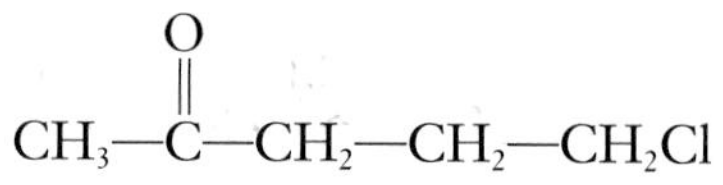

$$\mathrm{CH_3{-}\overset{\overset{\displaystyle O}{\|}}{C}{-}CH_2{-}CH_2{-}CH_2Cl}$$

Self-Test

1. Draw the structure of 4-methyl-3-hexanone.

2. What is the name of this compound? $\mathrm{CH_3\overset{\overset{\displaystyle O}{\|}}{C}\underset{\underset{\displaystyle CH_3}{|}}{C}HCH_2\overset{\overset{\displaystyle CH_3}{|}}{\underset{\underset{\displaystyle CH_3}{|}}{C}}CH_2CH_2CH_3}$

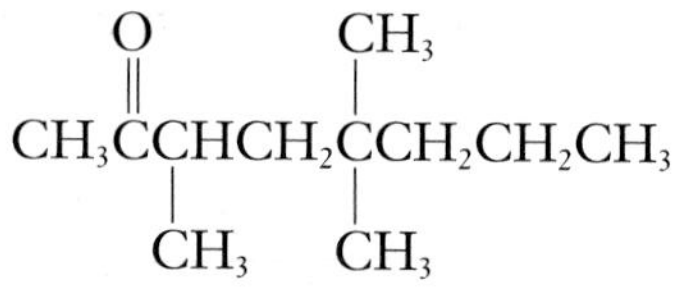

ANSWERS

1. 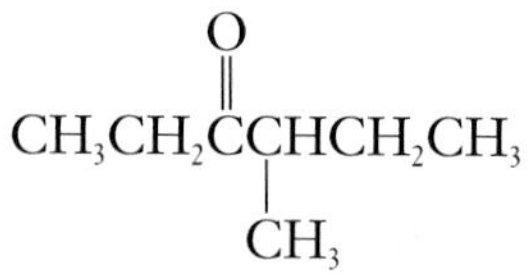 $\mathrm{CH_3CH_2\overset{\overset{\displaystyle O}{\|}}{C}\underset{\underset{\displaystyle CH_3}{|}}{C}HCH_2CH_3}$

2. 3,5,5-trimethyl-2-octanone

Physical Properties of Ketones

The carbonyl group makes ketones somewhat polar. This polarity makes the smallest ketones water-soluble, and they are usually also soluble in organic solvents. The smallest ketones, like acetone and ethyl methyl ketone, are common solvents. In larger ketones, the nonpolar alkyl groups and rings have more effect on solubility than the carbonyl group. Therefore larger ketones are water-insoluble but soluble in nonpolar solvents.

The smallest ketones are liquids, and larger ones tend to be solids.

Reactions of Ketones

Reduction

Ketones can be reduced to secondary alcohols:

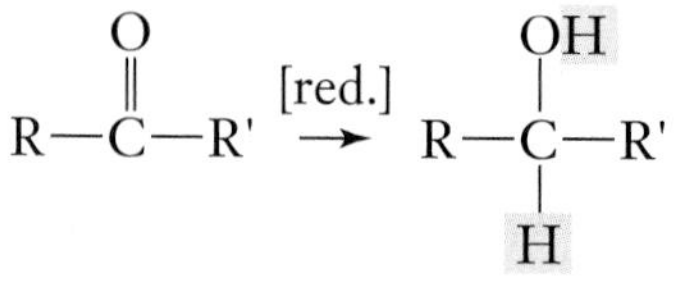

$$\underset{\text{Ketone}}{\mathrm{R{-}\overset{\overset{\displaystyle O}{\|}}{C}{-}R'}} \xrightarrow{[\text{red.}]} \underset{2^\circ\text{ alcohol}}{\mathrm{R{-}\overset{\overset{\displaystyle OH}{|}}{\underset{\underset{\displaystyle H}{|}}{C}}{-}R'}}$$

Animals use fat to store energy. During the synthesis of fats, ketones are reduced to alcohols. The synthesis of fats and their role in living systems is examined in Chapters 13 and 17. *(Art Wolfe/Tony Stone Images)*

Each atom of the carbonyl group gets a hydrogen atom during the reduction. Reduction of 2-butanone yields 2-butanol:

$$CH_3CH_2\overset{\overset{\displaystyle O}{\|}}{C}CH_3 \xrightarrow{[red.]} CH_3CH_2\underset{\underset{\displaystyle H}{|}}{\overset{\overset{\displaystyle OH}{|}}{C}}CH_3$$

2-Butanone 2-butanol

The reduction of ketones can serve as a synthesis for secondary alcohols, but this reaction is more important to us because it occurs in several reactions of the body.

Oxidation

Ketones cannot be easily oxidized, although like most other organic compounds, they are completely oxidized during combustion.

Formation of Hemiacetals and Acetals

Like aldehydes, ketones react with alcohols to form hemiacetals:

$$R-\underset{\underset{\displaystyle O}{\|}}{C}-R' + HO-R'' \overset{H^+}{\rightleftharpoons} R-\underset{\underset{\displaystyle OH}{|}}{\overset{\overset{\displaystyle OR''}{|}}{C}}-R'$$

Ketone Alcohol Hemiacetal

2-Butanone and 1-propanol would react to yield this product:

$$CH_3-\underset{\underset{\displaystyle O}{\|}}{C}-CH_2CH_3 + HOCH_2CH_2CH_3 \overset{H^+}{\rightleftharpoons} CH_3-\underset{\underset{\displaystyle OH}{|}}{\overset{\overset{\displaystyle OCH_2CH_2CH_3}{|}}{C}}-CH_2CH_3$$

2-Butanone 1-propanol A hemiacetal (its name is 2-propoxy-2-butanol)

The hemiacetals that are formed from ketones can react with an alcohol to yield an acetal:

$$R-\underset{\underset{\displaystyle OH}{|}}{\overset{\overset{\displaystyle OR''}{|}}{C}}-R' + HO-R''' \overset{H^+}{\rightleftharpoons} R-\underset{\underset{\displaystyle OR'''}{|}}{\overset{\overset{\displaystyle OR''}{|}}{C}}-R' + H_2O$$

Hemiacetal Alcohol Acetal

The hemiacetal formed from 2-butanone and 1-propanol reacts with a second 1-propanol molecule in this reaction:

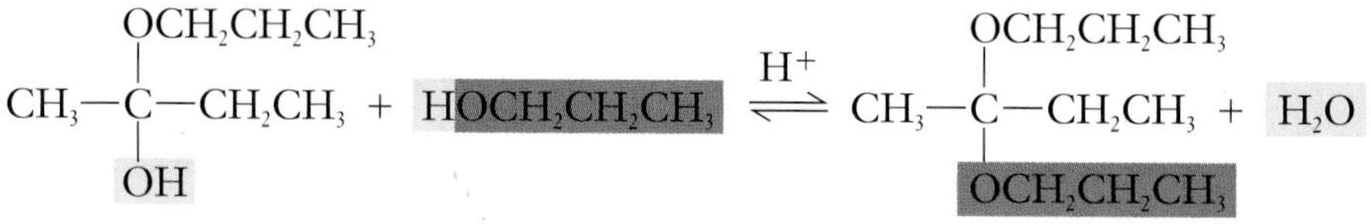

$$\underset{\text{Hemiacetal}}{CH_3-\overset{\overset{\displaystyle OCH_2CH_2CH_3}{|}}{\underset{\underset{\displaystyle OH}{|}}{C}}-CH_2CH_3} + \underset{\text{1-propanol}}{HOCH_2CH_2CH_3} \overset{H^+}{\rightleftharpoons} \underset{\text{Acetal}}{CH_3-\overset{\overset{\displaystyle OCH_2CH_2CH_3}{|}}{\underset{\underset{\displaystyle OCH_2CH_2CH_3}{|}}{C}}-CH_2CH_3} + H_2O$$

Take a moment to compare the acetals formed from aldehydes with those formed from ketones. Those formed from aldehydes have a hydrogen atom on the carbon that bears the two oxygen atoms. An acetal formed from a ketone has no hydrogen atom on this carbon atom.

Example 11.5 Reactions of Ketones

1. Predict the products for the **(a)** reduction, and **(b)** oxidation of 2-pentanone.
2. Draw the structure of the acetal that forms from 2-butanone and two moles of methanol.

SOLUTIONS

1. **(a)** Reduction of ketones yields a secondary alcohol. The carbonyl group at carbon 2 of the ketone becomes the hydroxyl group of the secondary alcohol, and so the product is 2-pentanol.

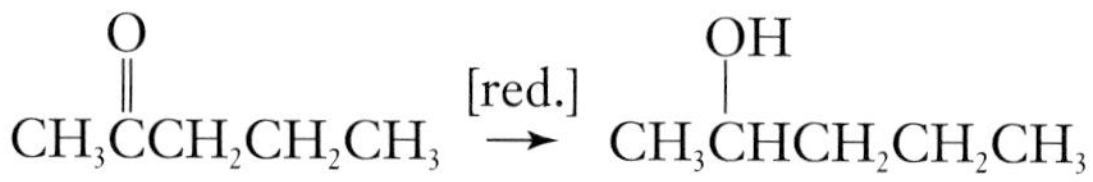

$$CH_3\overset{\overset{\displaystyle O}{\|}}{C}CH_2CH_2CH_3 \xrightarrow{[red.]} CH_3\overset{\overset{\displaystyle OH}{|}}{C}HCH_2CH_2CH_3$$

   (b) Ketones do not readily undergo oxidation, and so no reaction occurs.
2. The oxygen atom of each of the alcohol molecules would attach to the carbonyl carbon of the ketone. The ketone molecule would lose the oxygen from the carbonyl group, and the hydroxyl group of each of the alcohol molecules would lose their hydrogen atoms. The product would be

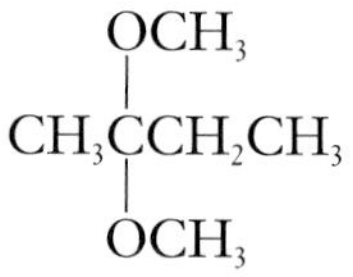

$$CH_3\overset{\overset{\displaystyle OCH_3}{|}}{\underset{\underset{\displaystyle OCH_3}{|}}{C}}CH_2CH_3$$

Self-Test

1. What ketone could be used to make 2-methyl-3-pentanol?
2. What is the structure of the hemiacetal made from 2-butanone and ethanol?

ANSWERS

1. 2-methyl-3-pentanone
2. $CH_3\overset{\overset{\displaystyle OCH_2CH_3}{|}}{\underset{\underset{\displaystyle OH}{|}}{C}}CH_2CH_3$

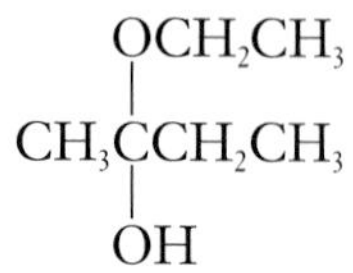

11.4 Carboxylic Acids

The most common organic acids are the **carboxylic acids.** A carboxylic acid contains the **carboxyl group,** which is a carbonyl group that has a hydroxyl group attached to it. The fourth bond to the carbon atom of the carbonyl group can be to a hydrogen atom, to an alkyl group, or to an aromatic ring:

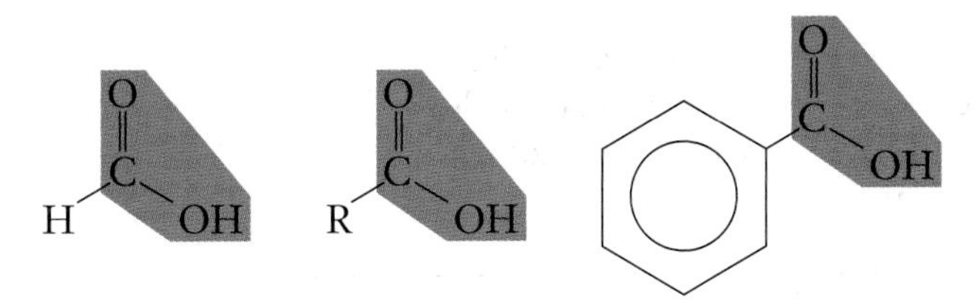

The condensed structural formula for the carboxyl group is drawn this way: —COOH .

Acids are hydrogen-ion donors, and in carboxylic acids it is the hydrogen on the hydroxyl group that is donated:

$$RCOOH \rightarrow RCOO^- + H^+$$

Remember that acids are hydrogen ion donors.

Carboxylic acids are common organic compounds, and you will see many more of them as you continue your studies of biochemistry and health care.

Naming Carboxylic Acids

The IUPAC rules for carboxylic acids are very similar to those for aldehydes and ketones. For a carboxylic acid, the suffix *-oic acid* replaces the *-e* in the corresponding alkane name. The carboxyl carbon is carbon 1 of the chain, and the rules for naming substituents are identical to those already given (Table 11.5).

The IUPAC name for the simplest aromatic carboxylic acid is benzoic acid:

Benzoic acid

Other aromatic carboxylic acids are named by placing the substituent name in front of *-benzoic acid.*

Table 11.5 IUPAC Rules for Naming Carboxylic Acids

Rule One	The longest continuous chain of carbon atoms that includes the carboxyl group is the parent. Combine the parent name and *an* with the suffix *-oic acid.*
Rule Two	Number the chain to make the carboxyl carbon carbon number 1.
Rule Three	The rules for substituents are the same as those for alkanes.
Rule Four	The rules for prefixes, commas, and hyphens are the same as for alkanes.

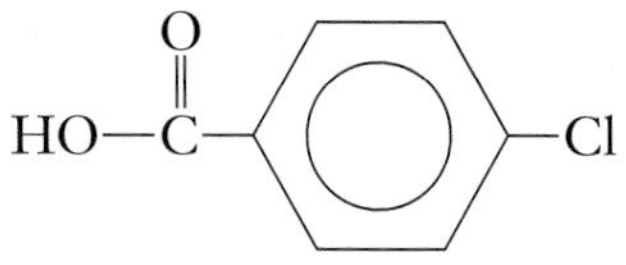

$$HO-\overset{\overset{O}{\|}}{C}-C_6H_4-Cl$$

4-chlorobenzoic acid
(p-chlorobenzoic acid)

Use the same location and punctuation rules you use for other aromatic compounds.

Some of the smaller carboxylic acids are identified by common names derived from the roots given for aldehydes in Table 11.3 and the suffix *-ic acid:*

$HCOOH$	CH_3COOH	CH_3CH_2COOH	$CH_3CH_2CH_2COOH$
Formic acid	Acetic acid	Propionic acid	Butyric acid

In these names, the locations of substituents are indicated with Greek letters:

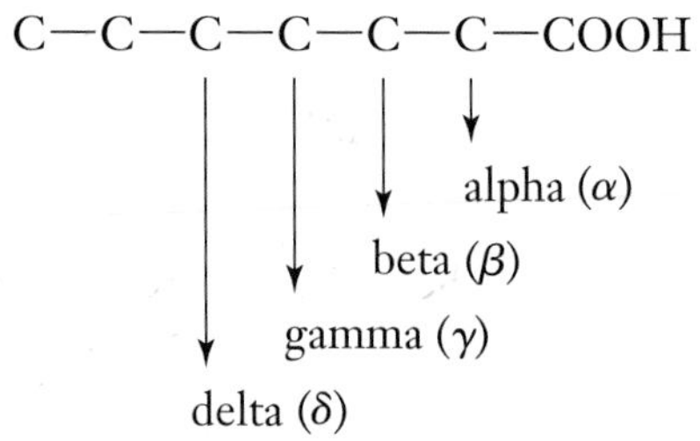

$$C-C-C-C-C-C-COOH$$

alpha (α)
beta (β)
gamma (γ)
delta (δ)

The acid named 3-hydroxybutanoic acid using IUPAC rules is called β-hydroxybutyric acid in common nomenclature:

$$CH_3\overset{\overset{OH}{|}}{C}HCH_2COOH$$

Some acids have common names that provide no information about structure. Lactic acid and pyruvic acid are examples you will see in biochemistry:

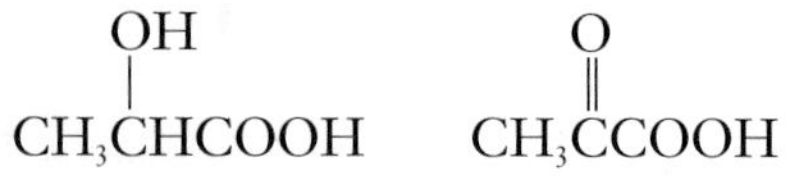

$$CH_3\overset{\overset{OH}{|}}{C}HCOOH \qquad CH_3\overset{\overset{O}{\|}}{C}COOH$$

Lactic acid   Pyruvic acid

Salicylic acid is an aromatic carboxylic acid that is the active form of aspirin:

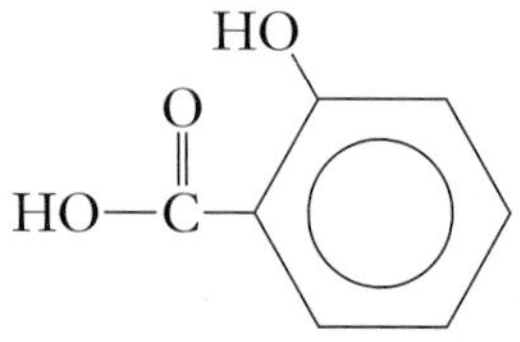

$$HO-\overset{\overset{O}{\|}}{C}-C_6H_4(OH)$$

Salicylic acid

Most commonly encountered dicarboxylic and tricarboxylic acids are usually identified by common names. Oxalic acid, succinic acid, citric acid, and isocitric acid are examples.

$$\text{HOOC—COOH} \qquad \text{HOOCCH}_2\text{CH}_2\text{COOH}$$

Oxalic acid   Succinic acid

$$\begin{array}{c} \text{CH}_2\text{COOH} \\ | \\ \text{HO—C—COOH} \\ | \\ \text{CH}_2\text{COOH} \end{array} \qquad \begin{array}{c} \text{CH}_2\text{COOH} \\ | \\ \text{H—C—COOH} \\ | \\ \text{HO—CH—COOH} \end{array}$$

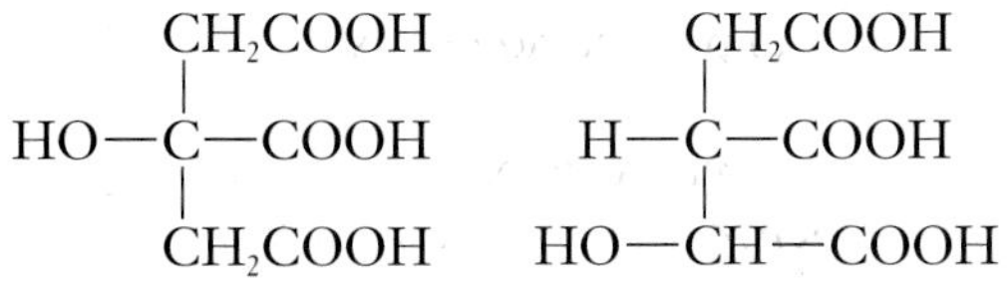

Citric acid   Isocitric acid

The leaves of the rhubarb plant contain poisonous oxalic acid. *(© Ann Reilly/PHOTO NATS)*

Example 11.6 Names and Structures of Carboxylic Acids

1. What are the IUPAC and common names of CH_3CHCH_2COOH? (Cl on the second carbon)

$$\begin{array}{l} \text{CH}_3\text{CHCH}_2\text{COOH} \\ \quad\;\; | \\ \quad\;\; \text{Cl} \end{array}$$

2. What is the IUPAC name for lactic acid?

SOLUTIONS

1. There are four carbon atoms in the chain. In IUPAC nomenclature, the parent is therefore *-butan-* and the suffix is *-oic acid: -butanoic acid.* The carboxyl carbon atom is carbon 1, and so the chloro group is on carbon 3: 3-chlorobutanoic acid. In common nomenclature, this four-carbon carboxylic acid is butyric acid. The chloro group is on the beta carbon, and so the name is β-chlorobutyric acid.
2. A look at the structure of lactic acid shows it to be a three-carbon carboxylic acid, and so the parent name is propanoic acid. The hydroxyl group is on carbon 2, and so the correct IUPAC name is 2-hydroxypropanoic acid.

Self-Test

1. What is the structure of 3-bromopentanoic acid?
2. What is the name of this compound? (Reminder: the $—NH_2$ group is the amino group.)

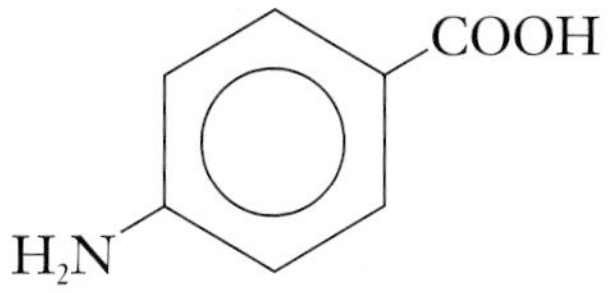

$$H_2N—C_6H_4—COOH \text{ (para)}$$

ANSWERS

1. 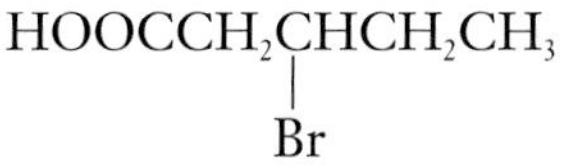 $$\begin{array}{l} \text{HOOCCH}_2\text{CHCH}_2\text{CH}_3 \\ \qquad\qquad\;\; | \\ \qquad\qquad\;\; \text{Br} \end{array}$$
2. 4-aminobenzoic acid (*p*-aminobenzoic acid). This compound, sometimes called PABA, is a good absorber of uv light, and is used in sunscreen lotions.

Naming Carboxylate Anions

Carboxylic acids, like other acids, donate a hydrogen ion in water:

$$\underset{\text{Carboxylic acid}}{R{-}COOH} + H_2O \rightleftharpoons \underset{\text{Carboxylate anion}}{R{-}COO^-} + H_3O^+$$

The anion formed when a carboxylic acid donates a hydrogen ion is called a **carboxylate anion.** To name a carboxylate anion, drop the *-ic acid* suffix from the carboxylic acid name and replace it with *-ate.* For instance, propanoic acid becomes propanoate, $CH_3CH_2COO^-$, and acetic acid becomes acetate, CH_3COO^-. Salts that contain a carboxylate anion are named like binary ionic compounds. The cation is named first, then comes the anion (the carboxylate) name. Sodium acetate is the correct name for the sodium salt formed from acetic acid:

$$\underset{\text{Sodium acetate}}{CH_3COO^-\ Na^+}$$

Sodium benzoate is used as a preservative in many foods and drinks, including some baked goods. *(Charles D. Winters)*

Example 11.7 The Names and Structures of Carboxylate Anions

1. What is the name for the carboxylate anion formed from benzoic acid?
2. What is the name of the salt formed from calcium ion and the carboxylate anion of propionic acid?

SOLUTIONS

1. The name is derived from the carboxylic acid name. Drop the *-ic acid* suffix to get *benzo-*. Now add the suffix *-ate:* benzoate.
2. This compound is named as a binary ionic compound. The cation is named first, then the carboxylate anion is named second. Propionic acid becomes propionate, and the name is calcium propionate.

Self-Test

1. What is the carboxylate anion of salicylic acid?
2. If potassium is the cation associated with the carboxylate anion of salicylic acid, what is the name of the salt?

ANSWERS

1. Salicylate anion
2. Potassium salicylate

Physical Properties of Carboxylic Acids

Both the carbonyl group and the hydroxyl group are polar, and so it should be no surprise that the carboxyl group is polar. In addition, carboxylic acids form hydrogen bonds. As a result of these relatively strong intermolecular interactions, all carboxylic acids are liquids or solids at room temperature. Carboxylic acids have boiling points higher than those of alcohols and aldehydes of the same size (Figure 11.3). Two carboxylic acid molecules (acetic acid is shown here) hydrogen-bond to each other to form a complex of two molecules (known as a dimer):

$$\begin{array}{c} \text{Hydrogen bond} \\ \downarrow \\ \begin{array}{ccc} & \text{OH}\cdots\text{O}=\text{C}-\text{CH}_3 \\ & |\qquad\quad\ \ | \\ \text{CH}_3- & \text{C}=\text{O}\cdots\text{HO} \end{array} \\ \uparrow \\ \text{Hydrogen bond} \end{array}$$

Because this dimer has twice the mass of a single molecule, a higher temperature is needed to boil carboxylic acids than would be needed for similarly sized aldehydes or alcohols. The smaller carboxylic acids are liquids, but decanoic acid (ten carbons) and longer ones are solids. Benzoic acid is a solid, as are many other aromatic carboxylic acids.

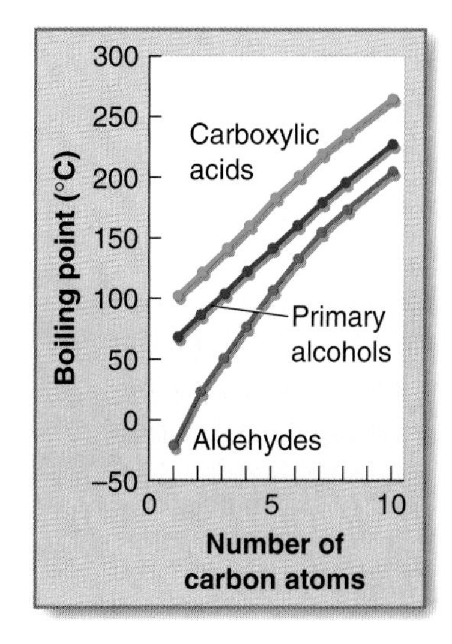

Figure 11.3 Carboxylic acids have boiling points higher than alcohols and aldehydes of similar size.

Connections

Acidosis—An Overabundance of Organic Acids

The body normally produces a number of carboxylic acids, but on some occasions too much acid is produced and the normal buffering capacity of blood is exceeded. When this occurs, *acidosis*, the abnormal lowering of blood pH, results. *Lactic acidosis* occurs when lactic acid exceeds normal ranges. This can occur during extreme exercise, hyperventilation, and in some diseases. *Diabetic acidosis* is caused by excess β-hydroxybutyric acid and acetoacetic acid. These compounds, which along with acetone are called the ketone bodies, are normally produced and maintained at low concentrations to provide fuel to the body. During a diabetic crisis, and during extreme fasting, these compounds accumulate (ketosis) and may cause acidosis.

$$\underset{\text{Lactic acid}}{CH_3-\overset{\displaystyle OH}{\overset{|}{C}H}-COOH} \qquad \underset{\beta\text{-hydroxybutyric acid}}{CH_3-\overset{\displaystyle OH}{\overset{|}{C}H}-CH_2-COOH} \qquad \underset{\text{Acetoacetic acid}}{CH_3-\overset{\displaystyle O}{\overset{\|}{C}}-CH_2-COOH}$$

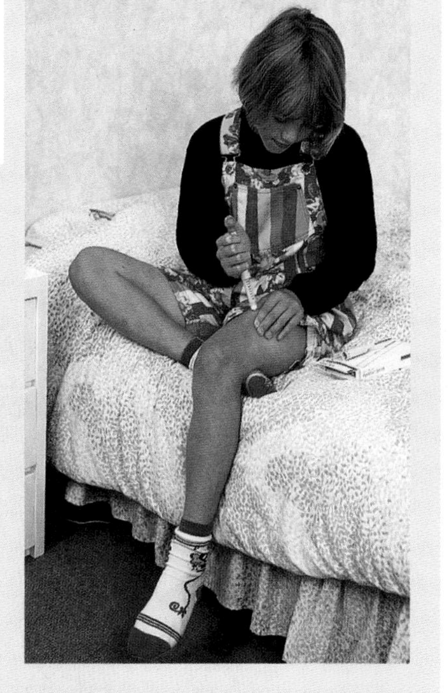

This diabetic is administering insulin, which will prevent excessive production of the carboxylic acids that are known as ketone bodies. Excess ketone bodies in the blood cause acidosis. *(Mark Clarke/Science Photo Library/Photo Researchers, Inc.)*

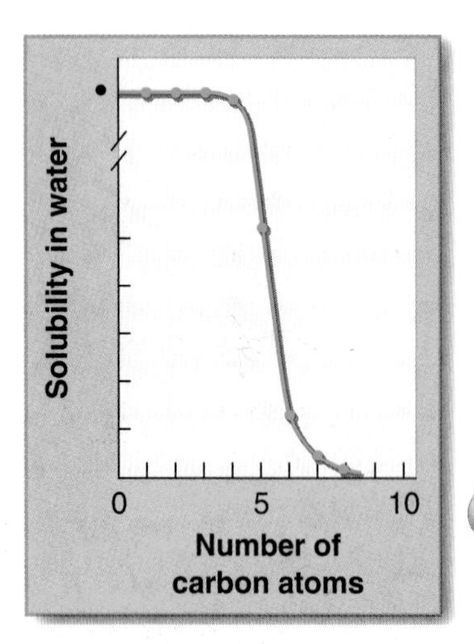

Figure 11.4 The solubility of carboxylic acids in water decreases with chain length because the increasing nonpolarity of the alkyl group or aromatic ring outweighs the polarity of the carboxyl group.

The smaller carboxylic acids are water-soluble, but those with more than five carbon atoms and most aromatic carboxylic acids are relatively insoluble (Figure 11.4). The solubility of the smaller acids is explained by their hydrogen-bonding to water molecules, whereas larger carboxylic acids are water-insoluble because of the nonpolarity of the relatively large alkyl group or aromatic ring. Carboxylic acids are soluble in many nonpolar solvents. Carboxylic acids appear to be soluble in basic aqueous solutions. This is because these acids react with the base to form carboxylate anions, which, like many other ions, are water-soluble. Actually, the apparent solubility of larger carboxylates requires some additional explanation that will be given in Chapter 13.

Reactions of Carboxylic Acids

The Acidity of Carboxylic Acids

Carboxylic acids are weak acids (Chapter 7), which means they ionize only partially when dissolved in water. When aqueous solutions of carboxylic acids are formed, the solutions are acidic, but most of the acid molecules remain as the carboxylic acid. In contrast, in neutral or basic solutions (when the pH is raised) most or all of the acid molecules have lost their hydrogen ion, and the carboxylate anion predominates. At the normal pH values found in the body (pH = 7.4), most carboxylic acids exist as carboxylate anions.

Acid–Base Reactions

When placed in water, some carboxylic acid molecules donate their hydrogen ions to water molecules, which act as a base:

$$\underset{\text{Carboxylic acid}}{RCOOH} + \underset{\text{Base}}{H_2O} \rightleftharpoons \underset{\text{Carboxylate anion}}{RCOO^-} + \underset{\text{Hydronium ion}}{H{-}OH_2^+}$$

The carboxylate anions can accept a hydrogen ion from an acid (such as hydronium ion). Because of this, the reaction of carboxylic acids with a weak base like water is shown as a reversible reaction.

Carboxylic acids also react with strong bases such as sodium hydroxide or potassium hydroxide to yield carboxylate salts:

$$\underset{\text{Carboxylic acid}}{RCOOH} + NaOH\,(aq) \rightarrow \underset{\text{Sodium salt of the carboxylate anion}}{RCOO^-\ Na^+\,(aq)} + H_2O$$

The reaction of acetic acid with potassium hydroxide provides a specific example:

$$\underset{\text{Acetic acid}}{CH_3COOH\,(l)} + \underset{\text{Potassium hydroxide}}{KOH\,(aq)} \rightarrow \underset{\text{Potassium acetate}}{CH_3COO^-\ K^+\,(aq)} + H_2O\,(l)$$

Esterification

Carboxylic acids react with alcohols to form esters. These reactions are catalyzed by acid, and water is a by-product:

$$\underset{\text{Carboxylic acid}}{\mathrm{R\overset{O}{\overset{\|}{C}}OH}} + \underset{\text{Alcohol}}{\mathrm{HOR'}} \overset{H^+}{\rightleftharpoons} \underset{\text{Ester}}{\mathrm{R\overset{O}{\overset{\|}{C}}OR'}} + H_2O$$

When acetic acid and ethanol react, they form ethyl acetate:

$$\underset{\text{Acetic acid}}{\mathrm{CH_3\overset{O}{\overset{\|}{C}}OH}} + \underset{\text{Ethanol}}{\mathrm{HOCH_2CH_3}} \overset{H^+}{\rightleftharpoons} \underset{\text{Ethyl acetate}}{\mathrm{CH_3\overset{O}{\overset{\|}{C}}OCH_2CH_3}} + H_2O$$

Ethyl acetate-containing products include some nail polishes, model airplane glues, and nail polish removers. *(Charles D. Winters)*

Formation of Amides

Amides are compounds formed from carboxylic acids and either ammonia or an amine. The reactants cannot simply be mixed together to form the amide because amines and ammonia, being basic, would react with the carboxylic acid in an acid–base reaction. Other synthetic routes must be used. In a commonly used reaction, ammonia (or an amine) reacts with an acid chloride (a reactive derivative of a carboxylic acid) to form an amide:

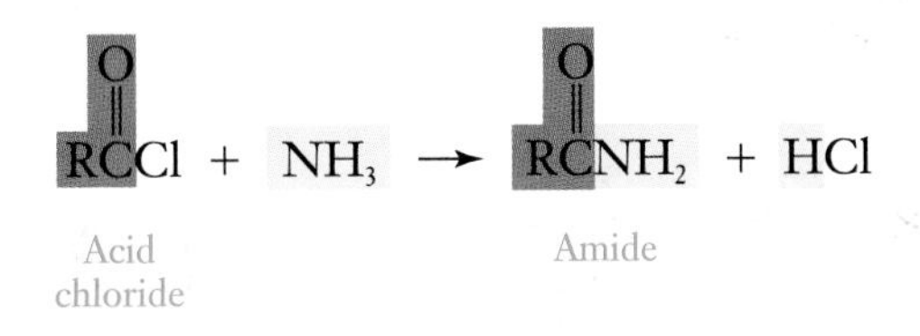

Nylon is both a polymer and an amide. Nylon is typically formed by the reaction between a diamine (it contains two amino groups) and a diacid chloride (it contains two acid chloride functional groups):

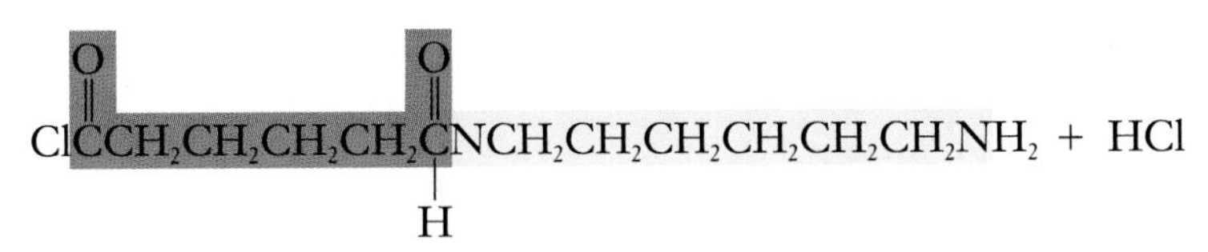

The amide that is formed still has a free amino group and a free acid chloride group. These groups can react with another molecule of the diamine at the acid chloride end of the molecule and with another diacid chloride molecule at the

Figure 11.5 The formation of Nylon 66 in the laboratory. An organic solvent containing the diacid chloride is layered over an aqueous solution of the diamine. The two compounds react with each other where the layers meet. The nylon that is formed is being pulled from the beaker and wrapped around a rod. *(Charles D. Winters)*

amino end. These reactions repeat, and the result is a polymer that consists of alternating units, one that comes from the diacid and the other that comes from the diamine. The product that is formed from the two molecules shown in Figure 11.5 is called Nylon 66.

Example 11.8 Reactions of Carboxylic Acids

1. What carboxylic acid and alcohol were used to make the ester

$$CH_3\overset{\overset{\displaystyle O}{\|}}{C}OCH_2CH_2CH_3\ ?$$

2. What is the structure of the amide formed in the reaction

$$CH_3CH_2CH_2\overset{\overset{\displaystyle O}{\|}}{C}{-}Cl + H_2NCH_2CH_3 \longrightarrow ?$$

SOLUTIONS

1. Both the carbonyl group and the methyl group attached to it came from the carboxylic acid. The two-carbon carboxylic acid is ethanoic acid (acetic acid). The noncarbonyl oxygen atom and propyl group came from the alcohol, which is 1-propanol.
2. The —Cl of the acid chloride is lost and replaced by the nitrogen atom of the amine. The nitrogen atom loses a hydrogen atom, leaving this structure:

$$CH_3CH_2CH_2\overset{\overset{\displaystyle O}{\|}}{C}{-}\underset{\underset{\displaystyle H}{|}}{N}CH_2CH_3$$

Self-Test

What is the structure of the ester formed in the reaction

$$HO{-}\overset{\overset{\displaystyle O}{\|}}{C}{-}C_6H_5 + HOCH_3 \overset{H^+}{\rightleftharpoons} ?$$

ANSWER

$$H_3C{-}O{-}\overset{\overset{\displaystyle O}{\|}}{C}{-}C_6H_5$$

11.5 Esters, Amides, and Other Derivatives of Carboxylic Acids

Esters

Several classes of carbonyl compounds are considered to be *derivatives* of carboxylic acids because carboxylic acids are used to make them. An **ester** is a derivative formed from a carboxylic acid and an alcohol or phenol. The generalized formula for an ester is

From the carboxylic acid → $R-\overset{\overset{\large O}{\|}}{C}-O-R'$ ← from the alcohol (or phenol)

↑ Ester bond

The condensed formula for an ester is RCOOR' , where R and R' are alkyl or aryl groups.

Many natural and synthetic compounds are esters. Aspirin is an ester made from acetic acid—the carboxylic acid, and salicylic acid—the phenol.

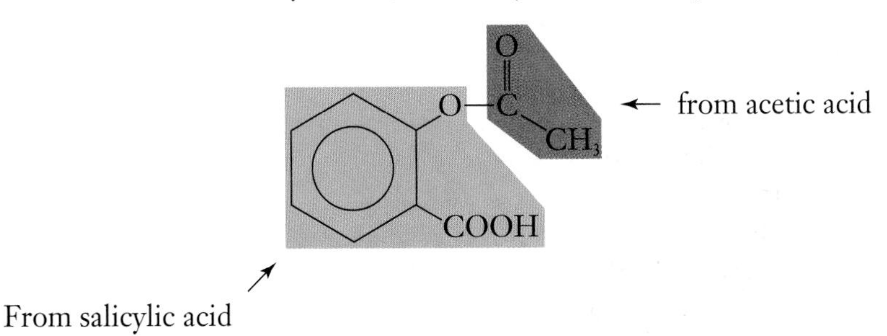

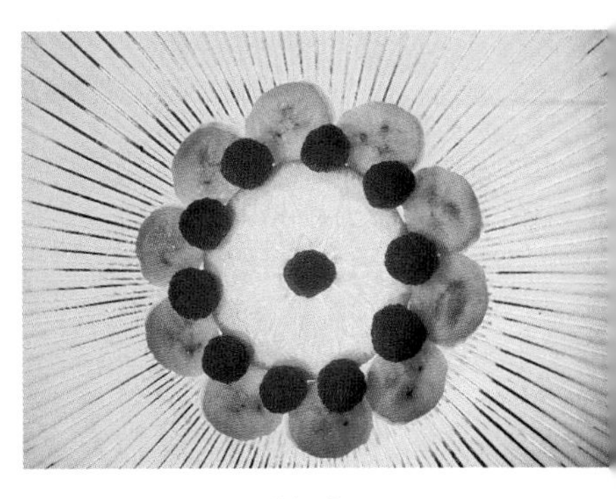

Esters are responsible for many common smells such as those of raspberries, bananas, and oranges. Isoamyl acetate is the principal component of the smell of bananas. *(Charles D. Winters)*

The fats and oils of the body and diet are esters. You will see examples of esters over and over again as you continue your studies.

Naming Esters

Both the IUPAC and common name of an ester are derived from the names of the carboxylic acid and the alcohol found in the ester. Consider the ester name *methyl ethanoate* as an example. The word *ethanoate* is the carboxylate anion name of ethanoic acid. The word *methyl* is the alkyl part of the alcohol methanol. Combining them yields *methyl ethanoate*, the name of the ester made from methanol and ethanoic acid. If the ester is made from phenol, then the word *phenyl* is used to indicate the presence of phenol. Common ester names are similar to the IUPAC name, but use the common carboxylate anion name of the carboxylic acid. The common name of methyl ethanoate is methyl acetate.

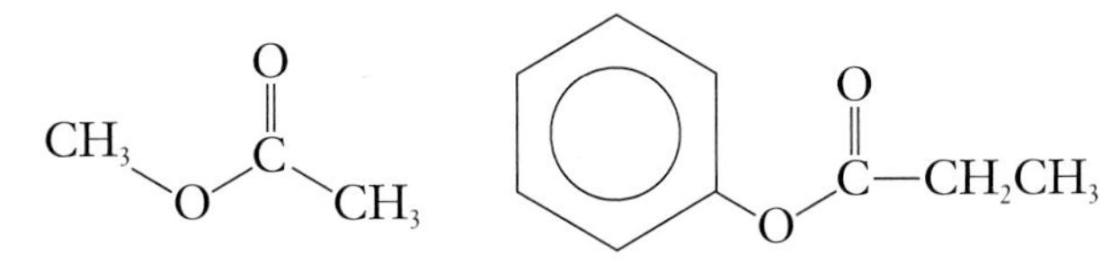

IUPAC: methyl ethanoate
Common: methyl acetate

IUPAC: phenyl propanoate
Common: phenyl propionate

Example 11.9 Names and Structures of Esters

Phenyl ethanoate is the ester formed from which carboxylic acid and alcohol?

SOLUTION

In the name of an ester, the alcohol is named using the *-yl* suffix, and then the carboxylic acid is named as its carboxylate anion. The alcohol in phenyl ethanoate is therefore phenol, and the carboxylic acid is ethanoic acid.

Self-Test

1. What is the name of $CH_3CH_2CH_2O\overset{\overset{O}{\|}}{C}CH_2CH_2CH_3$?
2. What is the structure of phenyl benzoate?

ANSWERS

1. Propyl butanoate
2. 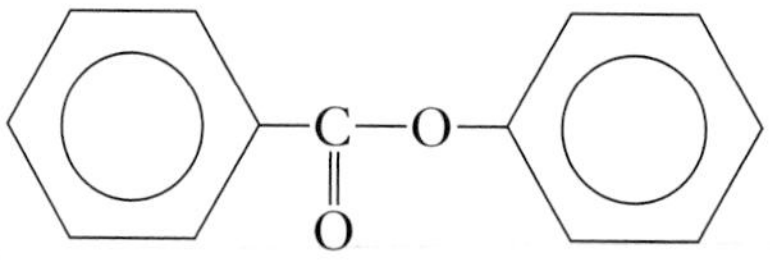 $C_6H_5-\underset{\underset{O}{\|}}{C}-O-C_6H_5$

Properties of Esters

The carbonyl group and other oxygen atom make an ester slightly more polar than hydrocarbons but less so than aldehydes. Esters are generally water-insoluble but they are soluble in organic solvents. The smaller esters are liquids at room temperature, and larger ones are solids.

Reactions of Esters

Esters undergo *hydrolysis*, which is cleavage of an ester by the addition of a water molecule to yield a carboxylic acid and an alcohol:

$$\underset{\text{Ester}}{R-\overset{\overset{O}{\|}}{C}-O-R'} + H_2O \xrightarrow{H^+} \underset{\text{Carboxylic acid}}{R-\overset{\overset{O}{\|}}{C}-OH} + \underset{\text{Alcohol}}{HOR'}$$

These hydrolyses are catalyzed (promoted) by either acids or bases. The reaction for the acid hydrolysis of isopropyl acetate, for example, is

$$\underset{\text{Isopropyl acetate}}{CH_3COOCH(CH_3)_2} + H_2O \xrightarrow{H^+} \underset{\text{Acetic acid}}{CH_3COOH} + \underset{\text{Isopropyl alcohol}}{HOCH(CH_3)_2}$$

Base hydrolysis of esters is also called *saponification*, and this important reaction will be discussed further in Chapter 13.

Connections
Aspirin

Early peoples used many plants for medicinal purposes. Among those used for pain relief were willow and birch bark and wintergreen leaves. Today we know these plant materials contain salicylic acid or related compounds and that these salicylates provided the pain relief. Pure salicylic acid is a poor choice as a drug, however, because it damages the stomach lining.

Aspirin is an ester in which the phenolic hydroxyl of salicylic acid has reacted with acetic acid (its synthesis is shown in Chapter 4). Aspirin has much less effect on the stomach, and when ingested it is hydrolyzed to salicylic acid, which is responsible for the observed medicinal effects. Aspirin was developed and first marketed in the 1890s and has become one of the most commonly used drugs in the world. Many billions of tablets are consumed annually as an analgesic (pain killer), as an antiinflammatory agent, and as an inhibitor of blood clotting.

Aspirin is not a risk-free drug, however. Gram quantities can be fatal to a child, and allergic reactions and stomach ulcers are possible side effects. In addition, children with flu-like symptoms should not be given aspirin because Reye's syndrome, which is a potentially fatal condition, could develop.

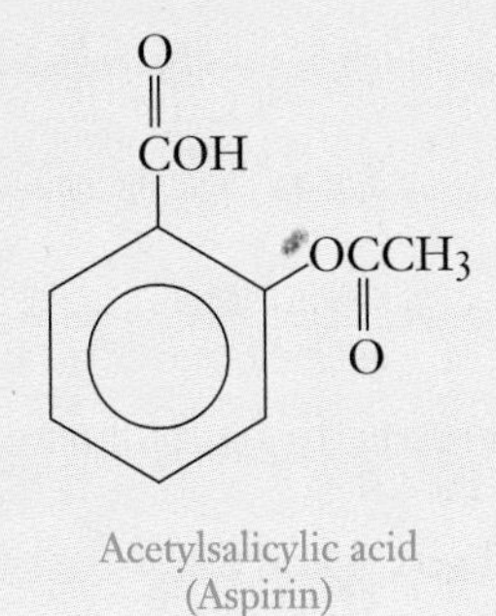

Acetylsalicylic acid (Aspirin)

Willow bark was the first source of salicylic acid, a compound with analgesic properties. Today, aspirin is synthesized from salicylic acid. *(Charles D. Winters)*

Amides

Amides are organic compounds that are formed from a carboxylic acid and either ammonia or an amine. The general structures for these two types of amides are

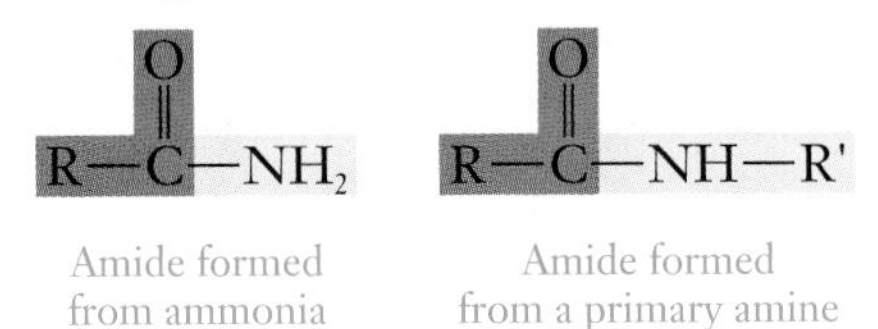

The condensed structural formula for an amide made from ammonia is $RCONH_2$. There are numerous examples of natural and synthetic amides, including nylon. Proteins are natural amides. Proteins are so important, an entire chapter of this textbook is devoted to them (Chapter 14).

Naming Amides

When naming an amide formed from ammonia, the parent name of the carboxylic acid is used (drop the *-oic acid* suffix) with the suffix *-amide.* Thus the amide made from propanoic acid and ammonia is called propanamide:

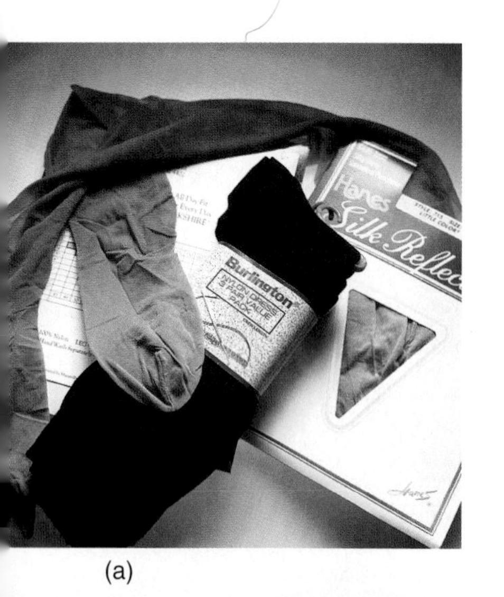

(a)

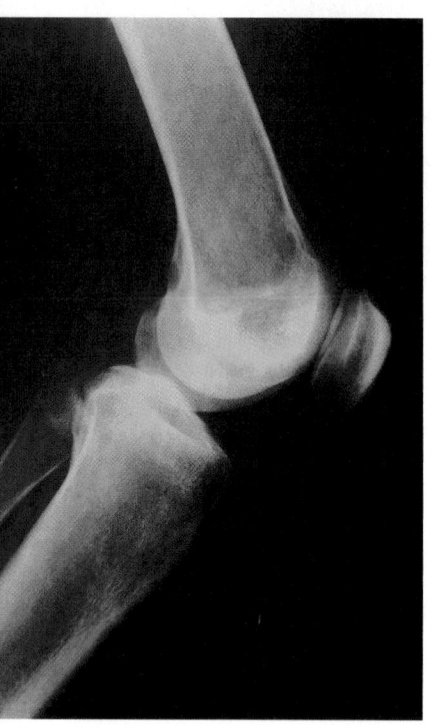

(b)

(a) The nylon in these fabrics is a polyamide. **(b)** The protein collagen is a principal component of bone and connective tissues. Proteins are natural polyamides. *(a, Charles D. Winters, b, Nicholas Veasey/Tony Stone Images)*

$$CH_3CH_2\overset{\overset{\displaystyle O}{\|}}{C}\text{—}NH_2$$

Propanamide

For an amide formed from an amine, an alkyl name is used to designate the alkyl part of the amine. The amide formed from methylamine and ethanoic acid is called *N*-methylethanamide:

$$CH_3\overset{\overset{\displaystyle O}{\|}}{C}\text{—}\overset{\overset{\displaystyle H}{|}}{N}\text{—}CH_3$$

N-methylethanamide

The prefix *N-methyl-* identifies the amine, the parent name *-ethan-* indicates the carboxylic acid, and the suffix *-amide* indicates that the compound is an amide.

The two smallest amides have common names that are frequently used. These names are derived from formic acid and acetic acid:

$$H\overset{\overset{\displaystyle O}{\|}}{C}\text{—}NH_2 \qquad CH_3\overset{\overset{\displaystyle O}{\|}}{C}\text{—}NH_2$$

Formamide (Methanamide) — Acetamide (Ethanamide)

Example 11.10 Names and Structures of Amides

1. What is the structure of butanamide?
2. What is the name of $C_6H_5\overset{\overset{\displaystyle O}{\|}}{C}\text{—}NHCH_2CH_3$? (benzene ring)

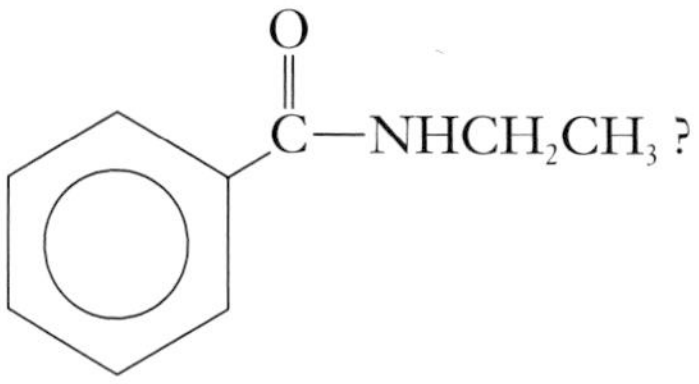

SOLUTIONS

1. The suffix *-amide* tells us this compound is derived from a carboxylic acid and ammonia. The root *butan-* indicates the carboxylic acid is butanoic acid. Combining this carboxylic acid and ammonia into an amide yields this structure:

$$CH_3CH_2CH_2\underset{\underset{\displaystyle O}{\|}}{C}NH_2$$

2. The carboxylic acid part of the amide is benzoic acid, thus the parent amide is *benzamide.* This amide has an ethyl group attached to a nitrogen atom, which is indicated by the prefix *N-ethyl.* Combining the prefix and parent yields *N-ethylbenzamide.*

Self-Test

1. What is the name of $H\underset{\underset{\displaystyle O}{\|}}{C}NHCH_3$?

2. What is the structure of *N*-isopropylpropanamide?

ANSWERS

1. *N-methylmethanamide* (*N*-methylformamide)
2. $CH_3CH_2\underset{\underset{\displaystyle O}{\|}}{C}NHCH(CH_3)_2$

Properties of Amides

Amides are relatively nonpolar, but hydrogen bonds do form between the oxygen atom of the carbonyl group of one amide and one of the hydrogen atoms on the nitrogen atom of another one.

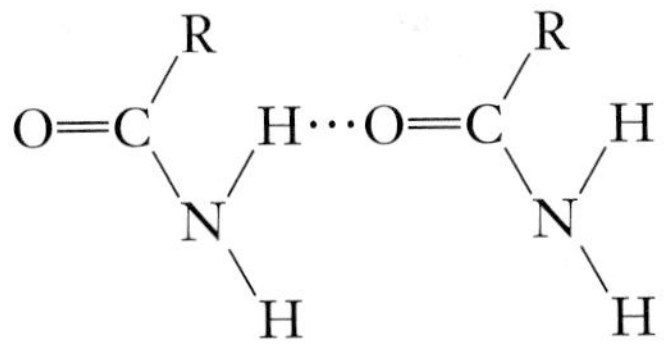

```
       R            R
      /            /
 O=C     H···O=C      H
    \   /         \   /
     N             N
      \             \
       H             H
```

Because of this relatively strong interaction, most amides are solids at room temperature. Amides are generally soluble in organic solvents, but only the smallest amides are water-soluble.

Reactions of Amides

An amide can be hydrolyzed to the parent carboxylic acid and ammonia/amine:

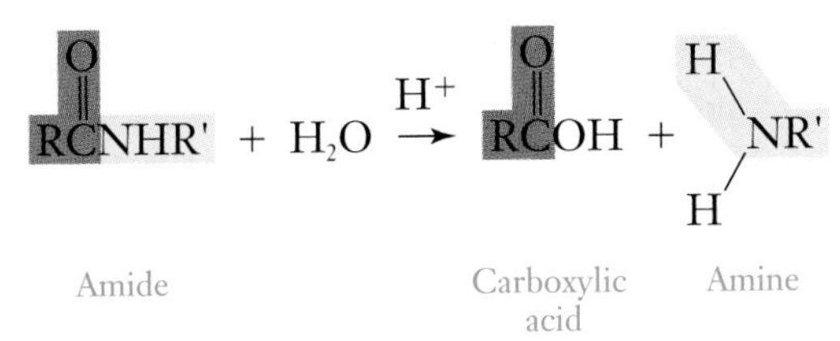

This is a very important reaction in biochemistry. Proteins are polymers of amino acids linked by amide (peptide) bonds. During digestion, these bonds are hydrolyzed to yield free amino acids, which are then absorbed (see Chapter 17).

Acid Chlorides, Acid Anhydrides, and Thiol Esters

Three more derivatives of carboxylic acids warrant attention. *Acid chlorides* consist of a chlorine atom attached to a carbonyl group. An acid chloride has the general structure

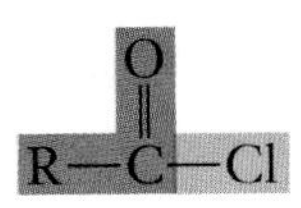

Acid chloride

Acid chlorides are quite reactive and are often used to introduce a carboxylic acid into an organic molecule. Acid chlorides are often identified by common names. By far the most common acid chloride is acetyl chloride:

$$CH_3-\overset{\overset{\displaystyle O}{\|}}{C}-Cl$$

Acetyl chloride

Other acid halides such as acetyl bromide exist, but they are not commonly encountered.

Anhydrous means without water, and an *acid anhydride* (also simply called an anhydride) is formed when two carboxylic acid molecules combine and lose a water molecule:

$$R-\overset{\overset{\displaystyle O}{\|}}{C}-OH + HO-\overset{\overset{\displaystyle O}{\|}}{C}-R \rightarrow R-\overset{\overset{\displaystyle O}{\|}}{C}-O-\overset{\overset{\displaystyle O}{\|}}{C}-R + H_2O$$

Carboxylic acid   Carboxylic acid   Acid anhydride

Some anhydrides, such as acetic anhydride, are used in organic synthesis to make esters or amides.

A third group of compounds derived from carboxylic acids are the *thiol esters.* These compounds are formed from carboxylic acids and thiols (Chapter 10). These compounds are represented with the general structure

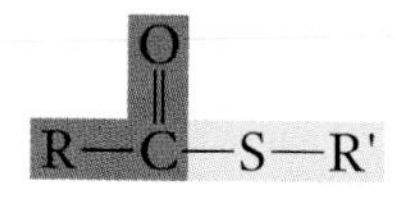

Thiol ester

The yellow part of this thiol ester comes from the thiol, and the rest comes from the carboxylic acid. Thiol esters have an important role in many of the reactions of the body, including synthesis and breakdown of fats.

CHAPTER SUMMARY

The **carbonyl group** consists of a carbon atom double-bonded to an oxygen atom. **Aldehydes** have a hydrogen atom bound to the carbon atom of a carbonyl group. In IUPAC nomenclature aldehydes are identified by the *-al* suffix. Aldehydes are more polar than hydrocarbons but less polar than alcohols. Aldehydes can be oxidized or reduced and react with one alcohol molecule to form a **hemiacetal** or two alcohol molecules to form an **acetal.** A **ketone** has a carbon atom bound to each of the two remaining bonds on the carbon atom of a carbonyl group. Ketones are recognized by the *-one* suffix in their name and are less polar than aldehydes. Ketones can be reduced, and like aldehydes react with alcohols to form hemiacetals and acetals. The **carboxyl group** has a hydroxyl group bound to the carbon atom of the carbonyl group, and this is the functional group of **carboxylic acids.** These acids are recognized by the *-oic acid* suffix in their name. The hydrogen atom of the carboxyl group is acidic, and the anion formed when a carboxylic acid loses its acidic hydrogen is called a **carboxylate anion.** These anions have the *-oate* suffix in their name. Carboxylic acids are more polar than alcohols, and carboxylate anions are ionic. Carboxylic acids are weak acids that react in acid–base reactions and exist as carboxylate anions at normal body pH. These acids react with several different molecules to form carboxylic acid derivatives. Several classes of organic compounds are made (derived) from carboxylic acids. An **ester** is made from a carboxylic acid and an alcohol, and an ester name includes information about both the alcohol and the carboxylic acid. Esters are relatively nonpolar. Amides are derived from a carboxylic acid and either an amine or ammonia. **Amide** names include the suffix *-amide.* Amides are relatively nonpolar but form hydrogen bonds. Amides and esters can be hydrolyzed to carboxylic acids and amines (ammonia) or an alcohol, respectively. If the hydroxyl group of a carboxylic acid is replaced by a chlorine atom, then the compound is an acid chloride. Acid anhydrides are made by dehydration of two carboxylic acids. Thiol esters are formed from a carboxylic acid and a thiol.

Aldehydes and Ketones

Oxidation (aldehydes only)

$$\text{Aldehyde} \xrightarrow{[ox.]} \text{carboxylic acid}$$

$$CH_3CHO \xrightarrow{[ox.]} CH_3COOH$$

Reduction

$$\text{Aldehyde} \xrightarrow{[red.]} 1° \text{ alcohol}$$

$$CH_3CHO \xrightarrow{[red.]} CH_3CH_2OH$$

$$\text{Ketone} \xrightarrow{[red.]} 2° \text{ alcohol}$$

$$CH_3\overset{\overset{O}{\|}}{C}CH_3 \xrightarrow{[red.]} CH_3\overset{\overset{OH}{|}}{C}HCH_3$$

Formation of hemiacetals and acetals

$$\text{Aldehyde} + \text{alcohol} \underset{}{\overset{H^+}{\rightleftharpoons}} \text{hemiacetal}$$

$$CH_3CHO + CH_3CH_2OH \overset{H^+}{\rightleftharpoons} CH_3\underset{\underset{H}{|}}{\overset{\overset{OH}{|}}{C}}OCH_2CH_3$$

$$\text{Ketone} + \text{alcohol} \overset{H^+}{\rightleftharpoons} \text{hemiacetal}$$

$$CH_3\overset{\overset{O}{\|}}{C}CH_3 + CH_3CH_2OH \overset{H^+}{\rightleftharpoons} CH_3\underset{\underset{CH_3}{|}}{\overset{\overset{OH}{|}}{C}}OCH_2CH_3$$

$$\text{Aldehyde} + 2 \text{ alcohol} \overset{H^+}{\rightleftharpoons} \text{acetal} + \text{water}$$

$$CH_3CHO + 2CH_3CH_2OH \overset{H^+}{\rightleftharpoons} CH_3\underset{\underset{H}{|}}{\overset{\overset{OCH_2CH_3}{|}}{C}}OCH_2CH_3 + H_2O$$

$$\text{Ketone} + 2 \text{ alcohol} \overset{H^+}{\rightleftharpoons} \text{acetal} + \text{water}$$

$$CH_3\overset{\overset{O}{\|}}{C}CH_3 + 2CH_3CH_2OH \overset{H^+}{\rightleftharpoons} CH_3\underset{\underset{CH_3}{|}}{\overset{\overset{OCH_2CH_3}{|}}{C}}OCH_2CH_3 + H_2O$$

Carboxylic acids

Acidity

$$\text{Carboxylic acid} + \text{strong base} \rightarrow \text{salt} + \text{water}$$

$$CH_3COOH + NaOH \rightarrow CH_3COO^-Na^+ + H_2O$$

Formation of Esters

$$\text{Carboxylic acid} + \text{alcohol} \overset{H^+}{\rightleftharpoons} \text{ester} + \text{water}$$

$$CH_3COOH + CH_3CH_2OH \overset{H^+}{\rightleftharpoons} CH_3COOCH_2CH_3 + H_2O$$

Formation of Amides

$$\text{Acid chloride} + \text{amine} \rightarrow \text{amide} + \text{hydrogen chloride}$$

$$CH_3COCl + CH_3CH_2NH_2 \rightarrow CH_3CONHCH_2CH_3 + HCl$$

Esters and Amides

Hydrolysis

$$\text{Ester} + \text{water} \xrightarrow{H^+} \text{carboxylic acid} + \text{alcohol}$$

$$CH_3COOCH_2CH_3 + H_2O \xrightarrow{H^+} CH_3COOH + CH_3CH_2OH$$

$$\text{Amide} + \text{water} \xrightarrow{H^+} \text{carboxylic acid} + \text{amine}$$

$$CH_3CONHCH_2CH_3 + H_2O \xrightarrow{H^+} CH_3COOH + CH_3CH_2NH_2$$

KEY TERMS

Section 11.1
carbonyl group

Section 11.2
acetal
aldehyde
hemiacetal

Section 11.3
ketone

Section 11.4
carboxylate anion
carboxyl group
carboxylic acid

Section 11.5
amide
ester

EXERCISES

The Carbonyl Group (*Section 11.1*)

1. Draw the structure of the carbonyl group.

2. List the common classes of organic compounds that possess a carbonyl group. Provide an example of a compound for each group.

3. Using either structures or words, show the difference between the structure of an aldehyde and a ketone.

Aldehydes (*Section 11.2*)

4. Provide the IUPAC name for each of these aldehydes.

(a) $HCCH_2CH_2CH_3$ (with $=O$ on the first C)

(b) CH_3CHCH_2CHCH (with CH_3 on C-4, Br on C-2, and $=O$ on C-1)

(c) 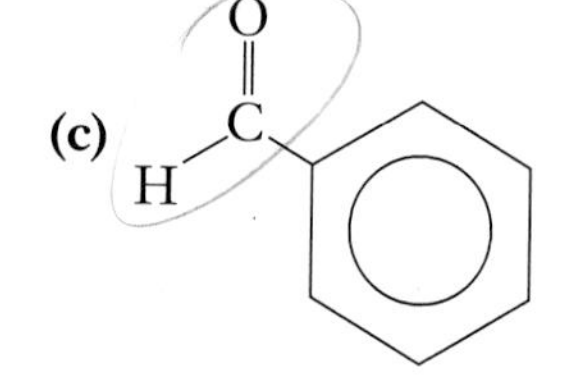

5. What are the IUPAC names of these compounds?

(a) $CH_3CHCH_2CHCH_2CH$ (with CH_2CH_3 on the second C, CH_3 on the fourth C, and $=O$ on the last C)

(b) $CH_3CH_2CHCHCHO$ (with Cl on the third C and Cl on the fourth C)

(c) 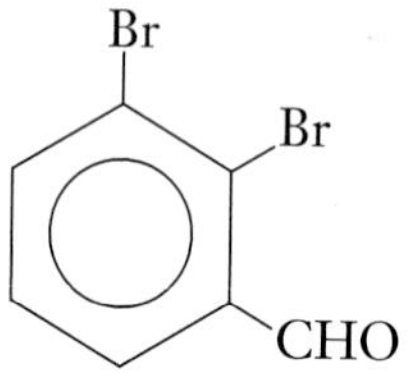 Benzene ring bearing Br, Br, and CHO on adjacent positions

6. State both the IUPAC and common name for each of these aldehydes.

(a) $HCHO$
(b) CH_3CHO
(c) CH_3CH_2CHO

7. Provide both the IUPAC and common name for each of these compounds.

(a) CH_3CHCHO (with Br on the middle C)

(b) $HOCH_2CH_2CH_2CHO$

(c) 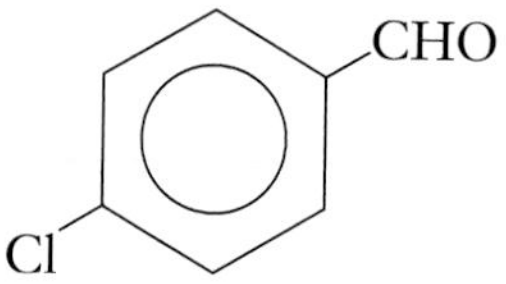 Benzene ring with CHO and Cl in para positions

8. Provide the structure for each of these aldehydes.
(a) Pentanal
(b) 2-ethylbutanal
(c) 3,5-dimethylhexanal
(d) 2-bromo-4-methylheptanal
(e) 2,6-dichlorobenzaldehyde

9. Draw the structures of these aldehydes.
(a) Butyraldehyde
(b) α-chloropropionaldehyde
(c) 3-bromo-2-hydroxyoctanal
(d) 9-phenylnonanal
(e) *o*-hydroxybenzaldehyde

10. Would hexanal dissolve more readily in water or ethanol? Support your answer.

11. Draw the structures and name the products of these reduction reactions:

(a) $CH_3CH_2CH_2CHO + H_2 \xrightarrow{Ni} ?$

(b) $CH_3CCHO + H_2 \xrightarrow{Pt} ?$ (with two CH_3 groups on the second C)

12. Draw the structures and name the products of these reduction reactions:

(a) $CH_3CHO + H_2 \xrightarrow{Ni} ?$

(b) $C_6H_5CHO + H_2 \xrightarrow{Pt} ?$

(Product name is benzyl alcohol)

13. Which aldehyde, when oxidized, would yield this product: $CH_3(CH_2)_7COOH$?

14. Aldehydes can be oxidized to produce carboxylic acids. Predict and draw the carboxylic acids made from these aldehydes.
(a) Acetaldehyde
(b) 2-methylbutanal

15. Predict and draw the structures of the carboxylic acids formed by oxidation of these aldehydes.
(a) Benzaldehyde
(b) 5-phenylpentanal

16. A positive result for the Tollen's test yields a striking result. Provide an equation summarizing the reaction of an aldehyde with silver ion (Ag^+) during a Tollen's test.

17. Which of the following would yield a positive test when treated with Tollen's reagent?
(a) CH_3CHO
(b) Acetone
(c) Propanamide
(d) Butyraldehyde

18. Which of the following would yield a positive test when treated with Benedict's solution?
(a) Propionaldehyde
(b) Cyclohexanone
(c) Benzoic acid
(d) Glucose

19. Molecules of what two classes of organic compounds are needed to produce a hemiacetal?

20. What is the difference between a hemiacetal and an acetal? Provide a sketch of a molecule of each formed from ethanal and ethanol to support your answer.

21. Provide the structure of the hemiacetal formed in each of these reactions.

(a) $CH_3CHO + CH_3OH \xrightleftharpoons{H^+} ?$

(b) $CH_3CHO + CH_3CH_2CH_2OH \xrightleftharpoons{H^+} ?$

22. What is the structure of the hemiacetal formed in each of these reactions?

(a) $CH_3CH_2CHO + CH_3OH \xrightleftharpoons{H^+} ?$

(b) $CH_3CHO + CH_3CH_2OH \xrightleftharpoons{H^+} ?$

***Ketones** (Section 11.3)*

23. Write the IUPAC names for these ketones.

(a) $CH_3\overset{O}{\overset{\|}{C}}CH_2CH_2CH_3$

(b) $CH_3\overset{CH_3}{\overset{|}{C}}HCH_2\underset{\underset{CH_3}{|}}{C}H\overset{O}{\overset{\|}{C}}CH_3$

(c) $C_6H_5\overset{O}{\overset{\|}{C}}CH_3$

(d) $ClCH_2\overset{O}{\overset{\|}{C}}CH_3$

24. Provide the IUPAC names for these compounds.

(a) $C_6H_5\overset{O}{\overset{\|}{C}}C_6H_5$

(b) $CH_3-\underset{\underset{CH_3}{|}}{\overset{H_3C}{\overset{|}{C}}}-\overset{O}{\overset{\|}{C}}-\underset{\underset{CH_3}{|}}{\overset{CH_3}{\overset{|}{C}}}-CH_3$

(c) $ClCH_2CH_2\overset{O}{\overset{\|}{C}}CH_3$

(d) $CH_3\underset{\underset{CH_3}{|}}{C}H\overset{O}{\overset{\|}{C}}CH_2CH_2CHBr_2$

25. What is the common name for each of these ketones?

(a) $CH_3CH_2CH_2\overset{O}{\overset{\|}{C}}CH_3$

(b) $CH_3CH_2\overset{O}{\overset{\|}{C}}CH_2CH_3$

26. Each of these ketones can be identified by what common name?

(a) $CH_3CH_2\overset{O}{\overset{\|}{C}}CH_2CH_2CH_3$

(b) $C_6H_5-\overset{O}{\overset{\|}{C}}-CH_2CH_2CH_2CH_3$

27. Draw the structures of these ketones.
(a) Acetone
(b) 4,4-dichloro-2-hexanone
(c) 1,3-dibromopropanone

28. What are the structures of these ketones?
(a) 6-methyl-3-octanone
(b) 7-ethyl-2-methyl-3-nonanone
(c) Isopropyl phenyl ketone

29. This name is incorrect: 5-octanone. Identify the problem or problems.

30. Why are the smallest ketones water-soluble and larger ketones water-insoluble?

31. When a ketone is reduced what product is formed?

32. Predict and draw the structure of the product of each of these reactions.

(a) $CH_3CH_2CH_2CH_2\overset{O}{\overset{\|}{C}}CH_2CH_3 \xrightarrow{[red.]} ?$

(b) $CH_3\overset{O}{\overset{\|}{C}}-C_6H_5 \xrightarrow{[red.]} ?$

33. Predict and draw the structure of the product of each of these reactions.

(a) $CH_3\overset{O}{\overset{\|}{C}}CH(CH_3)_2 \xrightarrow{[red.]} ?$

(b) $CH_3CH_2\overset{O}{\overset{\|}{C}}-C_5H_9 \xrightarrow{[red.]} ?$ (cyclopentyl)

34. What is the structure of the product of this reduction reaction?

$$CH_3CH_2\overset{O}{\overset{\|}{C}}CH_2\overset{CH_3}{\overset{|}{C}}HCH_3 + H_2 \xrightarrow{Ni} ?$$

35. What ketone would you use to make isopropyl alcohol by a reduction reaction?

36. Draw the structure of the hemiacetal that forms from the reaction of one mole of cyclohexyl methyl ketone and one mole of ethanol.

37. This acetal is formed from one mole of what ketone and two moles of what alcohol?

$$CH_3-\underset{OCH_2CH_3}{\underset{|}{\overset{OCH_2CH_3}{\overset{|}{C}}}}-CH_3$$

Carboxylic Acids (Section 11.4)

38. Draw the carboxyl group. This group is composed of what two other groups?

39. What part of the carboxyl group donates a hydrogen ion?

40. State the IUPAC names for these carboxylic acids.
(a) Cl_3CCOOH

(b) $HOOCCH_2\overset{OH}{\overset{|}{C}}HCH_2CH_3$

(c)

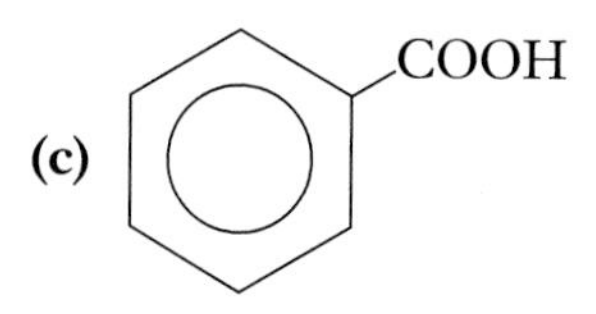

(d) $NH_2CH_2CH_2CH_2COOH$

41. What is the IUPAC name for each of these carboxylic acids?

(a) $CH_3CH_2\overset{OH}{\overset{|}{C}}HCH_2COOH$

(b)

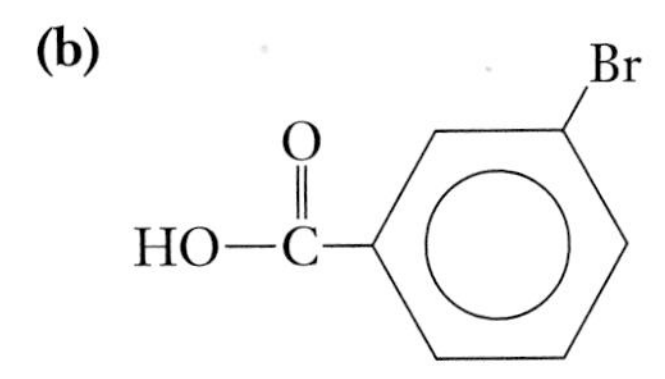

(c) $HOOCC(CH_3)_2CH_3$ (a central C carrying two CH_3 groups, one above and one below, and bonded to HOOC and CH_3)

(d) $CH_3CH_2CHCH_2CH_3$ with a COOH group on the central CH

42. Draw the structure for each of these carboxylic acids.
    (a) Acetic acid
    (b) 2-methylpropanoic acid
    (c) 2,4,6-trichlorobenzoic acid
    (d) Pyruvic acid
43. Provide the structure for each of these carboxylic acids:
    (a) Salicylic acid
    (b) *p*-hydroxybenzoic acid
    (c) γ-iodobutyric acid
    (d) Citric acid
44. What name is given to the anion of a carboxylic acid that has donated its hydrogen ion?
45. State both the IUPAC and common names of these carboxylate anions.
    (a) CH_3COO^-
    (b) Benzene ring bonded to C(=O)–O^-
    (c) $^-OOCCH_2CH_2CH_3$
46. Provide both the IUPAC and common names for these carboxylate anions.
    (a) $HCOO^-$
    (b) $^-OOCCH_2CH_3$
    (c) $CH_3CHCH_2COO^-$ with an OH group on the CH
47. Name the sodium salts that contain the carboxylate anions in Exercise 45.
48. Name the calcium salts that contain the carboxylate anions in Exercise 46.
49. Draw the structures of these carboxylate ions:
    (a) 3-bromobenzoate ion
    (b) 2,5-dimethylhexanoate ion
50. Draw the structures of these carboxylate ions:
    (a) 2-chloroethanoate ion
    (b) *o*-hydroxybenzoate ion
51. Complete these reactions giving both the structure and name of the product.
    (a) $CH_3CH_2COOH(aq) + NaOH(aq) \longrightarrow ?$
    (b) $(CH_3)_2CHCH_2CH_2CH_2COOH + KOH(aq) \longrightarrow ?$
    (c) Benzene ring–COOH + NaOH(aq) $\longrightarrow$?
52. Complete the reactions between these carboxylic acids and a base.
    (a) $CH_3COOH + NH_3(aq) \longrightarrow ?$
    (b) $HCOOH + KOH(aq) \longrightarrow ?$
    (c) Salicylic acid + excess NaOH(aq) $\longrightarrow$?
53. What is the carboxylic acid and alcohol needed to make each of these esters:
    (a) $HC(=O)OCH_2CH_3$
    (b) Benzene ring–C(=O)–O–CH_3
54. What is the carboxylic acid and alcohol needed to make each of these esters:
    (a) $CH_3CH_2CH_2C(=O)OCH(CH_3)_2$
    (b) $CH_3OC(=O)C(=O)CH_3$
55. What amide forms in each of these reactions?
    (a) $CH_3C(=O)Cl + NH_3 \longrightarrow ?$

(b) $C_6H_5-\overset{O}{\overset{\|}{C}}Cl + CH_3NH_2 \rightarrow$?

56. What reactants would you use to make this amide?

$CH_3CH_2CH_2\overset{O}{\overset{\|}{C}}NHCH_2-C_6H_5$

Esters, Amides, and Other Derivatives of Carboxylic Acids (Section 11.5)

57. An ester forms when what two kinds of organic compounds react?

58. Name these esters:

(a) $CH_3CH_2CH_2\overset{O}{\overset{\|}{C}}OCH_3$

(b) $CH_3-\overset{O}{\overset{\|}{C}}-O-C_6H_5$

(c) $CH_3\overset{CH_3}{\overset{|}{C}}HO\overset{O}{\overset{\|}{C}}CH_2CH_2\overset{Br}{\overset{|}{C}}HCH_3$

(d) $CH_3\overset{O}{\overset{\|}{C}}OCH_2CH_3$ (Both IUPAC and common)

59. Name these esters:

(a) $CH_3\overset{O}{\overset{\|}{C}}O-C_5H_9$ (cyclopentyl)

(b) $CH_3CH_2CH_2CH_2CH_2O\overset{O}{\overset{\|}{C}}CH_2CH_3$

(c) $H\overset{O}{\overset{\|}{C}}OCH_2CH_2CH_3$

(d) $C_6H_5-\overset{O}{\overset{\|}{C}}-O-CH_3$

60. What is the IUPAC name for each of the esters in Exercise 53?

61. What is the IUPAC name for each of the esters in Exercise 54?

62. Draw the structures of these esters:
(a) Butyl formate
(b) Propyl decanoate
(c) Ethyl lactate
(d) 2-methylpentyl ethanoate

63. Draw the structures of these esters:
(a) Methyl butyrate
(b) Cyclopentyl propionate
(c) Isopropyl benzoate
(d) Methyl salicylate

64. Draw the structure of the functional group found in an amide.

65. Provide the names for these amides.

(a) $H\overset{O}{\overset{\|}{C}}NH_2$

(b) $C_6H_5-\overset{O}{\overset{\|}{C}}-NH_2$

66. Provide the names for these amides.

(a) $NH_2\overset{O}{\overset{\|}{C}}CH_2CH_2CH_2CH_3$

(b) $CH_3NH\overset{O}{\overset{\|}{C}}CH_3$ (Hint: The naming of substituents on the nitrogen atom of an amide is the same as for amines.)

67. Sketch the structure of these amides.
(a) Acetamide
(b) 3-methylbutanamide

68. Draw the structure of these amides.
(a) Propionamide
(b) N-ethylethanamide (Hint: The naming of substituents on the nitrogen atom of an amide is the same as for amines.)

69. Predict the products formed when these amides are hydrolyzed.

(a) Propanamide
(b) Benzamide
(c) Nylon

70. Predict the products formed when these amides are hydrolyzed.

(a) $CH_3C(=O)NHCH_3$

(b) $CH_3CH_2CH_2CH_2C(=O)NH_2$

(c) 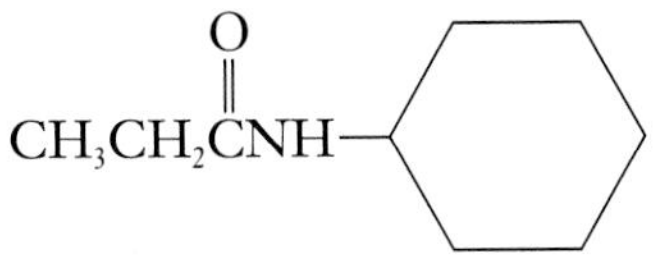 $CH_3CH_2C(=O)NH$–cyclohexyl

71. Draw the structures of the organic product formed in these reactions:

(a) $CH_3CH_2C(=O)Cl + NH_3 \rightarrow ?$

(b) $CH_3C(=O)Cl + NH(CH_3)_2 \rightarrow ?$

72. Draw the structures of the organic product formed in these reactions:

(a) 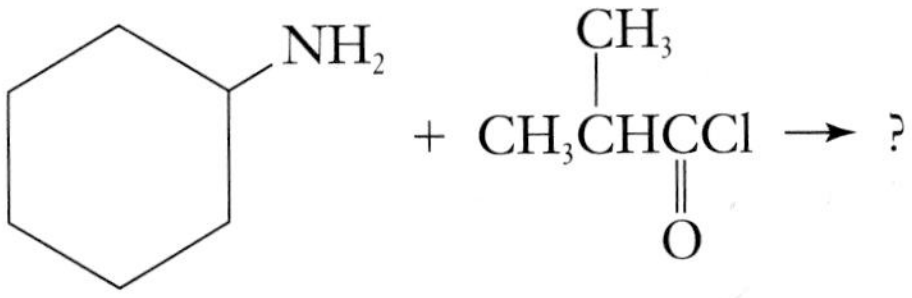 Cyclohexyl–NH_2 $+ CH_3CH(CH_3)C(=O)Cl \rightarrow ?$

(b) 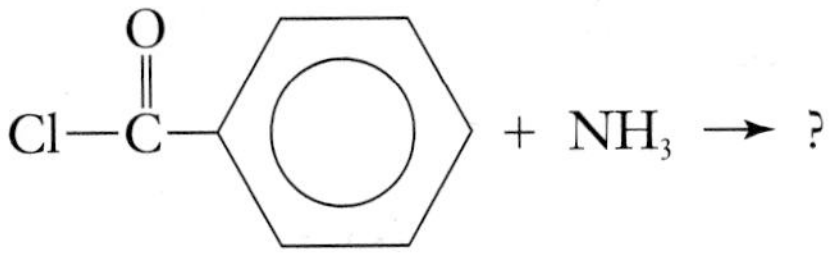 $Cl-C(=O)-$benzene ring $+ NH_3 \rightarrow ?$

73. The acid chlorides contain what functional group?

74. Draw the functional group of the acid anhydrides.

75. Compare the structure of a thiol ester to that of an ester.

76. Identify the general use of acid chlorides in the chemistry laboratory.

General Questions

77. Did an aldehyde or ketone react with two molecules of methanol to yield this acetal?

$$CH_3CH_2CH(OCH_3)-OCH_3$$

78. Name the aldehyde or ketone you identified in Exercise 77.

79. What is the structure of the acetal formed in each of these reactions?

(a) $CH_3CH_2CHO + 2CH_3OH \overset{H^+}{\rightleftharpoons} ?$

(b) $CH_3C(=O)CH_2CH_3 + 2CH_3OH \overset{H^+}{\rightleftharpoons} ?$

80. If an amine and a carboxylic acid are mixed together, what kind of reaction will occur?

81. These compounds belong to what class of carbonyl-containing compounds? What is the IUPAC name of each of them?
(a) $HCOOCH_2CH_2Br$
(b) $ClCH_2CH_2CH_2CHO$
(c) $CH_3CH_2COCH_2CH_3$
(d) $HOOCCH_2CH_2CH_3$

82. These compounds belong to what class of carbonyl-containing compounds? What is the IUPAC name of each of them?
(a) $CH_3CH_2CH_2COOCH_2CH_3$
(b) $HOOCCH_2CH_2COOH$
(c) $HCONHCH_3$
(d) $CH_3COCH_2CH_2CH_2CH_3$

Challenge Problems

83. Acidosis exists whenever blood pH is below its normal range. Alkalosis exists when blood is above its normal pH range. Using the library, the Internet, and other sources available to you, identify the chemical factors that contribute to alkalosis.

84. Oil of wintergreen contains methyl salicylate, a topical analgesic. Write a balanced equation for the synthesis of this compound. What structural relationship exists between the active compound of oil of wintergreen and aspirin?

Carbohydrates

12

SUGARS AND MACROMOLECULES THAT PROVIDE ENERGY AND STRUCTURE TO LIVING ORGANISMS.

BIOLOGICAL CHEMISTRY, or *biochemistry*, is the study of the chemistry of living organisms. This is not new chemistry for you; rather, it is the application of general and organic chemistry to living things. Biochemistry studies *biomolecules*, such as carbohydrates, lipids, proteins, and nucleic acids, and it also studies the reactions of the body and the catalysts that control them. We begin our study of biochemistry with a look at the carbohydrates.

Outline

12.1 Classes and Names of Carbohydrates

Identify a compound as a carbohydrate and distinguish among monosaccharides, disaccharides, and polysaccharides.

12.2 Monosaccharides

Recognize these common monosaccharides: glyceraldehyde, glucose, ribose, deoxyribose, fructose, and galactose.

12.3 Properties and Reactions of Sugars

Understand the formation of cyclic sugars and the glycosidic bond.

12.4 Oligosaccharides

Identify the common disaccharides and briefly describe their biological origin.

12.5 Polysaccharides

Recognize the common polysaccharides and describe their biological roles.

Breads, grains, and pasta are sources of carbohydrates.
(Charles D. Winters)

12.1 Classes and Names of Carbohydrates

We correctly perceive paper, cotton, crab shells, nectar, and blood as quite different substances. Yet all of these materials are related by composition—they all contain the sugar glucose. The glucose is not in the same form in these substances, though. Blood has glucose (blood sugar) in solution. Nectar contains sucrose (common table sugar), which contains glucose. Paper and cotton contain cellulose, which is a *macromolecule* (a very large molecule) that contains only glucose. Crab shells contain chitin, a macromolecule that is essentially a modified cellulose. Glucose is a widely distributed and abundant compound. Glucose is a carbohydrate.

Carbohydrates are polyhydroxy aldehydes or ketones, or derivatives of these compounds.

All of these materials contain glucose or modified glucose. *(Charles D. Winters)*

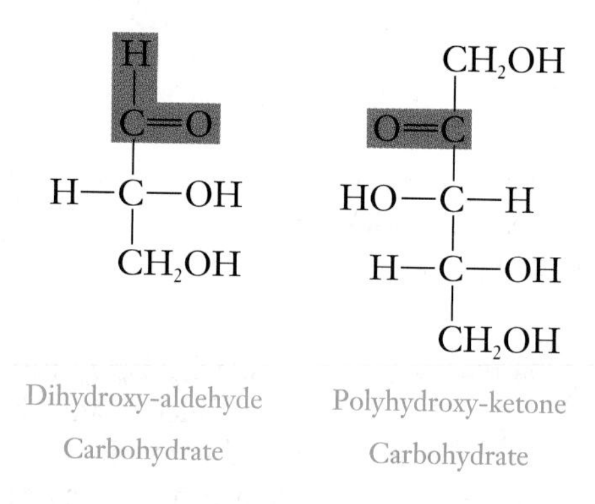

Although *poly-* normally means many, as used here it means two or more groups. Sugars, starch, and cellulose are common examples of carbohydrates.

The word *carbohydrate* comes from the early observation that some carbohydrates have formulas that can be written as $C_n(H_2O)_n$. In other words, they appeared to be "hydrates of carbon". An observation that seems to confirm this perception involves the reaction of a sugar and concentrated sulfuric acid. The sugar is converted to a black residue that is nearly pure carbon, and water is liberated as steam (Figure 12.1):

$$\underset{\text{Sugar}}{C_n(H_2O)_n(s)} \xrightarrow[\text{(Dehydrating agent)}]{H_2SO_4} \underset{\text{Carbon}}{nC(s)} + \underset{\text{Steam}}{n(H_2O)(g)}$$

Early chemists named these compounds for this property, and the name carbohydrate has been retained even though we now recognize that carbohydrates are the polyhydroxy-aldehydes and -ketones found in foods, plants, and animals.

The names of many carbohydrates contain the suffix **-ose.** Gluc*ose*, sucr*ose*, and cellul*ose* are clearly identified as carbohydrates. Some carbohydrates do not follow this simple rule of nomenclature—starch, glycogen, and chitin are examples. Carbohydrates can be named by IUPAC rules, but generally common names are used because they are simpler and are universally understood. The six examples cited in this paragraph are all common names.

Carbohydrates vary in size, with the larger ones made up of two or more smaller carbohydrate molecules. The smallest carbohydrates, the **monosaccha-**

Figure 12.1 Sugars react with concentrated sulfuric acid to yield carbon and steam. Most of the steam escapes, leaving a black residue of carbon. This reaction supports the name *hydrates of carbon*. *(Charles D. Winters)*

rides or *simple sugars*, have the general formula $C_nH_{2n}O_n$. For example, glucose is a monosaccharide with the formula $C_6H_{12}O_6$. Glucose and the other monosaccharides are the basic units found in larger carbohydrates.

Oligosaccharides are carbohydrates that contain between two and ten monosaccharides covalently linked together. Sucrose, which is both table sugar and the sugar found in nectar, contains two monosaccharides, glucose and fructose. Oligosaccharides are often identified by the number of monosaccharides present in the molecule. For instance, the *di*saccharide lactose contains *two* monosaccharides, and the *tetra*saccharide maltotetrose contains *four* monosaccharides. Most of the oligosaccharides discussed in this textbook are disaccharides.

Cellulose, starch, and chitin are very large carbohydrates. These substances are classified as **polysaccharides,** where *poly-* means many. Thus polysaccharides contain many monosaccharide units. For our purposes, a carbohydrate is classified as a polysaccharide if it contains more than ten monosaccharide units.

Self-Test

Classes and Names of Carbohydrates

1. How many monosaccharide units are present in a trisaccharide?
2. What name is given to a carbohydrate that contains many monosaccharide units?
3. How many monosaccharides could be in an oligosaccharide?

ANSWERS

1. Three
2. Polysaccharide
3. Two to ten

12.2 Monosaccharides

Monosaccharides are sugars that are found both free in nature and as components of other compounds. Monosaccharides vary from one another (1) by the number of carbon atoms that are present in the chain and (2) by the functional group present. The number of carbon atoms in a monosaccharide is indicated by these prefixes: *tri-* (for three carbons), *tetr-* (four carbons), *pent-* (five carbons), *hex-* (six carbons), and *hept-* (seven carbons). This numerical prefix is combined with the suffix *-ose* to yield a name that indicates the size of a monosaccharide. A *triose* is a monosaccharide containing three carbon atoms, and a *pentose* contains five carbon atoms (Table 12.1).

When the functional group in a sugar is used for classification, sugars containing an *ald*ehyde group are **aldoses,** and those containing a *ket*o group are **ketoses.**

Table 12.1 Some Representative Monosaccharides

COMMON NAME	MOLECULAR FORMULA	STRUCTURE	BY CHAIN LENGTH	CLASSIFICATION BY FUNCTIONAL GROUP	BY BOTH VARIABLES
Glyceraldehyde	$C_3H_6O_3$	H | 1 C=O | H—2 C—OH | 3 CH_2OH	Triose	Aldose	Aldotriose
Ribose	$C_5H_{10}O_5$	H | 1 C=O | H—2 C—OH | H—3 C—OH | H—4 C—OH | 5 CH_2OH	Pentose	Aldose	Aldopentose
Fructose	$C_6H_{12}O_6$	1 CH_2OH | O=2 C | HO—3 C—H | H—4 C—OH | H—5 C—OH | 6 CH_2OH	Hexose	Ketose	Ketohexose

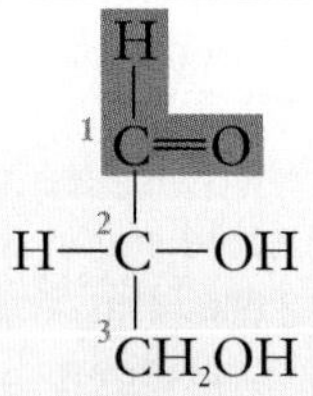

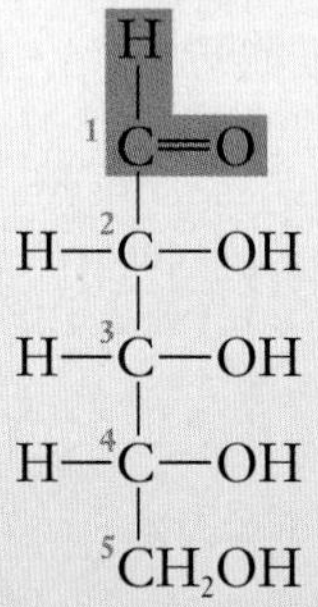

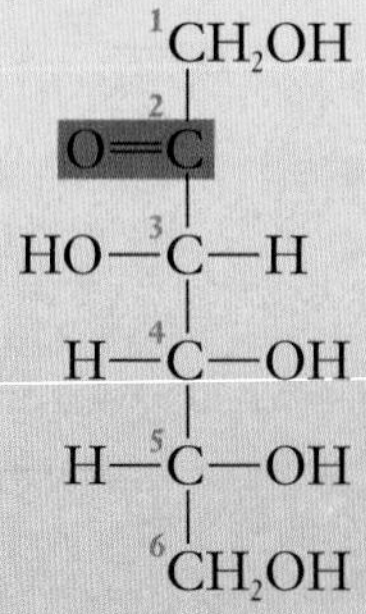

When a sugar is classified by both variables, by functional group and chain length, the *aldo-* or *keto-* prefix is placed before the prefix for chain length: ketohexose is an example (Table 12.1).

Example 12.1 Classification of Monosaccharides

Classify this monosaccharide first with respect to the number of carbon atoms in the chain, then by the functional group present, and finally by both chain length and functional group.

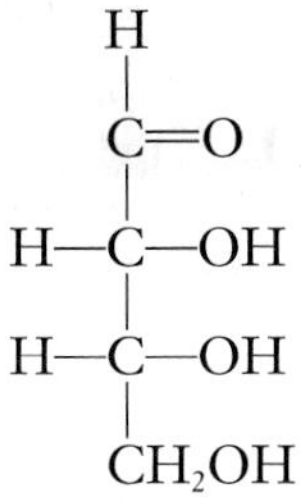

```
       H
       |
       C=O
       |
    H—C—OH
       |
    H—C—OH
       |
       CH2OH
```

SOLUTION

Because this molecule contains four carbon atoms, *tetr-* is the correct prefix, and this sugar is a tetrose. The molecule contains an aldehyde group, and so it is an aldose. These two designations can be combined to yield aldotetrose.

Self-Test

Give the combined classification name for

```
(a)   H              (b)  CH2OH            (c)  CH2OH
      |                   |                     |
      C=O                 C=O                   C=O
      |                   |                     |
   H—C—OH              H—C—OH                H—C—OH
      |                   |                     |
      CH2OH               CH2OH              H—C—OH
                                                |
                                                CH2OH
```

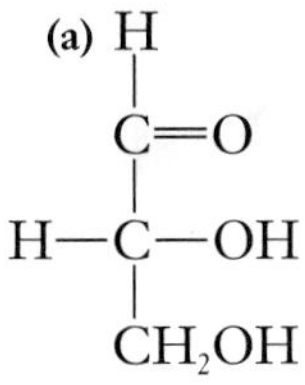

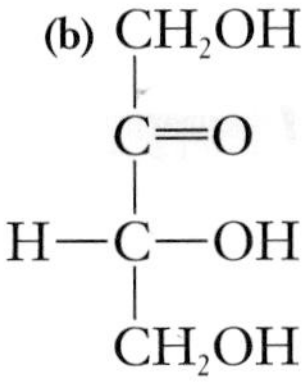

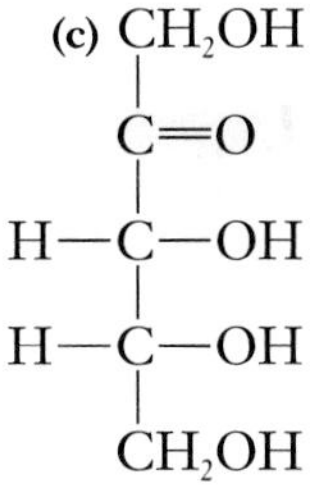

ANSWERS:

(a) Aldotriose **(b)** Ketotetrose **(c)** Ketopentose

Stereochemistry of Monosaccharides: D- and L- Families

Monosaccharides, like many other biological compounds, possess one or more chiral centers. Recall that a chiral center is an atom having four different substituents attached to it (see Chapter 8). The presence of a chiral center means two stereoisomers—two enantiomers—can exist, and they are mirror images of each other. Consider the aldotriose glyceraldehyde, which is the smallest monosaccharide containing a chiral center. Two stereoisomers of glyceraldehyde exist and they are distinguished from each other by using the prefixes **D-** and **L-** to designate stereochemistry:

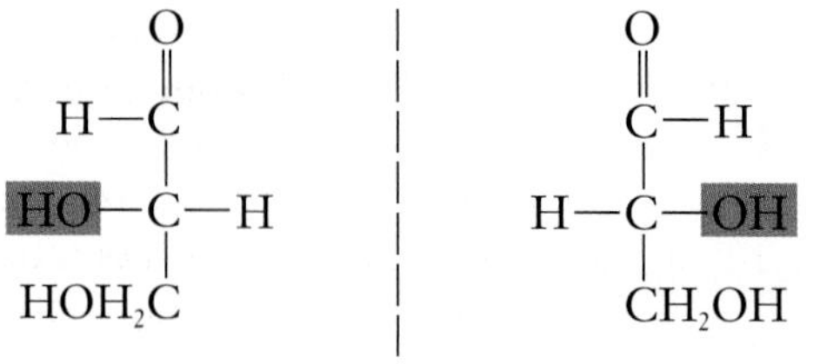

```
        O                    O
        ‖                    ‖
     H—C                    C—H
        |                    |
    HO—C—H               H—C—OH
        |                    |
   HOH2C                   CH2OH
```

L-Glyceraldehyde | D-Glyceraldehyde

Mirror

By convention, when glyceraldehyde is drawn with the carbonyl group at the top (as drawn here), the enantiomer with the shaded —OH group on the left side of the molecule is called L-glyceraldehyde. The other enantiomer—the one with the shaded —OH group on the right side of the molecule—is designated D-glyceraldehyde. The L- and D- are derived from the Latin words *levo-* (left) and *dextro-* (right).

The stereochemistry of the glyceraldehydes is used to assign all other carbohydrates as D- or L-. If more than one chiral center is present in the molecule, the one most distant from the carbonyl group is used to determine stereochemistry. For example, the middle four carbon atoms of D-glucose and L-glucose are chiral centers, but the highlighted hydroxyl group on the last chiral center is used to determine their D- or L-designation:

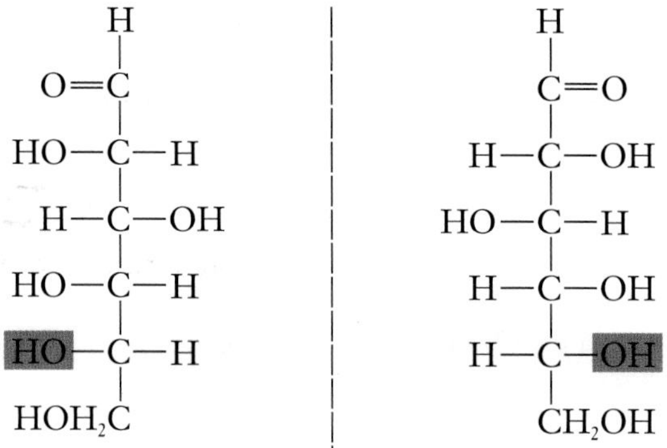

```
         H                    H
         |                    |
       O=C                    C=O
         |                    |
     HO—C—H               H—C—OH
         |                    |
      H—C—OH             HO—C—H
         |                    |
     HO—C—H               H—C—OH
         |                    |
     HO—C—H               H—C—OH
         |                    |
    HOH2C                   CH2OH
```

Mirror

L-Glucose | D-Glucose

Note that L-glucose is the mirror image (the enantiomer) of D-glucose. Most naturally occurring sugars belong to the D- family.

Example 12.2 Stereochemical Families of Monosaccharides

This monosaccharide belongs to which stereochemical family?

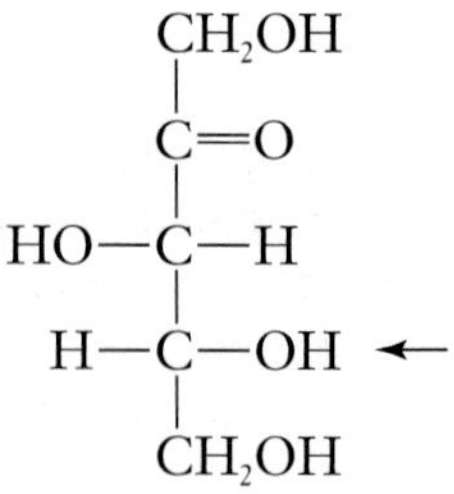

```
        CH2OH
         |
         C=O
         |
     HO—C—H
         |
      H—C—OH  ←
         |
        CH2OH
```

SOLUTION

Although this simple sugar is a ketose, the rules remain the same. When determining the stereochemistry of these sugars, draw the molecule with the carbonyl group near the top, as done in this example. Now use the chiral center most distant from the carbonyl group (the one with the arrow pointing toward it) to determine which stereoisomer you have. The hydroxyl group is on the right, and so this is the D- stereoisomer.

Self-Test

1. This monosaccharide belongs to what stereochemical family?

$$\begin{array}{c} CH_2OH \\ | \\ C{=}O \\ | \\ HO{-}C{-}H \\ | \\ CH_2OH \end{array}$$

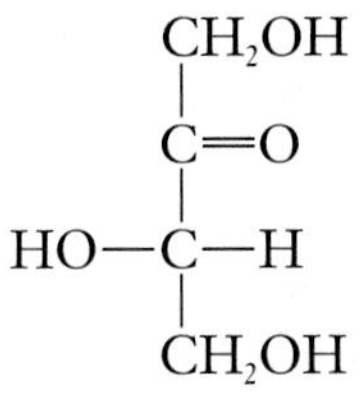

2. The three sugars in Table 12.1 belong to what stereochemical family?

ANSWERS

1. L- 2. All are D-.

Cyclic Forms of Monosaccharides

Hemiacetals (Chapter 11) form when an alcohol reacts with an aldehyde or ketone:

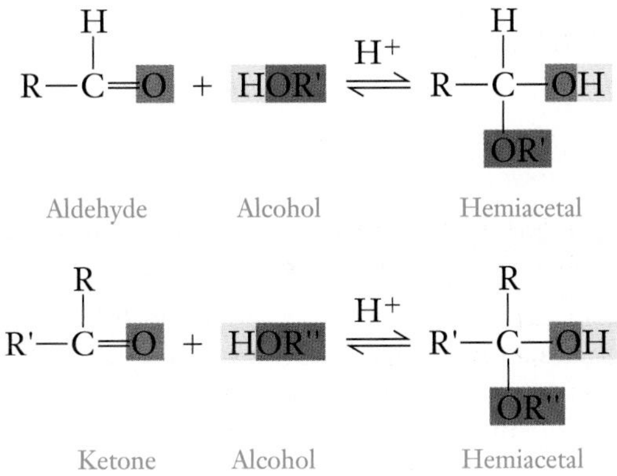

$$\underset{\text{Aldehyde}}{R{-}\overset{\displaystyle H}{\overset{|}{C}}{=}O} + \underset{\text{Alcohol}}{HOR'} \overset{H^+}{\rightleftharpoons} \underset{\text{Hemiacetal}}{R{-}\overset{\displaystyle H}{\overset{|}{\underset{\underset{\displaystyle OR'}{|}}{C}}}{-}OH}$$

$$\underset{\text{Ketone}}{R'{-}\overset{\displaystyle R}{\overset{|}{C}}{=}O} + \underset{\text{Alcohol}}{HOR''} \overset{H^+}{\rightleftharpoons} \underset{\text{Hemiacetal}}{R'{-}\overset{\displaystyle R}{\overset{|}{\underset{\underset{\displaystyle OR''}{|}}{C}}}{-}OH}$$

Because sugars contain both hydroxyl groups, which are the functional groups of alcohols, and an aldehyde or ketone group, *intramolecular* hemiacetals are possible, and in some sugars, quite likely. Pentoses and hexoses form intramolecular hemiacetals that are rings with five or six atoms. Ribose and other aldopentoses form intramolecular, five-atom-ring, hemiacetals through the reaction of the hydroxyl group on carbon 4 with carbon 1:

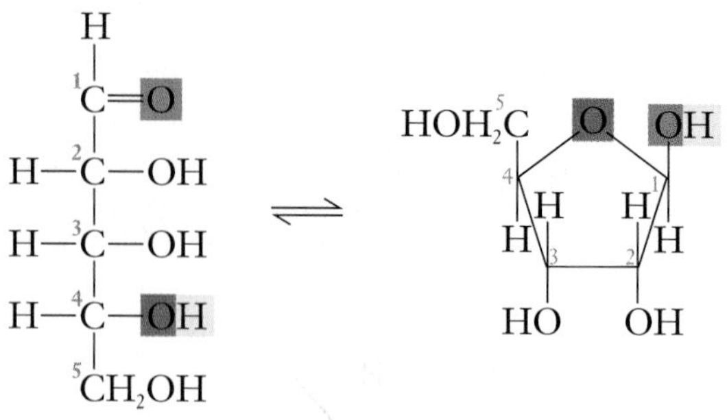

D-Ribose ⇌ D-Ribose

Monosaccharides containing rings with five atoms are called **furanoses** because their ring resembles the ring in the cyclic ether *furan:*

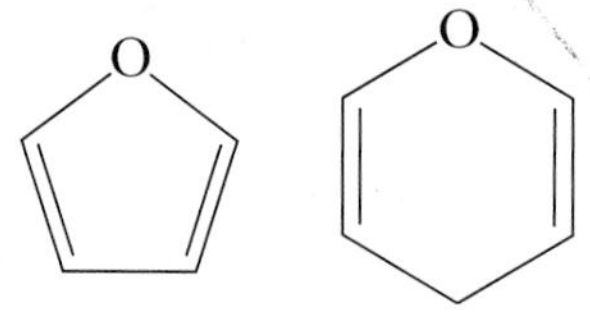

Furan Pyran

Glucose forms an intramolecular hemiacetal that has a ring with six atoms:

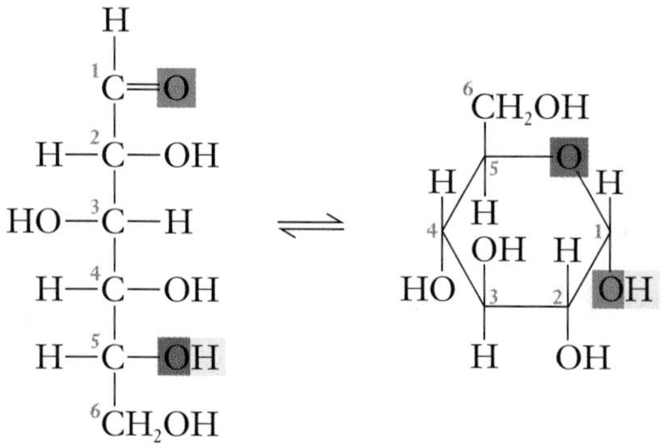

D-Glucose ⇌ D-Glucose

A monosaccharide containing a six-atom ring is called a **pyranose** because the ring resembles the ring of the cyclic ether *pyran.*

A sugar containing fewer than five carbon atoms does not form a ring but instead remains in the linear form.

The introduction of a ring into a sugar results in the formation of two cyclic isomers. D-Glucose, for instance, forms two different pyranoses, as Figure 12.2 shows. Both isomers form through a reversible reaction between the carbonyl group and the hydroxyl group on carbon 5. Notice in Figure 12.2 that the products of the reaction differ with respect to the orientation of the hydroxyl group on carbon atom number 1. The oxygen atom of the hydroxyl group on carbon 5 can attack *either* side of the carbonyl carbon, yielding two isomers.

The carbon atom that had been the carbonyl carbon before ring formation is called the **anomeric carbon,** and the two isomers that form are called **anomers.** In the ring form, the anomeric carbon is bonded to a hydroxyl group that can be

α-D-glucopyranose (α-D-glucose) D-glucose β-D-glucopyranose (β-D-glucose)

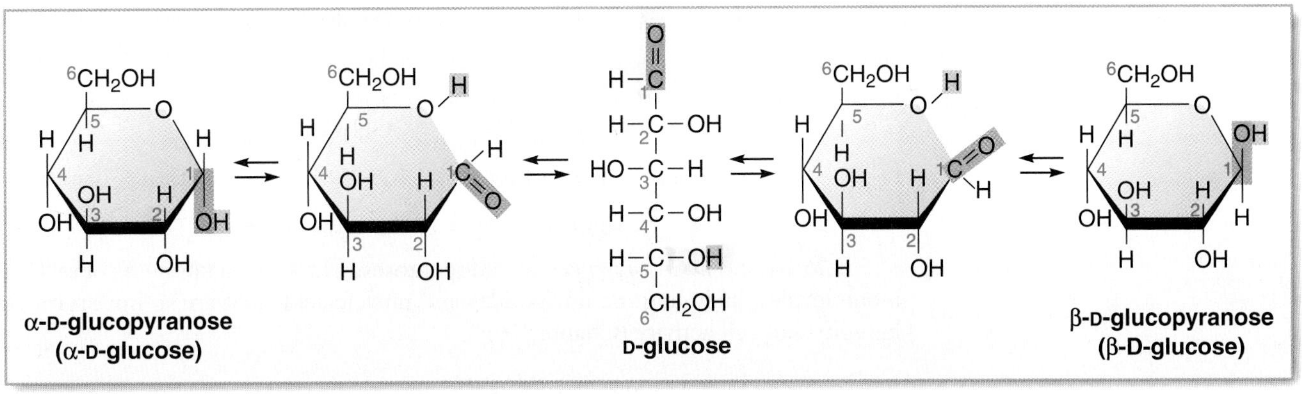

Figure 12.2 The formation of α-D-glucose and β-D-glucose from the open-chain form of glucose. Both cyclic forms exist because the hydroxyl group on carbon 5 can add to either side of the carbonyl carbon.

drawn either above the ring, on the same side as the —CH_2OH (up), or below the ring (down). If the anomeric hydroxyl group is drawn up, the anomer is the beta (β) anomer; the down orientation is called the alpha (α) anomer.

The anomers of glucose are formally called α-D-glucopyranose and β-D-glucopyranose, but are commonly referred to as α-D-glucose and β-D-glucose, respectively. Because the reaction that produces the ring forms of sugars is reversible, these rings can revert back to their chain forms.

Self-Test

Cyclic Forms of Monosaccharides

1. Identify the rings in deoxyribose, fructose, and galactose as either furans or pyrans.

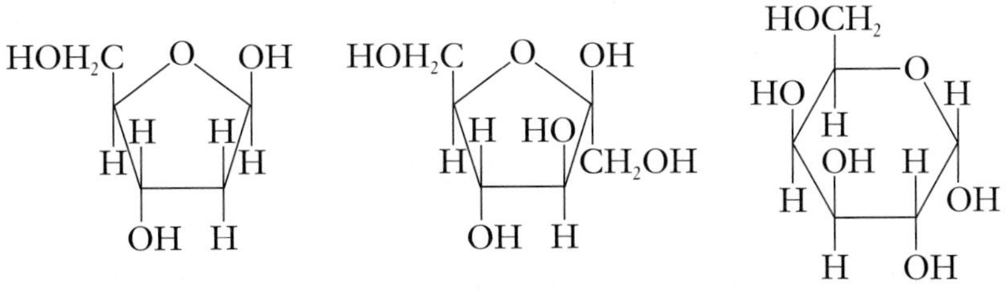

Deoxyribose Fructose Galactose

2. Which sugars do not exist as cyclic forms?

ANSWERS

1. Deoxyribose, fructose—furan; galactose—pyran
2. Sugars containing three or four carbon atoms

Some Common and Important Monosaccharides

The two smallest monosaccharides, *D-glyceraldehyde* and *dihydroxyacetone*, are trioses. These sugars are not common in the free form, but their derivatives are important in some energy-producing reactions of the body.

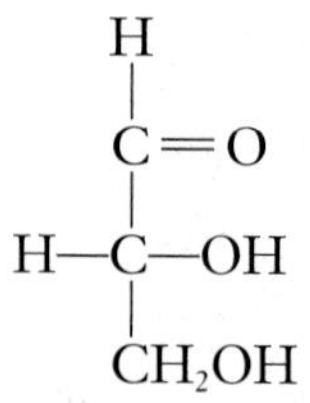

H
|
C=O
|
H—C—OH
|
CH_2OH

D-Glyceraldehyde

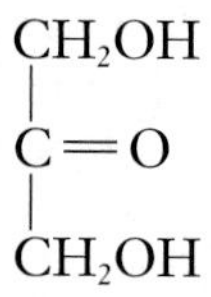

CH_2OH
|
C=O
|
CH_2OH

Dihydroxyacetone

D-Ribose and *D-2-deoxyribose* are aldopentoses. They are components of several biomolecules, including the nucleotides and nucleic acids that are so important in heredity and cell activity (Chapter 16).

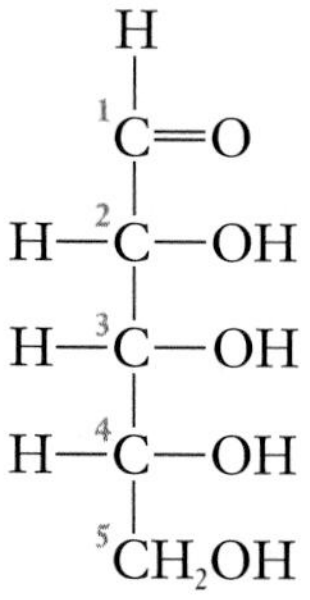

H
|
^{1}C=O
|
H—^{2}C—OH
|
H—^{3}C—OH
|
H—^{4}C—OH
|
$^5CH_2OH$

D-Ribose

H
|
^{1}C=O
|
H—^{2}C—H ← no oxygen atom on carbon 2
|
H—^{3}C—OH
|
H—^{4}C—OH
|
$^5CH_2OH$

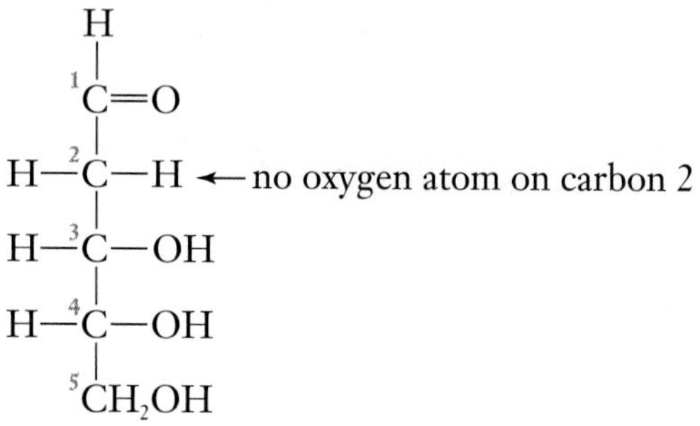

D-2-Deoxyribose

Glucose is one of several important monosaccharides that are hexoses. It is found as a component of many other carbohydrates, but its role as blood sugar makes it of great importance to the health sciences. Blood circulates glucose throughout the body, making this sugar available to all cells. The normal range for blood glucose in humans is from 60mg/100mL of blood to about 100mg/100mL of blood. If the concentration is consistently above this normal range, the condition known as *hyperglycemia* exists. This condition is seen in the disease diabetes mellitus. If concentrations persist below the normal range, then the person has *hypoglycemia* (low blood sugar).

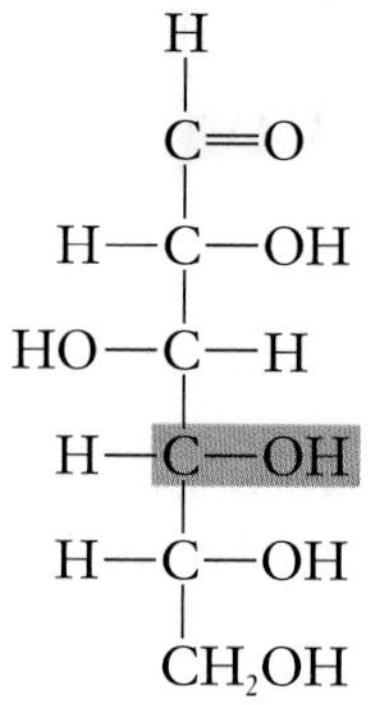

H
|
C=O
|
H—C—OH
|
HO—C—H
|
H—C—OH
|
H—C—OH
|
CH_2OH

D-Glucose

Fructose, along with glucose, is found in honey. Derivatives of fructose are important in energy metabolism. Commercially, fructose is now used as a sweetener in the form of high-fructose syrups. Fructose is 1.7 times sweeter than table sugar (sucrose), therefore less fructose is needed for the same degree of sweetness. Be-

Bees gather nectar from flowers and hydrolyze the sucrose in it to glucose and fructose. Honey is a thick mixture of glucose and fructose. *(Thomas C. Boyden/Dembinksy Photo Associates)*

cause less is needed to get a particular sweetness, cost is reduced, and fewer Calories are added to the food.

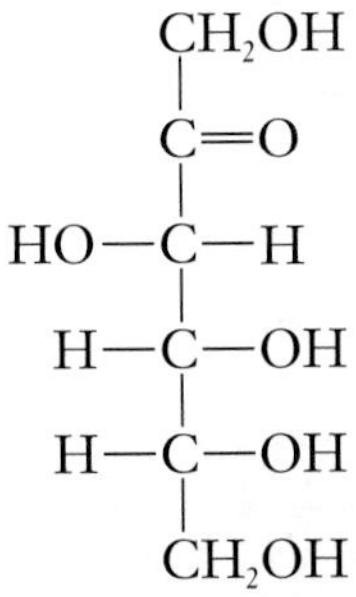

$$
\begin{array}{rcl}
 & CH_2OH & \\
 & | & \\
 & C{=}O & \\
 & | & \\
HO- & C & -H \\
 & | & \\
H- & C & -OH \\
 & | & \\
H- & C & -OH \\
 & | & \\
 & CH_2OH &
\end{array}
$$

D-Fructose

Galactose is a third important hexose. It is a component of lactose (milk sugar), and it is found as a component in a number of larger molecules in the body. Several genetic diseases such as galactosemia, Fabry's disease, and Tay-Sachs disease involve impaired metabolism of galactose or galactose-containing compounds. Lactose intolerance, a common problem for many adults, is discussed in Chapter 17.

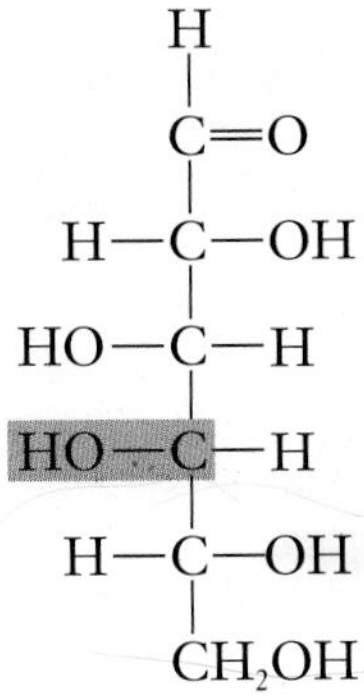

$$
\begin{array}{rcl}
 & H & \\
 & | & \\
 & C{=}O & \\
 & | & \\
H- & C & -OH \\
 & | & \\
HO- & C & -H \\
 & | & \\
HO- & C & -H \\
 & | & \\
H- & C & -OH \\
 & | & \\
 & CH_2OH &
\end{array}
$$

D-Galactose

The structures of galactose and glucose are very similar; compare the orientation of the hydroxyl group on carbon 4 of galactose (highlighted) to the same hydroxyl group in glucose.

Connections

Artificial Sweeteners

Many of us have turned to artificial sweeteners to satisfy our "sweet tooth" and avoid the Calories found in sugar, honey, and other sweeteners. Artificial sweeteners available now or in the past include saccharin, cyclamates, and aspartame.

Saccharin, discovered in 1879, is the oldest known low-calorie sweetener. Saccharin became very widely used, making it *the* artificial sweetener until the early 1960s, when the introduction of the cyclamates provided another option. The cyclamates quickly replaced saccharin in a portion of the market because cyclamates lack the aftertaste associated with saccharin. Both of these sweeteners are nonnutritive—that is, the body does not use them either for energy or to build other molecules.

In the late 1960s, however, cyclamates were shown to cause cancer in laboratory animals and were removed from use as sweeteners. Later saccharin was shown to cause bladder cancer in laboratory animals, and since 1978, a cancer-causing warning label appears on every product containing saccharin.

In 1981, a new compound—aspartame—was approved as a low-calorie sweetener. In exhaustive testing to meet FDA approval, no apparent health hazards were found for aspartame, although some people continue to express concerns about its safety. Aspartame was quickly and widely accepted, in part because it lacks the aftertaste of saccharin. Aspartame, like sugars, is a nutritive sweetener—that is, it can be used by the body for energy. Although aspartame has about the same caloric value per gram as sucrose, it is 180 times sweeter and thus much less is need to obtain the same degree of sweetness.

$$\mathrm{HOOC{-}CH_2{-}\underset{\displaystyle NH_2}{CH}{-}\overset{\displaystyle O}{\overset{\|}{C}}{-}NH{-}\underset{\displaystyle CH_2C_6H_5}{CH}{-}\underset{\displaystyle O}{\underset{\|}{C}}{-}OCH_3}$$

Aspartame

(a)

(b)

(a) This is the structure of aspartame, which is a sweetener widely used in a variety of foods and beverages. **(b)** Check the labels of your diet and low-calorie foods to see how many of them contain aspartame. *(Charles D. Winters)*

SWEETENERS	SWEETNESS RELATIVE TO TABLE SUGAR (SUCROSE)
Glucose (blood sugar)	0.5
Fructose (a sugar in honey)	1.7
Sucrose (table sugar)	1.0
Cyclamate	30
Saccharin	450
Aspartame	180

12.3 Properties and Reactions of Sugars

Solubility

The common sugars, which are mono- and disaccharides, are water-soluble because they form hydrogen bonds with water molecules. In contrast, the very large (polymeric) carbohydrates, such as starch, are not water-soluble. They are too big to dissolve and form dispersions instead (Chapter 6).

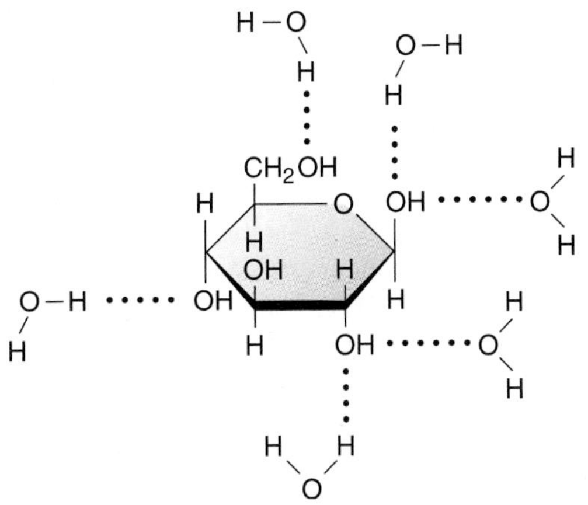

Sugar molecules are water-soluble because they hydrogen bond to water molecules.

Optical Properties of Sugars

Sugars contain one or more chiral centers. Compounds containing chiral centers are *optically active* because they rotate a special type of light called plane-polarized light. One of a pair of stereoisomers rotates this light clockwise and the other rotates it counterclockwise by the same amount. Clockwise rotation is designated by a (+) symbol, and counterclockwise rotation is specified by a (−). D-Glyceraldehyde and D-glucose rotate light to the right (+), and L-glyceraldehyde and L-glucose rotate light to the left (−). It is simply chance that these two D-monosaccharides both rotate light to the right. There is *no* relationship between stereochemical family (D− or L−) and the rotation of light (+ or −). D-Fructose, for example, rotates light to the left (−). The D− and L− classification is related to the arrangement of atoms about a single chiral center in the molecule. It is not related to optical activity.

Oxidation-Reduction Reactions of Sugars

The aldehyde group of aldoses is easily oxidized, and in aqueous solutions, ketoses can slowly rearrange to aldoses. Thus all monosaccharides are *reducing agents,* which means they cause other substances to be reduced as the carbonyl group of the aldose is oxidized. For this reason, the monosaccharides are classified as **reducing sugars.** The oxidation of glucose yields the carboxylic acid gluconic acid:

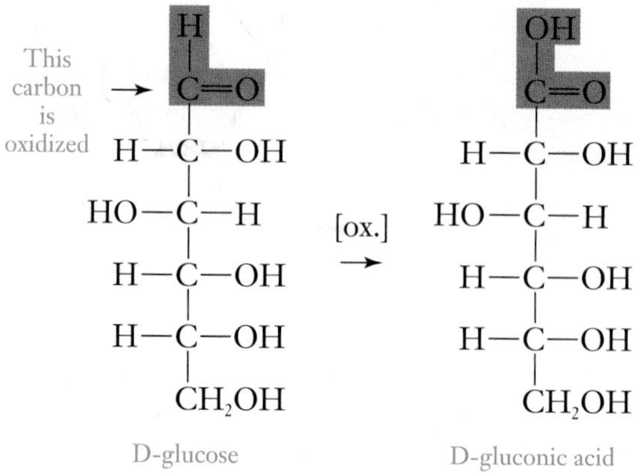

This carbon is oxidized → H–C=O

H–C–OH
HO–C–H
H–C–OH
H–C–OH
CH_2OH

D-glucose

[ox.] →

OH–C=O
H–C–OH
HO–C–H
H–C–OH
H–C–OH
CH_2OH

D-gluconic acid

The reducing property of monosaccharides is sometimes used to identify carbohydrates and to determine their concentrations. Several convenient analyses for sugars are available, including the Benedict's and Fehling's tests (see Chapter 11). The basis for these tests is that any reducing sugar in a solution—of urine, for example—reduces some metal ion. The progress of the reaction is measured by color changes that occur as metal ions are reduced (Figure 12.3).

The carbonyl group of monosaccharides can also be reduced. Several derivatives of sugars found in living systems are either oxidized or reduced sugars.

Glycosidic Bonds

Hemiacetals react with alcohols to form acetals:

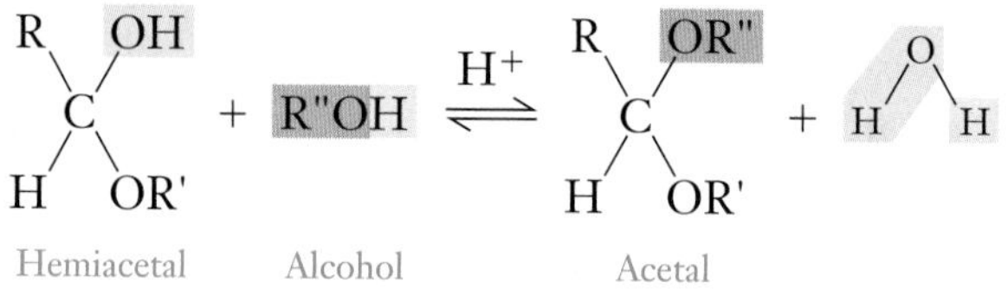

R(OH)C(H)(OR') + R"OH $\overset{H^+}{\rightleftharpoons}$ R(OR")C(H)(OR') + H_2O

Hemiacetal · Alcohol · Acetal

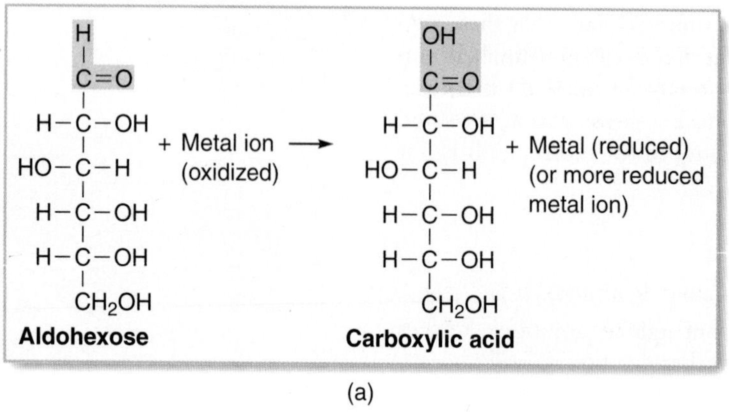

H–C=O
H–C–OH
HO–C–H
H–C–OH
H–C–OH
CH_2OH
Aldohexose

+ Metal ion (oxidized) →

OH–C=O
H–C–OH
HO–C–H
H–C–OH
H–C–OH
CH_2OH
Carboxylic acid

+ Metal (reduced) (or more reduced metal ion)

(a)

Color

Percentage glucose 0.5% ⟷ 2% (or greater)

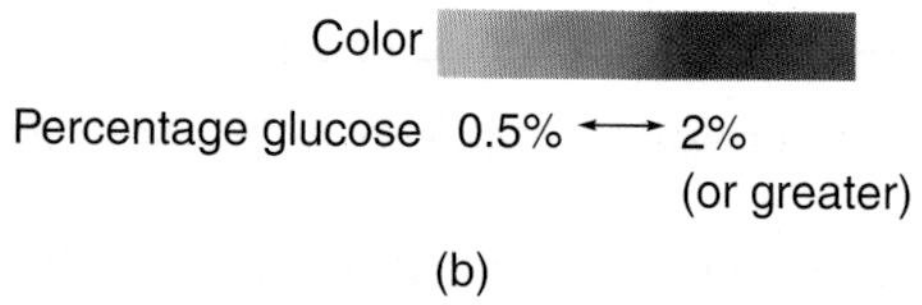

(b)

Figure 12.3 **(a)** The oxidation of an aldose is accompanied by the reduction of a metal ion. **(b)** The strips used to test for glucose in urine are based on this chemistry, and the amount of glucose present in a urine sample can be estimated by the color of the test strip.

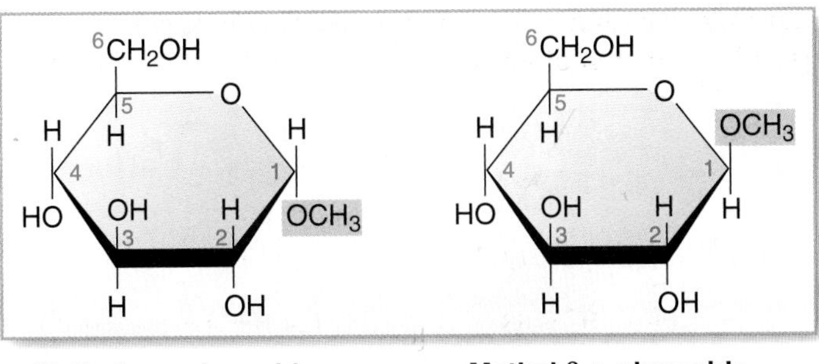

Figure 12.4 These glycosides were formed when glucose reacted with methanol. The name indicates which alcohol and sugar formed the glycoside. Note that two isomers are possible because the sugar can exist as either of two anomers.

Hemiacetals formed from ketones also form acetals by the same reaction. Because cyclic sugars are hemiacetals, they also react with an alcohol to form the carbohydrate equivalent of an acetal, which is called a **glycoside:**

$$\text{D-Glucose} + CH_3CH_2OH \underset{}{\overset{H^+}{\rightleftharpoons}} \text{Glycoside (ethyl β-D-glucoside)} + H\text{—}O\text{—}H$$

D-Glucose   Ethanol

Glycosidic bond

Glycoside (ethyl β-D-glucoside)

The covalent bond between the anomeric carbon atom of the cyclic sugar and the oxygen atom of the alcohol is called a **glycosidic bond.** The bond is designated as either alpha or beta, depending on the orientation of the oxygen atom; up is β and down is α.

The name of the sugar in the glycoside is used to name the glycoside. If the sugar is glucose, the glycoside is named as a *glucoside*. Fructose is the sugar in a *fructoside*, and galactoside is the sugar in a *galactoside*. Figure 12.4 shows two glycosides.

Self-Test

Properties and Reactions of Sugars reducing   Non-reducing

1. **(a)** Classify glucose, lactose, and sucrose as reducing or nonreducing sugars. (Refer to the next section on oligosaccharides to determine the structures of lactose and sucrose.)

   (b) Why is each sugar in **(a)** reducing or nonreducing?

glucose is a hemiacetal   sucrose no hemiacetal
lactose contains a hemiacetal   - does not
- yield free carbonyl group

2. Do the following glycosides have an alpha or a beta linkage?

(a) [structure: HOH_2C, O, H, H, H, H, OCH_3, OH, OH] (b) [structure: $HOCH_2$, O, HO, OCH_2CH_3, H, OH, H, H, H, H, OH]

3. What kind of glycoside is shown in 2(a) and 2(b)?

ANSWERS

1. **(a)** Glucose, lactose—reducing; sucrose—nonreducing **(b)** Because glucose is a hemiacetal and lactose contains a hemiacetal, they can both react to yield a free carbonyl group. Sucrose does not have a hemiacetal, thus no free carbonyl group is available. 2.**(a)** Alpha **(b)** Beta **3.** Riboside, galactoside

12.4 Oligosaccharides

When a glycosidic bond forms between any hydroxyl group of one monosaccharide and a hemiacetal of another one, the product is an oligosaccharide. For example, the hydroxyl group on carbon 4 of one glucose can react with the anomeric carbon of another glucose to yield this disaccharide:

D-Glucose + D-Glucose ⇌ Maltose + H_2O

The glycosidic bonds in oligosaccharides are described by the orientation of the oxygen atom (α or β) and by the atoms that are participating in the bond. In the disaccharide formed above, the bond is called an $\alpha(1 \longrightarrow 4)$ bond because the glycosidic bond involves the anomeric carbon of the first glucose (carbon 1) and the oxygen on carbon 4 of the second glucose.

In biochemistry, the units (monomers) within a larger molecule (polymer) are referred to as **residues.** Thus we say that the residues in an oligosaccharide or polysaccharide are monosaccharides.

Disaccharides are by far the most common oligosaccharides, and there are three common ones that you should know about. *Lactose,* which is milk sugar, is an important source of energy for young mammals. This sugar makes up about 5% of cow's milk, but human milk contains somewhat more. Lactose consists of the monosaccharides galactose and glucose linked by a $\beta(1 \longrightarrow 4)$ glycosidic bond [Fig-

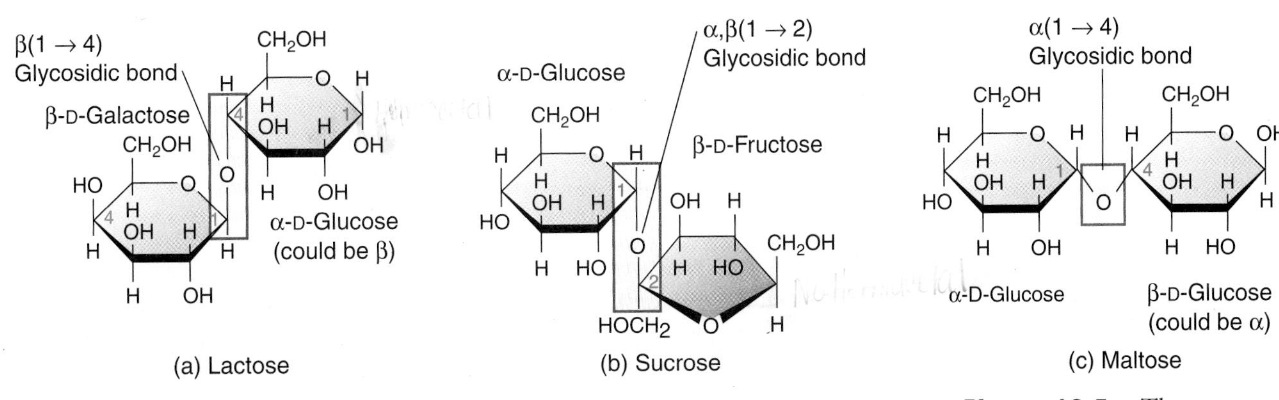

Figure 12.5 Three important disaccharides, **(a)** lactose, **(b)** sucrose, and **(c)** maltose, commonly found in foods or the digestive tract. Take a moment to identify the monosaccharides and glycosidic bonds found in each.

ure 12.5(a)]. Do you see what $\beta(1 \rightarrow 4)$ means? The oxygen of the glycosidic bond is in the beta position, and the bond involves carbon one of the galactose and carbon four of the glucose. Note that the anomeric carbon atom of the glucose residue in lactose does not participate in the glycosidic bond. Because of this, the hemiacetal of the glucose can reduce metal ions, thus lactose is a reducing sugar.

Sucrose is a disaccharide found in plants, where it may be stored or transported. In addition, sucrose is found in the nectar of flowers, where it attracts and rewards pollinators. Sucrose contains fructose and glucose linked by an $\alpha,\beta(1 \rightarrow 2)$ glycosidic bond [Figure 12.5(b)]. Note that in sucrose the anomeric carbons of both fructose and glucose are involved in the glycosidic bond. This is stated in the bond name, and because neither monosaccharide has a free hemiacetal, neither could reduce metal ions. Unlike the aldoses and lactose, sucrose is not a reducing sugar.

Maltose is a disaccharide made up of two glucose units linked by an $\alpha(1 \rightarrow 4)$ glycosidic bond [Figure 12.5(c)]. Maltose is obtained from the malt of germinating grain, and it is also present in the digestive tract during the digestion of starch. Maltose is a degradation product of larger carbohydrates such as starch. Malt, which is obtained from germinating barley, is used to make beer. The sugars in malt, including maltose, serve as the energy source for the yeast during fermentation.

Although many oligosaccharides play a role in the normal structure and function of the body, only a few are mentioned in this textbook. You will see some of these oligosaccharides when you study lipids and proteins in Chapters 13 and 14.

Sucrose in nectar attracts pollinators to flowers. *(Anthony Mercieca/Dembinsky Photo Associates)*

Self-Test

Oligosaccharides

1. Name the simple sugars found in (a) lactose (b) maltose (c) sucrose.
2. Explain the designation "$\beta(1 \rightarrow 4)$" used to describe the glycosidic bond of lactose.

ANSWERS

1. **(a)** Galactose, glucose **(b)** Glucose **(c)** Fructose, glucose 2. The oxygen of the anomeric carbon of galactose is in the beta position and is bonded to carbon 4 of glucose.

Connections

ABO Blood Type

What is your blood type? Many people know their blood type, and if you do, you might answer type O, A, B, or AB. What do these terms mean? They refer to the ABO blood group, one of several blood groups discovered by Karl Landsteiner for which he received the 1930 Nobel Prize for physiology/medicine. Today we know that the blood groups are related to substances found on the surface of blood cells.

Human red blood cells (erythrocytes) have many proteins and lipids on the outside surface of the cell membrane, some of which have oligosaccharides attached to them. One of the lipids found is a sphingolipid (Chapter 13), which can have one of three different oligosaccharides bound to it. It is the oligosaccharides bound to this lipid molecule that determine the ABO blood type for an individual. The oligosaccharide exists as one of three variations called antigens. The oligosaccharide in antigen O is a pentasaccharide ending in galactose. Antigen A has the same pentasaccharide plus an N-acetylgalactosamine residue attached to the terminal galactose. Antigen B has the same pentasaccharide with an additional galactose attached to the terminal galactose residue. People with type O blood have only the O antigen on the surface of their red blood cells. People with AB blood have both the A and B antigens on their erythrocytes. A person with type A blood has only antigen A or both antigen A and antigen O, and a person with type B has only antigen B or both antigen B and antigen O. (By the way, if you have had genetics you should be able to explain this pattern.)

The ABO blood group must be considered when donor blood is selected for a patient in need of a blood transfusion because some ABO blood types are incompatible with others. The compatibility within the ABO blood group is summarized in the table.

ABO Blood Group Compatibility

BLOOD TYPE	CAN DONATE BLOOD TO	CAN RECEIVE BLOOD FROM
A	A and AB	A and O
B	B and AB	B and O
AB	AB	A, B, AB, and O
O	A, B, AB, and O	O

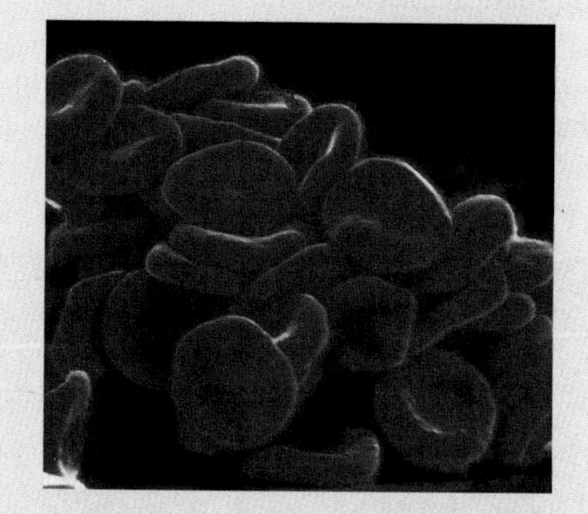

The ABO blood types are due to oligosaccharides found at the surface of red blood cells. *(CNRI/Science Photo Library/Photo Researchers, Inc.)*

12.5 Polysaccharides

Polysaccharides are macromolecules made up of monosaccharides linked by glycosidic bonds. Starch, glycogen, cellulose, and chitin are common examples of polysaccharides. Although each is a polymer of glucose or a glucose derivative, they have structural differences that result in distinctly different properties.

Starch and Glycogen: Storage Polysaccharides

The major storage polysaccharide in animals is glycogen, and that in plants is starch. **Starch** and **glycogen** are similar molecules that contain many D-glucose molecules linked by $\alpha(1 \rightarrow 4)$ glycosidic bonds and, in some cases, $\alpha(1 \rightarrow 6)$ bonds. Plants use starches to store energy, and these substances are the **complex carbohydrates** of our diet. There are two types of polymers found in starches, amylose and amylopectin, which differ in size and shape. **Amylose** is a linear polymer—that is, it has all of its glucose residues connected in one long chain by $\alpha(1 \rightarrow 4)$ glycosidic bonds [Figure 12.6(a)]. **Amylopectin** is larger than amylose and is branched. Some of the glucose residues contain both $\alpha(1 \rightarrow 4)$ and $\alpha(1 \rightarrow 6)$ glycosidic bonds, and a branch occurs at each $\alpha(1 \rightarrow 6)$ bond. The difference in size, from several hundred glucose residues for amylose to several thousand for amylopectin, accounts for the difference in their behavior in water. Neither is truly soluble in water because both are too big to form true solutions. Instead, they form dispersions (Chapter 6) in hot water. Because it is the smaller of the two forms of starch, amylose forms dispersions more readily. In water, the $\alpha(1 \rightarrow 4)$ bonds of these starch molecules are relatively exposed to water molecules and the bonds can be readily cleaved by hydrolysis.

Typical cornstarch produced in the United States is about 25% amylose and 75% amylopectin.

Glycogen, like amylopectin, is a polymer of $\alpha(1 \rightarrow 4)$-linked subunits of glucose with $\alpha(1 \rightarrow 6)$-linked branches [Figure 12.6(b)]. Glycogen is more branched and compact than amylopectin. The branching in glycogen has physiological significance. In liver and muscle cells, a specific, highly regulated enzyme named

Grains, tubers, and other plant products are important sources of starches. *(Charles D. Winters)*

glycogen phosphorylase catalyzes the release of individual phosphorylated glucose molecules from the ends of the chains of glycogen. These molecules can be used by the cell for energy or other needs. Because glycogen is branched, many phosphorylated glucose molecules can be released at any moment. In contrast, a linear polysaccharide like amylose has only one or two ends where the enzyme can release individual molecules from the polymer. Clearly, the highly branched nature of glycogen allows for its rapid breakdown.

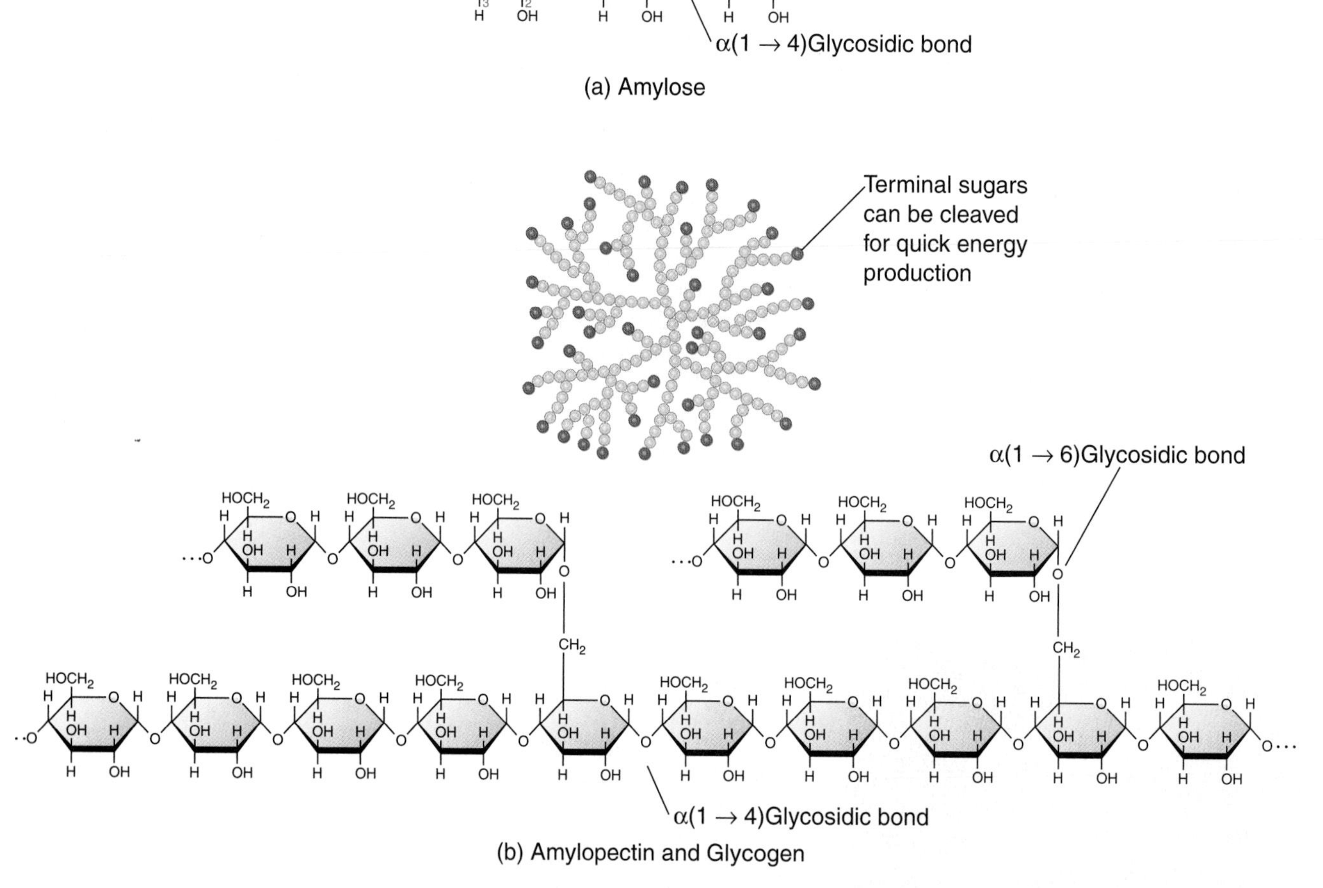

Figure 12.6 A comparison of amylose to amylopectin and glycogen. **(a)** Amylose is a linear polymer composed only of glucose residues linked by $\alpha(1 \rightarrow 4)$ bonds. The chain of glucose residues is normally coiled into a helix. **(b)** Amylopectin (in plants) and glycogen (in animals) consist of glucose residues linked by both $\alpha(1 \rightarrow 4)$ and $\alpha(1 \rightarrow 6)$ glycosidic bonds. The $\alpha(1 \rightarrow 6)$ bonds provide the branching seen in these molecules. In amylopectin, branching occurs every 24–30 glucose residues. In glycogen, it is every 8–12 residues. In vertebrates, terminal sugars in glycogen (shown in blue) can be cleaved quickly, yielding a glucose derivative for quick energy production.

Cellulose and Chitin—Structural Polysaccharides

Cellulose and **chitin,** which have roles in some structures of plants and animals, are polymers containing $\beta(1 \rightarrow 4)$ glycosidic bonds. This is in contrast to starches and glycogen that contain $\alpha(1 \rightarrow 4)$ bonds. In general, $\alpha(1 \rightarrow 4)$ glycosidic bonds are more easily cleaved than $\beta(1 \rightarrow 4)$ bonds because the former are more exposed. Figure 12.7(a) shows the structure of a cellulose molecule, and Figure 12.7(b) shows a cellulose fiber, which consists of numerous cellulose molecules hydrogen-bonded to each other. Note the extensive hydrogen bonding that extends up, down, and sideways. As a consequence, few of the $\beta(1 \rightarrow 4)$ bonds are exposed, and the rate of cellulose breakdown is slower than that of starches. This difference in breakdown rate makes sense—structural materials should be relatively resistant to

β(1 → 4) Glycosidic bond

(a) Cellulose

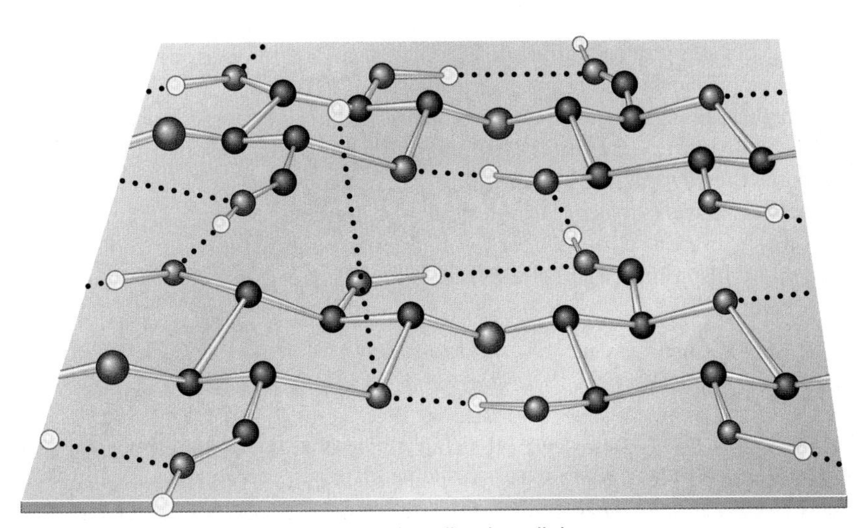

(b) Hydrogen bonding in cellulose

Figure 12.7 **(a)** Cellulose, like amylose, is a linear polymer but contains $\beta(1 \rightarrow 4)$ bonds rather than $\alpha(1 \rightarrow 4)$ bonds. **(b)** Hydrogen bonds between hydroxyl groups hold a cellulose molecule to other cellulose molecules that are above, below, and beside it. Because cellulose molecules are usually buried by other ones, breakdown of cellulose is usually slow. The rings in **(b)** are drawn in the chair form to show the alignment of the atoms in the hydrogen bonds.

Connections

Carbohydrates and Nutrition

The carbohydrates can be classified into three nutritional categories—sugars, complex carbohydrates, and fiber. Dietary sugars include the monosaccharides such as glucose and fructose, and the oligosaccharides such as the disaccharides sucrose and lactose. Through digestion, the oligosacchrides are broken down to monosaccharides, which are absorbed and used by the body.

Complex carbohydrates and fiber are polysaccharides. The complex carbohydrates are principally starches that are broken down by digestion to monosaccharides. Fiber consists primarily of polysaccharides such as cellulose that cannot be digested. Although indigestible, fiber appears to serve a role in human health; some studies indicate a correlation between intestinal cancer and a low-fiber diet.

The monosaccharides supplied by the diet and digestion are a major source of energy for the body. Four Calories (4 kcal) are available from each gram of digestible dietary carbohydrate. In addition, monosaccharides are also used by the body to make a variety of biomolecules because sugars are found as components of all of the major classes of biomolecules of the body.

There are many dietary sources of carbohydrates. Generally foods from plants are higher in carbohydrates than foods derived from animals. Roughly 40 to 50% of the Calories consumed in the North American diet are derived from carbohydrates. In many cultures the percentage is somewhat higher, up to around 90%. Many nutritionists believe a larger proportion of our energy needs should come from dietary carbohydrate rather than fat and protein, and that this increase should include more complex carbohydrates and natural sugars and fewer processed sugars.

Fruits, vegetables, and whole-grain products are among the foods and beverages that provide carbohydrates. *(Charles D. Winters)*

breakdown, whereas energy-storage molecules should readily be converted to smaller, more easily used units.

Cellulose is a polymer of glucose. Chitin is similar, but in it the hydroxyl group on carbon 2 of each glucose molecule is replaced by an amide group (Figure 12.8). Chitin is the polysaccharide found in the exoskeleton of crabs, shrimp, insects, and other arthropods. Cellulose is the principal structural polysaccharide of plants found especially in the stems, stalks, trunks, and woody portions. Because plants are so abundant, they synthesize about 100 pounds of cellulose daily for every person on Earth. This material has great potential as a source of energy and chemicals.

Wood is primarily cellulose fibers embedded in a highly polymerized substance called *lignin.* This complex of cellulose and lignin is a strong, enduring substance

Cellulose is the principal structural component of plants, giving rigidity and strength to cell walls. Plants provide many things for us, including materials, energy, and a beautiful environment. *(Darrell Gulin/Dembinsky Photo Associates)*

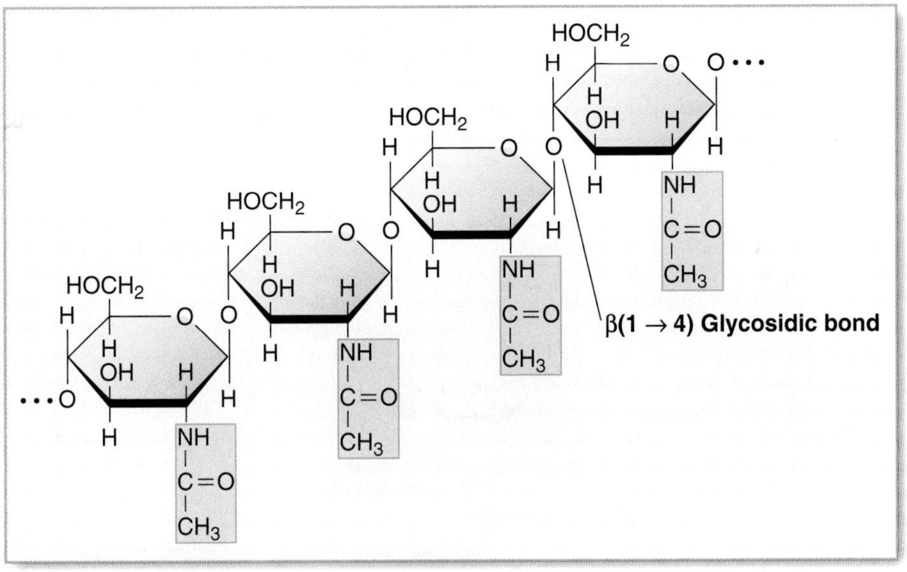

Chitin

Figure 12.8 Chitin is a structural polysaccharide like cellulose, but its composition differs from cellulose. The residues of cellulose are glucose, whereas chitin has N-acetylglucosamine residues. Chitin gives rigidity and strength to the exoskeletons of arthropods—insects, lobsters, and crabs, for example.

that is ideal for many structural uses. Steel-reinforced concrete mimics the structure of wood because the steel rods are analogous to the cellulose fibers and the concrete is like the lignin. Similarly, modern fishing rods consist of fiberglass or graphite embedded in resins.

Starch—$\alpha(1 \rightarrow 4)$ bonds—and cellulose—$\beta(1 \rightarrow 4)$ bonds—differ only in the orientation of the glycosidic linkage between the glucose residues, yet humans efficiently digest starch in food but cannot use cellulose. In general, mammals and other higher animals do not possess the enzymes necessary to cleave cellulose to glucose. However, grazing animals, such as sheep and cattle (ungulates), possess intestinal microflora that make these enzymes. Cellulose hydrolysis is catalyzed by microbial enzymes, and the products are thus available to the microflora and the host animal. Thus cattle and related organisms can digest many plants not suitable for other animals. Dietary cellulose, however, is not a completely inert, wasted portion of the diet for humans and other animals. Cellulose is a major part of the indigestible, tasteless **fiber** present in our diet as roughage. Research indicates that there may be health benefits derived from a high-fiber diet.

Other Polysaccharides—Hyaluronic Acid and Heparin

The polysaccharides discussed so far have repeating units consisting of a single sugar or a single-sugar derivative. Several important polysaccharides have a more complex repeating unit. **Hyaluronic acid** is found in the connective tissue of higher animals. It is part of the viscous material around bone joints that absorbs shock and acts as a lubricant for the bone surfaces. Figure 12.9 shows the glycosidic bonding and the *two*-sugar repeating unit of hyaluronic acid, as well as the large complexes that hyaluronic acid forms in joints.

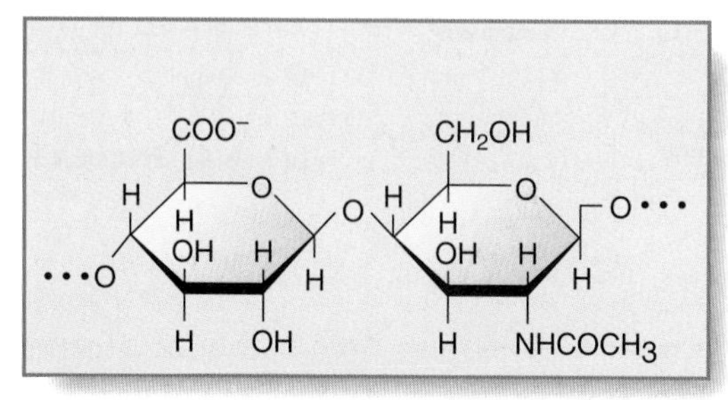

(a) Hyaluronic acid

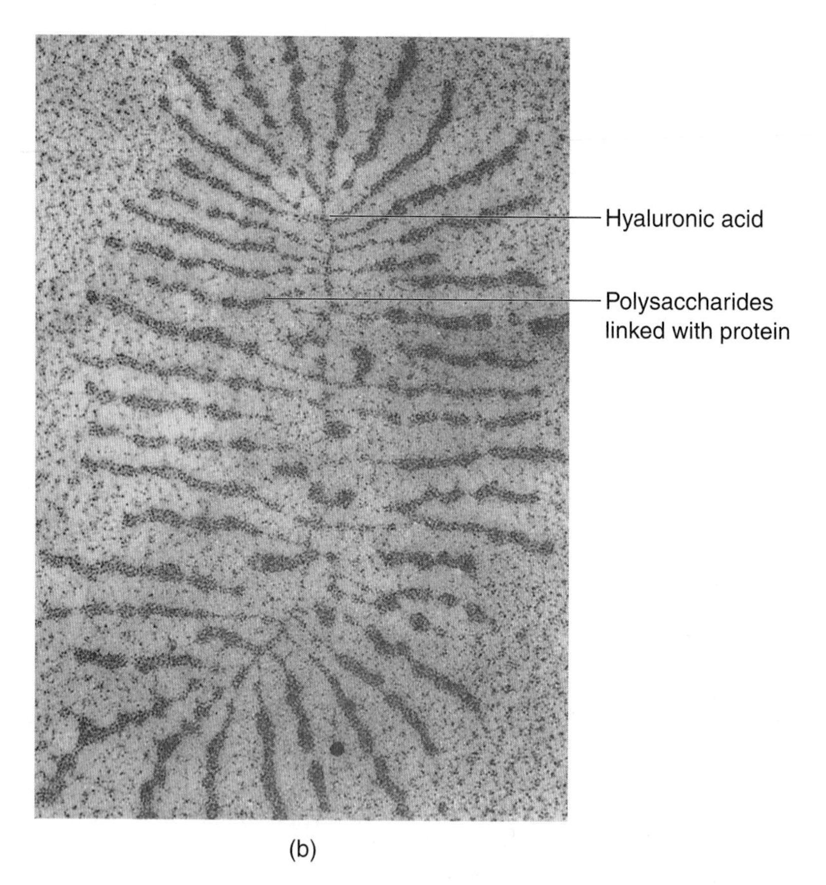

(b)

Figure 12.9 **(a)** This disaccharide is the repeating unit in hyaluronic acid, which is a polysaccharide found in connective tissues. It provides the lubricant for joints. Hyaluronic acid is associated with proteins and other polysaccharides to form proteoglycans shown in this electron micrograph **(b)**. *[b, From Ruckwalter, J.A., and Rosenberg, L.* Collagen and Related Research, *3, 489–504 (1983).]*

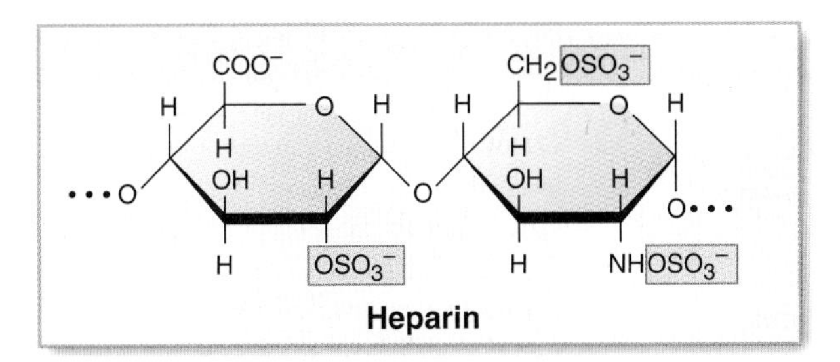

Figure 12.10 The structure of a short segment of heparin. Note the repeating unit of heparin and the sulfate groups. Purified heparin is added to blood samples donated for transfusion to prevent clotting.

Heparin is a polysaccharide produced by some cells of the circulatory system. It is a powerful anticoagulant used in a number of medical applications in which the chance of blood clotting must be reduced or eliminated. Heparin is common to most animals and is a linear polysaccharide. It has a unique two-sugar repeating unit and is characterized by a large number of sulfate groups (Figure 12.10).

CHAPTER SUMMARY

Carbohydrates are polyhydroxy-aldehydes or -ketones or derivatives of them, and many can be recognized by the suffix ***-ose*** in their name. Carbohydrates are classified by size with the **monosaccharides,** the simple sugars, the smallest carbohydrates. Monosaccharides can be classified by the carbonyl group they contain—**aldose** or **ketose**—or by the number of carbon atoms in the molecule—triose—or by both functional group and number of carbon atoms—aldohexose. Larger carbohydrates contain two or more monosaccharides linked by **glycosidic bonds.** Most monosaccharides belong to the D-stereochemical family, and pentoses and hexoses exist in cyclic forms called **furanoses** and **pyranoses.** Monosaccharides are water-soluble, rotate plane-polarized light, can be oxidized or reduced, and form glycosidic bonds. Glycosidic bonds link the two to ten monosaccharide units found in **oligosaccharides** and the many monosaccharide units found in **polysaccharides. Disaccharides** such as maltose, lactose, and sucrose are the most common oligosaccharides, though others play important roles in membranes and other cellular structures. Some polysaccharides such as the **starches—amylose** and **amylopectin—** and **glycogen** store glucose molecules in living organisms. The highly branched nature of glycogen allows for rapid mobilization of glucose. Other polysaccharides play a role in the structure of living organisms—**cellulose** in wood and **chitin** in arthropod shells are examples. **Hyaluronic acid** serves as a viscous lubricant found in joints, and **heparin** is an anticoagulant in blood.

KEY TERMS

Section 12.1
carbohydrate
monosaccharide
oligosaccharide
-ose
polysaccharide

Section 12.2
aldose
anomeric carbon
anomers
D- and L- families
furanose
ketose
pyranose

Section 12.3
glycoside
glycosidic bond
reducing sugar

Section 12.4
disaccharide
residue

Section 12.5
amylopectin
amylose
cellulose
chitin
complex carbohydrates
fiber
glycogen
heparin
hyaluronic acid
starch

EXERCISES

Classes and Names of Carbohydrates (Section 12.1)

1. Define the terms *monosaccharide*, *oligosaccharide*, and *polysaccharide*.
2. Identify the functional groups found in carbohydrates. State the definition of a carbohydrate and compare your answer to this definition.
3. Explain the historical origin of the word *carbohydrate*.

Monosaccharides (Section 12.2)

4. Explain the structural information contained in these terms.
   (a) Ketopentose **(b)** Aldotetrose
   (c) Ketohexose
5. Explain the structural information contained in these terms.
   (a) Aldotriose **(b)** Ketoheptose
   (c) Aldopentose
6. Draw the structure of the simplest aldose and ketose.
7. Name the smallest aldose and ketose.
8. What distinguishes an L-sugar from a D-sugar?
9. Identify each of the following sugars by functional group, chain length, and D- or L-family.

(a)
$$\begin{array}{c} \text{H} \\ | \\ \text{C=O} \\ | \\ \text{H—C—OH} \\ | \\ \text{HO—C—H} \\ | \\ \text{CH}_2\text{OH} \end{array}$$

(b)
$$\begin{array}{c} \text{CH}_2\text{OH} \\ | \\ \text{C=O} \\ | \\ \text{HO—C—H} \\ | \\ \text{H—C—OH} \\ | \\ \text{CH}_2\text{OH} \end{array}$$

10. Identify each of the following sugars by functional group, chain length, and D- or L-family.

(a)
$$\begin{array}{c} \text{H} \\ | \\ \text{C=O} \\ | \\ \text{H—C—OH} \\ | \\ \text{HO—C—H} \\ | \\ \text{HO—C—H} \\ | \\ \text{H—C—OH} \\ | \\ \text{CH}_2\text{OH} \end{array}$$

(b)
$$\begin{array}{c} \text{CH}_2\text{OH} \\ | \\ \text{C=O} \\ | \\ \text{CH}_2\text{OH} \end{array}$$

11. Draw the open-chain structure of
    (a) An aldopentose
    (b) A ketohexose
    (c) An aldotetrose in the L-stereochemical family
12. Draw the open-chain structure of
    (a) An aldohexose
    (b) An aldoheptose
    (c) A ketotetrose in the D-stereochemical family
13. You have a drawing of an open-chain D-aldohexose in front of you (or you can flip back a few pages to view the structure). Identify the minimal structural change(s) you would have to make to convert it into an L-aldohexose.
14. Furnish the names of these monosaccharides.

(a)
$$\begin{array}{c} \text{H} \\ | \\ \text{C=O} \\ | \\ \text{H—C—OH} \\ | \\ \text{HO—C—H} \\ | \\ \text{H—C—OH} \\ | \\ \text{H—C—OH} \\ | \\ \text{CH}_2\text{OH} \end{array}$$

(b)
$$\begin{array}{c} \text{CH}_2\text{OH} \\ | \\ \text{C=O} \\ | \\ \text{HO—C—H} \\ | \\ \text{H—C—OH} \\ | \\ \text{H—C—OH} \\ | \\ \text{CH}_2\text{OH} \end{array}$$

15. What are the names of these monosaccharides?

(a)
$$\begin{array}{c} \text{H} \\ | \\ \text{C=O} \\ | \\ \text{H—C—OH} \\ | \\ \text{H—C—OH} \\ | \\ \text{H—C—OH} \\ | \\ \text{CH}_2\text{OH} \end{array}$$

(b)
$$\begin{array}{c} \text{H} \\ | \\ \text{C=O} \\ | \\ \text{HO—C—H} \\ | \\ \text{CH}_2\text{OH} \end{array}$$

16. Identify the functional groups on the carbon atoms of glucose and fructose (use the open-chain forms).
17. Classify these carbohydrates as monosaccharides, oligosaccharides, or polysaccharides.
    (a) Sucrose **(b)** Galactose
    (c) Maltose **(d)** Cellulose
18. List the similarities and differences between glucose and fructose.
19. List the similarities and differences between ribose and 2-deoxyribose.

20. Which of the following sugars is a monosaccharide and a furanose?
(a) Glucose (b) Galactose
(c) Fructose (d) Sucrose

21. Identify by number the carbon atom of an aldopentose that is the anomeric carbon atom in the cyclic hemiacetal. Which carbon atom is the anomeric carbon atom in a cyclic ketohexose?

22. Indicate which of these structures are cyclic hemiacetals, and in the hemiacetals identify the carbon atom that is the anomeric carbon atom.

(a) (b)

23. Indicate which of these structures are cyclic hemiacetals, and in the hemiacetals identify the carbon atom that is the anomeric carbon atom.

(a) (b)

24. How is an alpha anomer recognized?

25. Describe the furanose and pyranose forms of monosaccharides.

26. Explain how hemiacetal formation and the cyclic forms of sugars are related.

Properties and Reactions of Sugars (Section 12.3)

27. Monosaccharides and disaccharides are water-soluble. Why?

28. A carbohydrate is shown to be an aldotetrose with two chiral centers. It is shown to be in the D-stereochemical family. Draw the possible structure(s) for this compound.

29. How many chiral centers are present in D-fructose? How many stereoisomers exist for ketohexose?

30. D-Ribose is levorotatory but D-glucose is dextrorotatory. How can this be?

31. A reducing sugar will
(a) React with Benedict's reagent
(b) Have fewer Calories
(c) Always be a ketose
(d) None of these

32. Describe the chemistry involved in the designation "reducing sugar".

33. Provide an explanation for the observation that both galactose and lactose are reducing sugars.

34. Sucrose is not a reducing sugar. Why?

35. Fructose is not an aldose, yet it is classified as a reducing sugar. How can this be?

36. Identify each of these glycosides by the sugar they contain, and indicate whether the glycosidic linkage is alpha or beta:

(a)

(b)

(c)

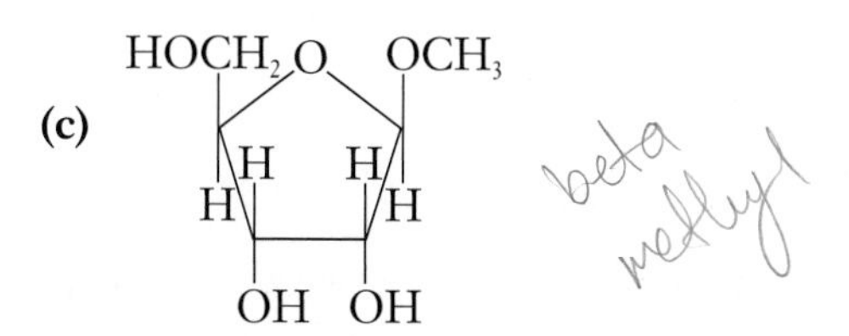

37. Identify each of these glycosides by the sugar they contain, and indicate whether the glycosidic linkage is alpha or beta:

(a)

(b)
$HOCH_2$
O
H H
H
OH H
OH OCH_3
H OH

(c)
$HOCH_2$ O OCH_2CH_3
H H
H H
HO H

Oligosaccharides *(Section 12.4)*

38. Name these disaccharides.

(a)
$HOCH_2$ $HOCH_2$
O O
H H H H
H H
OH H OH H
HO O OH
H OH H OH

(b)
$HOCH_2$
O
H OH
$HOCH_2$ H
OH H
O
HO O H
H
OH H H OH
H H
H OH

(c)
$HOCH_2$
O
H H
H OH H
OH H
HO O CH_2OH
H HO
H HO
$HOCH_2$ O H

39. Common table sugar is
(a) Glucose (b) Fructose
(c) Sucrose (d) Maltose

40. Honey is a mixture composed of an aldose and a ketose. Identify these sugars.

41. Hydrolysis of the glycosidic bond in disaccharides yields two simple sugars. What is (are) the simple sugar(s) obtained from the hydrolysis of the following?
(a) Sucrose (b) Maltose (c) Lactose

42. Identify the glycosidic bonds found in the disaccharides in Exercise 41.

43. Draw one of the two structures of maltose.
(a) Circle any hemiacetal that is present.
(b) Circle any acetal that is present.
(c) Identify the glycosidic bond in maltose.

44. What factor determines whether a carbohydrate is classified as a mono-, oligo-, or polysaccharide?

Polysaccharides *(Section 12.5)*

45. Identify the polysaccharides that yield only D-glucose when hydrolyzed.

46. Pinpoint the structural difference between amylose and cellulose.

47. What is the structural difference between amylose and amylopectin?

48. Describe the similarities and differences of cellulose and chitin.

49. Describe the physiological role of glycogen and explain how its structure contributes to this role.

50. Explain why glycogen can be broken down rapidly in liver or muscle cells.

51. Why can cattle digest cellulose and humans cannot?

52. Where is hyaluronic acid found and what is its role?

53. Provide a brief explanation of the structure and physiological role of heparin.

General Questions

54. Match the term on the left with the description on the right.

(A) Amylose	(a) Ketose
(B) Glycogen	(b) Table sugar
(C) Fructose	(c) A polymer of glucose with beta glycosidic linkages
(D) Sucrose	(d) A polymer of glucose stored in the body
(E) Triose	(e) A three-carbon sugar
(F) Cellulose	(f) A form of starch

55. For each of the following words, give an explanation, definition, or example that demonstrates your understanding of the word.
(a) Carbohydrate **(b)** Pentose
(c) Aldose **(d)** Ketose
(e) Ketohexose **(f)** L-family
(g) Furanose **(h)** Anomeric carbon
(i) Anomers **(j)** Glycosidic bond
(k) Monosaccharide **(l)** Disaccharide
(m) Polysaccharide **(n)** Glycogen
(o) Cellulose **(p)** Amylose
(q) Reducing sugar **(r)** Lactose

56. Draw structures for each of the following:
(a) D-aldopentose
(b) L-glyceraldehyde
(c) α-D-glucopyranose
(d) The isomers of "ketotetrose"

57. Which of the following carbohydrates are classified as reducing sugars?
(a) Galactose **(b)** Fructose
(c) Starch **(d)** Glycogen
(e) Sucrose **(f)** Lactose

Challenge Exercises

58. When cellulose is partially hydrolyzed, cellobiose—a disaccharide—is one of the products. What is the composition of cellobiose? What glycosidic bond is present in this disaccharide?

59. A children's breakfast cereal contains 11 g of sugar and 15 g of complex carbohydrate per 1-oz serving. **(a)** Determine the number of calories provided by these carbohydrates. **(b)** What percentage is provided by complex carbohydrates?

60. Many distance runners practice "carbo loading" (eating carbohydrate-rich foods) for one to three days preceding a race. What do they hope to accomplish?

Lipids

13

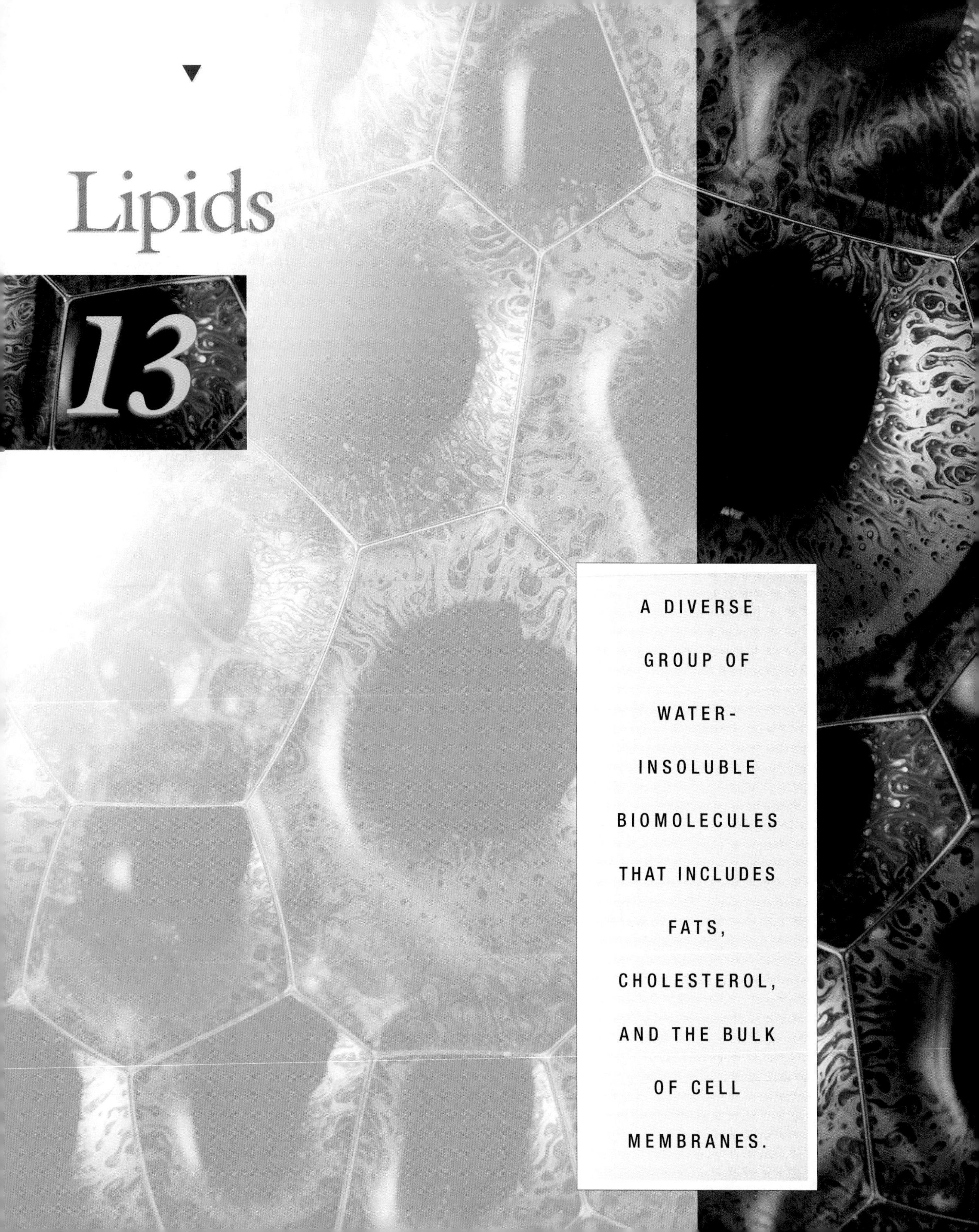

A DIVERSE GROUP OF WATER-INSOLUBLE BIOMOLECULES THAT INCLUDES FATS, CHOLESTEROL, AND THE BULK OF CELL MEMBRANES.

LIPIDS ARE the biological molecules that are insoluble in water but soluble in organic solvents. Unlike carbohydrates, lipids are not composed of similar, chemically related structural units. Instead, they vary greatly in both structure and function. Lipids are classified as either simple or complex. Complex lipids include the triacylglycerols, glycolipids, sphingolipids, and a variety of phosphate-containing lipids called phospholipids. Simple lipids include the steroids, prostaglandins, and leukotrienes. Triacylgylcerols (fats and oils) are involved in energy storage and use, and the other complex lipids are common components of membranes, as is the simple lipid cholesterol. Some lipids are hormones, some play protective roles, and others aid in digestion and nutrient transport.

Outline

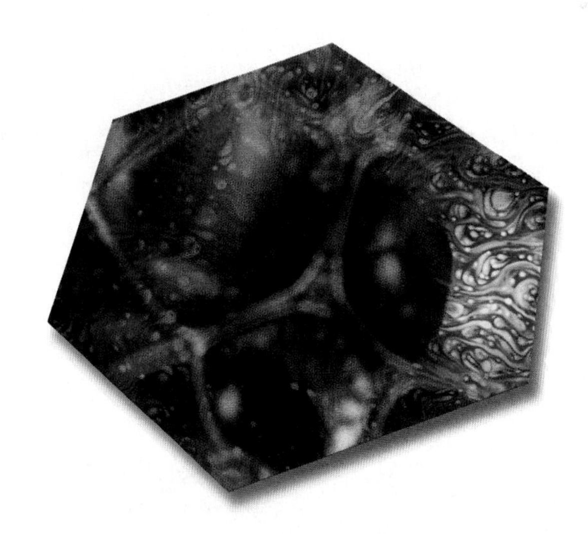

Soap bubbles consist of soap and water molecules arranged in a sheet-like structure that resembles cell membranes. ***(Dr. Jeremy Burgess/ Science Photo Library/ Photo Researchers, Inc.)***

13.1 Fatty Acids

Complex lipids contain long-chain carboxylic acids called **fatty acids,** which **simple lipids** lack [Figure 13.1(a)]. The fatty acids in complex lipids typically have 10 to 24 carbon atoms in the carbon chain. Fatty acids nearly always contain an even number of carbon atoms (they are made from two-carbon pieces), and they are almost never branched. These compounds are rarely found free in the cells of living organisms. In fact, free fatty acids are somewhat toxic to cells.

Fatty acids usually contain zero, one, two, or three carbon–carbon double bonds, and on this basis are classified as saturated fatty acids or unsaturated fatty acids. **Saturated fatty acids** contain no carbon–carbon double or triple bonds. The term *saturated* means these fatty acids do not accept any more hydrogen—that is, they are saturated with hydrogen. **Unsaturated fatty acids** contain one or more carbon–carbon double bonds. (Carbon–carbon triple bonds are usually not found in fatty acids.) If the acid contains just one double bond, it is called a *monounsat-*

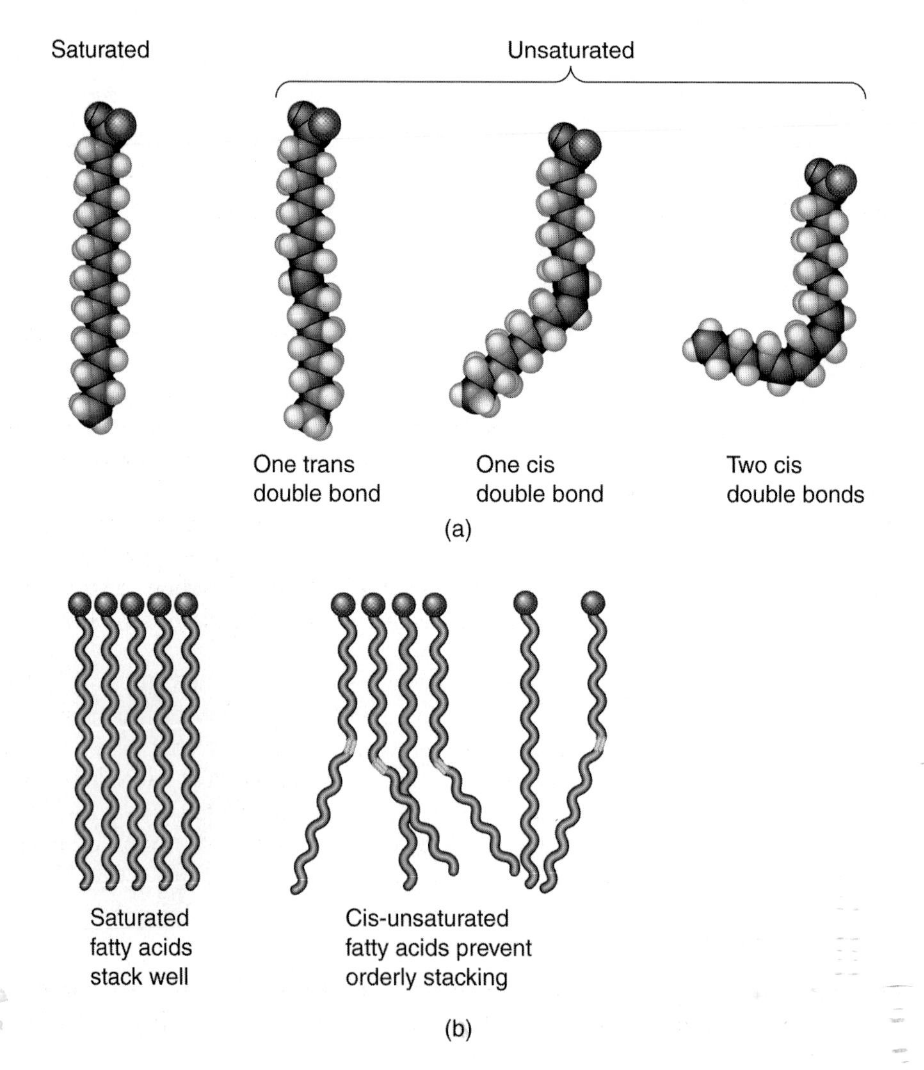

Figure 13.1 **(a)** Space-filling models of some saturated and unsaturated fatty acids. *Cis*-double bonds introduce a "kink" that changes the shape of the molecule. **(b)** *Cis*-Unsaturated fatty acids do not stack well, thus they are liquids at room temperature. Saturated fatty acids are solids.

Table 13.1 Some Common Fatty Acids

NUMBER OF CARBON ATOMS: NUMBER OF DOUBLE BONDS	COMMON NAME	STRUCTURE
Saturated		
14:0	Myristic acid	$CH_3(CH_2)_{12}COOH$
16:0	Palmitic acid	$CH_3(CH_2)_{14}COOH$
18:0	Stearic acid	$CH_3(CH_2)_{16}COOH$
Unsaturated		
16:1	Palmitoleic acid	$CH_3(CH_2)_5-CH=CH-(CH_2)_7COOH$
18:1	Oleic acid	$CH_3(CH_2)_7-CH=CH-(CH_2)_7COOH$
18:2	Linoleic acid	$CH_3(CH_2)_4-CH=CH-CH_2-CH=CH-(CH_2)_7COOH$
18:3	Linolenic acid	$CH_3CH_2-CH=CH-CH_2-CH=CH-CH_2-CH=CH-(CH_2)_7COOH$

urated fatty acid. If two or more double bonds are present, it is a *polyunsaturated* fatty acid.

Table 13.1 shows the names and structures of some common fatty acids. All these acids can be named according to IUPAC rules for carboxylic acids, but more frequently common names are used because IUPAC names are often too long for common use. Some fatty acids were named for the source from which they were first isolated. Palmitic acid, for example, is the saturated fatty acid typical of palm oil. A convenient shorthand notation for fatty acids consists of two numbers separated by a colon. The first number indicates the number of carbon atoms in the molecule, and the second number expresses the number of carbon–carbon double bonds.

In organic chemistry you learned that the orientation and position of double bonds should be expressed in a compound's name because several isomers of each compound are possible. Unfortunately, the common names of fatty acids include none of this information, but the carbon–carbon double bonds in most fatty acids are *cis*-. The location of the double bond in monounsaturated fatty acids is usually between carbon atoms 9 and 10, and the position of the double bonds in linoleic and linolenic acids are typical of polyunsaturated fatty acids.

Physical Properties

The water solubility of fatty acids varies with pH. Fatty acids lose their carboxylate hydrogen ion between pH 4 and 5. Below a pH of about 4, they exist in the protonated form as a carboxylic acid; above about pH 5 the unprotonated carboxylate form dominates.

$$\underset{\substack{\text{Carboxylic acid}\\ \text{(Stearic acid)}\\ \text{Predominates below pH} \approx 4\\ \text{Insoluble in water}}}{CH_3(CH_2)_{16}COOH} \rightleftharpoons \underset{\substack{\text{Carboxylate anion}\\ \text{(Stearate)}\\ \text{Predominates above pH} \approx 5\\ \text{Soluble in water}}}{CH_3(CH_2)_{16}COO^- + H^+}$$

These two forms have different solubilities in water. The carboxyl group of the carboxylic acid is polar, but the large nonpolar alkyl group makes fatty acids insoluble in water and soluble in nonpolar organic solvents. The carboxylate salt of a fatty acid is ionic. Ionic compounds are generally insoluble in nonpolar organic solvents, but they are soluble in water because the charged group interacts with water molecules.

Molecules that possess significant polar and nonpolar ends are called **amphipathic molecules.** The carboxylate form of a carboxylic acid is amphipathic:

$$\underbrace{CH_3CH_2CH_2CH_2CH_2CH_2CH_2CH_2CH_2CH_2CH_2CH_2CH_2CH_2CH_2}_{\text{Nonpolar}}\underbrace{COO^-}_{\text{Polar}}$$

When amphipathic molecules such as carboxylates of fatty acids are mixed with water, they appear to dissolve or suspend in water. True solutions are not formed, however; instead micelles or other large complexes form (see the last section in this chapter).

Both the melting points and boiling points of fatty acids increase with increasing size (Table 13.2). In addition, the melting points of unsaturated fatty acids are also affected by the presence of *cis*- carbon–carbon double bonds. These *cis*- double bonds alter the shape of the molecule by introducing "kinks" (bends) that reduce the ability of the molecules to "stack" in an orderly fashion in the solid state [Figure 13.1(b)]. As a result, the melting points of unsaturated fatty acids are lower than the melting points of saturated fatty acids of the same size. The same effect is seen in molecules that contain fatty acids. Complex lipids that contain appreciable amounts of unsaturated fatty acids are liquids at room temperature, whereas those containing saturated fatty acids are solids.

Table 13.2 Melting Points of Some Fatty Acids

FATTY ACID		MELTING POINT (°C)
Saturated fatty acids		
Myristic acid	14:0	58
Palmitic acid	16:0	64
Stearic acid	18:0	69
Unsaturated fatty acids		
Palmitoleic acid	16:1	−1
Oleic acid	18:1	16
Linoleic acid	18:2	−5
Linolenic acid	18:3	−11

Chemical Properties

The chemical properties of fatty acids arise from the presence of the carboxyl group and carbon–carbon double bonds. The carboxyl group undergoes the typical reactions described in Chapter 11, and in complex lipids, fatty acids are typically esterified to alcohols.

The carbon–carbon double bonds of unsaturated fatty acids undergo the typical addition reactions described in Chapter 9. In hydration, water adds to these double bonds to introduce a hydroxyl group into the molecule. This addition is important in the metabolic reactions where fatty acids are broken down to produce energy (Chapter 17):

$$\underset{\text{Unsaturated fatty acid}}{\mathrm{RCH{=}CH(CH_2)_nCOOH}} + \mathrm{H_2O} \rightarrow \underset{\text{Hydroxy-fatty acid}}{\mathrm{R}\overset{\mathrm{OH}}{\overset{|}{\mathrm{CH}}}{-}\overset{\mathrm{H}}{\overset{|}{\mathrm{C}}}\mathrm{H(CH_2)_nCOOH}}$$

In hydrogenation, hydrogen adds to these double bonds:

$$\underset{\text{Unsaturated fatty acid}}{\mathrm{RCH{=}CH(CH_2)_nCOOH}} + \mathrm{H_2} \rightarrow \underset{\text{Saturated fatty acid}}{\mathrm{R}\overset{\mathrm{H}}{\overset{|}{\mathrm{CH}}}{-}\overset{\mathrm{H}}{\overset{|}{\mathrm{C}}}\mathrm{H(CH_2)_nCOOH}}$$

Several important reactions of the body involve hydrogenation, but this reaction is also important in the food industry. When you get a chance, read the label on some food packages. Look for the term *partially hydrogenated*, which is used to describe some oils. This term indicates that some of the carbon–carbon double bonds in the fatty acids of these oils have had hydrogen added to them.

A third addition reaction of importance is halogenation, the addition of halogen to the double bond:

$$\underset{\text{Unsaturated fatty acid}}{\mathrm{RCH{=}CH(CH_2)_nCOOH}} + \mathrm{X_2} \dashrightarrow \underset{\text{Halogenated-fatty acid}}{\mathrm{R}\overset{\mathrm{X}}{\overset{|}{\mathrm{CH}}}{-}\overset{\mathrm{X}}{\overset{|}{\mathrm{C}}}\mathrm{H(CH_2)_nCOOH}}$$

This reaction does not occur in living organisms, but it is used in lipid analysis as a test called the iodine number. The amount of iodine that adds to a lipid sample indicates the degree of unsaturation in the lipid molecule.

One other reaction of unsaturated fatty acids is important in understanding the properties of fats and oils: Carbon–carbon double bonds can be oxidized by oxygen. This reaction cleaves the double bond, yielding products that contain an aldehyde group and products that contain a carboxyl group. If the product molecules are small enough, they are volatile (readily vaporize).

$$\underset{\text{Unsaturated fatty acid}}{\mathrm{CH_3(CH_2)_mCH = CH(CH_2)_nCOOH}} + \mathrm{O_2} \rightarrow$$

$$\underset{\text{Volatile aldehydes and acids}}{\mathrm{CH_3(CH_2)_mCHO + OHC(CH_2)_nCOOH}}$$

These products contain partially hydrogenated vegetable oils. *(Charles D. Winters)*

The smell of rancid butter is caused by butanoic (butyric) acid formed by this reaction.

Self-Test

Fatty Acids

1. Name the following fatty acids:
   (a) The saturated fatty acid containing 18 carbon atoms.
   (b) The unsaturated fatty acid containing 16 carbon atoms and one double bond.
   (c) The fatty acid designated by 18:3.
2. The double bonds in most fatty acids have which configuration, *cis*- or *trans*-?
3. (a) What is the appropriate name for the carboxylate anion of oleic acid?
   (b) Which form of oleic acid predominates at pH 7?
4. Which fatty acid and which addition reaction would yield this product:

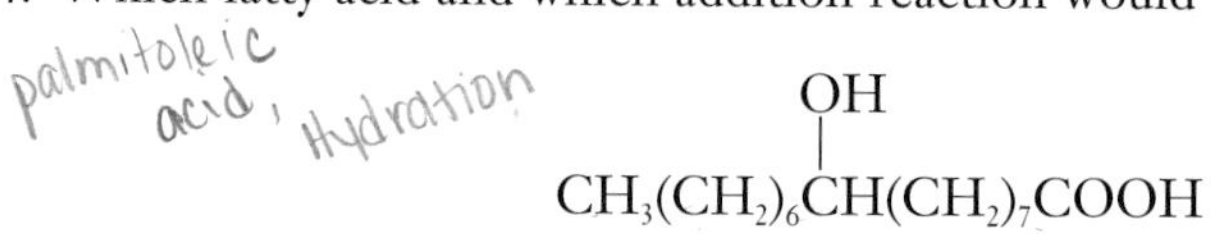

$$CH_3(CH_2)_6CH(OH)(CH_2)_7COOH$$

ANSWERS

1. (a) Stearic acid (b) Palmitoleic acid (c) Linolenic acid 2. *Cis*- 3. (a) Oleate (b) Oleate (carboxylate) form 4. Palmitoleic acid; hydration

13.2 Triacylglycerols

Triacylglycerols are nonpolar complex lipids that serve as energy reserves. These lipids are triesters formed from three fatty acids esterified to the three hydroxyl groups of glycerol (Figure 13.2). All triacylglycerols contain glycerol, but the fatty acids may differ from one triacylglycerol to another. (During certain reactions in the body, only one or two fatty acids may be bonded to a glycerol instead of three. These molecules are called *monoacylglycerols* and *diacylglycerols*, respectively. These *mono-* and *di-* compounds are found in low concentrations in living systems.)

Triacylglycerols are known by several synonyms. *Triglycerides* is an older name that means the same thing: three fatty acids esterified to glycerol. **Fats** are triacylglycerols that are solid at room temperature. Lard (from pork) and tallow (from beef) are fats. The triacylglycerols of most animals are fats. **Oils** are triacylglycerols that are liquids at room temperature. Corn oil, cotton seed oil, and sunflower oil are examples. Many plant triacylglycerols are oils. (The word "oil" is also used for such liquids as crude oil, mineral or paraffin oil, and motor oil. These materials are hydrocarbons, not triacylglycerols.)

Fats are solids at room temperature and oils are liquids. *(Charles D. Winters)*

Physical and Chemical Properties

Why are some triacylglycerols solids at room temperature (fats) and others are liquids (oils)? This difference is largely due to the degree of unsaturation in the fatty acids. Fats contain primarily saturated fatty acids, whereas oils contain a larger percentage of unsaturated fatty acids. Thus the physical state of a triacylglycerol at room temperature provides a clue to the degree of unsaturation in its fatty acids.

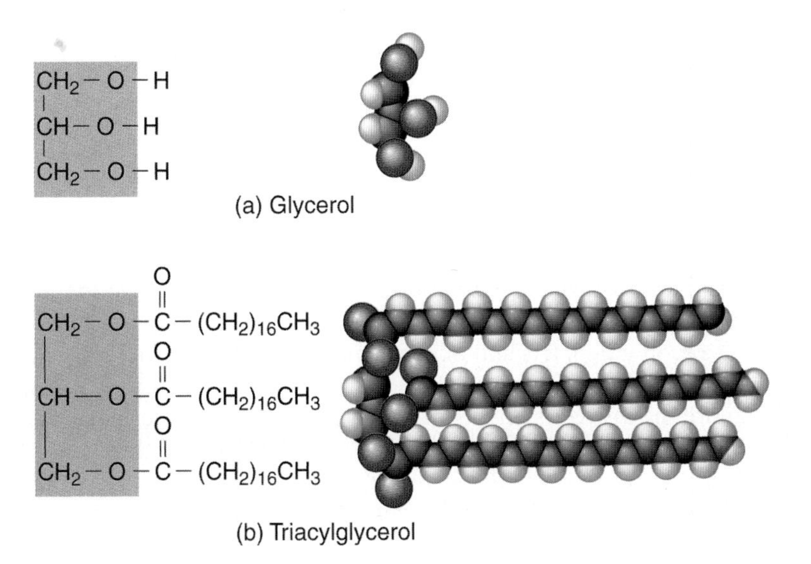

Figure 13.2 Glycerol and a triacylglycerol. **(a)** Glycerol is a trihydroxy alcohol. **(b)** A triacylglycerol has three fatty acids bound to glycerol (highlighted) through ester bonds.

There is a convenient test for estimating the amount of unsaturation in a sample of fat or oil. A sample of triacylglycerols is titrated with the halogen iodine, which readily adds to carbon–carbon double bonds:

$$\ldots CH_2CH_2CH{=}CHCH_2CH_2\ldots + \underset{\text{(Reddish brown)}}{I_2} \dashrightarrow \underset{\text{(Colorless)}}{\ldots CH_2CH_2\overset{\displaystyle I}{\overset{|}{C}}H{-}\overset{\displaystyle I}{\overset{|}{C}}HCH_2CH_2\ldots}$$

When the iodine color in the titrant persists, all of the carbon–carbon bonds have reacted. The more double bonds present in the sample, the more iodine required to complete the titration. The *iodine number* is defined as the grams of iodine required to react with all of the carbon–carbon bonds in 100 g of sample. A high iodine number indicates a high degree of unsaturation. A low number means the triacylglycerol is relatively saturated. Lard has an iodine number around 65 to 70, whereas corn oil has an iodine number around 125 to 130.

The presence of carbon–carbon double bonds makes triacylglycerols susceptible to oxidation when they are exposed to air. They can readily become **rancid,** which means they smell and taste bad because of the presence of volatile organic acids and aldehydes that have formed via oxidation of double bonds. Even a small amount of oxidative cleavage can make an oil or fat inedible. The food industry often does one or two things to reduce the chance of a food becoming rancid. They either (a) add antioxidants such as BHA (butylated hydroxyanisole) or BHT (butylated hydroxytoluene) that react with the oxidizing agent to prevent its reacting with double bonds, or (b) partially hydrogenate an oil to reduce the number of double bonds present. Partial or complete hydrogenation increases shelf life, but the degree of saturation in dietary oils and fats is correlated to coronary disease (see Section 13.4). Is extended shelf life gained at the expense of good nutrition?

Partial hydrogenation also changes the triacylglycerol from a liquid to a semisolid, and complete hydrogenation changes it to a hard solid.

Fats and oils are esters, thus they can be saponified (hydrolyzed in base). (Acid-catalyzed hydrolysis is described in Chapter 11.)

$$\text{Triacylglycerol} + 3NaOH \longrightarrow \text{glycerol} + 3 \text{ fatty acid carboxylate anions} + 3Na^+$$

Large quantities of fats and oils are saponified annually to make soaps, which are the salts of fatty acid carboxylate anions.

Biological Roles

Triacylglycerols are energy reserves for animals and many plants. In general, the fewer the number of oxygen atoms in an organic molecule, the more energy the molecule contains. Thus on a per-gram basis, fats and oils store more energy than sugars, starches, and glycogen because these carbohydrates contain numerous oxygen atoms not present in fats and oils. The same argument can be made for proteins. As a rough guide, fats and oils contain about 9 Cal of energy per gram, and carbohydrates and proteins about 4 Cal per gram. Thus on a mass basis, fats and oils are more than twice as efficient for energy storage. Furthermore, fats and oils are stored "dry" (anhydrous) in the fat cells of the body, but glycogen is hydrated by numerous water molecules. Therefore, 1 g of hydrated glycogen in the body contains far less energy than 1 g of anhydrous body fat. Most people have a significant amount of body fat (10 to 30 percent for the average North American). The energy stored in fat can provide usable energy for several months.

Triacylglycerols are broken down in the body during metabolism, and most of the glycerol and fatty acids produced in the breakdown are used for energy production. However, some of the fatty acids are used as building blocks for other molecules. Two of these, the polyunsaturated linoleic and linolenic acids, are *essential fatty acids.* They are essential in the diet because the body cannot synthesize them, and yet they are used to synthesize other necessary biomolecules. These two fatty acids cannot be made because animals cannot introduce carbon–carbon double bonds into the appropriate positions in monounsaturated fatty acids to convert them to polyunsaturated fatty acids.

Large amounts of stored fat are common in mammals that hibernate and in those that live in cold climates (Figure 13.3). This stored fat serves two useful purposes: (1) It provides energy during hibernation or fast, and (2) it acts as a source of insulation against cold. Marine mammals use fat (blubber) as insulation against the cold polar waters, for buoyancy, and as an energy source during the breeding and pup-rearing seasons. Hibernating animals use it as both insulation and as a food source. Subcutaneous (under the skin) fat in humans appears to serve an insulating role. Slender people may feel cold in cool weather whereas heavier people are more comfortable. Yet when it is hot, the better insulated individual is uncomfortable.

Figure 13.3 Hibernating animals and marine mammals rely on stored fat for energy and insulation. Some marine mammals have several inches of blubber for insulation against cold waters.

Connections

Fat Substitutes

Many people want to reduce their weight, caloric intake, or intake of fat for better health. These people often choose a diet containing fewer high-fat foods to accomplish their goal. The amounts of high-fat foods can be reduced by eating less of them, or some high-fat foods can be replaced by foods containing *fat substitutes.* Fat substitutes have fewer total calories than the fats they replace, and also have fewer or no Calories from fat.

To function as a fat substitute a substance must mimic the culinary properties of fats. Current fat substitutes are based on carbohydrates, proteins, or fats. Carbohydrate- and protein-based fat substitutes provide both fewer calories and the same texture and taste as the fat-based products they replace. These fat substitutes are used in prepared foods such as desserts, luncheon meats, and salad dressings. They work well in these foods, but cannot be used for fried foods because the high temperatures used in frying would destroy them.

The fat substitutes used in fried foods are derived from fats. An example is the product marketed as Olestra®. This product is a sucrose polyester consisting of a sucrose molecule with six to eight fatty acids esterified to the hydroxyl groups on the sugar. Olestra has many of the properties of fats—taste and heat stability are examples—but cannot be digested or absorbed in the gastrointestinal tract and thus provides no Calories. Olestra was approved by the Food and Drug Administration (FDA), but some people complain that it gives them distress of the lower gastrointestinal tract, and others are concerned that large amounts of Olestra in a diet could reduce uptake of fat-soluble nutrients such as vitamins A and D.

These products contain Olestra, a fat substitute. *(George Semple)*

Fatty tissue may also serve a protective role. Fatty layers around the internal organs of vertebrates serve as a cushion to protect the organs from injury.

Example 13.1 Calculating Energy Derived from Fat in Foods

A 12-oz chocolate milk shake contains 10 g of protein, 72 g of carbohydrate, and 9 g of fat. Determine **(a)** the number of Calories in the shake and **(b)** the percentage of Calories that come from fat. Remember that there are 9 Cal/g of fat and 4 Cal/g of carbohydrate or protein.

SOLUTION

(a) To obtain the number of Calories, first use dimensional analysis to determine how many Calories are contributed by each group of nutrient and then sum the Calories:

$$10 \text{ g protein} \times 4 \text{ Cal/g protein} + 72 \text{ g carbohydrate} \times 4 \text{ Cal/g carbohydrate} + 9 \text{ g fat} \times 9 \text{ Cal/g fat} = 40 \text{ Cal} + 288 \text{ Cal} + 81 \text{ Cal} = 409 \text{ Cal}$$

(b) To determine the percentage of the energy that comes from fat, divide the energy from fat by the total energy and multiply by 100:

$$\frac{81 \text{ Cal fat energy}}{409 \text{ Cal total energy}} \times 100 = 19.8\% = 20\%$$

Self-Test

1. What are the components of a triacylglycerol?
2. What structural factor greatly affects whether a triacylglycerol is a fat or an oil?
3. Fats and oils serve three biological roles. What are those roles?
4. A sample of food contains 8.7 g of protein, 43.3 g of carbohydrate, and 10.1 g of fat. (a) Determine the number of Calories in this food, and (b) determine the percentage of energy from fat.

ANSWERS

1. Three fatty acids molecules and a glycerol molecule 2. The degree of unsaturation in the fatty acids 3. Energy storage, insulation, protection of internal organs 4. 300 Cal; 30%

13.3 Complex Lipids of Membranes and Blood

Some complex lipids are structural components of cell **membranes,** which are the sheet-like bilayered structures that surround cells and divide the cell interior. Some are also found in the circulatory system, where they play a role in normal transport of other lipids.

The most common membrane lipids are those that contain phosphorus, the *phospholipids.* Many phospholipids are **phosphoacylglycerols,** compounds structurally related to the triacylglycerols. A molecule consisting of glycerol with fatty acids esterified to two of the hydroxyl groups and phosphoric acid esterified to the third is called a **phosphatidic acid** [Figure 13.4(a)]. When an additional alcohol is bonded to the phosphate of the phosphatidic acid, the molecule is a phosphoacylglycerol. If choline is the additional molecule bound to the phosphate, as shown in Figure 13.4(b), the compound is phosphatidylcholine (also known as lecithin), which is the most abundant phospholipid. If ethanolamine is bound, the phosphoacylglycerol is phosphatidylethanolamine [Figure 13.4(c)], which is also known as cephalin. Other derivatives of phosphatidic acids include phosphatidylserine and phosphatidylinositol.

Triacylglycerols are considered nonpolar lipids even though they contain the slightly polar ester groups. In contrast, the phosphoacylglycerols are **polar lipids,** because they contain a polar head group consisting of the phosphate group and the alcohol attached to it. The rest of the phosphoacylglycerol (the alkyl part of the two fatty acids) is nonpolar and called the nonpolar tail. Thus phosphoacylglycerols, like soaps, are amphipathic molecules because they contain both a polar region (the head) and a nonpolar region [the tail; Figure 13.4(d)].

The **sphingolipids** are another group of complex lipids found in membranes. These lipids contain (1) the base *sphingosine* [Figure 13.5(a)], (2) a fatty acid, and (3) one or more other molecules. The fatty acid is bonded to the amino group of sphingosine through an amide bond. The other molecules are polar molecules such as sugars or phosphate-choline. These polar groups are bonded to the sphingosine

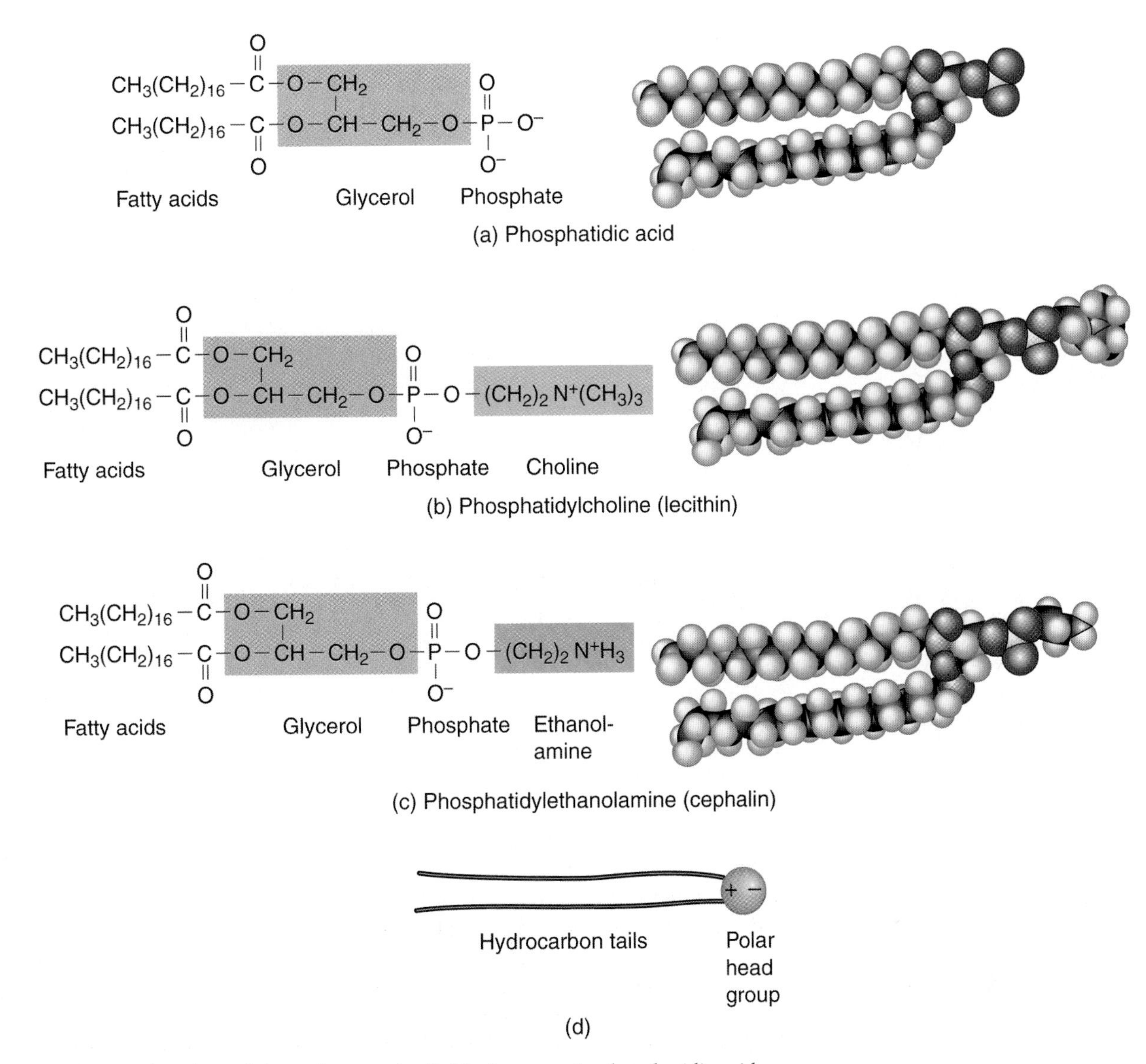

Figure 13.4 Phosphoacylglycerols are polar lipids that contain phosphatidic acids. **(a)** Phosphatidic acids contain glycerol, two fatty acids and a phosphate group. **(b)** Phosphatidylcholine (lecithin) is the most common phosphoacylglycerol. It contains a phosphatidic acid and a molecule of choline. **(c)** Phosphatidylethanolamine (cephalin) contains a phosphatidic acid and ethanolamine. **(d)** Most polar lipids can be represented by a polar-head group with two hydrocarbon tails.

through an ester bond to a hydroxyl group. The sphingolipids differ in composition from the phosphoacylglycerols, but these two groups of lipids resemble one another structurally, with both being amphipathic. This structural similarity is why they have similar roles in membranes.

There are two major types of sphingolipids—*sphingomyelin*, which contain phosphate and choline [Figure 13.5(b)], and *cerebrosides*, which contain sugars.

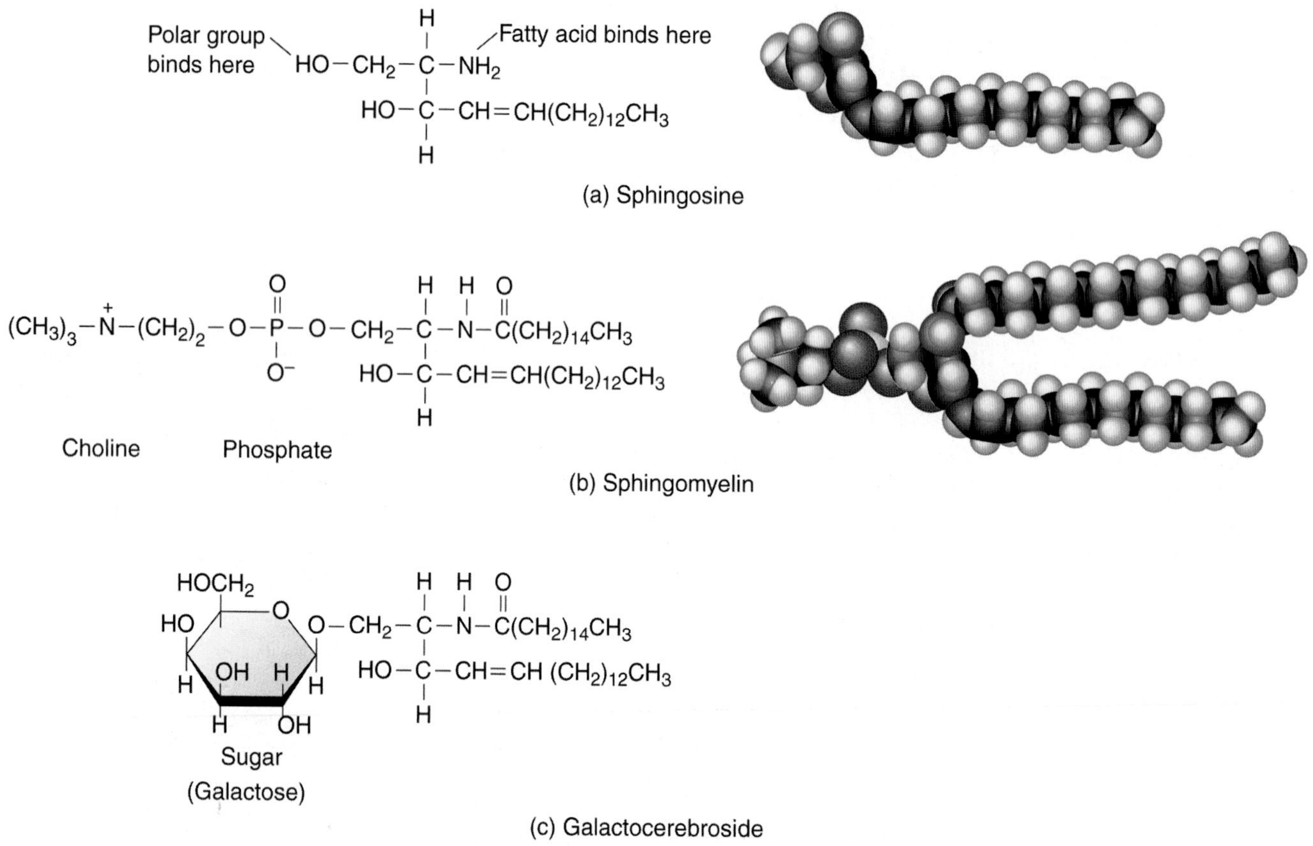

Polar group binds here

Fatty acid binds here

$HO-CH_2-CH(NH_2)-CH(OH)-CH=CH(CH_2)_{12}CH_3$

(a) Sphingosine

$(CH_3)_3-\overset{+}{N}-(CH_2)_2-O-P(=O)(O^-)-O-CH_2-CH(NH-C(=O)(CH_2)_{14}CH_3)-CH(OH)-CH=CH(CH_2)_{12}CH_3$

Choline

Phosphate

(b) Sphingomyelin

$HOCH_2$, HO, OH, H, OH — $O-CH_2-CH(NH-C(=O)(CH_2)_{14}CH_3)-CH(OH)-CH=CH(CH_2)_{12}CH_3$

Sugar (Galactose)

(c) Galactocerebroside

Figure 13.5 Sphingosine and two sphingolipids. **(a)** Sphingosine. **(b)** Sphingomyelin contains sphingosine, a fatty acid, a phosphate group, and a molecule of choline. Compare the shape of sphingomyelin to that of phosphatidylcholine in Figure 13.4(b). **(c)** This cerebroside contains sphingosine, a fatty acid, and galactose.

In your studies you may encounter the term **glycolipid.** This term is used to describe a lipid that has a carbohydrate as a component. Phosphatidyl inositol and the cerebrosides are examples of glycolipids.

Self-Test

Complex Lipids of Membranes

1. What are the components of phosphatidic acid?
2. Glycolipids contain what kinds of components?
3. What makes a molecule amphipathic?

ANSWERS

1. Two fatty acids, glycerol, one phosphate 2. Carbohydrates and lipids 3. Having both a polar and a nonpolar end

13.4 Simple Lipids

Unlike the complex lipids, the simple lipids do not contain fatty acids linked by ester bonds. Therefore base hydrolysis (saponification) has no effect on their structure. This section introduces several common simple lipids—steroids, prostaglandins, and leukotrienes.

Steroids

Steroids are compounds that contain the fused-ring system commonly called the *steroid nucleus* (Figure 13.6). Although the steroids are related to each other by structure, they carry out a wide variety of functions in the body. Regardless of their function, however, you can always recognize a steroid by its steroid nucleus.

The most abundant steroid in animals is **cholesterol** (Figure 13.6). It is found in membranes, and, like other membrane lipids, is amphipathic. The steroid nucleus is nonpolar, and the hydroxyl group is polar. The irregular shape of cholesterol does not fit well in the membrane, just as unsaturated fatty acids do not stack well. As a result, cholesterol reduces the rigidity of a membrane, keeping the membrane fluid. Cholesterol is also important because many other steroids are synthesized from it.

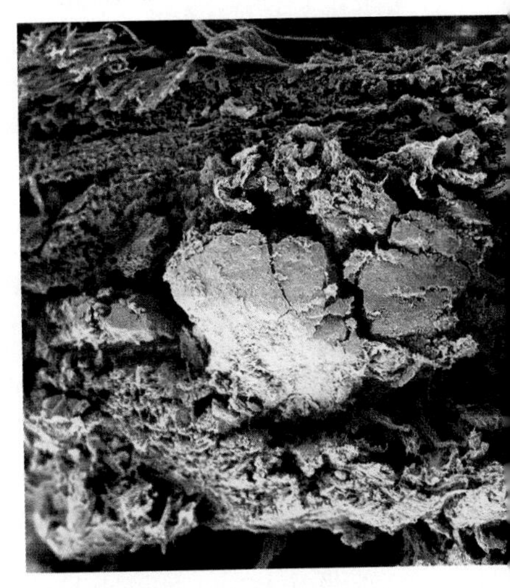

Atherosclerosis is a hardening and thickening of the arterial walls. *(Meckes/Ottawa/Photo Researchers, Inc.)*

Cholesterol is obtained from the diet and is also made in the body, primarily by the liver. The circulatory system transports cholesterol to the rest of the body. In some people, deposits containing cholesterol build up on the interior surface of arteries, a condition known as *atherosclerosis.* These deposits reduce the diameter of the arteries that supply the heart and other vital organs. This reduction means these vessels are more easily blocked, perhaps by a small blood clot that would normally pass through. If no blood passes, that portion of the heart or other organ fed by the blocked artery is deprived of oxygen and nutrients and may die. Surgical techniques (angioplasty) can be used to reduce constrictions in arteries, but this is useful only before the artery is totally clogged and while the heart and other organ

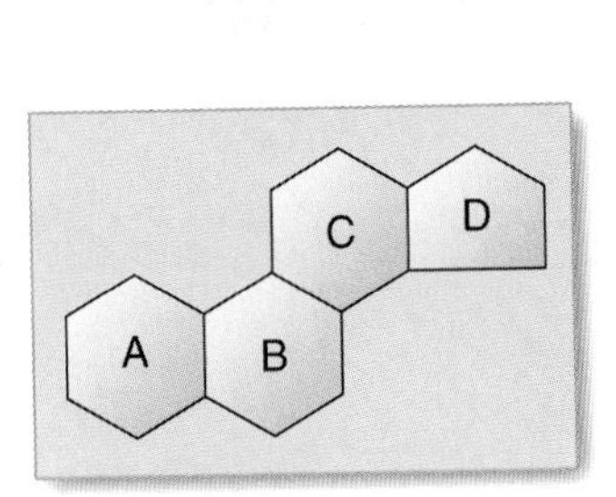

Steroid nucleus

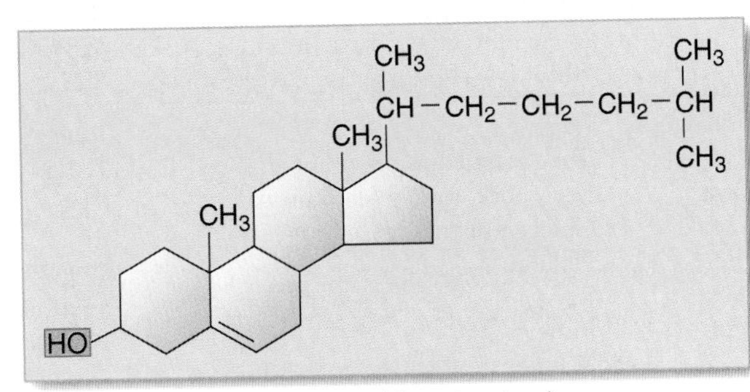

Cholesterol (ring structure)

Cholesterol (space-filling model)

Figure 13.6 All steroids possess the steroid nucleus. Cholesterol, the most abundant steroid, is shown as both a flat ring structure and in the more realistic space-filling form. The —OH group gives cholesterol some polarity.

tissue is still alive. Although the specific role of cholesterol in this disease process is not fully understood, it is known that a relationship exists between blood cholesterol levels and atherosclerosis. In general, the higher the concentration of blood cholesterol, the higher the probability for heart attack.

Another group of steroids found in the body are the *steroid hormones*, which like all hormones are produced in one part of the body and then circulated in blood to the rest of the body, where they bind to specific target cells and exert influence over

Connections

Correlations Between Lipids and Health

Several correlations exist between lipids and good health. First, there is a correlation between obesity—defined as 30% or greater body fat—and cardiovascular disease. Because fats and oils are energy-rich, restricting the intake of these substances can help reduce weight and lower the risk of cardiovascular disease. However, it is neither easy nor desirable to eliminate all fat from the diet. A small amount of the polyunsaturated oils should be eaten regularly to provide the essential fatty acids.

A second correlation between lipids and good health is related to the degree of saturation in dietary fats and oils. In general, increased proportions of saturated fats increase the risk of cardiovascular disease. If the ratio of saturated fats to polyunsaturated fats is greater than 1, the risk of heart disease is increased. If the ratio is below 1, the risk is decreased. Because foods from animal sources are higher in saturated fats and those from plants are higher in unsaturated fats, a reduction in foods from animal sources and an increase in consumption of foods from plants will normally decrease this ratio.

A third correlation exists between blood cholesterol levels and cardiovascular disease. Many of us are tempted to conclude that high blood cholesterol and dietary cholesterol are linked, but studies do not yet support a correlation between dietary cholesterol and the development of heart disease in all humans. People with known heart conditions should reduce their blood cholesterol levels through medication and diet, but for the general public it has not been shown that reducing dietary cholesterol will *by itself* reduce the risk of heart attacks. This is not to say that dietary cholesterol might not be a factor in heart disease, but at this time the link has not yet been established.

Foods from animal sources contain cholesterol but foods from plants do not. *(Charles D. Winters)*

The Amount of Cholesterol in Some Common Foods

FOOD	SERVING SIZE	MILLIGRAMS OF CHOLESTEROL
Egg	1 egg	250
Tuna (canned in oil, drained)	184 g	116
Beef, pork, turkey (dark meat)	84 g	67
Chicken or turkey (light meat)	84 g	55
Halibut	84 g	55
Salmon	84 g	40
Butter	1 Tbsp	35
Hot dog	1 hot dog	34
Milk	1 cup	34
Cheddar cheese	28 g	28
Ice cream	1/2 cup	27
Skim milk	1 cup	5

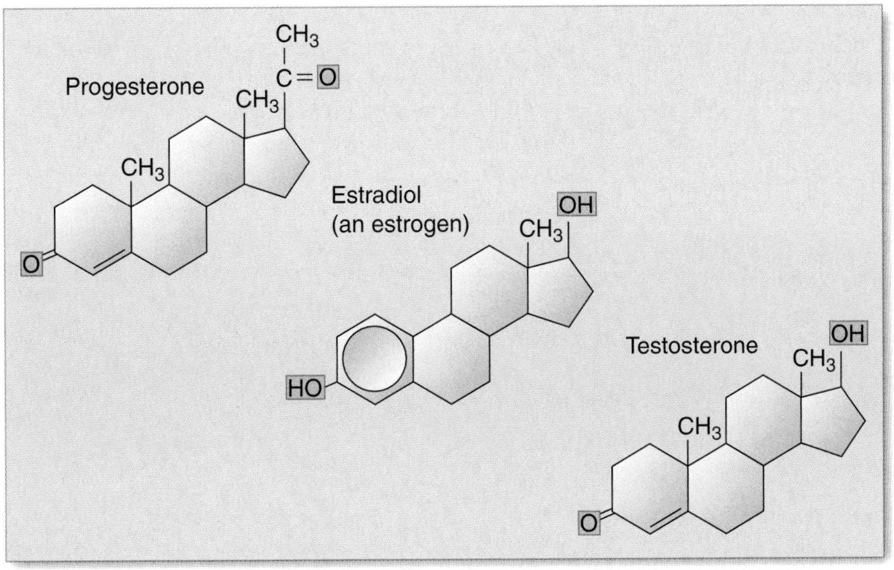

Figure 13.7 The common human sex hormones.

the activity in these cells. Some steroid hormones are involved in salt and water regulation, others are involved in the body's response to stress, and still others play a role in sexual function and in the development and maintenance of secondary sexual characteristics.

Sex hormones are steroid hormones that determine the secondary sexual characteristics of males and females. Although both sexes carry some of each of the hormone types, a careful balance must be maintained to provide normal growth and development. The male hormones, the *androgens* (testosterone), are quite similar in structure to a female hormone, *progesterone*. The other common female hormones, the *estrogens*, contain an aromatic ring in the fused ring system (Figure 13.7).

The *adrenocorticoid hormones* are another important group of steroid hormones. These hormones are produced by the cortex of the adrenal gland (thus their name). They play a central role in regulating some parts of metabolism. The most common member of this class is *hydrocortisone* (also known as *cortisol*), a powerful anti-inflammatory agent (Figure 13.8). Hydrocortisone, and related compounds such as cortisone and prednisolone, are used to treat some skin inflammations, asthma, and a variety of inflammatory diseases including arthritis. Although these compounds have highly beneficial properties, prolonged therapeutic use can result in problems, including excessive breakdown of protein (reflected in muscle deterioration, weakness, and excessive excretion of nitrogen-containing waste products), sodium retention, calcium excretion, and increased susceptibility to infection.

Bile salts are a group of steroids that play a role in digestion. These compounds are synthesized from cholesterol in the liver and then stored in the gallbladder. During digestion, bile salts enter the small intestine and break down (emulsify) fat globules, which aids in digestion and absorption of dietary lipids [Figure 13.9(a)]. Bile salts are amphipathic because they have polar groups projecting from one face of the steroid ring system, whereas the other face is nonpolar

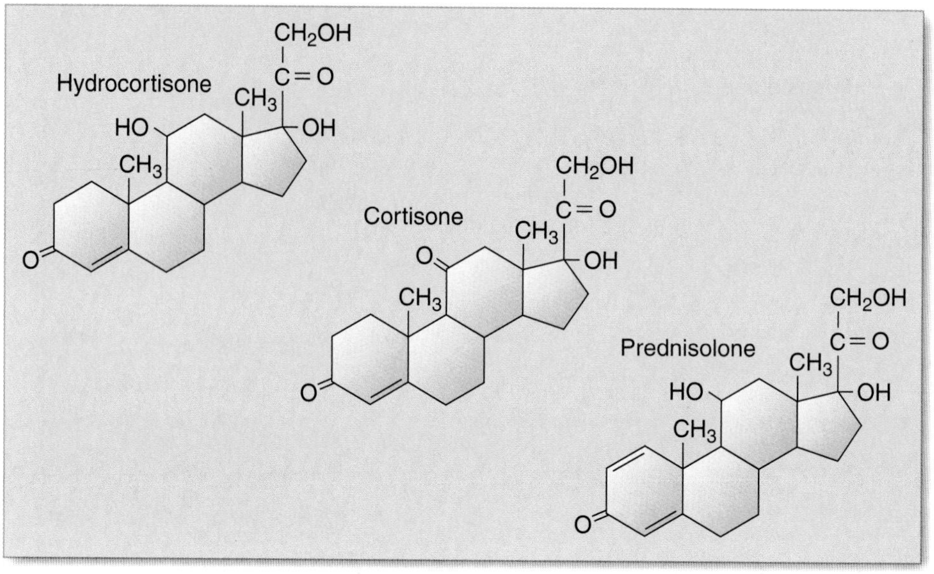

Figure 13.8 Hydrocortisone, cortisone, and the synthetic analog prednisolone.

[Figure 13.9(b)]. The nonpolar face binds to dietary fats and oils, and the polar face is exposed to the water in the intestine. The rate of lipid digestion is much faster in these bile-salt complexes than in the fat globules, because the complexes have a much greater surface area where digestive enzymes can interact with dietary lipids. Two typical bile salts, *taurocholate* and *glycocholate*, are shown in Figure 13.9.

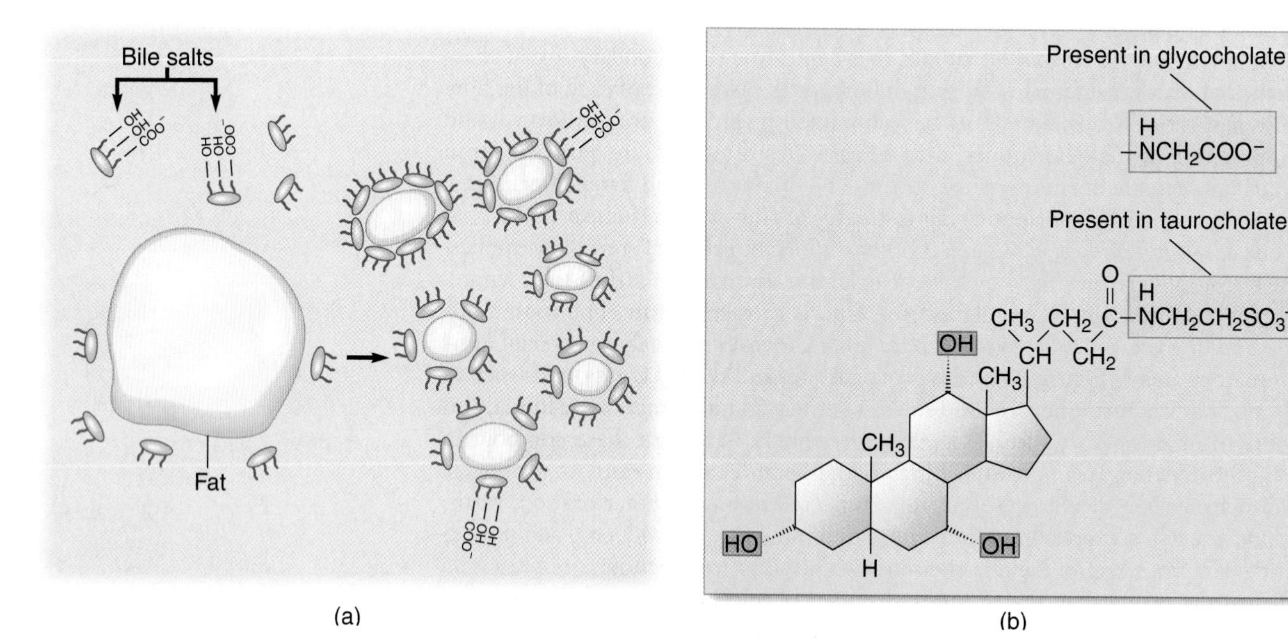

Figure 13.9 **(a)** Bile salts break down fat globules into smaller complexes of lipid and bile salts, where digestion occurs faster. **(b)** The common bile salts glycocholate and taurocholate both contain a steroid nucleus with polar hydroxyl groups (red) located on the same face, but have different polar groups on the right side of the molecule.

Anabolic Steroids in Athletics

Some people thrive on competition, and athletics provides an outlet. Sometimes, however, internal drive or external pressure to win forces a dilemma: Win at all costs, or play the game by the rules? Serious competitors train hard, many eat balanced diets, most conscientiously obtain enough rest, and nearly all abstain from tobacco and other substances that reduce performance. Some take vitamin and mineral supplements because they believe doing so will enhance their performance. Some athletes, however, overstep the boundary between acceptable and unacceptable by using performance-enhancing drugs. Within this group of drugs are the anabolic steroids. These compounds are either natural steroid hormones or more commonly synthetic versions of them.

Anabolic refers to reactions that build or synthesize. Anabolic hormones stimulate the body's reactions related to synthesis and growth. The male sex hormone testosterone and its synthetic mimics are examples of anabolic hormones that stimulate increased muscle mass. Some athletes use anabolic steroids to stimulate muscle growth to obtain a competitive edge, but is this action to enhance performance acceptable? Most people (and the governing bodies of sports) do not think so. It provides an unfair edge. Furthermore, there are a number of known or suspected side effects that are potentially quite harmful: liver tumors, aggressive–impulsive behavior, testicular atrophy in males, and the development of masculine traits in females. The use of performance-enhancing steroids involves great health risks.

Anabolic steroids enhance muscle building but cause serious side effects. *(Larime Photographic/Dembinsky Photo Associates)*

Prostaglandins and Leukotrienes

Though found in the body only in trace amounts, the simple lipids known as prostaglandins and leukotrienes play a role in regulation of body function. These lipids are synthesized from polyunsaturated fatty acids. The **prostaglandins** were first isolated from secretions of the male reproductive tract and were thought to originate in the prostate gland. It is now known that they are widespread in both sexes. The structure of a common prostaglandin (PGE_2) is shown in Figure 13.10(a). Prostaglandins are present in very small amounts in many tissues of the body and have pronounced regulatory properties, including a role in the physiology of fever and pain. Aspirin is the world's most common analgesic (pain killer) and antipyretic (fever reducer). Although aspirin has been used for more than 100 years, until recently no one understood how pain and inflammation are affected by it. Now we know that aspirin interferes with the body's ability to synthesize prostaglandins.

One of the more recently discovered groups of lipids is the **leukotrienes,** which play a role in allergic and inflammation responses. Because some of these

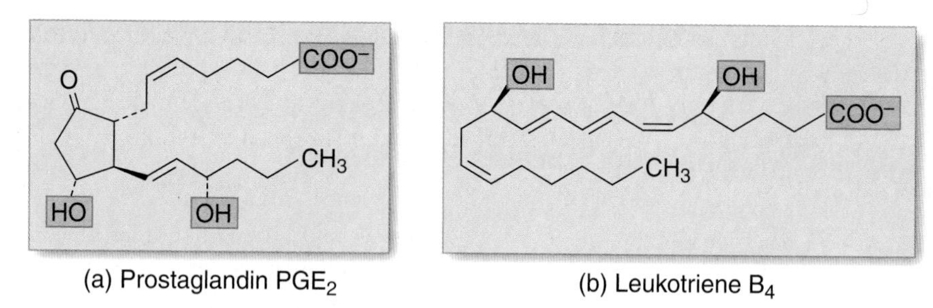

(a) Prostaglandin PGE_2 (b) Leukotriene B_4

Figure 13.10 Prostaglandins and leukotrienes. **(a)** Prostaglandin PGE_2 is a typical prostaglandin. **(b)** Leukotriene B_4 is a representative leukotriene.

compounds were isolated from leukocytes (white blood cells), their name reflects this origin as well as the unsaturated nature of the compounds. The structure of a common leukotriene is shown in Figure 13.10(b).

Self-Test

Simple Lipids

1. Steroids are identified by the presence of one structural feature. Name and draw it.
2. Cholesterol serves what roles in living organisms?
3. Prostaglandins and leukotrienes are synthesized from which compounds?

ANSWERS

1. The steroid nucleus:

2. Component of membranes; used to make other steroids
3. Polyunsaturated fatty acids

13.5 Membranes and Related Structures

Polar lipids are important components of cell membranes (see the earlier section on lipids of membranes and blood), but to understand the role of polar lipids in membrane structure, we must first examine related structures called micelles, bilayers, and liposomes.

In water, some amphipathic molecules (including soaps) can aggregate to form *micelles*, which are spherical structures that have a nonpolar, hydrophobic environment in the core, and a polar, aqueous environment on the surface. Micelles have the long nonpolar tails of the amphipathic molecules clustered together in the core and the polar-head groups on the surface (Figure 13.11). The bile salt–lipid complexes formed in the small intestine are micelles.

Why do amphipathic molecules form micelles in water? One useful model says the strong hydrogen bonding between water molecules excludes the nonpolar tails of the amphipathic molecules from water, whereas the polar-head groups can interact with the water molecules. As a result, the clustering of water molecules and polar-head groups forces the nonpolar tails away from water, where they aggregate together in the core of the micelle. In the core the nonpolar tails bond with each other through London forces. The interactions (bonding) between the nonpolar tails clustered inside the micelle are referred to as **hydrophobic interactions.** Such interactions play a very important role in biological membranes and protein structure. The term *hydrophobic interaction* implies that water-hating, nonpolar molecules prefer to be together, but the reality is they have been squeezed together by polar water molecules.

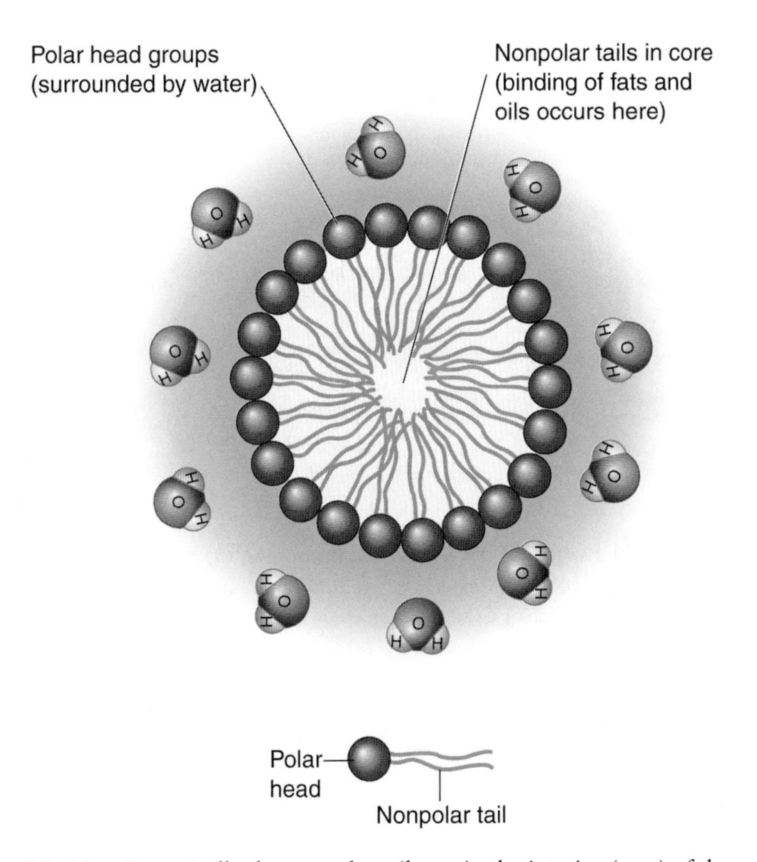

Figure 13.11 In a micelle the nonpolar tails are in the interior (core) of the micelle and the polar-head groups are in contact with the polar solvent.

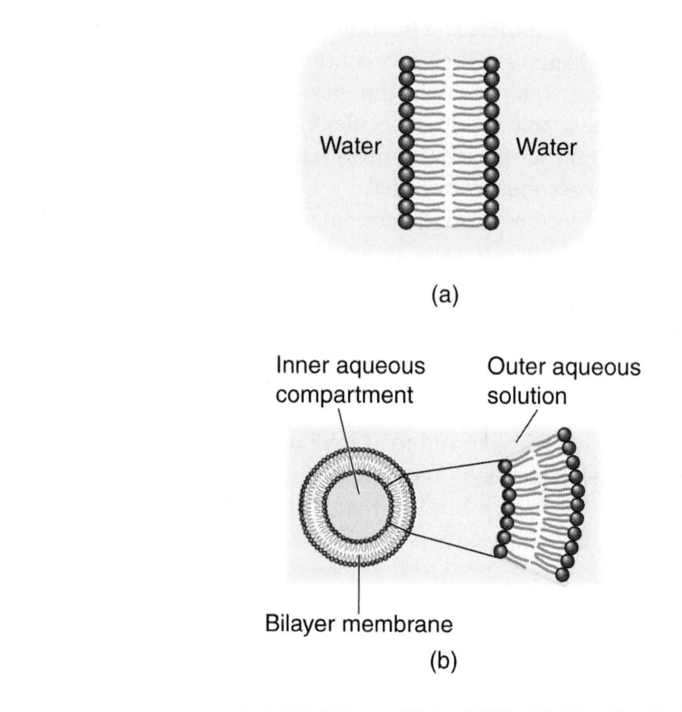

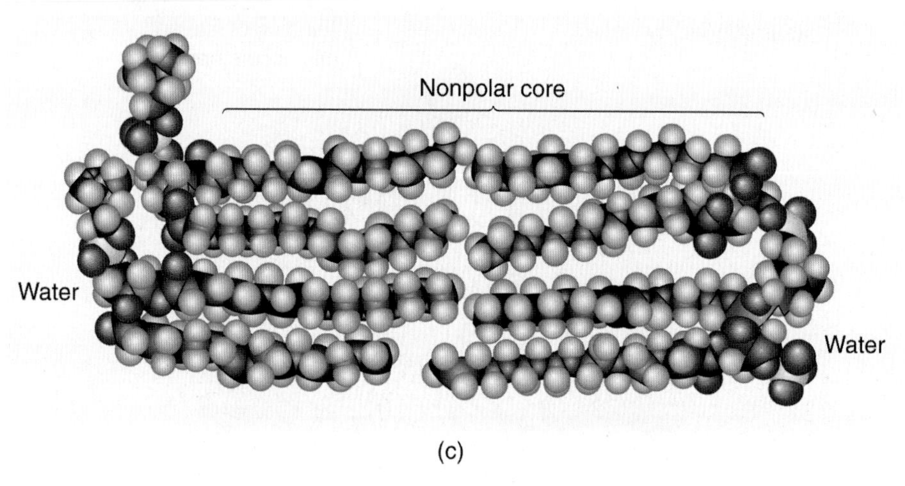

Figure 13.12 The structure of bilayers and liposomes. **(a)** Bilayers consist of polar lipids that have their nonpolar regions aggregated together, and their polar-head groups facing the aqueous solution. **(b)** Liposomes are spherical structures that are bilayers folded around a core of water or other polar solvent. **(c)** Space-filling models of some polar lipids in a bilayer.

A **bilayer** is another structure that forms when some amphipathic molecules are put in water [Figure 13.12(a)]. The nonpolar tails in the bilayer project into the core. This puts the polar-head groups on the exterior surfaces.

Bilayers can close back on themselves to form a continuous bilayer surrounding a core of water. These structures are called **liposomes** [Figure 13.12(b)] and can be prepared as relatively stable structures. They show some promise for improved delivery of drugs because the outer polar surface of the liposome keeps them in suspension in blood and other body fluids, and the core carries an aqueous solution or suspension of the drug.

Micelles, bilayers, and liposomes have served another important role in biology as models for biological membranes. As we learned earlier in the chapter, membranes are bilayered structures that separate a cell from its external environment and divide the cell into compartments. The most abundant components of membranes are phospholipids, glycolipids, and cholesterol. Like bilayers, membranes have a nonpolar core that is hydrophobic and two surfaces that are polar. The polar ends of the polar lipids include positive and negative charges that attract each other. These attractions significantly stabilize membranes and liposomes derived from them.

Proteins are another major component of membranes. These macromolecules are discussed in detail in Chapter 14, but for now picture them as large molecules that may have both polar and nonpolar regions on their surface. Membrane proteins are imbedded in the lipids of a membrane. Their nonpolar portions are in contact with the nonpolar tails of the lipids in the core of the membrane, and their polar portions face outward to bond with water. Hydrophobic interactions hold the nonpolar portion of the protein in the core, and various polar interactions with water and solutes keep the polar part on the surface.

The current model for cellular membranes is called the *fluid mosaic model* because the proteins form a mosaic in the membrane (Figure 13.13). Protein and lipid molecules are free to diffuse around in the membrane (i.e., they are fluid) because only weak, noncovalent bonding holds them in the membrane.

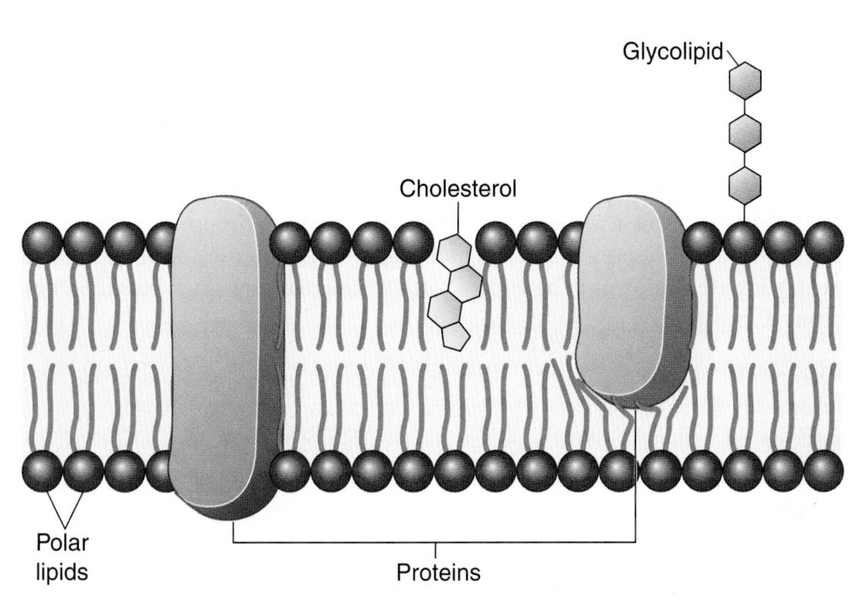

Figure 13.13 The fluid mosaic model says membranes consist of proteins and polar lipids that interact noncovalently in the nonpolar core and at the polar surface. Because the molecules are not covalently bound, they are free to diffuse throughout the membrane and thus the membrane is fluid.

Membranes are stable biological structures, which means that their lipid components and some of their proteins are not easily removed. This is because ionic and polar attractions at the membrane surface must be broken before such removal can occur, and a nonpolar tail or surface must be brought into contact with water molecules or other polar molecules and ions.

Transport Across Biological Membranes

Lipid bilayers and membranes are selectively permeable barriers that restrict the movement of many ions and molecules. Polar molecules cannot pass through the nonpolar core of a cell membrane. Nonpolar molecules, however, such as molecular oxygen or carbon dioxide, do pass through a membrane because they dissolve in the nonpolar core and readily cross. The transport of these molecules across a membrane appears to be by **simple diffusion** [Figure 13.14(a)], which means they move from a region of higher concentration to lower concentration, like the dye molecules of a drop of dye put into water.

How do polar molecules and ions cross a membrane? Their transport involves some of the membrane proteins. For example, some proteins are involved in the transport of glucose molecules, whereas others are involved in the transport of sodium ions. Proteins aid passage of materials through a membrane by one of two ways: facilitated diffusion or active transport. In **facilitated diffusion** proteins act as polar channels that allow ions or polar molecules to pass through the membrane [Figure 13.14(b)]. The molecules or ions are never exposed to a nonpolar environment as they move from one aqueous solution, through the polar channel, and into another aqueous environment. Just as in simple diffusion, movement is from a region of higher concentration to one of lower concentration. It occurs quickly because the proteins allow passage through the nonpolar core of the membrane. Facilitated diffusion does not require the input of energy.

Active transport of ions and polar molecules across a membrane requires energy. The transporting protein in the membrane uses energy to move molecules *against* a concentration gradient (from low concentration to high concentration). We can think of active transport as being like a pump. The proteins of the transport system pump the ion or molecule across the membrane, using energy to do the work [Figure 13.14(c)]. The energy used in active transport is provided by metabolism (Chapter 17).

Self-Test

Membranes

1. What are the two principal classes of compounds found in membranes?
2. Describe the interaction (bonding) between a protein molecule in a membrane and the molecules surrounding it.
3. What name is given to the energy-requiring process that passes material through a membrane?

ANSWERS

1. Polar lipids and proteins 2. The nonpolar part of the protein molecule takes part in hydrophobic interactions with the nonpolar tails of the lipids in the core, and the polar part interacts with the polar-head groups of the lipids and with water surrounding the membrane through dipole–dipole and ionic interactions, and hydrogen bonds. 3. Active transport

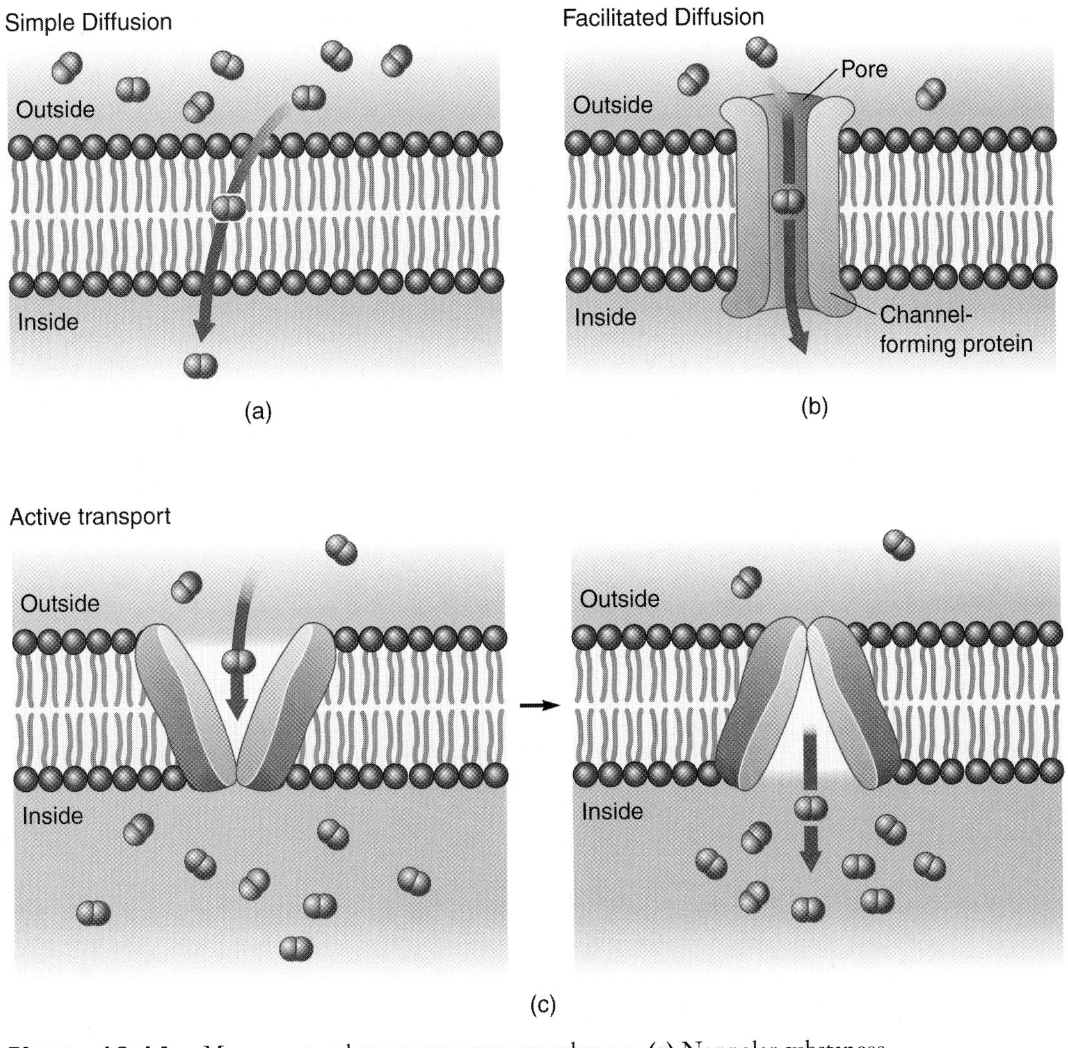

Figure 13.14 Movement and transport across membranes. **(a)** Nonpolar substances cross membranes by simple diffusion. **(b)** Some polar substances pass through polar channels that are made up of proteins. This is facilitated diffusion. **(c)** Some substances are transported across membranes in an energy-requiring process. This is active transport.

Connections

The Fat-Soluble Vitamins

The fat-soluble vitamins—A, D, E, and K—are lipids found in a variety of foods that are stored in the lipids of the body. Vitamin A was the first human vitamin to be discovered. Some vitamin A, as retinal, serves a role in the detection of light in the retina of the eye. As a result, a deficiency of vitamin A may manifest itself as a form of night blindness. This vitamin is also involved in the maintenance of normal skin and other surface tissues, in growth, and perhaps in membrane function.

Toxicity from vitamin A overdose can occur because excess vitamin A is stored in body fat and is not excreted. Most overdoses involve dietary supplements or medications; rarely does the diet provide toxic quantities. (Large daily doses of liver could cause it, but who among us eats large amounts of liver regularly?)

Vitamin D, which exists in several forms, is active in the body as dihydroxycholecalciferol (calcitriol). This compound is involved in calcium uptake and use, and proper bone mineralization will not occur unless adequate amounts are present. The disease rickets may result from vitamin D deficiency in children.

Vitamin D is sometimes called the sunshine vitamin because ultraviolet light in the skin converts a steroid to this vitamin. As long as people are exposed to adequate amounts of sunlight, there should be little or no need for supplementary sources of vitamin D. Few natural sources of this vitamin are available. Liver, fatty fishes, egg yolk, and butter provide some. A more common source is the vitamin D added to dairy products and some other foods. This is the major source of dietary vitamin D in North America. Strict vegetarians should have no deficiency unless they avoid sunlight.

Vitamin E is a collection of compounds known as *tocopherols*. The tocopherols are antioxidants; they reduce the rate of oxidation of double bonds, and they appear to function in this role in cell membranes. In addition, vitamin E is involved in forming various cells and tissues. Deficiencies in humans are rare, but in laboratory animals several deficiencies have been

Vitamin A

Vitamin D

Lipids are nonpolar and thus insoluble in body fluids such as blood. **Lipoproteins,** which like biological membranes are complexes of lipids and proteins, transport lipids in blood. Chylomicrons are lipoproteins that transport dietary fats and cholesterol from the intestine to the tissues of the body. Low-density lipoproteins (LDLs) transport cholesterol and other lipids from the liver to the cells of the

seen or induced by deficient diets. The symptoms of the deficiencies include sterility, muscle atrophy, and nervous disorders.

Vitamin K, the final fat-soluble vitamin to be considered, plays a role in the synthesis of prothrombin, which is necessary for normal blood clotting. Without vitamin K, inadequate amounts of prothrombin are formed, and normal clotting may not occur. Because vitamin K is readily available in leafy vegetables and is formed by intestinal bacteria, human deficiencies of this vitamin are uncommon.

These foods are good sources of the fat-soluble vitamins. *(Charles D. Winters)*

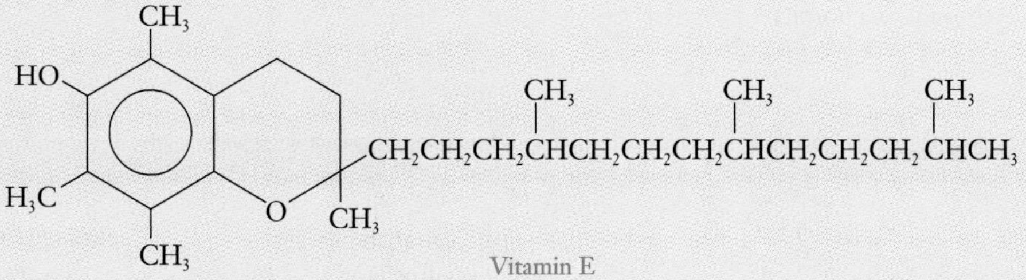

Vitamin E

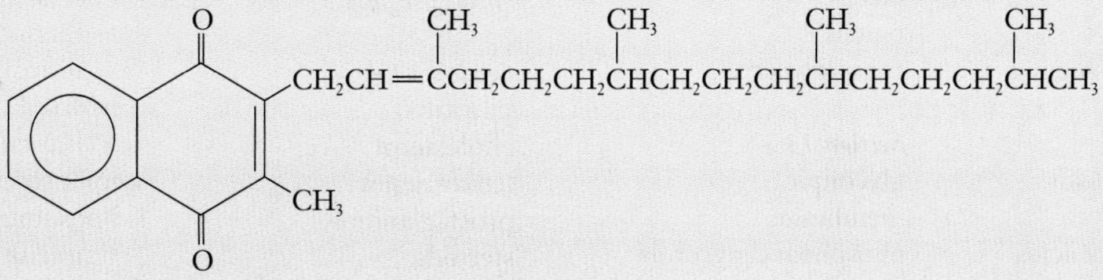

Vitamin K

body, where they are absorbed and used. High-density lipoproteins (HDLs) transport cholesterol from body cells back to the liver. Because high ratios of LDL to HDL have been correlated with cardiovascular disease, the cholesterol in LDL is sometimes referred to as *bad cholesterol*, whereas that in HDL is sometimes referred to as *good cholesterol*.

CHAPTER SUMMARY

Lipids are the water-insoluble molecules of the body that are soluble in organic solvents. Lipids are classified as **complex lipids**—those that contain fatty acids—and **simple lipids,** which do not. **Fatty acids** are unbranched, long-chain carboxylic acids containing an even number of carbon atoms. **Saturated fatty acids** contain no carbon–carbon double bonds, and **unsaturated fatty acids** contain one or more of these double bonds or, rarely, triple bonds. Fatty acids are water-insoluble but react with bases to yield carboxylate anions, which are **amphipathic molecules** that appear to be water-soluble. Fatty acids form esters, and if unsaturated, react with hydrogen, water, or halogens. Unsaturated fatty acids can also be oxidized to aldehydes or other products. **Triacylglycerols** are triesters of glycerol and three fatty acids that store energy and provide insulation and protection for the body. These nonpolar complex lipids are also called fats and oils or triglycerides. **Fats** contain a larger proportion of saturated fatty acids and are solids at room temperature. Oils are more unsaturated and are liquids at room temperature. **Oils,** and to a lesser extent fats, become **rancid** if oxidation of double bonds occurs. Polar complex lipids are found in membranes and blood. **Phosphoacylglycerols** consist of a glycerol, two fatty acids, a phosphate group, and an alcohol. Phosphatidyl choline (lecithin) and phosphatidyl ethanolamine are the most common phosphoacylglycerols. **Sphingolipids** consist of sphingosine, a fatty acid, and a polar-head group. Both sphingolipids and phosphoacylglycerols are amphipathic molecules. Simple lipids include the **steroids, prostaglandins,** and **leukotrienes.** Steroids include **cholesterol,** which has a structural role in membranes, **bile salts,** which are involved in digestion, and other steroids that have numerous roles in the body. Prostaglandins and leukotrienes are involved in regulating cellular activity. Micelles are aggregates of carboxylate anions or other amphipathic molecules and have the nonpolar tails of the carboxylates in the core and the polar head groups at the micelle surface. **Bilayers** and **liposomes** are aggregates of amphipathic molecules that have a bilayer arrangement and serve as models for biological membranes. Cell **membranes** consist of polar complex lipids and proteins bound noncovalently, thus the membrane is fluid. Non–energy-requiring transport across membranes can be accomplished by **simple diffusion** or by **diffusion that is facilitated** by one or more proteins. **Active transport** across membranes requires energy. **Lipoproteins** transport cholesterol and other lipids in blood.

KEY TERMS

Introduction
lipids

Section 13.1
amphipathic molecules
complex lipids
fatty acids
saturated fatty acids
simple lipids
unsaturated fatty acids

Section 13.2
fats
oils
rancid
triacylglycerols

Section 13.3
glycolipid
membrane
phosphoacylglycerols
phosphatidic acid
polar lipids
sphingolipids

Section 13.4
bile salts
cholesterol
leukotrienes
prostaglandins
steroids

Section 13.5
active transport
bilayer
facilitated diffusion
hydrophobic interactions
lipoproteins
liposomes
simple diffusion

EXERCISES

Introduction

1. The compounds discussed in this chapter are lipids. What is a common characteristic of all lipids?
2. Is petroleum a lipid?
3. List the common complex lipids introduced in this chapter.
4. List the simple lipids discussed in this chapter.
5. Which of the following solvents may be suitable for extracting (dissolving) lipids from biological tissues?
   (a) Chloroform
   (b) Ethanol
   (c) Saturated NaCl(aq)
   (d) Toluene
   (e) Water

Fatty Acids (Section 13.1)

6. Describe the difference between a simple and a complex lipid.

7. Fatty acids are carboxylic acids. What distinguishes them from other carboxylic acids?

8. Which of the following groups are likely to be found in a fatty acid?

(a) C=C (b) C≡C (c) —$CH(CH_3)_2$ (d) —COOH (e) —OH

9. Fatty acids are sometimes classified as saturated, monounsaturated, or polyunsaturated. What do these terms mean?

10. Which of the fatty acids described in this textbook are saturated? Which are monounsaturated? Which are polyunsaturated?

11. Describe the effect a *cis*- double bond has on the shape of a fatty acid molecule.

12. The melting points of fatty acids and compounds containing them are affected by *cis*- double bonds. Identify this effect and explain it.

13. Name the common unsaturated fatty acids that contain 18 carbon atoms. Write their structures showing the correct stereochemistry.

14. Draw the structure of both palmitoleic acid and stearic acid and point out any structural differences.

15. Draw the structure of the fatty acid that is designated 16:0.

16. Write the equation for the reaction of palmitoleic acid with these reagents:

(a) Water (b) Hydrogen (c) Iodine

17. Write a balanced equation for the hydrogenation of these fatty acids:

(a) Oleic acid (b) Palmitic acid

(c) Linolenic acid

Triacylglycerols (Section 13.2)

18. Draw the structure of a triacylglycerol that contains palmitoleic, stearic, and linoleic acids.

19. Indicate a change you could make in the structure of the triacylglycerol in Exercise 18 to produce a different isomer.

20. Provide three synonyms for triacylglycerol.

21. Fats and oils share what structural features?

22. How does an oil differ structurally from a fat?

23. A food label lists an oil as "partially hydrogenated". What does this mean?

24. Oils stored for long periods of time may become rancid. Explain the chemical changes that occur as this happens.

25. What can be done to reduce the chances of rancidity in fats and oils found in foods?

26. There are two main factors that make fats more efficient for energy storage than glycogen. What are these factors?

27. Saponification of a triacylglycerol will yield what products?

28. Some liquid soaps contain primarily the salts of unsaturated fatty acids, whereas solid soaps contain primarily salts of saturated fatty acids. What would you use to make a liquid soap? A solid soap?

29. If you hydrogenate an oil what do you get as the product?

30. How many moles of hydrogen are needed to completely hydrogenate a mole of triacylglycerol containing equal parts of oleic acid, linoleic acid, and linolenic acid?

31. Define an essential fatty acid. Which fatty acids are essential?

32. List and briefly discuss the correlations between dietary lipids and cardiovascular disease.

Complex Lipids of Membranes and Blood (Section 13.3)

33. The basic unit found in phosphoacylglycerols is phosphatidic acid. Draw the structure of a phosphatidic acid.

34. Indicate the structural changes you would have to make to the structure in Exercise 33 to get a

(a) Cephalin (b) Phosphatidylcholine

35. List the components of the phospholipid lecithin.

36. What is an amphipathic molecule? What cell structures contain this type of molecule?

37. Identify the structural differences between a triacylglycerol and a phosphoacylglycerol.

38. Phospholipids are found in membranes, but triacylglycerols are not. Why?

39. Describe in your own words or sketch the structural differences between a cephalin and a lecithin.

40. What is a sphingolipid?

41. Identify both this complex lipid and its components:

$$\begin{array}{l} CH_3(CH_2)_{14}COOCH_2 \\ \qquad\qquad\qquad\quad | \\ CH_3(CH_2)_{16}COOCH \\ \qquad\qquad\qquad\quad | \qquad O \\ \qquad\qquad\qquad\quad | \qquad \| \\ \qquad\qquad\qquad\quad CH_2OPOCH_2CH_2NH_3^+ \\ \qquad\qquad\qquad\qquad\quad | \\ \qquad\qquad\qquad\qquad\quad O^- \end{array}$$

42. Identify both this complex lipid and its components:

$$CH_3(CH_2)_{14}CONHCHCH_2O\overset{O}{\overset{\|}{P}}OCH_2CH_2N^+(CH_3)_3$$

(with O^- on P, and the CH carbon bonded to $CH_3(CH_2)_{12}CH{=}CHCHOH$)

Simple Lipids (Section 13.4)

43. Identify the steroid found in animal membranes.
44. Draw the structure of cholesterol and highlight the structural feature it has in common with other steroids.
45. Explain the relationship between blood cholesterol and atherosclerosis.
46. Name the principal sex hormones found in humans. What is their general function?
47. There has been an increased awareness of athletes taking anabolic steroids (those resembling testosterone). Why would an athlete want to take these?
48. How do the estrogens vary in structure from most other steroids?
49. Hydrocortisone is a steroid. What is its biological role?
50. Name the two common bile salts. What is their biological role?
51. What is a prostaglandin, and what roles does it play in our well-being?
52. How does aspirin reduce inflammation and pain?
53. The steroids found in the body are synthesized from ________.
    (a) Cholesterol (c) Cortisone
    (b) Testosterone (d) Bile acids
54. Which of the following is a male sex hormone?
    (a) Progesterone (c) Estrogen
    (b) Androgens (d) Hydrocortisone

Membranes and Related Structures (Section 13.5)

55. Draw a small piece of a micelle to show how amphipathic molecules are aligned in the structure.
56. Compare the polarities of the core and surface of a micelle.
57. Provide the term used to describe the interaction of nonpolar molecules or nonpolar parts of molecules when they are in an aqueous environment.
58. Draw a short piece of a lipid bilayer. Label the parts of the molecules with respect to polarity.
59. Draw part of a liposome. What potential use might the pharmaceutical industry have for liposomes?
60. Identify the two types of biological molecules found in biological membranes.
61. Describe the fluid mosaic model for biological membranes.
62. Compare simple diffusion, facilitated diffusion, and active transport.
63. Which of the following are found in membranes?
    (a) Free fatty acids (c) Phospholipids
    (b) Triacylglycerols (d) Prednisolone

General Questions

64. Match the term on the left with the description on the right.
    (A) Triacylglycerols (a) Made from polyunsaturated fatty acid
    (B) Phospholipid (b) Female sex hormone
    (C) Prostaglandin (c) Fats and oils
    (D) Progesterone (d) Constituent of membrane
65. For each of the following words or phrases, provide a definition, description, or example to show that you understand it.
    (a) Lipid (i) Prostaglandin
    (b) Triacylglycerol (j) Steroid
    (c) Unsaturated fatty acid (k) Estrogen
    (d) Fat (l) Testosterone
    (e) Phosphoacylglycerol (m) Bile salts
    (f) Facilitated transport (n) Cholesterol
    (g) Phosphatidyl choline (o) Hydrocortisone
    (h) Glycolipid
66. Much of the chemical behavior of compounds can be directly related to the structures of the compounds. For each of the following pairs of compounds, describe the similarities and differences.
    (a) Oil–fat
    (b) Fat–phospholipid
    (c) Estrogen–progesterone
    (d) Prostaglandin–leukotriene

Challenge Exercises

67. A 50-kg person with 20% body fat begins a fast. Daily activity consumes 2000 Calories per day during the fast. If we assume that all of the energy used

comes from metabolism of fat (not strictly true) and that all 20% of the body fat is available (also not strictly true), how long will body fat sustain energy needs?

68. The role of dietary cholesterol in heart disease is widely discussed. First use the Internet to gather information about this topic, then compare the views and opinions that are expressed in these sources. What sources of information could you use next to clarify what you have learned from the Internet?

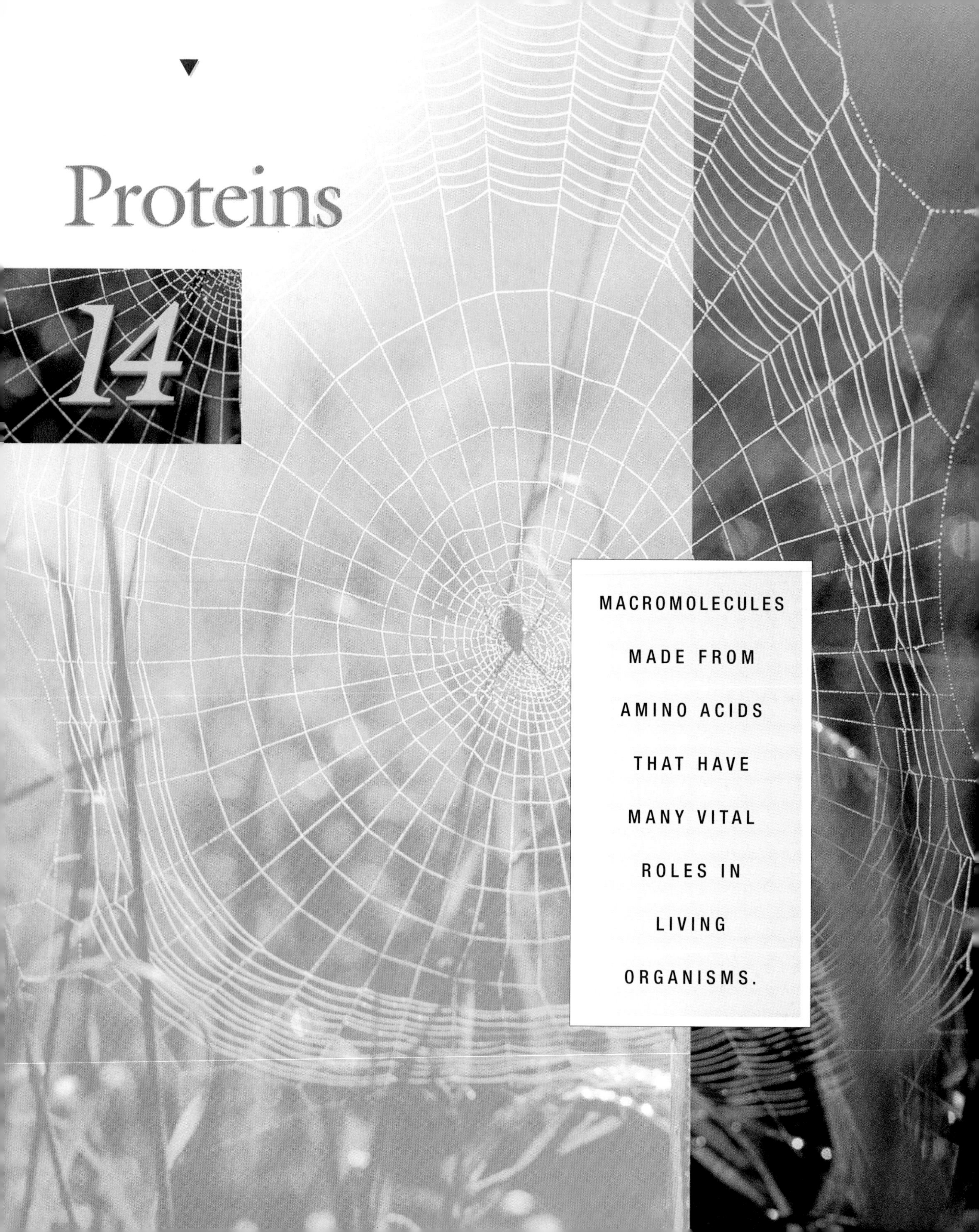

Proteins

14

MACROMOLECULES MADE FROM AMINO ACIDS THAT HAVE MANY VITAL ROLES IN LIVING ORGANISMS.

ABOUT HALF of the dry mass of cells is made up of proteins, which makes these important macromolecules the single most abundant class of biomolecules. The name protein comes from the Greek word *proteios*, which means of the first rank or importance. Proteins are indeed important, having many fundamental roles in life. Proteins are an integral part of the structure of cells, and they carry out and control many cellular activities.

Outline

14.1 Alpha Amino Acids
Recognize the 20 common amino acids and classify them as nonpolar, polar, acidic, or basic.

14.2 The Peptide Bond
Describe the formation of a peptide bond and give the definition of a peptide.

14.3 Protein Structure
Distinguish among primary, secondary, tertiary, and quaternary protein structure and explain the forces that are responsible for them.

14.4 Some Functions and Properties of Proteins
List some common functions of proteins and explain the forces and interactions involved in protein solubility, denaturation, and electrophoresis.

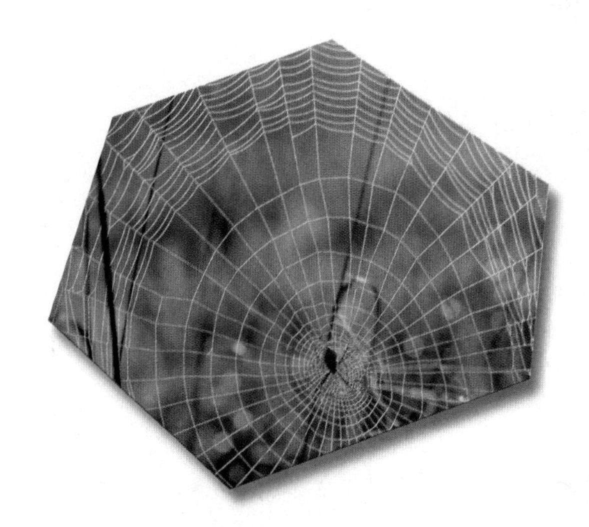

The silk of a spider web consists of fibers of the protein fibroin glued together by the protein sericin. ***(Paul Boisvert, FPG International)***

14.1 Alpha Amino Acids

Proteins are macromolecules, but more specifically they are polymers of smaller molecules called amino acids. Our study of proteins begins, therefore, with a survey of these amino acids.

Amino Acid Structure

Amino acids are organic compounds that contain both an amino group and a carboxyl group. In *α-amino acids*, both groups are attached to the same carbon atom, which is the α-carbon (alpha carbon) in common nomenclature.

COOH ← Carboxyl group

Amino group → $NH_2—C_\alpha—H$ ← α-carbon

R ← Side chain

The α-carbon atom of α-amino acids is bonded to four substituents. Three of these are the same in all amino acids: carboxyl group, amino group, and hydrogen atom. The fourth substituent is different in each amino acid and is represented by R-. There are 20 amino acids commonly found in proteins.

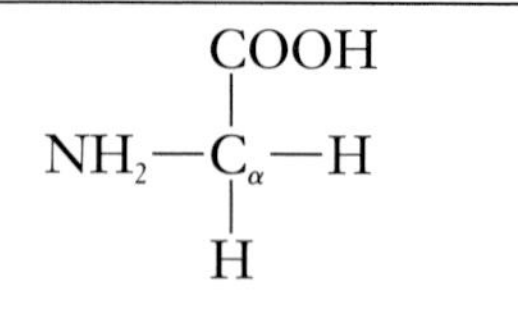

Glycine has no chiral center because there are two hydrogen atoms attached to the α-carbon atom.

Except for glycine, which has two hydrogen atoms (R = H) bonded to the α-carbon, the α-carbon of the amino acids is a chiral center (Chapter 8), and a pair of stereoisomers (enantiomers) exist. Figure 14.1 shows these stereoisomers. They are designated D- and L-, which indicates that they are in the same stereochemical families as the D- and L-glyceraldehydes (Chapter 12). When the molecules are drawn with the carboxyl group at the top and with the side chain pointing downward, the L-amino acid has the amino group on the left and the D-amino acid has the amino group on the right. The amino acids found in proteins are L-amino acids. D-amino acids are rare but do occur naturally. One example is their presence in some bacterial cell walls.

With the exception of proline, the α-amino acids are identical except for the side chain (Table 14.1). It is the structure of these side chains that distinguishes the

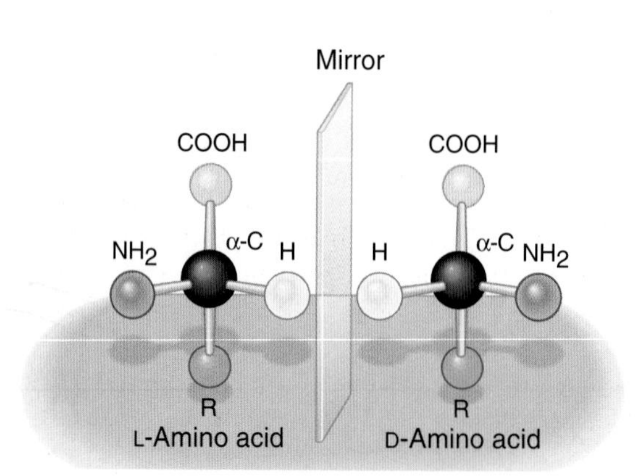

Figure 14.1 General formula for an α-amino acid. Both the amino group and the carboxyl group are attached to the same carbon, called the α-carbon. These compounds possess a chiral center, thus a pair of enantiomers exists. The two shown here are aligned as mirror images of each other. The amino acids in proteins are L-amino acids.

amino acids from each other. It is worthwhile to examine them briefly and classify them into general groups. We will group the amino acids according to the polarity of the side chain. (Remember, the amino group and carboxyl group are polar or ionic; this grouping refers just to the side chain.)

The *nonpolar amino acids* contain a nonpolar side chain, usually an alkyl group or aromatic ring. These amino acids play an important role in determining and maintaining protein structure. The nonpolar amino acid known as *proline* is unusual because the end of its side chain is bonded to the α-amino group, forming a ring (and also forming a 2° amine). The resulting ring structure gives proline a special role in protein structure because the ring often forces a protein to fold or bend. *Glycine* also deserves special mention because its "side chain" is a hydrogen atom, which of course is much smaller than the other side chains. This means glycine can fit into a protein where there is very little space.

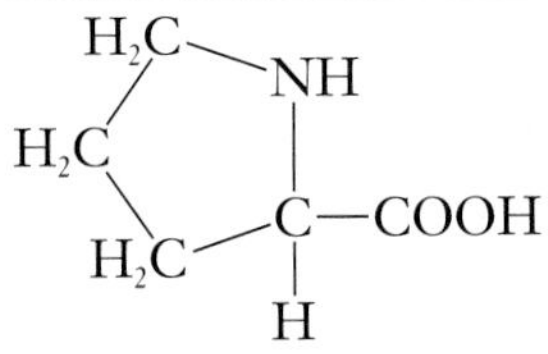

H_2C—NH; H_2C; H_2C—C—COOH; H

Proline

In *polar amino acids* the side chain contains one or more polar groups. These amino acids also play vital roles in protein structure, but some of them play special roles in protein function as well. The polar amino acids are divided into three subgroups: neutral, acidic, and basic. The *neutral polar amino acids* contain side chains that are polar but not usually ionic. The *acidic amino acids* contain a carboxyl group in the side chain (this is *in addition* to the one attached to the α-carbon atom). The *basic amino acids* contain one or more nitrogen atoms in the side chain, which makes them basic like ammonia.

Acid–Base Properties of Amino Acids

Amino acids are both acids and bases because they possess an α-amino group and an α-carboxyl group. Like all other acids and bases, these groups take part in acid–base reactions. In this case, however, the reaction is *intramolecular* because the carboxyl group donates its hydrogen ion to the amino group in the same molecule:

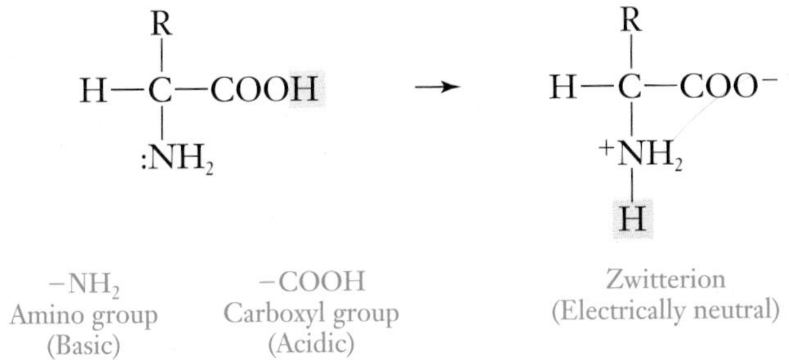

$$\mathrm{H{-}\underset{:NH_2}{\overset{R}{C}}{-}COOH} \rightarrow \mathrm{H{-}\underset{{}^{+}NH_2{-}H}{\overset{R}{C}}{-}COO^-}$$

$-NH_2$ Amino group (Basic) — $-COOH$ Carboxyl group (Acidic) — Zwitterion (Electrically neutral)

The product of this intramolecular reaction is a dipolar ion called a **zwitterion** (from the German word *zwei* for two). It is electrically neutral because the opposite charges cancel. Because amino acids are zwitterions, they are water-soluble and have high melting points, just like most other ionic compounds. (Because amino acids are shown sometimes as zwitterions and sometimes in the un-ionized form, you should learn to recognize both.)

Molecules that possess both acidic and basic properties are called **amphoteric molecules.** In the presence of a base, an amino acid behaves as an acid because

(text continues on page 430)

Table 14.1 The 20 Common Amino Acids

NAME	ABBREVIATION	SIDE-CHAIN STRUCTURE
Nonpolar Side Chain		
Alanine	Ala	$CH_3-CH(NH_2)-COOH$
Glycine	Gly	$H-CH(NH_2)-COOH$
Isoleucine	Ile	$CH_3-CH_2-CH(CH_3)-CH(NH_2)-COOH$
Leucine	Leu	$CH_3-CH(CH_3)-CH_2-CH(NH_2)-COOH$
Methionine	Met	$CH_3-S-CH_2-CH_2-CH(NH_2)-COOH$
Phenylalanine	Phe	$C_6H_5-CH_2-CH(NH_2)-COOH$
Proline	Pro	CH_2 ring: $CH_2-CH_2-CH_2-NH-CH(COOH)$ (cyclic)
Tryptophan	Trp	indole (N–H) $-CH_2-CH(NH_2)-COOH$

Table 14.1 *continued*

NAME	ABBREVIATION	SIDE-CHAIN STRUCTURE
Nonpolar Side Chain (continued)		
Valine	Val	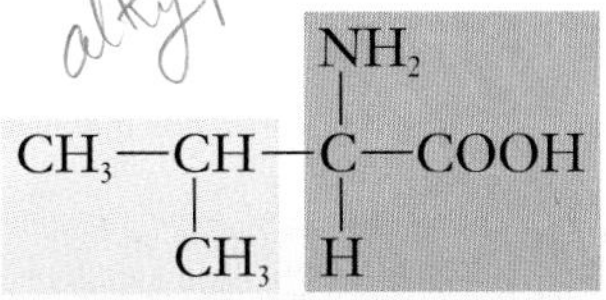 $CH_3-CH(CH_3)-C(NH_2)(H)-COOH$
Polar, Neutral Side Chain		
Asparagine	Asn	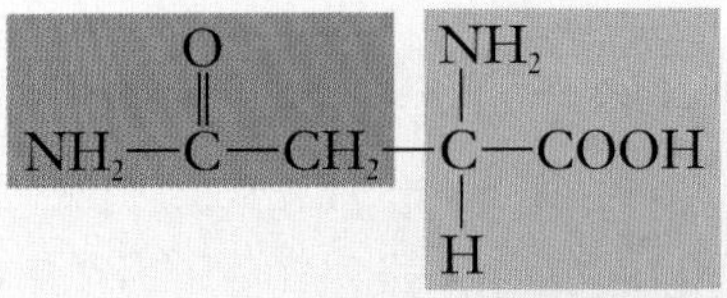 $NH_2-C(=O)-CH_2-C(NH_2)(H)-COOH$
Cysteine	Cys	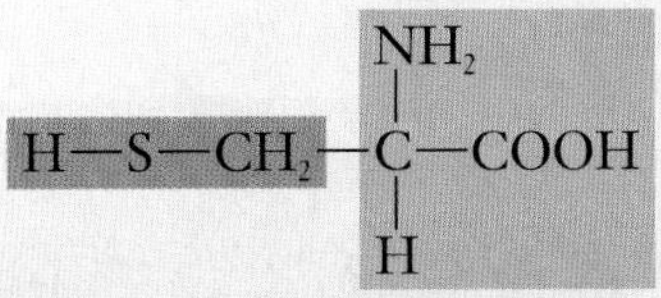 $H-S-CH_2-C(NH_2)(H)-COOH$
Glutamine	Gln	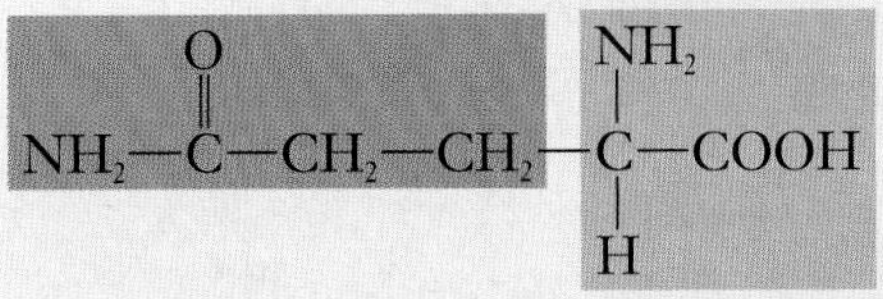 $NH_2-C(=O)-CH_2-CH_2-C(NH_2)(H)-COOH$
Serine	Ser	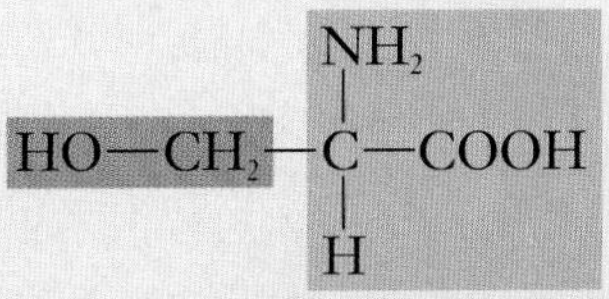 $HO-CH_2-C(NH_2)(H)-COOH$
Threonine	Thr	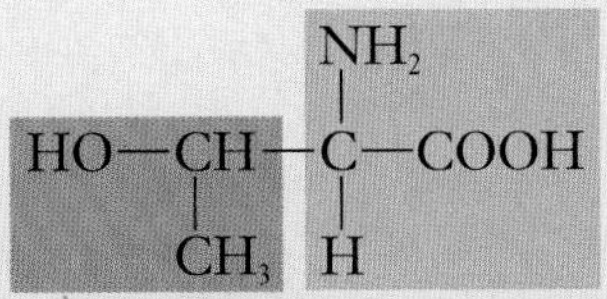 $HO-CH(CH_3)-C(NH_2)(H)-COOH$
Tyrosine	Tyr	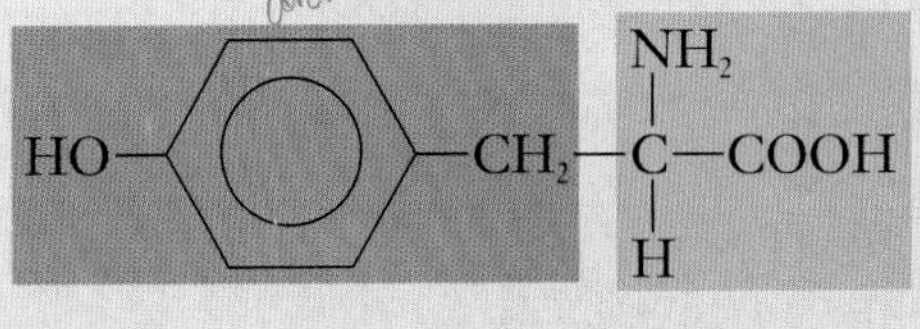 $HO-C_6H_4-CH_2-C(NH_2)(H)-COOH$
Polar, Acidic Side Chain		
Aspartic acid	Asp	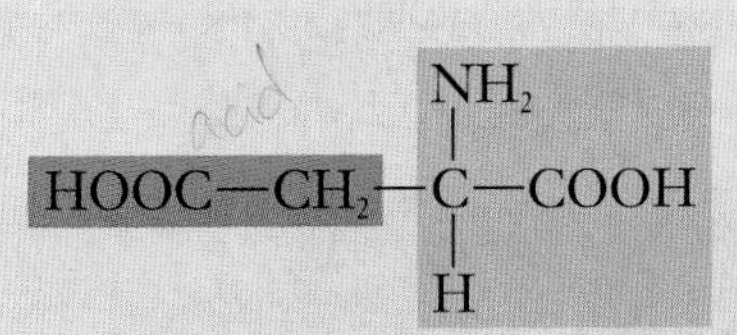 $HOOC-CH_2-C(NH_2)(H)-COOH$

continues

Table 14.1 *continued*

NAME	ABBREVIATION	SIDE-CHAIN STRUCTURE
Polar, Acidic Side Chain (continued)		
Glutamic acid	Glu	$HOOC{-}CH_2{-}CH_2{-}\overset{NH_2}{\underset{H}{C}}{-}COOH$
Polar, Basic Side Chain		
Arginine	Arg	$NH_2{-}\overset{NH}{\overset{\Vert}{C}}{-}NH{-}CH_2{-}CH_2{-}CH_2{-}\overset{NH_2}{\underset{H}{C}}{-}COOH$
Histidine	His	(imidazole ring: N, N–H) $-CH_2{-}\overset{NH_2}{\underset{H}{C}}{-}COOH$
Lysine	Lys	$NH_2{-}CH_2{-}CH_2{-}CH_2{-}CH_2{-}\overset{NH_2}{\underset{H}{C}}{-}COOH$

it gives up a proton from the $-NH_3^+$ group (Figure 14.2). In an acidic solution, an amino acid functions as a base because the carboxylate group accepts a hydrogen ion.

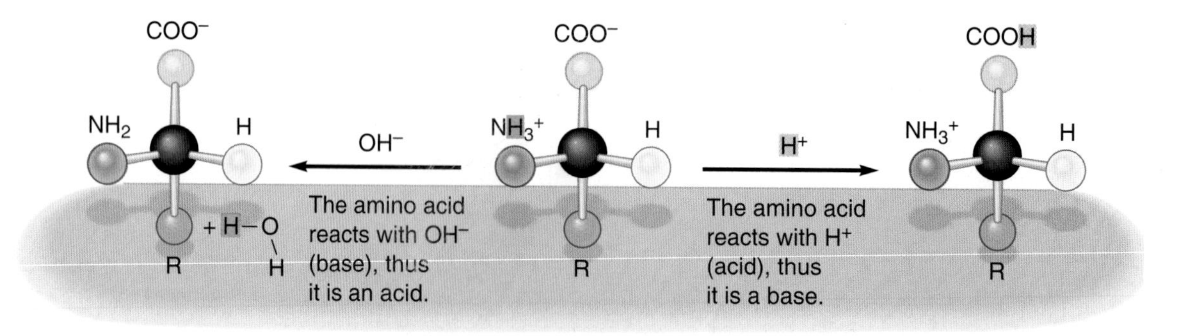

Figure 14.2 Amino acids are amphoteric, which means they can act as both acids and bases. The protonated amino group acts as an acid, and the carboxylate group acts as a base.

Charged Properties of Amino Acids

In aqueous solutions near pH 7, many of the zwitterions of the amino acids are neutral. At other pH values, however, these amino acids carry a net electrical charge. In an acidic solution (pH <7) the zwitterion gains a hydrogen ion, and its charge changes from 0 to 1+:

$$\mathrm{H{-}\underset{\underset{NH_3^+}{|}}{\overset{\overset{R}{|}}{C}}{-}COO^-} + \mathrm{H^+} \longrightarrow \mathrm{H{-}\underset{\underset{NH_3^+}{|}}{\overset{\overset{R}{|}}{C}}{-}COOH}$$

Neutral charge = 0 | Charge = 1+ | (Low pH form) Charge = 1+

Similarly, a loss of a hydrogen ion in basic solutions (pH > 7) alters the zwitterion charge from 0 to 1−:

$$\mathrm{H{-}\underset{\underset{\underset{\underset{H}{|}}{HNH^+}}{|}}{\overset{\overset{R}{|}}{C}}{-}COO^-} + \mathrm{OH^-} \longrightarrow \mathrm{H{-}\underset{\underset{HNH}{|}}{\overset{\overset{R}{|}}{C}}{-}COO^-} + \mathrm{H{-}OH}$$

Neutral charge = 0 | Charge = 1− | (High pH form) Charge = 1−

Amino acids are neutral at only one pH value. The pH value at which an amino acid is neutral—in other words, where it has no net charge—is the **isoelectric point (pI).**

Some amino acid side chains contain groups that can also act as acids and bases. These groups contribute to the pI of the amino acid, but in proteins they are also the groups that contribute most to the protein's acid–base properties. This is because the α-amino and α-carboxyl groups of amino acids in proteins are not free; instead they are involved in covalent bonds. In contrast, the side chains are usually free and thus can act as acids or bases.

Example 14.1 Amino Acids

Alanine is a nonpolar amino acid, and serine is a polar one. Yet these two amino acids have similar solubilities in water. Why?

SOLUTION

Although the amino acids are classified by the polarities of the side chain (alanine has a nonpolar methyl group and serine has a polar hydroxyl group) it is the polarity of the entire molecule that determines solubility. Because both these amino acids normally exist as zwitterions, they are water-soluble like many ionic compounds.

Self-Test

1. What is the name of the amino acid that has the side chain

$$-\underset{\displaystyle CH_3}{\underset{|}{C}H}CH_2CH_3?$$

In this amino acid, what other groups are attached to the α-carbon?

2. Identify each of the following amino acids as neutral, acidic, or basic:

(a)
$$\begin{array}{c} COOH \\ | \\ NH_2-C-H \\ | \\ CH_2OH \end{array}$$

(b)
$$\begin{array}{c} COOH \\ | \\ NH_2-C-H \\ | \\ (CH_2)_4NH_2 \end{array}$$

(c)
$$\begin{array}{c} COOH \\ | \\ NH_2-C-H \\ | \\ CH_2SH \end{array}$$

(d)
$$\begin{array}{c} COOH \\ | \\ NH_2-C-H \\ | \\ CH_2COOH \end{array}$$

3. Give the product(s) for each of the following reactions:

(a)
$$\begin{array}{c} H \\ | \\ CH_3-C-COO^- \\ | \\ NH_3^+ \end{array} + H^+ \rightarrow \quad \begin{array}{c} CH_3-C-COOH \\ | \\ NH_3^+ \end{array}$$

(b)
$$\begin{array}{c} H \\ | \\ CH_3-C-COO^- \\ | \\ NH_3^+ \end{array} + OH^- \rightarrow \quad \begin{array}{c} CH_3-C-COO^- \\ | \\ NH_2 \end{array} + H_2O$$

ANSWERS

1. Isoleucine; an amino group, a carboxyl group, and a H atom. 2. **(a)** Neutral **(b)** Basic **(c)** Neutral **(d)** Acidic

3. **(a)**
$$\begin{array}{c} H \\ | \\ CH_3-C-COOH \\ | \\ NH_3^+ \end{array}$$

(b)
$$\begin{array}{c} H \\ | \\ CH_3-C-COO^- \\ | \\ NH_2 \end{array} + H_2O$$

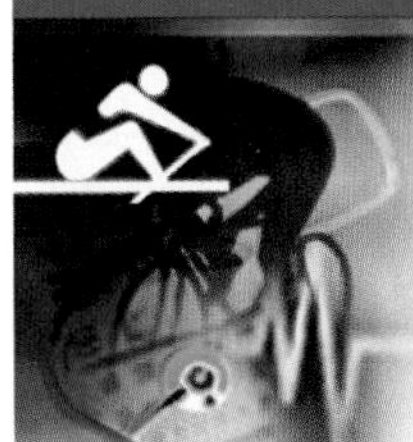

Connections

Proteins and Amino Acids in Nutrition

We need all 20 of the amino acids to synthesize the many vital proteins in our bodies. If we lack even one of them, protein synthesis cannot occur. Our diet normally provides adequate amounts of these amino acids, but we can also synthesize some of them. The amino acids that we cannot synthesize are called **essential amino acids** because they must be present in our diet.

Dietary proteins are the body's primary source of amino acids. During digestion, dietary proteins are broken down to amino acids, which are then absorbed and used by the body. If the amounts of amino acids are in excess of the body's needs for protein synthesis, the excess amino acids are broken down to yield energy. The amino acids from a gram of protein yield 4 Calories (kcal) of energy, about the same as carbohydrates but less than fats and oils.

When assessing a food as a source of protein two factors must be considered: (1) the total amount of protein in the food, and (2) the amino acid composition of the proteins in the food. Even if a food contains adequate amounts of total protein, that food would be a poor sole source of protein if one or more essential amino acids were in inadequate amounts. Meats and dairy products are particularly good sources of dietary protein, because these foods contain large amounts of protein that contain all of the essential amino acids. In contrast, many foods from plants have adequate amounts of protein but are often low in one or more of the essential amino acids. A vegetarian diet is not necessarily a diet with inadequate amounts of essential amino acids, however. If a vegetarian eats several protein-containing foods from plants at each meal, adequate amounts of the essential amino acids will be eaten. For example, the traditional corn, beans, and rice diet of Mexico and the American Southwest provides adequate amounts of the essential amino acids.

These foods are good sources of protein. *(Charles D. Winters)*

The Essential Amino Acids for Humans

Arginine[a]	Methionine
Histidine	Phenylalanine
Isoleucine	Threonine
Leucine	Tryptophan
Lysine	Valine

[a]Arginine is required for growth in children but is not required by adults.

14.2 The Peptide Bond

As we learned in the previous section, the macromolecules we know as proteins are polymers of amino acids. **Peptides** are oligomers and polymers of amino acids that have a molar mass that is generally less than 5000 amu. **Proteins** are larger polymers, those having a molar mass greater than 5000 amu. Because the boundary between peptides and proteins is based on mass, there is no fixed limit to the *number* of amino acids in a peptide because the different amino acids have different molar masses. Peptides generally contain from two to a few dozen amino acids.

Amino acids are covalently bonded to each other in proteins and peptides:

$$^{+}H_3N-\underset{R_1}{\overset{H}{C}}-\overset{O}{\overset{\|}{C}}-O^- + {}^{+}H_3N-\underset{R_2}{\overset{H}{C}}-\overset{O}{\overset{\|}{C}}-O^- \rightarrow {}^{+}H_3N-\underset{R_1}{\overset{H}{C}}-\overset{O}{\overset{\|}{C}}-\underset{H}{N}-\underset{R_2}{\overset{H}{C}}-\overset{O}{\overset{\|}{C}}-O^- + H_2O$$

Amino acid   Amino acid   Peptide (Peptide bond)

In organic chemistry, the bond formed between the carbon atom of a carbonyl group and the nitrogen atom of an amino group is called an amide bond (Chapter 11). When this bond involves amino acids, it is called a peptide bond. **Peptide bonds** connect amino acids in peptides and proteins.

Remember, an "activated" carboxyl group is needed to form a peptide bond; a carboxylate anion or carboxyl group will not react with an amine or ammonium salt.

The formation of a peptide bond between two amino acids leaves one of the amino acids in the newly formed peptide with a free carboxyl group and the other with a free amino group. Thus each of these amino acids can now form another peptide bond to yield a peptide containing four amino acid residues. One terminal amino acid of the four-residue peptide has a free amino group, and the other terminal amino acid has a free carboxyl group, thus two more peptide bonds can be formed between these ends and free amino acids (Figure 14.3). In this way several to many amino acids can be linked together to form peptides and proteins.

The reaction that forms a peptide bond is readily reversible in a hydrolysis reaction (cleavage by water):

$$H_3N^{+}-\underset{R_1}{\overset{H}{C}}-\overset{O}{\overset{\|}{C}}-\underset{H}{N}-\underset{R_2}{\overset{H}{C}}-\overset{O}{\overset{\|}{C}}-O^- + H_2O \rightarrow H_3N^{+}-\underset{R_1}{\overset{H}{C}}-\overset{O}{\overset{\|}{C}}-O^- + H_3N^{+}-\underset{R_2}{\overset{H}{C}}-\overset{O}{\overset{\|}{C}}-O^-$$

Peptide   Amino acid$_1$ Amino acid$_2$

Both reactions occur in the body. Peptide bonds are formed when the body synthesizes proteins from amino acids, as occurs during growth and during normal replacement of proteins. In the reverse reaction, peptide bonds in proteins are broken down by hydrolysis. Dietary proteins are cleaved in the digestive tract, and cellular proteins are hydrolyzed and replaced routinely during the normal course of cellular activity.

Structure and Nomenclature of Peptides

Peptides are sometimes identified by the number of amino acids in them. A peptide containing two amino acids is called a *dipeptide.* One with three is a *tripeptide;* four amino acids are present in a *tetrapeptide. Oligo*peptides contain several amino acids. For our purposes, ten amino acids is the upper limit for an oligopeptide. *Poly*peptides contain more than ten amino acids. (This boundary between oligopeptides and polypeptides is arbitrary.)

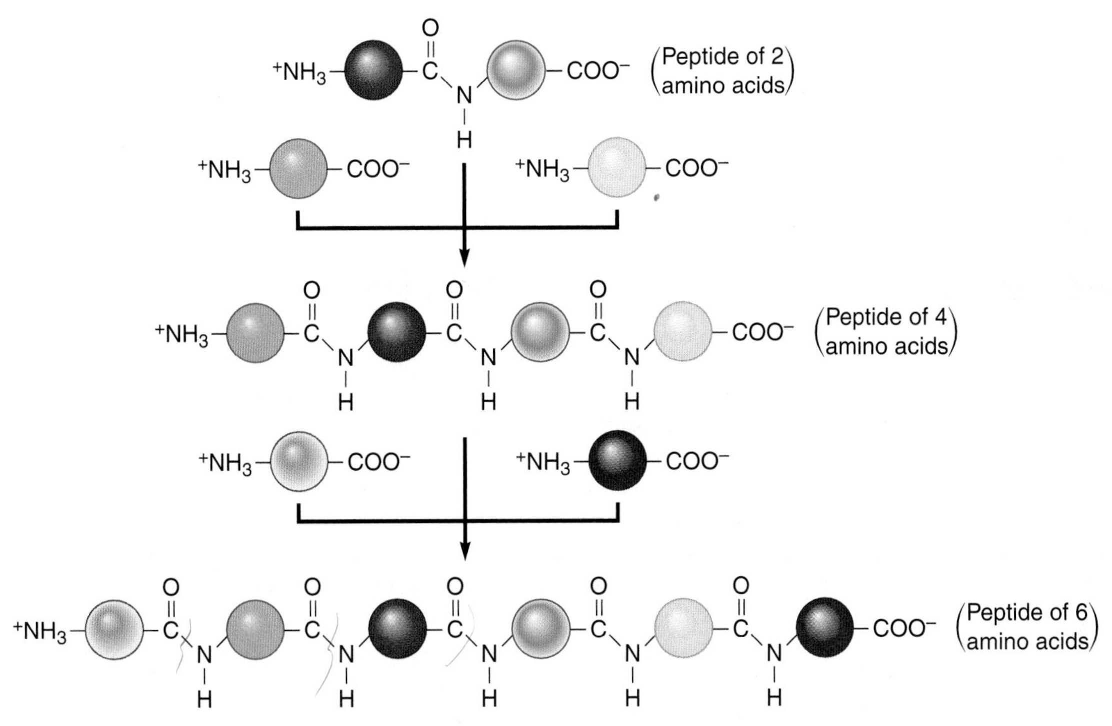

Figure 14.3 Because amino acids are bifunctional, the amino group and the carboxyl group can each react with other amino acids to form peptide bonds. As a consequence, amino acids can be bonded together to form oligomers and polymers of amino acids. The colored circles represent the rest of the amino acid.

Peptides are sometimes named systematically. The amino acid that has the free amino group (it is called the N-terminal amino acid) is named first, and then all the other amino acids are named in order. The last amino acid in the name is the one that has the free carboxyl group (it is called the C-terminal amino acid). For all amino acids in the name except the last one, the suffix *-yl* replaces the suffix in the amino acid name (the C-terminal amino acid retains its full name). For example, the systematic name for the tripeptide shown in Figure 14.4(a) is alanylmethionylglycine.

If more than a few amino acids are present in a peptide, the systematic name becomes large and inconvenient. For this reason, most peptides obtained from nature have a common name that is used more often than the systematic name. The peptide shown in Figure 14.4(b), for instance, is commonly called leu-enkephalin, but its systematic name is tyrosinylglycylglycylphenylalanylleucine.

Biologically Important Peptides

Many naturally occurring peptides are hormones, and it is worth while to look at a few of the more common ones. Oxytocin [Figure 14.4(d)] is a nonapeptide that causes contraction of uterine and other smooth muscles. It is produced by the body to initiate the contractions that accompany labor, and is used in clinical settings to

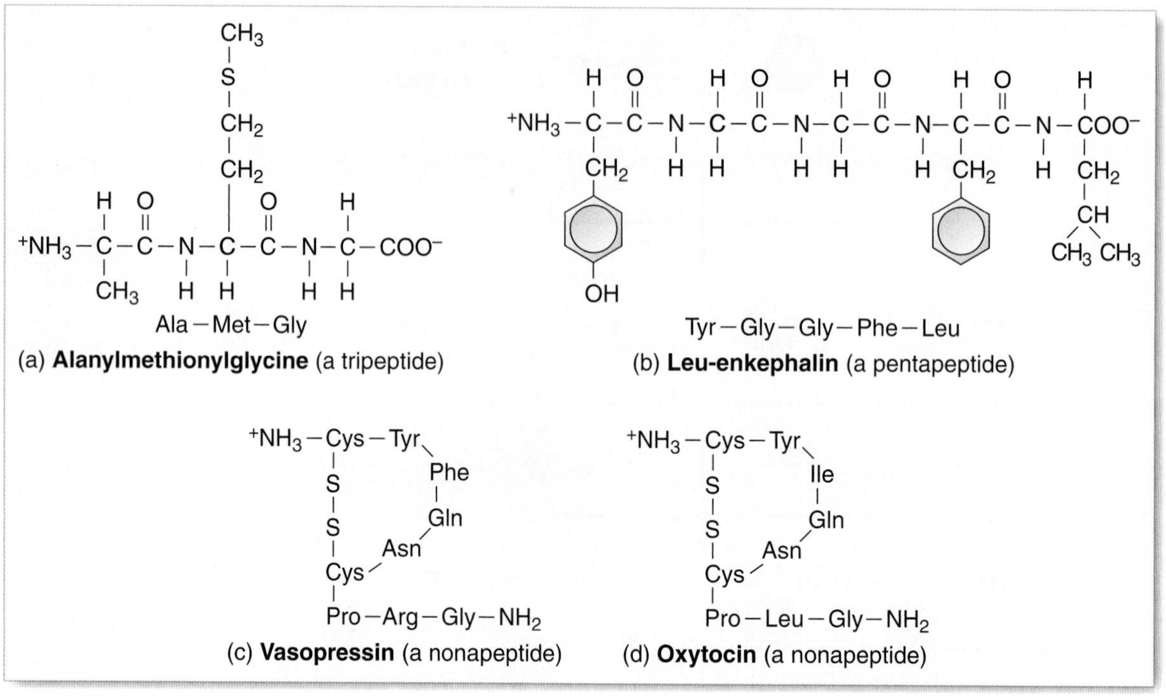

Ala—Met—Gly

(a) **Alanylmethionylglycine** (a tripeptide)

Tyr—Gly—Gly—Phe—Leu

(b) **Leu-enkephalin** (a pentapeptide)

$^{+}NH_3$—Cys—Tyr—Phe—Gln—Asn—Cys—Pro—Arg—Gly—NH_2 (Cys–S–S–Cys)

(c) **Vasopressin** (a nonapeptide)

$^{+}NH_3$—Cys—Tyr—Ile—Gln—Asn—Cys—Pro—Leu—Gly—NH_2 (Cys–S–S–Cys)

(d) **Oxytocin** (a nonapeptide)

Figure 14.4 Some peptides. The entire structures of alanylmethionylglycine **(a)**, and leu-enkephalin **(b)** are shown. Vasopressin **(c)** and oxytocin **(d)** are represented by the three-letter abbreviations of their amino acids. Note the structural similarity between vasopressin and oxytocin, even though these two peptides cause quite different physiological responses. Gly-NH_2 indicates that the carboxyl group of glycine is bonded to ammonia via an amide bond.

induce labor. Vasopressin [Figure 14.4(c)] is very similar in structure to oxytocin, but it is involved in water retention and blood pressure regulation. Several peptides, including the endorphins [Figure 14.4(b)], are involved in pain and pleasure sensation. Morphine and heroin appear to mimic the effects of endorphins and enkephalins. These peptides are also the agent responsible for the "runner's high" that follows intense aerobic exercise.

Example 14.2 Peptide Bond and Peptides

Give the systematic name for the peptide NH_2—Ser—Lys—Val—Tyr—COOH.

SOLUTION

The N-terminal amino acid is named first, and all amino acids except for the C-terminal one have their suffix replaced by *-yl.* Thus serine becomes seryl, and the next two amino acids become lysyl and valyl. Finally, the full name of the C-terminal amino acid is used: seryllysylvalyltyrosine.

Self-Test

1. The peptide bond is the same as which bond of organic chemistry?

2. Examine oxytocin in Figure 14.4(d). Which amino acid is N-terminal? Name the C-terminal amino acid in leu-enkephalin [Figure 14.4(b)].
3. What is the distinction between a peptide and a protein?

ANSWERS

1. Amide bond 2. Cysteine; leucine 3. Size. If the compound has a molar mass greater than 5000 amu, it is a protein; if the molar mass is less than 5000 amu, it is a peptide.

14.3 Protein Structure

Many proteins consist only of a single polypeptide chain, but others have two or more. In addition, some proteins contain one or more other components. Most proteins are very large; many have molar masses greater than 50,000 amu, and some have masses exceeding a million amu.

The location of each amino acid in a protein is an important part of the protein structure. To distinguish the amino acid residues of the polypeptide chain from each other, each residue is numbered. The N-terminal amino acid occupies the first residue, the next amino acid is the second residue, and so on until the C-terminal amino acid, which occupies the last residue.

The number of proteins that can be made from the 20 common amino acids is very large. Let's make a few assumptions and calculations to get a feeling for the number of possibilities. Let's assume a protein has 100 residues (a rather small protein). Assume that any of the 20 amino acids can occupy any residue (this is the case for real proteins). The number of possible amino acid sequences is 20 times itself one hundred times, or 20^{100}. This is about 1×10^{130}! (Compare this huge number with the estimated number of atoms in the entire universe, which is around 10^{80}, give or take a few orders of magnitude.)

And remember, this number of amino acid sequences (10^{130}) is for only 100 amino acid residues. In reality, any number of amino acids from around 50 to as many as hundreds could be present in a protein. The number of possible proteins that could be made from the 20 amino acids is truly astronomical. In reality, the number of proteins found in living organisms is much smaller than the numbers in these calculations. It is estimated that the number of different proteins present in all organisms is around a trillion (10^{12}).

Example 14.3 Number of Nonapeptides

How many nonapeptides can be made from the 20 amino acids?

SOLUTION

Any of the 20 amino acids can occupy any of the nine residues in a nonapeptide. The answer is calculated this way: amino $acid_1 \times$ amino $acid_2 \ldots \times \ldots$ amino $acid_9 = (20)_1 \times (20)_2 \ldots \times \ldots (20)_9 = (20)^9 = 5.1 \times 10^{11}$ nonapeptides, roughly 100 times more than the number of humans on Earth. (The subscripts by the number 20 indicate each residue.)

Self-Test

Leu-enkephalin is a pentapeptide. How many pentapeptides can be made from the 20 amino acids?

ANSWER

3.2×10^6

Protein Conformation

Each protein has its own unique amino acid sequence. How are these amino acids and their atoms arranged in space? Free rotation about single bonds allows for an infinite number of conformations even in small molecules. Because there are many single bonds in a protein, an infinite number of possible conformations exist. Yet biochemists are quite sure that all molecules of a protein exist in one or at most a very few conformations.

The conformation in which a protein most frequently exists is called its **native conformation** (Figure 14.5). In this conformation the protein has its normal (native) function. Why does each protein have only one conformation? In general, the native conformation of a protein is its most stable, lowest energy conformation. The protein is held in that shape by specific forces (bonding) within the molecule. The idea of a unique, most-stable conformation for a molecule should not be new or surprising. In Chapter 8 it was shown that ethane (C_2H_6) has an infinite number of possible conformations, but one is the most stable. The same idea holds for proteins, but they are of course, much larger molecules. If its native conformation is changed, a protein loses its native function or activity. This process is called **denaturation** (Figure 14.6).

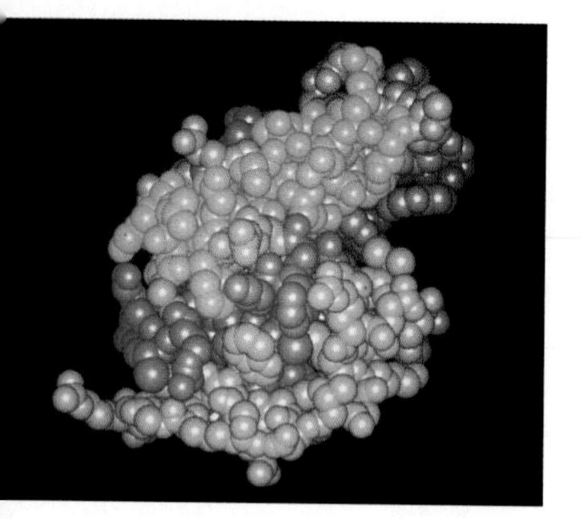

Figure 14.5 This computer-drawn image represents the native conformation of the protein lysozyme from hen egg whites. Each protein has its own unique shape. *(Charles Grisham)*

Because protein structure is critical to protein function, much of the rest of this chapter deals with protein structure, which can be studied at four levels, as Figure 14.7 illustrates.

Primary Structure

The **primary structure** of a protein is its amino acid sequence. The amino acid sequences of many proteins are now known, and each protein appears to have its own

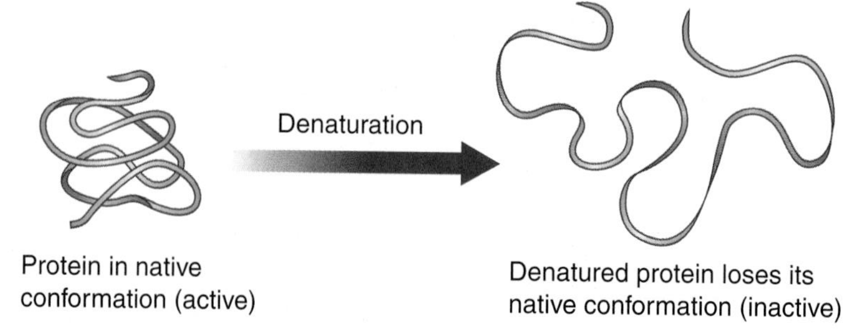

Figure 14.6 Denaturation changes a protein's shape and causes loss of function.

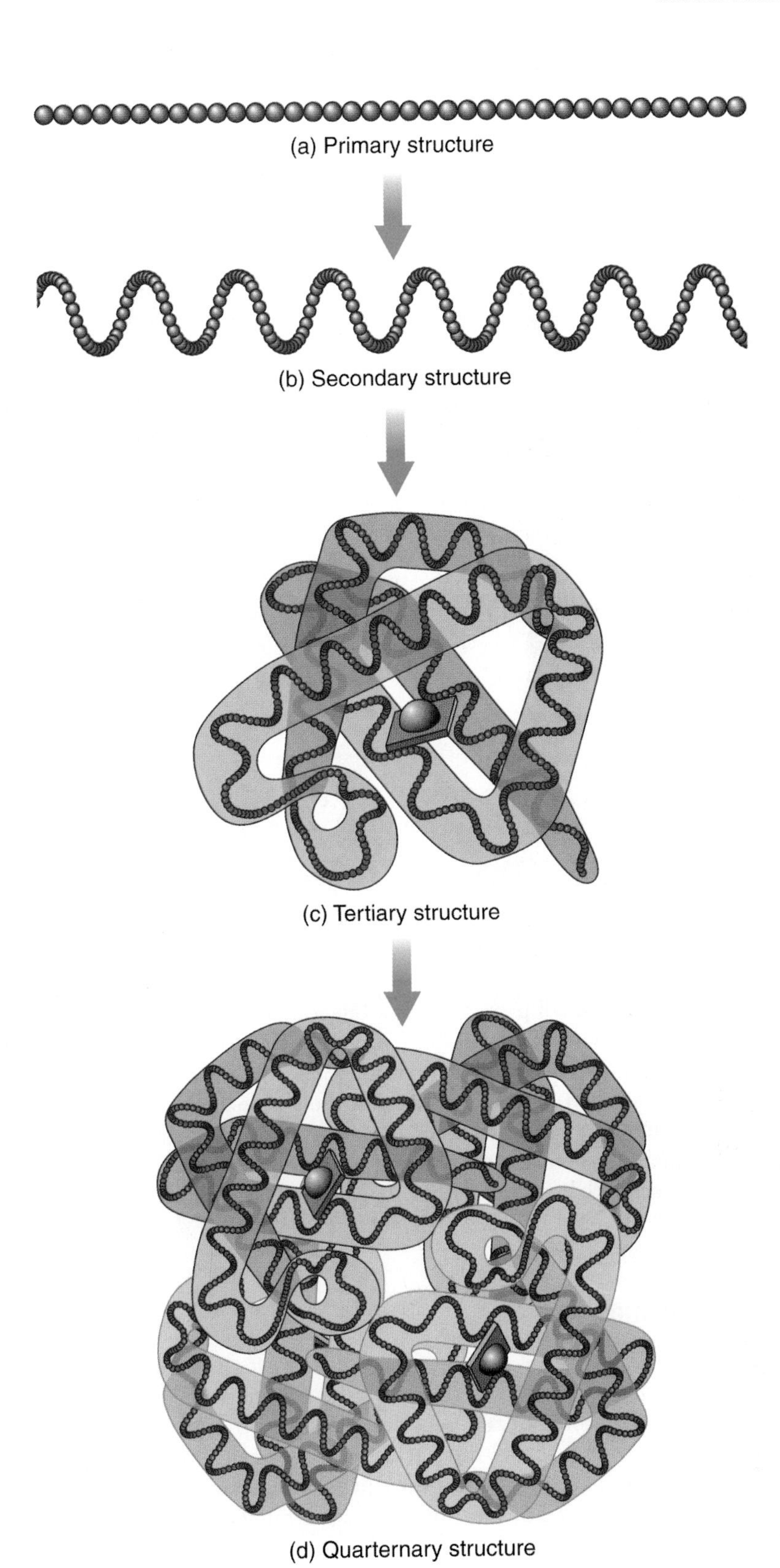

Figure 14.7 The four levels of protein structure.

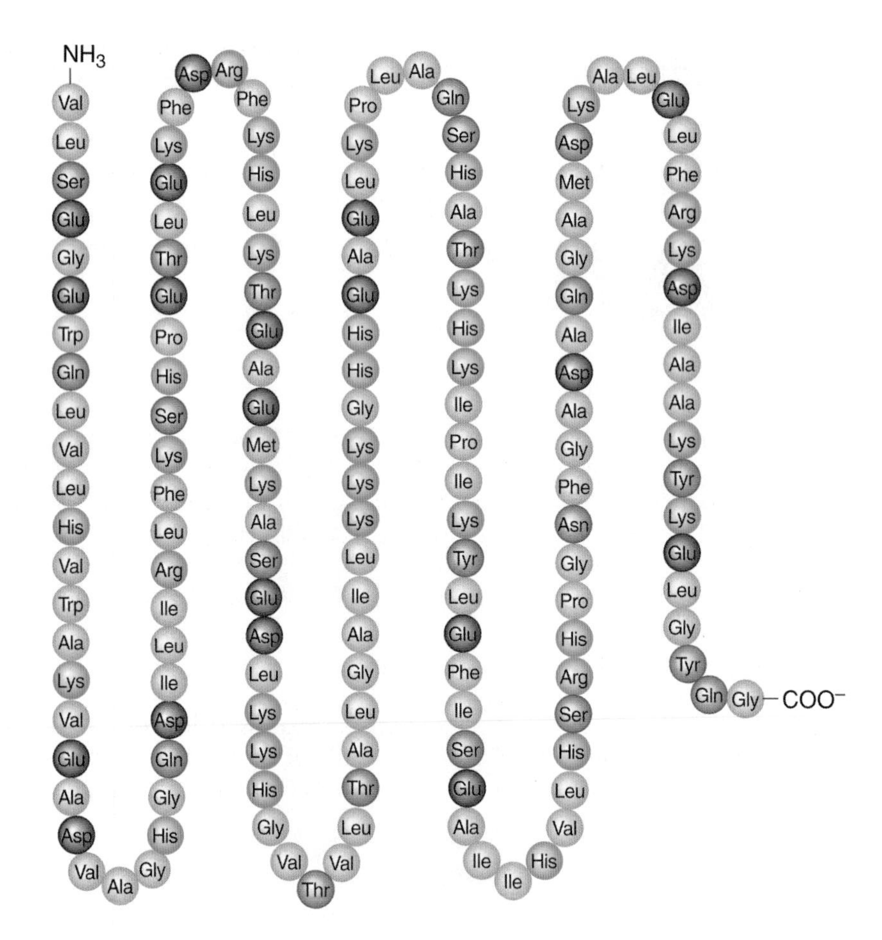

Figure 14.8 The primary structure of myoglobin from sperm whale. This protein is made up of 153 amino acids in this specific sequence. Myoglobin does not look like this; the actual shape of myoglobin is determined by three levels of structure: primary, secondary, and tertiary.

unique sequence. The small protein myoglobin contains 153 amino acids linked in the specific order shown in Figure 14.8. Ultimately, the primary structure of a protein determines its other levels of structure because different amino acids occupy different residues in each protein. As a result, the amino acid at any given residue interacts with other amino acid residues and with water differently in each protein. A protein's primary structure determines the unique combination of interactions between amino acids that occurs within that protein, and this in turn results in a unique shape for the protein. *Each protein has its own unique shape because it has its own unique amino acid sequence.*

Secondary Structure

The **secondary structure** of a protein involves the arrangement of the atoms of the polypeptide chain, which is sometimes called the molecule's backbone. This chain consists of the atoms of the peptide bonds plus the α-carbon atom of the amino acids. Figure 14.9 shows the backbone of a small piece of a polypeptide chain. Many proteins have portions of their peptide chain arranged into secondary structure, and a few have all or most of the chain arranged in this way.

There are two common types of secondary structure: (1) α-helix and (2) β-pleated sheet. In an **α-helix,** the atoms of the backbone are arranged in a

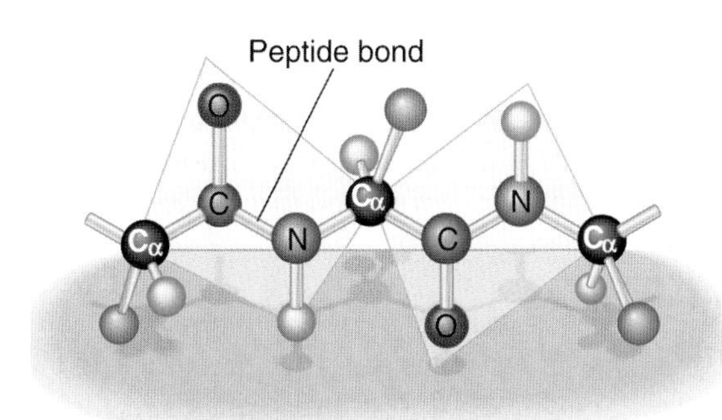

Figure 14.9 The backbone of the polypeptide chain includes the atoms of the peptide bond, C=O, NH, and the α-carbon, ignoring the R-group (green) and the hydrogen atom on the α-carbon atom.

helix, which has a spiral shape [Figure 14.10(a)]. In an α-helix, hydrogen bonds exist between atoms of peptide bonds that are four residues apart in the chain. This bonding holds the polypeptide chain in the helical shape [Figure 14.10(b)]. The hydrogen bonds form between a nitrogen-bound hydrogen atom in one peptide bond and the carbonyl oxygen atom in another peptide bond. Collectively these hydrogen bonds are strong enough to hold the polypeptide chain in the α-helix conformation.

It appears that an α-helix will form in a protein as long as no forces prevent the hydrogen bonds from forming. Some proteins have nearly the whole polypeptide chain arranged into an α-helix, whereas other proteins have little or none of their chain in this conformation. What factors prevent an entire chain from assuming the α-helix conformation? One factor is the presence of the amino acid proline.

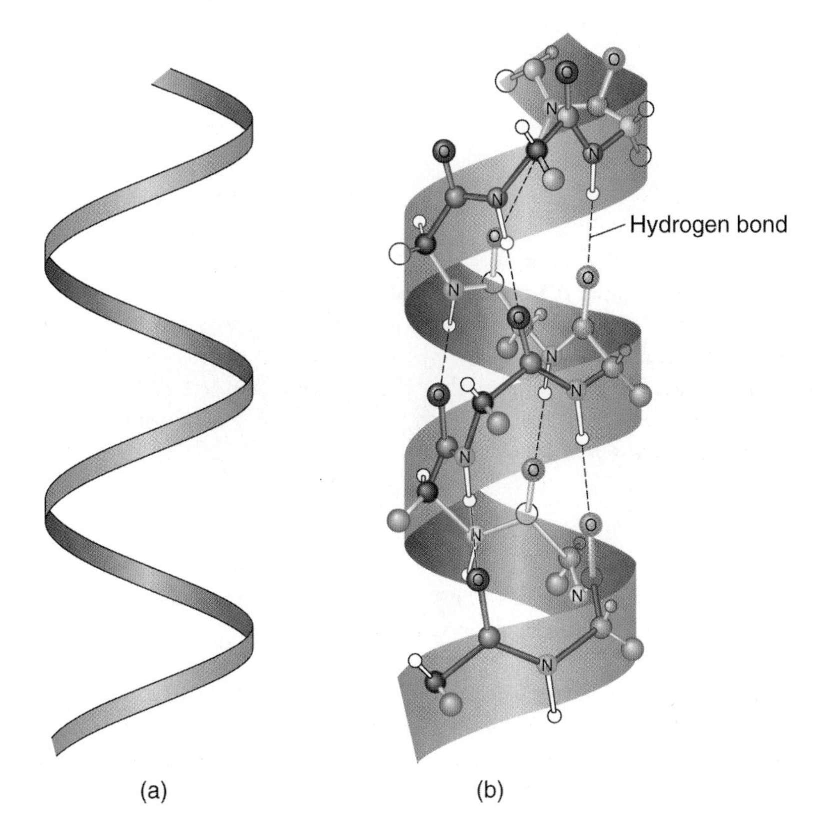

Figure 14.10 α-Helix is one type of secondary structure. **(a)** Coiled springs and slinkies are common objects with spiral or helical structures. **(b)** The atoms of the polypeptide chain are arranged into a spiral and are held there by hydrogen bonds between the carbonyl oxygen in one residue and the hydrogen on the nitrogen atom four residues away.

When it is in a chain, proline's unique shape introduces a "kink" into the chain, which prevents a helix at that residue. Another factor is amino acids with charged side chains. For example, like-charged side chains (arginine and lysine are examples) will repel each other and prevent formation of the helix. In addition, because of their shape, the side chains of asparagine, leucine, serine, and threonine will prevent helix formation if they occur frequently in the chain.

The second common type of protein secondary structure is the ***β*-pleated sheet** (Figure 14.11). This structure also results from hydrogen bonding between atoms of the peptide bond, but the bonds form between atoms that are either (a) in distant portions of the same chain or (b) in different polypeptide chains in those proteins that contain more than one chain. (Remember, in an *α*-helix the hydrogen bonding is between atoms that are four residues apart in the same chain.)

Proteins that are fiber-like are called **fibrous proteins** and are often water-insoluble. In *α*-fibrous proteins the polypeptide chains are arranged into an *α*-helix yielding a long, cylindrical protein; the structural proteins found in skin or hair are α-fibrous proteins. In *β*-fibrous proteins the polypeptide chain is arranged into *β*-pleated sheet; fibroin, the protein in threads of silk, is an example.

Tertiary Structure

Most proteins are not fiber-like, nor are their structures entirely *α*-helix or *β*-pleated sheet, although there may be some regions of polypeptide chain that are in *α*-helix or *β*-pleated sheet. Instead, these proteins are usually compact and roughly spher-

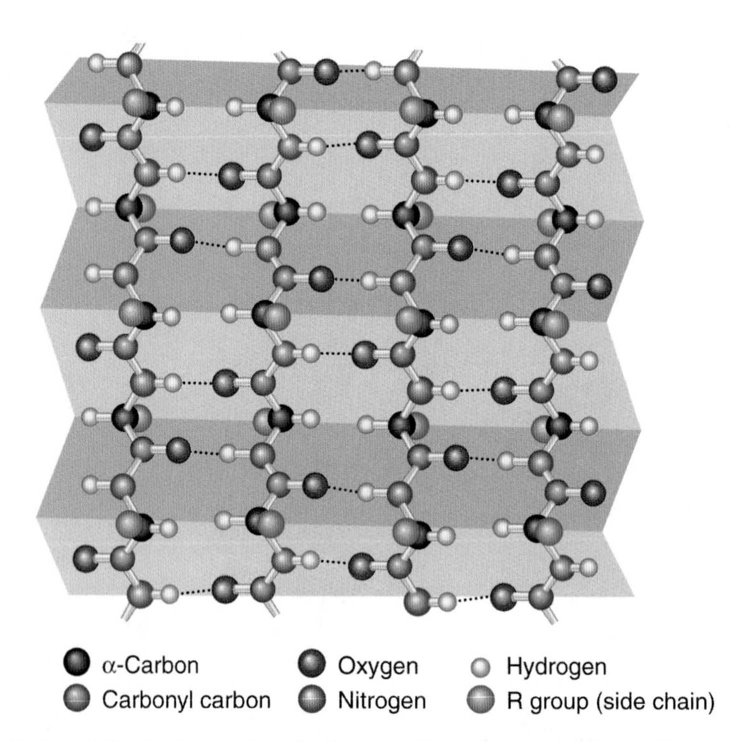

Figure 14.11 The *β*-pleated sheet is the second common type of secondary structure in proteins. In β-pleated sheets two or more polypeptide chains (or distant portions of the same chain) are held together by hydrogen bonds between atoms of the polypeptide backbones.

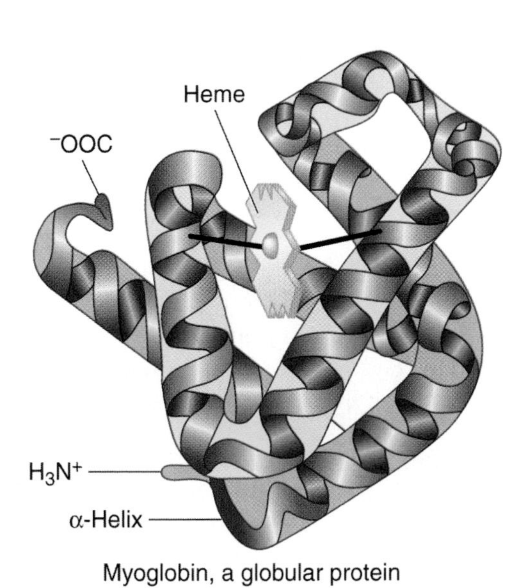

Figure 14.12 The tertiary structure of myoglobin. Much of the polypeptide chain of this protein exists as α-helix. Only the atoms of the polypeptide backbone are represented in the figure.

ical. These proteins are called **globular proteins.** Their shape is produced by the folding and bending of the peptide chain, and the result of this folding–bending is the **tertiary structure** of the protein (Figure 14.12).

There are several forces involved in tertiary structure, sharing one characteristic: They occur between atoms that may be many residues apart. This is why the compact folding is stable: Interactions between distant parts of the molecule maintain the tertiary structure. The specific forces involved include these noncovalent forces: hydrogen bonding, ionic and polar interactions, and hydrophobic interactions.

In most tertiary structures hydrogen bonds form either between atoms of two different side chains or between an atom of a side chain and an atom of the backbone [Figure 14.13(a)]. (Contrast this bonding with that in secondary structure where hydrogen bonding involves *only* atoms of peptide bonds.) These hydrogen bonds maintain the tertiary structure and stabilize the protein. Any one hydrogen bond is weak, of course, but *many* may be present in a globular protein.

Ionic and polar interactions result whenever a full or partial positive charge is close to a full or partial negative charge [Figure 14.13(b)]. In a compactly folded protein, several of these interactions may be present. Ionic interactions are usually strong, whereas dipole–dipole interactions are weaker—but again, collectively these interactions contribute to protein stability.

Hydrophobic interactions also contribute to protein tertiary structure [Figure 14.13(c)]. In globular proteins, nonpolar side chains are found in the interior of the molecule, where there are few or no water molecules. These nonpolar, hydrophobic groups are forced into the interior by water molecules that surround the protein. This clustering of nonpolar side chains is very much like the clustering of the hydrophobic tails of membrane lipids described in Chapter 13. The stability of many globular proteins seems to depend greatly on these hydrophobic interactions.

One type of covalent bond, the **disulfide bridge,** is important to tertiary structure. This bond forms by oxidation of two thiols in two cysteine residues

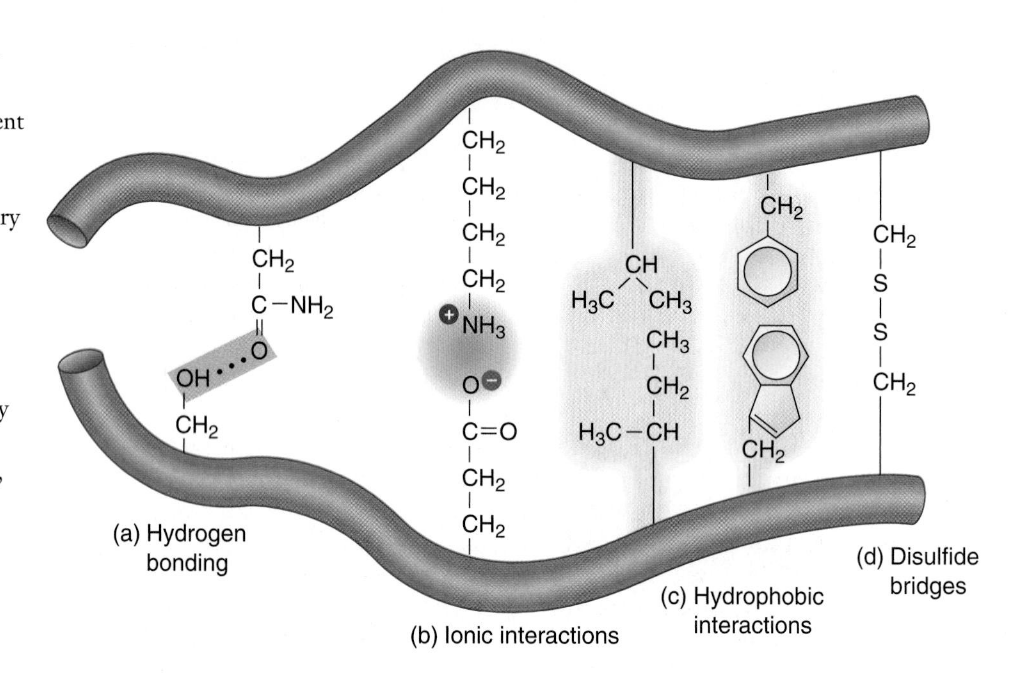

Figure 14.13 The tertiary structure of a protein is maintained by several noncovalent forces. **(a)** Hydrogen bonding, often between polar groups in side chains, contributes to tertiary structure. **(b)** Ionic and other polar interactions, like those between lysine and glutamate, are strong interactions. **(c)** Hydrophobic interactions involve regions that contain only nonpolar side groups of amino acids. In most globular proteins, the interior contains only nonpolar amino acids. **(d)** Disulfide bridges may form, and these covalently bound bridges stabilize the tertiary shape.

(Chapter 10), and these disulfide bridges in turn can be broken by a reduction reaction:

$$\underset{\text{Thiol}}{\text{cys—SH}} + \underset{\text{Thiol}}{\text{HS—cys}} \underset{[\text{red.}]}{\overset{[\text{ox.}]}{\rightleftharpoons}} \underset{\text{Disulfide}}{\text{cys—S—S—cys}}$$

When a protein first folds to its native conformation, one or more pairs of cysteine side chains may be brought next to each other. The sulfhydryl groups of these amino acids oxidize to form disulfide bonds [Figure 14.13(d)]. These covalent bonds are stronger than the noncovalent interactions discussed previously, and when present they significantly stabilize the protein in this tertiary conformation. Disulfide bridges are common in extracellular proteins that may require this added stability.

Example 14.4 Amino Acid Side Chains in Noncovalent Interactions

Consider the amino acids serine, lysine, and tryptophan. What types of interactions in a protein might involve their side chains?

SOLUTION

Serine might be involved in hydrogen bonding, lysine in an ionic interaction or hydrogen bonding, tryptophan in hydrophobic interactions.

Self-Test

Consider the amino acids leucine, cysteine, and glutamic acid. What types of interactions in a protein might involve their side chains?

ANSWERS

Leucine—hydrophobic interactions; cysteine—polar interactions; glutamic acid—ionic and hydrogen bonding

Quaternary Structure

Some proteins contain more than one polypeptide chain, and for these the term **quaternary structure** is used to explain the arrangement of the polypeptide chains in the protein. These proteins are called *oligomeric proteins*, which means they are oligomers of polypeptide chains, with each chain called a *subunit* or monomer of the protein. The oxygen-transport protein hemoglobin, for example, has four *subunits* (it is therefore said to be a tetramer), two identical ones designated α and two other identical ones designated β. The four subunits in hemoglobin have a very specific arrangement designated $\alpha_2\beta_2$ (Figure 14.14).

What forces hold subunits together in an oligomeric protein? The same forces involved in tertiary structure—hydrogen bonds, ionic and polar interactions, hydrophobic interactions, and in some cases disulfide bridges. In the dimeric protein shown in Figure 14.15, for example, the subunit on the left has a hydrophobic surface that matches with a hydrophobic surface of the one on the right. The left subunit possesses some positively charged side chains that are attracted to negative side chains on the other one. The subunit on the left has just the right groups in just the right orientation to form hydrogen bonds to groups on the face of the other subunit. Because these subunits match up, we say they are *complementary* to each other. Collectively, these forces hold the subunits together in a specific orientation. Disulfide bridges may form to further stabilize the structure.

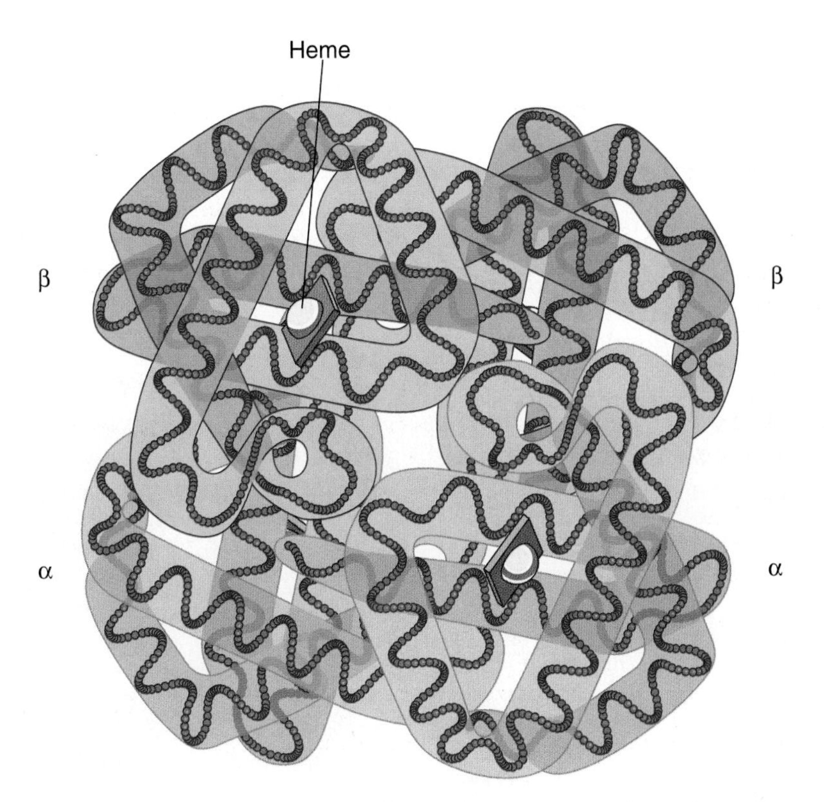

Figure 14.14 Quaternary structure of hemoglobin consists of four polypeptide chains, two called alpha subunits and two called beta. It contains four heme prosthetic groups.

Figure 14.15 Subunit interaction in a hypothetical dimeric protein. The two faces bind to each other through ionic and other polar bonding (green), hydrogen bonding (red) and hydrophobic interactions (yellow).

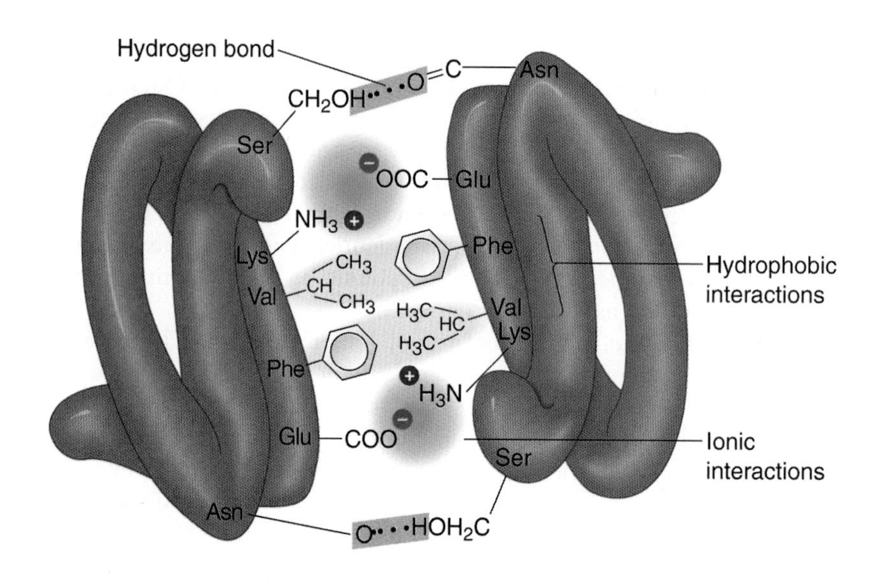

Connections

Sickle Cell Anemia

The structure of a protein determines the properties and function of the protein. This basic premise of biochemistry is illustrated by a comparison of normal hemoglobin molecules to the hemoglobin S molecules found in patients of sickle cell anemia. People with this condition usually live either in the malarial belt of the world or have family origins there. In the United States, most patients are African Americans, with a smaller number of patients of Mediterranean descent.

Hemoglobin molecules are normally highly soluble and tightly packed in the red blood cells that contain them. These normal hemoglobin molecules remain soluble as they first bind oxygen—to become oxyhemoglobin molecules—and then release oxygen becoming deoxyhemoglobin molecules. In contrast, hemoglobin S molecules have a tendency to become insoluble when they lose oxygen—when they become deoxyhemoglobin S molecules. If the concentration of deoxyhemoglobin S molecules becomes high in the red blood cell, the molecules aggregate together (become insoluble) to form long filaments in the red blood cell. These filaments push against the walls of the cell, changing its shape to sickle-shaped (see figure). Sickled red blood cells are fragile and easily broken, which explains the anemia found in these patients. Furthermore, these cells do not readily pass through capillaries, which results in impaired blood flow to tissues, which can lead to tissue destruction and, ultimately, death.

The decreased solubility of hemoglobin S has been traced to a single amino acid residue in the beta chain of hemoglobin. Hemoglobin S contains valine in the sixth residue of this chain, whereas normal hemoglobin contains glutamic acid. This change from an acidic residue to a nonpolar residue changes the molecule's solubility, resulting in altered cell shape that leads to a dangerous and debilitating disease.

Normal (rounded) and sickled red blood cells. *(Dr. Gopal Murti/Science Photo Library/Photo Researchers, Inc.)*

Other Components in Proteins

Proteins that contain only amino acids are called **simple proteins.** Other proteins, called **conjugated proteins,** contain one or more additional components, called prosthetic groups. A **prosthetic group** is a nonamino acid component of a protein. Hemoglobin contains the prosthetic group *heme,* which is the oxygen-binding part of the molecule (Figure 14.14). Carboxypeptidase, a protein involved in digestion, contains zinc ion (Zn^{2+}) as its prosthetic group. Conjugated proteins cannot carry out their functions if their prosthetic group is removed.

Macromolecular Complexes

Each protein has a unique shape because internal forces cause the protein molecule to assume its most stable conformation. Are these forces responsible for larger molecular complexes? Most definitely. Hydrogen bonding, ionic and polar interactions, and hydrophobic interactions all play a part in holding macromolecular complexes together. Ribosomes, for instance, are macromolecular complexes involved in protein synthesis (Chapter 16). They consist of many different proteins and a few large molecules of ribonucleic acid (RNA). Each of the proteins and RNAs of a ribosome are synthesized separately, and yet they bind together to form the subunits of a ribosome. This process, called *self-assembly,* occurs because the components are more stable when they are bound to other components than when they are free in solution. Self-assembly is a characteristic of all macromolecular complexes. Viruses are complexes of nucleic acid, either RNA or DNA, surrounded by a protein coat (Figure 14.16). The protein coat serves to protect the nucleic acid and is involved in entry and infection of certain host cells. Some viruses have a third layer—a lipid membrane—surrounding the protein coat.

RNA and DNA are discussed in Chapter 16.

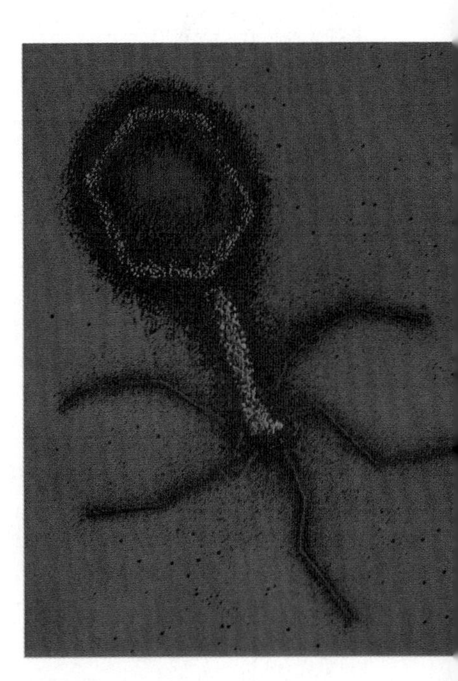

Figure 14.16 This bacteriophage viral particle consists of a nucleic acid surrounded by a protein coat. The particles self-assembled to yield the virus. A bacteriophage is a virus that infects bacteria. *(Lee D. Simon/Science Source/Photo Researchers, Inc.)*

Self-Test

Protein Structure

1. What is the primary structure of a protein?
2. What forces are responsible for maintaining tertiary structure in a protein?
3. A protein is broken down in the laboratory to yield amino acids and an organic molecule called pyridoxal. How would you classify this protein? In this context, what is pyridoxal called?

ANSWERS

1. Its amino acid sequence 2. Hydrogen bonding, ionic and polar interactions, hydrophobic interactions, and disulfide bridges 3. A conjugated protein, prosthetic group

14.4 Some Functions and Properties of Proteins

Function

The range of protein function is amazingly broad. Many serve as catalysts (enzymes), and some are structural components. Some are transporters, some have a

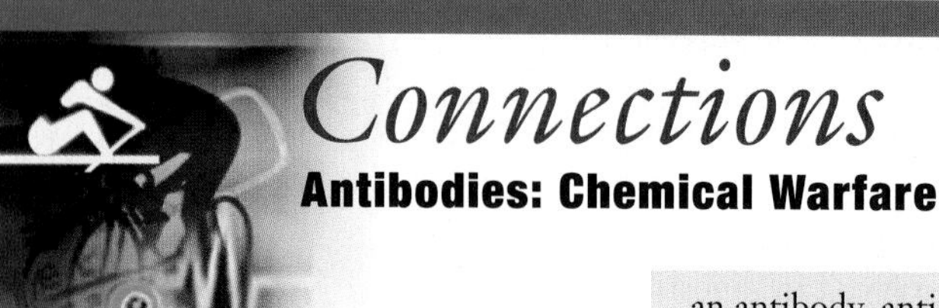

Connections

Antibodies: Chemical Warfare

We are constantly exposed to foreign particles such as viruses and bacteria. Our immune system is designed to recognize these invaders as foreign—as *nonself*—while also recognizing the substances normally found in our body as *self*. Among our immune system's defenses are the *antibodies*, proteins that bind to specific parts of a foreign particle (the antigen) forming an antibody–antigen complex. These complexes are in turn recognized by other components of the immune system—such as white blood cells—which then destroy the foreign body.

Antibodies do not normally treat the natural components of the body as antigens because the molecules of these components are identified by the immune system as *self* during embryonic development. Unfortunately, the body's ability to distinguish between self and nonself is not perfect, and an *autoimmune disease*, such as multiple sclerosis (MS), may result. In MS, the body produces antibodies that damage and destroy the lipid-rich membrane called the myelin sheath that covers the nerve fibers in the central nervous system. As multiple sites of the nervous system become damaged because of this disease, a variety of symptoms—including fatigue, weakness, vision problems, and coordination problems—result. Some other autoimmune diseases, the tissues or organs they affect, and their symptoms are listed in this table.

Some Autoimmune Diseases

DISEASE	AREAS AFFECTED	SYMPTOMS
Systemic lupus erythematosus	Connective tissue, joints, kidney	Facial skin rash, painful joints, fever, fatigue, kidney problems, weight loss
Type 1 diabetes	Insulin-producing cells in pancreas	Excessive urine production, blurred vision, weight loss, fatigue, irritability
Graves' disease	Thyroid	Weakness, irritability, heat intolerance, increased sweating, weight loss, insomnia
Rheumatoid arthritis	Joints	Crippling inflammation of the joints

protective role, some are involved in motion (muscles), some have a role in storage, still others are involved in regulating cell and body activities. Table 14.2 provides some examples of these functions.

Solubility

A protein molecule in an aqueous solution interacts with the water molecules and solutes that surround it. These interactions involve the polar and ionic side chains on the protein surface [Figure 14.17(a)] and play an important role in keeping the protein in solution.

Remember, amphoteric molecules can either give up or accept hydrogen ions.

Another important factor that affects protein solubility is the charge on the protein molecule. Proteins, like amino acids, are amphoteric molecules. At the pH values typical of a cell, the carboxyl groups in the side chains have lost their hy-

Table 14.2 Some Common Proteins and Their Functions

NAME	FUNCTION
Enzymes	
Amylase	Involved in digestion of starch
Lysozyme	Destroys cell walls of some bacteria
Hexokinase	Phosphorylates glucose in first step of glycolysis
Structural proteins	
Collagen	Found in fibrous connective tissue; most abundant protein in higher animals
α-keratin	Found in skin and hair
Transport proteins	
Hemoglobin	Transports oxygen in blood
Serum albumin	Transports fatty acids in blood
Protective proteins	
Antibodies	React with foreign bodies to aid in their removal
Fibrinogen and thrombin	Involved in blood clotting
Contractile proteins	
Actin	Found in thin muscle filaments
Myosin	Found in thick muscle filaments
Storage proteins	
Myoglobin	Stores oxygen in muscles
Ovalbumin	Stores amino acids for use by developing young in egg
Regulatory proteins	
Growth hormone	Involved in regulation of growth and metabolism
Insulin	Involved in use of nutrients
Macromolecular complexes	
Ribosomes	Protein synthesis
Cytoskeleton	Proteinaceous framework in cells

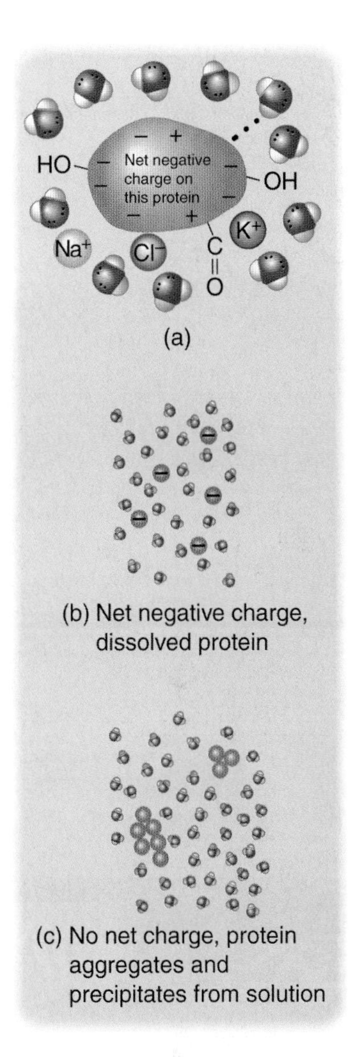

Figure 14.17 Protein–solvent and protein–protein interactions. **(a)** The polar and ionic groups on the surface of a dissolved protein molecule form noncovalent bonds with water and solute particles. **(b)** At this physiological pH, these protein molecules have a net negative charge, thus they repel each other and the protein remains in solution. **(c)** When the pH of the solution is adjusted so that the protein has no net charge, the protein molecules no longer repel each other. They aggregate and precipitate from solution.

drogen ions, leaving the negatively charged carboxylate anions, and the basic groups have gained hydrogen ions to become positively charged. These ionic groups are located mostly on the surface of the protein, which makes the surface a patchwork of charges [Figure 14.17(a)]. A protein with more negative than positive charges has a *net* or overall negative charge and one with more positive than negative charges has a net positive charge. Because all molecules of a given protein have the same net charge, they repel each other. This means the proteins remain in solution as distinct particles [Figure 14.17(b)].

The net charge on a protein changes with the pH of the solution. When a solution containing a dissolved protein becomes more acidic, some carboxylate groups accept some of the hydrogen ions to become neutral carboxyl groups:

$$—COO^- + H^+ \rightarrow —COOH$$

This reaction changes the net charge on the protein. Because there are now fewer carboxylate groups, the net charge becomes more positive. In basic solutions, positively charged side chains give up hydrogen ions to become neutral:

$$-NH_3^+ + OH^- \rightarrow -NH_2 + H_2O$$

Because there are now fewer ammonium groups, the net charge becomes more negative.

What effects do changes in pH have on the solubility of a protein? Changes in pH alter the net charge on a protein. At one pH, at the *isoelectric point*, the protein has no net charge and so the protein molecules no longer repel each other. Instead, they aggregate and precipitate (fall out) from solution [Figure 14.17(c)].

Denaturation of Proteins

Changes in pH can also change protein conformation—that is, they can denature a protein. Denaturation occurs whenever a protein changes from its native conformation to some other conformation (see Figure 14.6). Recall that the conformation of a protein may depend on ionic interactions. If the pH changes, certain side chains may accept or lose hydrogen ions. Ions may form or disappear. Stabilizing ionic interactions may be lost and destabilizing ionic interactions may appear. In general, proteins are denatured if the pH of the solution goes beyond a certain range.

Several other factors affect the conformation of a protein. Heat, changes in solvent, and heavy metals are all examples of denaturing agents. Heating a solution

Connections

Protein—Calorie Malnutrition

All forms of malnutrition are harmful, but inadequate amounts of protein in the diet are especially harmful to children. This is because a protein-deficient diet provides inadequate amounts of the amino acids needed for protein synthesis that is essential for normal childhood development. Lacking adequate protein, a child becomes physically or mentally impaired. The conditions produced by protein deficient diets are called *protein-calorie malnutrition* (PCM) and are seen most commonly in the children of some developing nations.

Kwashiorkor is a form of PCM in which caloric intake is adequate or close to adequate but dietary protein is too low, often because the diet relies heavily on starchy foods from plants with little or no meat. A child suffering from kwashiorkor typically has edema, an enlarged liver, and peeling skin (see figure). They often appear listless, sullen, and have little or no appetite. *Marasmus* is another form of PCM, in which the child's diet is too low in both calories and protein. This form of PCM results from starvation, and as a result the child's body appears more wasted away than in kwashiorkor and edema tends to be absent. In either form of PCM, continued severe deficiencies lead to death.

Children suffering from kwashiorkor. *(Charles Cecil/Visuals Unlimited)*

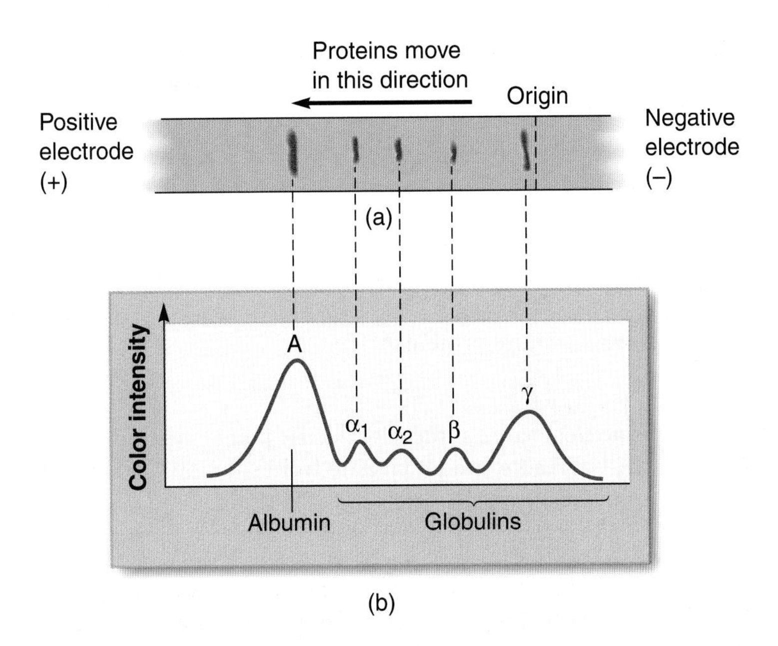

Figure 14.18 Blood serum proteins separated by electrophoresis. **(a)** Serum is placed onto an electrophoresis strip at the origin, and the strip is placed into an electric field. At the pH of this experiment, blood proteins have a net negative charge, and so are attracted to the positive electrode. Because the proteins have different charges and sizes, they migrate at different rates. **(b)** After separation, the strip is stained with a protein-specific dye and the amount of dye bound to the strip is proportional to the amount of protein. Albumin and γ-globulin are the most abundant proteins in this serum sample.

increases the thermal energy of the molecules in the solution. As the protein molecules gain energy, the noncovalent bonds in the molecule break. The native conformation is lost as the protein assumes other, more random shapes. With loss of native conformation comes loss of function and properties (denaturation). When you fry an egg, for example, the soluble proteins in the egg white become denatured. These denatured proteins are highly insoluble; thus a white "solid" forms from the clear, viscous egg white.

Proteins also denature in solvents or solutions that do not resemble either blood or the solution found in cells. Both organic solvents and salt solutions that are either very dilute or very concentrated denature many proteins. Similarly, some heavy metal ions are denaturing agents, because they break disulfide bridges.

Electrophoresis of Proteins

The charges on proteins can also be used to analyze a mixture of proteins. Protein mixtures must be separated in many medical and research applications. A powerful tool for these separations is **electrophoresis,** which is a technique that uses an applied electric field to cause charged particles (like proteins) to move. When a mixture of proteins is exposed to an electric field, the proteins migrate differently because they differ in both charge and size. This differential migration separates the proteins from each other. Figure 14.18 illustrates a common medical application, the analysis of serum proteins.

Example 14.5 Properties of Proteins

Egg white is a solution containing the protein ovalbumin in dilute salt water. When egg white is treated with ethanol, it turns from a transparent, viscous solution to a white, opaque solid. What has happened and why?

SOLUTION

The ovalbumin has denatured. This occurs when the alcohol mixing with the dilute salt water changes the solvent. Ovalbumin is not soluble in a mixed solvent containing water and alcohol, thus the protein denatures and precipitates.

Self-Test

1. Why are proteins least soluble in a solution that has a pH equal to the pI of the protein?
2. What factors can denature proteins?

ANSWERS

1. Proteins have no net charge at their isoelectric point, and thus do not repel each other. Instead they aggregate and precipitate from solution. 2. Heat, pH, changes in solvent, and some heavy metals

CHAPTER SUMMARY

Proteins are polymers of amino acids that have many diverse functions in the body. Each of the twenty α-amino acids found in proteins contain an amino group, carboxyl group, and a unique side chain. **Amino acids** belong to the L-stereochemical family and show the typical acid–base properties of carboxylic acids and amines. They typically exist as **zwitterions** and are **amphoteric.** The amino acids in **peptides** and proteins are linked by **peptide bonds,** with peptides having molar masses less than 5000 amu and proteins having molar masses greater than 5000 amu. A multitude of protein structures potentially exist and protein structure is described at four levels. A protein's **primary structure** is its amino acid sequence. **Secondary structure** involves hydrogen bonding between atoms of peptide bonds, and results in either **α-helix** or **β-pleated** sheet. The compact, tightly folded shape known as **tertiary structure** results from folding and bending of the peptide chain. Tertiary structure is stabilized by ionic and polar interactions, hydrogen bonding, hydrophobic interactions, and sometimes **disulfide bridges.** When two or more polypeptide chains are found in a protein the overall shape of the protein is its **quaternary structure,** which is stabilized by the same forces that stabilize tertiary structure. Proteins may contain one or more nonamino acid components called **prosthetic groups.** Macromolecular complexes self-assemble from proteins and other biomolecules, and are stabilized by the same interactions responsible for tertiary and quaternary structure. **Globular proteins** are usually water-soluble, and solubility changes with the pH of the solution. Proteins can be **denatured** by changes in pH or solvent, heat, or heavy metals. Proteins are charged particles that can be separated by **electrophoresis.**

KEY TERMS

Section 14.1
amino acid
amphoteric molecule
essential amino acid
isoelectric point (pI)
zwitterion

Section 14.2
peptide
peptide bond
protein

Section 14.3
α-helix
β-pleated sheet
conjugated protein
denaturation
disulfide bridge
fibrous proteins
globular proteins
native conformation
primary structure
prosthetic group
quaternary structure
secondary structure
simple protein
subunit
tertiary structure

Section 14.4
electrophoresis

EXERCISES

Alpha Amino Acids (*Section 14.1*)

1. Draw the structure of a generalized α-amino acid and label the three groups (substituents) that are always bonded to the α-carbon in these compounds.
2. Why are the 20 common amino acids called α-amino acids? Draw the structure of both an α-amino acid and a β-amino acid to illustrate the difference between them.
3. All of the common amino acids are ______.
   (a) In the D-stereochemical family
   (b) In the L-stereochemical family
   (c) Split evenly between the two stereochemical families
4. Define the term zwitterion.
5. Complete the structural skeleton below to obtain the structure of L-valine in its zwitterion form.

$$\begin{array}{c} COO^- \\ | \\ -C- \\ | \end{array}$$

6. Draw the zwitterion form of serine.
7. Examine the structure of threonine. How many chiral centers are present in this molecule? Which chiral center is compared to the glyceraldehydes to determine stereochemical family?
8. Sketch a molecule of alanine showing the stereochemistry that occurs in nature. What change in structure would yield the other stereoisomer?
9. Abbreviations are commonly used for amino acids, especially in protein structures. Which amino acids correspond to these abbreviations?
   (a) Phe (b) Cys (c) Glu (d) Lys
10. These abbreviations correspond to what amino acids?
   (a) Trp (b) Thr (c) Tyr (d) Arg
11. Identify the structural feature(s) that distinguish a polar neutral, an acidic, or a basic amino acid from a nonpolar amino acid.
12. List the nonpolar amino acids.
13. Proline is structurally distinct from the other amino acids. Identify proline's unique structural feature.
14. Draw the structures of a nonpolar and polar neutral amino acid in both the un-ionized and zwitterion forms.
15. Draw the structures of an acidic and basic amino acid in both the un-ionized and zwitterion forms.
16. What property must a compound possess to be classed as amphoteric?
17. Is there a pH value where all the functional groups of an α-amino acid are un-ionized? Why or why not?
18. Draw the structure of lysine showing the correct charge on the ionizable groups at pH 1, pH 7, pH 13.
19. Sketch the structure of aspartic acid showing the correct charge on the ionizable groups at pH 1, pH 7, pH 13.
20. An essential amino acid ______.
   (a) Is incorporated into all proteins
   (b) Is found in all food
   (c) Cannot be synthesized by the body
21. List the essential amino acids.
22. A zwitterion is not ______.
   (a) Soluble in hydrocarbons (b) Soluble in water
   (c) Electrically neutral (d) Polar

The Peptide Bond (*Section 14.2*)

23. Define a peptide bond and provide a sketch to illustrate your definition.
24. There are two different dipeptides containing glycine and alanine. Provide a sketch of each to illustrate how they differ.
25. Draw the structure of glutamylphenylalanylleucine.
26. Sketch the structure of valyltyrosylglycylasparagine.
27. Name this tripeptide:

$$^{+}H_3N-\underset{CH_3}{\overset{H}{C}}-\overset{O}{\overset{\|}{C}}-\underset{H}{\overset{}{N}}-\underset{H}{\overset{H}{C}}-\underset{O}{\underset{\|}{C}}-\overset{H}{N}-\underset{CH(CH_3)_2}{\overset{H}{C}}-COO^-$$

28. What is the name of this tripeptide?

$$^{+}H_3N-\underset{HOH_2C}{\overset{H}{C}}-\overset{O}{\overset{\|}{C}}-\overset{H}{N}-\underset{CH_2CH_2CH_2CH_2NH_3^+}{\overset{H}{C}}-\overset{O}{\overset{\|}{C}}-\overset{H}{N}-\underset{CH_2SH}{\overset{H}{C}}-COO^-$$

29. Which of the following is a peptide?
   (a) Hemoglobin (b) Casein
   (c) Myoglobin (d) Vasopressin

Protein Structure (*Section 14.3*)

30. Define primary structure in a protein.
31. What is the secondary structure of a protein?
32. Both α-helix and β-pleated sheet are examples of secondary structure in a protein. Explain the difference between them.
33. Define the tertiary structure of a protein.
34. What distinguishes a protein with quaternary structure from a protein lacking this structural feature?
35. Describe the forces that are responsible for primary protein structure.
36. Describe the forces that are responsible for secondary protein structure.
37. Describe the forces that are responsible for tertiary protein structure.
38. Describe the forces that are responsible for quaternary protein structure.
39. The α-helix of a protein is an example of ________.
    (a) Primary structure (b) Secondary structure
    (c) Tertiary structure (d) Quaternary structure
40. The shape of globular proteins is an example of ________.
    (a) Primary structure (b) Secondary structure
    (c) Tertiary structure (d) Quaternary structure
41. A β-pleated sheet is an example of ________.
    (a) Primary structure (b) Secondary structure
    (c) Tertiary structure (d) Quaternary structure
42. The structure of hemoglobin is an example of ________.
    (a) Primary structure (b) Secondary structure
    (c) Tertiary structure (d) Quaternary structure
43. In a protein in aqueous solution, the side chain of a leucine residue and the side chain of a valine residue are clustered together. This interaction is referred to as ________.
    (a) Disulfide bonding
    (b) Hydrogen bonding
    (c) Hydrophobic interactions
    (d) Ionic bonding
44. In a protein, the bonding of the hydroxyl hydrogen of serine to the side chain carbonyl oxygen of aspartate is referred to as ________.
    (a) Disulfide bonding
    (b) Hydrogen bonding
    (c) Hydrophobic interactions
    (d) Ionic bonding
45. Define a *disulfide bridge.* What role do disulfide bridges serve in the structure of a protein?
46. If two proteins have the same amino acid composition, does it necessarily follow that they have the same structure?
47. Describe the self-assembly of a virus or a cellular macromolecular complex.
48. Compare the composition of a simple protein and a conjugated protein.
49. Consider a globular protein that is water-soluble. Which of these amino acids are probably on the protein surface and which are probably in the core of the protein?
    (a) Alanine (e) Threonine
    (b) Glutamic acid (f) Valine
    (c) Leucine (g) Phenylalanine
    (d) Lysine
50. Why are the amino acids listed in Exercise 49 oriented on the surface or in the core?

Use the amino acids listed in Exercise 49 to answer Exercises 51 through 53.

51. Which of these amino acids could form ionic interactions when in a protein?
52. Which could form hydrogen bonds?
53. Which could be in hydrophobic interactions?
54. Define both protein native conformation and protein denaturation.
55. Which amino acids would contribute a positive charge to a protein in pH 7 solution? Which would contribute a negative charge?

Some Functions and Properties of Proteins (*Section 14.4*)

56. List the general classes of proteins and their functions that were discussed in this chapter.
57. Describe the specific function of these proteins.
    (a) Enzymes (e) Myoglobin
    (b) Antibodies (f) Lysozyme
    (c) Serum albumin (g) Insulin
    (d) Hemoglobin (h) Collagen
58. Which of the following proteins serves a defensive role?
    (a) Casein (b) Myoglobin
    (c) Lysozyme (d) Cytochrome c
59. A protein has no net charge at pH 6.13. What charge would it have at pH 7? At pH 9? At pH 5.5?
60. At which pH would the protein in Exercise 59 have the lowest solubility?
61. Explain what is meant by the term *isoelectric point.*

62. Examine Figure 14.18. Predict the movement of these proteins if the pH of the solution were made more acidic.

63. Changes in pH most affect what type of bonding in proteins?
(a) Hydrophobic interactions
(b) Ionic interactions
(c) Hydrogen bonding
(d) Disulfide bridges

64. Proteins that have been isolated from a biological source are kept in buffered solutions at cold temperatures. Why are these two conditions important?

General Exercises

65. A list of some words and phrases introduced in this chapter is presented below. For each, provide a description, definition, or example to show that you understand the word or phrase.
(a) α-amino acid
(b) L-amino acid
(c) Zwitterion
(d) Amphoteric
(e) Isoelectric point
(f) Dipeptide
(g) Peptide bond
(h) Protein
(i) Primary structure
(j) Secondary structure
(k) Tertiary structure
(l) Quaternary structure
(m) Hydrophobic interaction
(n) Disulfide bridge
(o) Denaturation

Challenge Exercises

66. An ounce of breakfast cereal contains 3 g of protein, 1 g of fat, and 23 g of carbohydrate. Estimate the Calories in an ounce of this cereal. (4 Cal/g of protein; 9 Cal/g of fat; 4 Cal/g of carbohydrate.)

67. Sickle cell anemia is caused by the change in a single amino acid in the β-chain of hemoglobin. Using this textbook and other sources, **(a)** identify this change, **(b)** explain how this change results in sickle cell anemia, and **(c)** list current medical treatments for sickle cell anemia.

Enzymes

15

PROTEINS THAT CATALYZE AND REGULATE THE REACTIONS OF THE BODY.

ENZYMES ARE proteins that catalyze virtually all of the reactions in the body. Catalysis is necessary because, without it, these reactions would either (1) be too slow or (2) would require high temperatures, highly reactive and concentrated reagents, and nonaqueous solvents. None of these conditions exist in the body or are needed because enzymes permit body chemistry to occur under the conditions that do exist within a cell.

Outline

15.1 An Introduction to Enzymes
Define an enzyme, and recognize an enzyme and its function from its name.

15.2 Enzyme Specificity
Describe the binding of substrate to an enzyme and explain why enzymes have specificity.

15.3 Enzyme Activity
List the factors that contribute to enzyme activity.

15.4 Rates of Enzyme-Catalyzed Reactions
Explain the effects of concentration, temperature, and pH on enzyme activity.

15.5 Regulation of Enzyme Activity
Differentiate among the types of enzyme regulation.

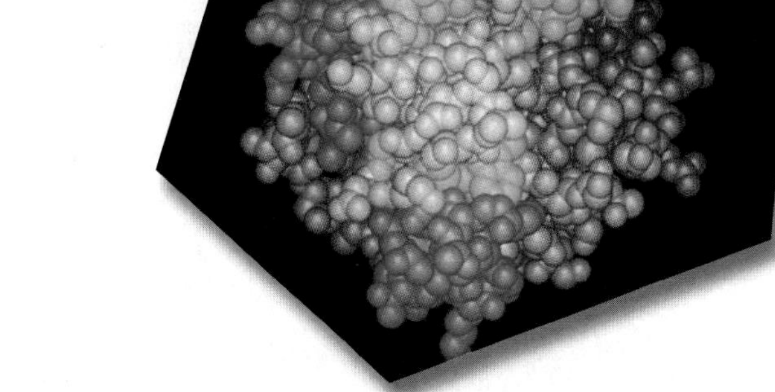

A computer-generated model of the enzyme, chymotrypsin. ***(Charles Grisham)***

15.1 An Introduction to Enzymes

Let's compare glucose use in the body to the burning of wood. Through several dozen enzyme-catalyzed steps, glucose is converted quickly and efficiently at 37° C to carbon dioxide, water, and energy. This same transformation occurs when wood is burned (remember, wood is principally cellulose, which is a polymer of glucose), but combustion occurs at temperatures that would destroy living tissues. Enzymes are efficient catalysts that speed up the chemical reactions that occur in the body at the modest temperatures and mild conditions that exist in cells. Enzymes, like all catalysts, speed up reactions by lowering the energy of activation for the reaction.

Enzymes are more than just efficient catalysts, however. They are usually specific for certain substances. Cells contain many hundreds of different compounds, and yet only one or a few of these compounds are affected by a particular enzyme. Furthermore, the activity of some enzymes varies with conditions in the cell. These changes in activity help regulate the body's chemical reactions and activities.

Enzyme Nomenclature

The suffix ***-ase*** in its name identifies a substance as an enzyme, although some older names use the suffix *-in*. Ure*ase*, RN*ase*, and hexokin*ase* are enzymes, as are tryps*in*, chymotryps*in*, and peps*in*. In addition, the names of some enzymes describe the reaction catalyzed by the enzyme. An *oxidase* catalyzes an oxidation reaction, and an *isomerase* catalyzes the isomerization of a reactant to one of its isomers. An enzyme name may also provide information about the reacting molecules, which are called

substrates. *Glucose oxidase* catalyzes the oxidation of its substrate glucose, and *hexokinase* catalyzes reactions in which the substrates are hexoses. Table 15.1 lists the six major classes of enzymes.

Cofactors and Coenzymes

Some enzymes consist only of amino acids—they are simple proteins—and require no other molecule or ion to be active. Others, however, require one or more inorganic ions—such as Mg^{2+} or Zn^{2+}—or organic molecules—such as heme or a niacin derivative—to be active. The required ions and organic molecules collectively are called **cofactors,** but a cofactor that is an organic compound is sometimes referred to as a **coenzyme.**

Table 15.1 The Major Enzyme Classes

CLASS	TYPE OF REACTION CATALYZED
1. Oxidoreductases	Oxidation-reduction
2. Transferases	Transfer one or more atoms from one substance to another
3. Hydrolases	Hydrolytic cleavage or the reverse
4. Lyases	Cleavage or the reverse, but not oxidation-reduction or hydrolysis
5. Isomerases	Intramolecular rearrangements (isomerizations)
6. Ligases	Energy requiring bond formation

An enzyme coupled with its cofactor is called a **holoenzyme;** an enzyme lacking its required cofactor is called an **apoenzyme.** In some cases, an enzyme and its cofactor are coupled only when the enzyme is involved in catalysis. In other cases, the enzyme and cofactor are permanently coupled and the cofactor is considered the *prosthetic* group of the enzyme. Apoenzymes and cofactors are catalytically inactive unless they are combined.

Self-Test

Enzymes

1. How do enzymes speed up the reactions of the body?
2. Use enzyme nomenclature to predict the function of triose phosphate isomerase.
3. What name is given to the nonamino acid portion of an enzyme that is required for activity?

ANSWERS

1. They lower the reaction's energy of activation. 2. This enzyme isomerizes triose phosphates. 3. Cofactor

15.2 Enzyme Specificity

Enzymes catalyze the conversion of specific substrate (reactant) molecules to product molecules. For some enzyme-catalyzed reactions a single substrate is converted to a single product; in other cases, two or more substrates or products are involved. An enzyme-catalyzed reaction involving a single substrate and product is represented by

$$\underset{\text{(Substrate)}}{S} \xrightarrow[]{\overset{\text{(Enzyme)}}{E}} \underset{\text{(Product)}}{P}$$

Initially the substrate or substrates bind noncovalently to one particular place on the enzyme molecule, which is called the **active site** of the enzyme, and forms an enzyme-substrate complex. An active site is either on the surface of the enzyme or in a groove or pocket on the enzyme surface (Figure 15.1). When its active site is occupied, the enzyme catalyzes the reaction converting the substrate(s) to product(s). Thus the active site has two distinct functions: to bind substrate and to catalyze the reaction.

Enzyme Specificity

The active site of an enzyme binds only one or a few of the many compounds in the cell. In other words, an enzyme shows **specificity** for substrate binding. Binding specificity is determined by two factors: (1) fit, which involves the size and shape of both the active site and the substrate (Figure 15.2), and (2) specific interactions between amino acid side chains at the active site and groups of atoms on the substrate. All other molecules in the cell either do not fit the active site or do not

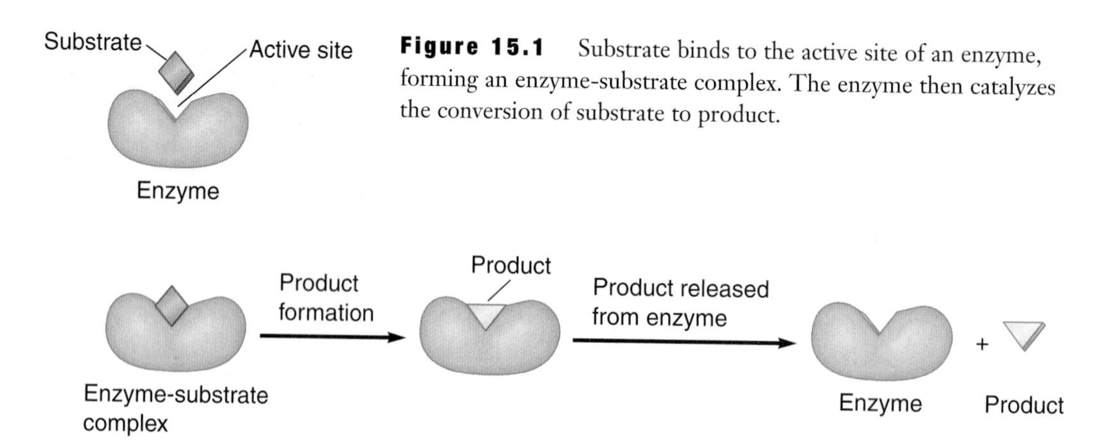

Figure 15.1 Substrate binds to the active site of an enzyme, forming an enzyme-substrate complex. The enzyme then catalyzes the conversion of substrate to product.

form bonds to side chains in the active site. Thus they cannot bind to the active site and they are not substrates of the enzyme.

Consider the enzyme lysozyme, which catalyzes the hydrolysis of glycosidic bonds in polysaccharides found in bacterial cell walls. The long, narrow active site of lysozyme accommodates the polysaccharide chain quite nicely, but if you think about it, many other molecules are also long like a polysaccharide. A polypeptide chain or a fatty acid of a triacylglycerol might also fit into the cleft. Lysozyme does not bind these molecules, however, which means that something more than size and shape contributes to specificity. Look now at Figure 15.3(a). This view of binding between lysozyme and its substrate shows specific hydrogen bonds between them. Not only are the sizes and shapes of the substrate and active site complementary, but the orientation of atoms on the substrate and active site permit hydrogen bonding between them. Fatty acids and polypeptides may fit into the active site, but they do not bond to it.

The interactions that occur between an enzyme and its substrate are the same interactions that are responsible for tertiary structure in proteins: hydrogen bond-

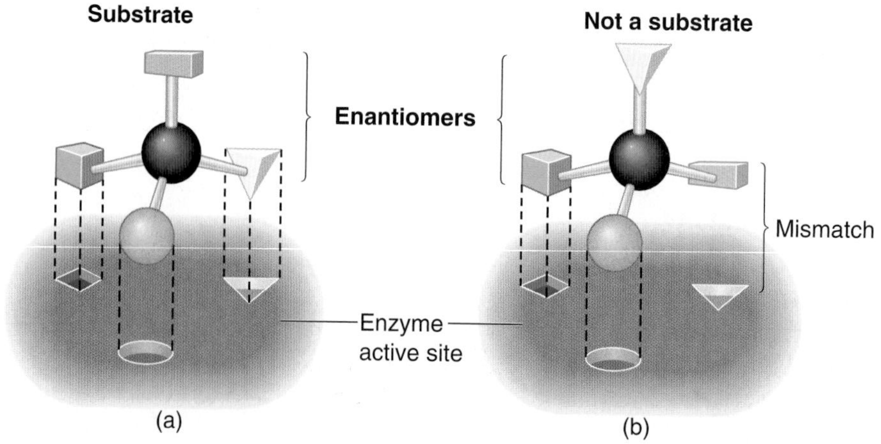

Figure 15.2 A substrate and an active site have complementary sizes and shapes that allow them to fit together. Even two very similar molecules—a pair of enantiomers (mirror images)—fit differently. One enantiomer fits and binds to the active site **(a),** and the other does not **(b).**

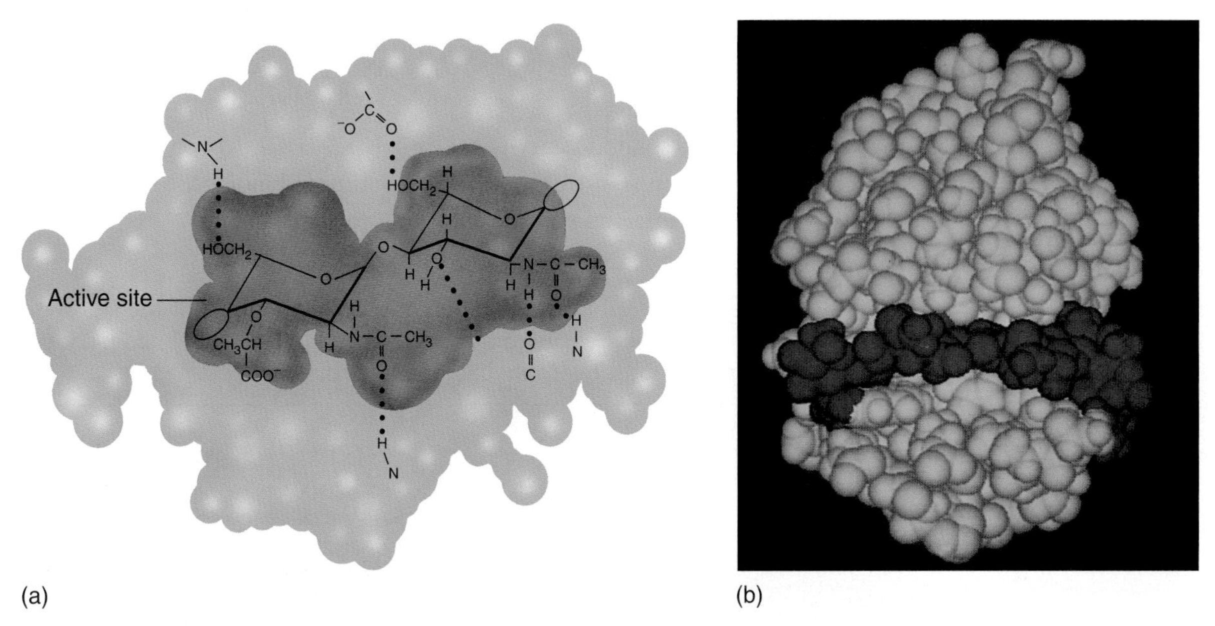

Figure 15.3 **(a)** When the polysaccharide substrate fits into the active site of lysozyme, hydrogen bonds form between atoms in the sugar residues of the polysaccharide and atoms in the enzyme. **(b)** A polysaccharide substrate is shown bound to the active site of this space-filling model of the enzyme lysozyme. *(National Institutes of Health, Science Photo Library/Photo Researchers, Inc.)*

ing, hydrophobic interactions, and ionic and other polar interactions. For instance, a lipid binds to the active site of an enzyme because the sizes and shapes are complementary and because the active site of the enzyme is lined with nonpolar amino acids that allow hydrophobic interactions with the lipid. A polysaccharide or polypeptide with similar size or shape could not bind to the same enzyme because they could not take part in these hydrophobic interactions.

This description of an active site corresponds to the **lock-and-key model** of enzyme-substrate interaction [Figure 15.4(a)]. Just as a key is complementary to the keyhole of a lock, the substrate fits into the active site of an enzyme. It is easy to picture this model, but for a variety of reasons it is now considered an oversimplification of enzyme-substrate interaction. The currently accepted view of these interactions is called the **induced-fit model,** which states that substrate-enzyme binding induces a change in the shape of one or, more likely, both of the molecules. When they are bound together, neither has the shape it had when free in solution [Figure 15.4(b)]. Either model explains enzyme-substrate binding, but the induced-fit model helps explain enzyme activity better than the older lock-and-key model.

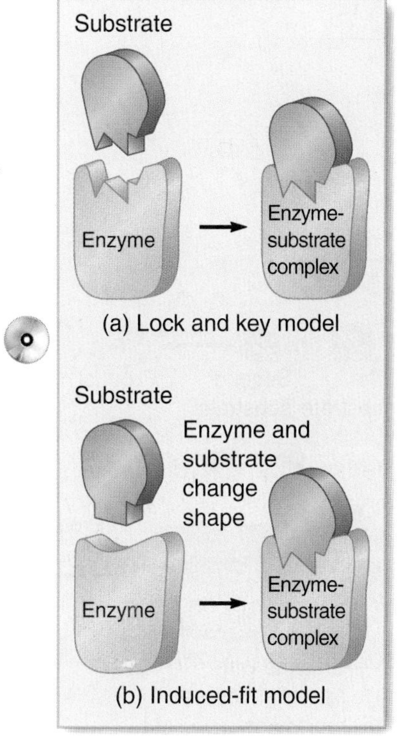

Figure 15.4 Models for enzyme-substrate binding. **(a)** In the lock-and-key model, the substrate and active site are complementary to each other before binding. They match like a lock and a key. **(b)** In the induced-fit model, the binding of substrate to enzyme causes changes in shape that make the substrate and active site complementary to each other.

Self-Test

Enzyme Specificity

1. Where do substrates bind to enzymes?
2. What two factors contribute to specificity of substrate binding?
3. What noncovalent interactions bind substrates to enzymes?

ANSWERS

1. At the active site. 2. Fit and specific bonding between atoms of the substrate and atoms on amino acid side chains at the active site. 3. Hydrogen bonding, ionic and other polar interactions, hydrophobic interactions.

15.3 Enzyme Activity

How does catalysis occur at the active site of an enzyme? Enzymes appear to use several factors in combination to enhance reaction rates: (1) proximity effects, (2) orientation effects, (3) acid–base catalysis, and (4) strain.

Proximity Effects

Enzymes help bring substrate molecules together—that is, they bring them into *proximity* with one another. Consider an enzyme that catalyzes this reaction:

$$\underset{\text{First substrate}}{S_1} + \underset{\text{Second substrate}}{S_2} \xrightarrow{E} \underset{\text{Product}}{P}$$

The enzyme binds both substrate molecules simultaneously, which brings these molecules into close proximity at the active site (Figure 15.5). Because the substrate molecules are close together, they react quickly to form product.

Orientation Effects

Enzymes also *orient* substrates to enhance the rate of reaction. When a substrate binds to the active site, it is held in a specific orientation relative to another sub-

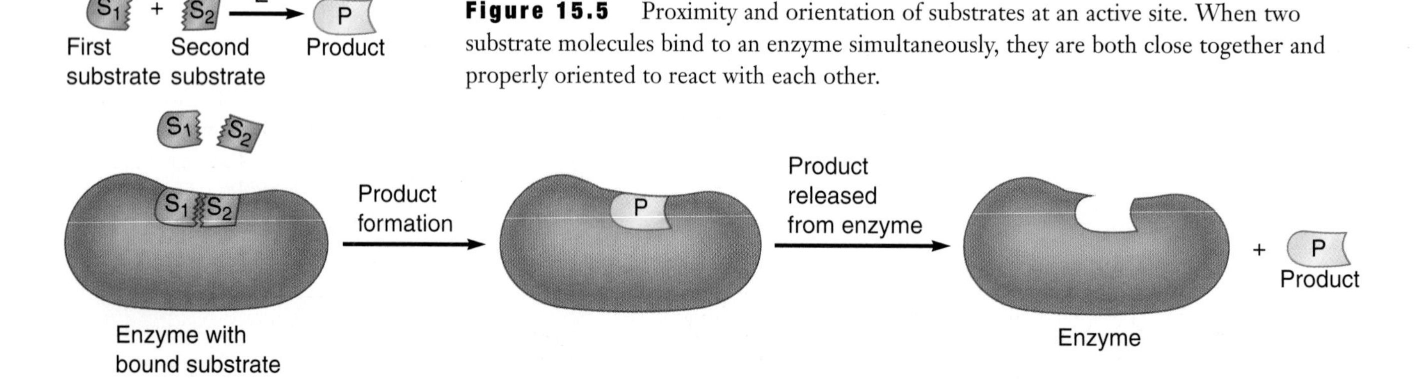

Figure 15.5 Proximity and orientation of substrates at an active site. When two substrate molecules bind to an enzyme simultaneously, they are both close together and properly oriented to react with each other.

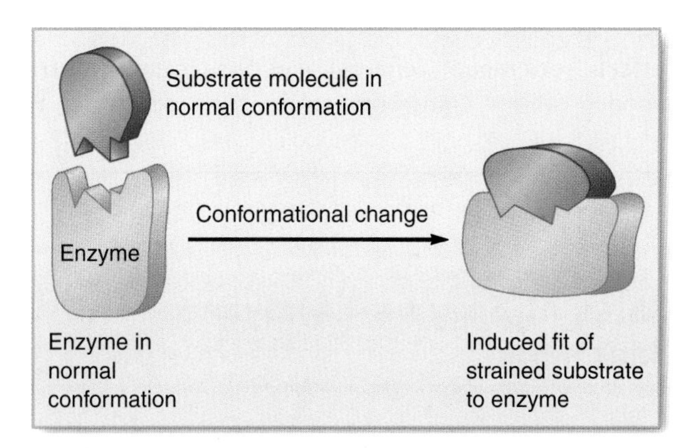

Figure 15.6 In the induced-fit model, the binding of substrate to enzyme results in conformational changes in these molecules. In its new conformation, the substrate is strained and therefore much more reactive.

strate molecule (Figure 15.5). In this orientation, a specific part of one substrate can react quickly with a specific part of the other substrate to form product.

Acid–Base Catalysis

Many enzymes can also act as acids and bases during catalysis. This speeds up the rate of reaction, just as acids and bases speed up many reactions in general and organic chemistry. Recall that certain amino acid side chains are either acidic or basic (Chapter 14), which means they can either donate or accept hydrogen ions. The acidic and basic groups at an active site have one enormous advantage over acids and bases in solution: The amino acids are both close and properly oriented to quickly give up or accept hydrogen ions from the substrate molecule.

Effect of Strain

A fourth factor influencing the rate of enzyme reactions is *strain* (stress). In the induced-fit model of substrate-enzyme binding, the enzyme and substrate undergo conformational changes when they bind. When a substrate is free in solution, it is generally in its lowest energy form, its most stable conformation. When the substrate binds to the enzyme, it is forced into another shape, a higher energy conformation (Figure 15.6). Because this enzyme-bound, strained-substrate molecule (ES*) is higher in energy than a free substrate molecule, its energy is closer to the energy of activation for the reaction than the unstrained substrate (Figure 15.7). As a result, it takes less energy to get from the substrate to product and the reaction occurs faster. On binding, many enzymes turn ordinary substrate molecules into strained, highly reactive species.

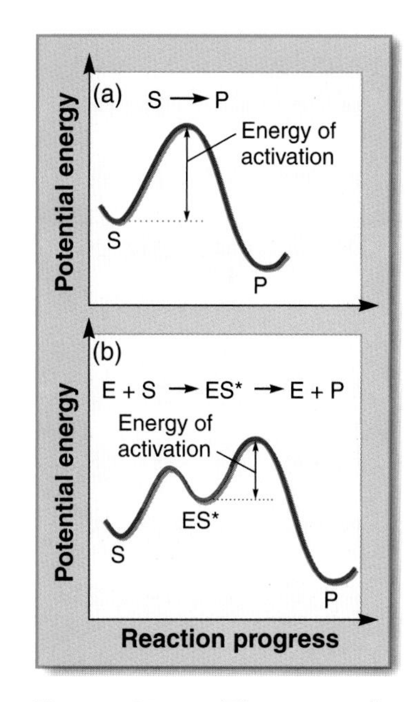

Figure 15.7 The energy of activation for a reaction **(a)** in the absence and **(b)** in the presence of an enzyme that induces strain in the substrate. In **(a)** the energy of activation is large. **(b)** When substrate binds to the enzyme, an enzyme-substrate complex (ES*) is formed. This complex is higher in energy than free substrate, thus the energy of activation is reduced. Because less energy is needed for the reaction to occur, the reaction is faster.

Self-Test

Enzyme Activity

1. List four factors that affect how enzymes catalyze reactions.
2. What advantage does the induced-fit model for enzyme-substrate binding have over the lock-and-key model?

ANSWERS

1. Proximity effects, orientation effects, acid–base catalysis, strain 2. The lock-and-key model cannot contribute strain to a substrate; the induced-fit model can.

15.4 Rates of Enzyme-Catalyzed Reactions

Biochemists use *velocity* as a synonym for *rate*.

Chemical reactions occur over a certain period of time, and the **reaction rate** is the speed at which the reaction occurs under a set of conditions. Under the same set of conditions, enzyme-catalyzed reactions are always much faster than uncatalyzed reactions. Whether reactions are enzyme-catalyzed or not, reaction rates are typically expressed either in terms of the amount of reactant (substrate) consumed per unit time or in terms of the amount of product formed per unit time. For example, consider the reaction $2H_2O \longrightarrow 2H_2 + O_2$. The rate of this reaction could be expressed as moles of water consumed per minute or as moles of hydrogen or oxygen produced per minute.

The rate of an enzyme-catalyzed reaction is affected by enzyme concentration, substrate concentration, temperature, pH, and enzyme inhibitors.

Effects of Enzyme Concentration

Usually in enzyme-catalyzed reactions there are many more substrate molecules than enzyme molecules. Under these conditions, if the concentration of the enzyme is increased, the rate of the reaction increases as well (Figure 15.8). This makes sense because at higher enzyme concentrations there are more enzyme molecules available for catalysis. If 1 enzyme molecule can convert 100 molecules of substrate to product in 1 second, then 2 enzyme molecules can make 200 molecules of product in the same time, which means the reaction rate has doubled. Thus if all other conditions are kept the same, the rate of a reaction is directly proportional to the amount of enzyme present.

Effects of Substrate Concentration

The rate of an enzyme-catalyzed reaction also varies with the concentration of substrate, but in an unusual way. For many reactions that are not enzyme catalyzed, the rate doubles when the concentration of the reactant doubles [Figure 15.9(a)]. For enzyme-catalyzed reactions, doubling the concentration of substrate appears to double the rate if the concentrations of substrate are low, whereas doubling high concentrations of substrate yields only a slight increase in reaction rate [Figure 15.9(b)]. At low substrate concentrations, only some of the enzyme molecules are bound to substrate. Thus only some of the enzyme molecules are catalyzing reactant to product at any moment in time. At these low concentrations, doubling the concentration of substrate results in doubling the number of enzyme molecules that have substrate bound to them [Figure 15.10(a)]. The rate doubles when the concentration of substrate doubles. In contrast, when concentrations of substrate are high, most of the active sites on the enzyme molecules are filled with substrate.

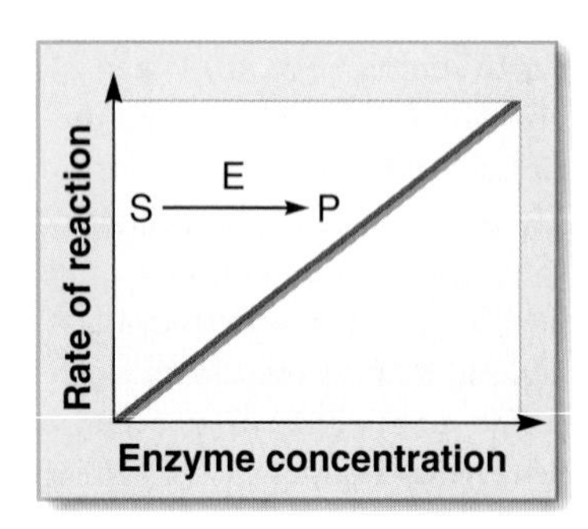

Figure 15.8 Increasing the concentration of an enzyme increases the rate of a reaction.

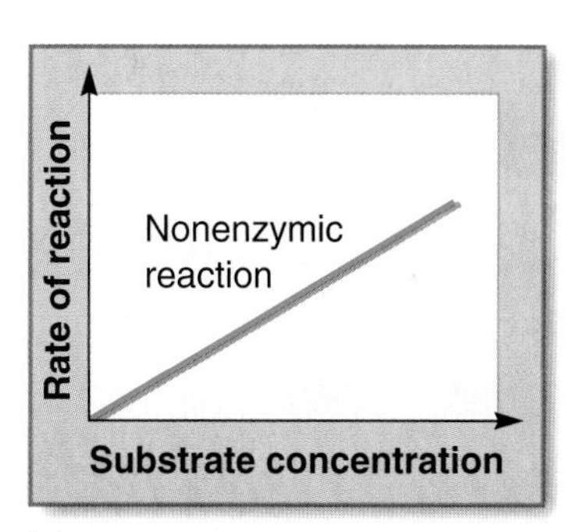

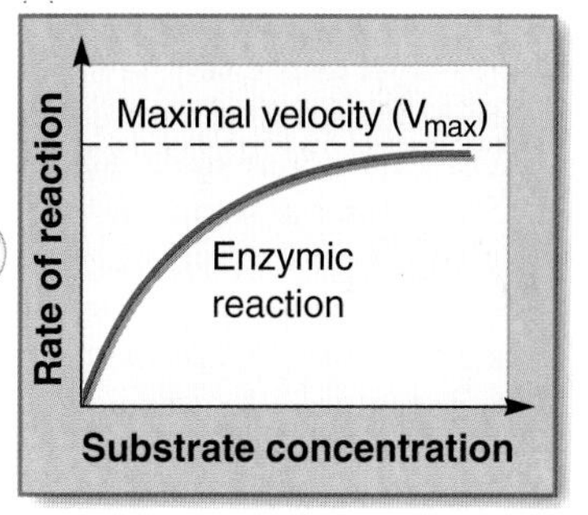

Figure 15.9 Increasing the concentration of a substrate increases both the rate of an uncatalyzed reaction and the rate of an enzyme-catalyzed reaction, but in quite different ways. At some point all of the enzyme is combined with substrate and the reaction reaches a maximal rate.

Doubling the concentration of substrate results only in a slightly larger proportion of enzyme molecules that have substrate bound [Figure 15.10(b)]. When enzymes are nearly saturated with substrate, doubling substrate concentration increases the rate only slightly. At high substrate concentrations, the rate (velocity) of the reaction approaches a maximal value, designated V_{max}. At very high concentrations of substrate, the reaction proceeds at its maximal rate, even if more substrate is added.

Effects of Temperature

In general, chemical reaction rates increase regularly with increasing temperature. This increase is due to the higher energy of the molecules at the higher temperatures. As a result of this increased energy, collisions between reactant molecules are more frequent, and a larger proportion of these collisions have enough energy to break old bonds and make new ones.

The rates of enzyme-catalyzed reactions also increase with temperature, but for these reactions the stability of the enzyme is also important. At lower temperatures,

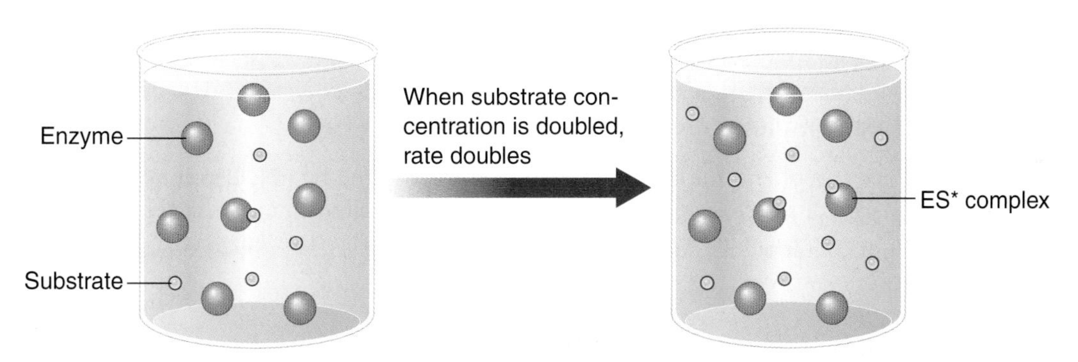

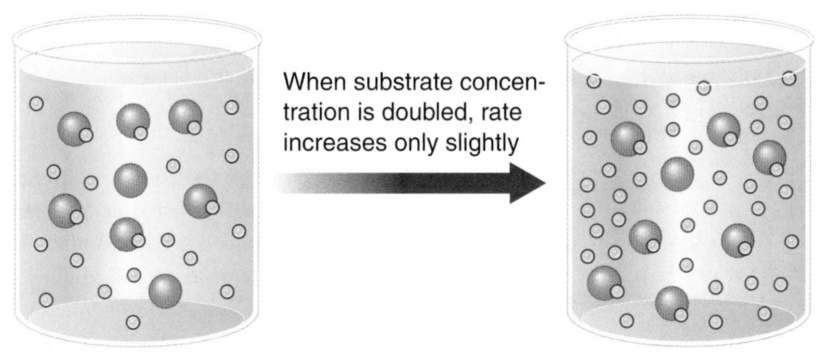

Figure 15.10 **(a)** At low substrate concentrations, doubling the concentration of substrate doubles the number of enzyme molecules that have substrate bound to them. **(b)** At high substrate concentrations, doubling the substrate concentration increases only slightly the number of enzyme molecules that have substrate bound.

We are *homeothermic* organisms, which means we maintain a more or less constant body temperature of 37°C (98.6°F). Occasionally, however, environmental stress or disease causes body temperature to deviate significantly from this normal range.

Hypothermia occurs whenever our core body temperature drops significantly below 37°C. At these decreased temperatures, life-sustaining biochemical reactions proceed more slowly, and as a result normal processes—including brain function—slow down. Mild hypothermia begins when body core temperature drops below 35°C and is characterized by uncontrollable shivering. As the temperature continues to decrease, disorientation and unconsciousness can occur. If core temperature approaches or drops below 30°C, death may result.

We have a *fever* whenever our core body temperature is significantly above 37°C. Fever results from some infections or exposure to hot and dehydrating conditions. When core body temperature exceeds 40°C, some proteins in the body may denature, and the loss of function of critical enzymes in the central nervous system may result in dysfunction and possibly death.

This person is being treated for hypothermia.

enzyme molecules are stable and so reaction rate increases with increasing temperature. At some upper temperature limit, however, the enzyme is denatured, which means that it loses its normal shape and normal function (Chapter 14). When this occurs, the enzyme no longer functions as a catalyst and the reaction rate drops drastically (Figure 15.11).

Effects of pH

The rates of enzyme-catalyzed reactions are affected by pH. Pepsin is a digestive enzyme of the stomach that works fastest in the acidic conditions (pH ≈ 2) found in the stomach. In contrast, the small-intestine digestive enzyme trypsin works

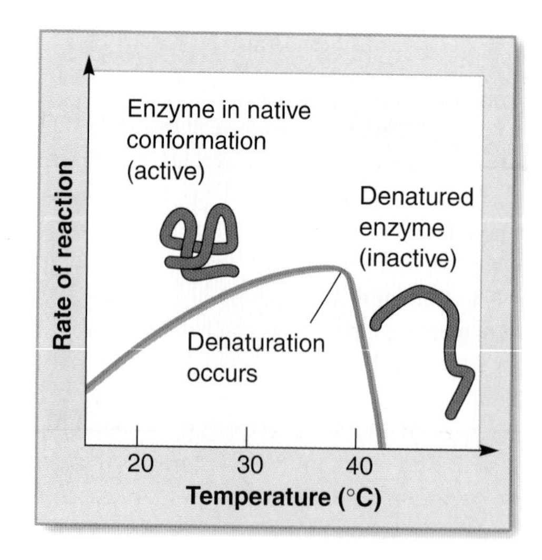

Figure 15.11 Increasing temperature speeds up an enzyme-catalyzed reaction until a temperature is reached that denatures the enzyme. At this temperature the rate of the reaction drops sharply.

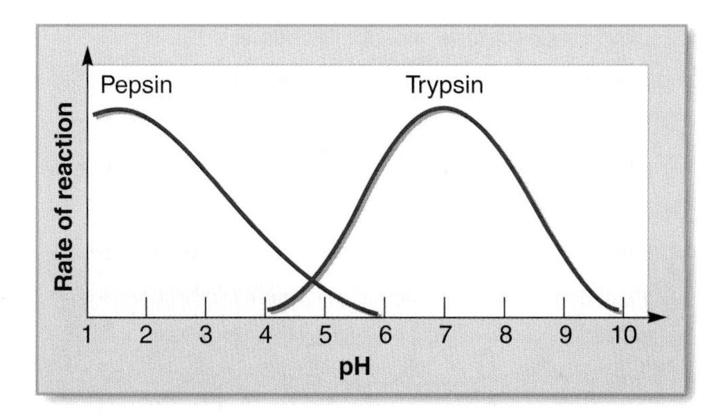

Figure 15.12 The effects of pH on enzyme activity. Pepsin, a digestive enzyme of the stomach, is most active under acidic conditions like those found in the stomach. Trypsin, a digestive enzyme of the small intestine, is most active at neutral to slightly alkaline pH, which is the pH range of the small intestine.

fastest at the neutral pH (pH ≈ 7) values of the small intestine (Figure 15.12). At other pH values, the rates of these enzyme-catalyzed reactions decrease. The effect of changing pH may simply be due to the denaturation of the protein, or it may be a bit more subtle. As noted in the previous section, the acid–base properties of some amino acid side chains are factors that contribute to enzyme catalysis. If the pH is too high or too low, these groups will be unprotonated or protonated too much of the time. The enzyme will function much less efficiently because it cannot donate or accept hydrogen ions as it normally would. It is only within a certain range of pH values that the acidic and basic side chains of the active site will be protonated or unprotonated an appropriate amount of time.

Changes in pH can also alter the substrate. As the pH changes, a group in the substrate may be protonated or deprotonated. The altered substrate may not bind to the enzyme, or it may react more slowly.

Effects of Enzyme Inhibitors

The rates of enzyme-catalyzed reactions can be decreased by substances called *inhibitors.* Inhibitors bind to enzymes and alter their catalytic activity. The inhibition may be permanent—that is, *irreversible*—or it may be *reversible* if activity returns when the inhibitor is removed.

Irreversible inhibition occurs whenever an **irreversible inhibitor** binds permanently to an enzyme, leaving the enzyme with little or no activity:

$$\underset{\text{(Active)}}{\text{Enzyme}} + \text{Inhibitor} \longrightarrow \underset{\text{(Inactive)}}{\text{Enzyme-inhibitor complex}}$$

The bond between inhibitor and enzyme is either a covalent bond or a very strong noncovalent interaction, but once it forms, the complex remains. The enzyme molecule still exists, but it lacks catalytic activity and makes no contribution to the cell. As a result, the effective concentration of the enzyme is reduced.

A group of phosphate-containing organic compounds provides an example of irreversible inhibition. This group includes the nerve gases and some insecticides such as malathion. The phosphate group in these compounds reacts readily with a hydroxyl group on a serine in the enzyme acetylcholinesterase (Figure 15.13). The result is a stable phosphoester bond. Once formed, the complex remains. Because this serine is an essential part of the active site, the enzyme molecule permanently loses its catalytic activity once the inhibitor binds. If a significant number of these enzyme molecules becomes inactive, normal muscle contraction and relaxation fails, and death from impaired breathing may result.

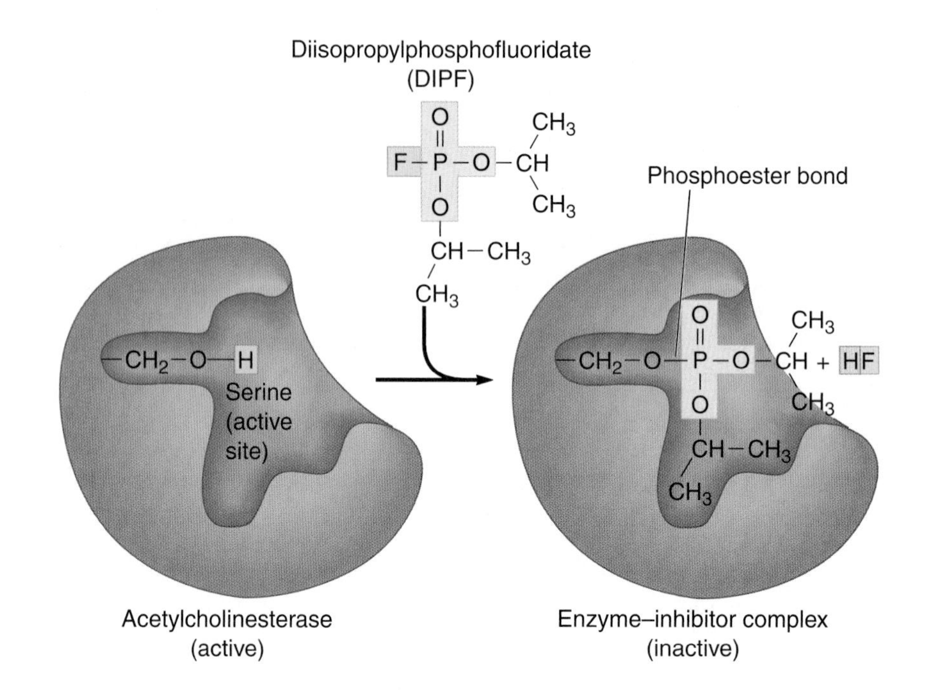

Figure 15.13 Some organophosphates—including nerve gases—are irreversible inhibitors of the enzyme acetylcholinesterase. They react irreversibly with a serine at the active site of the enzyme, forming an inactive enzyme-substrate complex.

Reversible inhibitors also bind to an enzyme molecule to reduce or eliminate enzymic activity, but only temporarily. Because a permanent complex is not formed, the inhibitor can be removed and activity restored:

$$\underset{\text{(Active)}}{\text{Enzyme}} + \text{Inhibitor} \rightleftharpoons \underset{\text{(Inactive)}}{\text{Enzyme-inhibitor complex}}$$

Reversible inhibitors bind to either the active site of the enzyme or elsewhere on the enzyme surface. If binding occurs at the active site, the substance is a **competitive inhibitor** because it competes with the substrate for the active site (Figure 15.14). While the inhibitor is bound to the active site, the substrate cannot bind. For that period of time, that enzyme molecule is catalytically inactive.

Competitive inhibitors have structures that resemble the structure of the substrate. These inhibitors form many or all of the noncovalent bonds that a substrate

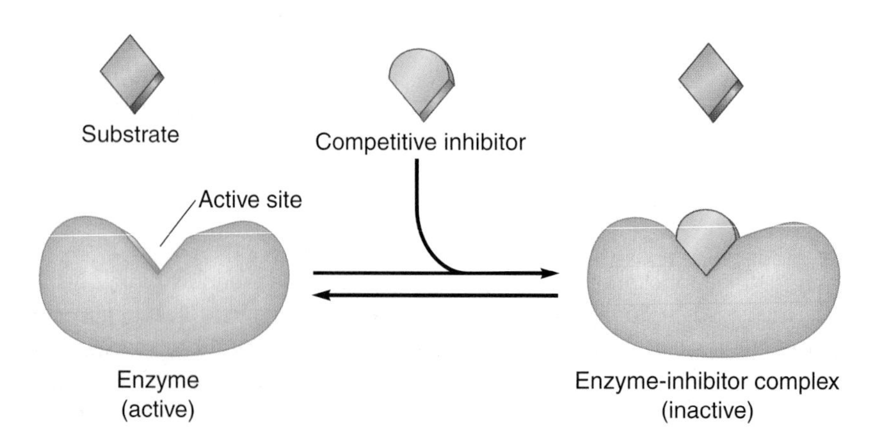

Figure 15.14 Competitive inhibition. A competitive inhibitor binds reversibly to the active site of an enzyme, preventing binding of the substrate.

would form with the active site. Unlike substrate, however, competitive inhibitors cannot react to yield a product. Competitive inhibitors bind to the active site, but no reaction occurs.

Sometimes competitive inhibitors of enzymes can be used to our advantage. *Sulfa drugs* are competitive inhibitors used to treat some bacterial infections. Bacteria use *p-aminobenzoic acid* to synthesize *folic acid*, which the bacteria need to grow. The sulfa drugs, including sulfanilamide, resemble *p*-aminobenzoic acid and bind to the active site of the enzyme dihydropteroate synthase (Figure 15.15). While sulfanilamide is bound to the active site, no *p*-aminobenzoic acid can bind and so no folic acid can be produced. If insufficient folic acid is produced, the growth of the bacterial population is inhibited. Sulfa drugs do not affect the patient's own cells because humans cannot make folic acid; we must obtain this vitamin in our diet.

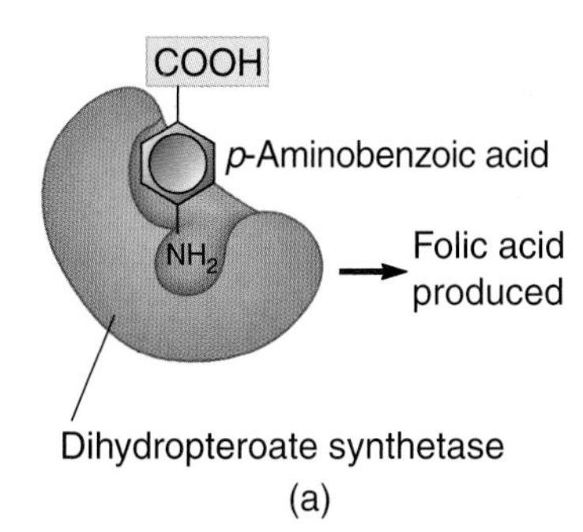

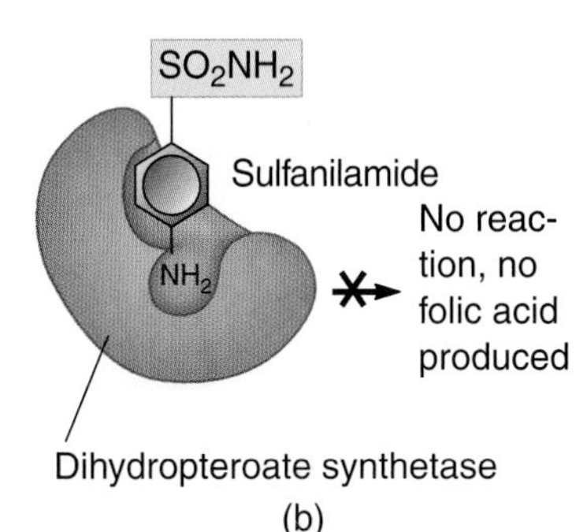

Figure 15.15 The sulfa drug sulfanilamide is a competitive inhibitor of a bacterial enzyme needed for folic acid synthesis. **(a)** The substrate *p*-aminobenzoic acid binds to the active site of this enzyme, which leads to the production of folic acid. **(b)** While the competitive inhibitor sulfanilamide is bound to the active site no folic acid is produced.

Example 15.1 Use of Competitive Inhibitors

Ethylene glycol, $HOCH_2CH_2OH$, is the main component of antifreeze, and in the body it is oxidized in an enzyme-catalyzed reaction to oxalic acid, HOOCCOOH, which is toxic. Ethanol, CH_3CH_2OH, is sometimes given to patients who have ingested ethylene glycol. Why?

SOLUTION

Note the similarity in structure between ethylene glycol and ethanol. Because they are similar, both bind to the active site of the first enzyme involved in the oxidation of ethylene glycol. When ethanol is bound, ethylene glycol cannot, and so its oxidation to oxalic acid is inhibited. While ethanol is present, the body excretes ethylene glycol rather than converting it to oxalic acid.

An interesting side point is that this first enzyme in ethylene glycol oxidation is alcohol dehydrogenase, which normally oxidizes ethanol. In this therapy, the normal substrate, ethanol, is used as a competitive inhibitor of a synthetic, unnatural substrate, ethylene glycol.

Self-Test

Succinate, $^-OOCCH_2CH_2COO^-$, is a substrate of the enzyme succinate dehydrogenase. Explain why malonate, $^-OOCCH_2COO^-$, is a competitive inhibitor of this enzyme.

ANSWER

Malonate resembles succinate closely. Each is the carboxylate of a dicarboxylic acid, and they differ in size only by one carbon (—CH_2—). Because of this similarity, malonate binds to the active site of the enzyme.

Some reversible inhibitors bind to the enzyme at a site other than the active site. These substances are called **noncompetitive inhibitors** because they do not compete with the substrate. Their binding to the enzyme causes a conformational change (Figure 15.16). In this new conformation the enzyme is either less active or inactive. In the new conformation, the enzyme may bind substrate either less efficiently or else not at all. Alternately, the new conformation may cause the catalytic groups of the active site to become misaligned, reducing the efficiency of catalysis.

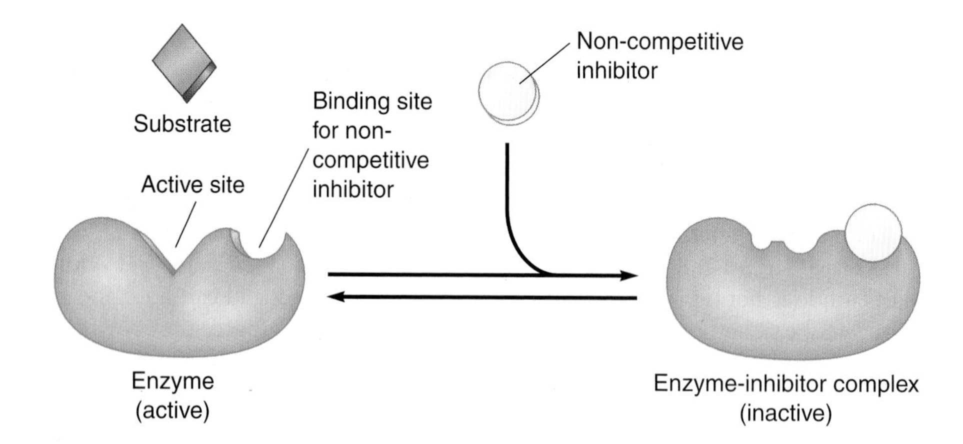

Figure 15.16
Noncompetitive inhibition. Noncompetitive inhibitors bind at a site other than the active site. This binding causes the enzyme to undergo a conformational change that reduces its activity. In this example, the conformational change prevents substrate from binding to the active site.

Self-Test

Rates of Enzyme-Catalyzed Reactions

1. Why is the rate of most enzyme-catalyzed reactions virtually zero at temperatures much above body temperature?
2. What is the main difference in the way competitive and noncompetitive inhibitors bind to enzymes?

ANSWERS

1. The enzyme is denatured at these temperatures and is therefore inactive.
2. Competitive inhibitors bind to the active site; noncompetitive inhibitors bind elsewhere on the enzyme.

15.5 Regulation of Enzyme Activity

The activity of enzymes is highly regulated. One way to change activity is to change the concentration of the enzyme in the cell—enzyme expression—which is discussed in Chapter 16. Enzyme activity can also be regulated in ways that do not change concentrations. Examples of this are (1) activation of zymogens, (2) reversible covalent modification of the enzyme, and (3) allosteric regulation.

Proenzymes

Cells synthesize some enzymes as *proenzymes*, which have no catalytic activity. **Zymogens** are proenzymes of protein-cleaving enzymes, many of which are involved in digestion and blood coagulation (see Connections, p. 472). Zymogens and other proenzymes can be recognized by the suffix *-ogen* or the prefix *pro-*. Trypsin*ogen* and *pro*elastase are examples.

Zymogens (and other proenzymes) gain activity when one or more peptide bonds are cleaved at some time following synthesis. This cleavage is followed by conformational changes that yield the active form of the protein. To understand the role of zymogens in the body, consider trypsin, a digestive enzyme that cleaves specific peptide bonds of ingested proteins. What prevents trypsin from breaking down the proteins of the pancreas, where it is made? This problem is prevented because trypsin is synthesized as the inactive zymogen, *trypsinogen*, which is released into the small intestine when partially digested food enters from the stomach. In the small intestine, trypsinogen is converted to trypsin by enzymes that are already there. These enzymes cleave a peptide bond near one end of the trypsinogen polypeptide chain. When freed of the small peptide, the molecule undergoes a conformational change to the active form, trypsin.

Covalent Modification of Enzymes

The activity of some enzymes can be reversibly increased or decreased by *covalent modification* of the enzyme molecules, which means some group is attached to the enzyme molecule by a covalent bond. An example is the phosphorylation–dephosphorylation of glycogen phosphorylase. The role of this enzyme is to cleave glucose molecules from glycogen and attach a phosphate group to them. This cleavage occurs on demand—that is, whenever more phosphorylated glucose is required by muscles or the liver. Glycogen phosphorylase with no phosphate groups covalently bound has little enzymic activity (Figure 15.17). When phosphorylated glucose is needed, a series of events is initiated that results in the attachment of phosphate groups to the glycogen phosphorylase. When it is phosphorylated, the enzyme is more active and rapidly forms phosphorylated glucose from glycogen. The enzyme remains active until dephosphorylation returns it to the less active form.

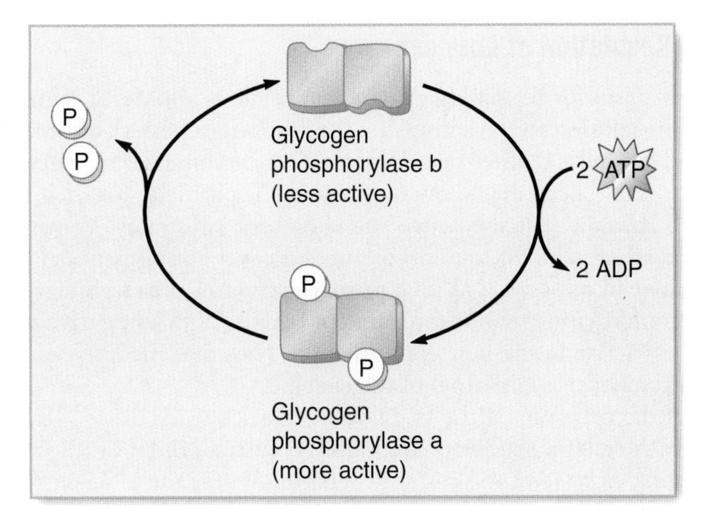

Figure 15.17 The reversible covalent modification of glycogen phosphorylase. When increased activity of glycogen phosphorylase is required, the enzyme is phosphorylated through a transfer of phosphate from ATP. (ATP is a high-energy molecule and is discussed in Chapters 16 and 17.) When the activity is not needed, phosphate is cleaved from the enzyme, which returns it to its less active form.

Connections
Blood Clotting

Normal blood clotting plays a vital role in preventing blood loss from injuries or surgery, but blood clots also cause some heart attacks. Blood normally contains appropriate concentrations of several proteins called clotting factors, which are zymogens. A cut or injury stimulates release of zymogen-activating factors that activate some zymogens, which in turn activate other ones. Step-by-step this process transforms increasing numbers of zymogens to active enzymes. This cascading event leads to the conversion of the soluble molecules of *fibrinogen* to insoluble molecules of *fibrin*, which aggregate to form the massive molecular plug that we call a clot. This process effectively closes the wound in all but life-threatening cuts or injuries. What began as the release of a small amount of activating factors in response to injury grows into a massive response that prevents excessive blood loss.

Other factors besides an open cut or wound may stimulate formation of blood clots. Bruises and impaired circulation are examples. Clots formed in this way may be confined to the cardiovascular system and can be carried from their site of origin to other parts of the body. If a clot becomes lodged in a blood vessel and blocks flow of blood beyond that point, cell death and loss of function will occur. If blood is blocked in an artery of the heart, a heart attack results. If the blockage occurs in the brain a stroke results.

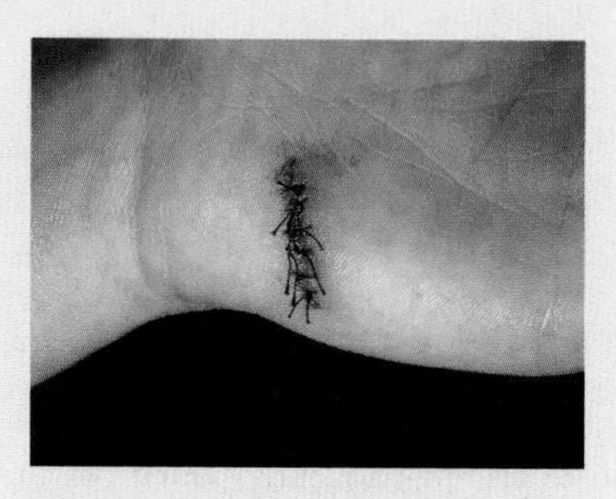

In the absence of normal blood clotting, a wound like this would be life-threatening. *(Alex Barel/Science Library/Photo Researchers Inc.)*

Allosteric Regulation of Enzymes

A third mechanism for regulating enzyme molecules is **allosteric regulation** (Figure 15.18). In addition to the active site, some enzymes possess one or more other sites—called allosteric ("other place") sites—that can bind certain small molecules found in the cell. These sites show the same high specificity for their small molecules, called *effectors*, that the active site shows for substrates. When an effector binds to the allosteric site, the enzyme undergoes a conformational change that causes a change in its activity. With a **positive effector,** the enzyme is more active in the new conformation than before effector bound. With a **negative effector,** the enzyme is less active in the new conformation. Negative effectors are, in essence, natural noncompetitive inhibitors of the enzyme.

Most of the reactions of the body occur in a sequence of steps called a *metabolic pathway*. Metabolic pathways are highly regulated, often by allosteric regulation of the first enzyme of the pathway. This allows for the efficient regulation of the entire pathway through a process called **feedback inhibition.** Consider a pathway consisting of three enzymes in which the first enzyme (E_1) converts substrate S to intermediate T, then the second enzyme (E_2) converts T into intermediate U, and finally the third enzyme (E_3) converts U into product P:

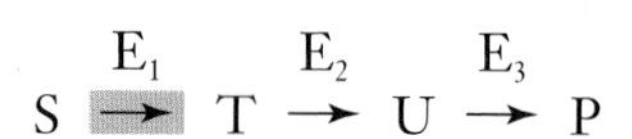

$$S \xrightarrow{E_1} T \xrightarrow{E_2} U \xrightarrow{E_3} P$$

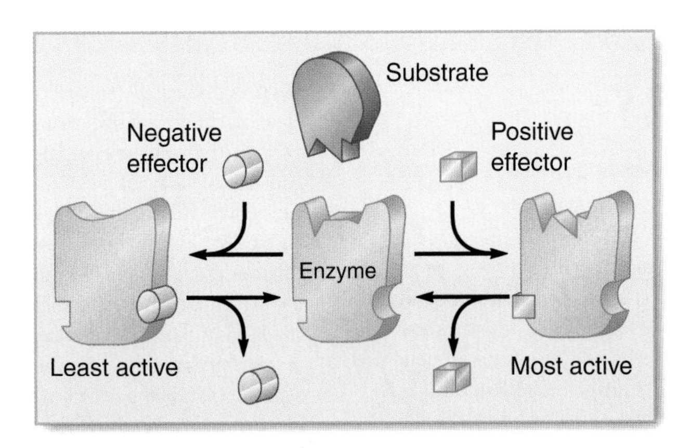

Figure 15.18 The regulation of allosteric enzymes by positive and negative effectors. Binding of a positive effector increases the activity of the enzyme, whereas binding of a negative effector decreases the activity. In this example, the effectors cause the enzyme to undergo conformational changes that affect binding of substrate.

Product P is a negative effector of the first enzyme (E_1) of the pathway. As S is converted to P, the concentration of P increases, resulting in P molecules binding to the allosteric sites of the E_1 molecules. This inhibits the E_1 molecules stopping the synthesis of molecules T, U, and P:

$$S \overset{E_1}{\nrightarrow} T \overset{E_2}{\rightarrow} U \overset{E_3}{\rightarrow} P$$

Thus, the accumulation of P is a message telling E_1 to shut down because the cell has enough P for its present needs.The inhibition of E_1 is reversible. As the concentration of P decreases in the cell, the P molecules bound to E_1 diffuse away from E_1. The E_1 molecules become active, and P is again formed. Allosteric control provides for rapid and sensitive regulation of metabolic activity.

Another example of regulation is seen in lactate dehydrogenase (LDH). LDH exists as **isozymes,** which are structurally similar enzymes that come from different tissues of the same organism. The isozymes of LDH differ in the subunits that are present in this tetrameric (four subunit) enzyme (Figure 15.19). M is the designation for the subunit that predominates in lactate dehydrogenase isolated from skeletal muscle, thus muscle LDH can be symbolized as M_4. H is the designation for the subunit that predominates in heart muscle, and so heart LDH is H_4. Other tissues usually contain LDH isozymes that consist of a mixture of these subunits. All the isozymes of LDH catalyze the reversible conversion of pyruvic acid to lactic acid.

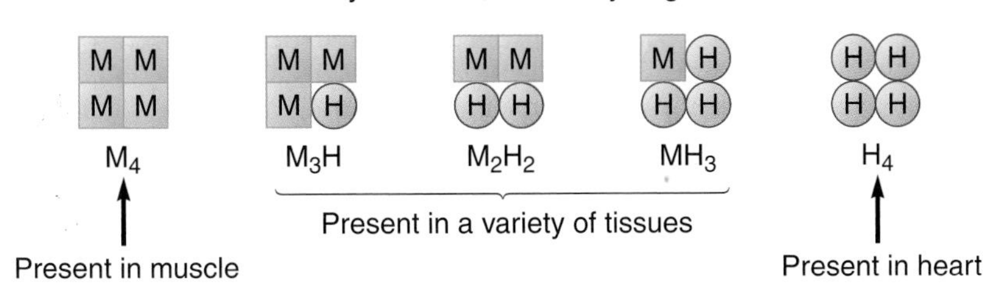

Figure 15.19 When the two kinds of subunits (M and H) of lactate dehydrogenase self-assemble to form the active tetramer, these combinations of subunits can result. Each combination is an isozyme.

Enzymes are essential to life, but they also play an increasingly direct role in medicine. Some of them are used in diagnosis (to determine the cause and nature of a disease) and others are used in treatment. Let's look at both the diagnosis and treatment of a heart attack as examples.

Suppose a person has intense chest pain and is admitted to a hospital. The patient's physician will immediately begin a diagnosis to determine the cause of the symptoms. A blood sample will be drawn from the patient, and the blood will be tested for the presence of heart proteins. Normally blood does not contain these proteins, but when a heart attack occurs, some heart muscle cells die and rupture. The contents of these broken cells, including proteins, enter the blood. If heart proteins are found in the patient's blood, then a heart attack is a very real possibility.

A different enzyme can be used to treat heart attack victims. Some heart attacks occur when a blood clot partially or completely blocks the flow of blood in an artery in the heart. The cells downstream from the blockage no longer receive oxygen and nutrients, and with time they begin to die with concurrent loss of function. The longer the duration of the blockage, the greater the probability of permanent injury or death. But if the clot can be removed or dissolved before significant permanent damage is done, the probability of death or permanent disability is significantly reduced.

A physician may administer *tissue-type plasminogen activator* (TPA) to a heart attack victim. This enzyme converts the zymogen plasminogen to plasmin, the enzyme that normally breaks down blood clots. Prompt treatment with TPA activates the normal clot-dissolving machinery of the body. If TPA is administered in time, the clot is dissolved and blood flow restored before a significant number of heart-muscle cells die.

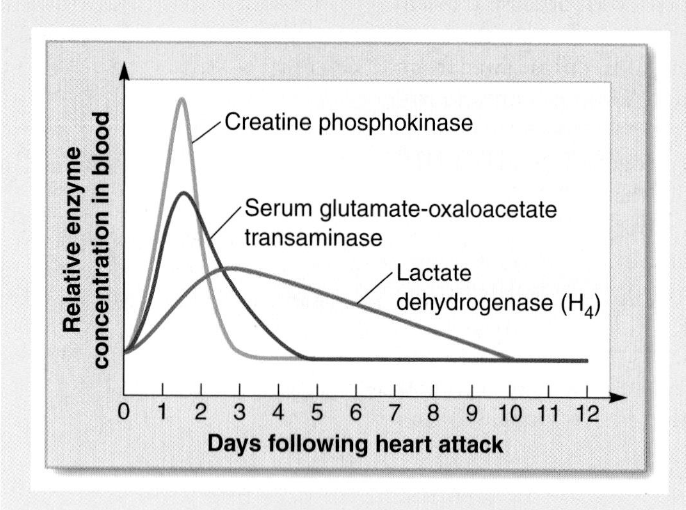

These three enzymes are normally absent from blood. During and after a heart attack, heart-muscle cells die and rupture, releasing their contents into blood. The presence of these heart-muscle enzymes in blood is indicative of a heart attack.

Although M_4 LDH and H_4 LDH catalyze the same reaction, they normally work in opposite directions. M_4 LDH normally converts pyruvate to lactate, which is just what is needed when muscle must, on occasion, function without oxygen (see anaerobic metabolism in Chapter 17). Conversely, the heart does not function without oxygen, and in heart tissue H_4 LDH converts lactate to pyruvate, thus preventing the build up of lactate in heart muscle.

Self-Test

Regulation of Enzyme Activity

1. Why are some digestive enzymes synthesized in an inactive (zymogen) form?
2. Explain how a thermostat works by feedback inhibition.

ANSWERS

1. If these enzymes were synthesized in an active form, they would hydrolyze the proteins found in the pancreas. 2. When a room gets too cold, an electrical contact in the thermostat closes, forming a complete electric circuit, which activates the heater. The heater heats the room until room temperature exceeds a set point in the thermostat. Now the electrical contact opens, breaking the circuit, and so the heater shuts down. Heat, the product of the heater, effectively shuts down its own production.

CHAPTER SUMMARY

Enzymes are proteins that catalyze the reactions of the body. Enzyme names typically end with the suffixes ***-ase*** or *-in*. Enzyme names often indicate the reaction catalyzed by the enzyme and the substrate of the enzyme. Some enzymes are simple proteins, others contain one or more **cofactors.** If a cofactor is an organic molecule, it is called a **coenzyme.** An enzyme with its cofactor is called a **holoenzyme;** one lacking its cofactor is an **apoenzyme.** One or more substrate molecules bind to an enzyme's active site through noncovalent interactions. Enzyme specificity is determined by how well a substrate fits in an active site and by the noncovalent interactions that form between the **substrate** and **active site.** The **lock-and-key model** of substrate-enzyme binding says the substrate fits the active site like a key fits a lock. The more generally accepted **induced-fit model** says that conformational changes occur for substrate and enzyme when the substrate binds. Enzyme activity depends on four factors: (1) proximity, (2) orientation, (3) acid–base catalysis, and (4) strain. The rates of enzyme-catalyzed reactions depend on the concentrations of both the enzyme and substrate. The rate of the reaction as a function of substrate concentration approaches a maximal value, V_{max}, because the enzyme molecules become saturated with substrate at high substrate concentrations. The rates of enzyme-catalyzed reactions increase with temperature until the enzyme denatures. The rates of enzyme-catalyzed reactions decrease outside a range of pH values because of protonation or deprotonation of groups on the enzyme or substrate. Enzymes can be inhibited irreversibly or reversibly. **Competitive inhibitors** bind at the active site of an enzyme; **noncompetitive inhibitors** bind at a site other than the active site. Regulation of preexisting enzyme molecules can occur by activation of inactive proteins called **zymogens,** by reversible covalent modification of the enzyme molecules, or by **allosteric regulation** in which **positive** or **negative effectors** bind to an allosteric site on the enzyme and alter its conformation and activity. The binding of a product molecule to the first enzyme of a pathway is an example of **feedback inhibition.**

KEY TERMS

Chapter Introduction
enzyme
holoenzyme
substrate

Section 15.1
apoenzyme
-ase
coenzyme
cofactor

Section 15.2
active site
induced-fit model
lock-and-key model

Section 15.4
competitive inhibitor
irreversible inhibitor
noncompetitive inhibitor
reaction rate (velocity)
reversible inhibitor

Section 15.5
allosteric regulation
feedback inhibition
isozymes
negative effector
positive effector
zymogen

EXERCISES

Introduction

1. Provide a definition for an enzyme and explain why enzymes are essential to living organisms.
2. To which group of biomolecules do the enzymes belong?

An Introduction to Enzymes (Section 15.1)

3. Which of the following substances are enzymes?
   (a) Trypsin (b) Galactose
   (c) Tryptophan (d) Acetylcholinesterase
   (e) Catalase (f) Sucrase
4. The names of the substances in Exercise 3 contain information about the substances. What information in these names did you use to determine whether they are enzymes?
5. Use the names of these enzymes to identify their substrate or substrates.
   (a) Xylulose reductase
   (b) UDP-glucuronosyl transferase
   (c) Xanthine oxidase
6. Use the name of the enzymes in Exercise 5 to determine the type of reaction they catalyze.
7. Use the names of these enzymes to identify their substrate or substrates.
   (a) Phosphohexose isomerase
   (b) Aspartate-ammonia ligase
   (c) Succinyl-CoA hydrolase
8. Use the name of the enzymes in Exercise 7 to determine the type of reaction they catalyze.
9. What is a substrate?
10. Provide a synonym for the prosthetic group of an enzyme.
11. What name is given to an enzyme that lacks its prosthetic group?
12. Some enzymes are active only when some organic molecule is also present. This organic compound is the enzyme's ___________ .

Enzyme Specificity (Section 15.2)

13. Provide a definition for the active site of an enzyme.
14. Compare the binding of substrate to an enzyme for the lock-and-key model and the induced-fit model.
15. Identify the types of bonds or interactions that bind a substrate to the active site of an enzyme.
16. It is said that a substrate is complementary to the active site of an enzyme. What factors make the active site and substrate complementary?
17. Fatty acid synthase complex, an enzyme involved in the synthesis of fatty acids, binds the fatty acid as it is being made. Explain why a polypeptide will not bind to the active site of this enzyme.
18. Why does amylose (a polysaccharide) not bind to the active site of the enzyme in Exercise 17?

Enzyme Activity (Section 15.3)

19. Explain the effect of an enzyme (or any catalyst) on the activation energy of a reaction.
20. How can the orientation of substrate molecules influence the activity of an enzyme?
21. Describe how the binding of two substrate molecules to an enzyme's active site enhances the activity of the enzyme. What term is used to describe this effect?
22. The side chains of some amino acids act as general acids or bases during catalysis. List the amino acids that can donate H^+ and those that can accept H^+ during catalysis.

Rates of Enzyme-Catalyzed Reactions (Section 15.4)

23. Sketch the rate (velocity) of an enzyme-catalyzed reaction as a function of enzyme concentration.
24. Sketch the rate of an enzyme-catalyzed reaction as substrate concentration is changed from zero to a high level.
25. Explain why the shape of the graph of rate of reaction versus enzyme concentration is different from the graph of rate of reaction versus substrate concentration.
26. Why does a graph of enzyme activity versus temperature increase in a linear fashion until a big drop is seen at some higher temperature?
27. Compare the activity of pepsin and trypsin in the pH range of 2 to 9. What is the physiological significance of these pH profiles?
28. Explain why a graph of the activity of an enzyme versus pH yields a bell-shaped curve.
29. Enzymes show varying rates with changes in pH and temperature. Are these factors normally important in the activity of the enzymes within the body?

30. Compare the binding of a reversible and irreversible enzyme inhibitor.
31. Compare the binding of a competitive and noncompetitive inhibitor.
32. What structural relationship exists between an enzyme's substrate and a competitive inhibitor of the enzyme?
33. How do sulfa drugs inhibit bacterial growth?
34. Succinate, $^{-}OOCCH_2CH_2COO^{-}$, is an intermediate in aerobic metabolism (see the discussion of the TCA cycle in Chapter 17). The enzyme succinate dehydrogenase oxidizes succinate to fumarate, $^{-}OOCCH{=}CHCOO^{-}$. This enzyme is inhibited by malonate, $^{-}OOCCH_2COO^{-}$. What kind of inhibition is this?

Regulation of Enzyme Activity (Section 15.5)

35. From its name, how can you recognize that a substance is a zymogen?
36. How are zymogens activated?
37. Why is trypsin not synthesized directly in the pancreas?
38. Give three examples of proteins that are synthesized in an inactive form.
39. How is glycogen phosphorylase activated?
40. What regulatory advantage does covalent modification of enzymes have over the process that leads to the activation of zymogens?
41. Identify the meaning of the word *allosteric* when it refers to enzymes.
42. Define allosteric regulation of enzyme activity and provide a real or hypothetical example as an illustration.
43. Define positive and negative effectors.
44. How are metabolic pathways regulated? Where in the pathway does this occur?
45. Provide a definition of feedback inhibition. With respect to enzyme catalyzed reactions, does this form of regulation require a positive or negative effector?
46. What name is given to different proteins that have the same catalytic activity?
47. Why are there different forms of lactate dehydrogenase (LDH)?

General Exercises

48. Explain how a study of enzyme concentrations in blood can be used to diagnose a heart attack.
49. Identify the role of tissue-type plasminogen activator in the treatment of heart attacks.

Challenge Exercises

50. You have isolated an enzyme from bacteria through a series of steps that removed all other proteins and small molecules. Your enzyme was active during the first steps but got less and less active as the purity increased. In its pure form, it has no activity. Curiously, full activity is restored if a small amount of the ground-up bacteria is added. Explain these observations.
51. A number of human genetic diseases have been traced to a missing or nonfunctional enzyme. Go to your library or use the Internet to identify three or more diseases with this cause, and identify the missing enzyme for each.

Heredity

16

MOLECULES OF DNA AND RNA CARRY INFORMATION AND DIRECT CELLULAR ACTIVITIES.

LIVING ORGANISMS are unique in that they alone have the ability to reproduce. "Like begets like" is a biblical phrase that shows this fact was known to ancient peoples. Yet variability is present within this constancy—cats beget cats, true, but the kittens differ from each other. We see this in humans, too. Sets of identical twins aside, people are distinctly different from one another in appearance. What is the molecular basis for these similarities and differences? Genetic (inherited) information is stored in deoxyribonucleic acid—a molecule better known as DNA. This chapter explains the composition, structure, and function of DNA and also examines the other molecules that are involved in the use and expression of genetic information.

Outline

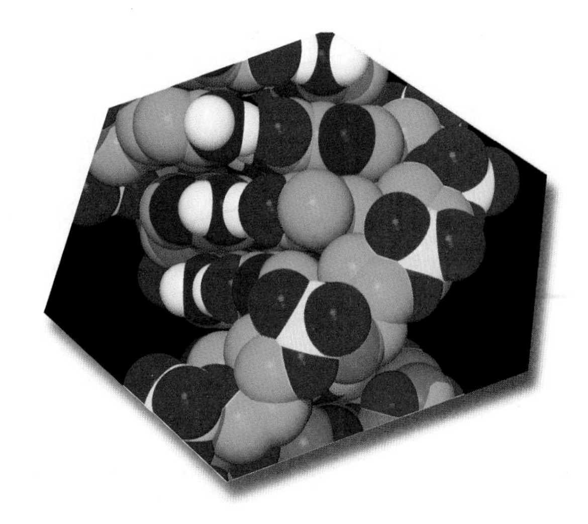

Genetic information is stored in molecules of DNA. ***(Professor K. Seddon and Dr. T. Evans/Science Photo Library/Photo Researchers, Inc.)***

16.1 Nucleotides

Nucleic acids: polymers of nucleotides

Nucleic acids are macromolecules that were first isolated from the *nuclei* of cells. Deoxyribonucleic acid (DNA) and the other nucleic acids are polymers of nucleotides, and so we begin our study of DNA with a discussion of these monomers. A **nucleotide** is a molecule that contains a nitrogenous (nitrogen-containing) base, a sugar, and one or more phosphate groups. Although there are many different nucleotides, the structure and composition of the most common one, adenosine triphosphate (ATP), is typical (Figure 16.1).

Nucleotide: sugar + nitrogenous base + phosphate(s)

A **nitrogenous base** is a molecule containing two or more nitrogen atoms in either a simple ring or a fused ring. There are five common nitrogenous bases found in the nucleotides of nucleic acids: *adenine*, *guanine*, *cytosine*, *thymine*, and *uracil* [Figure 16.2(a) and (b)]. Adenine and guanine contain the fused-ring system found in the compound *purine*, whereas the ring in cytosine, thymine, and uracil resembles the compound *pyrimidine*. Because of the ring or rings they contain, adenine and guanine are called **purines** and the other three nitrogenous bases are called **pyrimidines.**

The nitrogenous base of any nucleotide is bound to a sugar through one of the ring nitrogen atoms of the base. The sugars commonly found in nucleotides are ribose and 2-deoxyribose [Figure 16.2(c)]. The sugar in turn is bound to a phosphate group through a phosphoester bond (bond between a hydroxyl group and phosphoric acid). In biochemistry, a phosphate group is simply phosphoric acid as it exists at cellular pH values [Figure 16.2(d)].

Nucleoside: sugar + nitrogenous base

The names of nucleotides are based on the names of nucleosides. A **nucleoside** is a molecule made up of a sugar and a nitrogenous base, and the name of a nucleoside is based on these two components. The nucleoside containing the sugar deoxyribose and the nitrogenous base adenine is called deoxyadenosine; the one containing the sugar ribose and the nitrogenous base guanine is called guanosine (Figure 16.3). Thus the nucleotide made up of the nucleoside adenosine plus three phosphate groups is called adenosine triphosphate (look at Figure 16.1 again). Table 16.1 summarizes the nomenclature of some common nucleosides and nucleotides.

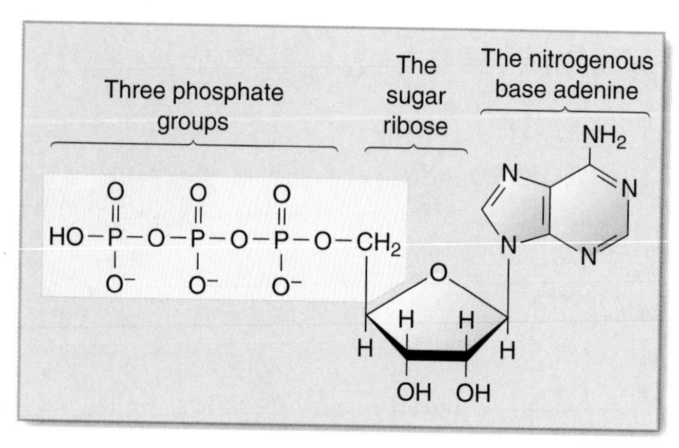

Figure 16.1 Adenosine triphosphate (ATP) is a typical nucleotide.

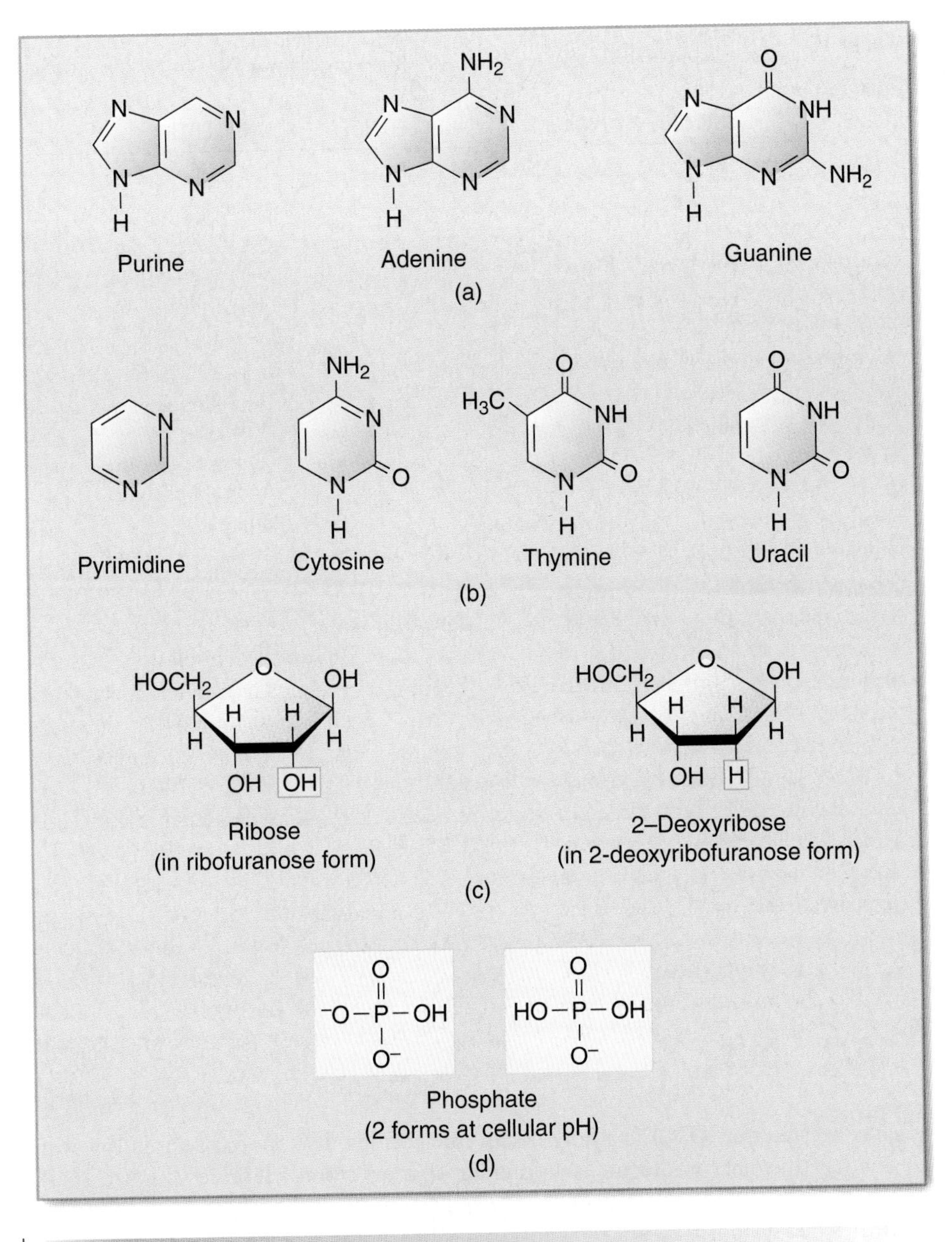

Figure 16.2 The components of nucleotides. **(a)** Purine and the nitrogenous bases adenine and guanine. **(b)** Pyrimidine and the nitrogenous bases cytosine, thymine, and uracil. **(c)** Ribose and 2-deoxyribose are the common sugars found in nucleotides. **(d)** Phosphate groups are phosphoric acid as it exists at cellular pH.

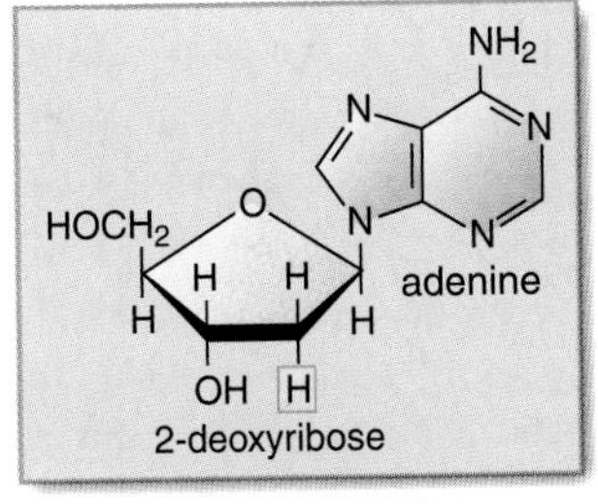

Deoxyadenosine

guanine
ribose

Guanosine

Figure 16.3 Nucleosides contain a nitrogenous base and a sugar.

Self-Test

Nucleotides

1. Name the nitrogenous bases that are **(a)** purines and **(b)** pyrimidines.
2. A compound that contains a nitrogenous base and a sugar belongs to what class of compounds?
3. Name the compound that contains the sugar ribose, the nitrogenous base uracil, and two phosphate groups.

ANSWERS

1. **(a)** Adenine, guanine **(b)** Cytosine, thymine, uracil 2. Nucleoside
3. Uridine diphosphate (UDP)

Table 16.1 Names of Common Nucleosides and Nucleotides

NITROGENOUS BASE	SYMBOL	SUGAR	NUCLEOSIDE	NUCLEOTIDE
Adenine	A	Deoxyribose	Deoxyadenosine	Deoxyadenosine monophosphate (dAMP)[a]
Guanine	G	Deoxyribose	Deoxyguanosine	Deoxyguanosine monophosphate (dGMP)
Cytosine	C	Deoxyribose	Deoxycytidine	Deoxycytidine monophosphate (dCMP)
Thymine	T	Deoxyribose	Deoxythymidine	Deoxythymidine monophosphate (dTMP)
Adenine	A	Ribose	Adenosine	Adenosine monophosphate (AMP)
Guanine	G	Ribose	Guanosine	Guanosine monophosphate (GMP)
Cytosine	C	Ribose	Cytidine	Cytidine monophosphate (CMP)
Uracil	U	Ribose	Uridine	Uridine monophosphate (UMP)

[a]The d indicates that the sugar in the nucleotide is *d*eoxyribose.

16.2 DNA

Earlier we said that DNA is a polymer of nucleotides. It is more precise, however, to say that this macromolecule is a polymer of deoxyribonucleotides (Figure 16.4), because **DNA** is an abbreviation for ***deoxyribo*****nucleic acid.** The name clearly identifies deoxyribose as the sugar found in DNA.

How are the nucleotides arranged in DNA? In other words, what is the structure of DNA? Often the structure of a biomolecule provides important clues about its function, and in the case of DNA, much of its function was learned once its structure became known.

The Sugar-Phosphate Backbone of DNA

The nucleotides in DNA and other nucleic acids are linked by phosphoester bonds between the sugar of one nucleotide and the phosphate group of another. Because each nucleotide in a nucleic acid contains both a sugar and a phosphate group, each nucleotide can form two of these phosphoester bonds. The result is a linear chain of alternating sugars and phosphate groups with a nitrogenous base sticking off of

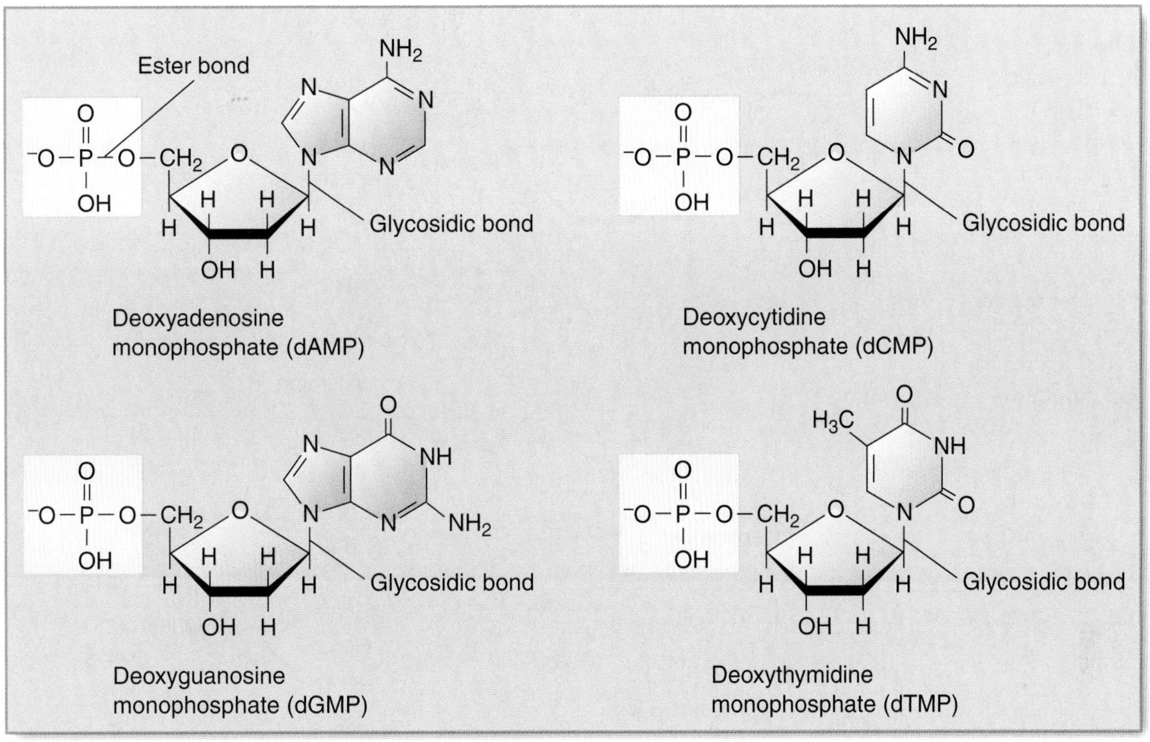

Figure 16.4 The deoxyribonucleotides found in DNA.

each sugar (Figure 16.5). The linear chain of alternating sugar and phosphate groups is called the *backbone* of the DNA molecule, and the whole thing—the backbone and attached nitrogenous bases—is called a *strand* of DNA. The end of the DNA strand that has a free hydroxyl group on carbon 5 of deoxyribose (5′ hydroxyl) is called the 5′ end of the strand; the 3′ end of the DNA strand has a free 3′ hydroxyl group on the deoxyribose.

The Double Helix

A complete DNA molecule is more than just a single strand, however. In the early 1950s, the British scientists Rosalind Franklin and Maurice Wilkins determined the dimensions of DNA. From these and other data, James Watson and Francis Crick proposed that DNA consists of two strands wrapped around each other to form a *double helix*. In the double helix, the phosphate–sugar backbones of the strands are on the outside of the helix and the nitrogenous bases project into its core (Figure 16.6).

What forces hold the two strands together in the double helix? They are held together (a) by hydrogen bonds between specific bases and (b) by nonspecific hydrophobic interactions. Individually, both types of interactions are weak, but there are thousands and thousands of hydrogen bonds present in a DNA molecule, and the hydrophobic interactions add more stability.

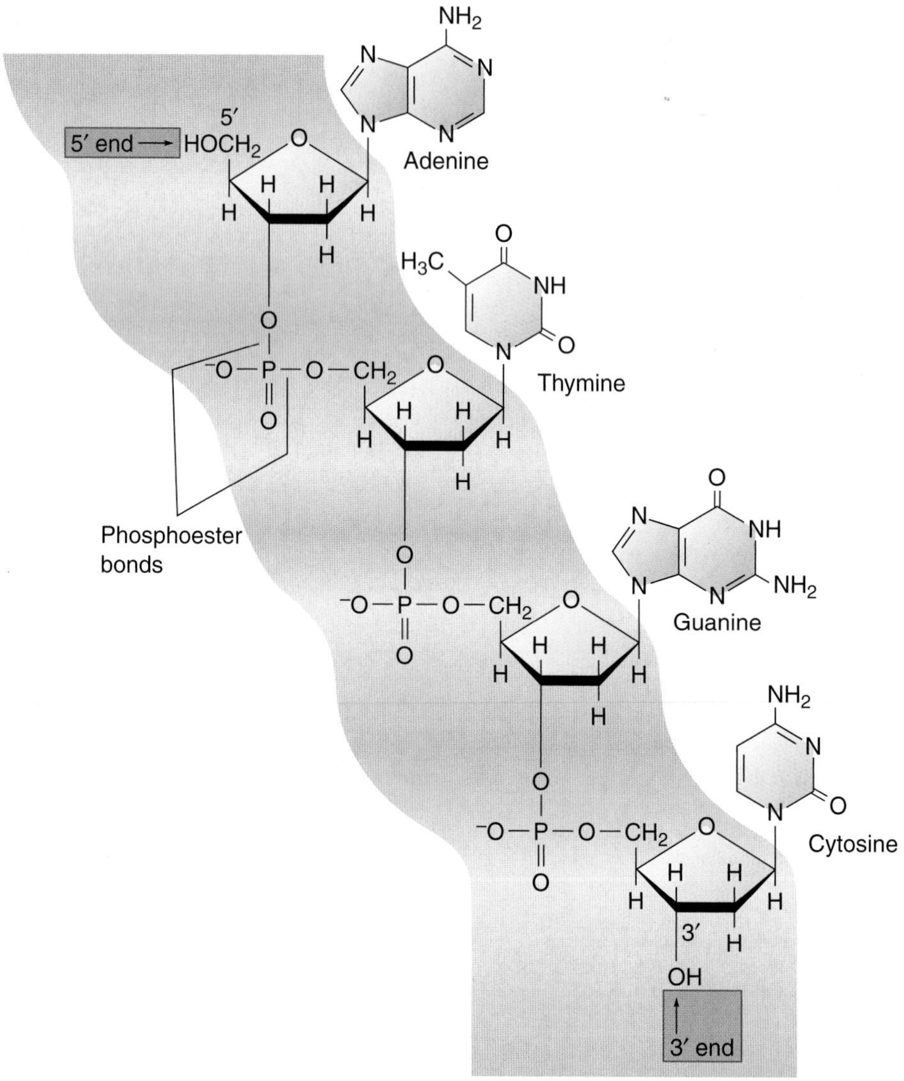

Figure 16.5 In DNA, deoxyribonucleotides are bound to each other through phosphoester bonds. This bonding pattern forms a linear polymer of DNA called a strand. The two ends of a strand are different from one another. One end has a free 5′ hydroxyl group on the sugar, and for this reason this end is called the 5′ end. The other end has a free 3′ hydroxyl group on the sugar, thus this is the 3′ end.

Let's take a closer look at the hydrogen bonding that occurs between strands. In DNA, adenine (A) pairs with thymine (T) through two hydrogen bonds, and guanine (G) pairs with cytosine (C) through three hydrogen bonds:

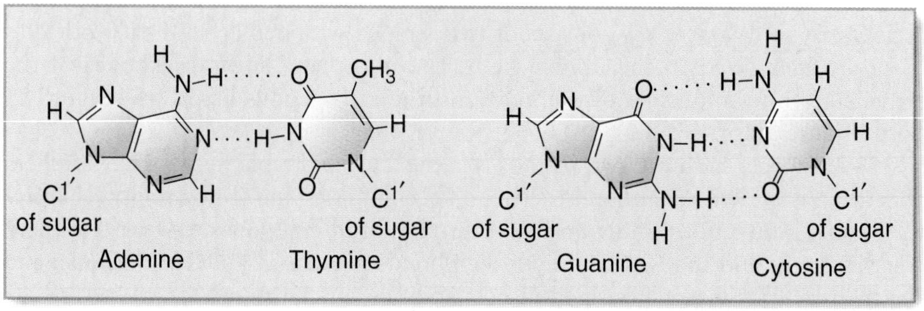

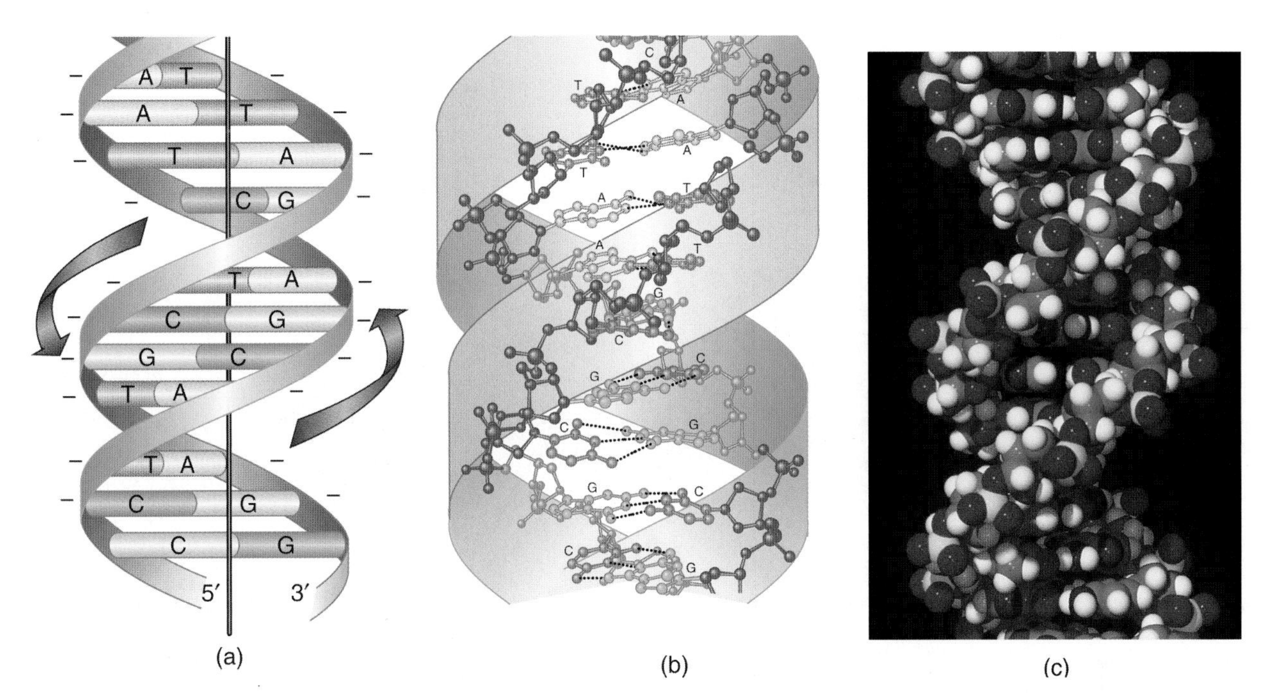

Figure 16.6 The double helix. **(a)** In this model the two sugar-phosphate backbones are represented by ribbons, and the bases are shown as sticks that meet in the core of the helix. The bases are labeled with their standard abbreviations. **(b)** The double helix is shown as a ball-and-stick model, with the hydrogen bonds between pairs of nitrogenous bases shown. **(c)** A molecular model of DNA, showing the double helix. *(Kenneth Eward/Biografx-Science Source/Photo Researchers, Inc.)*

Bases that pair with each other are called **complementary bases.** It is reasonable to ask why only certain bases pair with each other. These pairings occur because of the size and shape of DNA and because of the size and shape of nitrogenous bases. The width of the double helix core is just large enough to hold one purine and one pyrimidine. Two purines are too large, and two pyrimidines are too small. Thus the purine adenine (A) does not pair with the purine guanine (G), and the pyrimidine cytosine (C) does not pair with the pyrimidine thymine (T). Furthermore, in the core, adenine and cytosine do not have the right shapes to hydrogen bond effectively. The same is true for guanine and thymine. In DNA, therefore, the pairing of A with T and G with C is nearly absolute.

Need a memory aid? AT spells a word; G and C look very much alike.

Higher Levels of DNA Structure

As described so far, DNA consists of two long strands wrapped into a double helix. However, DNA is not found in this form in living cells. Bacterial DNA molecules are greater than 1 mm in length, and yet the typical bacterial cell is in the 1- to 10-μm range (less than 1/100 as long as the DNA). The bacterial DNA molecule is not just shoved into the cell like thread into a thimble; instead, it is arranged in an orderly, compact form in which the ends of the molecule are covalently linked to each other to form circular DNA (Figure 16.7). In addition, this circular DNA

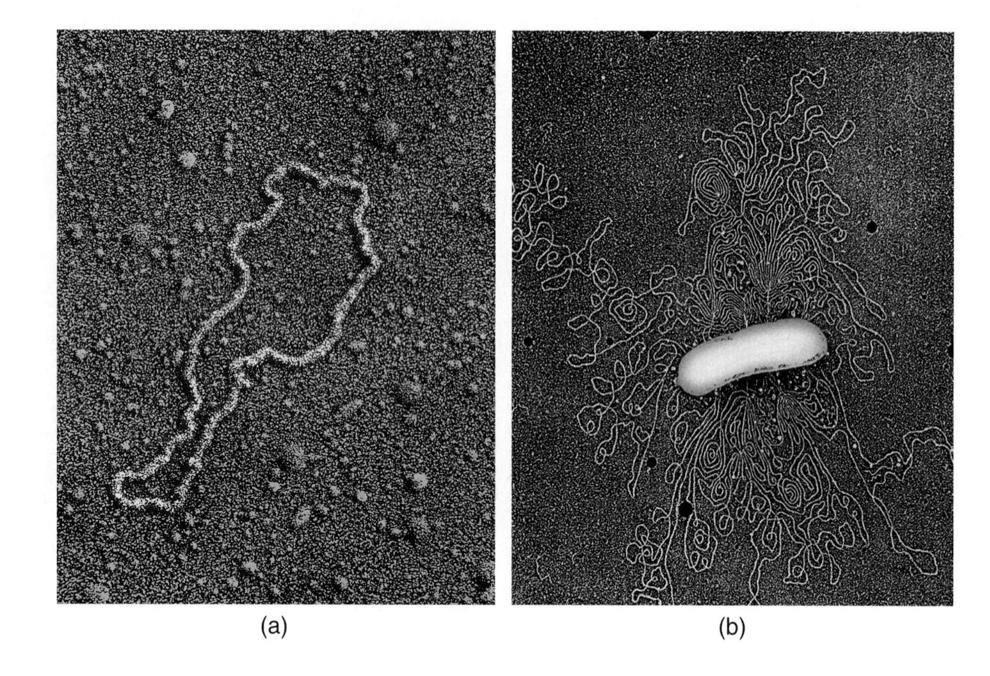

Figure 16.7 (a) Bacterial DNA is circular. (b) The supercoiled DNA spilling from this ruptured bacterial cell is much longer than the cell. *(a, Professor Stanley N. Cohen/Photo Researchers, Inc.; b, Dr. Gopal Murti/Science Photo Library/Photo Researchers, Inc.)*

molecule is supercoiled, which means it is wrapped around itself to yield a more compact structure. This arrangement allows the molecule to fit into the cell and allows the DNA to function normally.

The cells of higher organisms contain several to many long, linear DNA molecules. Even though these DNA molecules typically have a combined length of more than 1 m, the DNA is contained in a nucleus that is much less than 1 mm in diameter. The DNA molecules fit into the nucleus because they are complexed with histones to form nucleosomes that cluster to form **chromosomes** (Figure 16.8). Chromosomes are much more compact than linear DNA, and yet they still allow for all of the normal activities of DNA.

Replication of DNA

DNA stores the genetic information used to control the activities of a cell and, ultimately, the activities of a multicellular organism. In an organism, all cells contain the same genetic information, the same DNA. When body cells divide, each daughter cell gets a copy of DNA that is nearly always exactly like the DNA of the parent cell. Thus, with rare but important exceptions, the capabilities of daughter cells are identical to the parent cell from which they arose.

The structure of any molecule, including DNA, determines the properties and functions of that molecule. How can two identical DNA molecules (one for each daughter cell) arise from a single DNA molecule (that of the parent cell)? The answer to this question is found in DNA's structure. Because DNA contains complementary bases and is double-stranded, it follows that the two strands are also complementary. This is true because the base sequence of one strand is directly related to the base sequence of the other. Because the bases pair specifically—A with T and G with C—a sequence of bases on one strand *requires* a complementary set of bases on the other strand. If the bases A, T, C, and G are next to one another on one strand, then T, A, G, and C must be next to each other on the other strand.

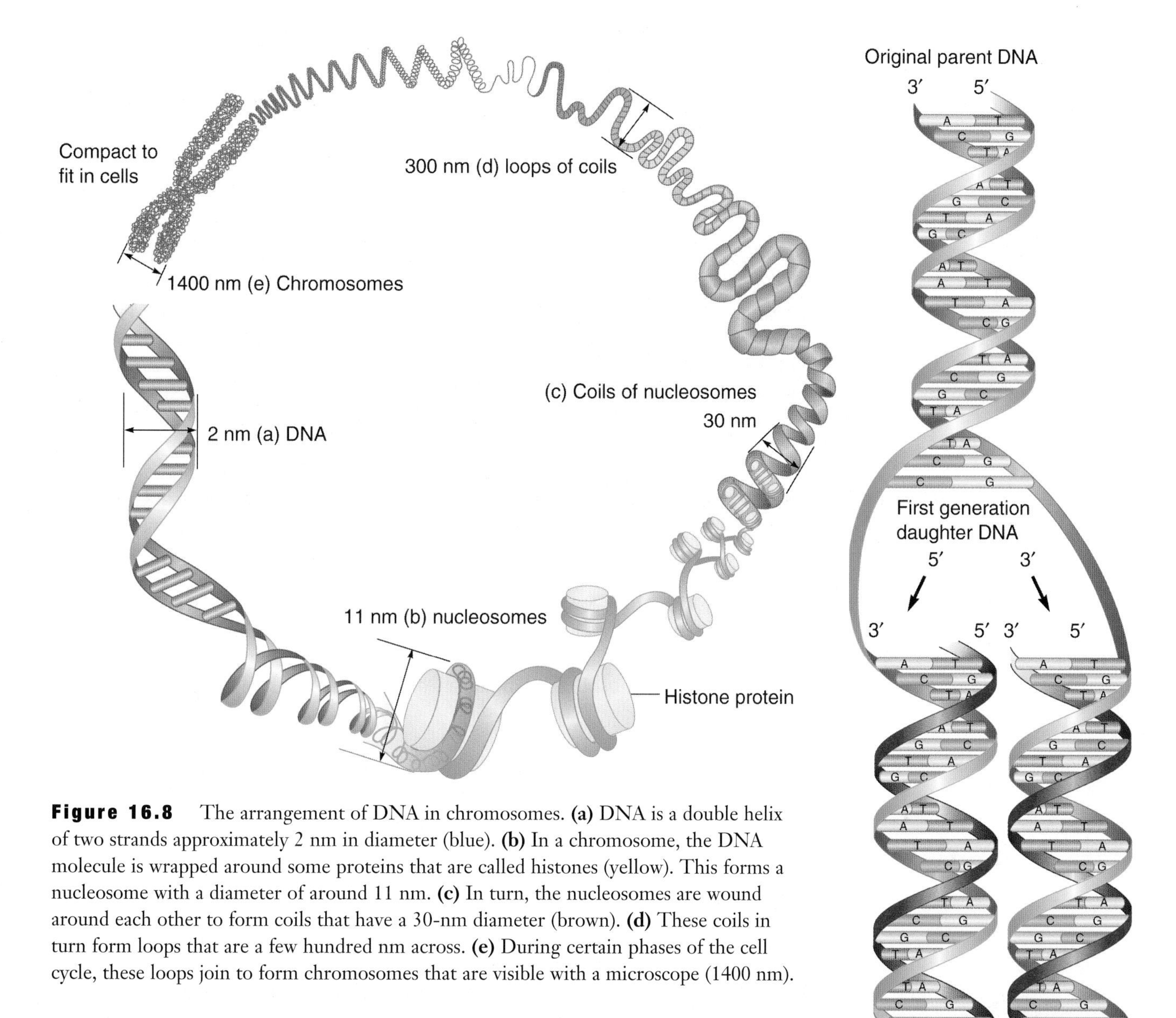

Figure 16.8 The arrangement of DNA in chromosomes. **(a)** DNA is a double helix of two strands approximately 2 nm in diameter (blue). **(b)** In a chromosome, the DNA molecule is wrapped around some proteins that are called histones (yellow). This forms a nucleosome with a diameter of around 11 nm. **(c)** In turn, the nucleosomes are wound around each other to form coils that have a 30-nm diameter (brown). **(d)** These coils in turn form loops that are a few hundred nm across. **(e)** During certain phases of the cell cycle, these loops join to form chromosomes that are visible with a microscope (1400 nm).

Figure 16.9 Semiconservative replication of DNA. During this process, each of the original strands of the parent DNA molecule (blue) is paired with a new strand (red) in the daughter molecules. Note that each base of the new strands is complementary to the corresponding base of the original strands.

When two strands of a DNA molecule are separated and a new strand is made from each of them, each new strand is complementary to the old strand with which it is now paired (Figure 16.9). This process is called *semiconservative* **replication,** because one strand of the original DNA molecule is conserved and passed on to each daughter DNA molecule. Through semiconservative replication, the two new DNA molecules are identical in composition and sequence to the parent DNA molecule from which they were made. Each daughter cell thus has the same genetic information that was in the parent cell.

Although semiconservative replication is basically the separation and duplication of the two strands of DNA, two factors make the process more complicated: (1) The DNA molecule is very large and usually associated with proteins, and (2) the replication must be very accurate to get every base paired with its complementary base.

Replication begins at a specific site on the DNA molecule called the *origin of replication.* In bacteria, there is only one origin of replication per DNA molecule. Higher organisms have several origins per chromosome to allow for rapid replication. At an origin of replication, the DNA strands are separated, and the enzyme DNA polymerase binds and catalyzes the synthesis of the new strands (Figure 16.10).

When a DNA polymerase binds to a strand of DNA (the template strand) it also binds two appropriate deoxyribonucleotides. That is, it binds the two nucleotides that are complementary to the two bases on the template strand that are in the active site of the polymerase (Figure 16.11). The enzyme then catalyzes the formation of a phosphoester bond between these deoxyribonucleotides. The polymerase then moves down the template strand by one more base (toward the 5′ end of the template strand) and again binds the deoxyribonucleotide that is complementary to the template-strand base that just entered the active site. The enzyme next forms a phosphoester bond between the new deoxyribonucleotide and the previous deoxyribonucleotide. The polymerase moves down the template strand one base at a time and binds the correct deoxyribonucleotide, adding it to the new strand. This process continues over and over until synthesis of that part of the new strand is complete.

Synthesis of the new strand is catalyzed by several DNA polymerase molecules (Figure 16.10). As a result, there are gaps on the new strand where one polymerase finished and another began. Another enzyme—polynucleotide ligase—catalyzes the formation of phosphoester bonds at these gaps to link them together to form the complete strand.

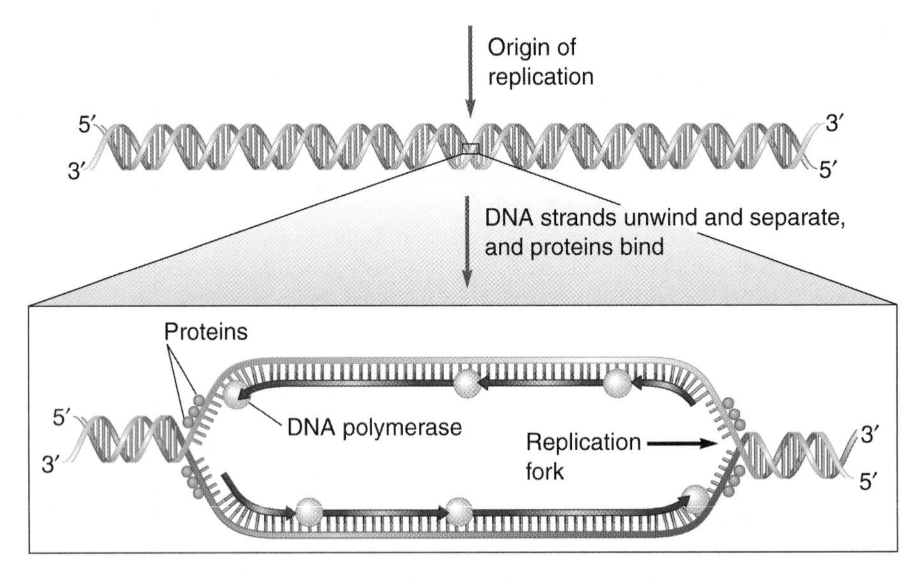

Figure 16.10 At an origin of replication, proteins (green) separate the DNA strands to yield two replication forks. At and near the forks, DNA polymerase molecules (yellow) bind to the DNA strands and catalyze the synthesis of a new DNA strand (red) complementary to the original strand (blue).

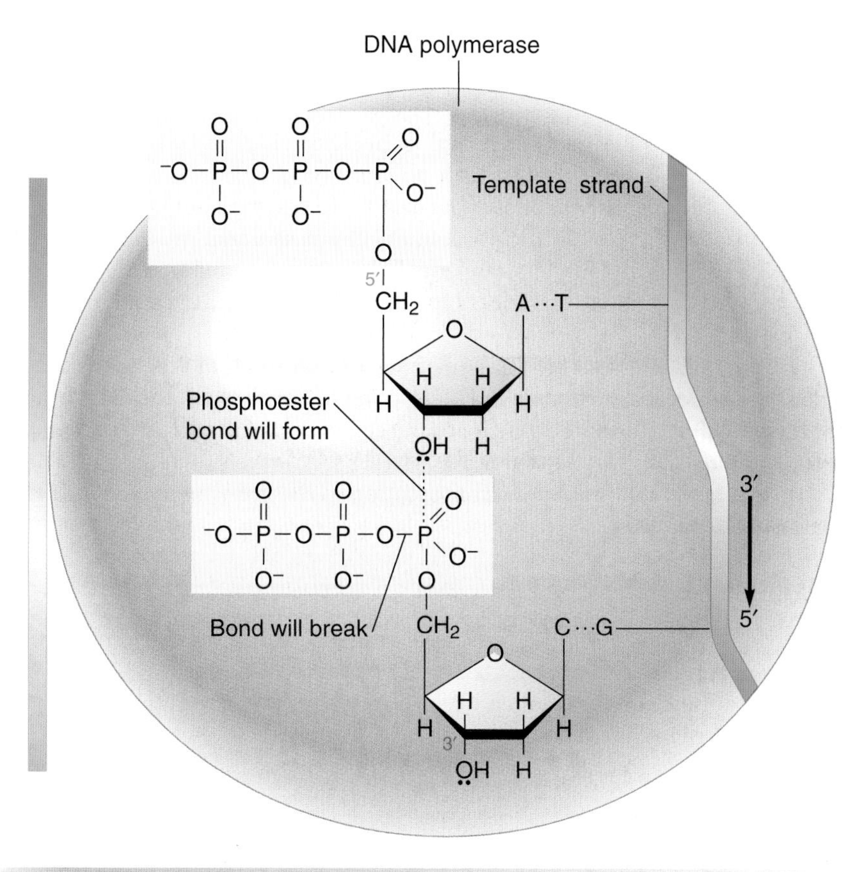

Figure 16.11 The role of DNA polymerase in semiconservative replication. The template strand of DNA and complementary deoxyribonucleotides bind to the active site of DNA polymerase.

Self-Test

DNA

1. What kind of bonds link deoxyribonucleotides in a strand of DNA?
2. A short segment of a DNA strand has the sequence AACTGGC. What is the sequence of the complementary strand?
3. What two types of interactions hold two DNA strands together?

ANSWERS

1. Phosphoester bonds 2. TTGACCG 3. Hydrogen bonds between complementary bases and hydrophobic interactions in the core of the double helix

After DNA replication, each daughter cell gets a complete DNA molecule that stores genetic information in that cell. But DNA does more than store information; it is also used to *express* genetic information. This expression occurs first through *transcription* to form RNA, which is then *translated* to make proteins. The flow of genetic information in cells can be summarized as

$$\text{DNA} \xrightarrow{\text{Replication}} \text{DNA} \xrightarrow{\text{Transcription}} \text{RNA} \xrightarrow{\text{Translation}} \text{Proteins}$$

We will study the processes of transcription and translation in the next few sections.

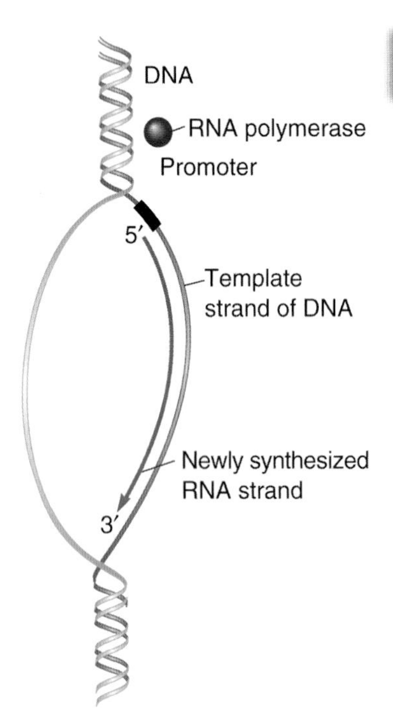

Figure 16.12 Transcription. RNA polymerase binds to the template strand of DNA at a promoter site and catalyzes the linking of ribonucleotides that are complementary to the bases of the DNA strand. Note that in transcription uracil (U) is the base complementary to adenine (A) rather than thymine (T).

16.3 RNA

Besides DNA, cells contain another kind of nucleic acid; this one called ***ribonu*cleic *a*cid,** RNA. This nucleic acid is a polymer of nucleotides that contain the sugar *ribose* rather than deoxyribose, but there are several additional differences between RNA and DNA. First, RNA contains the base uracil instead of thymine. Second, RNA is usually single-stranded. Third, it is a much smaller molecule than DNA. The structural differences between DNA and RNA are summarized in Table 16.2.

There are three classes of RNA. **Messenger RNA (mRNA)** functions as a carrier (messenger) of information from DNA to the rest of the cell. **Ribosomal RNA (rRNA)** is found in ribosomes, which are the cell components (organelles) responsible for protein synthesis. **Transfer RNAs (tRNA),** which function during protein synthesis, transfer amino acids from solution to the growing polypeptide chain during protein synthesis.

Transcription

RNA is made from DNA by the process of **transcription** using a strand of DNA as a *template* or guide. The enzyme that catalyzes this synthesis is RNA polymerase, an enzyme that functions very much like DNA polymerase, except it binds ribonucleotides rather than deoxyribonucleotides. RNA polymerase works by first binding to one strand of the DNA molecule—the template strand—at a sequence of bases called the *promoter*; then it binds and joins ribonucelotides that are complementary to the bases on the DNA template strand (Figure 16.12). One by one the ribonucleotides are bound and joined until a strand of RNA has been made that is complementary to that part of the DNA template strand. Because the bases in the RNA strand are complementary to those in the DNA strand, the information in the DNA strand has been *transcribed* and is now stored in the strand of RNA. It is because this RNA strand carries information that it is called messenger RNA (mRNA).

Transcription occurs in the cell nucleus, but the mRNA must be transported out of the nucleus to where it is used to make proteins through translation (Figure 16.13.) The organelles called *ribosomes* carry out translation and they do this in both the cytosol and in the membranous organelle called the *endoplasmic reticulum.*

Messenger RNAs, ribosomal RNAs, and transfer RNAs are all made by transcription. For each of them, transcription begins at a different promoter site on the DNA. Except for mRNA in bacteria, each product of transcription (this product is called the **primary transcript**) must be modified before it is functional. For an mRNA of a higher organism, the ends of the primary transcript are modified, and

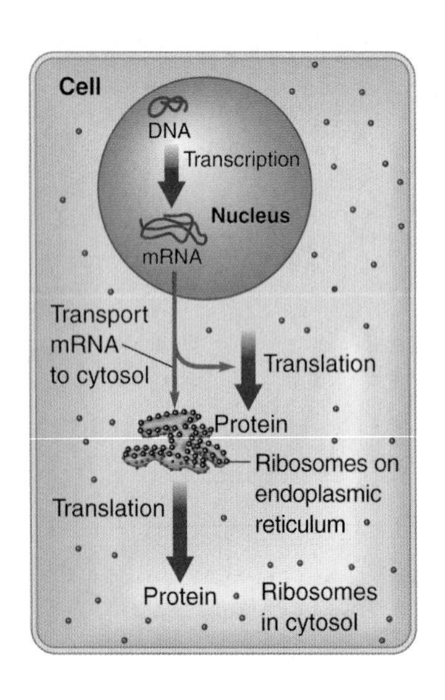

Figure 16.13 mRNA produced in the nucleus by transcription is transported from the nucleus to the cytosol and endoplasmic reticulum where it is translated by ribosomes to make proteins.

Table 16.2 The Structural Differences Between the Nucleic Acids DNA and RNA

NUCLEIC ACID	SUGAR	BASES	NUMBER OF STRANDS	SIZE
DNA	Deoxyribose	T, A, C, G	Two	Largest molecule in cell
RNA	Ribose	U, A, C, G	One	Macromolecules, but much smaller than DNA

nonfunctional pieces are removed from the strand. Ribosomal RNAs are formed by cutting a specific primary transcript into smaller pieces that are functional rRNA molecules. Transfer RNAs are also formed by cutting a specific primary transcript into smaller units, but in this case the cutting is only a first step. Some of the nucleotides in the units are then modified to yield functional tRNA.

Connections

Viruses

Viruses are the causative agent of some diseases. A virus is a tiny particle composed of a core of nucleic acid, a protein coat surrounding the core, and in some viruses a membranous envelope surrounding the protein coat. Although all cellular organisms use double-stranded DNA to store genetic information, viruses use single- or double-stranded DNA, or single- or double-stranded RNA. An example of a double-stranded RNA virus is the human immunodeficiency virus (HIV), which causes acquired immune deficiency syndrome (AIDS).

Viruses are inactive (dormant) outside their host cell. When a virus infects a host cell, it can enter another dormant stage or use some of its own enzymes and some host enzymes to make viral proteins and viral nucleic acids. These molecules then self-assemble to form mature viral particles that leave the host cell and infect other cells. The host cells for the HIV virus are certain white blood cells—T lymphocytes—that are involved in the body's immune response. As more and more T lymphocytes are infected and destroyed, the patient's normal immune response becomes compromised and can no longer protect against infectious diseases or some cancers.

Viral diseases are generally difficult to treat or cure for several reasons. First, viruses in a host spend most of their time within host cells where they avoid the white blood cells and antibodies of the immune system. Second, some viral genes for the protein coat mutate rapidly, which makes the protein coat unrecognizable to the immune system. Furthermore, viruses often use host enzymes for much of their transcription, translation, and replication, which means that any drug that affects these functions will damage or kill the host's cells along with the virus.

Research, however, sometimes leads to cures or partial control of viral disease. For example, during its replication cycle HIV uses a viral protease to cut large polypeptide chains into smaller functional enzymes and proteins. Researchers have found inhibitors of this enzyme, which interfere with HIV's replication cycle in patients. Unfortunately, the patient is not cured of HIV, but the progress of the disease is greatly slowed.

HIV viruses budding from a T-lymphocyte cell (pink). The protein coat is red, and the RNA core is green. *(NIBSV/Science Photo Library/Photo Researchers, Inc.)*

Self-Test

RNA

1. List four differences between DNA and RNA.
2. Which bases in a strand of DNA are complementary to the RNA segment -A---U---G---C-?
3. What enzyme is responsible for RNA synthesis?

ANSWERS

1. DNA contains the sugar deoxyribose, RNA contains ribose; DNA is double-stranded, RNA is single-stranded; DNA has the base thymine, RNA has the base uracil; DNA is the largest molecule in cell, RNA is much smaller. 2. TACG 3. RNA polymerase

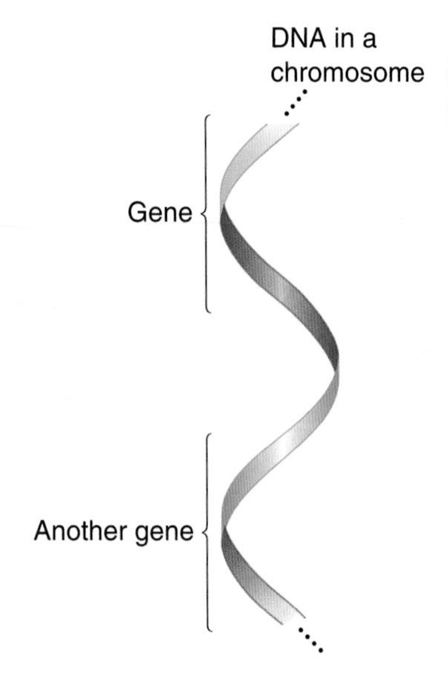

Figure 16.14 A gene is a part of a DNA molecule that codes for a particular polypeptide.

16.4 The Genetic Code

As mentioned earlier and as we shall see in more detail in a moment, translation is the process that uses information stored in mRNA to synthesize proteins. The information in mRNA comes from a **gene,** which is any particular portion of a DNA molecule that contains the information needed to synthesize a particular polypeptide chain (Figure 16.14). Information in a gene is stored in the unique sequence of bases in that part of the DNA, and transcription yields a mRNA with a unique, information-containing sequence of bases. This sequence, in turn, determines the amino acid sequence in the synthesized polypeptide chain.

How does the sequence of bases in a mRNA molecule provide the correct information for the synthesis of a protein? An analogy may be useful. The 26 letters in the English alphabet are used to make thousands and thousands of different words. Different combinations and different orders have different meanings, as CAT, RAT, and TAR show. Similarly, the four bases in mRNA—A, U, G, and C—are arranged in various combinations that are then used as signals that code for the amino acids that must go into a polypeptide chain during protein synthesis.

Codons are three-base sequences in mRNA that give rise to specific amino acids or signal start or stop.

How do the four bases in mRNA code for the 20 amino acids found in proteins? It cannot be one base for one amino acid (that would only give signals for four amino acids), nor will two bases for each amino acid work because there are only 16 combinations of four bases taken two at a time. It turns out that the bases in mRNA are grouped into *threes,* which yields 64 possible combinations (Table 16.3). Each three-*base* signal, or "word" (for example, AUC or GAG) is called a **codon.** If there are 64 codons and only 20 amino acids, do only 20 codons serve as signals for the 20 amino acids? Are the other codons meaningless, just as the letter combinations ATR and TRA are meaningless in English? Actually, all of the codons have meaning. Most amino acids have two or more codons. In addition, some codons serve as signals for starting and stopping protein synthesis.

Collectively, the roles of the codons are called the *genetic code.* All but 1 of the 64 codons code for just 1 amino acid or just 1 function. The exception is the codon AUG, which carries two messages: It is the codon for the amino acid methionine

Table 16.3 The Genetic Code

CODON	AMINO ACID	CODON	AMINO ACID	CODON	AMINO ACID	CODON	AMINO ACID
UUU	Phe	UCU	Ser	UAU	Tyr	UGU	Cys
UUC	Phe	UCC	Ser	UAC	Tyr	UGC	Cys
UUA	Leu	UCA	Ser	UAA	Stop	UGA	Stop
UUG	Leu	UCG	Ser	UAG	Stop	UGG	Trp
CUU	Leu	CCU	Pro	CAU	His	CGU	Arg
CUC	Leu	CCC	Pro	CAC	His	CGC	Arg
CUA	Leu	CCA	Pro	CAA	Gln	CGA	Arg
CUG	Leu	CCG	Pro	CAG	Gln	CGG	Arg
AUU	Ile	ACU	Thr	AAU	Asn	AGU	Ser
AUC	Ile	ACC	Thr	AAC	Asn	AGC	Ser
AUA	Ile	ACA	Thr	AAA	Lys	AGA	Arg
AUG	Met	ACG	Thr	AAG	Lys	AGG	Arg
GUU	Val	GCU	Ala	GAU	Asp	GGU	Gly
GUC	Val	GCC	Ala	GAC	Asp	GGC	Gly
GUA	Val	GCA	Ala	GAA	Glu	GGA	Gly
GUG	Val	GCG	Ala	GAG	Glu	GGG	Gly

Each codon either (1) codes for an amino acid or (2) serves as a start or stop signal. The start codon is shown in orange, and the stop codons are shown in blue.

and also signals a start. This codon always means methionine in protein synthesis, but in addition whenever it is preceded by some specific base sequences, it also means "start protein synthesis here".

The genetic code is universal. In all organisms, the same sequences of three bases code for the same amino acid.

Example 16.1

Which amino acids or functions correspond to these codons: GAA, CGC, UAG, AUG?

SOLUTION

Table 16.3 tells you that GAA corresponds to glutamic acid (Glu), CGC corresponds to arginine (Arg), and UAG is a "stop" codon. The dual-purpose codon AUG corresponds to the amino acid methionine and is also the "start" codon.

Self-Test

1. What is the function of the codons GCC, CUA, UGA?
2. **(a)** What are the codons for aspartic acid? **(b)** What codon(s) are for tryptophan?

ANSWERS

1. Code for alanine, leucine, and stop   2. **(a)** GAU, GAC   **(b)** UGG

16.5 Translation (Protein Synthesis)

The mRNA made by transcription carries the information needed to make a protein. After the mRNA is transported from the nucleus to the cytosol, it is used to make proteins by the process called **translation.** Translation involves three steps: (1) Initiation—the formation of a mRNA–ribosome–tRNA complex; (2) elongation—the sequential joining of amino acids to form the polypeptide chain; and (3) termination—the completion of protein synthesis and separation of the components of the complex formed in initiation.

Initiation

Ribosomes are the site of protein synthesis and exist as two separate subunits in the cell. In *initiation* an mRNA, the two subunits of a ribosome, and a specific tRNA join to form the initiation complex. The tRNA in this complex is called fmet-tRNA because it has a modified methionine bound to it. (Amino acids are bound to their specific tRNAs using ATP to provide the needed energy.) When the fmet-tRNA binds, it hydrogen bonds to the AUG codon of the mRNA by three bases on the tRNA. This three-base unit on the tRNA is called an **anticodon,** because the bases are complementary to those in the AUG codon on the mRNA [Figure 16.15(a)].

Elongation

Next a second tRNA with its amino acid (phenylalanine in this example) binds to the complex [Figure 16.15(b)]. The anticodon of this tRNA is hydrogen-bonded to the next codon on the mRNA, UUU, which is a codon for phenylalanine. Note that the two amino acids are now positioned next to each other in the ribosome. The tRNA molecules serve as adapters that ensure that the correct amino acids, corresponding to specific codons, are next to each other during protein synthesis. It is in this way that the correct polypeptide chain (in other words, the correct protein) is assembled. A protein in the ribosome now catalyzes the formation of a peptide bond between the two amino acids, leaving an empty tRNA and one with a peptide bound to it.

Next the ribosome is **translocated**—that is, it advances one codon along the mRNA. The empty tRNA falls from the complex, and the peptidyl-tRNA remains with its codon [Figure 16.15(c)]. Another tRNA with its amino acid binds to the next codon, a peptide bond forms, and translocation occurs again. One by one amino acids—bound to their tRNA—bind to the complex and are joined to form the growing polypeptide chain. Stated another way, the polypeptide chain is *elongated.*

Termination

When a stop codon on the mRNA enters the complex, translation terminates. This occurs because no amino acid is added and instead a ribosomal protein cleaves the polypeptide from the last tRNA and the complex falls apart [Figure 16.15(d)]. Often the polypeptide is modified following translation to yield the functional protein.

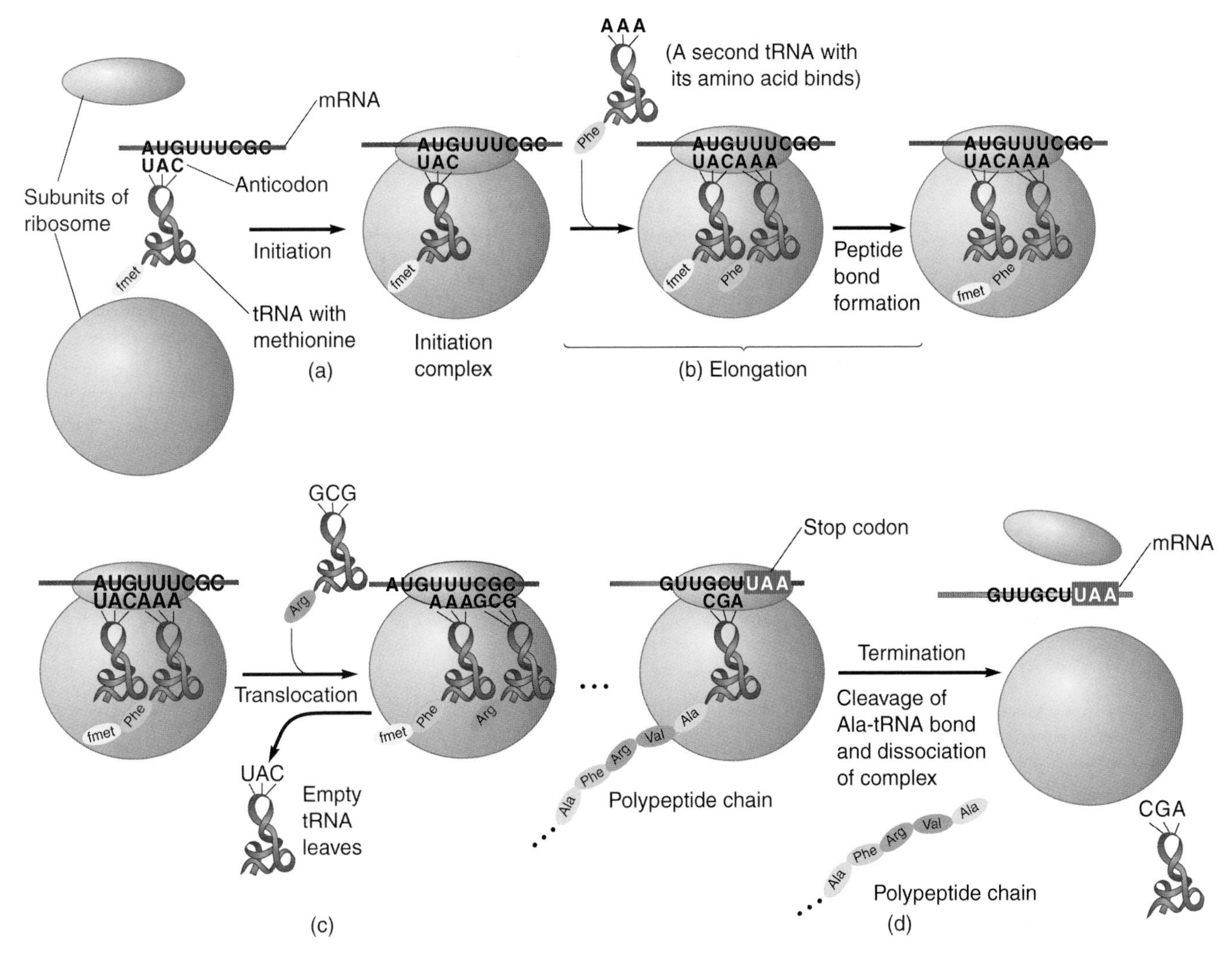

Figure 16.15 The process of translation. **(a)** Initiation. Translation starts with the formation of an initiation complex of mRNA, a ribosome, and a tRNA bound to modified methionine (fmet). **(b)** Elongation. A second tRNA with its amino acid binds, then a ribosomal enzyme forms a peptide bond between the amino acids, leaving a free tRNA and a tRNA bonded to a dipeptide. **(c)** Translocation. The ribosome and tRNA bearing the peptide move along the mRNA by one codon, the empty tRNA leaves, and elongation repeats. **(d)** Termination of translation. When a stop codon (green) is present in the ribosome-mRNA complex, an enzyme cleaves the bond between the tRNA and the polypeptide. The complex then dissociates, releasing a newly synthesized polypeptide chain into the cell.

Self-Test

Translation

1. What is the name for movement of a ribosome along an mRNA?
2. Which anticodon is complementary to the codon AGC?
3. What happens when a stop codon enters the ribosome-mRNA complex?

ANSWERS

1. Translocation 2. UCG 3. The polypeptide is cleaved from its tRNA and the complex dissociates.

16.6 Genetic Variability

Cats beget cats, and people beget people. Offspring are like their parents, but they are different as well. Clearly the individuals of a species share much genetic information, but just as clearly they have some differences. Where do these differences come from?

Ultimately these differences come from **mutations,** which are permanent changes in the DNA of a cell. Because genetic information is stored in the sequence of bases in DNA, any permanent change in the sequence is a mutation. Mutations change DNA, and these changes ultimately show up as altered proteins in the organism.

Mutagenesis is the process that leads to the formation of mutations. Mutations can arise spontaneously—that is, without apparent outside help—or they can be induced. Let's look at both types.

The young of a species are recognizable as members of that species, and yet differences are evident as well. *(Elyse Lewin/Image Bank)*

Example 16.2

Name the children shown in the margin.

SOLUTION:

Just kidding, but think about the genetic variability that leads to these differences in appearance.

Spontaneous Mutations

Replication is an important source of spontaneous mutations. Replication is a very accurate process, but it is not perfect. It is estimated that during replication an incorrect base is added less than once per one billion bases. This extremely low error rate is due to DNA polymerase. Unlike most other enzymes, DNA polymerase "proofreads" its product to make sure that the right base was inserted at the right place. After a phosphoester bond is formed, DNA polymerase checks again to see that the newest base is paired correctly to the complementary base on the parent strand. Of course, the polymerase cannot read anything, but it can sense correct geometric alignment, and if two bases are not hydrogen-bonded correctly, they will not have the correct shape. If the wrong base has been inserted, the DNA polymerase cleaves the phosphoester bond and tries again. Another nucleotide, presumably the correct one, diffuses in and the process is repeated. Because of proofreading, an error is rarely made in replication. As long as this proofreading catches the mistakes, no spontaneous mutation has occurred.

When proofreading fails, however, a spontaneous mutation will not necessarily occur, because the cell has additional protective measures. Cells possess DNA repair enzymes. These repair enzymes bind to DNA molecules and move along them. When they reach a point at which two paired bases are not complementary, the repair enzymes cut out the incorrect base, insert the proper one, and form new phosphoester bonds. The repair enzymes can generally tell the difference between the parent strand (which should be correct) and the daughter strand (where the mistake will probably be). This is because cells normally alter some of the bases of a DNA strand after the strand is synthesized. Older strands (like the parent strand) have these modifications, but new strands (daughter strands) lack them. Repair enzymes recognize the absence of modifications and work only on new strands—usually.

On very rare occasions, the repair enzymes cut out the wrong base (the one on the parent strand) and replace it with the complement of the *incorrect* base on the daughter strand. When this occurs, the wrong base pair is present in the DNA molecule. A different base pair has been substituted for the original pair, and all DNA molecules that arise from this one will have this change. A spontaneous **substitution mutation** has occurred.

What are the effects of a substitution mutation? The change of one base means that one codon has been altered [Figure 16.16(a)]. If the altered codon is one that codes for an amino acid, a different amino acid may be inserted in the polypeptide chain when new protein is being made by the cell. If this occurs, the protein will have one amino acid residue replaced (substituted) with another one. If the altered codon is a stop signal, translation will terminate when this codon enters the ribosome-mRNA complex, resulting in an incomplete protein.

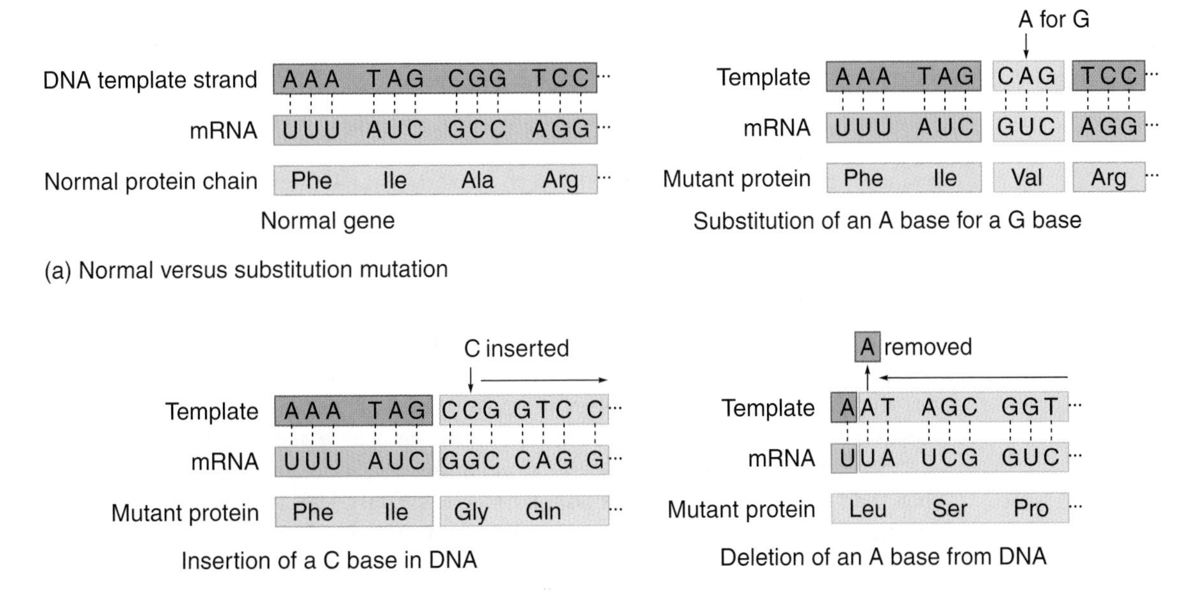

Figure 16.16 Changes in DNA structure. **(a)** Substitution of one base for another in DNA results in one altered codon in mRNA. This alteration usually means a different amino acid will be inserted into the protein. **(b)** Frameshift mutations result when a base is either inserted or deleted from a DNA strand. This type of mutation results in altered codons from that point in mRNA all the way to the end of the gene, so that many amino acids will be altered.

Induced Mutations

Mutations can be induced (caused) by a variety of agents called **mutagens.** Several forms of electromagnetic radiation are mutagenic, for example. Ultraviolet (UV) light and ionizing radiation, such as x-rays, may alter the structure of bases in DNA. If the repair enzymes don't fix this alteration properly, a permanent change results. There is a correlation between mutagenesis and *carcinogenesis*, the generation of cancer. The greater the exposure to UV light, the greater the chance of skin cancer, perhaps because one or more mutations in skin cells lead to uncontrolled growth of the cells.

A number of substances can cause mutations. Some of these mutagens chemically modify bases in DNA, and again the repair may not be accurate; the result is an induced substitution mutation. Other mutagens resemble a purine-pyrimidine pair, and because of this similarity can become wedged between base pairs in a DNA molecule. When this DNA molecule is replicated, a base pair could be inserted where it should not be or a pair could be left out. If this error is not corrected, an induced mutation has occurred. These mutations are called either *insertion* or *deletion* mutations. Collectively these are called **frameshift mutations,** because they shift the normal reading sequence of bases [Figure 16.16(b)]. Nearly all frameshift mutations result in defective proteins because many or most of the residues downstream from the frameshift will be affected.

Mutations—The Source of Genetic Variability

Mutations can occur in two types of cells: body cells (all cells of the body except reproductive cells), and sex cells (eggs and sperm and those cells that lead to them). The consequences of a mutation are quite different in these two cell types. A mutation in body-cell DNA can affect only the individual. The cell may divide to ultimately yield many cells carrying the same mutation, which could result in, say, a tumor, but the results are confined to that individual. If a change occurs in the DNA of a reproductive cell, however, the mutation can be transmitted to offspring. It is these sex-cell changes that ultimately yield differences in genetic information.

There is no unique sequence of DNA that can be called the DNA of a species. Instead, all the individuals of a species contain very similar, but nonidentical, DNA. This is why each kitten of a litter is unique, why each human is an individual distinctly different from all others. Unfortunately some of these genetic differences are the source of genetic diseases that are passed from generation to generation (Table 16.4).

Sometimes members of specific populations are more prone to certain disorders than are members of other populations. For example, Tay-Sachs disease, a disorder affecting lipid metabolism, is more common in Eastern European Jewish populations and their descendants than in other populations. To take another example, about 10% of the African American population carries a gene for sickle cell anemia (see Chapter 14 and the Connections that follows), and about 0.4% are afflicted with the disease.

Example 16.3

Two mutations have changed the DNA of an organism. Two mRNAs transcribed from this DNA have the codon changes GAC → GAA and UAC → UAA. What effect will this change have on the proteins synthesized from these mRNAs?

SOLUTION

Table 16.3 tells us that GAC codes for glutamate and GAA codes for aspartate. In this case, an acidic amino acid has substituted for another acidic amino acid. The protein is altered, but because the amino acids are very similar, the change in the protein may be small. In the second case, the table tells us that UAC codes for tyrosine, but UAA is a stop codon. Now protein synthesis will terminate at this codon. The protein will almost surely be nonfunctional.

Self-Test

To see how substitution, insertion, and deletion mutations can affect protein synthesis, consider this hypothetical base sequence in mRNA:

1 2 3 4 5 6 7 8 9 10 11 12 13 14
. . . C—G—A—C—U—G—G—A—U—A—A—U—C—C . . .

1. What is the corresponding amino acid sequence?
2. Replace $\overset{4}{\text{C}}$ with G. What is the new amino acid sequence?
3. Between $\overset{3}{\text{A}}$ and $\overset{4}{\text{C}}$, insert a U. What is the new amino acid sequence?
4. Delete $\overset{5}{\text{U}}$. What is the new amino acid sequence?

ANSWERS

1. Arginine-leucine-aspartate-asparagine. . . 2. Arginine-valine-aspartate-asparagine . . . 3. Arginine-serine-glycine-chain termination 4. Arginine-arginine-isoleucine-isoleucine . . .

Table 16.4 Some Human Genetic Diseases

DISORDER	CAUSE
Phenylketonuria (PKU)	An enzyme involved in converting phenylalanine to tyrosine is defective. Instead, phenylalanine is converted to phenylpyruvate and other phenylketones. (Phenylketonuria means phenylketones in the urine.) The accumulation of these phenylketones in a developing child causes neurological damage.
Galactosemia	An enzyme needed to change galactose to glucose is missing. The build-up of galactose is very serious in infants. Mental disorders as well as cataracts of the eye lens result.
Albinism	An enzyme needed for the conversion of tyrosine to a compound involved in skin color is missing. Albinism is the lack of this normal pigmentation.
Fabrey's disease	An enzyme needed to remove galactose from glycolipids is defective. This results in accumulation of the glycolipid in tissues. Death is often a result of kidney or cardiac failure due to this build-up.

Connections

Genetics of Sickle Cell Anemia

Sickle cell anemia is caused by red blood cells with an altered (sickled) shape resulting from the presence of hemoglobin S. Normal adult human hemoglobin has two alpha and two beta protein chains (Figure 14.14). Hemoglobin S differs from normal hemoglobin only in the beta chains and only in one residue: Glutamic acid in the sixth residue of normal hemoglobin is replaced by valine in hemoglobin S. This difference is the result of a substitution mutation in which a codon for glutamate changed to one for valine. The red blood cells of sickle cell anemics contain only hemoglobin S because both their genes for the beta chain of hemoglobin have the substitution mutation. Some people have only one gene for the mutated beta chain, and they are said to have sickle cell trait but not the disease.

Without modern health care, sickle cell anemia is lethal, usually resulting in death before childbearing age. Yet even though the disease is lethal, it is passed from generation to generation. For example, 4% of the people in Central Africa suffer from sickle cell anemia, and about 40% of them show the sickle cell trait. Why is the genetic information for hemoglobin S transmitted in such large numbers if people with only hemoglobin S died in the past before they had children?

Hemoglobin S in people with sickle cell trait provides a measure of protection from malaria, so that in malaria-ridden areas the transmission of these genes is advantageous. Because these people are less likely to die of malaria, they reach reproductive age and have children, some of whom also have sickle cell trait. In contrast, people with sickle cell anemia die young, and people with genes for only normal hemoglobin often die of malaria. Thus through people with sickle cell trait, the genes for hemoglobin S are maintained in human populations in malaria-infested regions of the world.

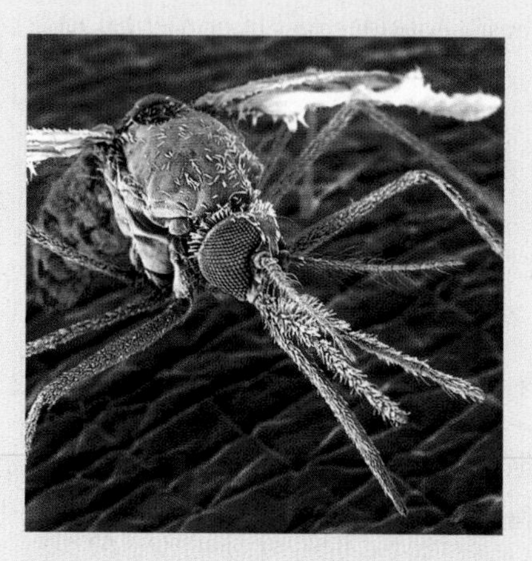

Malaria is transmitted to humans by the bite of female *Anopheles* mosquitoes. *(Oliver Meckes/Photo Researchers, Inc.)*

16.7 Regulation of Gene Activity

All body cells contain the same genetic information because they all contain the same DNA, and yet the cells of the body are not all identical to one another. Heart muscle cells have a different shape and function than liver cells. Neurons (nerve cells) resemble neither heart nor liver cells in either structure or function. Furthermore, the cells of an organism vary with time. Consider the changes that occur as a fertilized egg develops into an embryo, then a fetus, infant, child, and finally adult. How do these differences arise, and how are they maintained?

Cell structure and function result from the proteins that make up the cell. Each cell type is different because it has a different set of proteins. Some proteins, like DNA polymerase, are found in virtually all cells. Other proteins are found in only one type of cell. Hemoglobin, for instance, is found only in red blood cells and in the cells that develop into them. Furthermore, some cells can turn on or turn off the synthesis of some proteins in response to their environment.

How can cells contain different proteins when they all contain the same genetic information? Such differentiation is possible because cells regulate the synthesis of proteins by using only some of their genes at any given time. We'll concentrate our study of gene regulation on two examples from bacteria.

Induction of Gene Activity

Many bacteria can use a variety of organic compounds for energy. The intestinal bacterium *Escherichia coli (E. coli)* can grow on glucose, or on lactose, or on other compounds. When *E. coli* grow on glucose, though, the cells lack the enzymes needed to use lactose. (This makes sense; why waste resources making something you don't need?) When the cells are transferred to a medium that lacks glucose but has lactose, they begin making the enzymes needed to consume lactose. The presence of lactose has *induced* the synthesis of these enzymes. The process that turns on the synthesis of proteins when they are needed is called **induction.**

This is how lactose induction works (Figure 16.17). The genes for the lactose-using proteins are adjacent to each other on the DNA molecule. These genes are called *structural genes*, because they contain the information needed to make a protein. In Figure 16.17, a promoter is near one of the genes. (Remember, promoters are sites at which RNA polymerase binds to begin transcription.) If synthesis of the lactose-using proteins were not regulated, RNA polymerase would always bind here, transcription would yield mRNA, then the mRNA would be translated to yield the proteins. The synthesis of these proteins *is* regulated, however, because next to the promoter is a region called the *operator* site, that binds the protein called the *repressor*. The combination of promoter, operator, and structural genes is called an **operon.** This particular operon is the lactose operon, and it regulates the synthesis of lactose-using enzymes.

In the bacterium, the *lactose repressor* binds to the operator of the lactose operon [Figure 16.17(b)]. As long as the repressor is bound to the operator, RNA polymerase *cannot* bind properly to the promoter and thus transcription cannot occur. The repressor physically blocks the RNA polymerase. In the absence of lactose, the lactose repressor stays bound to the operator and the structural genes cannot be transcribed.

When some lactose enters the cell, a molecule of it binds to the repressor molecule. This changes the conformation of the repressor. In the new conformation, the repressor no longer binds to the operator, and the repressor-lactose complex leaves the DNA. Because the operator of the operon is now free, RNA polymerase molecules can bind to the promoter [Figure 16.17(c)], and transcription and translation occur to yield the enzymes needed for lactose use. Through induction, the bacterium makes these proteins only when they are needed, which is whenever there is some lactose around. Cellular resources are not wasted on unneeded proteins.

Repression of Gene Activity

Another way *E. coli* regulates protein synthesis is through **repression,** the *turning off* of protein synthesis when specific proteins are not needed. If the amino acid histidine is not available to *E. coli*, the bacterium makes it. If histidine is available, the bacterium stops synthesizing the enzymes needed for its synthesis. The presence of histidine *represses* its own synthesis.

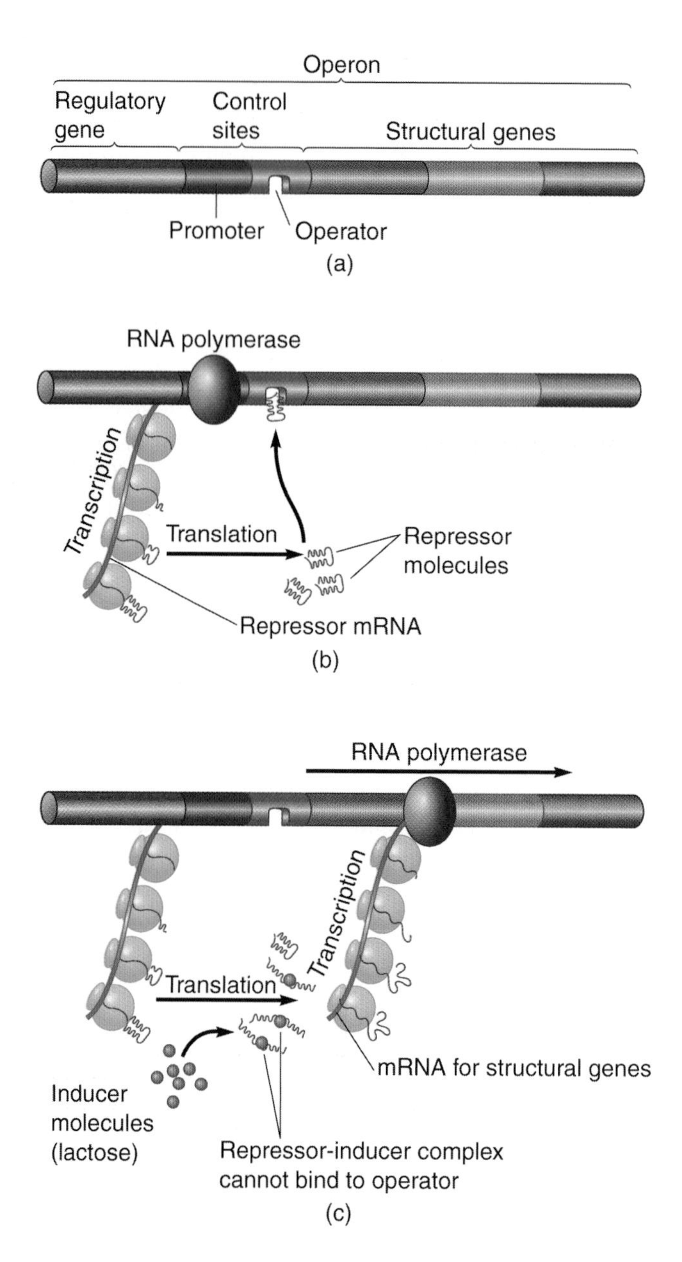

Figure 16.17 In *E. coli* bacteria, lactose in the culture medium induces the appearance of the enzymes that break down lactose. **(a)** The arrangement of the genes in the lactose operon. **(b)** When lactose is absent, the repressor binds to the operator, and this binding blocks RNA polymerase at the promoter site. As a result the structural genes needed to synthesize lactose-using enzymes cannot be transcribed. **(c)** When lactose is present, it binds to the repressor and alters its shape. The complex can no longer bind to the operator, and RNA polymerase now binds to the promoter and transcribes RNA.

Repression uses operons very similar to those of induction (Figure 16.18). The histidine operon has an operator, a promoter, the structural genes for the proteins, and a repressor. The main difference between lactose induction and histidine repression is the binding properties of the repressor. The histidine repressor does not bind to the operator by itself. It is a repressor-histidine complex that binds to the operator and thus prevents transcription. Anytime histidine is present, therefore, the bacterium does not make the histidine-synthesizing enzymes. When there is no histidine in the cell, there is no histidine-repressor complex to bind to the operator, and so RNA polymerase can bind and transcription and translation yield histidine-synthesizing enzymes.

Regulation in Higher Organisms

Regulation of gene activity also occurs in higher organisms. Genes or sets of genes are turned on or off as a fertilized egg develops into an infant, for example. Some genes are always turned on in all cells (DNA polymerase is one of many examples), some are transcribed only in one particular cell type (hemoglobin in red blood

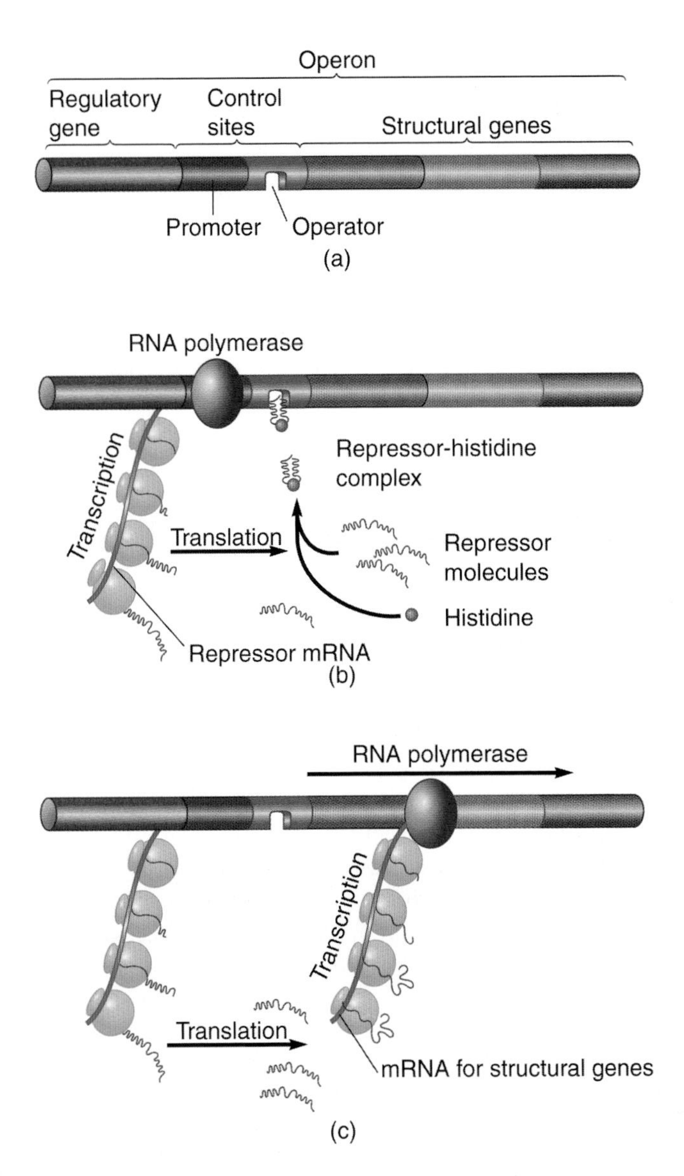

Figure 16.18 Histidine repression in *E. coli* bacteria. **(a)** The arrangement of genes in the histidine operon. **(b)** When histidine is present, it binds to the repressor and alters its shape. The repressor–histidine complex binds to the operator and prevents RNA polymerase from transcribing the structural genes. **(c)** When histidine is absent, the repressor cannot bind to the operator and RNA polymerase is free to transcribe the structural genes.

cells). Gene regulation in higher organisms is not well-understood, but researchers are sure it is not identical to the induction and repression just described. Some specific proteins do appear to bind to DNA and influence the transcription of genes, but the details of the process as we presently know it are not those of induction and repression. Gene regulation in higher organisms is of great importance in biochemistry and medicine. More research is needed to determine the details of this regulation and apply it to health care.

Self-Test

Regulation of Gene Activity

1. What is the role of the operator in an operon?
2. How does a repressor contribute to control of gene activity?
3. What are structural genes?

ANSWERS

1. The operator is the site on a strand of DNA where a repressor binds. 2. When the repressor is bound to the operator, RNA polymerase cannot bind properly to the promoter. 3. Genes whose products are proteins

16.8 Genetic Engineering

The information gathered by the Human Genome Project will contribute greatly to human genetic engineering.

During the past two decades, a technology has developed that can permanently alter DNA in very specific ways. Genes can be added to an organism, and the protein coded for in those genes will be synthesized. This process is called either *biotechnology* or *genetic engineering*.

Consider the production of human insulin, which is now used to treat diabetes. In the past, diabetics were given insulin obtained from slaughtered animals because there was no source of human insulin. Animal insulins are not identical to human insulin, and medical complications could result. Now genetically altered bacteria are used to make human insulin. These bacteria are grown in large numbers, and the insulin is isolated from them, just as some antibiotics are harvested from cultured microorganisms. The insulin is then processed and distributed to the medical community. Quantities of pure human insulin are now available at a reasonable cost.

The strategy for introducing a human gene into a bacterium is straightforward. The genetic information is first isolated from a human source and then introduced into a bacterium. Unfortunately, developing the process is much harder than developing the strategy. The first complication is that a primary transcript in a human cell (and all other higher cells) must be modified before it becomes an mRNA. Bacteria cannot make these modifications (they do not need them because the primary transcript in bacteria is already a functional mRNA). As a consequence, you cannot simply put a human gene into a bacterium, because the gene would not yield a functional mRNA for the protein you want to produce.

Two alternatives can be used to get around this problem. One, the gene could be synthesized in the laboratory. If the amino acid sequence of the protein is known, the codons needed to make the protein can be determined. Start and stop codons can be added, and perhaps a section at the beginning to act as a promoter. Each end of the synthesized gene is extended a few bases to provide a means for connecting the gene to the bacterial DNA.

The second method involves human mRNA for the desired protein. The mRNA is isolated, and then the enzyme *reverse transcriptase* is used to make a single strand of DNA that is complementary to the mRNA. The single-stranded DNA is converted to double-stranded DNA by DNA polymerase. This DNA is not the natural gene, but when transcribed later, it will yield a mRNA that is translated to yield the desired protein. Again, it is necessary to make modifications to the ends of this gene to assist insertion into the bacterial DNA.

The next step is to insert the gene into a plasmid (Figure 16.19). The gene is mixed with a small piece of bacterial DNA called a **plasmid,** which is a small circular DNA molecule. The plasmid DNA is cleaved by a *restriction endonuclease,* breaking the circle and leaving the resulting DNA with a short piece of single-stranded DNA at each end (Figure 16.19). These short pieces are complementary to the ends that were added to the synthetic gene. When copies of the gene are mixed with the cleaved plasmid DNA, some copies of the gene become part of the plasmids, held by hydrogen bonding between complementary bases. A ligase then forms phosphoester bonds between the pieces to yield an intact (closed-circle) plasmid that contains the gene.

The plasmid is now mixed with the bacterial species of choice and some of the plasmids, it is hoped, are taken into bacteria (Figure 16.19). The bacteria are cultured, and if the process has been successful, the gene will be transcribed to yield mRNA that is in turn translated into functional protein such as insulin. The insulin is harvested and processed for distribution and use.

Bioengineering someday may be used to alter the genes of human beings. The genetic makeup of microorganisms can readily be changed now, and it will soon be possible to alter DNA in humans. The technology holds great promise. Someday human cells could be genetically altered to replace the DNA responsible for a genetic disease, say, or to introduce functional DNA into a patient. A couple could have a child with little risk of a genetic disease, or a patient with a genetic disease could be spared its effects.

As the technology progresses, one can imagine that virtually all genetic traits could be altered. The potential for altering humankind is most impressive—and somewhat scary. When and how these changes come about will be determined ultimately by you and other members of society at large, not by bioengineers. Technology opens many doors, but the members of a society choose which ones we pass through.

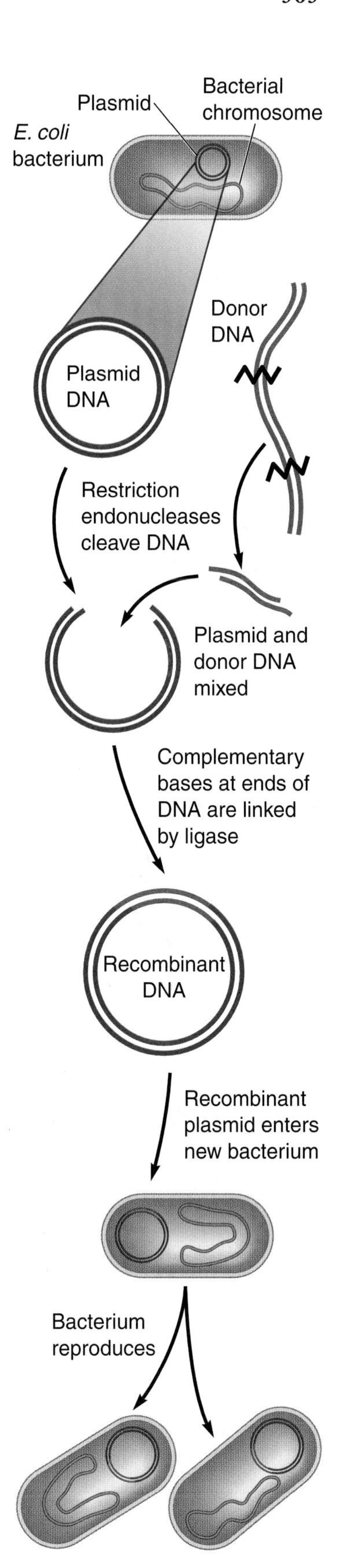

Figure 16.19 The insertion of DNA for a human protein into a bacterium. The DNA for the human protein is either synthesized in the laboratory or made from the mRNA for the protein. A plasmid is cleaved with a restriction endonuclease so that its ends match the DNA that is to be introduced. When these pieces are mixed, they first bind together through complementary bases at their ends and then are linked covalently by a ligase. The plasmid with its new DNA is then put into a bacterium and the bacterium is allowed to reproduce. All the offspring can produce the human protein.

Self-Test

Genetic Engineering

1. Why is human DNA nonfunctional in a bacterium?
2. What is the first enzyme used when a DNA strand is made from mRNA?
3. What type of enzyme is used to cut DNA into pieces? Why are these enzymes used?

ANSWERS

1. A primary transcript of human DNA must be modified to form functional mRNA, and bacteria cannot make these modifications. 2. Reverse transcriptase 3. Restriction endonucleases; they cut DNA, leaving short single-stranded ends that are then used to connect the DNA pieces to complementary single-stranded ends in plasmids.

CHAPTER SUMMARY

Genetic information is carried by DNA and provides constancy for species—like begets like—and variability among individuals of a species. **Nucleosides** are compounds containing a sugar and a **nitrogenous base; nucleotides** are nucleosides containing one or more phosphate groups. A strand of **deoxyribonucleic acid (DNA)** is a polymer containing deoxyribonucleotides linked by phosphoester bonds. A molecule of DNA consists of two DNA strands held together by hydrogen bonding between complementary bases on the two strands and by hydrophobic interactions. DNA molecules are organized into circular, supercoiled structures in bacteria and **chromosomes** in higher organisms. Cells make multiple copies of DNA through semiconservative **replication. Ribonucleic acid (RNA)** is a polymer of ribonucleotides, linked by phosphoester bonds and made by **transcription** of DNA. **Messenger RNA (mRNA)** carries genetic information that directs protein synthesis, **ribosomal RNA (rRNA)** is found in ribosomes, and **transfer RNA (tRNA)** carries amino acids to ribosomes for incorporation of the amino acids into the polypeptide chain. Information in messenger RNA is carried in three-base sequences called **codons** that code for a particular amino acid or serve as a start or stop signal. **Translation** is the process that uses information stored in messenger RNA to make proteins. There is variation in DNA structure (sequence) between individuals of a species. These differences arise by **mutations,** which may be spontaneous or induced. The expression of genetic information is regulated, thus gene products are formed only when appropriate or needed. **Induction** and **repression** in bacteria are examples of regulation of gene expression. Genetic engineering allows genetic information to be manipulated and provides significant promise for improving human life. With this promise comes the need for important decisions and choices.

KEY TERMS

Section 16.1
nitrogenous base
nucleoside
nucleotide
purine
pyrimidine

Section 16.2
chromosomes
complementary bases
deoxyribonucleic acid (DNA)
replication

Section 16.3
messenger RNA (mRNA)
primary transcript
ribonucleic acid (RNA)
ribosomal RNA (rRNA)
transcription
transfer RNA (tRNA)

Section 16.4
codon
gene

Section 16.5
anticodon
translation

Section 16.6
frameshift mutation
mutagen
mutation
substitution mutation

Section 16.7
induction
operon
repression

Section 16.8
plasmid

EXERCISES

Nucleotides (*Section 16.1*)

1. Name the sugars normally found in nucleic acids.
2. Compare the structure of the sugar found in DNA to the sugar found in RNA.
3. Which nitrogenous bases are purines? Which are pyrimidines?
4. What is the structural difference between ribose and deoxyribose?
5. Describe the structural similarities and difference between a nucleoside and a nucleotide.
6. Draw the structure of deoxyadenosine monophosphate and identify the three components of this nucleotide.
7. Draw the nucleoside containing uracil and ribose.
8. Name the nucleoside in Exercise 7.
9. Identify the compound that would be formed if three phosphates were attached to the nucleoside in Exercise 7.
10. Draw the compound containing cytosine, deoxyribose, and two phosphate groups.
11. Name the nucleoside in Exercise 10.
12. To what class of compounds would the compound formed in Exercise 10 belong?

DNA (*Section 16.2*)

13. Name the four nucleotides found in DNA.
14. The bonds that link the nucleotides in a strand of DNA are called ___________ .
15. The backbone of DNA is *not* composed of ___________ .
    (a) Sugars (b) Bases
    (c) Phosphoric acid
16. Describe the base pairing in DNA.
17. The two strands of DNA are not linked by covalent bonds, because covalently linked strands cannot be separated from each other when needed. What two types of bonds hold the two strands of DNA together?
18. Nucleotides are too big to draw easily, thus a single-letter code can be used to indicate the base in a nucleotide or nucleic acid. What base sequence is implied by this set of letters: ATTGC?
19. In DNA, the ratio of adenine to thymine and guanine to cytosine is 1 to 1. Why?
20. These are the base sequences of a part of one strand of DNA. What is the base sequence in the other strand?
    (a) . . . AGGCTA . . .
    (b) . . . CGTAG . . .
    (c) . . . GCCATA . . .
21. These are the base sequences of a part of one strand of DNA. What is the base sequence in the other strand?
    (a) . . . TGCTAC . . .
    (b) . . . GACCAT . . .
    (c) . . . CTGACG . . .
22. Name the enzyme that is responsible for the insertion of the correct nucleotide into the growing DNA strand during replication.
23. When replication is nearly complete, what enzyme connects the pieces of the synthesized DNA?
24. DNA is found in what structures in the nucleus?
25. Replication is described as semiconservative. What does this mean?

RNA (*Section 16.3*)

26. RNAs are divided into three groups. Identify these three groups of RNA and describe their functions.
27. What sugar is found in RNA?
28. Which nitrogenous bases are found in DNA and which are found in RNA?
29. Name the four nucleotides found in RNA.
30. What bases in RNA are complementary to the bases in DNA?
31. Which is larger, DNA or RNA?
32. During transcription, what serves as a template?
33. Transcription begins at what sites on DNA?
34. What enzyme is responsible for RNA synthesis?
35. Which is more accurate, replication or transcription? Why?
36. What is the primary transcript?

The Genetic Code (*Section 16.4*)

37. Provide the definition of a gene.
38. Why can't one or two bases be used to code for the amino acids in the genetic code?
39. If a segment of DNA has the base sequence, —A—C—G—G—T—A—C—T—G—, what will be the corresponding sequence on the mRNA? What amino acids would be coded?
40. If a segment of DNA has the base sequence, G—G—G—C—T—A—T—A—T, what will be the corresponding sequence on the mRNA? What amino acids would be coded?

41. Define the word *codon*. How many codons exist in the genetic code?
42. Why isn't the triplet T—A—C a codon for protein synthesis?
43. In what molecules are codons found?
44. Write a sequence of bases in mRNA that would code for the tripeptide phenylalanylprolylleucine.
45. Is the sequence you wrote in Exercise 44 the only base sequence that will code for this tripeptide? Explain.
46. A portion of the amino acid sequence for normal hemoglobin is

    Val—His—Leu—Thr—Pro—Glu—Glu—Lys

    The corresponding amino acid sequence for sickle cell hemoglobin S is

    Val—His—Leu—Thr—Pro—Val—Glu—Lys

    Write the *codon* change that has occurred.

Translation *(Section 16.5)*

47. Name the organelles that are the site of protein synthesis.
48. You have defined "codon" in Exercise 41. How is an *anticodon* related to a codon?
49. In what molecules are anticodons found?
50. Provide the anticodon for these codons.
    (a) AGC (b) CUA (c) GCU
51. Provide the anticodon for these codons.
    (a) UCA (b) ACG (c) CGU
52. Which nucleic acids are carriers of amino acids?
53. Describe the events that occur during the initiation step of translation.
54. Describe the events that occur during the translocation step of translation.
55. Describe the events that occur during termination of translation.
56. Are all proteins complete when translation terminates?

Genetic Variability *(Section 16.6)*

57. The synthesis of __________ is most directly affected by changes or errors in DNA.
    (a) Carbohydrates (b) Lipids
    (c) Proteins (d) Vitamins
58. DNA polymerases are said to have "proofreading" ability. What does this mean?
59. Define the word *mutagen* and provide common examples.
60. Provide a definition for *substitution mutation.*
61. Describe how substitution of a base in DNA can alter a protein.
62. A substitution has occurred in a three-base sequence in DNA that changes a codon as shown. What amino acid change occurs in the protein?
    (a) AUU ⟶ AGU
    (b) CAG ⟶ GAG
    (c) AGC ⟶ AGA
63. A substitution has occurred in a three-base sequence in DNA that changes a codon as shown. What amino acid change occurs in the protein?
    (a) UCU ⟶ ACU
    (b) GAC ⟶ GAA
    (c) ACC ⟶ AAC
64. If a substitution in DNA changes a codon from GGC to GGG, what change will occur in the protein?
65. Define the term *frameshift mutation.*

Regulation of Gene Activity *(Section 16.7)*

66. Provide the term that refers to the turning on of a set of genes in bacteria.
67. Give an example of a compound that turns on a set of genes.
68. Provide a definition of the term *operon.*
69. What is the role of an *operator* in an operon?
70. How is a *repressor* of an operon involved in gene expression?
71. Describe the effects on gene expression when a repressor binds to an operator.
72. What name is given to the turning off of a set of genes in bacteria?
73. Give an example of a compound that turns off a set of genes.

Genetic Engineering *(Section 16.8)*

74. When a gene is introduced into a bacterium it is often included in a carrier called a __________.
75. How are restriction endonucleases used in genetic engineering?
76. Name the enzyme that can synthesize DNA using RNA as a template.
77. Synthetic DNA could be made for a gene if the amino acid sequence of its protein were known. Would the base sequence of the synthetic gene resemble the actual base sequence of the gene? Why?
78. You have isolated a nucleic acid from a virus and have found the bases adenine, guanine, uracil,

and cytosine. What nucleic acid is present in this virus?

Challenge Exercises

79. There is a major effort (nearing completion) to determine the human genome (the genetic information for the human species). Would you expect the results to show a single unique sequence of bases for the human species? Why or why not?

80. The human immunodeficiency virus (HIV) is the virus that causes acquired immune deficiency syndrome (AIDS). Go to the library or Internet to find the current ways patients are treated for HIV. Describe briefly how each agent used in treatment affects the virus.

Metabolism

THE CHEMICAL REACTIONS OF THE BODY PROVIDE THE ENERGY AND MOLECULES NEEDED FOR LIFE.

THIS CHAPTER discusses the reactions of the body, which collectively are called metabolism. Some of these reactions produce energy as they break large molecules into smaller ones:

Large molecules such as fatty acids, sugars, amino acids
↓ Catabolism
Small molecules such as pyruvate, carbon dioxide, water

These reactions make up the branch of metabolism that is called **catabolism.** Glycolysis and the citric acid cycle are examples of catabolism. Another set of reactions consume energy as they build larger, more complex molecules from smaller ones:

Small molecules such as amino acids, nucleotides, acetate
↓ Anabolism
proteins, nucleic acids, fatty acids

These reactions make up the branch of metabolism that is called **anabolism.** Protein synthesis, replication and transcription, and fatty acid synthesis are examples of anabolism.

Outline

17.1 Digestion, Absorption, and Transport
Explain the processes of digestion, absorption, and transport.

17.2 Metabolic Pathways and Coupled Reactions
Recognize metabolic pathways and coupled reactions.

17.3 Energy in Metabolic Reactions
Explain the role of ATP, NAD^+, and NADPH in metabolism.

17.4 Carbohydrate Metabolism
Describe the origins and fates of the compounds involved in glycolysis, gluconeogenesis, and glycogen metabolism.

17.5 The Citric Acid Cycle
Trace the catabolism of acetyl-CoA through the citric acid cycle.

17.6 Electron Transport and Oxidative Phosphorylation
Explain how electron transport and oxidative phosphorylation produce ATP.

17.7 Lipid Metabolism
Describe the catabolism and synthesis of lipid molecules.

17.8 Metabolism of Nitrogen-Containing Compounds
Explain the catabolism of amino acids and the role of the urea cycle.

Metabolism produces the energy and molecules needed for this athlete's efforts. ***(Jim Cummins, FPG International)***

17.1 Digestion, Absorption, and Transport

In the gastrointestinal tract, the catabolic process known as **digestion** hydrolyses large food molecules to smaller ones. First, chewing mechanically reduces the size of food particles, and then enzymes catalyze the hydrolysis of bonds in molecules in the particles. After hydrolysis, the small product molecules are absorbed by the cells lining the small intestine and released into the blood for transport throughout the cardiovascular system. As we examine digestion, you may want to use Figure 17.1 as a review of the gastrointestinal tract.

Carbohydrate Digestion

Starch digestion begins in the mouth when salivary *amylase* catalyzes the hydrolysis of glycosidic bonds in starch molecules, yielding smaller polysaccharides, oligosaccharides, and some glucose molecules [Figure 17.2(a)]. This hydrolysis occurs only briefly because salivary amylase is denatured when it enters the acidic environment of the stomach. Dietary sugars (such as sucrose and lactose) and cellulose are not affected by amylase [Figure 17.2(b) and (c)].

Little digestion of carbohydrates occurs in the stomach because there are no gastric (stomach) enzymes for carbohydrates, and any salivary amylase mixed in with the food is soon denatured. When partially digested food enters the small intestine, pancreatic secretions neutralize stomach acid with bicarbonate ion. These secretions also contain several digestive enzymes for carbohydrates. Pancreatic amylase finishes the hydrolysis of the partially digested starches, yielding maltose and glucose. This maltose is hydrolyzed by the enzyme maltase to yield glucose

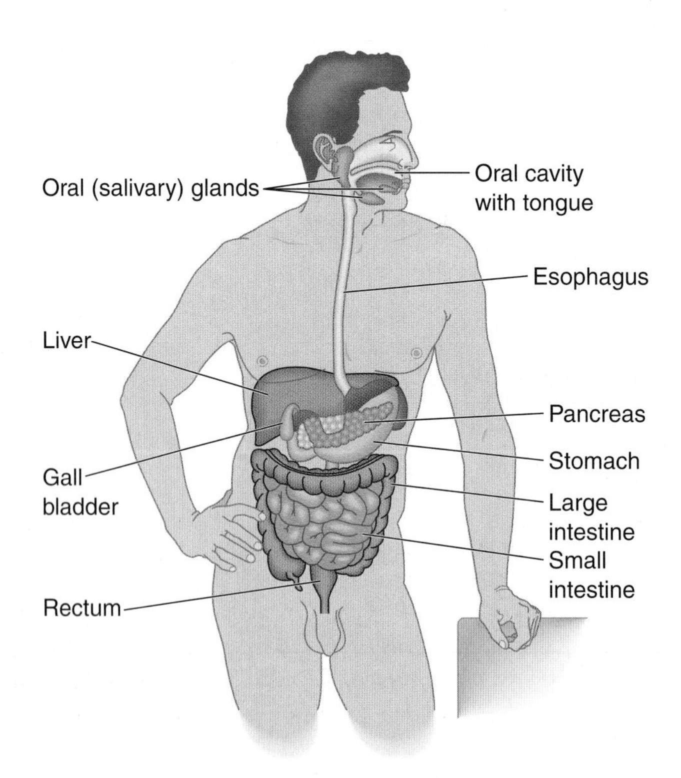

Figure 17.1 The gastrointestinal tract is responsible for the digestion and absorption of food.

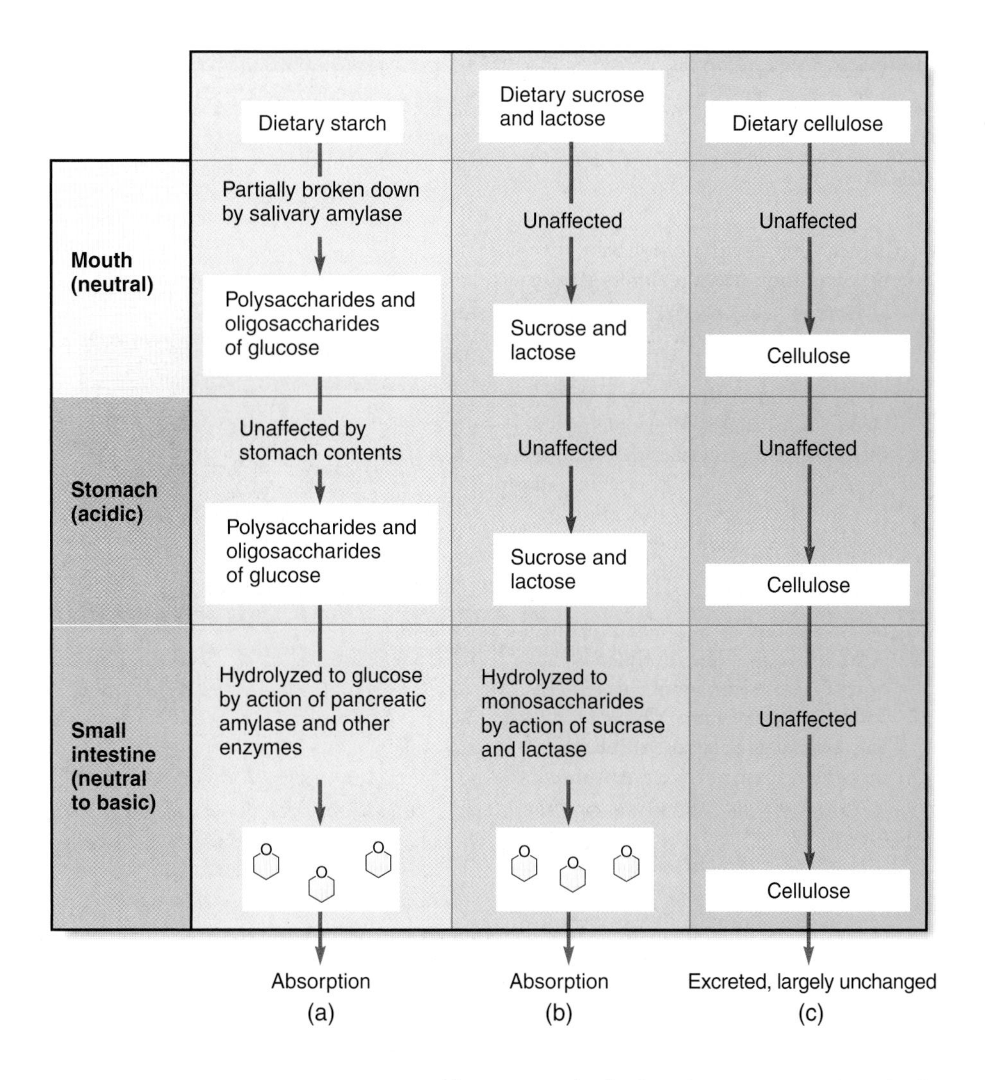

Figure 17.2 The digestion of the dietary carbohydrates **(a)** starches, **(b)** sugars (shown as sucrose and lactose) and **(c)** fiber (shown as cellulose).

[Figure 17.2(a)]. Dietary sucrose and lactose are hydrolyzed to monosaccharides by the enzymes sucrase and lactase, respectively [Figure 17.2(b)]. These simple sugars are absorbed by the cells lining the small intestine and then released into the blood stream.

Cellulose and other components of fiber are not digested by humans because they lack the enzymes needed to hydrolyze the β (1 ⟶ 4) glycosidic bonds in cellulose [Figure 17.2(c)].

Protein Digestion

Saliva contains no digestive enzymes for proteins; thus protein digestion does not begin until food enters the stomach. The acidic environment of the stomach denatures dietary proteins, which unfolds them and exposes their peptide bonds to hydrolysis. In addition, the stomach enzyme *pepsin* catalyzes the hydrolysis of some peptide bonds in proteins, yielding smaller polypeptides and oligopeptides (Figure 17.3). However, the hydrolysis of polypeptides is not completed in the stomach. When the partially digested food enters the small intestine, it is exposed to additional digestive enzymes.

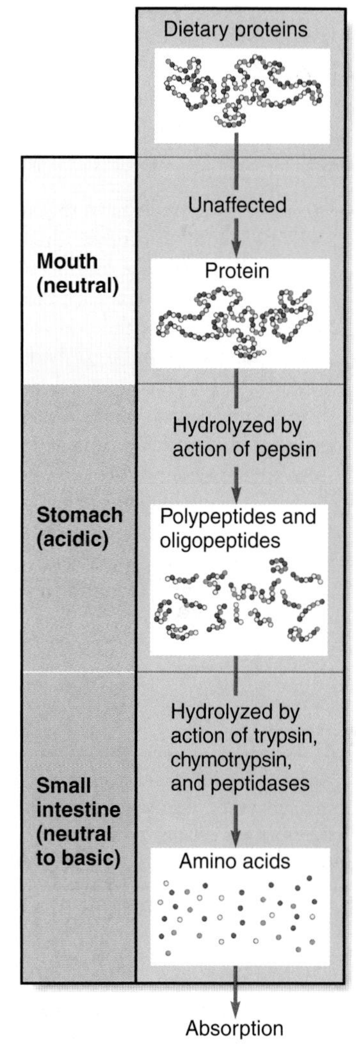

Figure 17.3 Digestion of dietary proteins occurs primarily in the stomach and small intestine.

Connections
Lactose Intolerance

The lactose in milk is an important nutrient for newborn babies and infants. A small fraction of newborn humans, however, cannot digest lactose because they have the relatively rare condition called *congenital lactose intolerance.* Furthermore, many children and adults have *acquired lactose intolerance*—a form of lactose intolerance that appears after infancy. What causes lactose intolerance and how is it recognized?

Young mammals have the intestinal enzyme lactase to catalyze the hydrolysis of dietary lactose to glucose and galactose. These sugars are then absorbed and used by the body to make other substances and to provide energy. Babies with congenital lactose intolerance cannot make lactase or make it in insufficient quantities. As a result, lactose is not broken down and remains in the digestive tract, where it alters normal water and nutrient absorption (because of increased osmotic pressure; Chapter 6), and leads to diarrhea and dehydration. A baby with this condition may die unless given an alternative diet that can be digested and absorbed properly.

When young mammals are weaned, most of them stop making lactase. This causes no problems for them because adult mammals do not generally ingest milk. Human children and adults are obvious exceptions, however, because lactose-containing dairy products are available to us throughout our lives. Many of us stop making lactase in significant quantities as we grow up. This results in an acquired lactose intolerance called *primary lactose intolerance.* Other individuals may acquire lactose intolerance through disease or intestinal surgery, and these people have *secondary lactose intolerance.*

Lactose-intolerant people vary greatly in the amounts of lactose they can tolerate, but anyone with lactose intolerance who eats more lactose than they can digest will show varying degrees of bloating, gas, and diarrhea after eating lactose-containing foods. They should avoid or eat smaller amounts of lactose-containing foods, or may benefit from using digestive aids containing lactase.

An understanding of lactose intolerance has led to the development of these products. *(Charles D. Winters)*

Zymogens are inactive forms of digestive enzymes. Can you guess why these enzymes exist as zymogens until needed?

If the bile duct becomes obstructed, bile does not enter the small intestine and lipid digestion and transport are greatly reduced. Surgery is often used to remove these obstructions.

Pancreatic secretions contain several zymogens that are activated to *proteases* and *peptidases,* the enzymes that catalyze the hydrolysis of proteins and peptides, respectively. Trypsin and chymotrypsin are the principal proteases and catalyze further hydrolysis of the polypeptide chains. Several peptidases complete protein breakdown by hydrolyzing the smaller peptides to amino acids. These amino acids are absorbed by intestinal cells and released to the cardiovascular system for distribution throughout the body.

Lipid Digestion

Digestion of dietary lipids does not begin until the small intestine, where they are mixed with secretions from the pancreas and gallbladder (Figure 17.4). The gallbladder releases bile, a secretion that contains *bile salts.* These amphipathic molecules form micelles (Chapter 13) with the water-insoluble lipids. The surface area of these micelles is very large, which increases contact between the lipids and their digestive enzymes.

Lipases catalyze the hydrolysis of ester bonds in triacylglycerols to yield fatty acids, glycerol, and monoacylglycerols. These products are absorbed into the intestinal cells that line the small intestine. In these cells, triacylglycerols are remade, then released into the bloodstream and transported throughout the body. Cholesterol, which is not affected by digestion, is also absorbed from the small intestine, then released into the blood stream.

Lipids are water-insoluble, but in blood they combine with specific proteins to form **lipoproteins,** which are transported throughout the body. Lipoproteins are like micelles; they possess a hydrophobic interior of lipid coated with a hydrophilic (water-loving) surface of protein molecules.

Self-Test

Digestion

1. Where does the digestion of starches occur?
2. Name the protease found in the stomach.
3. How do hydrophobic molecules such as triacylglycerols come into contact with the appropriate water-soluble digestive enzymes?

ANSWERS

1. Starch digestion begins in the mouth, but most of it occurs in the small intestine. 2. Pepsin 3. Dietary lipids and bile salts combine to form micelles. The large surface area of the micelles allows contact between the lipases and the triacylglycerols.

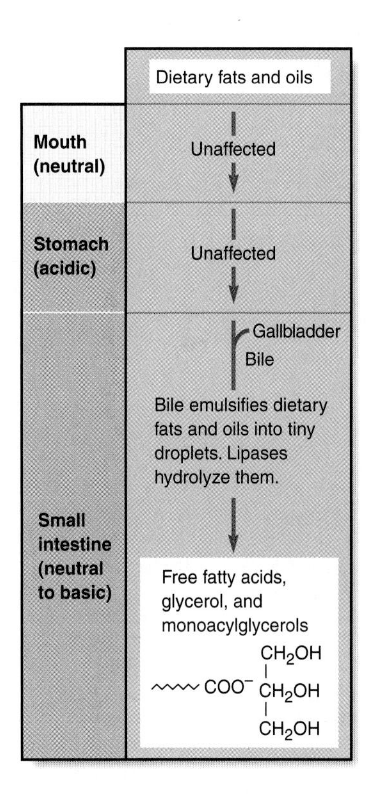

Figure 17.4 Digestion of fats and oils occurs in the small intestine through the action of both bile salts and lipases.

17.2 Metabolic Pathways and Coupled Reactions

Metabolic Pathways

The reactions of the body do not occur in an uncontrolled fashion, nor are they independent of each other. Many metabolic reactions are linked sequentially, with the product of one reaction becoming the reactant of the next one. A series of sequential reactions within the body is called a **metabolic pathway:**

$$A \xrightarrow{R_1} B \xrightarrow{R_2} C \xrightarrow{R_3} D \xrightarrow{R_4} E$$

In this diagram of a metabolic pathway, reactant A is converted to product B in reaction 1 (R_1); then B becomes the reactant in reaction 2 (R_2), the product of which is C, and so on until ultimately product E is formed. Usually each reaction in a metabolic pathway causes some modest change in its reactant, but the sum of these changes yields a final product that differs greatly from the original reactant. Many metabolic pathways have allosteric regulation that controls the rate of product formation (Chapter 15). You will see several important examples of metabolic pathways in this chapter.

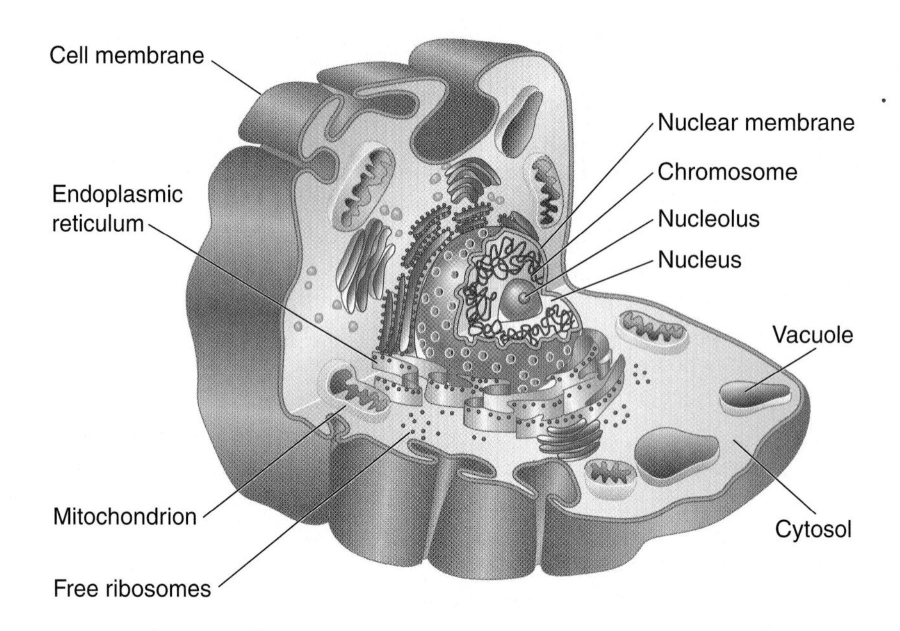

Figure 17.5 A representative animal cell.

Different parts of metabolism take place in different parts of the cell. As you study metabolism, use Figure 17.5 as a review of cell structure.

Coupled Reactions

Some of the reactions of the body occur concurrently with another reaction. These paired reactions are referred to as **coupled reactions.** Coupled reactions often involve energy. As one reaction releases energy, another gains some of the energy that is released. Some coupled reactions are oxidation–reduction reactions; as one molecule is oxidized in a reaction, another is reduced.

17.3 Energy in Metabolic Reactions

Energy is involved in many metabolic reactions. Some of the energy released by catabolic reactions is used in coupled reactions to make the nucleotide ***adenosine triphosphate (ATP).*** This molecule contains two phosphoanhydride bonds (Chapter 11), which are sometimes referred to as high-energy bonds. ATP is made by phosphorylating (adding a phosphate group to) ADP (Figure 17.6). The added phosphate group is bonded through a phosphoanhydride bond:

$$\text{ADP} + \text{P}_i \longrightarrow \text{ATP} + \text{H}_2\text{O} \qquad +7.3 \text{ kcal/mol}$$

P_i is a symbol for inorganic phosphate.

The +7.3 kcal/mol means there are 7.3 kcal more energy in a mole of ATP than in a mole of ADP—or stated another way, 7.3 kcal of energy has been added to a mole of molecules as ATP is made. The energy stored in ATP molecules is available for use by cells.

When cells need energy, ATP can be hydrolyzed to ADP and inorganic phosphate in a reaction that is the reverse of phosphorylation:

$$\text{ATP} + \text{H}_2\text{O} \longrightarrow \text{ADP} + \text{P}_i \qquad -7.3 \text{ kcal/mol}$$

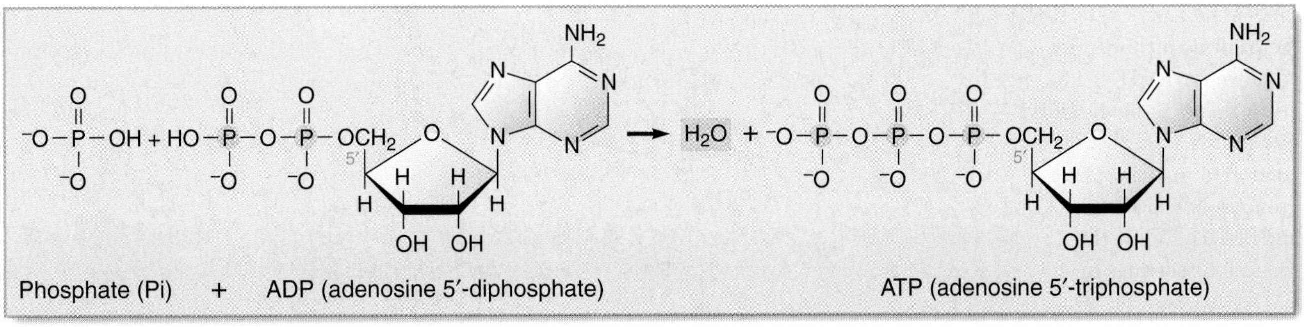

Phosphate (Pi) + ADP (adenosine 5′-diphosphate) → H_2O + ATP (adenosine 5′-triphosphate)

Figure 17.6 ATP stores chemical energy in high-energy bonds. ATP is formed from ADP by phosphorylation, which adds a second high-energy bond to the nucleotide.

The −7.3 kcal/mol means 7.3 kcal of energy are released when ATP is hydrolyzed, but this energy is not simply released as heat. Instead energy-yielding ATP hydrolysis is usually coupled to some anabolic reaction to provide the energy needed by that reaction. Cells store energy in several ways, but ATP is their principal energy-storing compound.

Catabolic reactions are generally oxidative. When some substrate molecules are oxidized, the coenzyme ***nicotinamide adenine dinucleotide (NAD⁺)*** is reduced to **NADH** [Figure 17.7(a)].

$$NAD^+ + 2e^- + H^+ \longrightarrow NADH \qquad +52.6 \text{ kcal/mol}$$

As with the synthesis of ATP, a significant amount (52.6 kcal/mol) of energy is stored in the NADH molecules. In some other catabolic reactions, the coenzyme *flavin adenine dinucleotide (FAD)* is reduced to $FADH_2$ [Figure 17.7(b)], again with storage of energy:

$$FAD + 2e^- + 2H^+ \longrightarrow FADH_2 \qquad +43.4 \text{ kcal/mol}$$

Both NADH and $FADH_2$ are readily reoxidized, yielding NAD^+ and FAD, respectively. During this oxidation, energy is released and used to make ATP.

Some anabolic reactions require both reduction and energy. These reactions use the coenzyme ***nicotinamide adenine dinucleotide phosphate (NADPH)*** as the electron donor, and as the reactant is reduced, the NADPH is oxidized to $NADP^+$:

$$NADPH \longrightarrow NADP^+ + 2e^- + H^+ \qquad -52.6 \text{ kcal/mol}$$

NADPH is reformed by reduction of **NADP⁺.** NADPH is very similar in structure to NADH [Figure 17.7(a)].

Figure 17.8 summarizes the role of some coenzymes in catabolism and anabolism.

Self-Test

Metabolic Pathways and Coupled Reactions

1. What is the role of ATP in metabolism?
2. Compare the metabolic roles of NAD^+/NADH and $NADP^+$/NADPH

ANSWERS

1. ATP stores energy for use in metabolism. 2. NAD^+/NADH is used in catabolism and $NADP^+$/NADPH is used in anabolism.

Figure 17.7 These coenzymes are commonly involved in redox reactions. **(a)** NADH and NAD^+. (Only the oxidized nicotinamide is shown for NAD^+.) The arrow points to the hydroxyl group of NADH that is connected to a phosphate group in NADPH. **(b)** FAD and $FADH_2$. (Only the reduced fused-rings of $FADH_2$ are shown.)

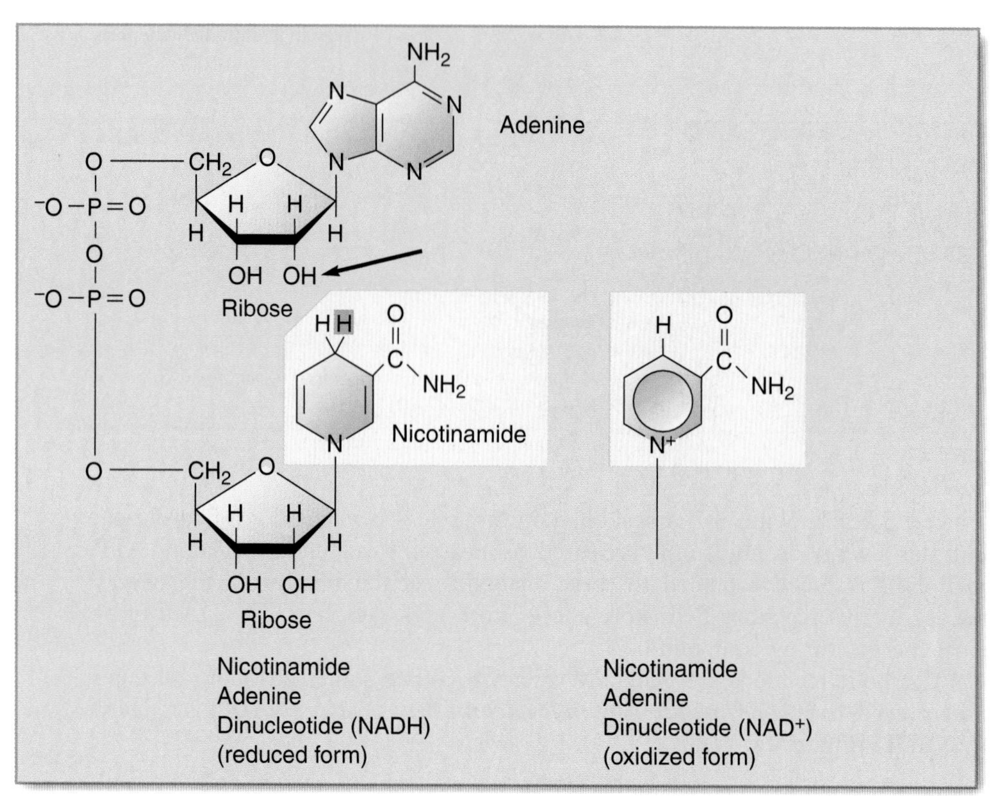

(a)

FAD

$FADH_2$

Flavin adenine dinucleotide, reduced form

Flavin adenine dinucleotide

Riboflavin

Adenine

Ribose

(b)

CATABOLISM
Converts larger molecules to smaller ones Yields energy : ADP → ATP Is oxidative : NAD^+ → NADH
ANABOLISM
Converts smaller molecules to larger ones Requires energy : ATP → ADP Is reductive : NADPH → $NADP^+$

Figure 17.8 A comparison of catabolism and anabolism. Through coupled reactions, the energy and electrons produced in catabolism are used to make ATP and NADH. Similarly, coupled reactions use ATP and NADPH to provide the energy and electrons needed by anabolism.

17.4 Carbohydrate Metabolism

Digestion yields glucose and smaller amounts of other monosaccharides. These molecules are generally used for energy production, but they can also be used to make other carbohydrates, glycolipids, and glycoproteins. Glucose is the most common sugar found in the diet and in blood; thus much of our discussion of carbohydrate metabolism focuses on this key molecule.

Glycolysis

Glycolysis is the principal pathway for catabolism of glucose and other simple sugars. These reactions occur in the cytosol (cytoplasm) of all cells and can be summarized with this word equation:

$$\text{Glucose} + 2\text{NAD}^+ + 2\text{ADP} + 2\text{P}_i \rightarrow 2\ \text{pyruvate} + 2\ \text{NADH} + 2\text{ATP} + \text{energy}$$

Like all other catabolic pathways, glycolysis breaks down a larger molecule—glucose—to smaller ones—pyruvate—with energy released in the process. Some of that energy is conserved in the ATP and NADH molecules formed. The specific steps of glycolysis are shown in Figure 17.9; as you read the summary that follows, refer to this figure for specific details. Each step is catalyzed by the enzyme shown in red.

During steps 1 through 3 of glycolysis there is an input of energy. During these steps, glucose is phosphorylated, then isomerized to fructose-6-phosphate, which is then phosphorylated to fructose-1,6-diphosphate. These steps consume two molecules of ATP for each molecule of glucose. In step 4, this phosphorylated hexose is cleaved into two phosphorylated trioses, which are the same size as pyruvate. The remaining reactions of glycolysis rearrange and oxidize these trioses to yield two molecules of pyruvate, with production of four molecules of ATP and two of NADH. The net production of ATP is two molecules of ATP per molecule of glucose: −2ATP, then +4ATP = +2 net ATP.

Fructose-1,6-diphosphate is also called fructose-1,6-bisphosphate.

Glucose is the principal compound that enters glycolysis, but it is not the only one. Figure 17.10 shows the entry of several other common carbohydrates.

Pyruvate Catabolism

The pyruvate produced by glycolysis has two catabolic fates, an *aerobic* one if oxygen is available in the cell and an *anaerobic* one if it is not. When oxygen is available, aerobic catabolism converts pyruvate to acetyl-CoA (Section 17.5). The acetyl-CoA is in turn oxidized to CO_2 and water in a process called cellular **respiration** (do not confuse cellular respiration with breathing, which is sometimes

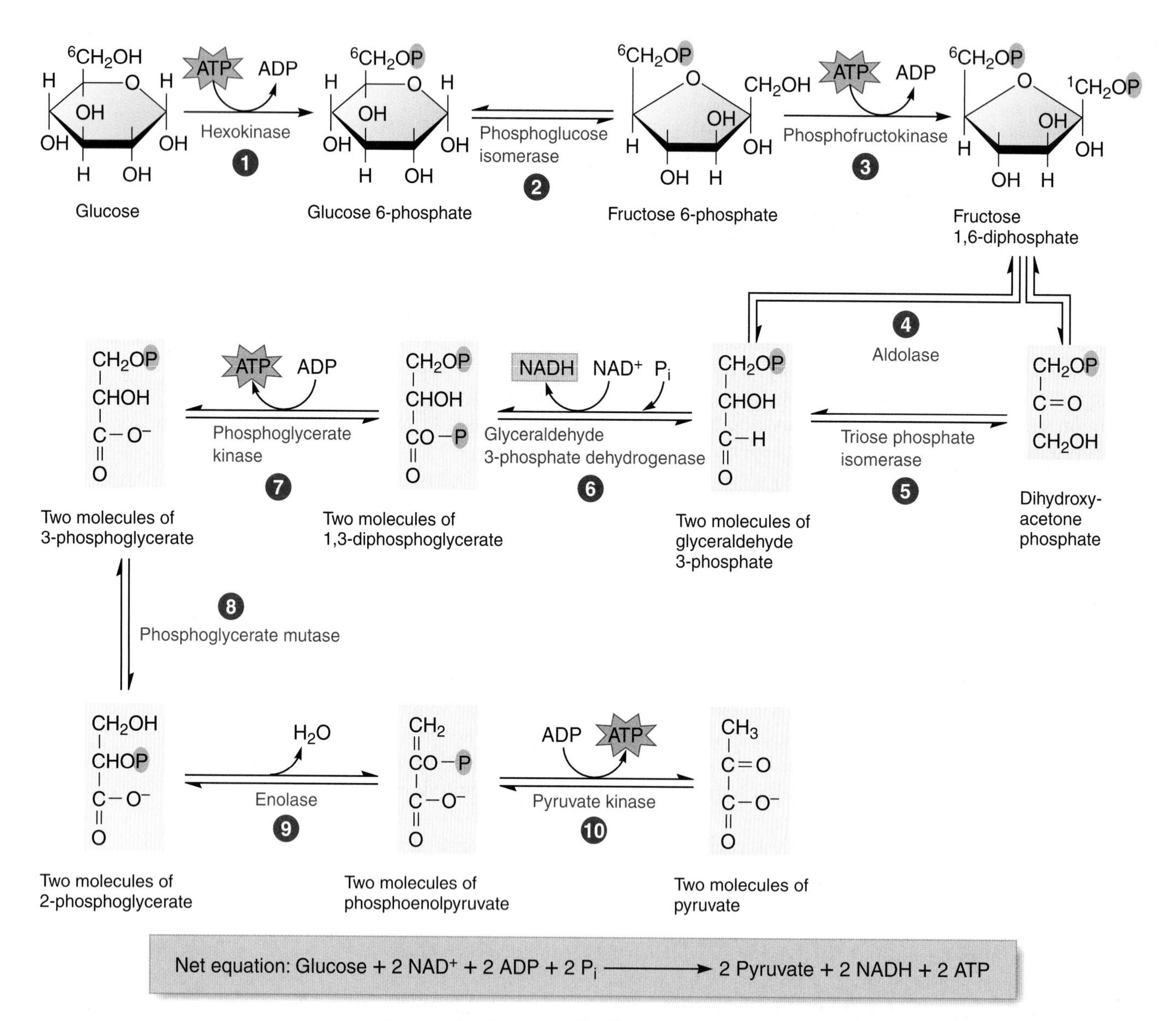

Figure 17.9 Virtually all living organisms have the glycolytic pathway. Glycolysis converts glucose to pyruvate and releases energy in the form of ATP as part of the process. Because glucose is broken into two trioses, all of the reactions from Step 6 on must be doubled, and this is reflected in the balanced reaction.

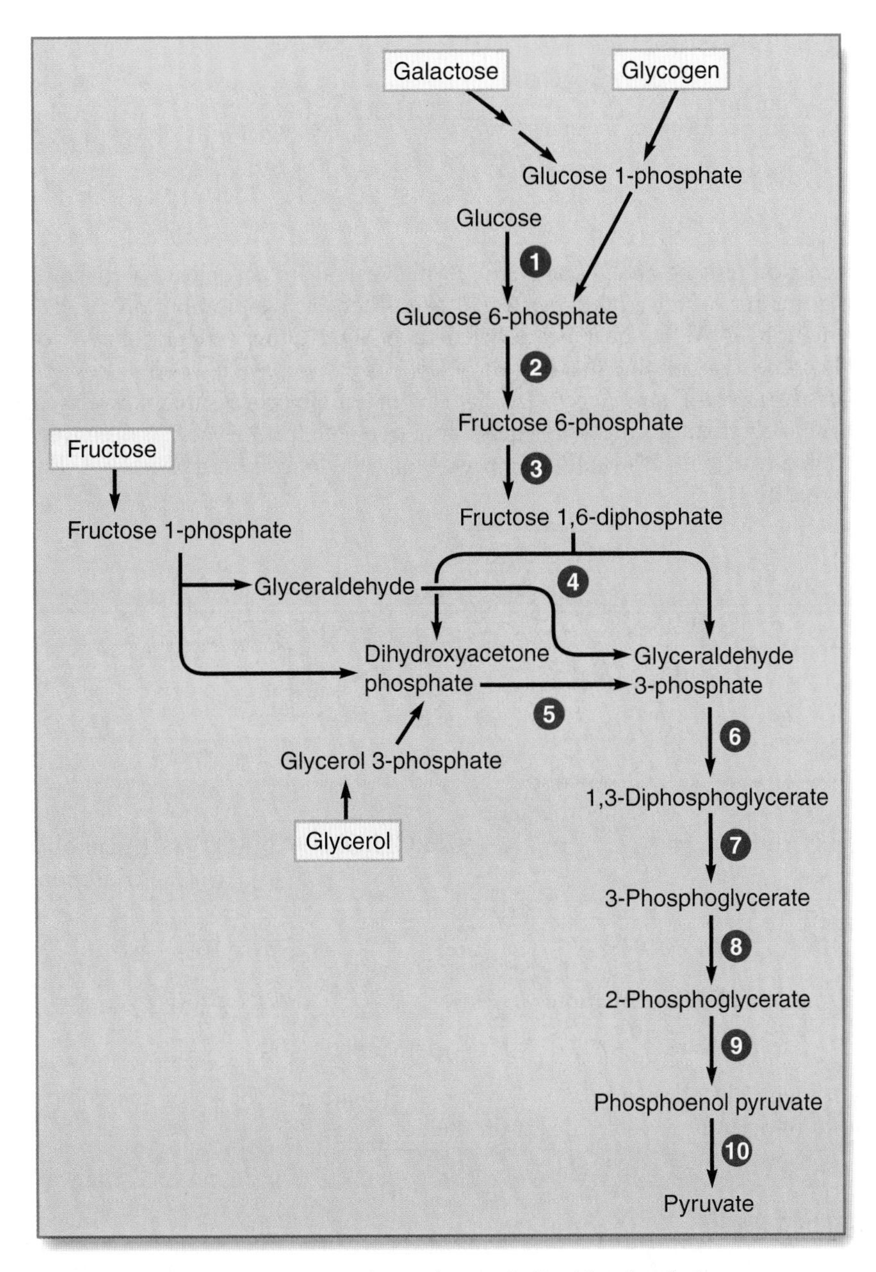

Figure 17.10 These compounds are also catabolized by glycolysis.

called respiration). If respiration occurs, the NADH produced by glycolysis is ultimately recycled to NAD^+ by the electron transport chain (Section 17.6).

During vigorous exercise, the amount of oxygen in our muscle cells is low. Under these conditions, the anaerobic process known as *lactate* **fermentation** occurs. In this process, pyruvate is reduced to lactate in a reaction using the NADH produced by glycolysis:

$$\underset{\text{Pyruvate}}{CH_3\overset{\overset{\large O}{\|}}{C}COO^-} + NADH + H^+ \xrightleftharpoons{\text{Lactate dehydrogenase}} \underset{\text{Lactate}}{CH_3\underset{\underset{\large OH}{|}}{C}HCOO^-} + NAD^+ \quad \text{(Anaerobic conditions)}$$

This reaction returns NADH to NAD^+, which can again participate in glycolysis.

No matter which route pyruvate takes—aerobic respiration or anaerobic fermentation—NADH must be recycled to NAD^+. Otherwise glycolysis would quickly come to a halt due to a lack of NAD^+.

Yeasts carry out *alcoholic fermentation*, an anaerobic process that converts sugars to alcohol (ethanol). In this fermentation, pyruvate from glycolysis is converted to acetaldehyde and carbon dioxide; then the acetaldehyde reacts with NADH to yield ethanol:

$$\underset{\text{Pyruvate}}{CH_3\overset{\overset{\large O}{\|}}{C}COO^-} \xrightleftharpoons{\text{Pyruvate decarboxylase}} \underset{\text{Acetaldehyde}}{CH_3\overset{\overset{\large O}{\|}}{C}H} + CO_2$$

$$\underset{\text{Acetaldehyde}}{CH_3\overset{\overset{\large O}{\|}}{C}H} + NADH + H^+ \xrightarrow{\text{Alcohol dehydrogenase}} \underset{\text{Ethanol}}{CH_3CH_2OH} + NAD^+ \quad \text{(Anaerobic conditions)}$$

Enzymes in yeast convert sugars from barley malt and grapes to the alcohol in beer and wine, respectively. The CO_2 produced makes the bubbles in beer and sparking wines.

NET for alcoholic fermentation:

$$\text{pyruvate} + NADH + H^+ \longrightarrow \text{ethanol} + CO_2 + NAD^+$$

Although glycolysis produces two net ATP, neither aerobic nor anaerobic catabolism of pyruvate involves direct production of ATP.

Gluconeogenesis

Although food normally provides large amounts of glucose to the body, there are times when the body must make its own. **Gluconeogenesis** ("new synthesis of glucose") is the anabolic pathway that carries out this synthesis. Gluconeogenesis is not simply the reverse of glycolysis, because each pathway has several unique enzymes that catalyze specific irreversible steps.

In gluconeogenesis, glucose is synthesized from smaller metabolic intermediates produced by catabolism (Figure 17.11). Glucose can be made from three sources: (1) lactate, (2) the glycogenic amino acids, and (3) glycerol. Lactate is produced by fermentation in anaerobic muscle. The *glycogenic amino acids* include all amino acids except leucine and lysine. Glycerol is produced in modest amounts by catabolism of fats and oils. Once the glucose is made, the body can use it for energy or to make all the other monosaccharides it requires.

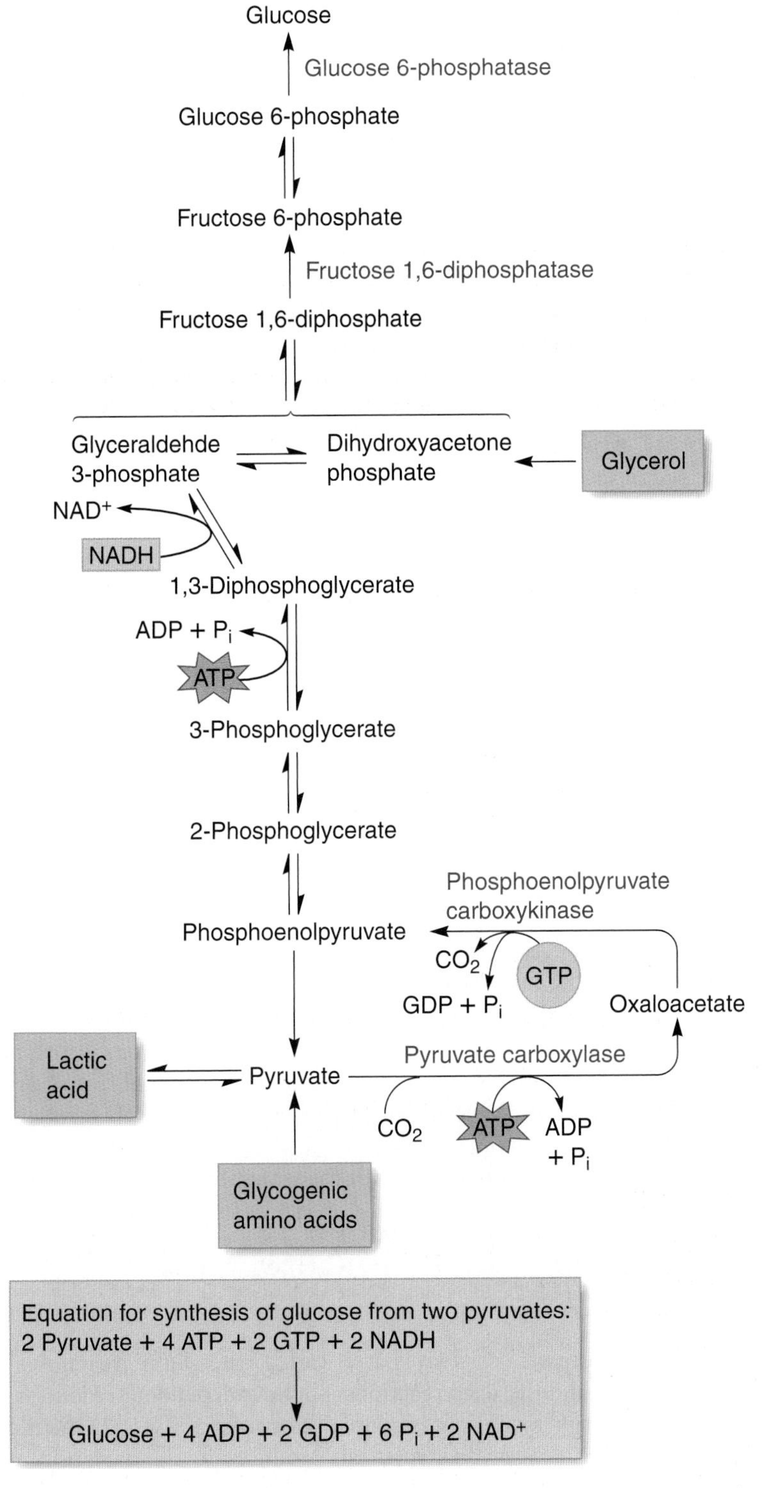

Figure 17.11 Gluconeogenesis synthesizes glucose from several other metabolites using many of the enzymes of glycolysis. The enzymes that differ from those of glycolysis are shown. The balanced equation shows the amounts of coenzymes needed per glucose; the reactions show the amount needed per substrate molecule for that reaction.

Glycogen Metabolism

Glycogen (Chapter 12) is a polymer of glucose that is used to store glucose in muscle and liver cells. When the body needs glucose, a glucose molecule is cleaved from glycogen and attached to inorganic phosphate:

$$\text{Glycogen}_{(n\ \text{glucose residues})} + \text{phosphate} \xrightarrow{\text{Glycogen phosphorylase}} \text{glycogen}_{(n-1\ \text{glucose residues})} + \text{glucose-1-phosphate}$$

The glucose-1-phosphate is converted to glucose-6-phosphate, which can be catabolized by glycolysis. Glycogen breakdown is a highly regulated process. The hormone epinephrin stimulates this process in muscle, and the hormone glucagon stimulates it in the liver.

Synthesis of glycogen generally occurs when glucose is available. **Glycogenesis,** the synthesis of glycogen, is catalyzed by glycogen synthetase and involves the transfer of a glucose molecule from the complex uridine diphosphate-glucose (UDP-glucose) to the growing glycogen chain:

$$\text{Glycogen}_{(n\ \text{glucose residues})} + \text{UDP-glucose} \xrightarrow{\text{Glycogen synthetase}} \text{glycogen}_{(n+1\ \text{glucose residues})} + \text{UDP}$$

The energy required to add a glucose residue to glycogen is provided by the cleavage of the bond between UDP and glucose. The synthesis and breakdown of glycogen are coordinated; when one occurs, the other is inactive.

Plants, through photosynthesis, use solar energy to make the biomolecules that sustain life. *(Dwight R. Kuhn)*

Photosynthesis

Humans and many other organisms are *heterotrophic,* which means they must feed on other animals or plants to get the molecules they need. Plants and some microorganisms are *autotrophic,* which means they can take simple inorganic molecules like water and carbon dioxide and use them plus light energy to synthesize organic compounds. This process is called **photosynthesis,** and life on Earth is almost totally dependent on it. Without photosynthetic organisms, heterotrophic organisms, including humans, would perish.

Photosynthesis can be summarized with this equation:

$$6CO_2 + 6H_2O \xrightarrow{\text{Light}} \underset{\text{Glucose}}{C_6H_{12}O_6} + 6O_2$$

Photosynthesis consists of two parts called the light reactions—light-dependent reactions—and the dark reactions—light-independent reactions. In the light reactions the energy in sunlight is coupled to two processes: (1) the reduction of $NADP^+$ to NADPH using hydrogen atoms produced by the oxidation of water, and (2) the production of ATP from ADP and inorganic phosphate (Figure 17.12). The water consumed in photosynthesis and the oxygen produced by it come from these light reactions. The NADPH and ATP produced by the light reactions are then used in the dark reactions to reduce the carbon atoms in carbon dioxide molecules and to connect them to form larger organic molecules (Figure 17.12). Although glucose is usually shown as the product of photosynthesis, all of the organic molecules in a plant are formed from carbon atoms originating in carbon dioxide.

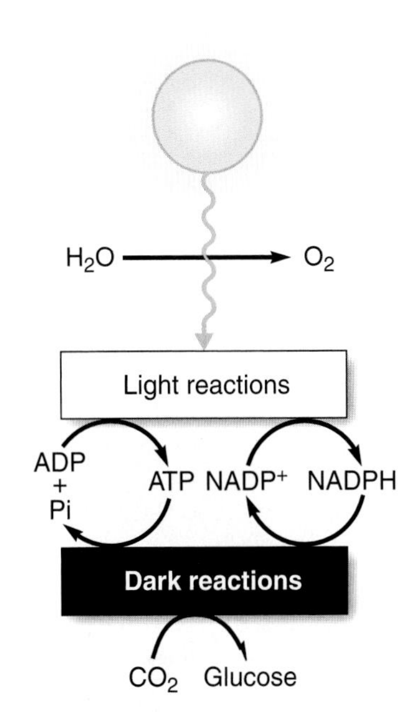

Figure 17.12 The NADPH and ATP produced by the light reactions of photosynthesis are used to make glucose from carbon dioxide.

Self-Test

Carbohydrate Catabolism

1. Compare the fate of pyruvate in human aerobic and anaerobic catabolism. What are the names of these two processes?
2. Pyruvate can be fermented to what two products?
3. Compare the gross and net production of ATP during glycolysis.

ANSWERS

1. During aerobic conditions pyruvate is converted to acetyl-CoA, which is catabolized further; this is part of respiration. During anaerobic conditions pyruvate is converted to lactate; this is fermentation. 2. Lactate in many organisms and ethanol in yeast. 3. Gross: four ATP molecules for each glucose molecule; net: two ATP molecules are consumed in the first three steps of glycolysis, and so the net product is two ATP molecules.

17.5 The Citric Acid Cycle

In glycolysis, a molecule of glucose gives a net yield of two molecules of ATP. The energy locked up in the high-energy bonds of ATP is only 2% of the energy stored in a glucose molecule. Most of the energy remains in the pyruvate and NADH molecules, thus additional catabolism is needed to gain more useful energy.

Pyruvate produced in glycolysis diffuses from the cytosol into the matrix of a mitochondrion (Figure 17.13), where pyruvate dehydrogenase catalyzes the reaction that forms acetyl-CoA:

$$\underset{\text{Pyruvate}}{CH_3\overset{\overset{\displaystyle O}{\|}}{C}COO^-} + CoA + NAD^+ \xrightarrow{\text{Pyruvate dehydrogenase}} \underset{\text{Acetyl-CoA}}{CH_3\overset{\overset{\displaystyle O}{\|}}{C}\text{—}CoA} + CO_2 + NADH$$

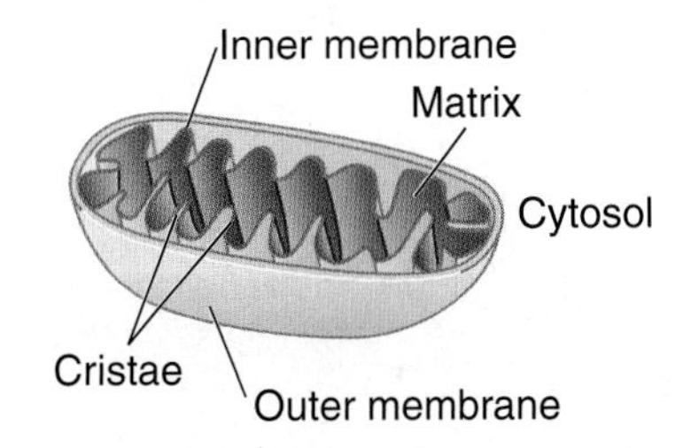

Figure 17.13 The structure of a mitochondrion. The pyruvate dehydrogenase complex is found in the matrix as are the enzymes of the citric acid cycle. The inner membrane is highly folded, forming cristae and contains the components of the electron transport chain and ATP synthetase.

This acetyl-CoA enters the **citric acid cycle,** a cyclic metabolic pathway that oxidizes acetyl-CoA to carbon dioxide (Figure 17.14).

In the first step of the citric acid cycle, the acetyl group of acetyl-CoA combines with a molecule of oxaloacetate, which is the product of the last reaction of this cyclic pathway. This reaction forms a molecule of citrate—thus the name citric acid cycle. Citrate is then isomerized to isocitrate (step 2), which is oxidized and cleaved to yield α-ketoglutarate and CO_2 (step 3). During this oxidation, a molecule of NAD^+ is reduced to NADH. In step 4, α-ketoglutarate is oxidized to succinyl-CoA and CO_2, and another NAD^+ molecule is reduced to NADH. The rest of the cycle returns the succinyl-CoA to oxaloacetate.

In step 5, succinyl-CoA is converted to succinate and CoA. Energy released in this reaction is used to make guanosine triphosphate (GTP)—an energy storing molecule similar to ATP—from GDP and phosphate. Succinate is next oxidized to

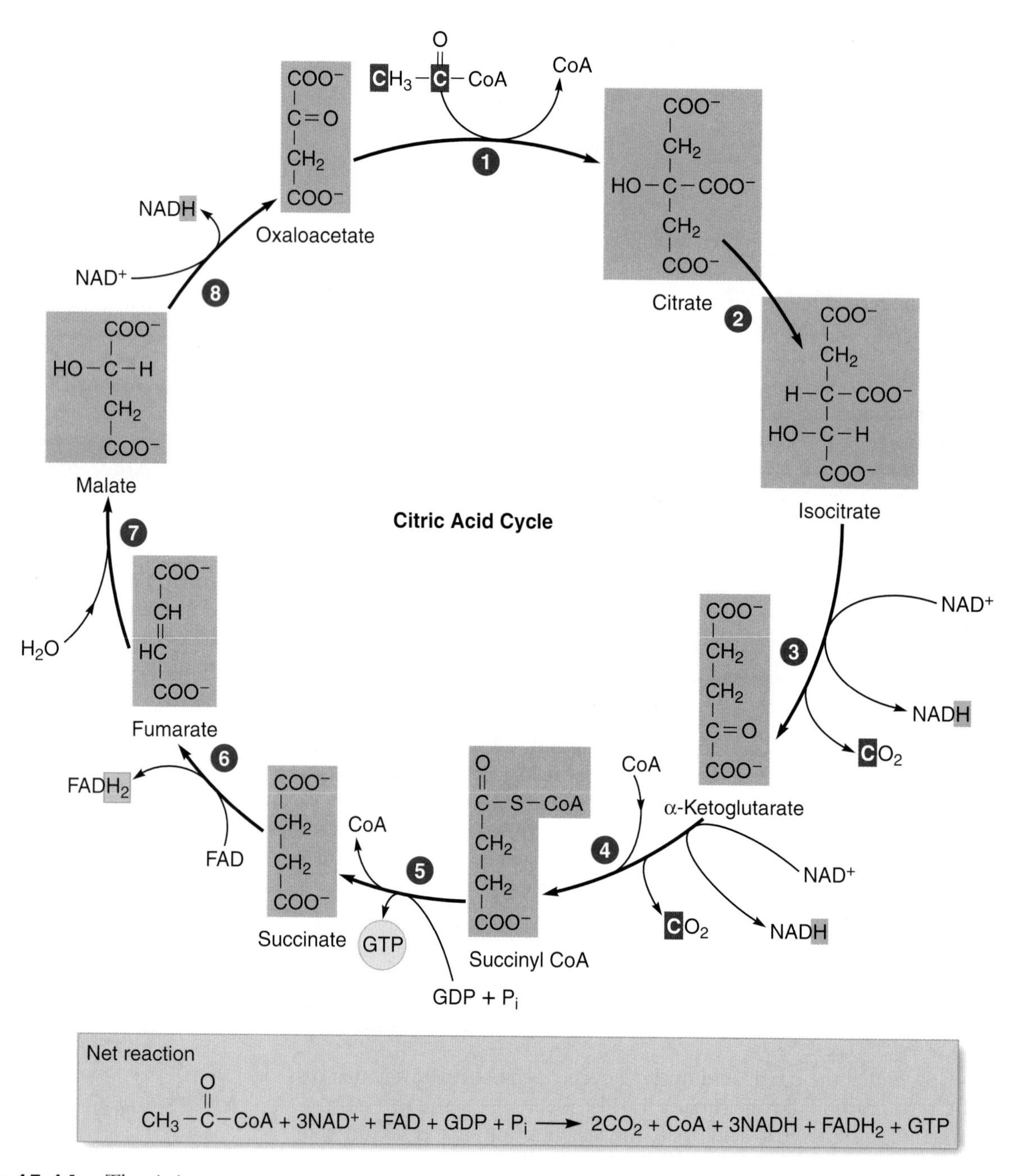

Figure 17.14 The citric acid cycle oxidizes the acetyl group of acetyl-CoA to carbon dioxide. Energy is conserved in molecules of NADH, $FADH_2$, and GTP. The acetyl-CoA is derived from the catabolism of glucose, fatty acids, and some amino acids.

fumarate. During this oxidation, flavin adenine dinucleotide (FAD) is reduced to $FADH_2$ (step 6). In step 7, a molecule of water is added to fumarate to yield malate (hydration of the carbon–carbon double bond). In step 8, malate is oxidized to oxaloacetate, with NAD^+ reduced to NADH concurrently. Because oxaloacetate is the starting compound of the cycle, the cycle is now complete.

This equation summarizes the citric acid cycle:

$$\text{Acetyl-CoA} + 3\text{NAD}^+ + \text{FAD} + \text{GDP} + \text{P}_i \longrightarrow 2\text{CO}_2 + \text{CoA} + 3\text{NADH} + \text{FADH}_2 + \text{GTP}$$

Little useful energy comes directly from this cycle because only one ATP equivalent (the GTP) is produced per acetyl-CoA. Because a molecule of glucose yields two molecules of pyruvate in glycolysis, and therefore two molecules of acetyl-CoA, two equivalents of ATP as GTP are produced in the citric acid cycle per glucose. The NADH and $FADH_2$ produced in the cycle and elsewhere, however, contain large amounts of energy that can be used to make ATP, and so it is to this formation of ATP that we next turn.

Self-Test

Citric Acid Cycle

1. How many molecules of NADH are produced in one turn of the citric acid cycle? How many turns of the citric acid cycle would occur for the acetyl-CoA that ultimately comes from one glucose molecule?
2. Which reactions produce NADH in the citric acid cycle?
3. No ATP is formed in the citric acid cycle, but when the total ATP production is tallied, the cycle is given credit for one ATP per acetyl-CoA. Why? How many ATP are credited per glucose molecule?

ANSWERS

1. Three per turn (per acetyl-CoA); two turns per glucose molecule 2. The oxidations of isocitrate, α-ketoglutarate, and malate 3. One GTP molecule is formed, and it is equivalent in energy to an ATP molecule; two ATP per glucose molecule

17.6 Electron Transport and Oxidative Phosphorylation

Aerobic catabolism of glucose generally yields 36 ATP molecules per glucose molecule. A net of two ATPs is produced in glycolysis, and the two GTP molecules produced in the citric acid cycle are equivalent to two ATP; these two sources therefore account for four ATPs. Where do the remaining 32 ATP come from? They are made using the energy stored in NADH and $FADH_2$, and the **chemiosmotic theory** provides an explanation of how this energy is used to make ATP. According to this theory, energy released by the oxidation of NADH and $FADH_2$ is used to maintain a gradient (concentration difference) of hydrogen ions between the inside and outside of the mitochondrion. This concentration difference contains stored (potential) energy that is used to make ATP.

The Electron-Transport Chain

The purpose of the electron transport chain is to provide energy for the synthesis of ATP and to reoxidize NADH and $FADH_2$. The **electron-transport chain** consists of several proteins and other molecules located in the inner mitochondrial membrane (Figure 17.15). These molecules alternately accept and donate electrons, essentially forming a chain that passes two electrons from NADH and $FADH_2$ to O_2. The NADH and $FADH_2$ are oxidized to NAD^+ and FAD, and oxygen, the electrons, and two hydrogen ions become water [Figure 17.15(a)].

Considerable energy is released as the electrons flow through the electron-transport chain from NADH and $FADH_2$ to oxygen. Some of this energy is used to pump protons (H^+) from the mitochondrial matrix to the outside of the mitochondrion [Figure 17.15(a)]. For each pair of electrons that flows from an NADH molecule to molecular oxygen, three sets of protons are pumped out [Figure 17.15(a)]. This pumping of protons increases the concentration of protons outside the mitochondrial membrane and decreases it inside. We say that a gradient has been established that is both chemical and electrical in nature. Any time a substance has a higher concentration in one region than another, a *chemical gradient* (a difference in concentration) exists. If the substance is also electrically charged, the way a proton is, then an *electrical gradient* (a difference in electrical charge) also exists. Chemical and electrical gradients possess potential energy (it is the energy of a chemical gradient that is responsible for the diffusion of particles from a region of higher concentration to a region of lower concentration). The chemical energy originally stored in the reduced state of NADH is now stored in a chemical and electrical gradient resulting from the pumping of protons.

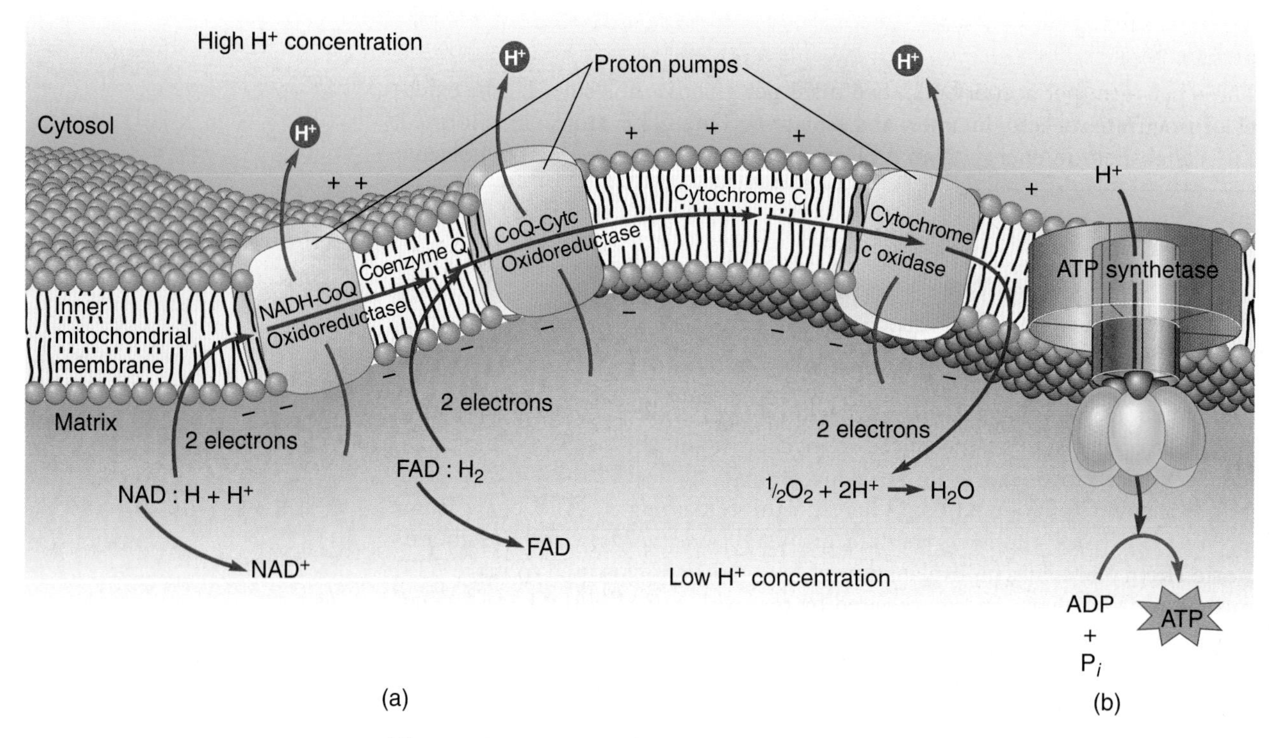

Figure 17.15 **(a)** The electron-transport chain passes electrons from NADH and $FADH_2$ through a series of electron carriers to oxygen, and uses the energy released to pump protons out of the mitochondrion. **(b)** ATP synthetase uses energy stored in the proton gradient to make ATP. As the protons pass through the channel in the enzyme, the energy released is used to make ATP.

The oxidation of $FADH_2$ to FAD by oxygen yields less energy than the oxidation of NADH. As a result, two sets of protons are pumped during this oxidation instead of the three sets when NADH is oxidized. Figure 17.15(a) shows the electrons from $FADH_2$ entering the electron-transport chain at a different point than the electrons from NADH. The energy from the oxidation of NADH drives three proton pumps, and the energy from the oxidation of $FADH_2$ drives only two of the pumps.

Oxidative Phosphorylation

The electron-transport chain establishes and maintains a gradient of protons (H^+) across the inner mitochondrial membrane as protons are pumped during the oxidation of NADH and $FADH_2$. The energy stored in this gradient is used to make ATP by the processes called **oxidative phosphorylation.** This process is catalyzed by the enzyme *ATP synthetase* located in the inner mitochondrial membrane:

$$\text{ADP} + \text{P}_i + 7.3\ \text{kcal/mol ADP} \xrightarrow{\text{ATP synthetase}} \text{ATP} + \text{H}_2\text{O}$$

Because the synthesis of ATP requires +7.3 kcal/mol, the synthesis of ATP must be coupled to an energy-yielding process. This coupled process is the movement of protons from the cytosol through a channel in ATP synthetase into the mitochondrion. (Remember, the gradient of protons has stored energy.) As these protons pass through the channel, some of the energy given up by their passage is used to bind phosphate to ADP to make ATP [Figure 17.15(b)].

The protons used to make ATP were pumped out of the mitochondrion during the oxidation of NADH and $FADH_2$. It is estimated that the oxidation of a mole of NADH pumps enough protons to make three moles of ATP, and the oxidation of a mole of $FADH_2$ pumps enough protons to make two moles of ATP.

Energy Yield from Glucose

From the aerobic catabolism of glucose that we have just discussed, 36 molecules of ATP are formed from each glucose molecule. Let's count them step by step. You can refer to Figure 17.16 and Table 17.1 as we go along. Glycolysis produces 4 ATP directly and also 2 NADH, which would yield 6 ATP through electron transport and oxidative phosphorylation. However, 2 ATP are consumed to activate glucose and two more ATP are lost when the NADH electrons are transferred into the mitochondrion to make $FADH_2$. This leaves a net yield of 6 ATP from glycolysis.

The oxidation of pyruvate by pyruvate dehydrogenase yields 1 NADH for each pyruvate. Because two pyruvate molecules are produced per glucose molecule this step yields 2 NADH per glucose molecule. Each NADH yields 3 ATP through oxidative phosphorylation, and so 6 ATP are produced in this step.

The citric acid cycle oxidizes the acetyl group of the two acetyl-CoA molecules produced by pyruvate dehydrogenase. For each acetyl-CoA that passes through the citric acid cycle, 3 NADH, 1 $FADH_2$ and 1 GTP are produced. Because two acetyl-CoA are formed per glucose molecule, 6 NADH, 2 $FADH_2$, and 2 GTP are produced by the cycle for each glucose molecule. Converting these to ATPs yields 18, 4, and 2 respectively, giving a net yield of 24 ATP from the citric acid cycle. Summing the net production from glycolysis (6), from the pyruvate dehydrogenase reaction (6), and from the citric acid cycle (24) yields a net ATP production of 36 ATP per glucose molecule.

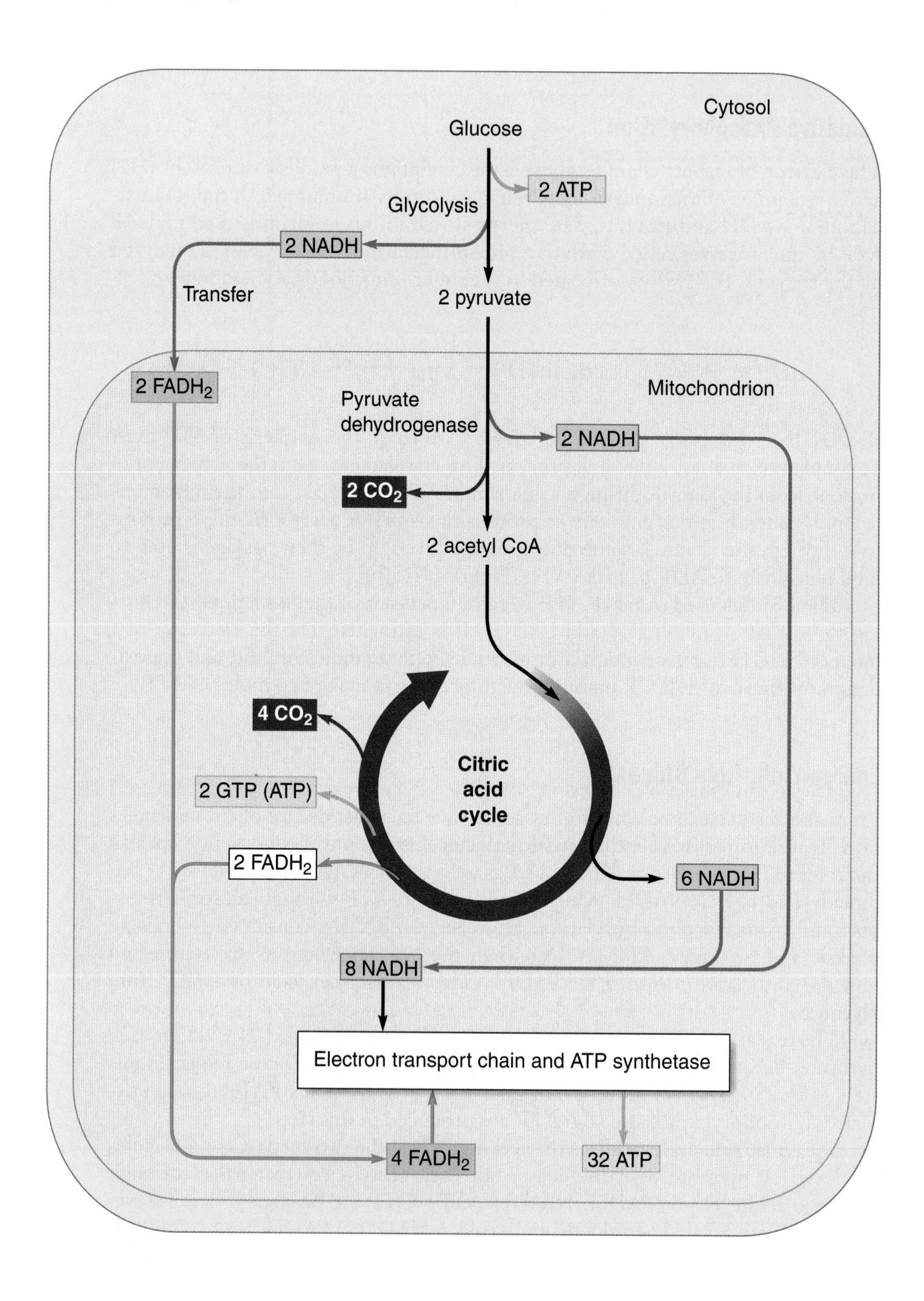

Figure 17.16 The production of 36 ATP from complete oxidation of glucose.

Table 17.1 Production of ATP During Complete Oxidation of Glucose

ATP from glycolysis	
ATP from 1,3-diphosphoglycerate	2 ATP
ATP from phosphoenolpyruvate	2 ATP
Oxidation of glyceraldehyde-3-phosphate (2 NADH)	6 ATP
Glucose activation	−2 ATP
Transfer of electrons from NADH into mitochondria	−2 ATP
Net	6 ATP
ATP from pyruvate dehydrogenase	
Oxidation of pyruvate (2 NADH)	6 ATP
Net	6 ATP
ATP from citric acid cycle	
Oxidation of isocitrate (2 NADH)	6 ATP
Oxidation of α-ketoglutarate (2 NADH)	6 ATP
GTP from succinyl-CoA	2 ATP
Oxidation of succinate (2 $FADH_2$)	4 ATP
Oxidation of malate (2 NADH)	6 ATP
Net	24 ATP
Total from complete oxidation of glucose	36 ATP

Overall: Glucose + $6O_2$ + 36 ADP + $36P_i \rightarrow 6CO_2 + 6H_2O$ + 36 ATP

Self-Test

Electron Transport and Oxidative Phosphorylation

1. Account for the 12 ATP produced from each acetyl-CoA that passes through the citric acid cycle.
2. What is the source of the energy used to pump protons from inside the mitochondrion to outside?

ANSWERS

1. One GTP is produced, equivalent to one ATP in energy. Three NADH are produced, and when one NADH is oxidized by oxygen, enough energy is released to synthesize three ATP. One $FADH_2$ is produced, and oxidation of this molecule by oxygen yields two ATP. Total: $1 + (3 \times 3) + 2 = 12$. 2. Oxidation of NADH and $FADH_2$

17.7 Lipid Metabolism

There are many metabolic pathways for the numerous lipids in the diet and the body. This section examines a few of these pathways to give you some insight into fatty acids, ketone bodies, and cholesterol.

Fatty Acid Catabolism

As we saw in an earlier section, digestion and absorption of lipids results in triacylglycerols bound to lipoproteins in blood. Most of these triacylglycerols are absorbed by adipose (fat) cells and stored until the body needs them for energy. When energy is needed, lipases in the adipose cells catalyze the hydrolysis of triacylglycerols, and the resulting fatty acids are released to the blood, where they bind to the blood protein serum albumin. These bound fatty acids circulate in blood until absorbed by the cells of the body.

Once inside a cell, a fatty acid reacts with coenzyme A to form fatty acyl-CoA. This reaction is an energy-requiring reaction that is coupled to ATP hydrolysis to AMP and pyrophosphate (PPi). Because two phosphate bonds are ultimately broken when ATP is converted to AMP, the energy of two high-energy bonds is released during the hydrolysis and used to form the fatty acyl-CoA:

$$^{-}O-\overset{\overset{O}{\|}}{\underset{\underset{OH}{|}}{P}}-O-\overset{\overset{O}{\|}}{\underset{\underset{OH}{|}}{P}}-O^{-}$$

Pyrophosphate

$$\text{Fatty acid} + \text{CoA} + \text{ATP} + H_2O \longrightarrow \text{fatty acyl-CoA} + \text{AMP} + \text{PPi} \qquad \text{(Fatty acid activation)}$$

Further catabolism of the fatty acid occurs inside mitochondria, but fatty acyl-CoA molecules cannot pass through the mitochondrial membrane. Transport through the membrane is accomplished in this way: (1) The fatty acyl-CoA binds to the surface of a mitochondrion where the fatty acid is removed from the CoA; (2) the fatty acid binds to a molecule of carnitine, and the fatty acyl-carnitine crosses the membrane; and (3) the fatty acid is cleaved from the carnitine and reattached to another CoA molecule inside the mitochondrion.

Inside the mitochondrion, the fatty acyl group is oxidized to CO_2 and water. This oxidation is called **beta oxidation** because it is the beta carbon of the acyl group that is oxidized. It yields a smaller fatty acid, an acetyl-CoA, an NADH, and a $FADH_2$. Like carbohydrate catabolism, beta oxidation yields acetyl-CoA and reduced coenzymes.

$$\underset{\beta}{RCH_2}-\underset{\alpha}{CH_2}\overset{\overset{O}{\|}}{C}-CoA \longrightarrow \longrightarrow NADH + FADH_2 + \underset{\beta}{R\overset{\overset{O}{\|}}{C}}-\underset{\alpha}{CH_2}\overset{\overset{O}{\|}}{C}-CoA \longrightarrow \longrightarrow R\overset{\overset{O}{\|}}{C}-CoA + CH_3\overset{\overset{O}{\|}}{C}-CoA$$

Although all fatty acids can undergo beta oxidation, palmitic acid (16:0) bound to coenzyme A (palmitoyl-CoA) is used to illustrate the steps of beta oxidation (Figure 17.17). In the first step, palmitoyl-CoA is *oxidized*, and FAD is reduced to $FADH_2$. Next a water molecule is added to the double bond formed in step 1. This *hydration* yields an alcohol having the hydroxyl group on the beta carbon. Now the β-hydroxyl group is *oxidized* to a β-keto group. This oxidation is accompanied by the reduction of NAD^+ to NADH. This β-keto intermediate has the same number of carbon atoms as the starting acyl-CoA, but now the beta carbon has been oxidized. In the fourth step of beta oxidation, the β-keto intermediate is *cleaved* into two smaller molecules. The bond between the alpha and beta carbon atoms is broken, and the beta carbon atom is transferred to a molecule of coenzyme A to form myristoyl-CoA, the CoA derivative of myristic acid (14:0). The other molecule is acetyl-CoA. The myristoyl-CoA is a compound identical to the starting material (palmitoyl-CoA) except that it has two fewer carbon atoms. These four steps of beta oxidation can be repeated over and over again. Each cycle yields another $FADH_2$ and NADH and removes another acetyl-CoA. From palmitoyl-CoA, eight

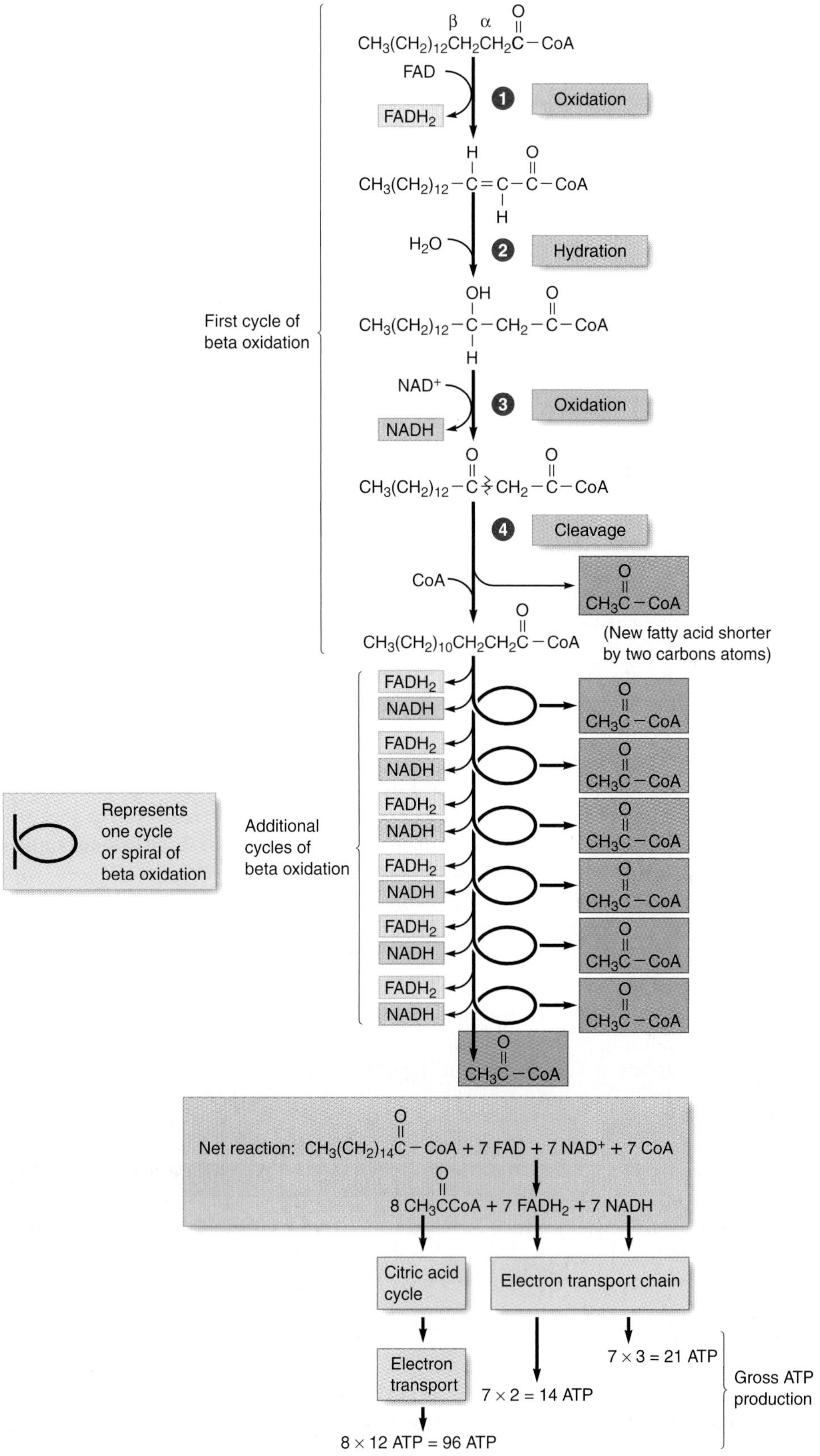

Figure 17.17 Beta oxidation of palmitic acid. Note that the number of acetyl-CoA produced (eight) is one more than the number of $FADH_2$ and NADH (seven each).

First cycle of beta oxidation

$CH_3(CH_2)_{12}CH_2CH_2C(=O)-CoA$ (β, α)

FAD → $FADH_2$ — 1 Oxidation

$CH_3(CH_2)_{12}-CH=CH-C(=O)-CoA$

H_2O — 2 Hydration

$CH_3(CH_2)_{12}-CH(OH)-CH_2-C(=O)-CoA$

NAD^+ → NADH — 3 Oxidation

$CH_3(CH_2)_{12}-C(=O)-CH_2-C(=O)-CoA$

4 Cleavage

CoA → $CH_3C(=O)-CoA$

$CH_3(CH_2)_{10}CH_2CH_2C(=O)-CoA$ (New fatty acid shorter by two carbons atoms)

Additional cycles of beta oxidation

$FADH_2$, NADH → $CH_3C(=O)-CoA$ (×6)

$CH_3C(=O)-CoA$

Represents one cycle or spiral of beta oxidation

Net reaction: $CH_3(CH_2)_{14}C(=O)-CoA + 7\ FAD + 7\ NAD^+ + 7\ CoA \rightarrow 8\ CH_3C(=O)CoA + 7\ FADH_2 + 7\ NADH$

Citric acid cycle → Electron transport → 8×12 ATP = 96 ATP

Electron transport chain → $7 \times 2 = 14$ ATP; $7 \times 3 = 21$ ATP

Gross ATP production

molecules of acetyl-CoA are produced along with seven molecules each of $FADH_2$ and NADH.

Unsaturated fatty acids are also oxidized by beta oxidation, but the product yield is different. Each double bond in a fatty acid reduces the yield of $FADH_2$ by one. This is because an $FADH_2$ is produced whenever a carbon–carbon double bond is formed during beta oxidation, and if the bond already exists in the molecule, then the reaction does not occur.

How much energy is gained from oxidation of fatty acids? Consider oxidation of palmitic acid as an example. This 16-carbon fatty acid is cleaved to eight acetyl-CoA molecules, each of which yields 12 ATP when catabolized in the citric acid cycle (Table 17.2). Beta oxidation of palmitic acid also yields 7NADH and $7FADH_2$, which yield ATP when oxidized (Table 17.2). There is a cost associated with beta oxidation, however: the ATP that is broken down to an AMP when the fatty acid is originally attached to coenzyme A. This hydrolysis reduces the net yield of ATP by two (Table 17.2).

Fatty Acid Synthesis

Triacylglycerols are used to store energy in the body and are made whenever energy intake exceeds immediate needs. The glycerol needed for the synthesis of triacylglycerols comes from the reduction of dihydroxyacetone phosphate formed in glycolysis, and the needed fatty acids come from the diet or **lipogenesis.** This process synthesizes fatty acids, with palmitic acid the major product (Figure 17.18). Lipogenesis occurs in the cytosol and is catalyzed by a complex of proteins called *fatty acid synthetase.* In the first step of this process, in an energy-requiring reaction, CO_2 is added to an acetyl-CoA to form malonyl-CoA. Next the malonyl group of the malonyl-CoA and the acetyl group of an acetyl-CoA are transferred to acyl carrier protein (ACP). These groups condense to form a β-keto intermediate similar to the one formed in beta oxidation. This intermediate is then reduced with NADPH (Section 17.2) to form a β-hydroxy intermediate, which is dehydrated to an unsaturated intermediate. This unsaturated intermediate is then reduced with a second NADPH. The result is a four-carbon carboxylic acid bound to the synthetase. Now the cycle repeats again and again until palmitic acid is formed.

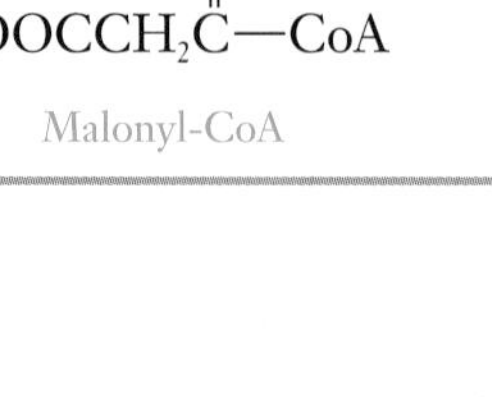

Malonyl-CoA

While palmitic acid is synthesized as described previously, other fatty acids are made from palmitic acid. An additional enzyme system can elongate palmitic acid two carbon atoms at a time, again using acetyl-CoA to provide the carbon atoms.

Table 17.2 Energy Yield (in ATP) from Palmitic Acid

8 acetyl-CoA × 12 ATP/acetyl-CoA =	96 ATP
7 NADH × 3 ATP/NADH =	21 ATP
7 $FADH_2$ × 2 ATP/$FADH_2$ =	14 ATP
Gross production of ATP from palmitic acid =	131 ATP
Activation of fatty acid for entry into beta oxidation =	−2 ATP
Net production of ATP from palmitic acid =	129 ATP

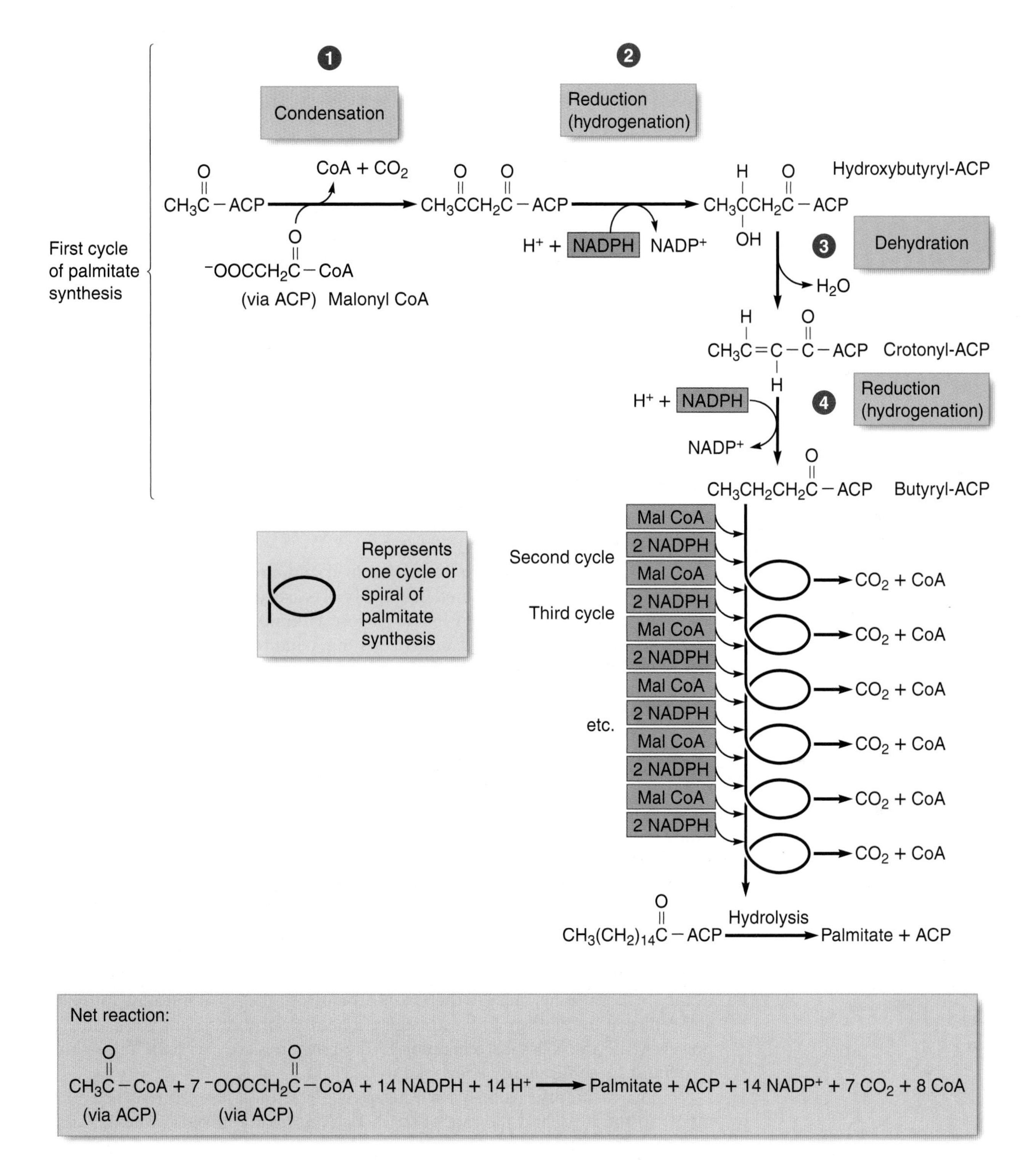

Figure 17.18 Fatty acid biosynthesis. Note that this anabolic pathway uses NADPH for reductions.

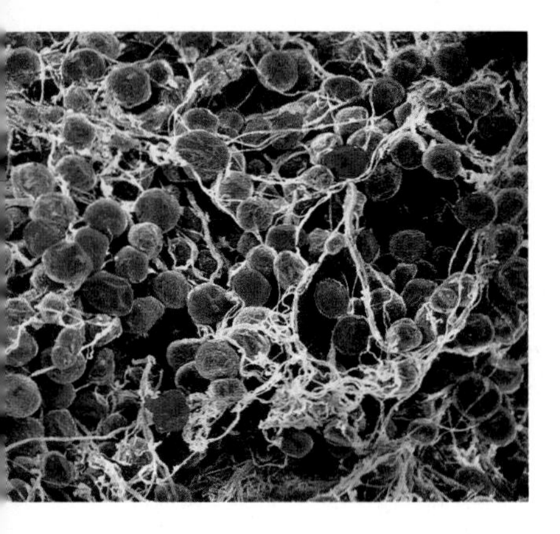

These rounded cells are called adipocytes (fat cells). Fat not used up in metabolic processes is housed in these cells. The connective tissue fibers have a supporting function for the fat cells. *(Professor P. Motta/Science Photo Library/Photo Researchers, Inc.)*

The steps are very similar to those of the pathway that uses fatty acid synthetase. Still another enzyme system introduces carbon–carbon double bonds to yield unsaturated fatty acids. This system in animals cannot introduce double bonds beyond carbon 10 of a chain, however, and for this reason linoleic acid and linolenic acid cannot be synthesized by animals.

After their synthesis, fatty acids are incorporated into complex lipids, including triacylglycerols that are stored in adipose tissue until needed.

Ketone Bodies

Ketone bodies, which are soluble molecules found in blood, are derived from fatty acids:

$$CH_3\overset{\overset{O}{\|}}{C}CH_2\overset{\overset{O}{\|}}{C}OH \qquad CH_3\overset{\overset{OH}{|}}{C}HCH_2\overset{\overset{O}{\|}}{C}OH \qquad CH_3\overset{\overset{O}{\|}}{C}CH_3$$

Acetoacetic acid — β-hydroxybutyric acid — Acetone

The ketone bodies are synthesized in the liver and circulate throughout the body at modest concentrations in blood. Ketone bodies are normally used by cells for energy, but during fasting, when blood glucose is limited, their concentrations increase markedly. Under starvation or fasting conditions, the concentrations of ketone bodies may be high enough that the odor of acetone can be detected in the breath. This also occurs in diabetic crisis, where synthesis of ketone bodies may be excessive. The other two ketone bodies are acids and lose their hydrogen ions to become their conjugate bases, acetoacetate and 3-hydroxybutyrate. At high concentrations, these ketone bodies in the blood (*ketosis*) may cause *acidosis* which is a below-normal blood pH. Extreme acidosis leads to coma and death.

Ketone Bodies in Diabetic Ketosis

	BLOOD CONCENTRATION MG/100 ML	URINARY EXCRETION MG/24 HR
Normal	<3	≤125
Untreated diabetic	90	5000

Cholesterol Metabolism

The catabolism of cholesterol deserves comment because cholesterol has such an apparent role in health. Small amounts of cholesterol are converted to steroid hormones, and some is lost in the digestive tract. The principal route for catabolism of cholesterol, however, perhaps three quarters of it, is its conversion to bile salts in the liver. These compounds are stored in the gallbladder and emulsify dietary lipid in the small intestine, and are then reabsorbed from the small intestine into blood and returned to the liver and gallbladder. The reabsorption is not 100% efficient, however, and the bile salts remaining in the intestines are excreted. Thus cholesterol must be continuously converted to bile salts.

Humans and other animals synthesize cholesterol. This synthesis occurs primarily in the liver, and the rate of synthesis is normally linked to the amount of cholesterol ingested. Patients with high blood cholesterol often have a diet designed to minimize cholesterol intake, and they take drugs to inhibit cholesterol synthesis. Hydroxymethylglutaryl-CoA (HMG-CoA) is an intermediate in cholesterol synthesis. The drug lovostatin resembles HMG-CoA and is a competitive inhibitor of a key enzyme in cholesterol synthesis.

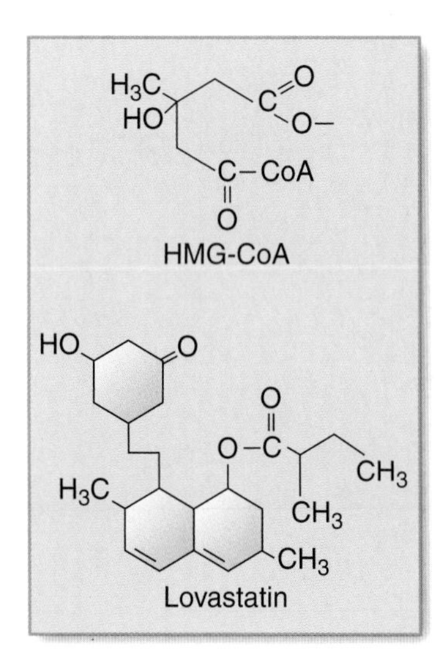

Connections

Tay-Sachs Disease

One of the better-known genetic disorders associated with lipid metabolism is Tay-Sachs disease. Gangliosides, which are glycolipids derived from sphingosine, are common in brain tissue. They are usually made and broken down, but a person with Tay-Sachs disease lacks the enzyme hexosaminidase A that normally cleaves some sugar residues from the gangliosides. Because of this genetic defect, the ganglioside G_{m2} accumulates in brain tissues, and this buildup results in several severe abnormalities. Development is normal for the first four to eight months of the infant's life, then the symptoms of blindness, deafness, and paralysis appear. Death usually occurs before the child reaches the age of 5.

Accumulated gangliosides caused this enlargement and degeneration in nerve cells in the brain in a Tay-Sachs disease patient. *(IMS Creative/Custom Medical Stock Photo)*

Self-Test

Lipid Metabolism

1. Determine the number of ATP molecules produced during beta oxidation of stearic acid (18:0). (Table 17.2 will serve as a useful guide.)
2. What is the cause of acidosis in some patients with diabetes?

ANSWERS

1. Stearic acid, as stearoyl-CoA, yields 9 acetyl-CoA, 8 NADH, and 8 $FADH_2$. These yield 108, 24, and 16 ATP, respectively, for a total gross production of 148 ATP. Because two ATP are required to activate the fatty acid, the net production is 146 ATP. 2. High concentrations of acetoacetic acid and β-hydroxybutyric acid in the blood.

17.8 Metabolism of Nitrogen-Containing Compounds

Most of the nitrogen in the body is found in amino acids and in nitrogenous bases. This section examines several key aspects of nitrogen metabolism.

Amino Acid Metabolism

Amino acids derived from dietary protein are absorbed by intestinal cells, released into the blood, and transported throughout the body. Some of them are used to synthesize specific proteins needed by the body, but most diets provide more amino acids than are needed for this synthesis. Although small amounts are used to make other molecules, most of the extra amino acids are catabolized to produce energy.

Catabolism of amino acids occurs primarily in the liver. A common step in this catabolism is **transamination,** a process that removes the amino group from an amino acid and transfers it to α-ketoglutarate:

$$\underset{\text{Amino acid}}{\mathrm{RC(NH_3^+)(H)COO^-}} + \underset{\alpha\text{-Ketoglutarate}}{\mathrm{^-OOCCH_2CH_2C(=O)COO^-}} \xrightarrow{\text{Transaminase}} \underset{\alpha\text{-Keto acid}}{\mathrm{RC(=O)COO^-}} + \underset{\text{Glutamate}}{\mathrm{^-OOCCH_2CH_2C(NH_3^+)(H)COO^-}}$$

There are several transaminases in liver cells that catalyze this reaction, and each uses α-ketoglutarate as the amino group acceptor. Thus transamination collects the amino groups from all catabolized amino acids and transfers them to α-ketoglutarate molecules to form glutamate molecules. The carbon skeletons of the amino acids become keto-acids that are catabolized, by individual pathways, to smaller molecules such as acetyl-CoA and citric acid cycle intermediates that enter catabolic pathways that you already know (Figure 17.19). Catabolism of α-keto acids yields the energy we associate with dietary protein.

All is well so far, but there is only a little α-ketoglutarate in a cell, and so it must be recycled. This is accomplished in two ways. Some glutamate molecules undergo oxidative **deamination** to yield α-ketoglutarate and ammonia:

$$\underset{\text{Glutamate}}{\mathrm{^-OOCCH_2CH_2C(NH_3^+)(H)COO^-}} + \mathrm{NAD^+} + \mathrm{H_2O} \xrightarrow{\text{Glutamate dehydrogenase}} \underset{\alpha\text{-Ketoglutarate}}{\mathrm{^-OOCCH_2CH_2C(=O)COO^-}} + \mathrm{NADH} + \mathrm{NH_4^+}$$

Some α-ketoglutarate is thus made available for transamination, but now ammonia (as ammonium ion) has been produced. Ammonia is toxic if its concentration builds up, but humans and many other animals use the urea cycle to convert ammonia to urea (Figure 17.19), which is much less toxic.

The remaining glutamate (that which was not deaminated) is returned to α-ketoglutarate via transamination during the synthesis of urea (Figure 17.19). Note that both of the nitrogen atoms in urea come from glutamate—one via deamination to ammonium ion and the other via transamination. In the urea cycle, these nitrogen atoms (represented in this equation as ammonium ions) are combined with carbon dioxide to form urea:

$$\mathrm{2NH_4^+} + \mathrm{CO_2} \longrightarrow \underset{\text{Urea}}{\mathrm{H_2NC(=O)NH_2}} + \mathrm{H_2O} + \mathrm{2H^+}$$

Plants and many bacteria make all of the amino acids. Mammals and other heterotrophs can make only some of them. Those that cannot be synthesized are the *essential amino acids* (see Chapter 14). For heterotrophs, it is not normally a problem to have lost this synthetic ability because amino acids are usually present in the foods they eat.

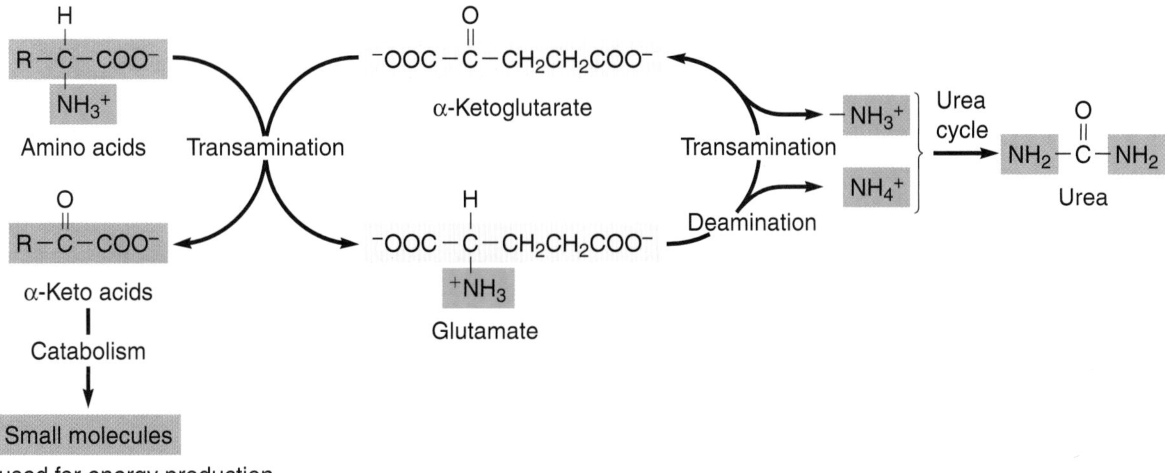

Figure 17.19 The carbon skeletons of the amino acids (tan) are broken down into several smaller metabolites. The amino groups of amino acids (blue) are used to make urea.

Connections

Phenylketonuria

Have you ever noticed the warning label for phenylketonurics on containers of food and beverage that contain NutraSweet®? NutraSweet is a brand name for aspartame, an artificial sweetener that is a dipeptide containing aspartic acid and phenylalanine. Phenylketonuria (PKU) is a genetic disease characterized by the absence of the enzyme phenylalanine hydroxylase. As a result of the deficiency, PKU patients cannot metabolize phenylalanine normally, and any phenylalanine above that needed for protein synthesis accumulates in the body. If the phenylalanine exceeds a threshold concentration, some other enzymes of the body convert it to phenylketo acids, such as phenylpyruvic acid. These compounds appear first in the blood and then in the urine (phenylketonuria literally means "phenylketones in the urine"). Unfortunately, a significant concentration of phenylketo acids in the blood tends to disrupt the normal development of the nervous system, leading to mental retardation and other abnormalities.

Because of the effect of phenylketo acids on development, the diagnosis of PKU in newborns is a routine analysis. If the diagnosis is positive, the infant must be given a diet that contains just enough phenylalanine to meet the needs of protein synthesis. If the level of phenylalanine is held carefully, there is no significant buildup of phenylketo acids in the blood and normal development can occur. The adult phenylketonuric should also restrict the intake of phenylalanine, and the labeling of products containing NutraSweet helps those patients monitor their phenylalanine uptake. A pregnant woman with PKU must be especially careful about her diet because phenylketo acids can cross the placenta to the fetus.

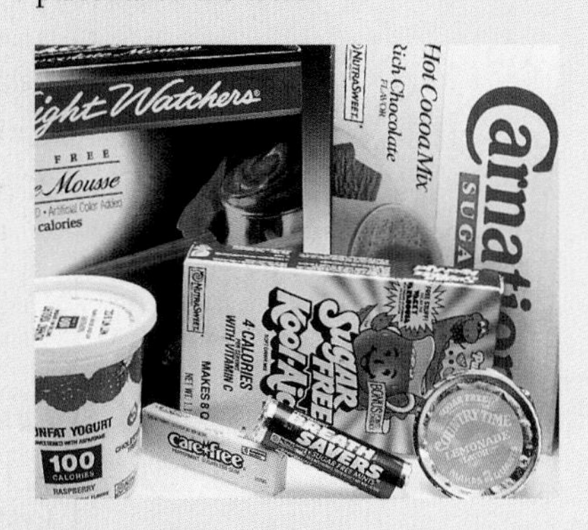

These products contain aspartame. *(Charles D. Winters)*

Metabolism of Nucleotides

Nucleotides and nucleic acids are never abundant in the diet, and as a result their catabolism provides no significant amount of energy or smaller molecules. On the other hand, some dietary molecules are very important in nucleotide anabolism. Dietary niacin is a vitamin required to make nicotinamide adenine dinucleotide (NAD^+), and the vitamin riboflavin is needed to make flavin adenine dinucleotide (FAD). Given the needed components, nucleotides can be synthesized by the body. Each nucleotide has its own biosynthetic pathway that is too detailed for our purposes. However, it is worthwhile to examine a part of one pathway to see an application of biochemistry in health care.

Deoxythymidine monophosphate (dTMP) is formed from deoxyuridine monophosphate (dUMP) by the transfer of a methyl group from a coenzyme called methylene tetrahydrofolate (methylene THF):

$$\text{dUMP} + \text{methylene THF} \xrightarrow{\text{Thymidylate synthase}} \text{dTMP} + \text{DHF}$$

In the body, the anticancer drug fluorouracil is altered to form an irreversible inhibitor of thymidylate synthase. When the proper amount of fluorouracil is present, little dTMP can be formed because most of the enzyme molecules are inactive. Other compounds, such as methotrexate, prevent the recycling of the coenzyme DHF to methylene THF. If the concentration of methylene THF is low, little dTMP is formed.

Why do these compounds work as anticancer drugs? Cancer cells divide more rapidly than most normal cells. Cell division requires DNA replication, which requires the deoxynucleotides. Inhibition of dTMP synthesis kills rapidly dividing cells. Unfortunately, these treatments also affect some normal cells, especially those that divide rapidly. Hair cells, cells lining the gastrointestinal tract, and the cells that divide to form blood cells are examples of rapidly dividing cells. Hair loss, gastrointestinal problems, and suppressed immunity are consequently all side effects of chemotherapy.

Self-Test

Metabolism of Nitrogen-Containing Compounds

1. What happens to the keto-acids produced by transamination?
2. What happens to the amino groups of amino acids after they are removed by transamination?
3. There are actually few major differences between a normal cell and a cancer cell. What difference is exploited by many anticancer drugs?

ANSWERS

1. Each keto-acid is catabolized by a separate pathway. 2. The amino groups are first collected in glutamate molecules, then removed by oxidative deamination and transamination and used to form urea. 3. The rate of growth of cancer cells is usually faster than that of normal cells.

CHAPTER SUMMARY

Nearly all of the reactions of the body are enzyme-catalyzed. **Metabolism** is the sum of all the chemical reactions in the body. **Catabolism** breaks larger molecules to smaller ones with release of energy, and **anabolism** uses energy to build larger molecules from smaller ones. A **metabolic pathway** is a regulated series of reactions that convert a reactant into a quite different product. Many reactions of the body are **coupled,** with an energy-yielding reaction providing the energy for an energy-requiring reaction. The nucleotide **ATP** stores energy in the body, and the nucleotides **NADH,** $FADH_2$, and **NADPH** are involved in oxidation-reduction reactions in the body. **Digestion** breaks down proteins, larger carbohydrates, and lipids, allowing amino acids, monosaccharides, and lipids to be absorbed and transported by the blood stream. Glucose and other carbohydrates are catabolized to pyruvate by **glycolysis.** Pyruvate is catabolized to lactate or ethanol under anaerobic conditions, or to acetyl-CoA under aerobic conditions. Acetyl-CoA is oxidized to carbon dioxide in the **citric acid cycle** with the production of GTP, NADH, and $FADH_2$. NADH and $FADH_2$ are oxidized by the **electron-transport chain,** which also establishes and maintains a proton gradient across the inner mitochondrial membrane. In **oxidative phosphorylation,** ATP synthetase uses a portion of the gradient's energy to phosphorylate ADP to form ATP. Fatty acids are **beta-oxidized** to acetyl-CoA, NADH, and $FADH_2$, and are synthesized by the process of **lipogenesis** using primarily acetyl-CoA and NADPH. Amino acids undergo **transamination,** yielding keto-acids that are catabolized to produce energy. The amino groups are stored temporarily in molecules of glutamate, and are then used to synthesize urea. Dietary nucleotides and nucleic acids do not contribute any significant amount of energy to the body, but several anabolic reactions for nucleotide synthesis are involved in anticancer therapy.

KEY TERMS

Introduction
anabolism
catabolism
metabolism

Section 17.1
digestion
lipoproteins

Section 17.2
coupled reactions
metabolic pathway

Section 17.3
adenosine triphosphate (ATP)
nicotinamide adenine dinucleotide phosphate ($NADPH/NADP^+$)
nicotinamide adenine dinucleotide ($NAD^+/NADH$)

Section 17.4
fermentation
gluconeogenesis
glycogenesis
glycolysis
photosynthesis
respiration

Section 17.5
citric acid cycle

Section 17.6
chemiosmotic theory
electron-transport chain
oxidative phosphorylation

Section 17.7
beta oxidation
lipogenesis

Section 17.8
deamination
transamination

EXERCISES

Introduction

1. Compare the roles of catabolism and anabolism.
2. Compare anabolism to catabolism with respect to size of reactants and products and whether the process is endothermic or exothermic.

Digestion, Absorption, and Transport (Section 17.1)

3. Describe the role of digestion in animals.
4. Only modest amounts of starch digestion occur before food enters the small intestine. Why?
5. Where are amylases produced and secreted? Why is more than one amylase needed?
6. Compare the roles of the enzymes maltase, sucrase, and lactase.
7. The pH of the stomach is much lower than other parts of the body. What is the function of this highly acidic environment?
8. What digestive enzymes catalyze the hydrolysis of peptide bonds?
9. What compounds complex with lipids in the small intestine to form micelles?

10. Why are dietary lipids incorporated into micelles in the small intestine?
11. Where does absorption of nutrients occur?
12. Why are lipoproteins needed for the transport of lipids in the blood?
13. Complete the following table.

FOOD CLASS	ENZYMES	SITES OF DIGESTION	PRODUCTS
Carbohydrates		Mouth and small intestine	
Proteins	Proteases and peptidases		
Lipids			Fatty acids, monoacylglycerols, and glycerol

Metabolic Pathways and Coupled Reactions (Section 17.2)

14. Provide a definition for *metabolic pathway.*
15. Give two or more examples of catabolic pathways and anabolic pathways.
16. Briefly describe the concept of coupled reactions.
17. After reviewing the metabolic pathways in this chapter, give two examples from them of coupled reactions that either conserve energy or couple oxidation–reduction.

Energy in Metabolic Reactions (Section 17.3)

18. ATP is a very common term in cell biology and biochemistry. What does the abbreviation ATP stand for, and what is its role in metabolism?
19. Identify the source of the energy used to make ATP from ADP and inorganic phosphate.
20. Could a reaction that requires 9.0 kcal/mol be coupled successfully to the hydrolysis of one mole of ATP? How could this reaction be coupled successfully to ATP hydrolysis?
21. Describe the role of NAD^+ and FAD in catabolism.
22. Explain the part that NADPH plays in anabolism.
23. In Chapter 15 you learned that enzymes differ in their specificity for substrates. How would enzymes distinguish between NAD^+/NADH and $NADP^+$/NADPH?

Carbohydrate Metabolism (Section 17.4)

24. **(a)** Look at the equation for glycolysis, then paraphrase the equation in your own words. **(b)** From the symbolic equation and your words, describe the role of glycolysis in the cell.
25. **(a)** Identify the general part of glycolysis that requires an input of energy. **(b)** What compound is the immediate source of this energy?
26. **(a)** Identify the part of glycolysis that is energy yielding. **(b)** What molecules store this energy?
27. In the absence of oxygen, glycolysis yields how many ATP per glucose?
28. Compounds other than glucose are catabolized by glycolysis. Where do they enter glycolysis?
29. If no or limited oxygen is present in muscle cells, the pyruvate produced by glycolysis is converted into what compound?
30. Is the reaction in Exercise 29 a fermentation or a respiration?
31. Compare the products of fermentation in yeast to the products of fermentation in animal cells.
32. Rapid anaerobic catabolism of glucose may lower cellular and blood pH. Why?
33. Provide a definition for *gluconeogenesis.* What is the role of this pathway in animals?
34. Make a list of the compounds that can serve as precursors to gluconeogenesis, and identify the product.
35. **(a)** What distinguishes an amino acid as *glycogenic?* **(b)** Identify the glycogenic amino acids.
36. What part of a triacylglycerol can be used for the synthesis of glucose? (Hint: Before you look up the answer in the textbook, break the word *triacylglycerol* into its parts.)
37. Formation of the glycosidic bonds in glycogen is endothermic. Identify the source of the energy used to form these bonds.
38. How do heterotrophic organisms obtain organic compounds? How do autotrophic organisms obtain organic compounds?

The Citric Acid Cycle (Section 17.5)

39. What percentage of the energy in glucose is conserved by the reactions of glycolysis? How is more of the energy originally present in glucose obtained?
40. **(a)** What reaction in the cell converts pyruvate into acetyl-CoA? **(b)** Where in the cell does this reaction occur?

41. What are the other products of the reaction in Exercise 40?
42. Ultimately, the carbon atoms of acetyl-CoA are converted into what molecule?
43. In what reaction of the citric acid cycle is GTP produced?
44. GTP is equivalent in energy to what molecule?
45. **(a)** List the reactions of the citric acid cycle that produce NADH. **(b)** Which one(s) produce $FADH_2$?

Electron Transport and Oxidative Phosphorylation *(Section 17.6)*

46. Describe the chemiosmotic theory.
47. Where does the oxidation of NADH and $FADH_2$ occur?
48. How are NADH and $FADH_2$ oxidized in mitochondria?
49. The electron transport chain removes electrons from (oxidizes) NADH and $FADH_2$. Identify the substance that is the ultimate acceptor of these electrons.
50. The acceptance and donation of electrons is only one function of the electron transport chain. What is the other one?
51. Explain why the gradient of protons (hydrogen ions) across the inner mitochondrial membrane is both chemical and electrical in nature.
52. Use your answer in Exercise 50 to explain why the oxidation of NADH yields 3 ATP but oxidation of $FADH_2$ yields only 2 ATP.
53. Explain what is meant by the term *oxidative phosphorylation.*
54. Energy stored in a proton gradient is used to drive the reaction that makes ATP. **(a)** What enzyme catalyzes this reaction? **(b)** Describe the location and spatial arrangement of this enzyme.
55. How many ATPs are ultimately produced when an acetyl-CoA passes through the citric acid cycle?
56. Oxidation of NADH normally yields enough energy to make 3 ATP, but the NADH formed in the cytosol by glycolysis are said to be equivalent to 2 ATP. Why?
57. Account for the 36 moles of ATP that are produced when a mole of glucose is completely oxidized.

Lipid Metabolism *(Section 17.7)*

58. Fatty acids are transported in blood to cells that use them for energy. **(a)** Explain where the fatty acids were stored and how they got into the blood. **(b)** Fatty acids are not water soluble; how are they transported in the blood?
59. Fatty acids are not water-soluble. How are they transported in cells and mitochondria?
60. Briefly describe the entry of fatty acids into the mitochondrion.
61. What is the energy cost associated with the activation of fatty acids?
62. Why is oxidation of fatty acids called beta oxidation?
63. Which coenzymes are used in beta oxidation? Which are used in fatty acid biosynthesis?
64. Calculate the net ATP yield from beta oxidation of lauric acid (14:0).
65. Why are the ATP yields from palmitic acid (16:0) and palmitoleic acid (16:1) different?
66. Provide both a definition for and a summary of *lipogenesis.*
67. Identify the fatty acid that is the product of fatty acid synthetase. How do humans make other saturated fatty acids?
68. How do animal cells make unsaturated fatty acids? Can they make linoleic acid and linolenic acid? Why?
69. Provide the name of the three ketone bodies, and identify their role in metabolism.
70. Compare the normal concentration of ketone bodies to those found in fasting organisms or diabetics in crisis.
71. Why is acidosis a potential problem in diabetics? Why can acetone sometimes be detected in a diabetic's breath?
72. How is cholesterol catabolized?
73. HMG-CoA is an intermediate in cholesterol biosynthesis. How has this knowledge been used to help patients with high blood cholesterol concentrations?

Metabolism of Nitrogen-Containing Compounds *(Section 17.8)*

74. State the two most common fates of dietary amino acids.
75. During amino acid catabolism, the amino groups of amino acids are transferred directly to what acceptor molecule?
76. When an amino acid loses its amino group it becomes a ____________-acid. What is the metabolic fate of these compounds?

77. **(a)** What compound is used for nitrogen excretion in humans? **(b)** Describe briefly how this compound is formed.
78. How do animals normally obtain amino acids?
79. Chemotherapy for cancer patients has several common side effects. What is the cause of these side effects?
80. Do anticancer drugs affect only cancer cells? Why or why not?

General Exercises

81. Aerobic catabolism of glucose yields acetyl-CoA. What other pathways yield this compound?

Challenge Exercises

82. Why do organisms have two very similar coenzymes—NAD^+/NADH and $NADP^+$/NADPH—for oxidation–reduction reactions rather than just one of them?

Appendix on Scientific Notation

Scientific notation (exponential notation) is a convenient way to express very large numbers, such as Avogadro's number—602,000,000,000,000,000,000,000—or to express very small numbers, such as the mass of an electron—0.0000000000000000000000000009109 g. Avogadro's number in scientific notation is 6.02×10^{23} and the mass of an electron is 9.109×10^{-28} g. There are no hard and fast rules for deciding when to use scientific notation, but numbers smaller than 0.01 or larger than 1000 are often shown in scientific notation.

Scientific notation expresses the value of a number in this format

$$N \times 10^n$$

where N is the *coefficient* and is a number between 1 and 10, and n is the *exponent (power)* and is a positive or negative integer. When the coefficient is multiplied times ten raised to the exponent, the answer has the same value as the number expressed in the conventional format:

$$2.7 \times 10^4 = 27,000$$

Let's take a look at exponents to get a better feel for scientific notation. For a positive exponent, the exponent represents the number of times 10 is multiplied by itself. The term 10^3 means

$$10^3 = 10 \times 10 \times 10 = 1000$$

A negative exponent represents the number of times 0.1 (1/10) is multiplied by itself. The term 10^{-4} means

$$10^{-4} = 0.1 \times 0.1 \times 0.1 \times 0.1 = 0.0001$$

When the exponential term is combined with the coefficient, the value of the number is obtained. Examples involving a positive and negative exponent are

$$5.3 \times 10^4 = 5.3 \times 10 \times 10 \times 10 \times 10 = 5.3 \times 10,000 = 53,000$$

and

$$1.39 \times 10^{-3} = 1.39 \times 0.1 \times 0.1 \times 0.1 = 1.39 \times 0.001 = 0.00139$$

Converting Conventional Numbers to Scientific Notation

To convert a number to scientific notation, first locate the decimal point in the number. If no decimal point is shown, then the decimal point lies to the right of the last digit. If the number is larger than 10, move the decimal point to the left until only one digit lies to the left of the decimal point. The number of places you move the decimal point will be the positive exponent in scientific notation. The coefficient will be expressed to the number of significant figures in the number (Chapter 1). For example, the number 113,000 would be converted to scientific notation this way:

$$113{,}000 = 1.13 \times 10^5$$

5 4 3 2 1
Places

Because the decimal point was moved five places to the left, the exponent is a positive 5.

If the number is smaller than 1, move the decimal point to the right until one digit lies to the left of the decimal point. The number of places you move the decimal point will be the negative exponent in scientific notation, and again, the coefficient will be expressed to the number of significant figures in the number.

$$0.00025 = 2.5 \times 10^{-4}$$

1 2 3 4
Places

Because the decimal point was moved four places to the right, the exponent is a negative 4.

Example Expressing Numbers in Scientific Notation

1. Write the number 10,400,000 in scientific notation.
2. Express 0.00003834 in scientific notation.

SOLUTIONS

1. The location of the decimal point in the number 10,400,000—although not shown—is to the right of the last zero. Move the decimal point to the left counting each place as you go:

7 6 5 4 3 2 1
Places

Because the decimal point was moved seven places to the left, the exponent will be a positive 7, and the exponential term is 10^7. There are three significant figures in the number, and so the coefficient is 1.04. The number 10,400,000 written in scientific notation is 1.04×10^7.

2. The decimal point in the number 0.00003834 is shown in the number. Move the decimal point to the right counting each place as you go:

$$0.00003834$$

1 2 3 4 5
Places

Because the decimal point was moved five places to the right, the exponent will be a negative 5, and the exponential term is 10^{-5}. There are four significant figures in the number, and so the coefficient is 3.834. The number 0.00003834 written in scientific notation is 3.834×10^{-5}.

Self-Test

Write these numbers in scientific notation: 1. 474,100 2. 0.027

ANSWERS

1. 4.741×10^5 2. 2.7×10^{-2}

Converting Scientific Notation to Conventional Numbers

To convert a number from scientific notation to a conventional number, use the exponent as a guide to moving the decimal point in the coefficient. For a positive exponent move the decimal point to the right a number of places equal to the exponent, adding zeros as needed; for a negative exponent move the decimal point to the left a number of places equal to the exponent, again adding zeros where needed.

$$2.91 \times 10^3 = 2910$$

Decimal point moved to the right three places

$$8.207 \times 10^{-4} = .0008207$$

Decimal point moved to the left four places

Example Converting Numbers in Scientific Notation to Conventional Numbers

Express these numbers in conventional format: 1. 3.8×10^2 2. 5.077×10^{-3}

SOLUTIONS

1. To convert 3.8×10^2 from scientific notation to conventional format, move the decimal point two places to the right adding zeros as needed:

$$3.8 \times 10^2 = 380$$

2. To convert 5.077×10^{-3} from scientific notation to conventional format, move the decimal point three places to the left adding zeros as needed:

$$5.077 \times 10^{-3} = .005077$$

Self-Test

Convert these numbers to conventional numbers: 1. 4.02×10^{-6} 2. 2.88×10^4

ANSWERS

1. 0.00000402 2. 28,800

Calculations Involving Scientific Notation

Calculations using numbers in scientific notation are most easily carried out with a scientific calculator. A number in scientific notation is entered in a calculator by

Brands of calculators use different symbols (and sometimes different order of entry) for keys. If the keys on your calculator differ from those shown here, consult your manual or talk to your instructor.

first pressing the number keys for the coefficient (press [+/−] to make the number negative), then pressing the [Exp] key, then entering the exponent (press [+/−] to make the exponent negative). To enter 7.04×10^7 press

To enter 3.4×10^{-5} press

To enter -2.66×10^{-4} press

[2] [.] [6] [6] [+/−] [Exp] [4] [+/−]

Carrying out an arithmetic operation such as multiplication or subtraction is straightforward on a calculator. Enter the first number in scientific notation, press the operator key, enter the second number, finish by pressing the equal key. To divide 2.9×10^7 by 3.8×10^4 press

[2] [.] [9] [Exp] [7] [÷] [3] [.] [8] [Exp] [4] [=]

and read the answer from the display: 763.15 . . . , which is 760 or 7.6×10^2 to the correct number of significant figures.

To subtract 4.91×10^{-3} from 1.82×10^{-2} press these keys

[1] [.] [8] [2] [Exp] [2] [+/−] [−]
[4] [.] [9] [1] [Exp] [3] [+/−] [=]

and read the answer from the display: 0.01329, which is 0.0133 or 1.33×10^{-2} to the correct number of significant figures.

Example Calculations Using Scientific Notation

1. Add 4.8×10^3 and 2.18×10^2.
2. Multiply 1.94×10^4 and 2.83×10^{-3}

SOLUTIONS

1. Use these key strokes to solve this addition:

[4] [.] [8] [Exp] [3] [+] [2] [.] [1] [8] [Exp] [2] [=]

The answer to the correct number of significant figures is 5.0×10^3.

2. The key strokes for this multiplication are

[1] [.] [9] [4] [Exp] [4] [×] [2] [.] [8] [3] [Exp] [3] [+/−] [=]

The answer to the correct number of significant figures is 54.9 (5.49×10^1)

Self-Test

1. Multiply 3.08×10^{-3} and 1.9×10^{7}.
2. Subtract 7.28×10^{4} from 5.9×10^{5}.

ANSWERS

1. 5.9×10^{4} 2. 5.2×10^{5}

Answers to Odd-Numbered Exercises

Chapter 1

1. (a) Gather data, form a hypothesis, test hypothesis
   (b) Journals and experiments would be primary sources of data; a hypothesis would be proposed that is consistent with all known data; the hypothesis would be tested by experimentation.
3. (a) A summarizing statement of observations
   (b) Hypotheses and theories are proposed explanations, not summaries.
5. Because an observation cannot prove a hypothesis correct, there is no number of times that can prove a hypothesis correct.
7. (a) A number and a unit (b) The unit is an indication of physical property and size, the number indicates the number of units. Both are needed to provide meaningful information in the measurement.
9. (a) Meter (b) Kilogram (c) Cubic meter (d) Second
11. (a) Gram, mass (b) Second, time (c) Liter, volume (d) Meter, length
13. Mass is a measure of the amount of matter; weight is a measure of the force of gravity on an object.
15. (a) 0.01 (b) 0.001 (c) 1000
17. (a) c (b) m (c) k
19. (a) Milliliter, volume (b) Nanosecond, time (c) Megagram, mass (d) Centimeter, length
21. (a) 1 (b) 1000
23. (a) 1 km (b) 1×10^6 μs (c) 1 Mg
25. 0.1 mm
27. (a) Precision is a measure of similarity in a set of measurements. (b) A fine watch is very precise; a sundial is relatively imprecise.
29. All of the numbers in a measurement, including the estimated place.
31. (a) 2 (b) 5 (c) 3
33. (a) 2 (b) 6 (c) 1 (d) 7 (e) 1
35. For multiplication and division, your answer will have the same number of significant figures as the measurement with the fewest number of significant figures. For addition and subtraction, your answer will have as many digits past the decimal point as the measurement with the fewest number of digits past the decimal point.
37. (a) 214 (b) 70,000 (c) 270 (d) 476
39. (a) 43,008 (b) 1.845 (c) −4.30 (d) 11.346
41. (a) 3.211×10^4 (b) 5.598×10^{-10} (c) 1.008×10^3 (d) 2.00×10^{-4}
43. (a) 177,000,000,000 (b) 29,200 (c) 0.00658 (d) 0.0000009477
45. (a) 1 s/1 × 10^9 ns and 1 × 10^9 ns/1 s
    (b) 1 kg/1000 g and 1000 g/1 kg
    (c) 1 L/1.057 qt and 1.057 qt/1 L
47. (a) 1 in. = 2.54 cm (b) 2.205 lb = 1 kg (c) 1000 m = 1 km
49. (a) 2.79 ft (b) 3.7 qt (c) 60.32 oz
51. (a) 8.07 m (b) 0.043 kg (c) 2310 mL
53. (a) 5.9 lb (b) 90 m (c) 39.5 qt
55. (a) 0.126 mL (b) 0.88 qt (c) 4.14×10^{-2} in.
57. 4.96 lb
59. 85 g
61. (a) 180 tablets (b) $30.60
63. (a) Absence of matter (b) Nowhere (c) No, but the density in space is extremely low.
65. Liquid
67. Definite shape and definite volume
69. (a) Physical process (b) Filter or allow grounds to settle (c) Yes, both are physical changes.
71. (a) Samples of matter that have constant composition
    (b) Many possibilities; examples include copper, carbon dioxide, and water.
73. (a) A substance containing two or more elements in fixed proportions (b) Many possibilities; examples include aspirin, baking soda, and sugar.
75. (a) O (b) Au (c) S (d) U (e) Na
77. (a) Potassium (b) Copper (c) Calcium (d) Nitrogen (e) Chlorine
79. Density is mass divided by volume; specific gravity is the density of an object/sample divided by the density of water.
81. (a) 10.5 g/mL (b) First, weigh the cube to get mass. Second, measure the length, width, and height of the cube, then multiply these measurements times each other to get volume. Third, divide the mass of the cube by the volume of the cube to get the density of the cube.
83. 168 g

85. The capacity to do work or cause a change
87. (a) Kinetic energy due to the motion of particles in matter (b) An increase in heat causes an increase in the motion of particles.
89. (a) 0.100 kJ, 23.9 cal, 0.0239 kcal, 0.0239 Cal
(b) Calorie
91. (a) English—°F, metric—°C, SI—kelvin
(b) °F (c) Yes; °C and kelvin
93. (a) 611.1°C (b) 39.4°C (c) −194°C
95. (a) 223°F (b) 40°F (c) −459.67°F
97. (a) 329 K (b) 536 K (c) 190 K
99. (a) 100°C (b) 556°C (c) −157°C
101. (a) The more precise an instrument, the greater the cost of production.
(b) You would need to know the degree of precision needed for the mass measurements in the laboratory.
103. 414.4 mL/prescription; to allow for spillage and waste, round to 420 or 425 mL.

Chapter 2

1. An atom of that element
3. Atoms of an element may vary in mass.
5. (a) B (b) Na (c) Al (d) Pt
7. (a) Sulfur (b) Tin (c) Bromine (d) Chromium
9. Proton, neutron, electron
11. (a) Proton (b) Electron (c) Neutron
13. Nucleons
15. $^1/_{12}$ of the mass of a C-12 atom; 1 amu ≈ 1.67×10^{-24} g
17. Number of protons in an atom of the element (and number of electrons in a neutral atom)
19. (a) 3 (b) 15 (c) 35 (d) 7 (e) 12 (f) 16
21. (a) Hydrogen (b) Oxygen
(c) Silver (d) Mercury
23. (a) 19 (b) 79 (c) 5
25. 30
27. The number of protons and neutrons (nucleons) in the nucleus of the isotope
29. 11, 23
31. (a) 46 neutrons, 35 protons
(b) 21 neutrons, 19 protons
(c) 10 neutrons, 8 protons
(d) 4 neutrons, 3 protons
33.

ELEMENT NAME	ELEMENT SYMBOL	ATOMIC NUMBER	MASS NUMBER	NUMBER OF PROTONS AND ELECTRONS	NUMBER OF NEUTRONS
Sulfur	S	16	32	16	16
Uranium	U	92	238	92	146
Chlorine	Cl	17	37	17	20
Potassium	K	19	41	19	22

35. Because the two isotopes are in approximately equal abundance, the atomic mass of bromine is approximately the average of the masses of the two isotopes.
37. (a) 30.97 amu (b) 22.99 amu (c) 9.01 amu
(d) 126.90 amu (e) 39.95 amu
39. (a) Tungsten (b) Boron (c) Nickel
41. The first shell
43. (a) 8 (b) 32 (c) 50
45. (a) 1st: 2, 2nd: 8, 3rd: 3 (b) 1st: 2, 2nd: 5
(c) 1st: 2, 2nd: 8, 3rd: 8, 4th: 2
47. (a) Hydrogen (b) Magnesium (c) Boron
49. In the outermost (valence) shell
51. (a) 1, 2 (b) 5, 2 (c) 2, 18
53. An electron moves from its present shell to one of higher energy (further from the nucleus).
55. Each line represents a transition of an electron from a higher energy shell to the second shell of the atom.
57. Groups or families
59. They have the same shell as their valence shell.
61. (a) 4th (b) 3rd (c) 2nd
63. (a) Sodium, magnesium, aluminum, silicon, phosphorus, sulfur, chlorine, argon
(b) Hydrogen, lithium, sodium, potassium, rubidium, cesium, francium
(c) Carbon, silicon, germanium, tin, lead
(d) Krypton
65. (a) 4 (b) 3 (c) 7
67. Group 7A elements: fluorine, chlorine, bromine, iodine, and astatine
69. Metals are located on the lower left portion of the periodic table.
71. Nonmetals are located in the upper right of the periodic table.
73. Metalloids are some of the elements located at the boundary between the metals and the nonmetals.
75. s, 2; p, 6; d, 10; f, 14
77. (a) He: $1s^2$ (b) B: $1s^2\ 2s^2\ 2p^1$
(c) Ar: $1s^2\ 2s^2\ 2p^6\ 3s^2\ 3p^6$
79. A region in space around the nucleus of an atom where electrons may be found
81. (a) 6 (b) 2 (c) 14 (d) 10
83. Radiation originating in the nuclei of atoms
85. Radioactive nuclides
87. 2
89. 112 hr
91. 0.156 g
93. The rate of carbon-14 decay of the sample is compared to the rate of decay from a sample of new plant material. From this comparison of rates, the age of the sample is estimated. Not very accurate. Because the rate of carbon-14 decay in new plant tissue is small, the differences caused by experimental error introduce considerable uncertainty.
95. Nuclear emission of beta particles (electrons)

97. (a) Polonium-218 (b) Bismuth-210
(c) Calcium-45 (d) Thorium-234
(e) Helium-3

99. Photographic film, Geiger counter, scintillation counter

101. Amount of radiation that results in the absorption of 1×10^{-2} J per kg of tissue

103. The amount of radiation (or toxic substance) that will kill 50% of the exposed population

105. The nucleus of a fissionable isotope, such as uranium-235, absorbs a neutron and then splits into two smaller daughter nuclei and several neutrons.

107. Huge amounts of energy to overcome the repulsion of the smaller nuclei

109. Because an ever-increasing number of neutrons are being produced by fission, the rate of fission increases very rapidly, leading to an explosion. This is the basis of an atomic bomb.

111. Production of electrical energy, production of nuclides used in health care, use of radiation in health care, irradiation of foods

113. 35.5 amu; yes, within the significant figures given in the exercise, the calculation is consistent with the periodic table.

115. 1×10^{-13} %

Chapter 3

1. 8; helium

3. Electron transfer—gain or loss of electrons—and sharing of electrons

5. (a) 2 in the 1st shell, 1 in 2nd shell; 1 valence electron
(b) 2 in the 1st shell, 2 in 2nd shell; 2 valence electron
(c) 2 in the 1st shell, 8 in 2nd shell, 8 in the 3rd shell, 1 in the 4th shell; 1 valence electron

7. (a) 1 (b) 6 (c) 7 (d) 8, satisfies octet rule
(e) 2 (f) 3

9. The nucleus and the core electrons

11. (a) Li· (b) :Ö: (c) ·Ċ· (d) ·Äl·

13. (a) K· (b) :Ṡ: (c) :C̈l: (d) :Är:
(e) ·Ca· (f) ·Ḃ·

15. An atom or group of atoms that has an electrical charge. The sodium atom is electrically neutral; the sodium ion has one few electrons and thus has a charge of plus one.

17. A negatively charged ion; many examples including S^{2-} and Br^{-}

19. As superscripts; the sign only for an ion with a charge of plus one or minus one, and a number followed by a sign for all other ions

21. (a) S^{2-}; anion (b) Ag^{+}; cation (c) Cr^{3+}; cation

23. A polyatomic ion is a group of atoms that have a charge; a monatomic ion is a single atom with a charge. They are polyatomic because two atoms—one carbon and one nitrogen—are present in the ion.

25. The atom gains one or more electrons to become an anion. :B̈r: + e^{-} → [:B̈r:]$^{-}$

27. (a) Gain; 2 (b) Lose; 2
(c) Lose; 1 (d) Gain; 2

29. (a) Minus 2 (b) Plus 2
(c) Plus 1 (d) Minus 2

31. (a) [:Ö:]$^{2-}$ (b) Mg^{2+} (c) Na^{+} (d) [:T̈e:]$^{2-}$

33. (a) Lithium ion, Li^{+}; sulfide ion, [:S̈:]$^{2-}$
(b) Calcium ion, Na^{+}; bromide ion, [:Ö:]$^{2-}$
(c) Aluminum ion, Mg^{2+}; chloride ion, [:F̈:]$^{-}$

35. One; Mg^{2+}, O^{2-}

37. The force of attraction between oppositely charged ions

39. (b)

41. A compound consisting of ions held together by ionic bonds

43. No. Because sulfur and oxygen are both nonmetals they form covalent bonds between atoms of them.

45. A substance that conducts electricity when in aqueous solution or melted; the salts found in the fluids of the body such as sodium chloride

47. A group of atoms held together by covalent bonds

59. (a) Ionic (b) Covalent (c) Covalent

51. Nonelectrolyte

53. They differ by the number of electron pairs in the bond; one pair in a single bond, two in a double bond and three in a triple bond.

55. A molecule containing two atoms

57. A separation of charge

59. The fluorine end; the phosphorus end F ←+ P

61. (a) Covalent (b) Ionic
(c) Covalent (d) Ionic

63. Upper right (nonmetals); lower left (metals)

65. (a) 1.8, ionic (b) 0.7, polar covalent
(c) 0, covalent (d) 1.0, polar covalent

67. The pairs of valence electrons around an atom stay as far apart as possible.

69. Pyramidal

71. No; it may contain polar bonds that cancel out, leaving the molecule nonpolar.

73. (a) Nonpolar (b) Polar (c) Polar

75. It means that nonpolar substances dissolve in nonpolar substances and polar and ionic substances dissolve in polar substances.

77. All three pairs would lie in the same plane as the boron atom and would be 120° apart.

Chapter 4

1. Identifies the elements present in the compound and identifies the number of atoms of each element present in the compound

3. It contains two atoms of oxygen and two atoms of hydrogen.

5. (a) One Mg^{2+}, one O^{2-} (b) Two Ag^+, one S^{2-}
(c) One Ba^{2+}, two I^- (d) One Cr^{3+}, three F^-
7. (a) One P, five Cl (b) One C, two S
(c) Two N, four H (d) Three H, one P, four O
9. Composed of a metal and a nonmetal;
Ionic: (a), (d), (e) Covalent: (b), (c)
11. (a) Aluminum ion (b) Bromide ion
(c) Nitride ion (d) Cesium ion
(e) Iron(III) ion
13. (a) Sodium iodide (b) Magnesium oxide
(c) Potassium nitride (d) Calcium bromide
15. (a) Iron(II) oxide (b) Copper(I) iodide
17. (a) Cupric sulfide (b) Ferrous bromide
(c) Mercuric chloride
19. (a) Cesium sulfide (b) Beryllium fluoride
(c) Mercury(II) bromide
21. (a) KBr (b) MgI_2 (c) Li_3P
23. (a) Cu_2S (b) HgF_2 (c) FeI_3
25. (a) Calcium hydroxide (b) Sodium sulfate
(c) Magnesium carbonate
27. (a) NaOH (b) SnF_2 (c) $Fe_3(PO_4)_2$
29. The atoms in a covalent compound are covalently bonded to form a molecule; the ions in an ionic compound are bonded by ionic bonds. Many possibilities exist, including water and glucose.
31. (a) Diphosphorus tetraiodide (b) Disulfur difluoride
(c) Boron trifluoride (d) Chlorine monofluoride
(e) Diselenium dichloride
33. (a) ICl (b) BP (c) SO_3 (d) SeS (e) N_2O_3
35. (a) SO (b) ICl_3 (c) IF_5 (d) SeO_2 (e) S_2O_7
37. (a), (d)
39. A concise symbolic representation of a chemical reaction
41. HgO(s)
43. 2 for HgO(s), 2 for Hg(l), 1 (but not written) for O_2(g)
45. $C_6H_{12}O_6$(s), O_2(g)
47. 1 (but not written) for $C_6H_{12}O_6$(s), 6 for O_2(g), 6 for CO_2(g), 6 for H_2O(g)
49. (a) Left side: 1 S, 4 O; right side: 1 S, 3 O; no
(b) Left side: 2 Al, 6 Br; right side: 2 Al, 6 Br; yes
(c) Left side: 2 C, 4 O; right side: 2 C, 4 O; yes
(d) Left side: 1 Al, 3 N, 10 O, 1 Na, 1 H; right side: 1 Al, 1 N, 6 O, 1 Na, 3 H; no
51. (a) Cl_2(g) becomes $2Cl_2$(g)
(b) HCl(g) becomes 2HCl(g)
(c) KNO_3(aq) becomes $2KNO_3$(aq)
53. (a) $2Na(s) + Cl_2(g) \rightarrow 2NaCl(s)$
(b) $2K(s) + 2H_2O(l) \rightarrow 2KOH(aq) + H_2(g)$
(c) $2LiOH(s) + CO_2(g) \rightarrow Li_2CO_3(s) + H_2O(l)$
55. (a) $2P(s) + 5Cl_2(g) \rightarrow 2PCl_5(s)$
(b) $Ba(OH)_2(aq) + 2HCl(aq) \rightarrow BaCl_2(aq) + 2H_2O(l)$
(c) $3Ca(NO_3)_2(aq) + 2K_3PO_4(aq) \rightarrow Ca_3(PO_4)_2(s) + 6KNO_3(aq)$
57. A mole is a unit corresponding to 6.02×10^{23} particles or objects; 6.02×10^{23} atoms; 6.02×10^{23} apples
59. 1.01×10^{24} atoms Sn
61. 22.4 mol S
63. 2.88×10^{23} molecules of water; 5.76×10^{23} atoms of hydrogen; 2.88×10^{23} atoms of oxygen
65. 0.166 mol glucose; 1.00 mol C; 2.00 mol H; 1.00 mol O
67. (a) 559 g U (b) 2.5×10^{-5} g Ag
(c) 87 g F_2
69. (a) 25.58 mol K (b) 1.3×10^{-7} mol Au
(c) 0.781 mol O_2
71. The mass of a mole of any substance; many examples exist including molar mass of iron = 55.85g/mol.
73. (a) 10.81 g/mol (b) 253.80 g/mol
(c) 149.12 g/mol (d) 27.03 g/mol
75. (a) 1.2 mol HCN (b) 0.220 mol $Cu(NO_3)_2$
(c) 1.97×10^{-4} mol KBr
77. (a) 2.8×10^2 g sucrose (b) 2.70 mg PbS
(c) 710 g H_2
79. 1.5 mol N_2, 4.5 mol H_2
81. 56.2 g NH_3
83. Yes; 10 mol $NaHCO_3$ yields 5 mol of Na_2CO_3, which is 530 g.
85. (a) Ammonium fluoride (b) Iron(II) sulfide
(c) Silicon monocarbide (d) Aluminum hydroxide
(e) Phosphorus pentabromide
(f) Beryllium bromide
87. (a) PbO (b) P_2O_4 (c) $BaCO_3$
(d) HgS (e) Si_3N_4 (f) $NaNO_3$
89. (a) Lithium chloride
(b) $2Li(s) + 2HCl(aq) \rightarrow 2LiCl(aq) + H_2(g)$
(c) 763 g LiCl
91. 121 g MgO

Chapter 5

1. Physical changes; no change in chemical composition occurs.
3. The boiling point of a substance at 1 atm of pressure; condensation
5. (a) Water, vaporizing (evaporating) (b) Oil or fat, freezing (solidifying) (c) Water, condensing
7. Heat of fusion
9. 6.0×10^3 cal
11. Heat of vaporization
13. 2.46×10^5 cal
15. (a) 1.83×10^3 cal (b) 2.99×10^3 cal
(c) 5.87×10^4 cal (d) 0.51 cal
17. The amount of heat needed to raise the temperature of 1 g of a substance 1°C.
19. 1.17×10^5 cal (117 kcal)
21. (a) 1.1×10^3 cal (b) 4.84×10^3 cal
(c) 1.55×10^4 cal
23. (a) Alcohol gained 5.04 cal.
(b) Lead lost 6.2×10^3 cal.
(c) Water gained 75.9 kcal.

25.

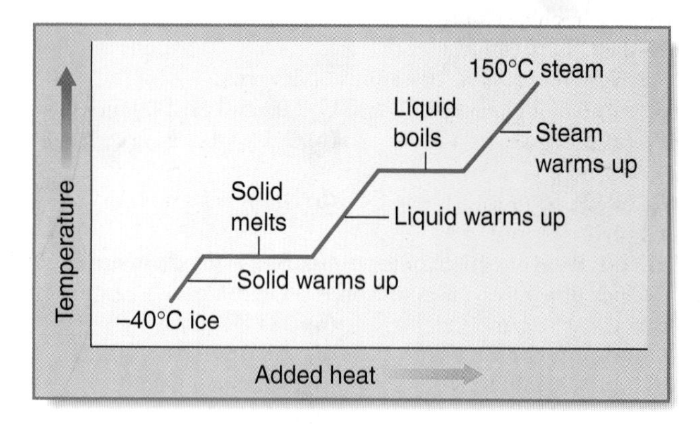

27. Gaseous; many examples including air, helium, and carbon dioxide
29. When moving gas particles strike the walls of their container, they exert a force on the wall, and a force per unit area is pressure. If the amount of gas were doubled twice as many collisions would occur with the wall, thus twice the pressure would be exerted.
31. Standard temperature, 0°C; standard pressure, 1 atm
33. (a) 746 torr (b) 0.982 atm (c) 29.4 in Hg
35. When no other factors change, the pressure (P) of a gas times the volume (V) of the gas equals a constant. No.
37. Pressure would increase by a factor of three; volume would double.
39. (a) 611 torr (b) 805 mL (c) 4.6 atm (d) 0.79 L
41. 10.9 L
43. When no other factors change, the volume of a gas divided by the temperature of the gas equals a constant.
45. The balloon would expand.
47. (a) 22°C (b) 93.6 mL (c) 62.8 L (d) 273 K
49. 29 L
51. Increase. Both an increase in temperature and a decrease in pressure will increase the volume of the gas in a balloon.
53. 4.4 L
55. 908 torr
57. 557 L
59. If no other factor changes, Avogadro's law states that the volume of a gas divided by the number of moles of a gas is a constant. Boyle's law and Charles's law relate the volume of a gas to pressure and temperature, respectively.
61. 65 L
63. PV = nRT
65. 1.5 atm
67. 0.0147 mol
69. The pressure exerted by a specific gas in a mixture of gases. Air is a mixture of gases that includes oxygen, and at STP has a pressure of 760 mm Hg; the partial pressure of oxygen in air at STP is 160 mm Hg.
71. Oxygen = 11 atm; nitrogen = 16 atm
73. Intermolecular forces
75. A bond between a very electronegative atom (such as O, N, or F) and a hydrogen atom bonded to a very electronegative atom; water
77. Solid
79. Dipole–dipole interactions, hydrogen bonds, London dispersion forces
81. The valence electrons of metal atoms are loosely held; these electrons can be "pushed" by a voltage to yield a current.
83. Liquid
85. 1.9×10^4 cal (19 kcal)
87. 1.10×10^{24} molecules of O_2; 2.20×10^{24} oxygen atoms

Chapter 6

1. Homogeneous mixture of two or more substances; many examples exist, including air and salt water.
3. The lesser component of the solution, the one being dissolved; salt
5. Iodine
7. Dipole–dipole interactions or hydrogen bonds
9. This interaction involves the attraction of the positive sodium ion to the negative end of the water molecule.
11. (a) Insoluble in water, soluble in benzene
(b) Soluble in water, insoluble in benzene
(c) Soluble in water, insoluble in benzene
(d) Insoluble in water, soluble in benzene
13. Unsaturated
15. Saturated
17. Molarity is a measure of concentration expressed in moles of solute dissolved in a liter of solution.; M = mol solute/liter of solution.
19. (a) 0.56 M (b) 0.20 M (c) 0.041 M
21. 0.062 M
23. (a) 0.30 mol (b) 0.69 mol (c) 0.114 mol
(d) 5.0×10^2 mol
25. 0.010 mol
27. Calculate the number of mol of KBr (0.025 mol) and NaCl (0.0625 mol) needed for the solutions. Use the molar mass of these compounds to convert these mole values to grams; 2.975 g KBr and 3.653 g NaCl. Weigh out the mass of each of these salts and place them in a 250-mL volumetric flask. Add enough water to dissolve the salts, then add water to bring the final volume to 250 mL.
29. (a) 2.1% $AgNO_3$ (b) 10.2% sucrose (c) 30.0% tin
31. 0.53 g
33. (a) 0.667% (b) 0.25% (c) 1.28%
35. (a) 23% (b) 14.8%
37. 28.1 mL
39. 0.21 L
41. 28 mL
43. The solubility of a gas in a liquid depends on the partial pressure of the gas (Henry's law). Because the partial pressure of oxygen is higher in the mixture than in air, more oxygen will dissolve in the blood of the patient.

45. The size of the solute particles are smaller in the solution than in a dispersion. Blood plasma
47. The particles suspended in a suspension are too large to remain in the liquid; they settle out under the influence of gravity, whereas the particles in a dispersion remain in the liquid indefinitely.
49. Solution
51. Could be either a solution or a dispersion. Shine a strong light through the liquid to see if there is a Tyndall effect.
53. An inadequate dose of any medication in the solid, an overdose of any medication in the solution; if not shaken, the dispensed medication has too much of the aqueous solution and too little of the solid.
55. The scattering of light by dispersed particles in a dispersion when a strong beam of light is passed through the mixture; a "sunbeam" in a sunlit room; any dispersion such as blood plasma or a protein solution would show the Tyndall effect.
57. Colligative properties are properties of solutions that depend only on the number of solute particles present in the solution. They all are influenced by the amount of solute added to the solution; the nature of the solute does not have an effect on them.
59. Freezing point is depressed (lowered); freezing point is depressed more.
61. The pressure that must be applied to a solution to prevent movement of water across a semipermeable membrane
63. Cell expands because water moves into cell (cell may rupture); cell shrinks (crenates) because water moves out of cell; no change
65. They would be the same in both solutions.
67. The can of diet soda would probably freeze, whereas the other two would not. The soda and beer have more solute particles (from sugar and alcohol, respectively) than the diet soda, thus the freezing point of their contents is lower.
69. The solution and blood have the same number of solute particles per volume.
71. 0.347 M $CaCl_2$; 0.347 M Ca^{2+}; 0.694 M Cl^-
73. 34.2 g
75. No; the dispersed particles and the solvent particles are both too small to be retained by the filter paper.

Chapter 7

1. An acid is a compound that can donate (give up) a hydrogen ion; many possibilities exist, including sulfuric acid, hydrochloric acid, and citric acid.
3. An ionic compound formed when an acid and base react; possibilities include sodium bicarbonate and sodium chloride.
5. The presence of one or more hydrogens at the beginning of the formula for the compound. **(a)**, **(d)**, and **(e)**
7. It is a base.
9. Ionization
11. $HCl(aq) + H_2O(l) \rightarrow H_3O^+(aq) + Cl^-(aq)$
13. Hydrochloric, nitric, and sulfuric acids
15. Weak acid
17. $HCHO_2(aq) \rightleftharpoons H^+(aq) + CHO_2^-(aq)$
19. Compounds that yield only small amounts of hydroxide ion in aqueous solutions; many possibilities exist, including sodium bicarbonate and ammonia.
21. Hydronium ion and hydroxide ion
23. Acidic
25. **(a)** Basic **(b)** Acidic **(c)** Basic **(d)** Neutral
27. 1.0×10^{-14}
29. **(a)** 1.90×10^{-3} M **(b)** 5.62×10^{-6} M **(c)** 2.55×10^{-10} M **(d)** 1.44×10^{-11} M
31. **(a)** 1.11×10^{-11} M **(b)** 4.03×10^{-6} M **(c)** 7.58×10^{-3} M **(d)** 1.86×10^{-9} M
33. **(a)** 4.394 **(b)** 9.1297 **(c)** 1.299 **(d)** 8.2337
35. **(a)** 1.5×10^{-5} M **(b)** 8.5×10^{-7} M **(c)** 3.9×10^{-13} M **(d)** 5.33×10^{-9} M
37. Acidic: **(a)** and **(c)**
Basic: **(b)** and **(d)**
39. Lemon juice, milk, blood, urine
41.

pH	[H⁺]	[OH⁻]	ACIDIC, NEUTRAL, OR BASIC?
6.41	3.9×10^{-7} M	2.6×10^{-8} M	Acidic
8.317	4.82×10^{-9} M	2.07×10^{-6} M	Basic
7.442	3.61×10^{-8} M	2.77×10^{-7} M	Basic

43. **(a)** $HBr(aq) + NaOH(aq) \rightarrow NaBr(aq) + H_2O(l)$
(b) $Ba(OH)_2(aq) + 2HNO_3(aq) \rightarrow Ba(NO_3)_2(aq) + 2H_2O(l)$
(c) $H_3PO_4(aq) + 3KOH(aq) \rightarrow K_3PO_4(aq) + 3H_2O(l)$
45. $Mg(OH)(s) + 2HCl(aq) \rightarrow MgCl_2(aq) + 2H_2O(l)^2$
47. A compound that has different colors at different pH values; a pH indicator can be selected that will change color at the pH where the titration is complete.
49. 8.51×10^{-2} M
51. **(a)** 1 **(b)** 2 **(c)** 1 **(d)** 2
53. A solution that resists changes in pH when either acid or base is added; this medication contains both aspirin and the components of a buffer. Because aspirin is acidic, the buffering components in the medication reduce the acidity of the aspirin.
55. **(a)**
57. Reactions in which the reactants gain or lose electrons; many examples exist, including the rusting of iron and the oxidation of glucose by the body.
59. Gain of electrons; the nonmetal gains one or more electrons to become an anion.
61. **(a)** Barium atoms are oxidized to Ba^{2+} ions; oxygen atoms are reduced to O^{2-} ions.

(b) Aluminum atoms are oxidized to Al^{3+} ions; bromine atoms are reduced to Br^- ions.
(c) Magnesium atoms are oxidized to Mg^{2+} ions; iodine atoms are reduced to I^- ions.

63. (a) Carbon is oxidized, oxygen is reduced.
(b) Carbon is reduced, hydrogen is oxidized
(c) Zinc is reduced, carbon is oxidized

65. The water will evaporate to form a saturated solution of the salt. The salt will then precipitate from solution.

67. The energy (heat) given up or taken in a reaction; usually expressed in J or cal per mole of reactant.

69. Rate depends on (1) the number of collisions, and (2) the energy of the collisions.

71. A catalyst speeds up a reaction but is not changed by the reaction.

73. At chemical equilibrium no net change occurs for a reversible reaction. The rates of the forward and backward reactions are the same.

75. When acid is added, the conjugate base in the buffer will react with the hydrogen ion from the acid until the system is again at equilibrium. Because of this reaction the concentration of hydrogen ion does not increase and the pH of the solution remains nearly the same.

77. Hypoventilation removes less carbon dioxide from the blood, thus it accumulates in the blood. The CO_2 reacts with water to form more carbonic acid, which ionizes to yield more hydrogen ion, making the blood more acidic than normal. Hyperventilation removes more carbon dioxide from the blood. Because less CO_2 is present, smaller amounts of carbonic acid are formed, less hydrogen ion is produced, and the blood pH is more alkaline than normal.

Chapter 8

1. The chemistry of carbon-containing compounds.

3. Organic; **(b)**, **(c)**; Inorganic: **(a)**, **(d)**

5. Inorganic; in general, inorganic compounds are water-soluble and are often ionic (or acids), whereas organic compounds are often water-insoluble and molecular.

7. Organic; organic compounds are molecular compounds and the interactions between molecules are usually weaker and more easily broken than those in ionic, inorganic compounds.

9. Organic

11. Carbon is in group 4A and thus has four electrons in its valence shell.

13. An organic compound containing only atoms of carbon and hydrogen; Many possible answers exist: Three are methane (natural gas), propane, ethene (ethylene).

15. Carbon-to-carbon single bonds and carbon-to-hydrogen single bonds

17. Tetrahedral

19. Propane

21. Propane:

```
    H  H  H
    |  |  |
 H—C—C—C—H
    |  |  |
    H  H  H
```

Pentane:

```
    H  H  H  H  H
    |  |  |  |  |
 H—C—C—C—C—C—H
    |  |  |  |  |
    H  H  H  H  H
```

23. (a)

```
    H  H  H
    |  |  |
 H—C—C—C—H
    |  |  |
    H  H  H
```

(b)

```
    H  H  H  H
    |  |  |  |
 H—C—C—C—C—H
    |  |  |  |
    H  H  |  H
        H—C—H
           |
           H
```

25. Ethane: CH_3CH_3
Nonane: $CH_3CH_2CH_2CH_2CH_2CH_2CH_2CH_2CH_3$

27. $CH_3CH_2CH_3$

29. Models provide three-dimensional images of molecules.

31. Wedges indicate a bond attached to a group that projects up out of the page, toward the viewer. A dashed line represents a bond to a group that projects below the plane of the page, away from the viewer.

33. Straight chain alkanes: **(a)**, **(c)**; branched alkane: **(b)**

35. Compounds having the same composition but with different structures

37. (a) Same compound (b) Isomers
(c) Different compounds

39. $CH_3CH_2CH_2CH_2CH_2CH_3$

```
CH3CHCH2CH2CH3
   |
   CH3
```

```
CH3CH2CHCH2CH3
      |
      CH3
```

```
   CH3
   |
CH3CCH2CH3
   |
   CH3
```

```
   CH3
   |
CH3CHCHCH3
      |
      CH3
```

41. Alkyl group names are derived from alkane names by dropping the *-ane* suffix of the alkane and replacing it with *-yl*. Prop*yl* is derived from prop*ane*.

43. Prefix: indicates number, location, and kinds of alkyl groups present in the molecule; parent: indicates longest continuous chain of carbon atoms; suffix: indicates the class of compound

45. **(a)** 4-ethyl-2-methylheptane
(b) 2,2,5-trimethylhexane
(c) 3,6-dimethylnonane
(d) 5-ethyl-2,2,5-trimethylheptane

47. **(a)**
$$\begin{array}{c} CH_3 \quad CH_2CH_2CH_3 \\ CH_3CHCHCHCH_2CH_2CH_2CH_2CH_3 \\ CH_2CH_3 \end{array}$$

(b)
$$\begin{array}{c} H_3C \quad CH_3 \\ CH_3C{-}CCH_2CH_2CH_2CH_3 \\ H_3C \quad CH_3 \end{array}$$

(c)
$$\begin{array}{c} CH_3 \\ CH_3CHCH_2CH_3 \end{array}$$

(d)
$$\begin{array}{c} H_3C \quad CH(CH_3)_2 \\ CH_3CH_2C{-}CHCH_2CH_2CH_2CH_3 \\ CH_3 \end{array}$$

49. An alkane containing one or more rings of carbon atoms.

51. **(a)** Methylcyclopentane; **(b)** Ethylcyclobutane;
(c) Propylcyclohexane; **(d)** Methylcyclooctane;
(e) 1,1-diethylcyclopentane
(f) 1-ethyl-4-methylcyclohexane

53. **(a)** 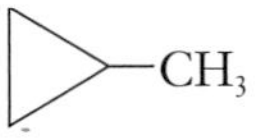 (cyclopropane ring)–CH_3

(b) 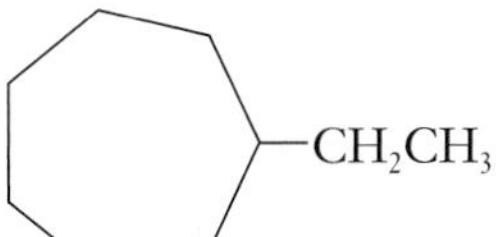 (cycloheptane ring)–CH_2CH_3

(c) 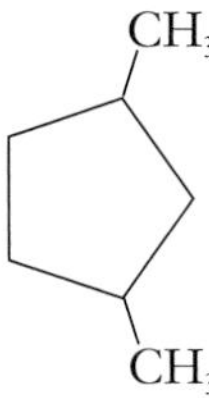 (cyclopentane ring with CH_3 and CH_3)

(d) 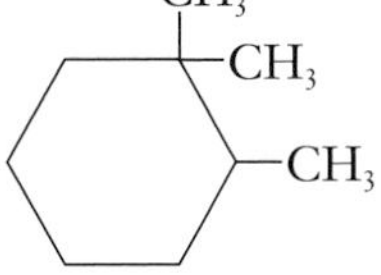 (cyclohexane ring with CH_3, CH_3 on one carbon and CH_3 on the adjacent carbon)

(e) 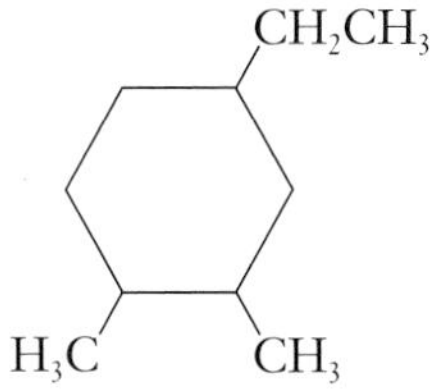 (cyclohexane ring with CH_2CH_3, H_3C and CH_3)

55. The numbering for the carbon atoms that have substituents should be the lowest possible numbers—1 and 2 in this case.

57. *cis*-1,3-dimethylcyclopentane (cyclopentane ring with CH_3 and CH_3)

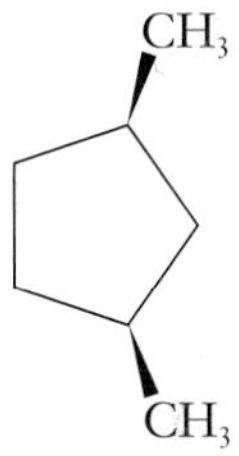

59. Because the carbon-to-carbon bond angles in alkanes are 109°, it is not possible for all six carbon atoms to lie in the same plane.

61. Chiral

63. **(b)**, **(c)**

65. **(b)**

67.
$$\begin{array}{c} CH_3 \\ CH_3CH_2CHCH_2CH_2CH_2CH_3 \end{array}$$

Carbon 3 is chiral.

69. Single bonds

71. Those between methane molecules are weaker than those between butane molecules. The larger the contact between the surface area of nonpolar molecules, the stronger the London dispersion forces.

73. No

75. Carbon dioxide and water

77. **(a)** $CH_3CH_2CH_3(g) + 5O_2(g) \rightarrow 3CO_2(g) + 4H_2O(g)$

(b) $2CH_3CH_2CH_2CH_2CH_2CH_3(l) + 19O_2(g) \rightarrow 12CO_2(g) + 14H_2O(g)$

(c) 2 (cyclopentane ring) $+ 15O_2(g) \rightarrow 10CO_2(g) + 10H_2O(g)$

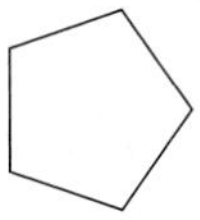

79. **(a)** $CH_3CH_3 + Br_2 \xrightarrow{\text{Light}} CH_3CH_2Br + HBr$

(b) $CH_4 + Cl_2 \xrightarrow{\text{Light}} CH_3Cl + HCl$

81. The structure of the compound

83. Alkynes: carbon-to-carbon triple bond; alkenes: carbon-to-carbon double bond; aromatic hydrocarbons: the aromatic ring

85. $C{-}N\langle$ (N bonded to C with two additional bonds)

87. (a) Aromatic ring, carboxyl group, ester
(b) Aromatic ring, hydroxyl group, amino group, carboxyl group

89. (a) 4.51×10^5 kJ
(b) $CH_3CH_2CH_3(g) + 5O_2(g) \rightarrow 3CO_2(g) + 4H_2O(g)$
(c) 59.9 lb

Chapter 9

1. A hydrocarbon containing fewer hydrogen atoms than an alkane with the same number of carbon atoms.
3. Alkenes, alkynes, aromatic hydrocarbons
5. (a) Petroleum refining; (b) used to make polymers for plastics and fibers, alcohol, antifreeze, and acetic acid.
7. Many possible answers: propene

9. (a) $CH_2{=}CHCH_2CH{=}CHCH_3$

(b)

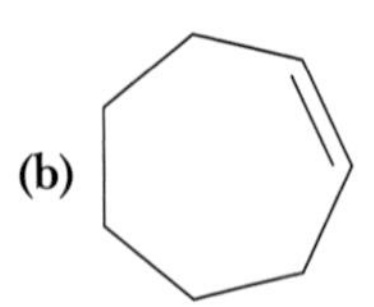

(c) $CH_2{=}CHCH_3$
(d) $CH_3CH_2CH{=}CHCH_2CH_2CH_3$

11. (a) 3-heptene (b) 4,5-dimethyl-1-hexene
(c) 2-methyl-3-propylcyclohexene

13. (a)

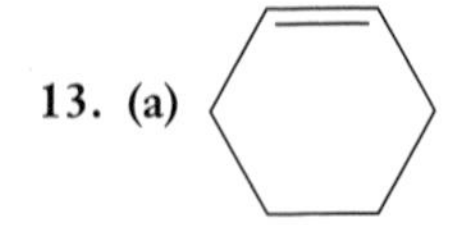

(b)

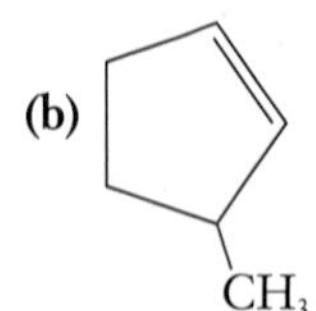

15. Single covalent bonds
17. The orientation of substituents attached to the two carbon atoms of the double bond; *cis* if similar substituents are on the same side of the double bond and *trans* if they are on opposite sides
19. No; because the two hydrogens on carbon 1 are identical

21. (a)
$$\begin{array}{c} H \\ | \\ CH_3CH_2C{=}CCH_2CH_2CH_3 \\ \phantom{CH_3CH_2C{=}}| \\ \phantom{CH_3CH_2C{=}}H \end{array}$$

(b)
$$\begin{array}{c} CH_3CH_2C{=}CCH_2CH_2CH_2CH_3 \\ |\quad| \\ H\quad H \end{array}$$

(c)
$$\begin{array}{c} CH_3 \\ | \\ CH_3C{=}CCHCH_3 \\ |\quad| \\ H\quad H \end{array}$$

(d)
$$\begin{array}{c} CH_3\quad H \\ |\qquad| \\ CH_3CCH_2C{=}CCH_2CH_2CH_2CH_3 \\ |\qquad\quad| \\ CH_3\qquad H \end{array}$$

23. The number for the position of the double bond must be the lowest possible; 2-butene is correct.
25. (a) *Trans*-9-methyl-4-decene
(b) *Cis*-4-methyl-2-pentene
(c) *Trans*-3,4-dimethyl-3-heptene
27. Ethyne (acetylene), $HC{\equiv}CH$; fuel for high-temperature torches
29. No; there are no geometric isomers for carbon–carbon triple bonds.
31. (a) Ethyne; acetylene
(b) 2-hexyne
(c) 4,4-dimethyl-2-pentyne
(d) 2,6,8-trimethyl-4-nonyne

33. (a) $CH_3CH_2C{\equiv}CCH_2CH_2CH_3$

(b)
$$\begin{array}{c} CH_3CHCH_2C{\equiv}CCH_2CH_2CH_2CH_3 \\ | \\ CH_3 \end{array}$$

(c)
$$\begin{array}{c} CH_3CH_2C{\equiv}CCHCH_2CH_2CH_2CH_2CH_3 \\ | \\ CH_2CH_3 \end{array}$$

(d)
$$\begin{array}{c} CH_3\quad CH_3 \\ |\qquad| \\ CH_3CC{\equiv}CCCH_3 \\ |\qquad| \\ CH_3\quad CH_3 \end{array}$$

35. London dispersion forces; the larger the molecules, the greater the London dispersion forces.
37. Addition, oxidation, polymerization
39. (a) $CH_3CH_2CH_2CH_3$

(b) $CH_3CH(CH_3)CH_2CH_3$

(c) $CH_3CH_2CH_2CH_3$

41. To the carbon atoms that were linked by the carbon–carbon double bond.

43. (a) $CH_3CHClCHClCH_2CH_3$

(b) $CH_3CH(CH_3)CH_2CH_3$

(c) $CH_2ClCHClCHClCH_2Cl$

45. When a hydrogen-containing asymmetric reagent adds to a carbon–carbon double bond, the hydrogen will add to the carbon atom of the double bond that has the most hydrogen atoms attached to it. No; both carbon atoms have the same number of hydrogen atoms.

47. (a) $CH_3CH_2CH{=}CH_2 + HCl \rightarrow CH_3CH_2CHClCH_3$

(b) $CH_3C(CH_3){=}CH_2 + HBr \rightarrow CH_3CBr(CH_3)CH_3$

(c) 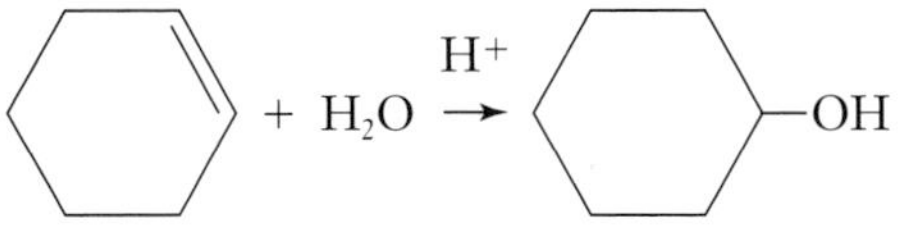 cyclohexene $+ H_2O \xrightarrow{H^+}$ cyclohexane—OH

49. (a) $CH_2{=}CHCH_3 \xrightarrow{\text{[mild oxidation]}} HOCH_2CH(OH)CH_3$

(b) $CH_2{=}CHCH_3 \xrightarrow{\text{[strong oxidation]}} H_2C{=}O + O{=}CHCH_3$

(c) $CH_3CH(CH_3)CH{=}C(CH_3)CH_3 \xrightarrow{\text{[strong oxidation]}}$ $CH_3CH(CH_3)CHO + O{=}C(CH_3)CH_3$

51. Many $CH_2{=}CH_2 \rightarrow \ldots—CH_2—CH_2—CH_2—CH_2—CH_2—CH_2—CH_2—CH_2—CH_2—CH_2—\ldots$

53. Aromatic ring

55. Electrons that are not localized between two atoms; they are delocalized throughout a portion or all of the molecule. Aromatic hydrocarbons

57. (a) 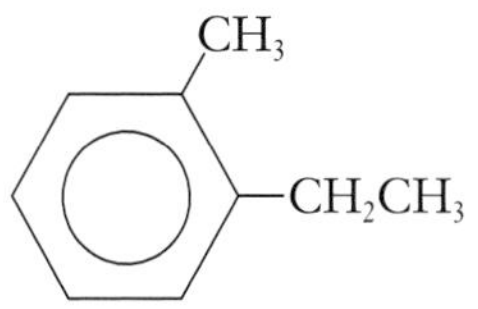 benzene ring with CH_3 and CH_2CH_3

(b) 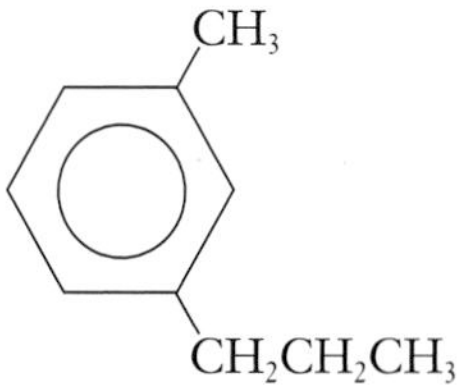 benzene ring with CH_3 and $CH_2CH_2CH_3$

(c) 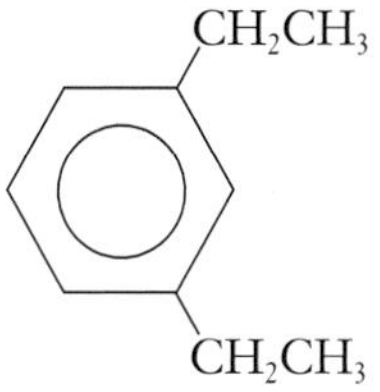 benzene ring with CH_2CH_3 and CH_2CH_3

59. (a) Propylbenzene
(b) 1,2-diethylbenzene (*o*-diethylbenzene)
(c) 2,4-diethyltoluene

61. (a) 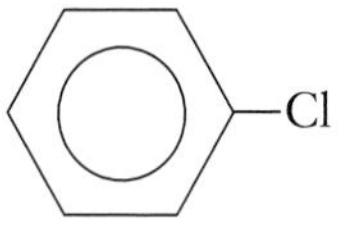 benzene ring—Cl **(b)** 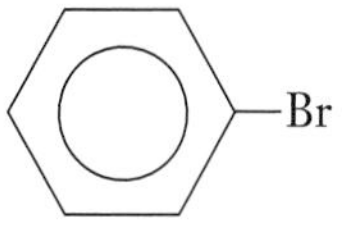 benzene ring—Br

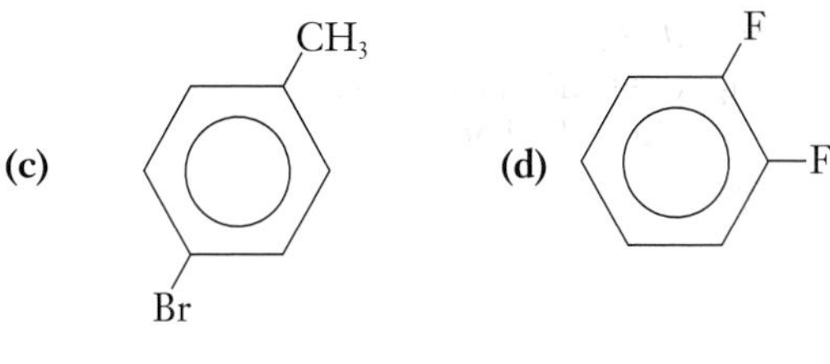

(e)

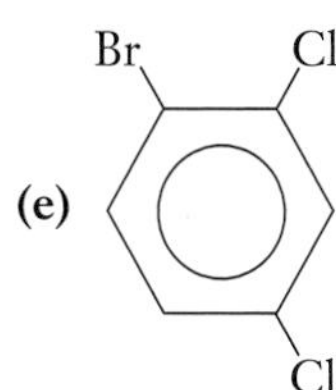

63. (a) 1-bromo-3-fluorobenzene (*m*-bromofluorobenzene)
(b) 2-nitrotoluene (*o*-nitrotoluene)
(c) 1,2-dibromo-3-nitrobenzene

65. Two or more attached (fused) rings; answer will vary—examples include naphthalene and anthracene.

67. Insoluble in water; generally soluble in nonpolar solvents

69. **(a)** (benzene) + SO_3 $\xrightarrow{H_2SO_4}$ (benzene)–SO_3H

(b) (benzene) + Cl_2 $\xrightarrow{FeCl_3}$ (benzene)–Cl + HCl

(c) (benzene) + $CH_3CHClCH_3$ $\xrightarrow{AlCl_3}$ (benzene)–CH(CH_3)(CH_3) + HCl

71. **(a)** $2HC{\equiv}CH + 5O_2 \rightarrow 4CO_2 + 2H_2O$
(b) 1.20×10^3 kcal; 5.00×10^3 kJ
(c) At STP, 215 L

Chapter 10

1. They contain one or more atoms of elements other than carbon and hydrogen.

3. A phenol has a hydroxyl group attached directly to a carbon atom of an aromatic ring; an alcohol has a hydroxyl group attached to a carbon atom that is not in an aromatic ring.

5. **(a)** HO–(cyclohexane)–CH_3

(b) $CH_3CH(CH_3)CH(OH)CH(CH_3)CH_2CH_2CH_2CH_3$ (OH on C-3; CH_3 groups on C-2 and C-4)

(c) $HOCH_2CH_2CH_2OH$

(d) $CH_3C(OH)(CH_3)CH_2CH_3$

(e) $HOCH_2CH_2CH_2CH_2CH_2CH_3$

7. **(a)** 2,4,4-trimethyl-2-pentanol
(b) 3-methyl-2-butanol
(c) 2,2,5-trimethylcyclohexanol
(d) 1,6-hexanediol
(e) 3-ethyl-3-pentanol

9. **(a)** 1° **(b)** 2° **(c)** 3° **(d)** 1°

11. **(a)** 4-chlorophenol (*p*-chlorophenol)
(b) 2-ethylphenol (*o*-ethylphenol)
(c) 2,4-dichlorophenol

13. **(a)** HO–(benzene)–NO_2 (para)

(b) (benzene) with OH and, ortho to it, $–CH_2CH_2CH_3$

(c) O_2N–(benzene)–OH, with a second NO_2 ortho to OH

(d) Cl–(benzene)–OH, with Cl on both carbons ortho to OH

15. These alcohols are small polar compounds; these alkanes are nonpolar.

17. Phenol is somewhat water soluble; larger phenols are generally insoluble.

19. Carbolic acid, phenol; too toxic and corrosive, and unpleasant to use

21. **(a)** $CH_3CH_2C(OH)(CH_2CH_3)CH_2CH_3 \xrightarrow{H^+} CH_3CH_2C(CH_2CH_3){=}CHCH_3 + H_2O$

(b) (cyclohexanol) $\xrightarrow{H^+}$ (cyclohexene) + H_2O

(c) (2-ethylcyclopentanol, OH and $–CH_2CH_3$) $\xrightarrow{H^+}$ (cyclopentene)–CH_2CH_3 + H_2O

23. **(a)** $CH_3CH_2CH_2CCH_3$ (with =O on the C)

(b) $CH_3CH_2CHCCH_3$ (with =O on the second C from the right and CH_3 on the CH)

(c) 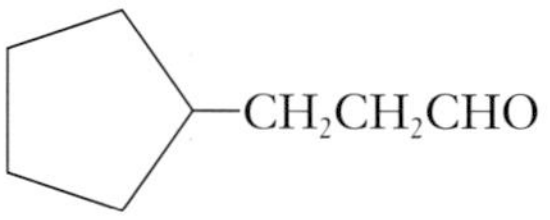 —CH_2CH_2CHO

(d) No reaction

25. 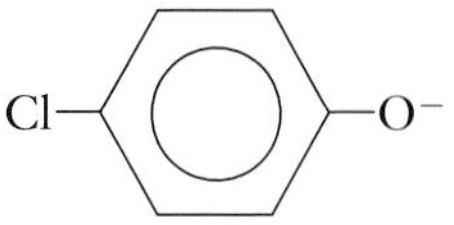 Cl— —O^- Na^+; sodium 4-chlorophenoxide

27. An oxygen atom single bonded to two carbon atoms

29. **(a)** Butyl methyl ether
(b) Ethyl propyl ether
(c) Cyclopentyl ethyl ether
(d) Dipropyl ether

31. **(a)** 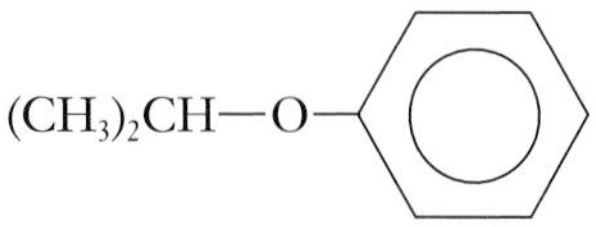 $(CH_3)_2CH$—O—

(b) 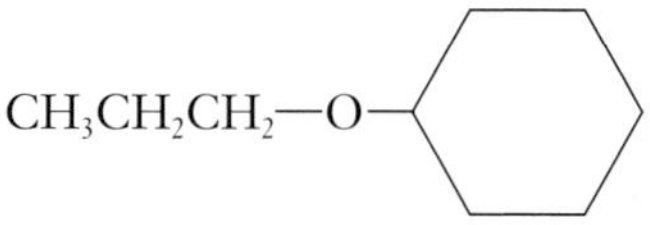 $CH_3CH_2CH_2$—O—

(c) 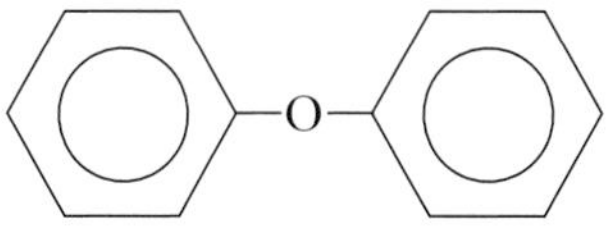 —O—

(d) $CH_3CH_2CH_2CH_2OCH(CH_3)_2$

33. Because ethers contain oxygen they can hydrogen bond with water molecules.

35. **(a)** Sulfhydryl group **(b)** Sulfhydryl—SH, hydroxyl—OH; because sulfur and oxygen are in the same group, group 6A, thiols and alcohols have somewhat similar properties.

37. **(a)** Methanethiol (methyl mercaptan)
(b) 3-pentanethiol **(c)** 4-ethyl-1-heptanethiol

39. **(a)** $CH_3CH_2CCH_2CH_3$ (with CH_3 and SH on the central C)

(b) $HSCH(CH_3)_2$

(c) 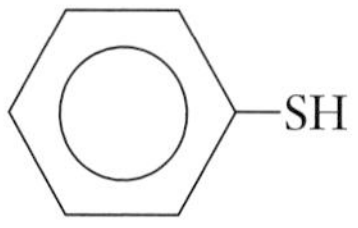 —SH

41. R—S—S—R $\xrightarrow{\text{[reduction]}}$ 2RSH

43. **(a)** 1° **(b)** 3° **(c)** 2°

45. The nitrogen atom in a quaternary ammonium salt is bonded to four carbon atoms and has a positive charge. In an amine the nitrogen atom is bonded to one, two, or three carbon atoms and is neutral.

47. A cyclic compound containing one or more noncarbon atoms in a ring

49. **(a)** 1-aminopropane (propylamine)
(b) Dicyclohexylamine
(c) N,N-diethylaniline

51. **(a)** N-ethylaniline
(b) Dimethylpentylamine
(c) Isopropylamine

53. **(a)** $CH_3CH_2CH_2N(CH_3)_2$

(b) 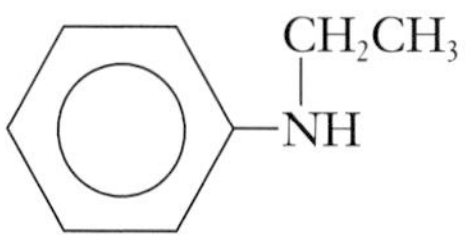 —NH (with CH_2CH_3 on the N)

(c) $CH_3CH_2NHCH(CH_3)_2$

55. Hydrogen bonding

57. The organic ammonium ions are weak acids and the ions of most other salts are not. (Ammonium ion is, of course, also a weak acid.)

59. **(a)** 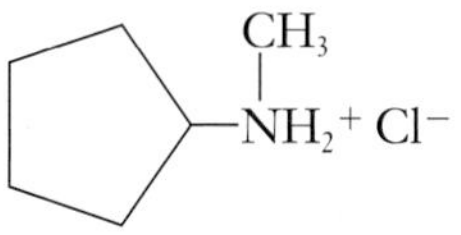 —NH_2^+ Cl^- (with CH_3 on the N)

(b) $2(CH_3)_3NH^+$ SO_4^{2-}

61. **(a)** Cyclohexylmethylammonium chloride
(b) Trimethylammonium sulfate

63. The halogen atom in an aryl halide is attached to a carbon atom of an aromatic ring. In an alkyl halide the halogen atom is attached to any carbon atom except one in an aromatic ring.

65. **(a)** 1,2-dichloroethane
(b) trichloromethane (chloroform)
(c) 1-bromo-2-methylcyclohexane

67. **(a)** 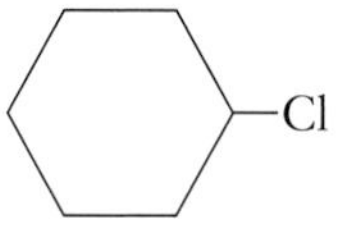 —Cl

(b) CI_4

(c) $CH_3CHCHCHCH_2CH_3$ (with CH_3 on C-2, F on C-3, CH_3 on C-4)

69. (a) 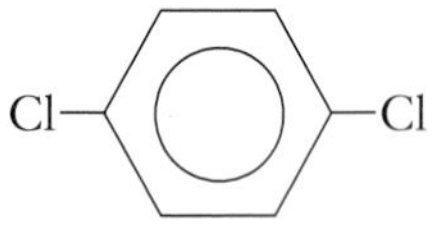 Cl — — Cl

(b)  Br, Br, Br

71. Generally insoluble in water and soluble in nonpolar organic solvents
73. They contain both an amino group and a carboxyl group. Serine and threonine are alcohols; tyrosine is a phenol; cysteine is a thiol.

Chapter 11

1. 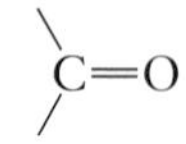 $\rangle C{=}O$

3. An aldehyde has a hydrogen atom bonded to the carbon of a carbonyl group; a ketone has two carbon atoms bonded to this carbon atom.
5. (a) 3,5-dimethylheptanal
   (b) 2,3-dichloropentanal
   (c) 2,3-dibromobenzaldehyde
7. (a) 2-bromopropanal (α-bromopropionaldehyde)
   (b) 4-hydroxybutanal (γ-hydroxybutyraldehyde)
   (c) 4-chlorobenzaldehyde (p-chlorobenzaldehyde)
9. (a) $CH_3CH_2CH_2CHO$

   (b) $CH_3CHClCHO$

   (c) $CH_3CH_2CH_2CH_2CH_2CHCHCHO$ (with Br on C-3 and OH on C-2)

   (d) 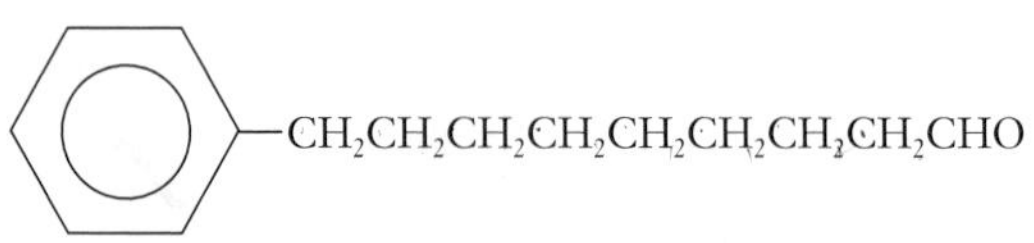 —$CH_2CH_2CH_2CH_2CH_2CH_2CH_2CH_2CHO$

   (e) 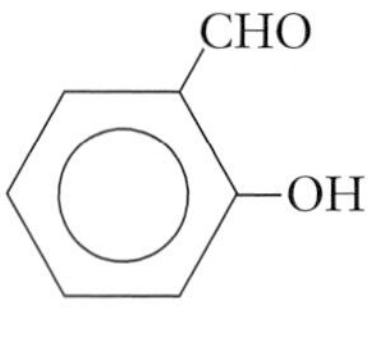 CHO, OH

11. (a) $CH_3CH_2CH_2CH_2OH$; 1-butanol

    (b) $CH_3C(CH_3)_2CH_2OH$; 2,2-dimethyl-1-propanol

13. Nonanal

15. (a) 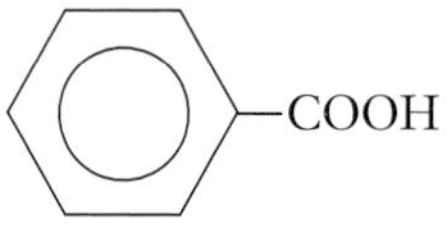 —COOH

    (b)  —$CH_2CH_2CH_2CH_2COOH$

17. (a), (d)
19. An alcohol and either an aldehyde or a ketone
21. (a) CH_3CHOH with OCH_3

    (b) CH_3CHOH with $OCH_2CH_2CH_3$

23. (a) 2-pentanone
    (b) 3,5-dimethyl-2-hexanone
    (c) Acetophenone
    (d) 1-chloropropanone
25. (a) Methyl propyl ketone
    (b) Diethyl ketone
27. (a) $CH_3\overset{O}{\overset{\|}{C}}CH_3$

    (b) $CH_3\overset{O}{\overset{\|}{C}}CH_2CCl_2CH_2CH_3$

    (c) $CH_2Br\overset{O}{\overset{\|}{C}}CH_2Br$

29. The carbonyl carbon should get the smallest possible number; 4-octanone is correct.

31. Secondary alcohol

33. **(a)** $CH_3CH(OH)CH(CH_3)_2$

(b) 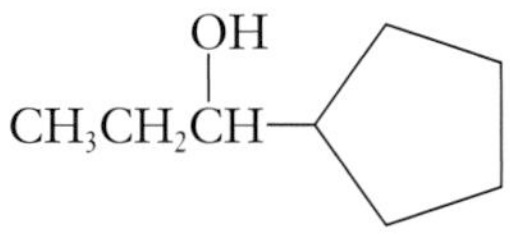 $CH_3CH_2CH(OH)$—(cyclopentyl)

35. Acetone

37. Acetone and ethanol

39. The hydroxyl group

41. **(a)** 3-hydroxypentanoic acid
(b) 3-bromobenzoic acid
(c) 2,2-dimethylpropanoic acid
(d) 2-ethylbutanoic acid

43. **(a)** 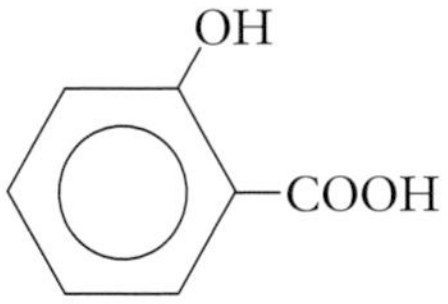 (benzene ring with OH and, on the adjacent carbon, COOH)

(b) 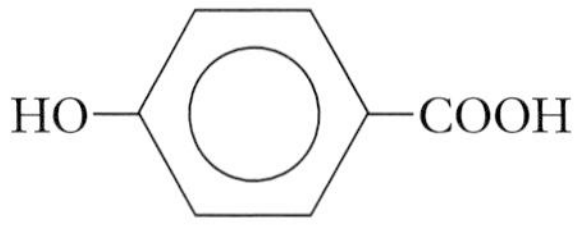 HO—(benzene ring)—COOH

(c) $ICH_2CH_2CH_2COOH$

(d) $HO—C(CH_2COOH)_2—COOH$

45. **(a)** Ethanoate (acetate) **(b)** Benzoate
(c) Butanoate (butyrate)

47. **(a)** Sodium ethanoate **(b)** Sodium benzoate
(c) Sodium butanoate

49. **(a)** 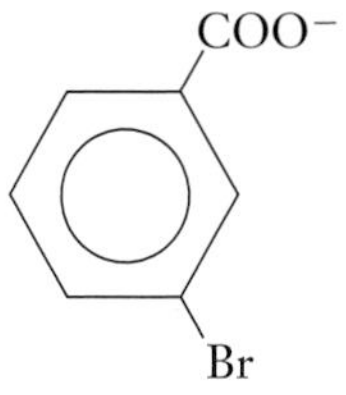 (benzene ring with COO^- and Br meta to each other)

(b) $CH_3CH(CH_3)CH_2CH_2CH(CH_3)COO^-$

51. **(a)** $CH_3CH_2COO^-$ Na^+; sodium propanoate
(b) $(CH_3)_2CHCH_2CH_2CH_2COO^-$ K^+; potassium 5-methylhexanoate
(c) 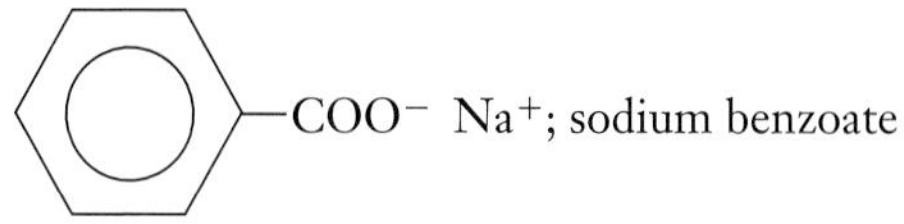 (benzene ring)—COO^- Na^+; sodium benzoate

53. **(a)** Methanoic acid (formic acid), ethanol
(b) Benzoic acid, methanol

55. **(a)** Ethanamide
(b) N-methylbenzamide

57. Carboxylic acid and alcohol

59. **(a)** Cyclopentyl ethanoate **(b)** Pentyl propanoate
(c) Propyl methanoate **(d)** Methyl benzoate

61. **(a)** Isopropyl butanoate **(b)** Methyl pyruvate

63. **(a)** $CH_3CH_2CH_2C(=O)OCH_3$

(b) 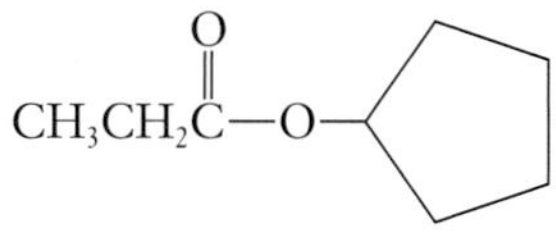 $CH_3CH_2C(=O)—O—$(cyclopentyl)

(c) 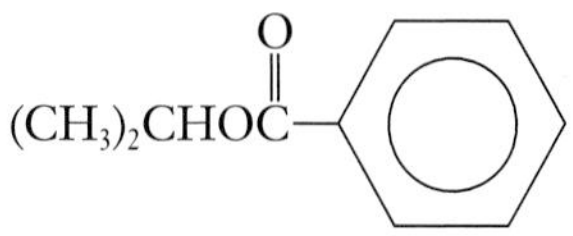 $(CH_3)_2CHOC(=O)—$(benzene ring)

(d) 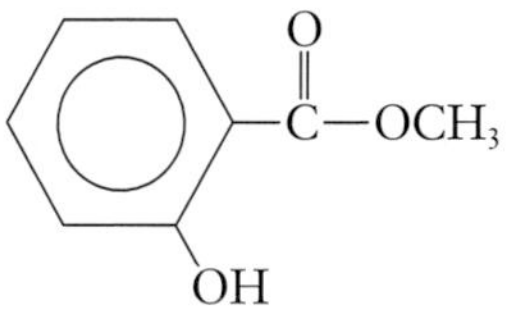 (benzene ring with OH and, on the adjacent carbon, $C(=O)—OCH_3$)

65. **(a)** Methanamide
(b) Benzamide

67. **(a)** $CH_3C(=O)NH_2$

(b) $CH_3CH(CH_3)CH_2C(=O)NH_2$

69. **(a)** Propanoic acid and ammonia
(b) Benzoic acid and ammonia
(c) A dicarboxylic acid and a diamine

71. **(a)** $CH_3CH_2C(=O)NH_2$

(b) 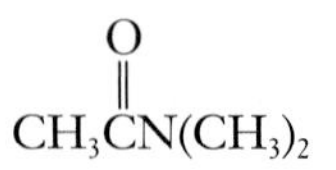 $CH_3\overset{O}{\overset{\|}{C}}N(CH_3)_2$

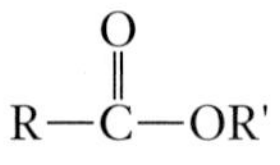

73. 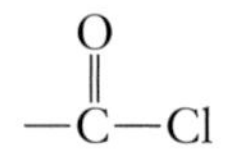 $-\overset{O}{\overset{\|}{C}}-Cl$

75. Ester: $R-\overset{O}{\overset{\|}{C}}-OR'$ Thiol ester: $R-\overset{O}{\overset{\|}{C}}-SR'$

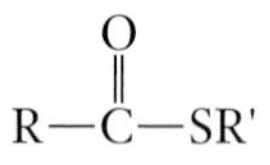

77. Aldehyde

79. (a) 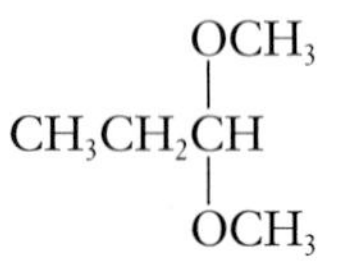 $CH_3CH_2\overset{OCH_3}{\overset{|}{\underset{\underset{OCH_3}{|}}{C}}}H$

(b) 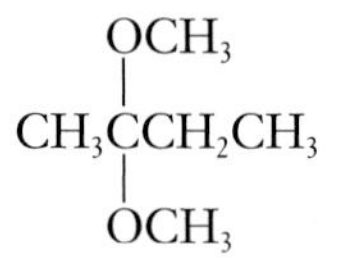 $CH_3\overset{OCH_3}{\overset{|}{\underset{\underset{OCH_3}{|}}{C}}}CH_2CH_3$

81. (a) Ester; 2-bromoethyl methanoate
(b) Aldehyde; 4-chlorobutanal
(c) Ketone; 3-pentanone
(d) Carboxylic acid; butanoic acid

83. Answers will vary.

Chapter 12

1. Monosaccharide: the smallest carbohydrates; the basic units found in larger carbohydrates
Oligosaccharide: carbohydrate made up of two to ten monosaccharides
Polysaccharide: carbohydrate containing many monosaccharides
3. Because many have formulas of this general form—$C_n(H_2O)_n$—they appear to be "hydrates" of carbon.
5. (a) A three-carbon sugar containing an aldehyde group
(b) A seven-carbon sugar containing a keto group
(c) A five-carbon sugar containing an aldehyde group
7. Glyceraldehyde and dihydroxyacetone
9. (a) L-aldotetrose (b) D-ketopentose
11. Several possibilities, including:

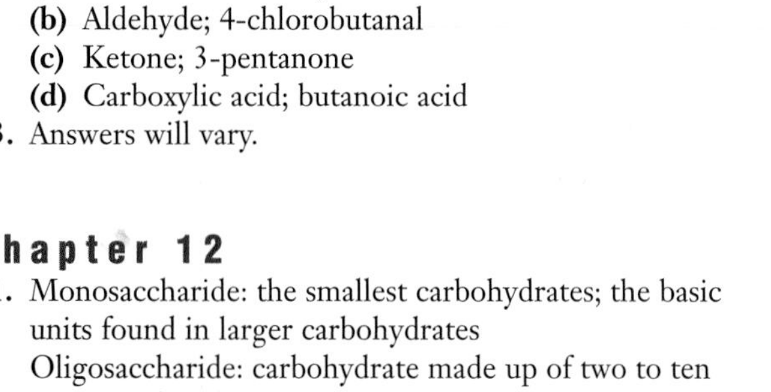

(a)
$$\begin{array}{c} CHO \\ | \\ H-C-OH \\ | \\ H-C-OH \\ | \\ H-C-OH \\ | \\ CH_2OH \end{array}$$

(b)
$$\begin{array}{c} CH_2OH \\ | \\ C{=}O \\ | \\ H-C-OH \\ | \\ H-C-OH \\ | \\ H-C-OH \\ | \\ CH_2OH \end{array}$$

(c)
$$\begin{array}{c} CHO \\ | \\ H-C-OH \\ | \\ HO-C-H \\ | \\ CH_2OH \end{array}$$

13. Interchange the hydrogen atom and hydroxyl group on carbon number five.
15. (a) D-ribose (b) L-glyceraldehyde
17. (a) Oligosaccharide (b) Monosaccharide
(c) Oligosaccharide (d) Polysaccharide
19. Both are aldopentoses that differ by a substituent on carbon 2. Ribose has a hydroxyl group, whereas 2-deoxyribose has a hydrogen atom instead.
21. Carbon number one; carbon number two
23. (a); Carbon number 2
25. These are the five-membered and six-membered ring forms, respectively, of the sugars that result from the formation of a hemiacetal.
27. They are quite polar because they contain several hydroxyl groups.
29. Three; eight
31. (a)
33. Each of them contain an anomeric carbon atom that can return to a free aldehyde.
35. Ketoses such as fructose can slowly isomerize to an aldose.
37. (a) Fructose, α (b) glucose, α
(c) 2-deoxyribose, β
39. (c)
41. (a) Glucose and fructose (b) Glucose
(c) Glucose and galactose
43.

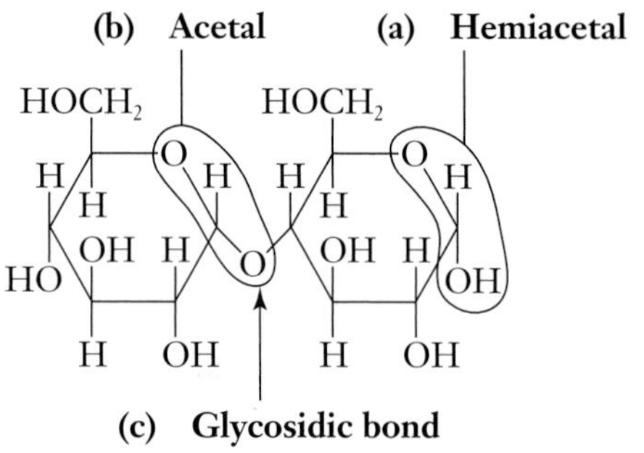

45. Amylose, amylopectin, glycogen, and cellulose
47. Amylose is a linear polymer of glucose, whereas amylopectin is a branched polymer of glucose because it contains some α 1⟶6 glycosidic bonds.
49. Glycogen stores glucose in the liver and muscles. Because glycogen is highly branched, it can readily be broken down to provide glucose as needed.
51. The stomachs of cattle contain microorganisms that produce cellulase, whereas humans' do not.
53. This polysaccharide consists of a repeating disaccharide containing sulfate groups. Heparin is an anticoagulant.
55. (a) Polyhydroxy aldehyde or ketone or a derivative of them
(b) Five-carbon sugar
(c) Sugar containing an aldehyde group
(d) Sugar containing a keto group

(e) Six-carbon sugar containing a keto group
(f) Stereochemical family related to the stereochemistry of L-glyceraldehyde
(g) Sugar in a five-member ring
(h) Carbon atom of a ring that had been the carbonyl carbon of the linear sugar
(i) Two sugars that are stereoisomers at the anomeric carbon atom
(j) Bond that links monosaccharides (and other molecules) in larger carbohydrates (and other polymers)
(k) Sugars that are the basic building blocks of the larger carbohydrates
(l) Carbohydrate containing two monosaccharides
(m) Carbohydrate containing many monosaccharides
(n) Storage polysaccharide for glucose found in animals
(o) Structural polysaccharide of glucose found in plants
(p) A linear starch molecule containing glucose
(q) Sugar that reduces metal ions in redox reactions
(r) Milk sugar—a disaccharide containing glucose and galactose

57. **(a)**; **(b)** weakly; **(f)**

59. **(a)** 104 Cal calculated, 100 Cal expressed to correct significant figures. **(b)** 58%, 60% to the correct number of significant figures

Chapter 13

1. Their solubility properties; insoluble in water and soluble in many organic solvents

3. Triacylglycerols, phosphoacylglycerols, sphingolipids

5. (a), (b), (d)

7. They have long hydrocarbon-like chains that make them water-insoluble.

9. Saturated: contains no carbon-carbon double bonds; monounsaturated: contains one carbon–carbon double bond; polyunsaturated: contains two or more carbon–carbon double bonds

11. It introduces a "kink," or bend into the chain.

13. Oleic acid:

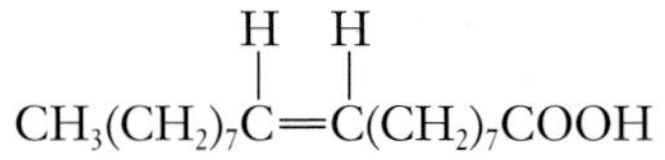

$$\mathrm{CH_3(CH_2)_7\overset{H}{C}{=}\overset{H}{C}(CH_2)_7COOH}$$

Linoleic acid:

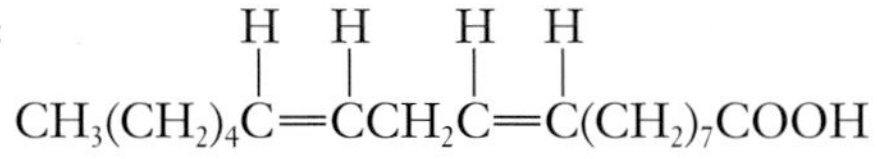

$$\mathrm{CH_3(CH_2)_4\overset{H}{C}{=}\overset{H}{C}CH_2\overset{H}{C}{=}\overset{H}{C}(CH_2)_7COOH}$$

Linolenic acid:

$$\mathrm{CH_3CH_2\overset{H}{C}{=}\overset{H}{C}CH_2\overset{H}{C}{=}\overset{H}{C}CH_2\overset{H}{C}{=}\overset{H}{C}(CH_2)_7COOH}$$

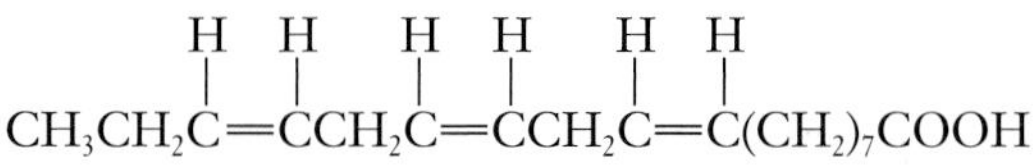

15. $\mathrm{CH_3CH_2CH_2CH_2CH_2CH_2CH_2CH_2CH_2CH_2CH_2CH_2CH_2CH_2CH_2COOH}$

17. **(a)** $\mathrm{CH_3(CH_2)_7CH{=}CH(CH_2)_7COOH + H_2 \xrightarrow{Pt} CH_3(CH_2)_{16}COOH}$

(b) No reaction

(c) $\mathrm{CH_3CH_2CH{=}CHCH_2CH{=}CHCH_2CH{=}CH(CH_2)_7COOH + 3H_2 \xrightarrow{Pt} CH_3(CH_2)_{16}COOH}$

19. Interchange the fatty acid on the middle hydroxyl group of glycerol with either of the other two fatty acids.

21. They are all triacylglycerols.

23. Some but not all of the double bonds present in the oil have been hydrogenated.

25. (1) Hydrogenate the oils to eliminate double bonds, (2) add preservatives such as BHT to prevent oxidation of double bonds.

27. Soap and glycerol

29. A fat

31. A fatty acid required in the diet because the body cannot make it; polyunsaturated ones such as linoleic and linolenic acids

33.

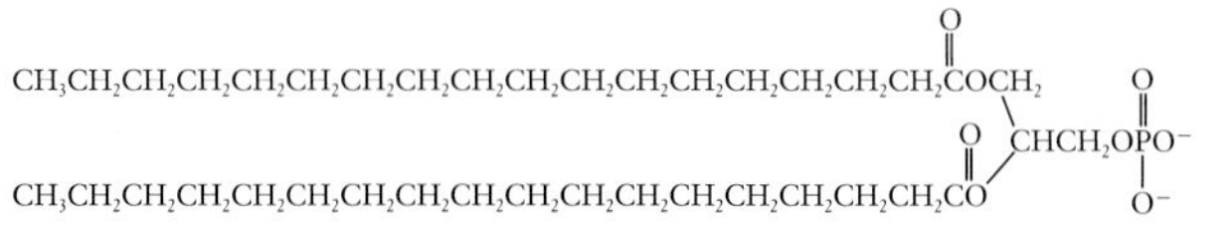

$$\mathrm{CH_3CH_2CH_2CH_2CH_2CH_2CH_2CH_2CH_2CH_2CH_2CH_2CH_2CH_2CH_2CH_2CH_2\overset{O}{\overset{\|}{C}}OCH_2}$$
$$\mathrm{\overset{\diagdown}{C}HCH_2O\overset{O}{\overset{\|}{P}}O^-\ (with\ O^-\ on\ P)}$$
$$\mathrm{CH_3CH_2CH_2CH_2CH_2CH_2CH_2CH_2CH_2CH_2CH_2CH_2CH_2CH_2CH_2CH_2CH_2\overset{O}{\overset{\|}{C}}O\diagup}$$

35. Two fatty acids, glycerol, phosphate, choline

37. A triacylglycerol has a fatty acid bound to the third hydroxyl group of glycerol; the phosphoacylglycerol has an alcohol-bearing phosphate group on this hydroxyl group.

39. Both contain a phosphatidic acid, but lecithin has choline bound to the phosphate and cephalin has ethanolamine.

41. Phosphatidyl ethanolamine; palmitic acid, stearic acid, glycerol, phosphate, ethanolamine

43. Cholesterol

45. There is a positive correlation between high blood cholesterol and atherosclerosis.

47. Anabolic steroids promote or aid development of muscle mass. Using these steroids makes the athlete stronger and faster.

49. Antiinflammatory agent.

51. Derivatives of unsaturated fatty acids that serve a variety of regulatory roles in the body

53. (a)

55. Compare your sketch to Figure 13.11.

57. Hydrophobic interactions

59. Compare your sketch to Figure 13.12. They may be used for drug delivery.

61. The proteins embedded in the lipid bilayer are free to move—fluid—and form a patchwork pattern—a mosaic—in the bilayer.

63. (c)
65. Many possibilities, including
(a) Water-insoluble compounds found in the body
(b) Lipids used for energy storage by the body
(c) Fatty acids containing one or more double bonds
(d) Triacylglycerol that is a solid at room temperature
(e) Polar lipid containing (minimally) two fatty acids, glycerol and phosphate
(f) Transport across a biological molecule that is helped by a protein in the membrane that acts as a channel
(g) Polar lipid containing a phosphatidic acid and choline
(h) A carbohydrate-containing lipid
(i) PGE_2
(j) A lipid containing the steroid nucleus
(k) Female sex hormone that is a steroid
(l) Male sex hormone that is a steroid
(m) Steroids that breakup dietary lipids into micelles to speed up the digestion of dietary lipids
(n) The most common steroid in animals
(o) An adrenocorticoid hormone that is antiinflammatory
67. 45 days

Chapter 14

1.

$$\text{Amino group} \rightarrow \begin{array}{c} \text{COOH} \leftarrow \text{carboxyl group} \\ | \\ H_2N-C-H \\ | \\ R \leftarrow \text{R group} \end{array}$$

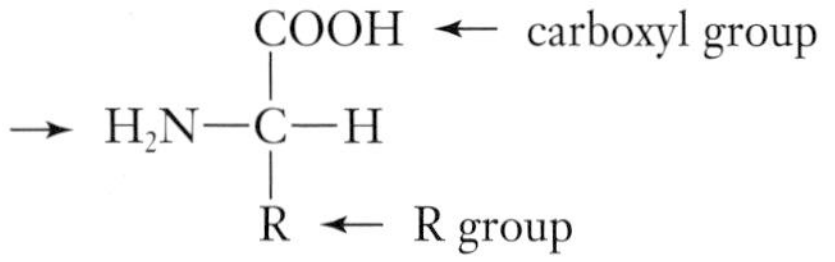

3. (b)

5.

$$\begin{array}{c} COO^- \\ | \\ H_3N^+-C-H \\ | \\ CH(CH_3)_2 \end{array}$$

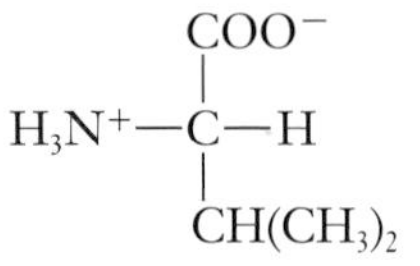

7. Two; the α-carbon
9. (a) Phenylalanine (b) Cysteine
(c) Glutamic acid (d) Lysine
11. The R group of a nonpolar amino acid is nonpolar; polar neutral R groups are uncharged but polar; an acidic R group has a negative charge at physiological pH; a basic one is positively charged at this pH.
13. Its amino group is covalently bonded to the R group forming a ring.
15. Several possibilities; the R group is shown un-ionized.
Glutamic acid:

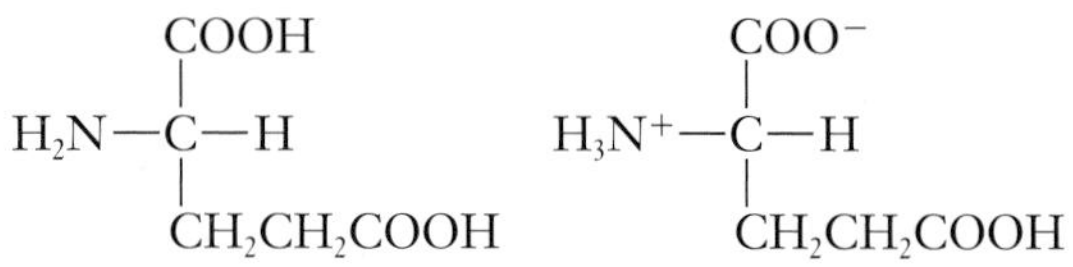

$$\begin{array}{c} COOH \\ | \\ H_2N-C-H \\ | \\ CH_2CH_2COOH \end{array} \qquad \begin{array}{c} COO^- \\ | \\ H_3N^+-C-H \\ | \\ CH_2CH_2COOH \end{array}$$

Lysine:

$$\begin{array}{c} COOH \\ | \\ H_2N-C-H \\ | \\ (CH_2)_4NH_2 \end{array} \qquad \begin{array}{c} COO^- \\ | \\ H_3N^+-C-H \\ | \\ (CH_2)_4NH_2 \end{array}$$

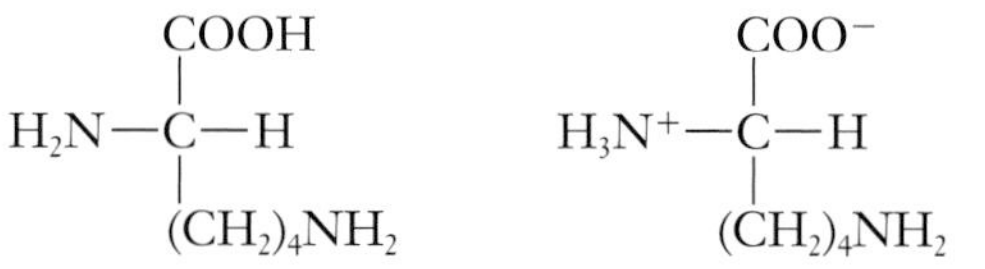

17. No; at all physiological pH values one or more of the functional groups will be ionized.

19.

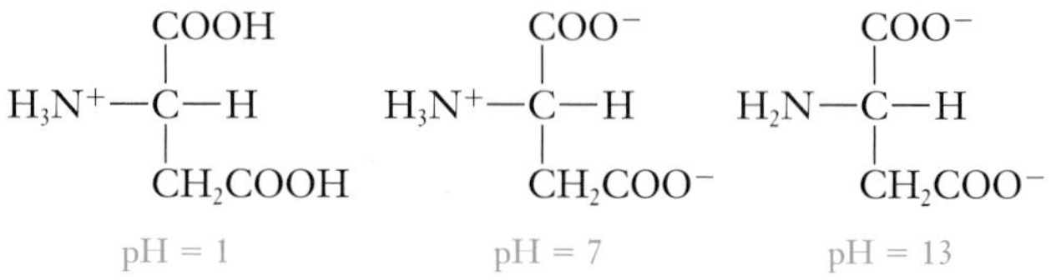

$$\begin{array}{c} COOH \\ | \\ H_3N^+-C-H \\ | \\ CH_2COOH \\ \text{pH} = 1 \end{array} \qquad \begin{array}{c} COO^- \\ | \\ H_3N^+-C-H \\ | \\ CH_2COO^- \\ \text{pH} = 7 \end{array} \qquad \begin{array}{c} COO^- \\ | \\ H_2N-C-H \\ | \\ CH_2COO^- \\ \text{pH} = 13 \end{array}$$

21. Arginine, histidine, isoleucine, leucine, lysine, methionine, phenylalanine, threonine, tryptophan, and valine.

23. The amide bond between two amino acids.

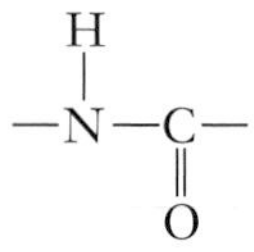

$$\begin{array}{c} H \\ | \\ -N-C- \\ \| \\ O \end{array}$$

25.

$$H_3N^+-\underset{^-OOCCH_2CH_2}{CH}-CO-NH-\underset{CH_2C_6H_5}{CH}-CO-NH-\underset{CH_2CH(CH_3)_2}{CH}-COO^-$$

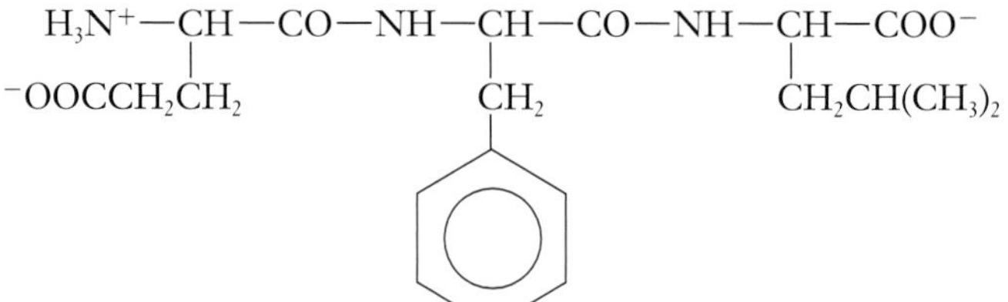

27. Alanylglycylvaline
29. (d)
31. The arrangement of the atoms of the polypeptide chain. This arrangement is maintained by hydrogen bonds between atoms involved in the peptide bonds.
33. The compact shape resulting from folding and bending of the polypeptide chain.
35. Peptide bonds—amide bonds between amino acids.
37. Hydrogen bonding, ionic and polar interactions, hydrophobic interactions; disulfide bridges, if present, help maintain tertiary structure.
39. (b)
41. (b)
43. (c)
45. The covalent bond between two sulfur atoms that forms when two thiols are oxidized; they help maintain tertiary structure.
47. Because the cellular components or the viral coat proteins are complementary, they bond noncovalently to each

other just like the subunits of an oligomeric protein. After aggregating, the subunits remain together because they are more stable in the complex than when free in solution.

49. Surface: **(b)**, **(d)**, **(e)** Core: **(a)**, **(c)**, **(f)**, **(g)**

51. **(b)** and **(d)**

53. **(a)**, **(c)**, **(f)**, **(g)**

55. Positive charge: lysine, arginine, histidine Negative charge: glutamate, aspartate

57. **(a)** Catalyze the reactions of the body
(b) Bind to and help remove foreign particles in the body
(c) Bind and transport fatty acids in blood
(d) Bind oxygen in lungs and transport it to the tissues of the body
(e) Binds and stores oxygen in muscle cells
(f) Catalyzes the breaking of bacterial cell walls
(g) A hormone that signals nutrients are available for use
(h) Structural protein in connective tissue

59. Negative at pH 7; negative at pH 9; positive at pH 5.5

61. The pH value where substances like amino acids and proteins have no net charge

63. **(b)**

65. Many possibilities, including
(a) A carboxylic acid with an amino group on the alpha carbon
(b) An alpha amino acid stereochemically related to L-glyceraldehyde
(c) Any molecule possessing two oppositely charged groups simultaneously
(d) Acting as either an acid or a base
(e) The pH value where an amino acid or protein has no net charge
(f) Peptide containing two amino acids
(g) The amide bond linking two amino acids
(h) Macromolecule consisting of many amino acids linked by peptide bonds.
(i) The amino acid sequence of a protein
(j) The arrangement of the atoms of the polypeptide chain of a protein
(k) The compact shape resulting from folding and bending of the polypeptide chain
(l) The shape resulting from aggregation of two or more subunits in an oligomeric protein
(m) The clustering of the nonpolar sidechains of the amino acids of a protein in an aqueous environment.
(n) The bond between two sulfur atoms formed by oxidation of two thiols.
(o) Loss of the native conformation of a protein.

67. The level of the answer will vary with the sources used. Portions of your answer can be compared to the Connections on sickle cell anemia in this chapter.

Chapter 15

1. Proteins that catalyze the reactions of the body; in the absence of enzymes the reactions of the body would be too slow to maintain life, or conditions would have to exist that would destroy cells.

3. **(a)**, **(d)**, **(e)**, **(f)**

5. **(a)** Xylulose **(b)** A UDP-glucuronoside (because you don't know all of this nomenclature, don't worry if your answer is somewhat different) **(c)** Xanthine

7. **(a)** Phosphohexose (such as a phosphorylated glucose) **(b)** Aspartate and ammonia **(c)** Succinyl-CoA

9. The small molecule that binds to the active site of an enzyme and is converted to product

11. Apoenzyme

13. The part of an enzyme that binds substrate(s) and catalyzes the conversion of substrate(s) to product(s)

15. Noncovalent ones such as hydrogen bonding, polar and ionic interactions, and hydrophobic interactions

17. The active site of an enzyme must be complementary to the substrate molecule. A polypeptide chain will not have both the correct shape and the right groups in the right places to bind to the active site of fatty acid synthetase.

19. It lowers the energy of activation.

21. Binding the two substrate molecules at the same time brings them close together so they can quickly react with each other. Proximity effect

23.

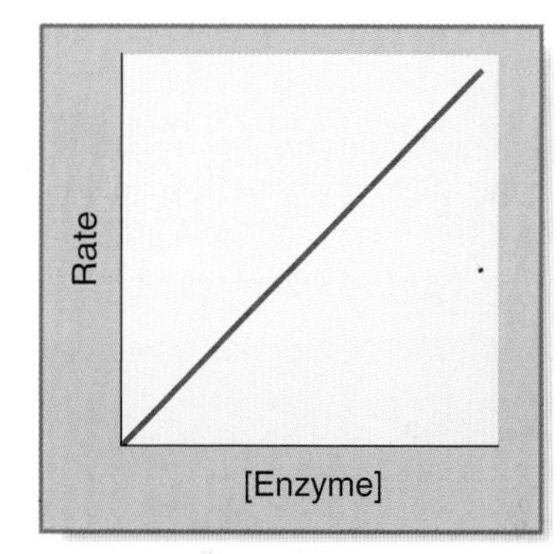

25. Even when more enzyme is added, each molecule of enzyme catalyzes the same number of reactions per unit time; thus a linear graph is the result. When more substrate molecules are added, the enzyme molecules become saturated with substrate; a curve results as the rate approaches a maximal value.

27. Pepsin is most active at acidic pH values around 2. Trypsin is most active around pH values of 7 to 8. Pepsin works in the stomach, where the normal pH is around 2; trypsin works in the small intestine, where the normal pH is around 7 to 8.

29. No, because the body normally maintains both pH and temperature in a narrow range of values where the enzymes are normally active.
31. A competitive inhibitor binds to the active site of an enzyme; a noncompetitive inhibitor binds to some other site on the enzyme.
33. These are competitive inhibitors of the enzyme dihydropteroate synthetase, and while they are bound no folic acid is synthesized, which stops bacterial growth.
35. By the suffix *-ogen* or the prefix *pro-*
37. Because it would catalyze the hydrolysis of the peptide bonds of the proteins found in the cells where it is made
39. By phosphorylation—by attaching several phosphate groups to the enzyme
41. Allosteric means "other site," so this means something will bind to a site other than the active site of the enzyme.
43. A positive effector binds to an allosteric site of an enzyme and causes the activity of the enzyme to increase. A negative effector binds to an allosteric site of an enzyme and causes the activity of the enzyme to decrease.
45. The product of a process inhibits the production of more of that product. A negative effector
47. Because in heart and muscle, the same reversible reaction must usually go in opposite directions, and the isozymes differ in their rates for the forward and reverse reactions.
49. This enzyme catalyzes the formation of plasmin, which breaks up blood clots that are blocking the flow of blood to tissues in the heart.
51. Answers will vary.

Chapter 16

1. Ribose and 2-deoxyribose
3. Adenine and guanine; cytosine, thymine, and uracil
5. A nucleoside consists of a nitrogenous base and a sugar; a nucleotide consists of a nitrogenous base, sugar, and one or more phosphates.
7. [structure: O, NH, O, N, $HOCH_2$, O, H, H, H, H, OH, OH]
9. Uridine triphosphate
11. Deoxycytidine diphosphate
13. Deoxyadenosine monophosphate, deoxyguanosine monophosphate, deoxycytidine monophosphate, and deoxytymidine monophosphate
15. (b)
17. Specific hydrogen bonds between the bases adenine and thymine and the bases guanine and cytosine; and hydrophobic interactions in the core of the DNA molecule
19. Because in double-stranded DNA each adenine on one strand is hydrogen-bonded to a thymine on the other, and each guanine on a strand is hydrogen-bonded to a cytosine on the other.
21. (a) ACGATG (b) CTGGTA (c) GACTGC
23. Polynucleotide ligase
25. When replication occurs, each new DNA molecule contains one complete old strand of DNA and one complete new strand of DNA.
27. Ribose
29. Adenosine monophosphate, guanosine monophosphate, cytidine monophosphate, and uridine monophosphate
31. DNA
33. Promoter sites
35. Replication; DNA polymerase possesses "proof-reading capability" not found in RNA polymerase.
37. The portion of a DNA molecule that contains the information for the synthesis of a polypeptide chain
39. UGCCAUGAC; cys, his, asp
41. A three-base signal or word found in mRNA that codes for an amino acid or signals start or stop for translation; 64
43. mRNA
45. No; several amino acids have more than one codon.
47. Ribosomes
49. Transfer RNAs (tRNAs)
51. (a) AGU (b) UGC (c) GCA
53. The two subunits of a ribosome, an mRNA and fmet-tRNA come together, with the tRNA hydrogen bonded to an AUG codon on the mRNA.
55. When a stop codon on the mRNA enters the ribosomal complex, the polypeptide is cleaved from the tRNA bound to it, and the complex falls apart.
57. (c)
59. An agent that causes a permanent change in DNA; it causes a mutation; uv light, ionizing radiations
61. The substitution of a base changes a codon. If the new codon signals for a different amino acid, then the base substitution will result in a polypeptide chain that has a new amino acid substituted for the original amino acid at that residue.
63. (a) Serine to threonine (b) Aspartate to glutamate (c) Threonine to asparagine
65. A mutation caused by the addition or deletion of a base in a DNA molecule; all codons from the point of deletion or addition will be changed because the "frame" on the mRNA has been "shifted."
67. Lactose
69. This is the site on the DNA molecule where the repressor protein binds.

71. Gene expression is turned off because RNA polymerase cannot get past the occupied operator, thus mRNA is no longer made.
73. Histidine
75. They are used to cut DNA to provide DNA with short, single-stranded DNA ends that are complementary to other DNA molecules cut with the same enzyme. When these two types of DNA are mixed, they join at the complementary ends to form a single DNA molecule.
77. No; because several amino acids have more than one codon.
79. No; if there were one unique sequence, we would all be genetically identical.

Chapter 17

1. Catabolism is often oxidative and breaks larger molecules into smaller ones with release of energy; anabolism is often reductive and makes larger molecules from smaller ones using energy.
3. Digestion breaks larger dietary molecules into smaller ones that can be absorbed from the gastrointestinal tract.
5. Salivary amylase is produced in the salivary glands and secreted into the mouth. Pancreatic amylases are made in the pancreas and secreted into the small intestine. Salivary amylase is denatured in the stomach before hydrolysis of starches is complete.
7. It causes denaturation of dietary proteins, which exposes peptide bonds to enzyme-catalyzed hydrolysis.
9. Bile salts
11. Primarily in the small intestine
13.

FOOD CLASS	ENZYMES	SITES OF DIGESTION	PRODUCTS
Carbohydrates	Amylases and sucrase, etc.	Mouth and small intestine	Monosaccharides
Proteins	Proteases and peptidases	Stomach and small intestine	Amino acids
Lipids	Lipases	Small intestine	Fatty acids, monoacylglycerols, and glycerol

15. Several possibilities, including catabolic: glycolysis, beta-oxidation; anabolic: protein synthesis, lipogenesis
17. Many possibilities, including: Step 10 in glycolysis conserves energy by forming ATP from ADP and the phosphate on phosphoenolpyruvate; Step 6 of glycolysis couples reduction of NAD^+ with oxidation of the aldehyde group in glyceraldehyde 3-phosphate.
19. It comes from some other coupled reaction or process that is exothermic.
21. Both NAD^+ and FAD act as oxidizing agents (acceptors of electrons) in catabolic reactions.
23. The active site of the appropriate anabolic enzymes is complementary to $NADP^+$/NADPH (NAD^+/NADH lacks the phosphate needed for binding). The active site of the appropriate catabolic enzymes is complementary to NAD^+/NADH (because $NADP^+$/NADPH has an additional phosphate, it cannot bind).
25. **(a)** The first part (the first 3 steps)
    (b) ATP.
27. Two
29. Lactate (lactic acid)
31. Yeast: ethanol and carbon dioxide; animal cells: lactic acid; both, NAD^+
33. The synthesis of glucose from glycerol, lactate, and the glycogenic amino acids; to make glucose when it is needed, for example when diet is unable to provide adequate amounts
35. **(a)** The carbons in it can be used to make glucose via neoglucogenesis. **(b)** All of them except leucine and lysine
37. The phosphoester bond (a high-energy bond) between glucose and UDP in uridine diphosphate-glucose is cleaved.
39. About 2%; by oxidation of pyruvate to carbon dioxide and water by pyruvate dehydrogenase, the citric acid cycle, and electron transport-oxidative phosphorylation.
41. Carbon dioxide and NADH
43. Reaction 5, the production of succinate from succinyl CoA.
45. **(a)** Reaction 3, oxidation of isocitrate; reaction 4, oxidation of α-ketoglutarate; reaction 8, oxidation of malate.
    (b) Reaction 6, oxidation of succinate
47. On the inner surface of the inner mitochondrial membrane
49. Molecular oxygen (O_2)
51. It is a chemical gradient because any substance including hydrogen ions forms a gradient when there is an unequal concentration across a membrane; it is an electrical gradient because hydrogen ions are charged, and if they are at unequal concentration, then the charge is unequal on the two sides of the membrane.
53. The phosphorylation of ADP to form ATP is coupled to the oxidation of NADH and $FADH_2$.
55. 12
57. See Table 17.1.
59. Covalently bound to a carrier molecule, CoA (Coenzyme A)
61. Two high energy bonds; $ATP \longrightarrow AMP + PP_i$
63. NAD^+ and FAD; NADPH
65. The double bond in palmitoleic acid reduces the production of $FADH_2$ by one. Because each $FADH_2$ is

equivalent to two ATP, the total ATP yield is reduced by two ATP.

67. Palmitic acid; we elongate palmitic acid by two carbon atoms using acetyl-CoA.

69. Acetoacetic acid, β-hydroxybutyric acid, acetone; They serve as fuel molecules circulating in the blood stream.

71. During diabetic crisis excess ketone bodies are formed that make the blood more acidic than normal. The blood concentration of acetone becomes high enough during a diabetic crisis to diffuse from the blood into the tissues lining the mouth and throat.

73. An enzyme competitive inhibitor that resembles HMG-CoA has been found. When taken by the patient, the inhibitor inhibits the synthesis of cholesterol.

75. α-Ketoglutarate

77. **(a)** Urea **(b)** In the urea cycle the amino groups from amino acids are combined with carbon dioxide to form urea.

79. Anticancer drugs affect all rapidly growing cells, including cancer cells. When rapidly growing normal cells are killed, effects such as hair loss and impaired immune responses appear.

81. Catabolism of both fatty acids and some amino acids

Glossary

A

absolute zero The zero temperature on the Kelvin scale; the temperature at which particles in matter are motionless.
accuracy An expression of how close a measurement is to the true value of the property being measured.
acetal An organic compound formed from a hemiacetal and an alcohol.
acid A compound that can donate (give up) a hydrogen ion.
active site The location on an enzyme where the substrate(s) binds and catalysis occurs.
active transport Transport across a membrane using energy to drive the movement of the molecules or ions.
adenosine triphosphate (ATP) Coenzyme (and nucleotide) that stores energy in living organisms.
alcohol An organic compound containing a hydroxyl group attached to a carbon atom of an alkyl group.
aldehyde An organic compound containing a carbonyl group that has a hydrogen atom attached to the carbon atom.
aldose A monosaccharide containing an aldehyde group.
alkane Hydrocarbon containing only carbon–carbon single bonds and carbon–hydrogen bonds.
alkene An unsaturated hydrocarbon containing one or more carbon–carbon double bonds.
alkyl group A substituent in a molecule that is formed from an alkane by removal of one hydrogen atom. Alkyl names are derived from the alkane they are formed from.
alkyl halide An organohalogen containing a halogen atom attached to an alkyl group.
alkyne An unsaturated hydrocarbon containing one or more carbon–carbon triple bonds.
allosteric regulation Regulation of enzyme activity involving the binding of effectors that either increase or decrease the enzyme's activity on binding.
alpha (α) helix Secondary protein structure in which the backbone of the polypeptide chain is arranged into a helix and held in place by hydrogen bonding between atoms of the peptide bonds.
alpha (α) radiation Radiation consisting of alpha particles (helium-4 nuclei).
amide An organic compound formed from a carboxylic acid and either ammonia or an amine.
amine An organic compound containing a nitrogen atom bonded to a carbon atom.
amino acids (α-amino acids) The 20 components of proteins consisting of a carboxyl group, α-amino group, and a side chain unique to each amino acid.
amino group The functional group of the amines.
ammonium salt An organic salt formed when an amine reacts with an acid.
amphipathic molecule A molecule containing both a significant polar region (polar head group) and a nonpolar region (nonpolar tail).
amphoteric molecule A molecule that possesses both acidic and basic properties.
amylopectin A starch consisting of branched molecules composed of glucose residues linked by $\alpha(1 \rightarrow 4)$ and $\alpha(1 \rightarrow 6)$ glycosidic bonds.
amylose A starch consisting of linear molecules composed of glucose linked by $\alpha(1 \rightarrow 4)$ glycosidic bonds.
anabolism The reactions of the body that consume energy as they build larger, more complex molecules from smaller ones.
anion A negatively charged ion.
anomeric carbon The carbon atom in the ring of a sugar that had been the carbonyl carbon before ring formation.
anomers The two isomers of the ring form of a sugar that result from formation of the ring.
anticodon A three-base sequence on a tRNA that is complementary to a codon in mRNA.
apoenzyme An enzyme lacking its required cofactor(s). An apoenzyme is catalytically inactive.
aromatic hydrocarbon An unsaturated hydrocarbon containing one or more aromatic rings.
aryl halide An organohalogen containing a halogen atom attached directly to an aromatic ring.
-ase A suffix used to identify a substance as an enzyme.
atom The smallest particle of an element that possesses the properties of that element.
atomic mass The average mass of an atom of an element.
atomic mass unit (amu) A unit defined as $\frac{1}{12}$ of the mass of a carbon-12 atom.
atomic number A number that expresses the number of protons found in any atom of an element.
Avogadro's law The volume of a gas divided by the moles of the gas are equal to a constant.

B

base A compound that can accept a hydrogen ion.
beta oxidation A catabolic process that breaks fatty acids down to acetyl-CoA with production of NADH and $FADH_2$.
beta (β) pleated sheet Secondary protein structure in which hydrogen bonding between atoms of the peptide

bonds hold different polypeptide chains together or hold distant portions of the same polypeptide chain together.
beta (β) radiation Radiation consisting of beta particles (electrons).
bilayer Two-layered structure of amphipathic molecules that has a nonpolar core and polar surfaces.
bile salts Amphipathic steroids that play a role in digestion by emulsifying dietary lipids to speed up their digestion.
binary covalent compound A covalent compound containing only atoms of two nonmetals.
binary ionic compound Ionic compound containing only two kinds of ions.
Boyle's law Pressure times volume of a gas is equal to a constant.
buffered solution (buffer) A solution that resists changes in pH when acid or base are added to it.

C

calorie The standard unit of energy in the metric system.
Calorie A unit of energy used in nutrition that is equal to 1 kilocalorie.
carbonyl group A functional group consisting of a carbon atom double-bonded to an oxygen atom.
carboxyl group (—COOH) The functional group of the carboxylic acids.
carboxylate anion The anion formed when a carboxylic acid gives up its hydrogen ion.
carboxylic acid An acidic organic compound containing a carboxyl group.
catabolism The reactions of the body that produce energy as they break large molecules into smaller ones.
catalyst Substance or mixture that speeds up the rate of a chemical reaction but is unchanged by the reaction.
cation A positively charged ion.
cellulose A structural polysaccharide of plants consisting of linear molecules composed of glucose linked by $\beta(1 \rightarrow 4)$ glycosidic bonds.
Celsius scale The temperature scale used in the metric system.
change in state A physical process involving energy that changes a sample of matter from one state to another.
Charles's law The volume of a gas divided by its temperature is equal to a constant.
chemical equation A concise symbolic representation of a chemical reaction.
chemical equilibrium A reversible reaction is at equilibrium when no net change occurs; the forward and the backward reaction occur at the same rate.
chemical reaction A change in a sample of matter that alters the composition of the sample.
chemical symbol An abbreviation of one or two letters used to represent an element.
chemiosmotic theory The energy released during oxidation of NADH and $FADH_2$ is used to pump protons from the mitochondrion, establishing an energy-rich gradient.
chemistry The study of matter and the changes that occur in matter.
chiral Term used to describe objects, such as human hands, that are nonsuperimposable mirror images.
chiral center An atom in the molecule (usually a carbon atom) that has four different substituents attached to it.
chitin A structural polysaccharide of arthropods consisting of linear molecules composed of N-acetylglucosamine linked by $\beta(1 \rightarrow 4)$ glycosidic bonds.
cholesterol The most abundant steroid in animals.
chromosomes Structures in the nucleus of higher cells consisting of DNA and proteins.
citric acid-cycle A cyclic catabolic pathway that oxidizes acetyl-CoA to carbon dioxide while producing reduced and phosphorylated coenzymes.
codon One of 64 three-base signals in mRNA that codes for an amino acid or signals start or stop for translation.
coefficient A whole number put in front of formulas or symbols of the substances in an equation to indicate the relative number of molecules, atoms, or ions that participate in the reaction.
coenzyme A cofactor that is an organic compound.
cofactor An ion or organic molecule needed by an enzyme to be catalytically active.
colligative properties Properties of solutions that depend only on the amount of solute in the solution, not on the nature of the solute.
combined gas law The product of the pressure and volume of a gas divided by the temperature of the gas equals a constant.
competitive inhibitor A reversible inhibitor that binds to the active site of an enzyme.
complementary bases Nitrogenous bases in nucleic acids that hydrogen bond to each other. Guanine pairs with cytosine, and adenine pairs with thymine in DNA, and uracil in RNA.
complex carbohydrates The starches in our diet.
complex lipids Lipids that contain one or more fatty acids bound to an alcohol, or less commonly, an amine.
compound A pure substance containing two or more elements in fixed proportions.
concentration A measure of the amount of solute present in a given quantity of a solution.
condensation The change from a gas to a liquid.
condensed structural formulas Structural formulas in which some or all of the lines that represent bonds are not shown.
conformation The infinite shapes of a molecule that depend on free rotation about single bonds in the molecule.
conjugated protein A protein that contains amino acids and one or more additional components called prosthetic groups.
conversion factor Ratios or fractions derived from definitions or equalities that are used in dimensional analysis.

core electrons All of the electrons located in the inner (nonvalence) shells of an atom.
coupled reactions Paired reactions that occur concurrently; often one is endothermic and the other exothermic.
covalent bond A bond resulting from the sharing of pairs of electrons by two atoms.
cubic centimeter A unit of volume equal to 1 milliliter.
curie (Ci) Unit of radioactivity equal to 3.7×10^{10} disintegrations per second (dps).
cycloalkane Alkane containing three or more carbon atoms bonded together to form a ring.

D

D- and L- stereochemical families Families of stereoisomers based on the chiral center in glyceraldehyde.
Dalton's law of partial pressures The total pressure of a mixture of gases is equal to the sum of the individual pressures of the gases.
data Specific factual information.
deamination Process that removes an amino group from a glutamate molecule yielding ammonia.
denaturation Process in which a protein or other biological macromolecule loses its native function or activity.
density A measure of the mass present in a unit volume of any substance.
deposition The change from a gas to a solid.
diagnosis Procedure used to determine both the cause of a medical condition and the course of treatment.
diatomic molecule A molecule containing two atoms.
digestion The catabolic process that breaks down large dietary molecules to smaller ones.
dimensional analysis A systematic approach for solving both conversions and calculations in which the number and unit of a measurement are multiplied by one or more conversion factors, yielding an answer that is correct for both the number and unit.
dipole Separation of charge; both bonds and molecules can exist as dipoles.
dipole–dipole interaction The force of attraction between oppositely charged ends of polar molecules.
disaccharide A sugar composed of two monosaccharides.
dispersion (colloid) A mixture with solute particles larger than those found in true solutions.
disulfide bridge Covalent bond in some proteins that forms by oxidation of two sulfhydryl groups in cysteine residues and helps maintain the tertiary structure of those proteins.
DNA (deoxyribonucleic acid) A macromolecule that is the carrier of genetic information in living organisms. Composed of nucleotides containing deoxyribose.
double bond A covalent bond involving two pairs of shared electrons.

E

electrolyte A compound that conducts electricity in aqueous solution or when melted.
electron A subatomic particle with a mass of 9.109×10^{-28} g and an electric charge of minus one.
electron transport chain A series of electron carriers that oxidizes NADH and $FADH_2$ and reduces oxygen; energy released during this process is used to pump protons out of the mitochondrion.
electronegativity A measure of the tendency of an atom to attract shared pairs of electrons to itself.
electrophoresis A process that uses an applied electric field to separate charged particles like proteins.
element A substance that cannot be broken down into any simpler substance.
enantiomers A pair of chiral molecules that are nonsuperimposable mirror images.
endothermic reaction A reaction that consumes (takes up) heat.
energy The capacity to cause a change; the ability to do work.
energy of activation The energy barrier that must be exceeded as a reactant goes to product in a reaction.
enthalpy of reaction The amount of energy given up or taken in by a reaction.
enzymes Proteins that catalyze virtually all of the reactions in the body.
equivalent The amount of an acid or base that produces one mole of hydrogen ions or one mole of hydroxide ions.
equivalent mass The mass of an acid or base that yields one mole of hydrogen ions or hydroxide ions.
essential amino acids The amino acids required in our diet because we cannot synthesize them.
ester An organic compound formed from a carboxylic acid and an alcohol or phenol.
ether An organic compound that contains an oxygen atom single-bonded to two carbon atoms.
exact numbers Numbers that have no uncertainty and that are assumed to have an unlimited number of significant figures.
exothermic reaction A reaction that produces heat.

F

facilitated diffusion Diffusion across a membrane through channels formed from protein molecules.
Fahrenheit scale The temperature scale used in the United States.
family See Group.
fats Triacylglycerols that are solid at room temperature.
fatty acids Long-chain, unbranched carboxylic acids containing an even number of carbon atoms.
feedback inhibition Process in which the end product of the process inhibits its own production. Many metabolic processes show feedback inhibition.
fermentation (lactate and alcoholic) Anaerobic catabolic processes that convert pyruvate into lactate and

ethanol respectively. NADH is concurrently recycled to NAD^+.
fiber The indigestible, tasteless component of our diet composed principally of cellulose and other plant polymers.
fibrous proteins Proteins that are fiber-like and often water-insoluble.
fission Process that splits a large nucleus into smaller nuclei with the release of enormous amounts of energy.
formula mass The total mass (in atomic mass units) of the ions in the formula of a ionic compound.
formulas (chemical formulas) Combinations of element symbols used to show the composition or structure of compounds.
frameshift mutation A mutation caused by the insertion or deletion of a base in the normal base sequence of DNA.
freezing The change from a liquid into a solid.
freezing point The temperature at which a liquid changes to a solid.
functional group An atom or groups of atoms that give a molecule its specific chemical and physical properties.
furanose A monosaccharide containing a five-membered ring.
fusion Process that combines (fuses) smaller nuclei into larger ones with release of enormous amounts of energy.

G

gamma (γ) radiation High-energy electromagnetic radiation similar to x-rays.
gas State of matter that has no definite volume or shape.
gene Any portion of a DNA molecule that contains the information needed to synthesize a particular polypeptide chain.
geometric isomers A term used for stereoisomers involving rings and double bonds. The prefixes *cis*- and *trans*- are used to distinguish these isomers from each other.
globular proteins Soluble proteins that are compact and roughly spherical. Most proteins are globular.
gluconeogenesis The anabolic process that makes glucose and other sugars from lactate, glycerol, and the glycogenic amino acids.
glycogen A storage polysaccharide of animals consisting of branched molecules composed of glucose residues linked by $\alpha(1 \rightarrow 4)$ and $\alpha(1 \rightarrow 6)$ glycosidic bonds.
glycogenesis The anabolic process that synthesizes glycogen.
glycolipid A lipid that has a carbohydrate as a component.
glycolysis The principal pathway for catabolism of glucose and other simple sugars.
glycoside Acetal formed from a sugar and an alcohol.
glycosidic bond The covalent bond linking a sugar to another sugar, an alcohol, or an amine.
gram The metric system standard unit for mass.
group (family) The elements found in a column in the periodic table.

H

half-life The time required for one half of a radioactive sample to decay.
halogenation An addition reaction in which the two halogen atoms of a molecule of a halogen add to the carbon atoms of a carbon–carbon double or triple bond.
heat Kinetic energy resulting from the motion or vibration of particles in matter.
heat of fusion The amount of energy (heat) a substance gains when it melts or loses when it freezes.
heat of vaporization The amount of energy (heat) that a substance gains when it vaporizes or loses when it condenses.
hemiacetal An organic compound formed from an aldehyde or ketone and an alcohol.
Henry's law When a gas is in contact with a liquid, the concentration of the gas in the liquid is directly proportional to the pressure of the gas.
heparin A polysaccharide produced by some cells of the circulatory system that is an anticoagulant.
hetero-atom An atom, other than a carbon or hydrogen atom, found in an organic compound.
holoenzyme An enzyme coupled with its cofactor(s). A holoenzyme is catalytically active.
hyaluronic acid A polysaccharide found in the connective tissue of higher animals that plays a role in protecting and lubricating joints.
hydration An addition reaction in which the atoms of a water molecule add to the carbon atoms of a carbon–carbon double or triple bond.
hydrocarbons Organic compounds that contain only carbon and hydrogen atoms.
hydrogen bond Force of attraction between a very electronegative atom (O, N, or F) and a hydrogen atom that is covalently bound to a very electronegative atom.
hydrogenation An addition reaction in which the two atoms of molecular hydrogen add to the carbon atoms of a carbon–carbon double or triple bond.
hydrohalogenation An addition reaction in which the two atoms of a hydrohalogen molecule add to the carbon atoms of a carbon–carbon double or triple bond.
hydronium ion (H_3O^+) The ion formed when a water molecule accepts a hydrogen ion.
hydrophobic interactions The noncovalent interactions (bonding) of nonpolar molecules or nonpolar parts of amphipathic molecules when they are in an aqueous environment.
hydroxyl group (—OH) The functional group of the alcohols and phenols.
hydroxylation An oxidation reaction of a carbon–carbon double bond that yields a product having a hydroxyl group on each of the carbons of the double bond.
hypothesis A statement that explains a known set of data.

I

ideal gas law Ideal gases show this relationship among these properties: PV = nRT.
induced-fit model A model for enzyme-substrate interaction that says the binding of the substrate to the active site induces a change in the shape of one or, more likely, both of the molecules.
induction A process in bacteria that turns on the synthesis of certain proteins when they are needed in the cell.
intermolecular forces The forces or bonds that exist between molecules in solids and liquids.
iodine number A number that represents the degree of unsaturation in a sample of fat or oil. It is defined as the grams of iodine required to react with all of the carbon–carbon double bonds in 100 g of sample.
ion An atom or group of atoms that has an electrical charge.
ionic bond The force of attraction between oppositely charged ions in a compound.
ionic compound A compound consisting of ions held together by ionic bonds.
ion-product constant of water (Kw) A constant—1.0×10^{-14} at 25° C—equal to the molar concentration of hydrogen ion times the molar concentration of hydroxide ion in water or any aqueous solution.
irreversible inhibitor A substance that binds permanently to an enzyme, causing the enzyme to permanently lose activity.
isoelectric point (pI) The pH value at which an amino acid or protein has no net charge.
isomers Two or more compounds that have the same composition but different structures.
isotopes Atoms of an element that contain different numbers of neutrons.
isozymes Structurally similar enzymes that come from different tissues of the same organism.

J

joule The standard SI unit for energy.

K

Kelvin scale The SI temperature scale.
ketone An organic compound containing a carbonyl group with two carbon atoms bonded to the carbon atom of the carbonyl group.
ketose A monosaccharide containing a keto group.
kilocalorie A unit of energy equal to 1000 calories.
kilogram The standard SI unit for mass.
kinetic energy Energy associated with motion.
kinetics The branch of chemistry that studies how fast reactions occur.

L

law A generalization or summary that is based on consistent experimental results or experiences.
law of conservation of matter In a chemical reaction matter is neither created nor destroyed.
le Chatelier's principle If a chemical equilibrium is disturbed, changes will occur to reestablish equilibrium.
leukotrienes Lipids synthesized from unsaturated fatty acids that play a role in allergic and inflammation responses.
Lewis structure (electron dot structure) A representation of an atom, ion, or molecule that uses dots to represent valence electrons.
lipids The biological molecules that are insoluble in water but soluble in organic solvents.
lipogenesis Anabolic process that synthesizes fatty acids, with palmitic acid the major product.
lipoproteins A macromolecular complex consisting of lipids and proteins; the lipoproteins of blood transport lipids throughout the cardiovascular system.
liposomes Structures formed from lipid bilayers that close back on themselves to form a continuous bilayer surrounding a core of water.
liquid State of matter that has a definite volume but no definite shape.
liter The standard metric unit for volume.
lock-and-key model A model for enzyme-substrate interaction that says the substrate and the active site are complementary like a lock and a key.
London dispersion force The force of attraction involving temporary interactions between a temporary dipole and an induced dipole.

M

Markovnikov's rule When an asymmetrical, hydrogen-containing reagent adds to a carbon–carbon double bond, the carbon atom of the double bond that has the most hydrogens gets the hydrogen atom of the adding molecule.
mass A measure of how much material (matter) is present in an object.
mass number Number used to indicate the number of neutrons and protons (nucleons) in an atom.
matter Anything that has mass and occupies space.
measurement An observation that provides numerical information about a subject or topic.
melting The change in state from a solid to a liquid.
melting point The temperature at which a solid melts.
membrane The sheet-like bilayered structures that surround cells and divide the cell interior.
messenger RNA (mRNA) RNA that functions as a carrier (messenger) of information from DNA to the rest of the cell.
metabolic pathway A series of sequential reactions within the body.
metabolism The sum of all the reactions of the body.

metal An element that is a good conductor of heat and electricity, that tends to be shiny, and is both malleable and ductile. Atoms of metals tend to lose electrons in chemical reactions.
metalloid An element possessing properties intermediate between those of metals and nonmetals.
meter The SI and metric standard unit for length.
milliliter A unit of volume equal to 1/1000th of a liter.
mixture A sample of matter with variable composition; a mixture consists of two or more substances that can be separated from each other by physical changes.
molar mass The mass of one mole of atoms, ions, or molecules. The molar mass of an element is simply the atomic mass expressed in grams.
molar volume The volume—22.4 L—occupied by 1 mol of any gas at STP.
molarity (M) A measure of concentration; a one-molar solution contains one mole of solute per one liter of solution.
mole A unit for counting particles or objects. One mole contains 6.022×10^{23} particles or objects.
molecular compound A compound consisting of molecules.
molecular formula Formula used to show the number of atoms of each element in a molecule of a molecular compound.
molecular mass The total mass (in amu) of the atoms in a molecule of a molecular compound.
molecular models Objects (models) used to show molecules in three dimensions.
molecule A group of atoms joined together by covalent bonds.
monatomic ion An ion containing only one atom.
monosaccharides (simple sugars) The smallest carbohydrates having the general formula $C_nH_{2n}O_n$.
mutagen Any agent that can cause a mutation.
mutation A permanent change in the sequence of bases in DNA.

N

native conformation The conformation of a biological macromolecule—such as a protein—in which the molecule normally exists and carries out its normal function.
negative effector A substance that binds to an allosteric enzyme causing the activity of the enzyme to decrease.
neutralization The acid–base reaction between a strong acid and a strong base.
neutron A subatomic particle with a mass of 1.675×10^{-24} g and no electric charge.
nicotinamide adenine dinucleotide A coenzyme used in many redox reactions of catabolism. NAD^+ is its oxidized form and NADH is its reduced form.
nicotinamide adenine dinucleotide phosphate A coenzyme used in redox reactions of anabolism. $NADP^+$ is its oxidized form and NADPH is its reduced form.
nitrogenous base A molecule found in nucleotides and nucleosides containing two or more nitrogen atoms in either a simple or fused ring.
noncompetitive inhibitor A reversible inhibitor that binds to an enzyme at a site other than the active site.
nonmetal An element that tends to be an electrical insulator, that is generally not shiny when solid, and is brittle. Atoms of nonmetals tend to gain or share electrons in chemical reactions.
normal boiling point The boiling point of a liquid substance at 1 atm of pressure.
nuclear radiation Radiation that originates in the nuclei of atoms.
nuclear transformation Process that converts one nucleus into another by high-energy bombardment with a particle.
nucleoside A compound containing a sugar and a nitrogenous base.
nucleotide A molecule containing a nitrogenous base, a sugar, and one or more phosphate groups.
nucleus The tiny core of an atom, containing the protons and neutrons of that atom.

O

octet rule The tendency for an atom to gain, lose, or share electrons to obtain eight electrons in their valence shell.
oils Triacylglycerols that are liquids at room temperature.
oligosaccharide A carbohydrate containing two to ten monosaccharides covalently linked by glycosidic bonds.
operon A regulatory unit on bacterial DNA that consists of a promoter, operator, and structural genes.
orbital A region in space around the nucleus of an atom where electrons can be found.
organic chemistry The study of carbon-containing compounds.
-ose A suffix used to identify a compound as a carbohydrate.
osmotic pressure The pressure that must be applied to a solution to prevent movement of water across a semipermeable membrane.
oxidation The loss of electrons by a reactant during a reaction.
oxidation-reduction (redox) reaction A reaction involving loss or gain of electrons by the reactants.
oxidative cleavage An oxidation reaction of a carbon–carbon double bond that cleaves the double bond, yielding aldehydes and ketones as products.
oxidative phosphorylation A process catalyzed by the enzyme ATP synthetase that phosphorylates ADP to make ATP using energy released as protons flow back into the mitochondrion.

P

partial pressure The pressure exerted by a gas in a mixture of gases.
peptide bond The amide bond that connects amino acids in peptides and proteins.

peptides Oligomers and polymers of amino acids that have a molar mass that is generally less than 5000 amu.
period A row of elements in a periodic table. The elements in a row have their valence electrons in the same shell.
periodic law A statement of the regular and periodic differences and similarities of the elements based on the arrangement of electrons in the atoms of the elements.
periodic table of the elements A table containing columns and rows of elements. The arrangement of the elements is based on their properties and arrangement of electrons.
pH scale A logarithmic measure of hydrogen ion concentration in aqueous solutions.
phenol An organic compound containing a hydroxyl group attached to a carbon atom of an aromatic ring.
phosphatidic acid A molecule consisting of glycerol, two fatty acids, and phosphate. Phosphatidic acids are the principle components of phosphoacylglycerols.
phosphoacylglycerols Polar complex lipids—consisting of two fatty acids, glycerol, phosphate, and an alcohol—that are components of membranes and lipoproteins.
photosynthesis The process used by autotrophic organisms to make glucose from carbon dioxide and water using solar energy.
plasmid A small circular DNA molecule found in some bacteria; can be used to introduce nonbacterial genes into a bacterium.
polar covalent bond A covalent bond that is polar because the bonded atoms differ significantly in electronegativity.
polar lipids Lipids such as phosphoacylglycerols and sphingolipids that contain a polar region called a polar head group. These lipids are amphipathic.
polyatomic ion An ion containing more than one atom.
polymerization A reaction in which small molecules called monomers are joined to form very large molecules called polymers.
polysaccharide A carbohydrate containing many monosaccharides covalently linked by glycosidic bonds.
positive effector A substance that binds to an allosteric enzyme, causing the activity of the enzyme to increase.
potential energy Energy that is stored energy.
precipitation reaction Reaction in which an insoluble product is formed.
precision An indicator of similarity in a set of measurements.
prefix A prefix is added to a unit to change the unit's value by a fixed amount.
pressure A measure of force applied to a given area.
primary amine An amine containing a nitrogen atom bonded to only one carbon atom.
primary structure The amino acid sequence of a protein.
primary transcript The RNA made directly by transcription; it is modified in various ways to form functional RNAs.
product A substance that is formed during a chemical reaction.
prostaglandins Simple lipids that play regulatory roles in the body and that are synthesized from polyunsaturated fatty acids.
prosthetic group The nonamino acid component of a conjugated protein.
protein Large polymers of amino acids that carry out a wide variety of roles in living organisms.
proton A subatomic particle with a mass of 1.673×10^{-24} g and an electric charge of plus one.
purine A nitrogenous base containing a fused ring resembling the ring of the compound purine.
pyranose A monosaccharide containing a six-membered ring.
pyrimidine A nitrogenous base containing a ring resembling the ring of the compound pyrimidine.

Q

quaternary ammonium salt An ionic organic compound containing a nitrogen atom bonded to four carbon atoms.
quaternary structure Protein structure that results from the arrangement of two or more polypeptide chains in the protein molecule.

R

radiation absorbed dose (rad) Unit expressing the amount of energy associated with radiation. Tissues exposed to 1 rad absorb 1×10^{-2} J of energy per kg of tissue.
radiation equivalent for man (rem) A unit used to express exposure to radiation. One rem is one rad multiplied by a factor called relative biological effectiveness, RBE.
radioactive decay Nuclear process in which a radioactive nuclide emits a particle to become an atom containing a different nucleus.
radioactive nuclides Atoms that have unstable nuclei that undergo radioactive decay.
rancid An adjective describing a bad-smelling sample of an oil or fat. The smell is due to the presence of volatile organic acids and aldehydes formed by oxidation of double bonds.
reactant A substance that reacts in a chemical reaction.
reaction rate The speed at which a reaction occurs under a set of conditions.
reducing sugar Sugar that causes other substances to be reduced as the carbonyl group of the sugar is oxidized.
reduction The gain of electrons by a reactant during a reaction.
replication (semiconservative replication) The process that forms two usually identical daughter DNA molecules from one DNA molecule. Each daughter DNA molecule gets one strand of the original DNA molecule and one new strand.

repression A process in bacteria that turns off the synthesis of certain proteins when they are not needed in the cell.
residue A unit (monomer) within a larger molecule (polymer); a location or position within a polymer.
respiration Aerobic catabolic process that converts small molecules such as pyruvate to carbon dioxide and water.
reversible inhibitor A substance that temporarily binds to an enzyme molecule, causing the enzyme to lose activity while the substance is bound.
ribosomal RNA (rRNA) The RNA found in ribosomes.
RNA (ribonucleic acid) A nucleic acid composed of nucleotides containing the sugar ribose.

S

salt An ionic compound formed when an acid and a base react.
saturated fatty acid A fatty acid containing no carbon–carbon double or triple bonds.
saturated solution A solution that contains all of a solute that will dissolve in the solvent under the conditions that exist.
scientific method A process that uses known information, statements of explanation, and experiments to gain understanding of phenomena.
scientific (exponential) notation A way of representing numbers that expresses them as a small number multiplied by a power of ten.
second The standard unit of time in the metric, SI, and English systems of units.
secondary amine An amine containing a nitrogen atom bonded to two carbon atoms.
secondary structure The structure of a protein involving the arrangement of the atoms of the polypeptide chain. This structure is maintained by hydrogen bonding between atoms of the peptide bonds.
shells (energy levels) Layers of electrons around the nucleus of an atom.
significant figures All of the numbers in a measurement including the estimated place.
simple diffusion The movement of molecules from a region of higher concentration to one of lower concentration.
simple lipids Lipids that do not contain a fatty acid as part of their structure.
simple protein A protein that contains only amino acids.
single bond A covalent bond involving one pair of shared electrons.
solid State of matter that has both definite shape and definite volume.
solubility The tendency of a substance to dissolve into another substance.
solute A substance in a solution that is dissolved in the solvent.
solution A homogeneous mixture of two or more pure substances.
solvent The most abundant component of a solution.
specific gravity The ratio of the density of a sample divided by the density of water.
specific heat The amount of energy (heat) required to raise the temperature of 1 g of a substance by 1°C.
sphingolipid Lipids containing the base sphingosine, a fatty acid, and one or more other molecules.
standard temperature and pressure (STP) Zero degrees Celsius and 1 atm of pressure.
starch Storage polysaccharides of plants composed of glucose residues.
states (states of matter) The three forms of matter that normally exist—gas, liquid, solid.
stereoisomers Compounds that have the same composition and the atoms bonded to the same atoms but that differ in the arrangement of atoms in three-dimensional space.
steroid A simple lipid containing the fused-ring system commonly called the steroid nucleus.
straight-chain alkanes (n-alkanes) Alkanes containing carbon atoms linked together in a continuous chain.
strong acid An acid in which all of the molecules donate their hydrogen ions when placed in water.
strong base A hydroxide-containing compound that dissociates in water to yield a high concentration of hydroxide ion.
structural formula A formula used to show how atoms are connected to each other in a molecule.
structural theory Model stating that the properties of a compound depend on the structure of the compound.
sublimation The change from a solid to a gas.
subshells A subset of a shell; subshells are labeled *s,p,d,* and *f,* and can hold up to 2, 6, 10 and 14 electrons respectively.
substance A sample of matter that always has constant composition. Substances cannot be separated into components by any physical changes.
substitution mutation A mutation that results in the replacement of a base in DNA with a different base.
substrate The reactant in an enzyme-catalyzed reaction.
sulfhydryl group (—SH) The functional group of the thiols.
suspension An unstable mixture of relatively large particles that are suspended in a liquid.
Systéme International (SI) A modification of the metric system of units that is used by scientists.

T

temperature A measure of the kinetic energy in matter that results from motion or vibration in the sample of matter.
tertiary amine An amine containing a nitrogen atom bonded to three carbon atoms.
tertiary structure The shape of a protein that results from the folding and bending of the polypeptide chain.
theory An explanation for a set of data that has gained considerable credibility and acceptance.

thiol An organic compound containing the sulfhydryl group.
titration A procedure in which volumes of a known solution are mixed with a volume of an unknown solution to determine the concentration of solute in the unknown solution.
transamination A process that removes the amino group from an amino acid and transfers it to α-ketoglutarate.
transcription A process that makes RNA using a strand of DNA as a template.
transfer RNA (tRNA) RNAs that transfer amino acids from solution to the growing polypeptide chain during protein synthesis.
translation The process that synthesizes proteins on ribosomes.
triacylglycerols Nonpolar complex lipids—composed of three fatty acids esterified to glycerol—that serve as energy reserves.
triple bond A covalent bond involving three pairs of shared electrons.
Tyndall effect The deflection of light as it passes through a dispersion.

U

unit The part of a measurement that specifies the physical property and size of the measurement.
universal gas constant (R) Constant used in the ideal gas law and other calculations; 0.0821 L atm/K mol.
unsaturated fatty acid A fatty acid containing one or more carbon–carbon double bonds.
unsaturated solution A solution containing less solute than would dissolve in the solvent under the existing conditions.

V

valence electrons The electrons located in the valence (outermost) shell in an atom.
valence shell The outermost shell of electrons in an atom.
vaporization The change from a liquid to a gas.

W

weak acid An acid in which only some of the molecules donate their hydrogen ions when placed in water.
weak base A compound that yields only small quantities of hydroxide ions when dissolved in water.
weight A measure of the gravitational pull on an object.

Z

zwitterion A dipolar ion in an amino acid formed by the intramolecular reaction of the carboxyl group and amino group.
zymogen Inactive forms of enzymes that are converted to the active form by cleavage of one or more peptide bonds.

Index

A

B

C

F

G

H

Q

R

T

U

V

W

X

Y

Z